K. Hunger (Editor)

Industrial Dyes

Chemistry, Properties, Applications

Further of Interest:

Herbst, W., Hunger, K.
Industrial Organic Pigments
Production, Properties, Applications
Third, Completely Revised Edition
2003
ISBN 3-527-30576-9

Buxbaum, G. (Ed.)
Industrial Inorganic Pigments
Second, Completely Revised Edition
1998
ISBN 3-527-28878-3

Smith, H. M. (Ed.)
High Performance Pigments
2002
ISBN 3-527-30204-2

Völz, H. G.
Industrial Color Testing
Fundamentals and Techniques
Second, Completely Revised Edition
2001
ISBN 3-527-30436-3

Freitag, W., Stoye, D. (Eds.)
Paints, Coatings and Solvents
Second, Completely Revised Edition
1998
ISBN 3-527-28863-5

Bieleman, J. (Ed.)
Additives for Coatings
2000
ISBN 3-527-29785-5

Klaus Hunger (Editor)

Industrial Dyes

Chemistry, Properties, Applications

Dr. Klaus Hunger (Editor)
Johann-Strauß-Str. 35
D-65779 Kelkheim
Germany
formerly Hoechst AG, Frankfurt, Germany

This book was carefully produced. Nevertheless, editor, authors and publisher do not warrant the information contained therein to be free of errors. Readers are advised to keep in mind that statements, data, illustrations, procedural details or other items may inadvertently be inaccurate.

Library of Congress Card No.: Applied for.

British Library Cataloguing-in-Publication Data: A catalogue record for this book is available from the British Library

Bibliographic information published by Die Deutsche Bibliothek
Die Deutsche Bibliothek lists this publication in the Deutsche Nationalbibliografie; detailed bibliographic data is available in the Internet at <http://dnb.ddb.de>

ISBN 3-527-30426-6

© 2003 WILEY-VCH Verlag GmbH & Co. KGaA, Weinheim

Printed on acid-free paper.

All rights reserved (including those of translation in other languages). No part of this book may be reproduced in any form – by photoprinting, microfilm, or any other means – nor transmitted or translated into machine language without written permission from the publishers. Registered names, trademarks, etc. used in this book, even when not specifically marked as such, are not to be considered unprotected by law.

Composition: Kühn & Weyh, 79111 Freiburg
Printing and Bookbinding: Druckhaus Darmstadt GmbH, Darmstadt

Printed in the Federal Republic of Germany.

Contents

Preface XXI

List of Contributors XXIII

1   Dyes, General Survey 1

1.1   Introduction   1

1.2   Classification Systems for Dyes   2

1.3   Classification of Dyes by Use or Application Method   3

1.4   Nomenclature of Dyes   6

1.5   Equipment and Manufacture   7

1.6   Economic Aspects   10

1.7   References   12

2   Important Chemical Chromophores of Dye Classes   13

        Introduction   13

2.1   Azo Chromophore   14
2.1.1   Introduction   14
2.1.2   General Synthesis   16
2.1.2.1 Diazo Components   16
2.1.2.2 Diazotization Methods   19
2.1.2.3 Coupling Components   20
2.1.2.4 Azo Coupling in Practice   28
2.1.3   Principal Properties   29
2.1.3.1 Tautomerism   29
2.1.3.2 Metallized Azo Dyes   32

Contents

2.1.3.3 Carbocyclic Azo Dyes   33
2.1.3.4 Heterocyclic Azo Dyes   34
2.1.4   References   35

2.2   Anthraquinone Chromophore   35
2.2.1   Introduction   35
2.2.2   General Synthesis   36
2.2.3   Principal Properties   36
2.2.3.1 Benzodifuranone Dyes   37
2.2.3.2 Polycyclic Aromatic Carbonyl Dyes   38
2.2.4   References   39

2.3   Indigoid Chromophore   40
2.3.1   Introduction   40
2.3.2   General Synthesis   40
2.3.3   Principal Properties   41
2.3.3.1 Color   41
2.3.3.2 Basic Chromophore   42
2.3.3.3 Solvatochromism   42
2.3.3.4 Redox System   43
2.3.4   References   43

2.4   Cationic Dyes as Chromophores   44
2.4.1   Introduction   44
2.4.2   General Synthesis   45
2.4.3   Chemical Structure and Classification   45
2.4.3.1 Dyes with Delocalized Charge   45
2.4.3.2 Dyes with Localized Charge   49
2.4.4   Principal Properties   52
2.4.4.1 Cationic Dyes for Synthetic Fibers   52
2.4.4.2 Cationic Dyes for Paper, Leather, and Other Substrates   53
2.4.5   References   55

2.5   Polymethine and Related Chromophores   56
2.5.1   Introduction   56
2.5.2   General Synthesis   57
2.5.3   Principal Properties and Classification   57
2.5.3.1 Azacarbocyanines   57
2.5.3.2 Hemicyanines   57
2.5.3.3 Diazahemicyanines   58
2.5.3.4 Styryl Dyes   58

2.6	Di- and Triarylcarbenium and Related Chromophores	59
2.6.1	Introduction	59
2.6.2	Chromophores	60
2.6.3	General Synthesis	62
2.6.4	Principal Properties	65
2.6.5	References	67
2.7	Phthalocyanine Chromophore	68
2.7.1	Introduction	68
2.7.2	General Synthesis	70
2.7.3	Principal Properties	72
2.7.4	Industrial Production	73
2.7.4.1	Copper Phthalocyanine	73
2.7.4.2	Phthalocyanine Derivatives	74
2.7.4.3	Pthalocyanine Sulfonic Acids and Sulfonyl Chlorides	75
2.7.5	References	76
2.8	Sulfur Compounds as Chromophores	78
2.8.1	Introduction	78
2.8.2	Chromophores	79
2.8.3	General Synthesis	79
2.8.3.1	Sulfur Bake and Polysulfide Bake Dyes	79
2.8.3.2	Polysulfide Melt Dyes	81
2.8.3.3	Pseudo Sulfur Dyes	83
2.8.4	Principal Properties	84
2.8.5	References	84
2.9	Metal Complexes as Chromophores	85
2.9.1	Introduction	85
2.9.2	Azo/Azomethine Complex Dyes	86
2.9.2.1	General Synthesis	86
2.9.2.2	Principal Properties	92
2.9.3	Formazan Dyes	97
2.9.3.1	Introduction	97
2.9.3.2	General Synthesis	98
2.9.3.3	Principal Properties	102
2.9.4	References	105
2.10	Fluorescent Dyes	107
2.11	Other Chromophores	109
2.11.1	Quinophthalone Dyes	109
2.11.2	Nitro and Nitroso Dyes	110
2.11.3	Stilbene Dyes	111
2.11.4	Formazan Dyes	111

VIII   Contents

2.11.5   Triphenodioxazine Dyes   112
2.11.6   References   112

3   Dye Classes For Principal Applications   113

3.1   Reactive Dyes   113
3.1.1   Introduction   113
3.1.2   Chemical Constitution of Reactive Systems   114
3.1.2.1   Mono-Anchor Dyes   114
3.1.2.2   Double-Anchor Dyes   117
3.1.2.3   Multiple-Anchor Dyes   118
3.1.3   Dye Classes (Chromogens) for Reactive Dyes   118
3.1.3.1   Azo Dyes   119
3.1.3.2   Metal-Complex Azo Dyes   119
3.1.3.3   Anthraquinone Dyes   119
3.1.3.4   Triphenodioxazine Dyes   120
3.1.3.5   Formazan Dyes   122
3.1.3.6   Phthalocyanine Dyes   122
3.1.4   Synthesis   123
3.1.5   Examples of Commercially Available Dyes   129
3.1.5.1   Azo Dyes   129
3.1.5.2   Metal Complex Azo Dyes   130
3.1.5.3   Formazan Dyes   130
3.1.5.4   Anthraquinone Dyes   130
3.1.5.5   Diphenodioxazine Dyes   131
3.1.6   Forms of Supply   131
3.1.7   References   132

3.2   Disperse Dyes   134
3.2.1   Introduction   134
3.2.2   Chemical Constitution   135
3.2.2.1   Azo Dyes   135
3.2.2.2   Anthraquinone Dyes   138
3.2.2.3   Other Chromophores   139
3.2.3   Synthesis   144
3.2.3.1   Monoazo Dyes   144
3.2.3.2   Disazo Dyes   145
3.2.3.3   Anthraquinone Dyes   145
3.2.3.4   Other Chromophores   145
3.2.4   Aftertreatment   145
3.2.5   Examples of Commercially Available Dyes   146
3.2.5.1   Monoazo Dyes   146
3.2.5.2   Disazo Dyes   154
3.2.5.3   Anthraquinone Dyes   155

3.2.5.4	Other Chromophores	155
3.2.6	References	156
3.3	Direct Dyes	158
3.3.1	Introduction	158
3.3.2	Chemical Constitution	159
3.3.2.1	Structural Characteristics	159
3.3.2.2	Precursors	159
3.3.2.3	Classification According to the Method of Application	161
3.3.3	Dye Classes	161
3.3.3.1	Monoazo Dyes	161
3.3.3.2	Disazo Dyes	162
3.3.3.3	Trisazo Dyes	164
3.3.3.4	Tetrakisazo Dyes	165
3.3.4	Synthesis	168
3.3.4.1	Monoazo Dyes	168
3.3.4.2	Disazo Dyes	169
3.3.4.3	Trisazo Dyes	170
3.3.4.4	Anthraquinone Direct Dyes	172
3.3.5	Direct Dyes with Aftertreatment	172
3.3.5.1	Aftertreatment with Cationic Auxiliaries	173
3.3.5.2	Aftertreatment with Formaldehyde	174
3.3.5.3	Diazotization Dyes	174
3.3.5.4	Aftertreatment with Metal Salts	175
3.3.6	Examples of Commercially Available Dyes	175
3.3.6.1	Monoazo Dyes	175
3.3.6.2	Disazo Dyes	176
3.3.6.3	Trisazo Dyes	176
3.3.6.4	Tetrakisazo Dyes	176
3.3.6.5	Condensation Dyes	177
3.3.6.6	Direct Dyes with a Urea Bridge	177
3.3.6.7	Triazinyl Dyes	177
3.3.7	References	178
3.4	Anthraquinone Dyes	178
3.4.1	Introduction	178
3.4.2	Chemical Constitution and Properties	180
3.4.2.1	Disperse Dyes	180
3.4.2.2	Vat Dyes	187
3.4.2.3	Acid Dyes	195
3.4.3	Synthesis	200
3.4.3.1	Introduction	200
3.4.3.2	Anthraquinonesulfonic Acids	200
3.4.3.3	Haloanthraquinones	201
3.4.3.4	Nitroanthraquinones	201
3.4.3.5	Aminoanthraquinones	201

3.4.3.6	Hydroxyanthraquinones 202
3.4.4	Examples of Commercially Available Dyes 202
3.4.5	References 203

3.5	Indigoid Dyes 204
3.5.1	Introduction 204
3.5.2	Chemical Constitution and Properties 205
3.5.2.1	Physical Properties 205
3.5.2.2	Chemical Properties 206
3.5.3	Synthesis 207
3.5.3.1	Chemical Synthesis 207
3.5.3.2	Biotechnological Synthesis 211
3.5.4	Commercially Available Dyes 213
3.5.4.1	Indigo, *C.I. Vat Blue 1*, 73000, [*482-89-3*] 213
3.5.4.2	Halogen Derivatives 213
3.5.4.3	Other Indigo Derivatives 214
3.5.5	References 214

3.6	Sulfur Dyes 215
3.6.1	Introduction 215
3.6.2	Chemical Constitution 216
3.6.3	Synthesis 216
3.6.3.1	Sulfur and Polysulfide Bake or Dry Fusion 216
3.6.3.2	Polysulfide Melt, Solvent Reflux, or Reflux Thionation Process 219
3.6.3.3	Indophenols 219
3.6.4	Modifications; Commercial Forms; Types of Sulfur Dyes 223
3.6.4.1	C.I. Sulphur Dyes 224
3.6.4.2	C.I. Leuco Sulphur Dyes 224
3.6.4.3	C.I. Solubilised Sulphur Dyes (Bunte Salts, S-Aryl Thiosulfate Dyes) 224
3.6.5	Pseudo (Synthetic) Sulfur Dyes 225
3.6.6	Commercially Available Dyes 226
3.6.7	References 226

3.7	Cationic Azo Dyes 227
3.7.1	Introduction 227
3.7.2	Chemical Constitution and Synthesis 227
3.7.2.1	Cationic Charge in the Coupling Component 227
3.7.2.2	Cationic Charge in the Diazo Component 237
3.7.2.3	Different Cationic Charges in Both the Coupling and the Diazo Component 244
3.7.2.4	Introduction of Cationic Substituents into Preformed Azo Dyes 245
3.7.2.5	Cationic Dyes with Sulfur or Phosphorus as Charge-Carrying Atoms 246
3.7.2.6	Dyes with Releasable Cationic Groups 247

3.7.3	Synthesis 248	
3.7.4	Examples of Commercially Available Dyes 248	
3.7.5	References 250	

3.8	Cationic Methine Dyes 254	
3.8.1	Introduction 254	
3.8.2	Chemical Constitution and Synthesis 254	
3.8.2.1	Streptocyanine Dyes 254	
3.8.2.2	Hemicyanine Dyes 255	
3.8.2.3	Higher Vinylogues of Hemicyanine Dyes 260	
3.8.2.4	Phenylogous Hemicyanine Dyes 261	
3.8.2.5	Cyanine Dyes 268	
3.8.3	Synthesis 271	
3.8.4	Examples of Commercially Available Dyes 272	
3.8.5	References 273	

3.9	Acid Dyes (Anionic Azo Dyes) 276	
3.9.1	Introduction 276	
3.9.2	Chemical Constitution and Synthesis 277	
3.9.2.1	Wool Dyes 278	
3.9.2.2	Polyamide Dyes 289	
3.9.2.3	Silk Dyes 291	
3.9.3	Examples of Commercially Available Dyes 291	
3.9.4	References 294	

3.10	Solvent Dyes 295	
3.10.1	Introduction 295	
3.10.2	Chemical Constitution and Application Properties 295	
3.10.2.1	Alcohol- and Ester-Soluble Dyes 295	
3.10.2.2	Fat- and Oil-Soluble Dyes 297	
3.10.2.3	Dyes Soluble in Polymers 298	
3.10.2.4	Solvent Dyes for Other Applications 299	
3.10.3	Examples of Commercially Available Dyes 300	
3.10.4	References 301	

3.11	Metal-Complex Dyes 302	
3.11.1	Introduction 302	
3.11.2	Chemical Constitution and Synthesis 304	
3.11.2.1	Chromium and Cobalt Complexes for Wool and Polyamides 304	
3.11.2.2	Metal Complexes for Cotton (see also Section 3.1) 311	
3.11.2.3	Metal Complexes for Leather (see also Section 5.1) 313	
3.11.2.4	Copper Complexes for Paper (see also Section 5.3) 315	
3.11.2.5	Metal-Complex Dyes for Polypropylene 316	
3.11.2.6	Formazan Dyes 316	
3.11.3	Miscellaneous Uses 319	

XII   Contents

3.11.3.1   Azo Metal-Complex Dyes   319
3.11.3.2   Formazan Dyes   324
3.11.4   References   327

3.12   Naphthoquinone and Benzoquinone Dyes   329
3.12.1   Introduction   329
3.12.2   Benzoquinone Dyes   330
3.12.3   1,4-Naphthoquinone Dyes   331
3.12.3.1   Simple 1,4-Naphthoquinones   331
3.12.3.2   Heteroannelated 1,4-Naphthoquinones   332
3.12.4   1,5-Naphthoquinones   335
3.12.5   References   338

4   Textile Dyeing   339

4.1   Introduction   339
4.1.1   Dyeing Technology   340
4.1.1.1   Principles of Dyeing   341
4.1.1.2   Bath Dyeing Technology   342
4.1.1.3   Continuous and Semicontinuous Dyeing   343
4.1.1.4   Printing   345
4.1.1.5   Dispensing Dyes and Chemicals   345
4.1.2   Standardization of Textile Dyes   346
4.1.3   Colorfastness of Textiles   348
4.1.4   Laboratory Dyeing Techniques   349

4.2   Reactive Dyes on Cellulose and Other Fibers   349
4.2.1   Fundamentals   350
4.2.2   Dyeing Techniques for Cellulose   353
4.2.3   Reactive Dyes on Wool, Silk and Polyamide Fibers   356
4.2.4   Reactive Dyes for Printing on Cellulose   357

4.3   Direct Dyes on Cellulosic Fibers   358
4.3.1   Dyeing Principle   358
4.3.2   Dyeing Parameters   359
4.3.3   Dyeing Techniques   360
4.3.4   Aftertreatment   361
4.3.5   Direct Dyes for Fiber Blends   361

4.4   Anthraquinone Vat Dyes on Cellulosic Fibers   362
4.4.1   Principles of Vat Dyeing   362
4.4.2   The Vat Dyeing Process   363
4.4.2.1   Vatting   363
4.4.2.2   Dye Absorption in the Exhaustion Process   364

Contents   XIII

4.4.2.3   Oxidation   364
4.4.2.4   Aftertreatment ("Soaping")   365
4.4.3   Dyeing Techniques   365
4.4.4   Vat Dyes for Fiber Blends   367

4.5   Leuco Esters of Vat Dyes on Cellulosic Fibers   367

4.6   Dyeing with Indigo (see 2.3, 3.5)   368
4.6.1   Dyeing Technique on Cotton   368
4.6.2   Indigo on Wool   369

4.7   Sulfur Dyes on Cellulosic Fibers   370
4.7.1   Types and Mode of Reaction   370
4.7.2   Additives to the Dye Bath   371
4.7.3   The Dyeing Process   372
4.7.4   Dyeing Techniques   373
4.7.5   Combination with Other Dyes   375

4.8   Azo (Naphtol AS) Dyes on Cellulosic Fibers   375
4.8.1   Application of Azo Dyes   375
4.8.2   Dyeing Processes on Cellulosic Fibers   376
4.8.3   Printing with Azo (Naphtol AS) Dyes on Cellulosic Fibers   377

4.9   Dyeing Cellulosic Fibers with Other Dye Classes   377
4.9.1   Mordant Dyes on Cellulosic Fibers   377
4.9.2   Acid Dyes on Plant Fibers   378
4.9.3   Basic Dyes on Cellulose   378
4.9.4   Oxidation Dyes on Cellulosic Fibers   378
4.9.5   Phthalogen Dyes on Cellulosic Fibers   379
4.9.6   Coupling and Diazotization Dyes on Cellulosic Fibers   379
4.9.7   Pigments and Mineral Dyes on Cellulose   380

4.10   Acid and Metal-Complex Dyes on Wool and Silk   381
4.10.1   Principles of Dyeing of Wool and Silk   381
4.10.2   Acid Dyes on Wool   382
4.10.3   Chrome Dyes on Wool   384
4.10.4   Metal-Complex Dyes on Wool   385
4.10.4.1 1:1 Metal-Complex Dyes   385
4.10.4.2 1:2 Metal-Complex Dyes   386

4.11   Acid and Metal-Complex Dyes on Polyamide   386
4.11.1   Chemical and Physical Structure of the Fiber   386
4.11.2   Interactions Between Dye and Fiber   387
4.11.3   Dyeing Processes with Different Classes of Dyes   388
4.11.3.1 Acid Dyes   388

XIV   Contents

4.11.3.2   1:2 Metal-Complex Dyes   390
4.11.4   Technology of Dyeing Polyamide   391

4.12   Disperse Dyes on Polyester and Other Man-Made Fibers   392
4.12.1   General Aspects   392
4.12.1.1   Dyeing in Aqueous Liquor   392
4.12.1.2   Thermosol Process   395
4.12.2   Dyeing Processes for Polyester Fibers with Disperse Dyes   396
4.12.2.1   Suitability of Disperse Dyes for Different Applications   396
4.12.2.2   Dyeing from Aqueous Dye Baths   397
4.12.2.3   Special Dyeing Processes   398
4.12.2.4   Continuous and Semicontinuous Dyeing Processes   399
4.12.2.5   Dyeing of PES Microfibers   400
4.12.2.6   Dyeing of Modified PES Fibers   401
4.12.2.7   Printing wth Disperse Dyes on Man-Made Fibers   401
4.12.3   Aftertreatment   403
4.12.4   Dyeing Blends Containing Polyester Fibers   403
4.12.4.1   Polyester–Cellulose Blends   403
4.12.4.2   Polyester–Wool Blends   407

4.13   Disperse Dyes on Other Fibers   409
4.13.1   Disperse Dyes on Cellulose Acetate   409
4.13.1.1   Dyeing Processes for Cellulose 2.5 Acetate   409
4.13.1.2   Dyeing Processes for Cellulose Triacetate   410
4.13.2   Disperse Dyes on Polyamide Fibers   410
4.13.3   Disperse Dyes on Other Fibers   411

4.14   Cationic Dyes on Acrylic Fibers   412
4.14.1   General Aspects   412
4.14.2   Cationic Dyes   412
4.14.3   Retarders and Auxiliaries   413
4.14.4   Exhaustion Process   414
4.14.5   Dyeing of Special Fiber Types with Cationic Dyes by the Exhaustion Process   416
4.14.6   Continuous Processes with Cationic Dyes   417
4.14.7   Cationic Dyes on Aramide Fibers   418
4.14.8   Cationic Dyes in Fiber Blends   419

4.15   References   421

5   **Nontextile Dyeing**   427

5.1   Leather Dyes   427
5.1.1   Introduction   427

5.1.2	Color Selection	428
5.1.2.1	Aniline Leather	428
5.1.2.2	Pigmented Leather	429
5.1.2.3	Colour Index	429
5.1.3	Natural and Mordant Dyes	431
5.1.3.1	Tanning Agents	431
5.1.3.2	Dyewood	432
5.1.3.3	Synthetic Mordant Dyes	433
5.1.4	Basic Dyes	433
5.1.4.1	Azine Dyes	433
5.1.4.2	Other Cationic (Basic) Dyes	434
5.1.5	Acid Dyes	434
5.1.5.1	Amphoteric Dyes	435
5.1.5.2	Anthraquinone Dyes	435
5.1.5.3	Low-Molecular Azo Dyes	436
5.1.5.4	Resorcinol Azo Dyes	437
5.1.5.5	Azo Metal-Complex Dyes	437
5.1.6	Direct Dyes	439
5.1.6.1	Condensation Dyes	439
5.1.6.2	Polyazo Dyes	440
5.1.6.3	Phthalocyanine Dyes	440
5.1.7	Sulfur Dyes	441
5.1.8	Reactive Dyes	442
5.1.9	Solvent Dyes	443
5.1.10	Pigments	444
5.1.11	References	445
5.2	Fur Dyes	446
5.2.1	Introduction	446
5.2.1.1	Origin of Fur	446
5.2.1.2	Animal Rights	447
5.2.1.3	Fur Hair and Classification	448
5.2.1.4	Fur Dressing	448
5.2.1.5	Fur Finishing	448
5.2.1.6	The Final Stage	449
5.2.1.7	Garments and Fashion	450
5.2.1.8	Labels	450
5.2.2	Fur Dyeing	451
5.2.2.1	History and Outlook	451
5.2.2.2	Color Selection and Colour Index	451
5.2.2.3	Vegetable Dyes	451
5.2.2.4	Oxidation Bases	452
5.2.2.5	Disperse Dyes	453
5.2.2.6	Acid and Direct Dyes	454
5.2.2.7	Metal-Complex Dyes	456

5.2.2.8	Other Synthetic Dyes 457
5.2.3	References 458

5.3	Paper Dyes 459
5.3.1	Introduction 459
5.3.2	Classification of Paper Dyes 460
5.3.3	Direct Dyes 460
5.3.3.1	Anionic Direct Dyes 461
5.3.3.2	Cationic Direct Dyes 466
5.3.4	Acid Dyes 469
5.3.5	Cationic (Basic) Dyes 470
5.3.6	Sulfur Dyes 471
5.3.7	Organic Pigments 472
5.3.8	Special Requirements for Paper Dyes 472
5.3.9	References 472

5.4	Hair Dyes 473
5.4.1	Bleaching 473
5.4.2	Dyeing with Oxidation Dyes 475
5.4.3	Dye Classes 479
5.4.3.1	Direct Dyes 479
5.4.3.2	Nitro Dyes 479
5.4.3.3	Cationic (Basic) Dyes (see Sections 3.7 and 3.8) 480
5.4.3.4	Anionic (Acid) Dyes (see Section 3.9) 480
5.4.3.5	Disperse Dyes (see Section 3.2) 480
5.4.3.6	Dyeing with Inorganic Compounds 480
5.4.3.7	Other Dyes 481
5.4.4	Product Forms 481
5.4.5	Dye-Removal Preparations 483
5.4.6	Testing Hair Dyes 483
5.4.7	References 484

5.5	Food Dyes 486
5.5.1	Introduction 486
5.5.1.1	Specifications 487
5.5.1.2	Uses and Individual Substances 487
5.5.2	Synthetic Dyes Approved for Coloring of Foodstuffs 487
5.5.3	Examples of Chemical Structures 489
5.5.4	Purity Requirements 491
5.5.5	Legal Aspects 492
5.5.5.1	Codex Alimentarius 492
5.5.5.2	EU and other European Countries 492
5.5.5.3	USA 493
5.5.5.4	Japan 494
5.5.6	References 494

5.6	Ink Dyes	495
5.6.1	Introduction	495
5.6.2	Application Principles	495
5.6.2.1	Ink-Jet Technology	495
5.6.2.2	Writing, Drawing and Marking Materials	497
5.6.3	Dye Classes	497
5.6.3.1	Dyes for Ink-Jet Application	497
5.6.3.2	Dyes for Writing, Drawing, and Marking	501
5.6.3.3	Fields of Application for Ink-Jet Printing	502
5.6.4	Inks	503
5.6.4.1	Ink-Jet Inks	503
5.6.4.2	Writing, Drawing, and Marking Inks	505
5.6.5	Properties of Ink-Jet Prints	507
5.6.6	References	508
5.7	Photographic Dyes	509
5.7.1	Cyanine Dyes	509
5.7.1.1	Introduction	509
5.7.1.2	Fundamental Aspects	509
5.7.1.3	Application of Sensitizing Dyes	511
5.7.1.4	Production of Sensitizing Dyes	512
5.7.1.5	Cyanine Dyes as Sensitizers	512
5.7.2	Merocyanine Dyes	515
5.7.3	Oxonol Dyes	516
5.7.4	Azomethine and Indoaniline Image Dyes	516
5.7.4.1	Introduction	516
5.7.4.2	Color Developers	517
5.7.4.3	Yellow Azomethine Dyes	517
5.7.4.4	Magenta Azomethine Dyes	518
5.7.4.5	Cyan Indoaniline Dyes	518
5.7.5	Azo Dyes	519
5.7.5.1	Diffusion-Transfer Imaging Systems	519
5.7.5.2	Silver Dye Bleach Processes	520
5.7.5.3	Color Masking	520
5.7.6	Metallized Dyes	521
5.7.7	Xanthene Dyes	521
5.7.8	Triarylmethane Dyes	522
5.7.9	Anthraquinone Dyes	523
5.7.10	References	523
5.8	Indicator Dyes	526
5.8.1	Introduction	526
5.8.2	General Principles	526
5.8.3	Classes of Indicators	527
5.8.3.1	ph Indicators	527

XVIII *Contents*

5.8.3.2   Redox Indicators   537
5.8.3.3   Metal Indicators   537
5.8.4   Indicator Papers   540
5.8.4.1   Bleeding Indicator Papers   540
5.8.4.2   Nonbleeding Indicator Papers   541
5.8.5   References   541

6   Functional Dyes   543

6.1   Introduction   543

6.2   Interactions of Functional Dyes   543

6.3   Functional Dyes by Application   545
6.3.1   Imaging   545
6.3.1.1   Laser Printing and Photocopying   545
6.3.1.2   Thermal Printing   551
6.3.1.3   Dyes for Ink-Jet Printing   555
6.3.1.4   Other Imaging Technologies   558
6.3.2   Invisible Imaging   559
6.3.2.1   Optical Data Storage   560
6.3.2.2   Other Technologies   564
6.3.3   Displays   566
6.3.3.1   Cathode Ray Tube   566
6.3.3.2   Liquid Crystal Displays   566
6.3.3.3   Organic Light-Emitting Devices   569
6.3.3.4   Electrochromic Displays   571
6.3.4   Electronic Materials   572
6.3.4.1   Organic Semiconductors   572
6.3.4.2   Solar Cells   573
6.3.4.3   Nonlinear Optical Dyes   574
6.3.4.4   Laser Dyes   576
6.3.5   Biomedical Applications   576
6.3.5.1   Fluorescent Sensors and Probes   577
6.3.5.2   Photodynamic Therapy   579

6.4   References   581

7   Optical Brighteners   585

7.1   Introduction   585
7.1.1   Physical Principles   586
7.1.2   Molecular Structure   588

7.1.3	History of Whitening	589
7.2	Chemistry of Technical Products	590
7.2.1	Carbocycles	590
7.2.1.1	Distyrylbenzenes	590
7.2.1.2	Distyrylbiphenyls	592
7.2.1.3	Divinylstilbenes	593
7.2.2	Triazinylaminostilbenes	593
7.2.3	Stilbenyl-2H-triazoles	596
7.2.3.1	Stilbenyl-2H-naphtho[1,2-d]triazoles	597
7.2.3.2	Bis(1,2,3-triazol-2-yl)stilbenes	597
7.2.4	Benzoxazoles	598
7.2.4.1	Stilbenylbenzoxazoles	598
7.2.4.2	Bis(benzoxazoles)	599
7.2.5	Furans, Benzo[b]furans, and Benzimidazoles	601
7.2.5.1	Bis(benzo[b]furan-2-yl)biphenyls	601
7.2.5.2	Cationic Benzimidazoles	602
7.2.6	1,3-Diphenyl-2-pyrazolines	605
7.2.7	Coumarins	607
7.2.8	Naphthalimides	608
7.2.9	1,3,5-Triazin-2-yl Derivatives	610
7.3	Commercial Forms and Brands	610
7.4	Uses	611
7.4.1	General Requirements	611
7.4.2	Textile Industry	611
7.4.3	Detergent Industry	613
7.4.4	Paper Industry	614
7.4.5	Plastics and Synthetic Fibers	614
7.4.6	Other Uses	615
7.5	Analytical Methods and Whiteness Assessment	615
7.5.1	Analytical Methods	615
7.5.2	Assessment of Whitening Effect	616
7.6	Environmental Aspects	616
7.7	References	617
8	**Health and Safety Aspects**	**625**
8.1	Introduction	625

XX *Contents*

8.2   Toxicology and Toxicity Assessments   626
8.2.1   Acute Toxicity   626
8.2.2   Sensitization   626
8.2.3   Mutagenicity   628
8.2.4   Carcinogenicity   629
8.2.4.1 Introduction   629
8.2.4.2 Metabolism of Azo Dyes   630

8.3   Environmental Assessment/Fate   633
8.3.1   Introduction   633
8.3.2   Treatment of Dye-Containing Wastewater   633

8.4   Legislation   634
8.4.1   Registration/Notification of New Substances   634
8.4.2   Principal Chemical Legislation also Relevant to Dyes   635
8.4.3   Special Regulations for Dyes (Colorants)   636
8.4.4   Material Safety Data Sheets   638

8.5   References   639
8.5.1   General References   639
8.5.2   Specific References   639

List of Examples of Commercially Available Dyes   643

Index   653

Preface

A great deal of single papers on color chemistry have been published over the years. Their subjects are preferably theoretical or physico-chemical considerations of dye chemistry. On the other hand, the knowledge of dye chemistry from a technical point of view is almost entirely concentrated in the chemical industry. Apart from lectures given at the very few conferences on color chemistry, the current state of industrial dye chemistry can only be extracted from the patent literature. There is little else being published on dyes, especially on industrial dyes and their applications.

This prompted us to write a reference book comprising the principal classes of industrially produced dyes with their syntheses, properties and main applications as well as a toxicological, ecological and legal survey of dyes.

Since the field of dyes has grown so big in recent decades, it cannot be comprehensively covered by one or only a few persons. The various authors are renowned experts in their different fields of dye chemistry, which guarantees most competent contributions to this book.

The book provides an overview of the present state of industrial dyes and is divided into the following chapters: After a general survey on dyes, important dye chromophores are covered in chapter 2, followed by classification of dyes for principal applications. Chapter 4 is devoted to textile dyeing with regard to the various dye classes. Non-textile dyeing is outlined in the next chapter. Because of their growing potential and interest, functional dyes are reviewed in a separate chapter. Optical brighteners make up chapter 7, and a general overview on the toxicology, ecology and legislation of dyes forms the final chapter 8.

The literature references relating to each chapter are positioned at their respective ends.

A comprehensive subject index should enable the reader to quickly find any sought item.

We omitted to use trade names, but tried to designate the dyes with the C.I. Colour Index Name, and where available, with the C.I. Constitution Number and with the CAS (Chemical Abstracts Service) Registry Number.

This book is intended for all those who are engaged and interested in the field of industrial dyes, especially chemists, engineers, application technicians, colorists and laboratory assistants throughout the dye industry and at universities and technical colleges.

I am very grateful to the authors for their dedicated cooperation for this book. My appreciation is also extended to the Wiley-VCH publishing company for its good collaboration, in particular to Ms. Karin Sora, for her support in numerous discussions throughout the project.

Kelkheim, November 2002 *Klaus Hunger*

List of Contributors

Editor

Dr. K. Hunger
Johann-Strauß-Str. 35
65779 Kelkheim/Ts.
Germany

formerly
Hoechst Aktiengesellschaft
Geschäftsbereich
Feinchemikalien und Farben
65926 Frankfurt am Main
Germany
Sections 2.1 (Gregory), 2.7, 3.1, 3.2, 3.3, 3.8, 3.9, 3.10, 3.12, 5.5, Chapter 8 (Sewekow)

Authors

Dr. W. Bauer
Masurenstr.6
63477 Maintal
Germany
Section 5.6

Dr. H. Berneth
Bayer AG
Farbenforschung SP-FE-PE
Geb. I1
51368 Leverkusen
Germany
Section 2.4

Dr. T. Clausen
Wella AG
Geschäftsleitung
Berliner Allee 65
64274 Darmstadt
Germany
Section 5.4

Dr. A. Engel
R & D PTD DyStar Textilfarben
GmbH & Co.
Deutschland KG
Kaiser-Wilhelm Allee
P.O. Box 100480
57304 Leverkusen
Germany
Section 3.7 (Hunger, Kunde)

Dr. M. Filosa
Polaroid Research Laboratories
Polaroid Corporation
Cambridge, MA 02139
USA
Section 5.7

Dr. P. Gregory
Avecia Research Center
Avecia Ltd.
Blackley Manchester M9 8ZS
UK
Sections 2.1 (Hunger), 2.2, 2.5, 2.6, 2.10, 2.11, Chapter 6

Dr. J. Griffiths
The University of Leeds
Dept. of Colour Chemistry
Leeds LS2 9JT
UK
Section 3.12 (Hunger)

Dr. R. Hamprecht
DyStar Textilfarben GmbH & Co.
Deutschland KG
BU D-F & E, Geb. 11
51304 Leverkusen
Germany
Section 3.2

Dr. C. Heid
Lauterbacher Str.10
60386 Frankfurt
Germany
Sections 2.8, 3.6

Dr. J. Kaschig
Ciba Speciality Chemicals
K-24.4.18
Klybeckstrasse
CH-4002 Basel
Switzerland
Chapter 7

Dr. K. Kunde
Bayer AG
SP-FE-PV 1
51368 Leverkusen
Section 3.7 (Engel, Hunger)

Dr. H. Leube
Grünberger Str. 6
67117 Limburgerhof
Germany
Chapter 4

Dr. W. Mennicke
Steglitzer Str. 8
51375 Leverkusen
Germany
Sections 2.9, 3.11

Dr. P. Miederer
BASF AG,
Abt.EFF/AI - D 306
67056 Ludwigshafen
Germany
Sections 2.3, 3.5

Dr. R. Pedrazzi
Clariant (Schweiz) AG
Head R&D Dystuffs
Rothausstr. 61
CH-4132 Muttenz
Switzerland
Section 5.3

Dr. A. G. Püntener
TLF France S. A.
4, rue de l'industrie
Boite Postale 310
F-68333 Huningue Cedex
France
Sections 5.1, 5.2

Dr. E. Ross
Merck KGaA
USF/ZRW
64271 Darmstadt
Germany
Section 5.8

Dr. Sewekow
Bayer AG, GB Farben Marketing
Allg.Technik
51368 Leverkusen
Germany
Chapter 8 (Hunger)

1 Dyes, General Survey

1.1 Introduction

The first synthetic dye, Mauveine, was discovered by Perkin in 1856. Hence the dyestuffs industry can rightly be described as mature. However, it remains a vibrant, challenging industry requiring a continuous stream of new products because of the quickly changing world in which we live. The early dyes industry saw the discovery of the principal dye chromogens (the basic arrangement of atoms responsible for the color of a dye). Indeed, apart from one or two notable exceptions, all the dye types used today were discovered in the 1800s [1]. The introduction of the synthetic fibers nylon, polyester, and polyacrylonitrile during the period 1930–1950, produced the next significant challenge. The discovery of reactive dyes in 1954 and their commercial launch in 1956 heralded a major breakthrough in the dyeing of cotton; intensive research into reactive dyes followed over the next two decades and, indeed, is still continuing today [1] (see Section 3.1). The oil crisis in the early 1970s, which resulted in a steep increase in the prices of raw materials for dyes, created a drive for more cost-effective dyes, both by improving the efficiency of the manufacturing processes and by replacing tinctorially weak chromogens, such as anthraquinone, with tinctorially stronger chromogens, such as (heterocyclic) azo and benzodifuranone. These themes are still important and ongoing, as are the current themes of product safety, quality, and protection of the environment. There is also considerable activity in dyes for high-tech applications, especially in the electronics and nonimpact printing industries (see Chapter 6).

The scale and growth of the dyes industry has been inextricably linked to that of the textile industry. World textile production has grown steadily to an estimated 35×10^6 t in 1990 [2,3]. The two most important textile fibers are cotton, the largest, and polyester. Consequently, dye manufacturers tend to concentrate their efforts on producing dyes for these two fibers. The estimated world production of dyes in 1990 was 1×10^6 t [2,3]. The figure is significantly smaller than that for textile fibers because a little dye goes a long way. For example, 1 t of dye is sufficient to color 42 000 suits [3].

The rapid growth in the high-tech uses of dyes, particularly in ink-jet printing, is beginning to make an impact. Although the *volumes* of hi-tech dyes will remain small in comparison to dyes for traditional applications, the *value* will be significant because of the higher price of these specialized dyes.

Perkin, an Englishman, working under a German professor, Hoffman, discovered the first synthetic dye, and even today the geographical focus of dye production lies in Germany (BASF, Dystar), England (Avecia), and Switzerland (Clariant, Ciba Specialties). Far Eastern countries, such as Japan, Korea, and Taiwan, as well as countries such as India, Brazil, and Mexico, also produce dyes.

1.2 Classification Systems for Dyes

Dyes may be classified according to chemical structure or by their usage or application method. The former approach is adopted by practicing dye chemists, who use terms such as azo dyes, anthraquinone dyes, and phthalocyanine dyes. The latter approach is used predominantly by the dye user, the dye technologist, who speaks of reactive dyes for cotton and disperse dyes for polyester. Very often, both terminologies are used, for example, an azo disperse dye for polyester and a phthalocyanine reactive dye for cotton.

Chemical Classification. The most appropriate system for the classification of dyes is by chemical structure, which has many advantages. First, it readily identifies dyes as belonging to a group that has characteristic properties, for example, azo dyes (strong, good all-round properties, cost-effective) and anthraquinone dyes (weak, expensive). Second, there are a manageable number of chemical groups (about a dozen). Most importantly, it is the classification used most widely by both the synthetic dye chemist and the dye technologist. Thus, both chemists and technologists can readily identify with phrases such as an azo yellow, an anthraquinone red, and a phthalocyanine blue.

The classification used in this chapter maintains the backbone of the Colour Index classification, but attempts to simplify and update it. This is done by showing the structural interrelationships of dyes that are assigned to separate classes by the Colour Index, and the classification is chosen to highlight some of the more recent discoveries in dye chemistry [4].

Usage Classification. It is advantageous to consider the classification of dyes by use or method of application before considering chemical structures in detail because of the dye nomenclature and jargon that arises from this system.

Classification by usage or application is the principal system adopted by the Colour Index [5]. Because the most important textile fibers are cotton and polyester, the most important dye types are those used for dyeing these two fibers, including polyester–cotton blends (see Chapter 4). Other textile fibers include nylon, polyacrylonitrile, and cellulose acetate.

1.3 Classification of Dyes by Use or Application Method

The classification of dyes according to their usage is summarised in Table 1.1, which is arranged according to the C.I. application classification. It shows the principal substrates, the methods of application, and the representative chemical types for each application class.

Although not shown in Table 1.1, dyes are also used in high-tech applications, such as in the medical, electronics, and especially the nonimpact printing industries. For example, they are used in electrophotography (photocopying and laser printing) in both the toner and the organic photoconductor, in ink-jet printing, and in direct and thermal transfer printing [6] (see Chapter 6). As in traditional applications, azo dyes predominate; phthalocyanine, anthraquinone, xanthene and triphenylmethane dyes are also used. These applications are currently low volume (tens of kilograms up to several hundred tonnes per annum) but high added value (hundreds of dollars to many thousand dollars per kilogram), with high growth rates (up to 60 %).

Reactive Dyes. These dyes form a covalent bond with the fiber, usually cotton, although they are used to a small extent on wool and nylon. This class of dyes, first introduced commercially in 1956 by ICI, made it possible to achieve extremely high washfastness properties by relatively simple dyeing methods. A marked advantage of reactive dyes over direct dyes is that their chemical structures are much simpler, their absorption spectra show narrower absorption bands, and the dyeings are brighter. The principal chemical classes of reactive dyes are azo (including metallized azo), triphendioxazine, phthalocyanine, formazan, and anthraquinone (see Section 3.1).

High-purity reactive dyes are used in the ink-jet printing of textiles, especially cotton.

Disperse Dyes. These are substantially water-insoluble nonionic dyes for application to hydrophobic fibers from aqueous dispersion. They are used predominantly on polyester and to a lesser extent on nylon, cellulose, cellulose acetate, and acrylic fibers. Thermal transfer printing and dye diffusion thermal transfer (D2T2) processes for electronic photography represent niche markets for selected members of this class (see Chapter 6).

Table 1.1 Usage classification of dyes

Class	Principal substrates	Method of application	Chemical types
Acid	nylon, wool, silk, paper, inks, and leather	usually from neutral to acidic dyebaths	azo(including premetallized), anthraquinone, triphenylmethane, azine, xanthene, nitro and nitroso
Azoic components and compositions	cotton, rayon, cellulose acetate and polyester	fiber impregnated with coupling component and treated with a solution of stabilized diazonium salt	azo
Basic	paper, polyacrylonitrile, modified nylon, polyester and inks	applied from acidic dyebaths	cyanine, hemicyanine, diazahemicyanine, diphenylmethane, triarylmethane, azo, azine, xanthene, acridine, oxazine, and anthraquinone
Direct	cotton, rayon, paper, leather and nylon	applied from neutral or slightly alkaline baths containing additional electrolyte	azo, phthalocyanine, stilbene, and oxazine
Disperse	polyester, polyamide, acetate, acrylic and plastics	fine aqueous dispersions often applied by high temperature/pressure or lower temperature carrier methods; dye may be padded on cloth and baked on or thermofixed	azo, anthraquinone, styryl, nitro, and benzodifuranone
Fluorescent brighteners	soaps and detergents, all fibers, oils, paints, and plastics	from solution, dispersion or suspension in a mass	stilbene, pyrazoles, coumarin, and naphthalimides
Food, drug, and cosmetic	foods, drugs, and cosmetics		azo, anthraquinone, carotenoid and triarylmethane
Mordant	wool, leather, and anodized aluminium	applied in conjunction with Cr salts	azo and anthraquinone
Oxidation bases	hair, fur, and cotton	aromatic amines and phenols oxidized on the substrate	aniline black and indeterminate structures
Reactive	cotton, wool, silk, and nylon	reactive site on dye reacts with functional group on fiber to bind dye covalently under influence of heat and pH (alkaline)	azo, anthraquinone, phthalocyanine, formazan, oxazine, and basic
Solvent	plastics, gasoline, varnishes, lacquers, stains, inks, fats, oils, and waxes	dissolution in the substrate	azo, triphenylmethane, anthraquinone, and phthalocyanine
Sulfur	cotton and rayon	aromatic substrate vatted with sodium sulfide and reoxidized to insoluble sulfur-containing products on fiber	indeterminate structures
Vat	cotton, rayon, and wool	water-insoluble dyes solubilized by reducing with sodium hydrogensulfite, then exhausted on fiber and reoxidized	anthraquinone (including polycyclic quinones) and indigoids

Direct Dyes. These water-soluble anionic dyes, when dyed from aqueous solution in the presence of electrolytes, are substantive to, i.e., have high affinity for, cellulosic fibers. Their principal use is the dyeing of cotton and regenerated cellulose, paper, leather, and, to a lesser extent, nylon. Most of the dyes in this class are polyazo compounds, along with some stilbenes, phthalocyanines, and oxazines. Aftertreatments, frequently applied to the dyed material to improve washfastness properties, include chelation with salts of metals (usually copper or chromium), and treatment with formaldehyde or a cationic dye-complexing resin.

Vat Dyes. These water-insoluble dyes are applied mainly to cellulosic fibers as soluble leuco salts after reduction in an alkaline bath, usually with sodium hydrogensulfite. Following exhaustion onto the fiber, the leuco forms are reoxidized to the insoluble keto forms and aftertreated, usually by soaping, to redevelop the crystal structure. The principal chemical classes of vat dyes are anthraquinone and indigoid.

Sulfur Dyes. These dyes are applied to cotton from an alkaline reducing bath with sodium sulfide as the reducing agent. Numerically this is a relatively small group of dyes. The low cost and good washfastness properties of the dyeings make this class important from an economic standpoint (see Section 3.6). However, they are under pressure from an environmental viewpoint.

Cationic (Basic) Dyes. These water-soluble cationic dyes are applied to paper, polyacrylonitrile (e.g. Dralon), modified nylons, and modified polyesters. Their original use was for silk, wool, and tannin-mordanted cotton when brightness of shade was more important than fastness to light and washing. Basic dyes are water-soluble and yield colored cations in solution. For this reason they are frequently referred to as cationic dyes. The principal chemical classes are diazahemicyanine, triarylmethane, cyanine, hemicyanine, thiazine, oxazine, and acridine. Some basic dyes show biological activity and are used in medicine as antiseptics.

Acid Dyes. These water-soluble anionic dyes are applied to nylon, wool, silk, and modified acrylics. They are also used to some extent for paper, leather, ink-jet printing, food, and cosmetics.

Solvent Dyes. These water-insoluble but solvent-soluble dyes are devoid of polar solubilizing groups such as sulfonic acid, carboxylic acid, or quaternary ammonium. They are used for coloring plastics, gasoline, oils, and waxes. The dyes are predominantly azo and anthraquinone, but phthalocyanine and triarylmethane dyes are also used.

1.4 Nomenclature of Dyes

Dyes are named either by their commercial trade name or by their Colour Index (C.I.) name. In the Colour Index [5] these are cross-referenced.

The commercial names of dyes are usually made up of three parts. The first is a trademark used by the particular manufacturer to designate both the manufacturer and the class of dye, the second is the color, and the third is a series of letters and numbers used as a code by the manufacturer to define more precisely the hue and also to indicate important properties of the dye. The code letters used by different manufacturers are not standardized. The most common letters used to designate hue are R for reddish, B for bluish, and G for greenish shades. Some of the more important letters used to denote the dyeings and fastness properties of dyes are W for washfastness and E for exhaust dyes. For solvent and disperse dyes, the heatfastness of the dye is denoted by letters A, B, C, or D, A being the lowest level of heatfastness, and D the highest. In reactive dyes for cotton, M denotes a warm- (ca 40 °C) dyeing dye, and H a hot-dyeing (ca 80 °C) dye.

There are instances in which one manufacturer may designate a bluish red dye as Red 4B and another manufacturer uses Violet 2R for the same dye. To resolve such a problem the manufacturers' pattern leaflets should be consulted.

These show actual dyed pieces of cloth (or other substrate) so the colors of the dyes in question can be compared directly in the actual application. Alternatively, colors can be specified in terms of color space coordinates. In the CIELAB system, which is is becoming the standard, the color of a dye is defined by L, a, and b coordinates.

The C.I. (*Colour Index*) name for a dye is derived from the application class to which the dye belongs, the color or hue of the dye and a sequential number, e.g., *C.I. Acid Yellow 3*, *C.I. Acid Red 266*, *C.I. Basic Blue 41*, and *C.I. Vat Black 7*. A five digit C.I. (Constitution) number is assigned to a dye when its chemical structure has been disclosed by the manufacturer. The following example illustrates these points:

Chemical structure:

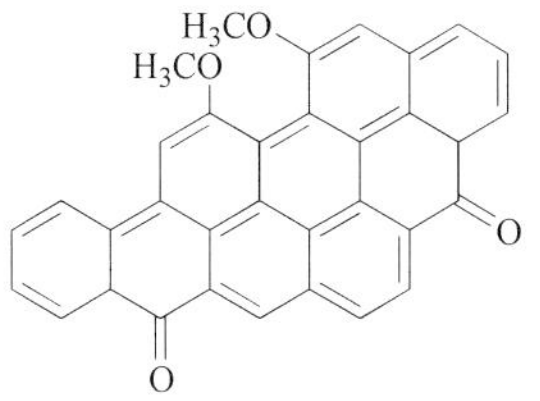

Molecular formula:	$C_{33}H_{20}O_4$
Chemical Abstracts name:	16,17-dimethoxydinaphthol[1,2,3-cd:3′,2′,1′-lm]perylene-5,10-dione
Trivial name:	jade green
C.I. name:	C.I. Vat Green 1
C.I. number:	C.I. 59825
Application class:	vat
Chemical class:	anthraquinone
CAS registry number:	[*128-58-5*]
Commercial names:	Solanthrene Green XBN, AVECIA
	Cibanone Brilliant Green, BF, 2BF, BFD,
	CIBA-GEIGY Indanthrene Brilliant Green, B, FB

It is important to recognise that although a dye has a C.I. number, the purity and precise chemical constitution will vary depending upon the name. Thus, the dye C.I. Direct Blue 99 from company A will not be identical to C.I. Direct Blue 99 from Company B.

1.5 Equipment and Manufacture

The basic steps of dye (and intermediate) manufacture are shown in Figure 1.1. There are usually several reaction steps or unit processes.

The reactor itself, in which the unit processes to produce the intermediates and dyes are carried out, is usually the focal point of the plant, but this does not mean that it is the most important part of the total manufacture, or that it absorbs most of the capital or operational costs. Operations subsequent to reaction are often referred to as workup stages. These vary from product to product, whereby intermediates (used without drying wherever practical) require less finishing operations than colorants.

The reactions for the production of intermediates and dyes are carried out in bomb-shaped reaction vessels made from cast iron, stainless steel, or steel lined with rubber, glass (enamel), brick, or carbon blocks. These vessels have capacities of 2–40 m^3 (ca 500–10 000 gallons) and are equipped with mechanical agitators, thermometers, or temperature recorders, condensers, pH probes, etc., depending

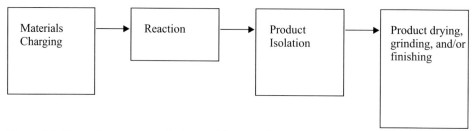

Figure 1.1. Operation sequence in dye and intermediate manufacture.

on the nature of the operation. Jackets or coils are used for heating and cooling by circulation of high-boiling fluids (e.g. hot oil or Dowtherm), steam, or hot water to raise the temperature, and air, cold water, or chilled brine to lower it. Unjacketed vessels are often used for reactions in aqueous solutions in which heating is effected by direct introduction of steam, and cooling by addition of ice or by heat exchangers. The reaction vessels normally span two or more floors in a plant to facilitate ease of operation.

Products are transferred from one piece of equipment to another by gravity flow, pumping, or blowing with air or inert gas. Solid products are separated from liquids in centrifuges, in filter boxes, on continuous belt filters, and, perhaps most frequently, in various designs of plate-and-frame or recessed-plate filter presses. The presses are dressed with cloths of cotton, Dynel, polypropylene, etc. Some provide separate channels for efficient washing, others have membranes for increasing the solids content of the presscake by pneumatic or hydraulic squeezing. The plates and frames are made of wood, cast iron, but more usually hard rubber, polyethylene, or polyester.

When possible, the intermediates are used for the subsequent manufacture of other intermediates or dyes without drying because of saving in energy costs and handling losses. This trend is also apparent with dyes. The use of membrane technology with techniques such as reverse osmosis to provide pure, usually aqueous, solutions of dyes has become much more prevalent. Dyes produced and sold as liquids are safer and easier to handle and save energy costs.

However, in some cases products, usually in the form of pastes discharged from a filter, must be dried. Even with optimization of physical form, the water content of pastes varies from product to product in the range of 20–80 %. Where drying is required, air or vacuum ovens (in which the product is spread on trays), rotary dryers, spray dryers, or less frequently drum dryers (flakers) are used. Spray dryers have become increasingly important. They require little labor and accomplish rapid drying by blowing concentrated slurries (e.g., reaction masses) through a small orifice into a large volume of hot air. Dyes, especially disperse dyes, that require wet grinding as the penultimate step are now often dried this way. In this case their final standardization, i.e., addition of desired amounts of auxiliary agents and solid diluents, is performed in the same operation.

The final stage in dye manufacture is grinding or milling. Dry grinding is usually carried out in impact mills (Atritor, KEK, or ST); considerable amounts of dust are generated and well-established methods are available to control this problem. Dry grinding is an inevitable consequence of oven drying, but more modern methods of drying, especially continuous drying, allow the production of materials that do not require a final comminution stage. The ball mill has been superseded by sand or bead mills. Wet milling has become increasingly important for pigments and disperse dyes. Many patented designs, particularly from Draiswerke GmbH and Gebrüder Netzsch, consist of vertical or horizontal cylinders equipped with high-speed agitators of various configurations with appropriate continuous feed and discharge arrangements. The advantages and disadvantages of vertical and horizontal types have been discussed [7]; these are so finely balanced as to lead to consideration of tilting versions to combine the advantages of both.

In the past the successful operation of batch processes depended mainly on the skill and accumulated experience of the operator. This operating experience was difficult to codify in a form that enabled full use to be made of it in developing new designs. The gradual evolution of better instrumentation, followed by the installation of sequence control systems, has enabled much more process data to be recorded, permitting maintenance of process variations within the minimum possible limits.

Full computerization of multiproduct batch plants is much more difficult than with single-product continuous units because the control parameters very fundamentally with respect to time. The first computerized azo [8] and intermediates [9] plants were bought on stream by ICI Organic Division (now Avecia) in the early 1970s, and have now been followed by many others. The additional cost (ca. 10 %) of computerization has been estimated to give a saving of 30–45 % in labor costs [10]. However, highly trained process operators and instrument engineers are required. Figure 1.2 shows the layout of a typical azo dye manufacturing plant.

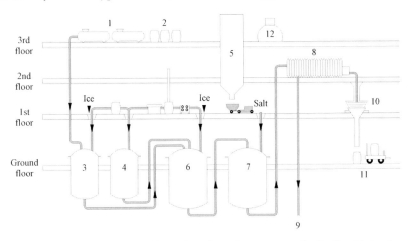

Figure 1.2 Layout of azo dye manufacturing plant. 1, storage tanks for liquid starting materials; 2, storage drums for solid starting materials; 3, diazotisation vessel; 4, coupling component vessel; 5, ice machine; 6, coupling vessel; 7, isolation vessel; 8, filter presses; 9, filtrate to waste liquor treatment plant; 10, dryers; 11, emptying of dyestuffs for feeding to the mill; 12, outgoing air purification plant.

1.6 Economic Aspects

In the early 1900s, about 85 % of world dye requirements were manufactured in Germany, with other European countries (Switzerland, UK, and France) accounting for a further 10 %. Eighty years and two world wars have seen dramatic changes in this pattern. Table 1.2 shows that, in weight terms, the Western European share of world production had dropped to 50 % by 1938 and to 40 % by 1974. It has declined even further with considerable production of commodity colorants moving to lower cost countries such as India, Taiwan, and China. However, since a large part of U.S. manufacture and some of others countries is based on Western European subsidiaries, their overall share remains at ca. 50 %. Since 1974 many of the national figures have not been published, but those that are available indicate that 1974 was the peak production year of the 1970s, with 1975 being the nadir. World recession caused a 20 % slump in production which has now been more than recovered.

Consequently, the figures of Table 1.2 are still applicable to the present day and show a decrease in Western European production and an increase in Eurasian production. More recent independent reviews [11, 12] indicate a slightly lower world output of 750 000 to 800 000 t/a. With the present state of the world economy in 2000 the growth rate for the traditional dye industries is likely to be around 3–4 %, which represents something like an additional 30 000 t/a. However, the growing high-tech industries such as ink-jet printing are now consuming significant amounts of dyes. Whilst the volumes are lower than for traditional dyes, the specialized nature of these new dyes mean they command much higher prices.

Table 1.2 World dye production trends, 10^3 t

Year	W. Europe	USA	Eurasia[a]	Japan	Others	Total
1938	110	37	35		28	210
1948		110				
1958	112	80	127		27	346
1966	191	130		49		
1974	300	138	200+	68	44	750+

[a] Eastern Europe, former USSR, and China.

For many years the major European chemical companies remained largely unchanged. Thus, chemical production, including dyes and drugs, were carried out in Germany by Bayer, BASF, and Hoechst; in Switzerland by Ciba-Geigy and Sandoz; and in the UK by ICI. These three countries represented the focus of the world's dyes (and pigments) production. The last decades have seen massive changes to the chemical industry, including the dyes industry. Companies have undergone major reorganizations, divestments, and acquisitions to focus on "core" activities, particularly life sciences and pharmaceuticals. Consequently, most of the dyes businesses have been divested, with mergers in certain cases. Table 1.3 shows the current situation in Western Europe.

Table 1.4, which shows the global manufacturers of colorants, highlights the growth of indigenous colorant producers in India and particularly the Far East.

Table 1.3 Major Western European dye procedures

Country	Original company	Current company
Germany	Bayer, Hoechst, BASF	→ Dystar
	(textile dyes) ↖	
UK	ICI → Zeneca →	Avecia
Switzerland	Sandoz (+ Hoechst Specialty Chemicals) →	Clariant
	Ciba Geigy →	Ciba Specialty Chemicals

Table 1.4 Global manufacturers of dyes.

Traditional	Indigenous	
Clariant	Everlight	Taiwan
Ciba Specialty Chemicals	Lucky Gold Star	Korea
Dystar	Daekwang	Korea
Avecia	Atul	India
	Crompton & Knowles	USA
	Milliken	USA
	Mitsubishi	Japan
	Sumitomo	Japan
	Nippon Kayaku	Japan
	Hodogaya	Japan
	Orient	Japan

1.7 References

General References

"Dyes and Dye Intermediates" in *Kirk-Othmer Encyclopedia of Chemical Technology*, Wiley, New York; in 1st ed., Vol. 5, pp. 327–354, by G. E. Goheen and J. Werner, General Aniline & Film Corp., and A. Merz, American Cyanamid Co.; in 2nd ed., Vol. 7, pp. 462–505, by D. W. Bannister and A. D. Olin, Toms River Chemical Corp.; in 3rd ed., Vol. 8, pp. 152–212, by D. W. Bannister, A. D. Olin, and H. A. Stingl, Toms River Chemical Corp.

Specific References

[1] P. F. Gordon, P. Gregory, *Organic Chemistry in Colour*, Springer-Verlag, Berlin, 1983.
[2] A. Calder, *Dyes in Non-Impact Printing*, IS and T's Seventh International Congress on Advances in Non-Impact Printing Technologies, Portland, Oregon, Oct., 1991.
[3] G. Booth, *The Manufacture of Organic Colorants and Intermediates*, Society of Dyers and Colourists, Bradford, UK, 1988.
[4] P. Gregory in *The Chemistry and Application of Dyes* (Eds.: D. R. Waring, G. Hallas), Plenum, New York, 1990, pp. 17–47.
[5] *Colour Index*, Vol. 4, 3rd ed., The Society of Dyers and Colourists, Bradford, UK, 1971.
[6] P. Gregory, *High Technology Applications of Organic Colorants*, Plenum, New York, 1991.
[7] K. Engels, *Farbe Lack* **85** (1979) 267.
[8] H. Pinkerton, P. R. Robinson, Institute of Electrical Engineers Conference Publication, No. 104, Oct. 1973.
[9] P. R. Robinson, J. M. Trappe, *Melliand Textilber.* **1975**, 557.
[10] S. Baruffaldi, *Chim. Ind.* **60** (1978) 213.
[11] O'Sullivan, *Chem. Eng. News* Feb. 26, 1979, 16.
[12] K.-H. Schündehütte, *DEFAZET Dtsch. Farben. Z.* **3** (1979) 202.

2 Important Chemical Chromophores of Dye Classes

Introduction

A principle focus of this book is the classification of dyes by chemical structure. This is certainly not the only possible classification scheme for dyes: ordering by application properties, e.g., naming according the substrate to be dyed is another alternative. Neither of these two categories can be used with the exclusion of the other one, and overlap is often inevitable. Nevertheless, for this book it was decided to make the chemical structure of dyes the main sorting system.

This chapter is devoted to the chemical chromophores of dyes, but the term "chromophore" is used here in a somewhat extended manner that also considers dye classes such those as based on cationic, di- and triarylcarbonium, and sulfur compounds, and metal complexes.

The two overriding trends in traditional colorants research for many years have been improved cost-effectiveness and increased technical excellence. Improved cost-effectiveness usually means replacing tinctorially weak dyes such as anthraquinones, until recently the second largest class after the azo dyes, with tinctorially stronger dyes such as heterocyclic azo dyes, triphendioxazines, and benzodifuranones. This theme will be pursued throughout this chapter, in which dyes are discussed by chemical structure.

During the last decade, the phenomenal rise in high-tech industries has fuelled the need for novel high-tech (functional) dyes having special properties. These hi-tech applications can bear higher costs than traditional dye applications, and this has facilitated the evaluation and use of more esoteric dyes (see Chapter 6).

2.1 Azo Chromophore

2.1.1 Introduction

The azo dyes are by far the most important class, accounting for over 50 % of all commercial dyes, and having been studied more than any other class. Azo dyes contain at least one azo group ($-N=N-$) but can contain two (disazo), three (trisazo), or, more rarely, four (tetrakisazo) or more (polyazo) azo groups. The azo group is attached to two groups, of which at least one, but more usually both, are aromatic. They exist in the *trans* form **1** in which the bond angle is ca. 120°, the nitrogen atoms are sp^2 hybridized, and the designation of A and E groups is consistent with C.I. usage [1].

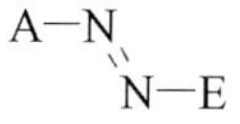

$$A-N=N-E$$

1

In monoazo dyes, the most important type, the A group often contains electron-accepting substituents, and the E group contains electron-donating substituents, particularly hydroxyl and amino groups. If the dyes contain only aromatic groups, such as benzene and naphthalene, they are known as carbocyclic azo dyes. If they contain one or more heterocyclic groups, the dyes are known as heterocyclic azo dyes. Examples of various azo dyes are shown in Figure 2.1. These illustrate the enormous structural variety possible in azo dyes, particularly with polyazo dyes.

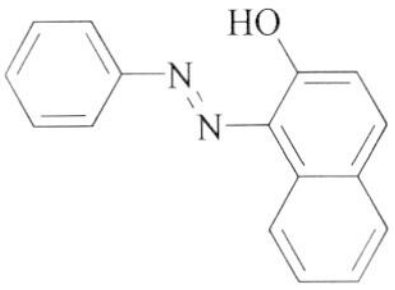

2

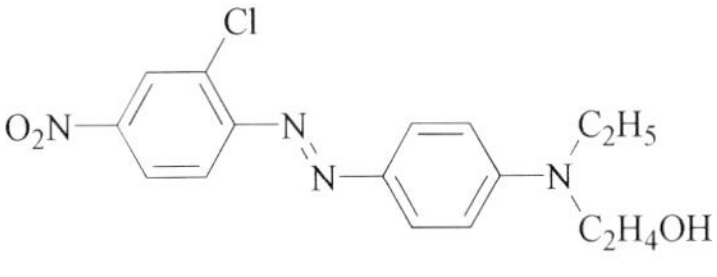

3

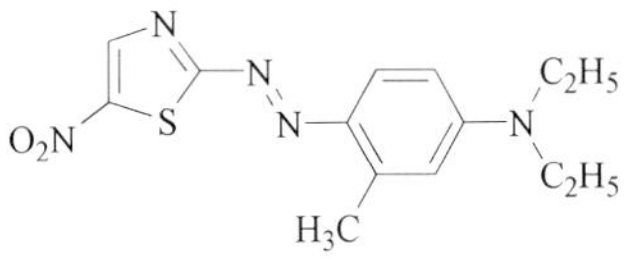

4

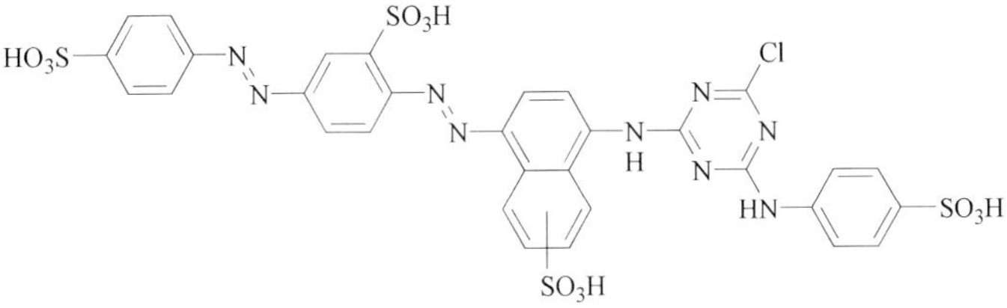

5

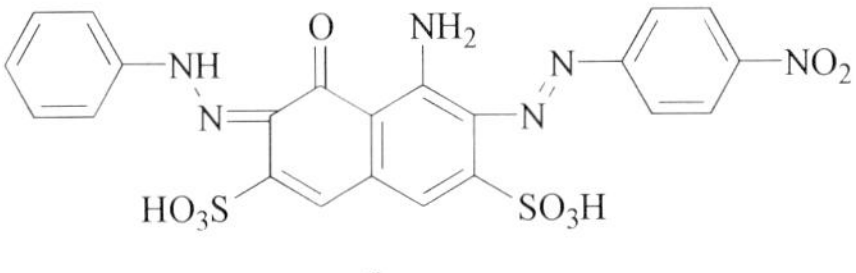

6

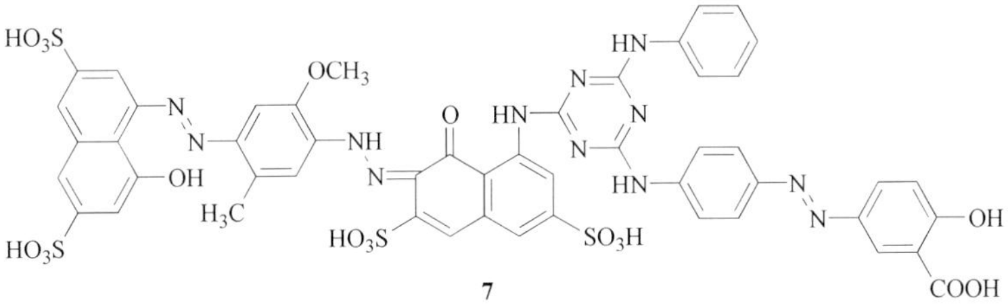

7

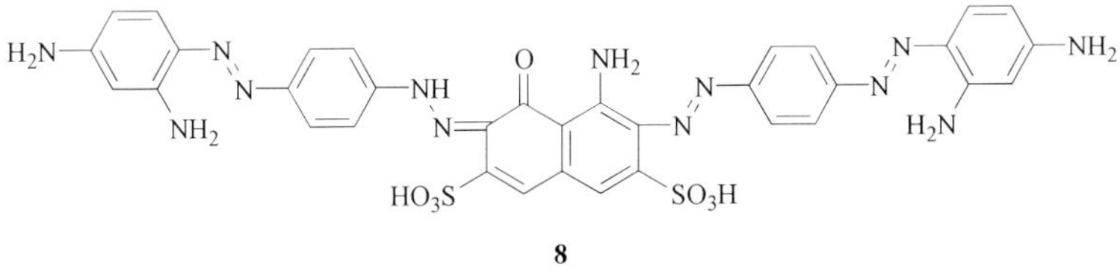

8

Figure 2.1 Azo dyes. *C.I. Solvent Yellow 14* (**2**), *C.I. Disperse Red 13* (**3**), *C.I. Disperse Blue* (**4**), *C.I. Reactive Brown 1* (**5**), *C.I. Acid Black 1* (**6**), *C.I. Direct Green 26* (**7**), *C.I. Direct Black 19* (**8**)

2.1.2 General Synthesis

Almost without exception, azo dyes are made by diazotization of a primary aromatic amine followed by coupling of the resultant diazonium salt with an electron-rich nucleophile. The diazotization reaction is carried out by treating the primary aromatic amine with nitrous acid, normally generated in situ with hydrochloric acid and sodium nitrite. The nitrous acid nitrosates the amine to generate the *N*-nitroso compound, which tautomerises to the diazo hydroxide [Eq. (1)].

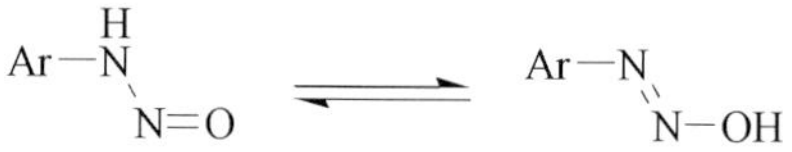

$$Ar-N(H)-N=O \rightleftharpoons Ar-N=N-OH \quad (1)$$

Protonation of the hydroxy group followed by the elimination of water generates the resonance-stabilised diazonium salt [Eq. (2)].

$$Ar-N=N^+ \longleftrightarrow Ar-N\equiv N^+ \quad (2)$$

For weakly basic amines, i.e., those containing several electron-withdrawing groups, nitrosyl sulfuric acid ($NO^+HSO_4^-$) is used as the nitrosating agent in sulfuric acid, which may be mixed with phosphoric, acetic, or propionic acid.

A diazonium salt is a weak electrophile and hence reacts only with highly electron rich species such as amino and hydroxy compounds. Even hydroxy compounds must be ionized for reaction to occur. Consequently, hydroxy compounds such as phenols and naphthols are coupled in an alkaline medium (pH≥pK_a of phenol or naphthol; typically pH 7–11), whereas aromatic amines such as *N,N*-dialkylamines are coupled in a slightly acid medium, typically pH 1–5. This provides optimum stability for the diazonium salt (stable in acid) without deactivating the nucleophile (protonation of the amine).

2.1.2.1 Diazo Components

Diazo components for the production of azo dyes can be divided into the following groups of aromatic amines:

Aniline and Substituted Anilines. Methyl, chloro, nitro, methoxyl, ethoxyl, phenoxyl, hydroxyl, carboxyl, carbalkoxyl, carbonamide, and sulfonic acid groups are primarily used as substituents.

Examples: Toluidines, nitroanilines, aminobenzoyl amides, and the sulfonic acids thereof.

Naphthylamines and Naphthylaminesulfonic Acids. The compounds shown in Figure 2.2, known by their common names, are examples of naphthylaminesulfonic acids.

2.1.2 General Synthesis

Many of these compounds can also be used as coupling components (see Section 2.2.2), as shown by the arrow indicating the coupling position.

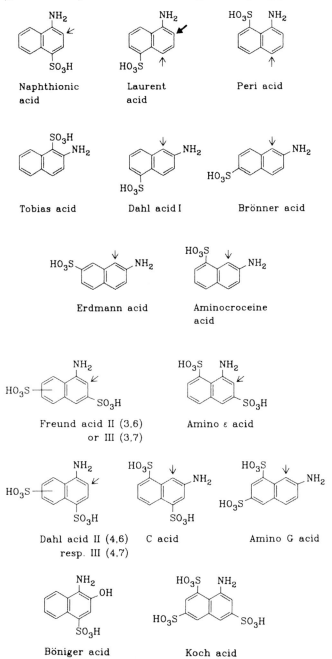

Figure 2.2

2.1 Azo Chromophore

Diamines H_2N-A-NH_2. The diamines can be divided into different groups:

a) A = phenylene or naphthylene group, also substituted, e.g., by methyl, chloro, nitro, methoxy, or sulfonic acid groups (e.g., **9–12**)

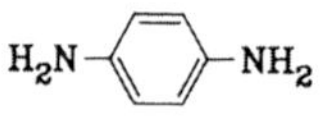

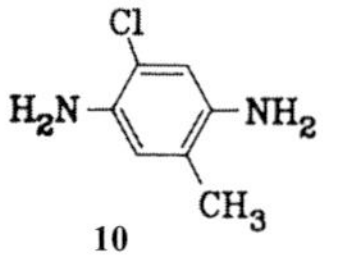

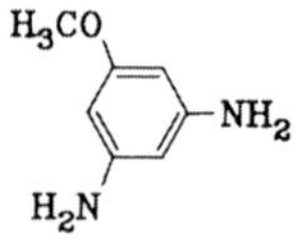

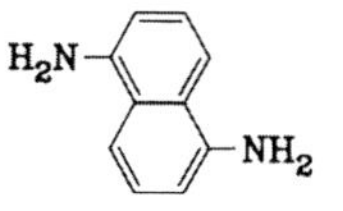

b) A = diphenyl group, substituted mainly by chlorine, carboxylic acid, or sulfonic acid groups (e.g., **13–16**).

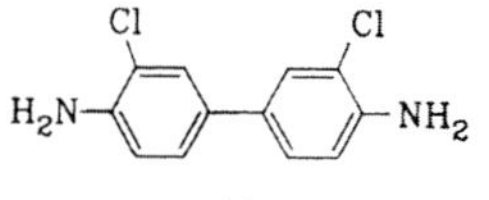

13 3,3'-Dichlorobenzidine

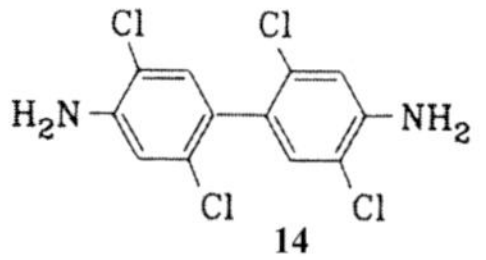

14 2,2',5,5"-Tetrachlorobenzidine

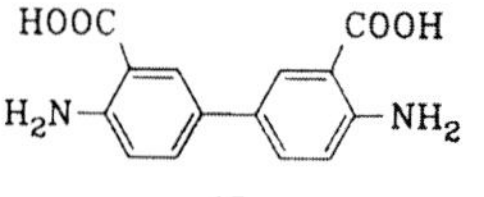

15 Benzidine-3,3'-dicarboxylic acid

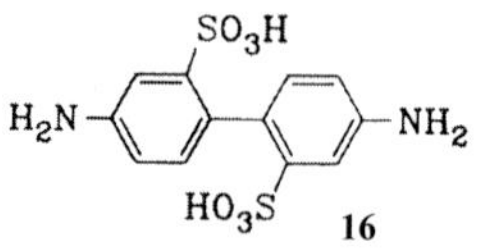

16 Benzidine-2,2'-disulfonic acid

c)

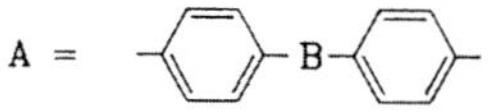

$$A = -\phi-B-\phi-$$

whereby the aromatic rings can also be substituted by methyl, chlorine, methoxy, carboxylic acid, or sulfonic acid groups; B = oxygen, sulfur, $-NH-$, $-SO_2-$, $-N=N-$, $-CH=CH-$, $-NHCO-$, $-NH-CO-HN-$ (e.g., **17–19**).

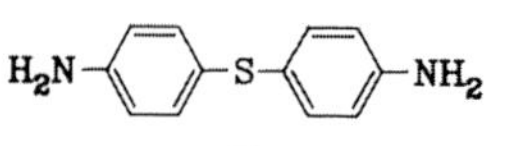

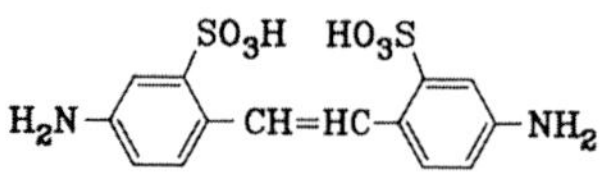

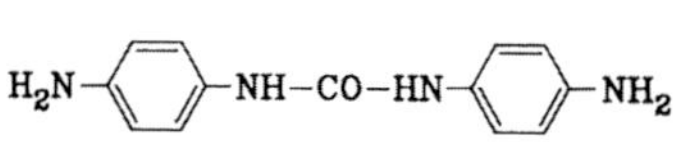

d) Heterocyclic amines and diamines (e.g., **20–25**)

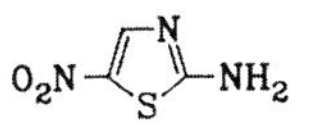

20, 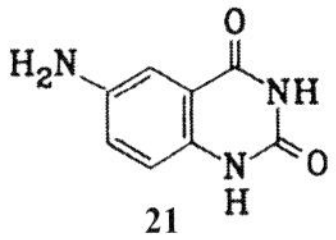 **21**, 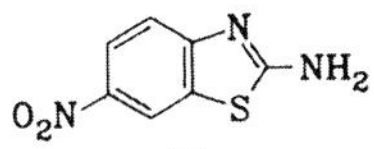 **22**, 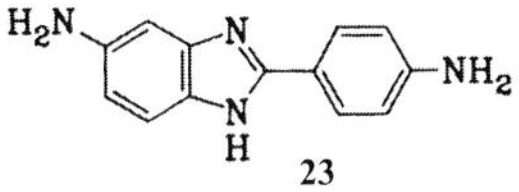 **23**, 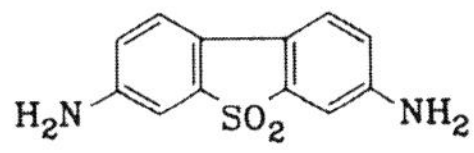 **24**, 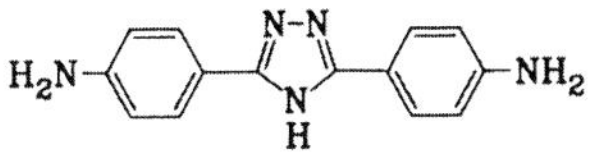 **25**

2.1.2.2 Diazotization Methods

Depending on the basicity and solubility of the amines being diazotized, the following diazotization methods find industrial use:

a) *Direct Diazotization.* The primary aromatic amine is dissolved or suspended in aqueous hydrochloric or sulfuric acid, and a concentrated aqueous sodium nitrite solution is added. An excess of 2.5–3 equivalents of acid per equivalent of amine is used. A temperature of 0–5 °C is maintained by adding ice.

b) *Indirect Diazotization.* Amines with sulfonic or carboxylic acid groups are often difficult to dissolve in dilute acid. Therefore, the amine is dissolved in water or a weak alkali, and the calculated amount of sodium nitrite solution is added to this amine solution which is stirred into the ice-cooled acid solution already in the vessel. The acid can also be added to the amine–nitrite mixture already at hand.

c) *Diazotization of Weakly Basic Amines.* Weakly basic amines are dissolved in concentrated sulfuric acid and diazotized with nitrosylsulfuric acid, which is easily prepared from solid sodium nitrite and concentrated sulfuric acid.

d) *Diazotization in Organic Solvents.* The water-insoluble or sparingly soluble amine is dissolved in glacial acetic acid or other organic solvents and, where necessary, diluted with water. After the addition of acid it is diazotized in the usual manner with sodium nitrite solution. Nitrosylsulfuric acid, nitrosyl chloride, alkyl nitrites, or nitrous gases can also be used instead of sodium nitrite. Temperature, pH, and the concentration of the diazotizing solution often have a considerable effect on the progress of diazotization. Physical properties (distribution, particle size) and the addition of emulsifiers and dispersing agents influence the diazotization of slightly soluble amines.

2.1 Azo Chromophore

Certain aromatic amines require special diazotizing processes:

1-Aminonaphthols, such as 1-amino-2-naphthol and several *1-aminonaphtholsulfonic acids*, such as 1-amino-2-naphthol-4-sulfonic acid, are oxidized to the respective quinone by nitrous acid. Diazotization can, however, take place under normal conditions in the presence of catalytic quantities of metal salts, such as copper or zinc salts.

o-Diamines, such as 1,2-phenylenediamine and 1,8-naphthylenediamine, undergo cyclization during the normal diazotization process to form triazoles:

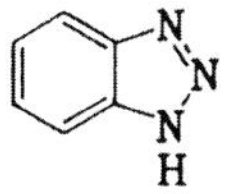

Benzotriazole 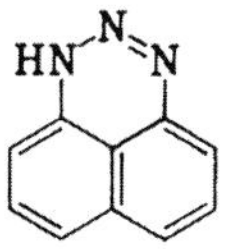   Naphtho[1,8]triazole
                    (Periazimine)

The desired bis-diazotization with *o-*, *m-*, and *p-*phenylenediamine is achieved in glacial acetic acid with nitrosylsulfuric acid. 1,8-Naphthylenediamine can be bis-diazotized in an excess of acid.

Diazonium compounds are generally stable only in aqueous solution at low temperatures. When heated, they frequently decompose by eliminating nitrogen to form the corresponding phenol. Some amines, however, can be diazotized at temperatures up to 40 °C. Metal ions also accelerate the decomposition of diazonium compounds. Therefore, diazotization is usually carried out in wooden vats or iron stirring vessels with an acid-proof lining or rubber coating.

2.1.2.3 Coupling Components

The most important coupling components can be divided into the following groups. Where there are several possibilities for coupling, the preferred coupling positions are marked by bold arrows and the other possible coupling positions by ordinary arrows.

Anilines, diaminobenzenes (e.g., **26–32**).

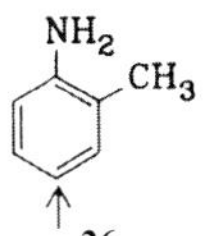

26: 2-methylaniline (o-toluidine)
27: 3-methylaniline (m-toluidine)
28: 2,6-dimethylaniline
29: 4-methyl-2-methylaniline (cresidine-type)
30: 1,2-diaminobenzene
31: 1,2-diamino-4-methylbenzene
32: 1,2-diamino-4,5-disulfonic acid benzene

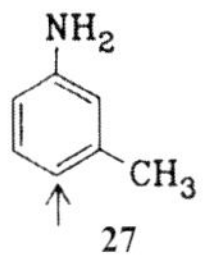

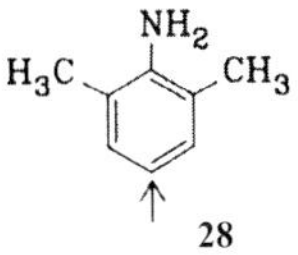

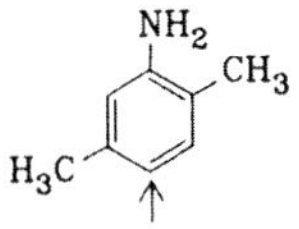

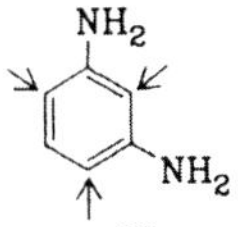

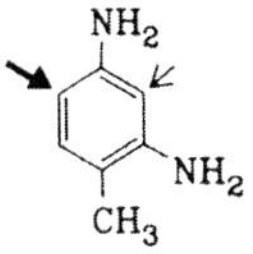

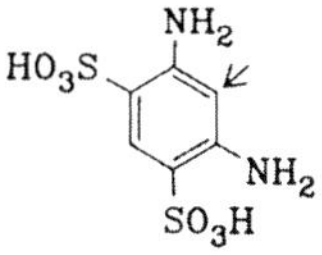

When treated with aniline, diazotized aniline (benzenediazonium chloride) yields, apart from a small quantity of 4-aminoazobenzene (**33**), diazoaminobenzene (**34**) as the main product:

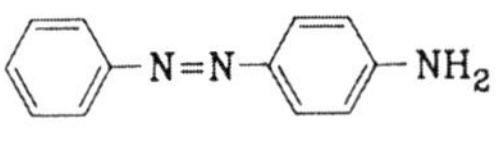

33: C$_6$H$_5$–N=N–C$_6$H$_4$–NH$_2$
34: C$_6$H$_5$–N=N–NH–C$_6$H$_5$

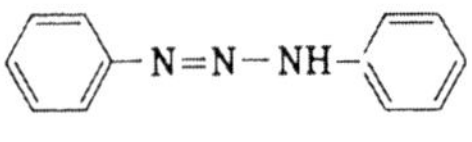

4-Aminoazobenzene (**33**) is obtained by coupling in a more strongly acid medium, but better still by heating **34** in aniline with addition of aniline hydrochloride.

Electron-donating substituents in the aniline, such as methyl or methoxy groups, especially at the *meta* position, promote coupling; the tendency to couple thus increases in the order aniline<*o*-toluidine<*m*-toluidine<*m*-anisidine<cresidine<1-amino-2,5-dimethoxybenzene (aminohydroquinone dimethyl ether) to the extent that without formation of the diazoamino compound, the last three bases are attached almost quantitatively in the desired position, i.e., at the 4-position relative to the amino group.

m-Phenylenediamines couple to form monoazo dyes (chrysoidines) or disazo dyes.

Naphthylamines, Naphthylaminesulfonic Acids. Some examples (**35–38**) are mentioned here; other compounds can be found among the diazo components (see Figure 2.2), because most naphthylaminesulfonic acids can be employed both for diazotization and for coupling purposes.

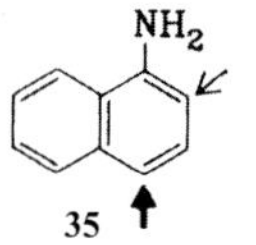

2.1 Azo Chromophore

35

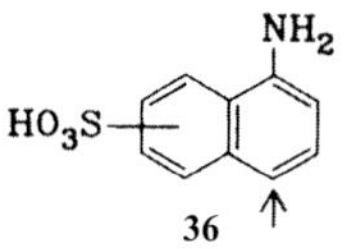

36 Cleve's acid 6 or 7

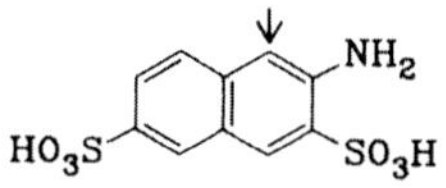

37 Amino R acid

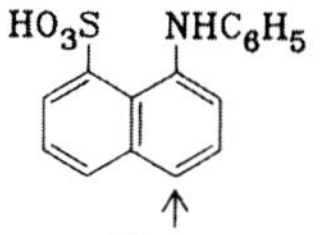

38 Phenyl peri acid

Whereas β-naphthylaminesulfonic acids always couple at the adjacent α position, in α-naphthylaminesulfonic acids, the coupling location is influenced by the position of the sulfonic acid group: 1,6-, 1,7-, and 1,8-naphthylaminesulfonic acids couple at the 4-position; 1,5-naphthylaminesulfonic acid (Laurent acid) only couples with very strong couplers (e.g., 2,4-dinitroaniline), mainly at the 4-position. With diazotized aniline, chloroaniline, or diazotized anilinesulfonic acids the 2-aminoazo dyes are obtained, but with diazotized nitroaniline, mixtures of the 2- and 4-coupling products form (Scheme 2.1).

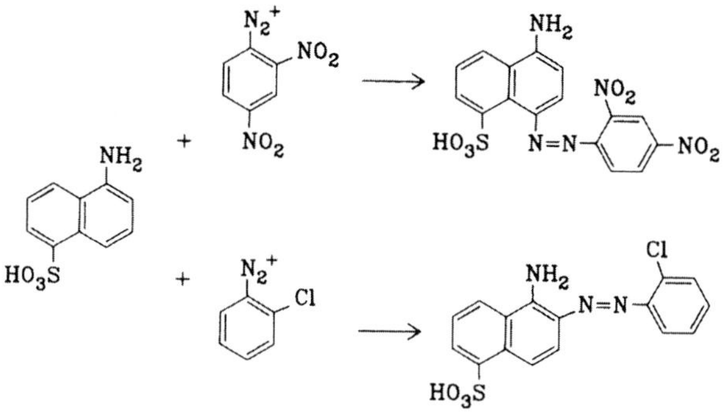

Scheme 2.1

Phenols, Naphthols (e.g., **39–47**). Phenols mainly couple at the 4-position, or at the 2-position if the 4-position is occupied. *p*-Hydroxybenzoic acid couples with elimination of CO_2; resorcinol couples twice: initially at the 4-position, and with a second equivalent of diazonium compound at the 2-position under acid conditions or at the 6-position under alkaline conditions. α-Naphthols mainly couple at the 4-position, in addition to which varying quantities of 2- and 2,4-coupling products are obtained, depending on the diazo component. β-Naphthol couples at the 1-position. Substituents in the 1-position, such as SO_3H, COOH, Cl, CH_2OH, or

CH$_2$NR$_2$, may be eliminated during the coupling process. 1-Methyl-2-naphthol does not form an azo dye.

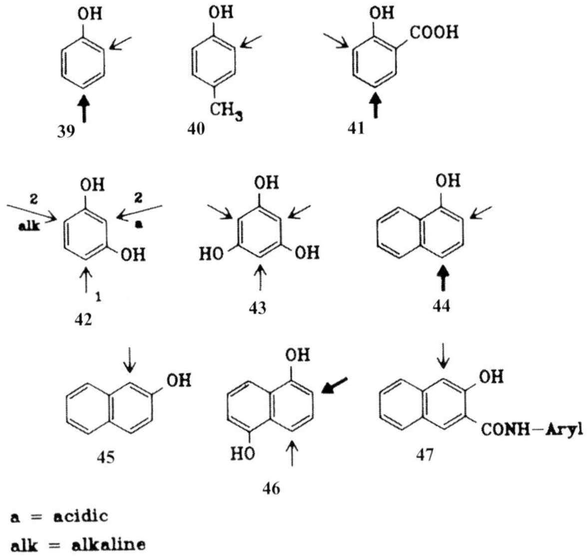

a = acidic
alk = alkaline

Coupling components in this series that are of particular industrial importance are 2-hydroxynaphthalene-3-carboxylic acid arylamides, in particular anilides, which also couple at the 1-position.

Phenols and naphthols generally couple more easily and rapidly than amines; phenol-3-sulfonic acid can be coupled, but not aniline-3-sulfonic acid.

The readiness of the phenols to couple increases with increasing number of hydroxyl groups, for example, in the following series:

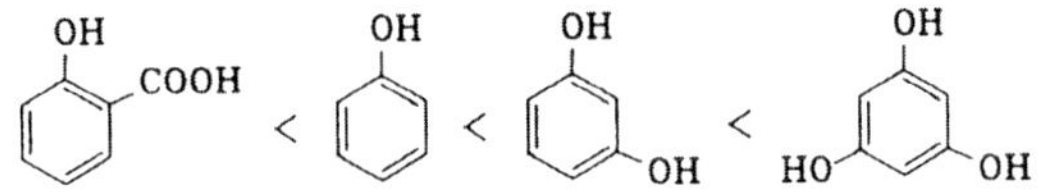

2.1 Azo Chromophore

Naphtholsulfonic Acids (Figure 2.3).

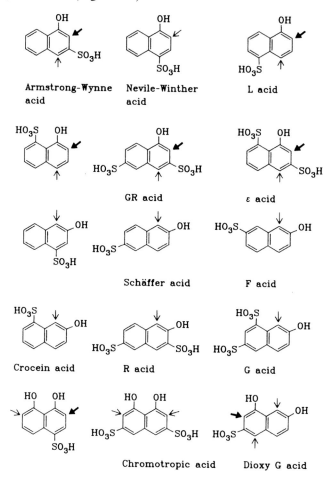

Figure 2.3

1-Naphtholsulfonic acids mainly couple at the 2-position. The 4-coupling products obtained as byproducts must be carefully removed from the azo dyes, because unlike the 2-substitution products, their shade changes as a function of the pH value (shade intensification with rising pH due to formation of phenolate or naphtholate resonance structures).

Aminophenols, Aminophenolsulfonic Acids, Aminonaphtholsulfonic Acids (e.g., **48–54**).

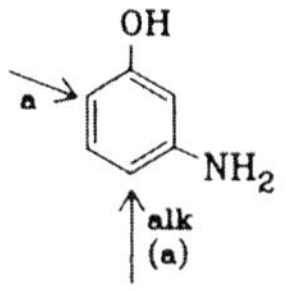

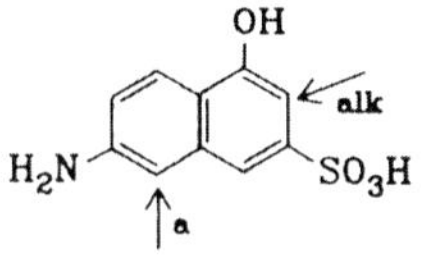

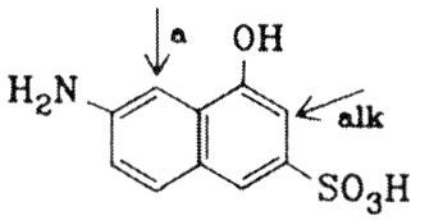

[Structures 48–54: 48 (m-aminophenol); 49 I acid; 50 γ acid; 51 Chicago acid SS; 52 M acid; 53 2 R acid; 54 2 Alkoxy Cleve acid (R = CH_3, C_2H_5)]

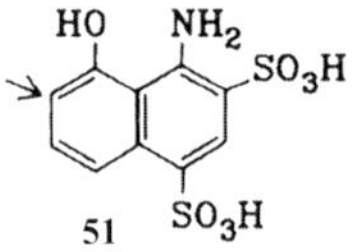

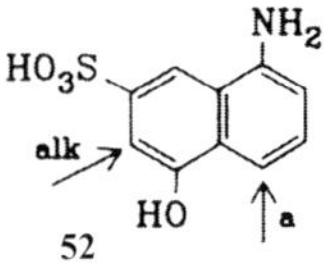

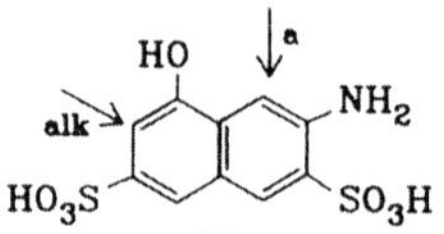

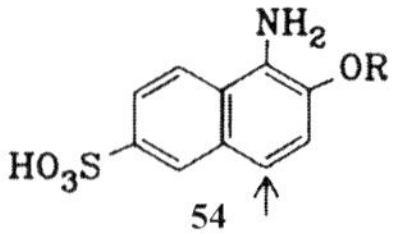

a = acid
alk = alkaline

In alkaline medium *m*-aminophenol couples at the position *para* to the hydroxyl group; in an acid solution *p*-hydroxy and *p*-aminoazo dyes are obtained.

In aminonaphtholsulfonic acids such as H, K, and S acids, orientation is strongly influenced by the pH of the coupling medium. In an acid medium the azo group enters the position *ortho* to the amino group, and in an alkaline medium, in the position *ortho* to the hydroxyl group. 1-Amino-8-naphtholsulfonic acids with free *ortho* positions can couple with two equivalents of diazonium compound, coupling initially taking place in the acid medium at the position *ortho* to amino group, followed by coupling in the alkaline range at the position *ortho* to hydroxy group. Aminonaphtholsulfonic acids (*o*-hydroxyazo dyes) initially coupled in an alkaline medium cannot normally undergo further coupling to form disazo dyes. However, if this initial coupling is followed by a metal-complexation step, further coupling is possible in the neutral or alkaline range.

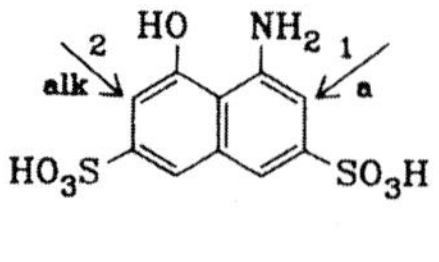

2.1 Azo Chromophore

H acid

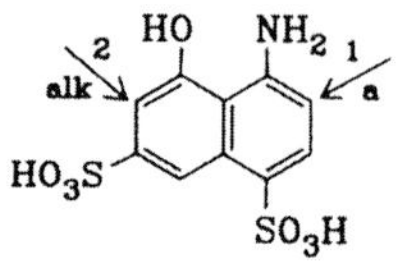

K acid

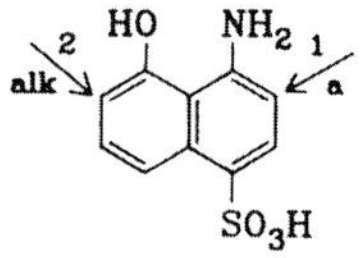

S acid
(Chicago acid S)

Compounds with Reactive Methylene Groups. In this group the coupling components with the greatest industrial importance are the *N*-acetoacetyl derivatives of aromatic amines (acetoacetarylides) $CH_3COCH_2CONHAr$. Anilines substituted by halogen, alkyl, alkoxy, nitro, and acylamino groups are most suitable (e.g., Napthol AS-IRG).

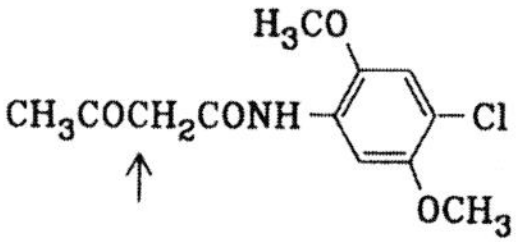

Naphtol AS–IRG

Heterocyclic Components. The most important compounds in this range are the 5-pyrazolones substituted at the 3-position.

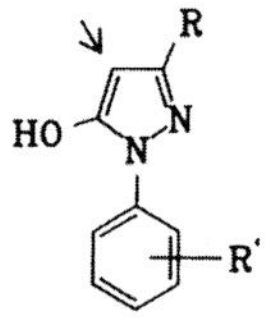

R = preferably CH_3, COOH, COOalkyl
R' = H, CH_3, Cl, SO_3H

Further examples are **(55)–(61)**.

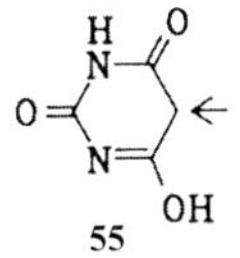

2.1.2 General Synthesis

55 Barbituric acid

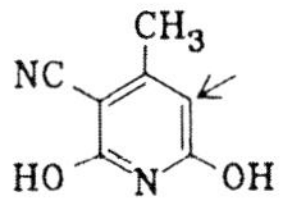

56 Pyridones

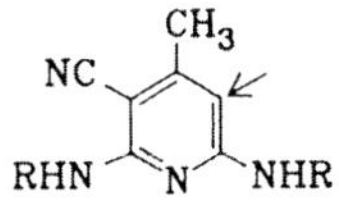

57 R = CH₃, CH₂CH₂OH

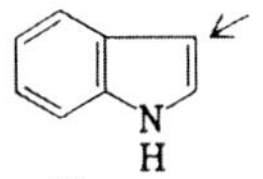

58

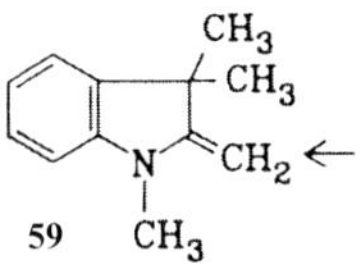

59 "Fischer Base"

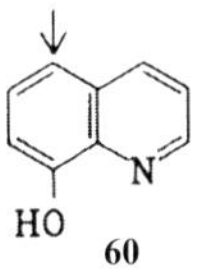

60 8-Hydroxyquinoline

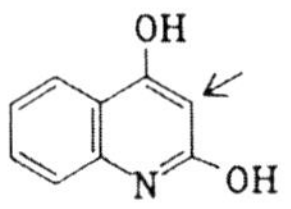

61 2,4-Dihydroxyquinoline

Coupling components containing both amino and hydroxyl groups, such as H acid (1-amino-8-naphthol-3,6-disulfonic acid) can be coupled stepwise (Scheme 2.2). Coupling is first carried out under acid conditions to effect azo formation in the amino-substituted ring. The pH is then raised to ionise the hydroxyl group (usually to pH≥7) to effect coupling in the naphtholate ring, with either the same or a different diazonium salt. Performing this process in the reverse order fails because the nucleophilicity of the amino group is insufficient to facilitate the second coupling step.

The unusual conditions needed to produce an azo dye, namely, strong acid plus nitrous acid for diazotization, the low temperatures necessary for the unstable diazonium salt to exist, and the presence of electron-rich amino or hydroxy compounds to effect coupling, means that azo dyes have no natural counterparts.

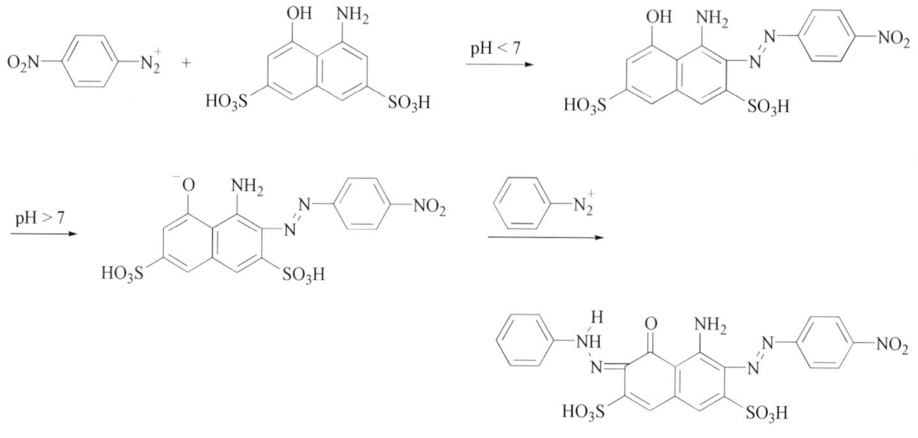

Scheme 2.2

2.1.2.4 Azo Coupling in Practice

The optimum coupling conditions depend on the nature of the diazo and coupling components used. The acidic diazonium salt solution is generally added to the solution of the coupling component. Because, in addition to acid derived from the diazotization process, additional acid is released during coupling, the optimum pH value must be adhered to by adding bases. Alkali metal hydroxides, sodium carbonate, sodium hydrogencarbonate, ammonia, calcium carbonate, magnesium oxide, etc., are used for this purpose. Sodium acetate or formate (weakly acid) or sodium phosphate (weakly alkaline) can also be added to buffer the acid. These are added to the reaction mixture before, during, or after combining the components. Because the pH value at the feed point of the diazonium salt solution is always different than that after thorough mixing has been completed, the type of stirrer and speed of stirring are also important in many instances. The sequence in which the two components are combined can also greatly influence the result.

The coupling reaction may be completed immediately after the components are mixed or after several hours. If the reaction requires a longer time, it is advisable to cool with ice and avoid exposure to bright light. To check whether excess diazonium compound is still present, a drop of reaction solution is spotted onto filter paper together with a component that couples readily (e.g., weakly alkaline H acid solution). If no coloration appears, coupling is complete. The presence of unconsumed coupling component can be determined by spotting with a solution of diazonium salt.

Attention must also be paid to the volume of coupling solution or suspension. With starting components of low solubility, the physical state is an important factor. To achieve complete reaction of diazo or coupling components with low solubility, it is often necessary to ensure that the reactants are distributed as finely as possible. This is carried out, for example, by adding dispersing agents to the diazo component or to the coupling mixture or by adding emulsifiers during acid precipitation of the coupling component prior to azo coupling. In each case, the most favorable reaction conditions must be precisely established and, because of the various influences that result in undesirable side effects, they must be carefully adhered to during manufacture.

2.1.3 Principal Properties

2.1.3.1 Tautomerism

In theory, azo dyes can undergo tautomerism: azo/hydrazone for hydroxyazo dyes; azo/imino for aminoazo dyes, and azonium/ammonium for protonated azo dyes. A more detailed account of azo dye trautomerism can be found elsewhere [2].

Azo/hydrazone tautomerism was discovered in 1884 [3]. The same orange dye was obtained by coupling benzenediazonium chloride with 1-naphthol and by condensing phenylhydrazine with 1,4-naphthoquinone. The expected products were the azo dye (**62**) (R=H) and the hydrazone (**63**) (R=H). It was correctly assumed that there was an equilibrium between the two forms, i.e., tautomerism.

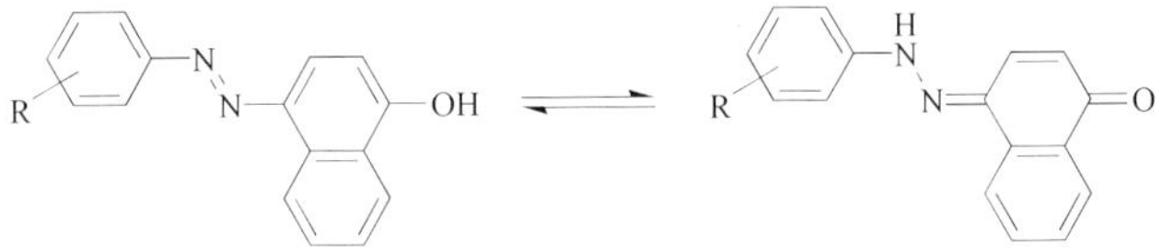

The discovery prompted extensive research into azo/hydrazone tautomerism, a phenomenon which is not only interesting but also extremely important as far as commercial azo dyes are concerned because the tautomers have different colors, different properties (e.g., lightfastness), different toxicological profiles, and, most importantly, different tinctorial strengths. Since the tinctorial strength of a dye primarily determines its cost-effectiveness, it is desirable that commercial azo dyes should exist in the strongest tautomeric form. This is the hydrazone form.

Hydroxyazo dyes vary in the proportion of tautomers present, from pure azo tautomer to mixtures of azo and hydrazone tautomers, to pure hydrazone tautomer. Almost all azophenol dyes (**64**) exist totally in the azo form, except for a few special cases [4].

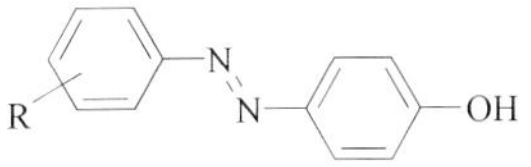

The energies of the azo and hydrazone forms of 4-phenylazo-1-naphthol dyes are similar, so both forms are present. The azo tautomers (**62**) are yellow (λ_{max} ca. 410 nm, ε_{max} ca. 20 000) and the hydrazone tautomers (**63**) orange (λ_{max} ca. 480 nm, ε_{max} ca. 40 000). The relative proportions of the tautomers are influenced by both solvent (Figure 2.4) and substituents (Figure 2.5).

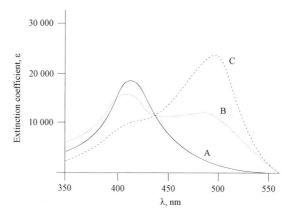

Figure 2.4 Effect of solvent on 4-phenylazo-1-naphthol absorption. A, pyridine; B, methanol; C, acetic acid.

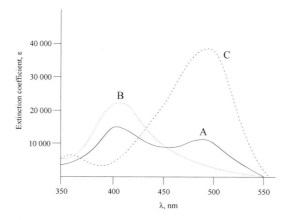

Figure 2.5 Electronic effect of substituents in 4-phenylazo-1-naphthol (**62**). A, R = H; B, R = p-OCH$_3$; C, R = p-NO$_2$.

The isomeric 2-phenylazo-1-naphthols (**65**) and 1-phenylazo-2-naphthols (**66**) exist more in the hydrazone form than the azo form, as shown by their UV spectra. Their λ_{max} values are each about 500 nm.

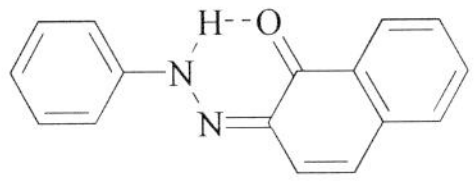

2.1.3 Principal Properties

65 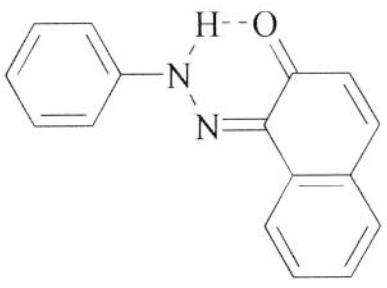 **66**

Important classes of dyes that exist totally in the hydrazone form are azopyrazolones (**67**), azopyridones (**68**), and azoacetoacetanilides (**69**).

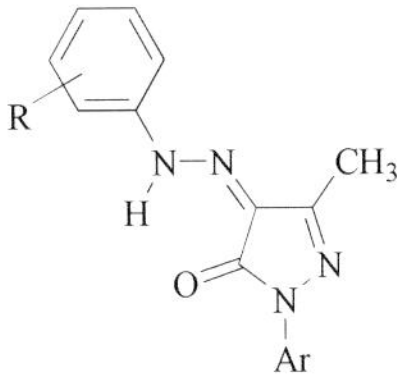

67 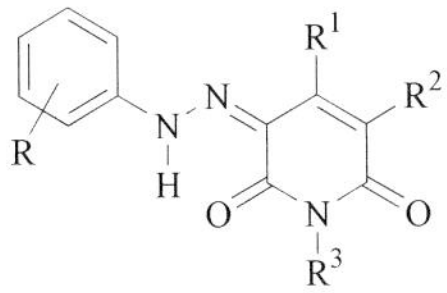 **68** 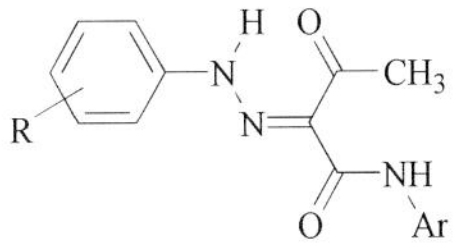 **69**

All aminoazo dyes exist exclusively as the azo form; there is no evidence for the imino form. Presumably, a key factor is the relative instability of the imino group.

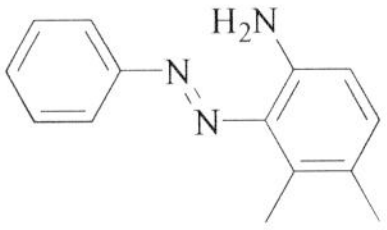

Azo 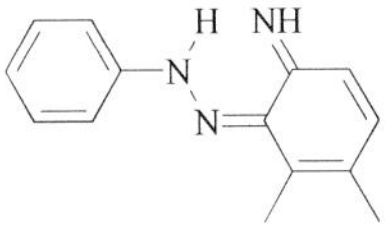 Imino

In disazo dyes derived from aminonaphthols, one group exists as a true azo group and one as a hydrazo group (**70**).

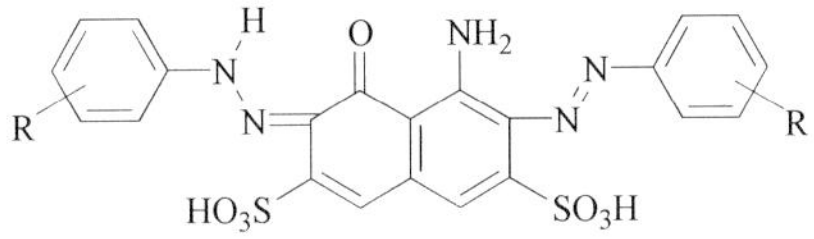

70

Aminoazo dyes undergo protonation at either the terminal nitrogen atom to give the essentially colorless ammonium tautomer (**71**) (λ_{max} ca. 325 nm),

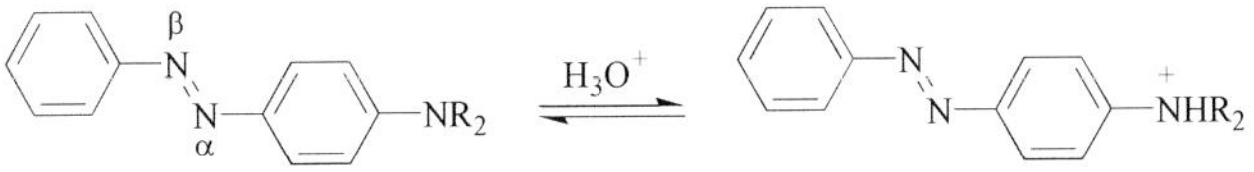

71

or at the β-nitrogen atom of the azo group to give a resonance-stabilized azonium tautomer (**73**), as shown for methyl orange (**72**). The azonium tautomer is brighter, generally more bathochromic and stronger (ε_{max} ca. 70 000) than the neutral azo dye (ε_{max} ca. 35 000). The azonium tautomers are related to diazahemicycanine dyes used for coloring polyacrylonitrile. The most familiar use of the protonation of azo dyes is in indicator dyes such as methyl orange (**72**) and methyl red (Scheme 2.3):

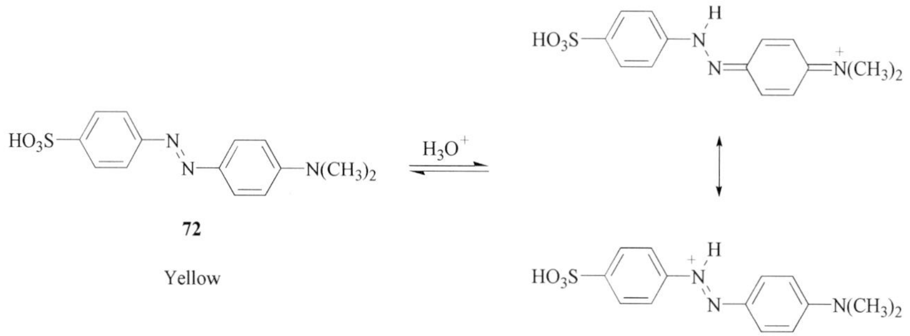

72

Yellow

73

Red

Scheme 2.3

2.1.3.2 Metallized Azo Dyes

The three metals of importance in azo dyes are copper, chromium, and cobalt. The most important copper dyes are the 1:1 planar copper(II) azo dye complexes (**74**).

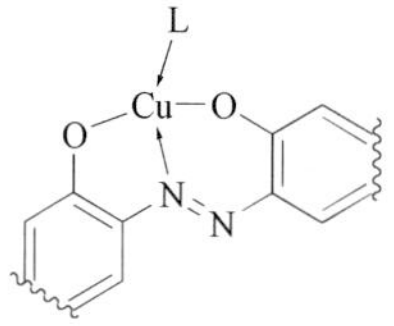

74

In contrast, chromium(III) and cobalt(III) form 2:1 dye:metal complexes that have nonplanar structures. Geometrical isomerism exists. The *o,o′*-dihydroxyazo dyes (**75**) form the Drew–Pfitzner or *mer* type (**76**) (A = C = O), whereas *o*-hydroxy-*o′*-carboxyazo dyes (**77**) form the Pfeiffer–Schetty or *fac* type (**78**) (A = CO_2 and C = O).

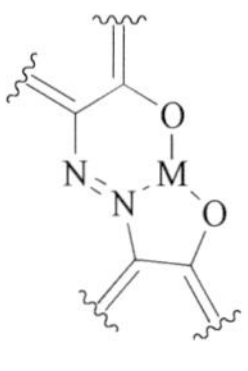

75

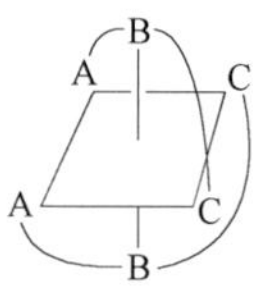

76

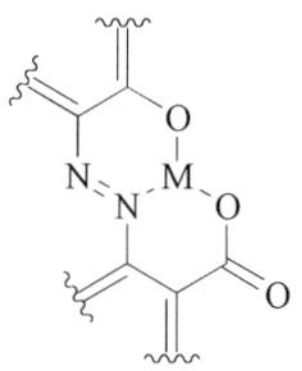

77

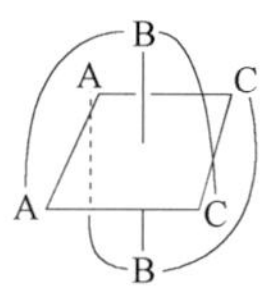

78

Metallization of dyes was originally carried out during the mordanting process to help fix the dye to substrate. Premetallized dyes are now used widely in various outlets to improve the properties of the dye, particularly lightfastness. However, this is at the expense of brightness, since metallized azo dyes are duller than non-metallized dyes.

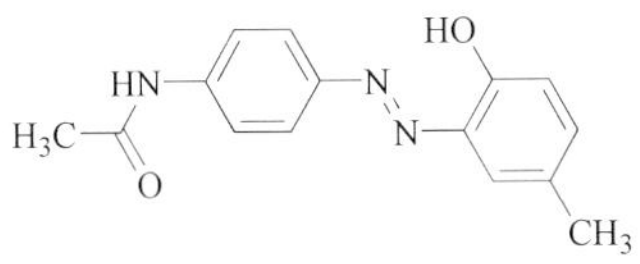

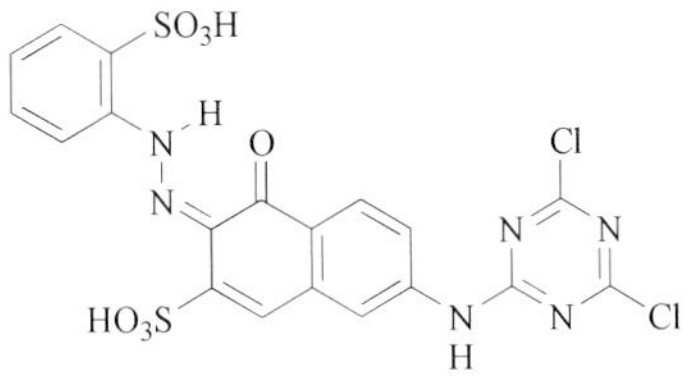

2.1.3.3 Carbocyclic Azo Dyes

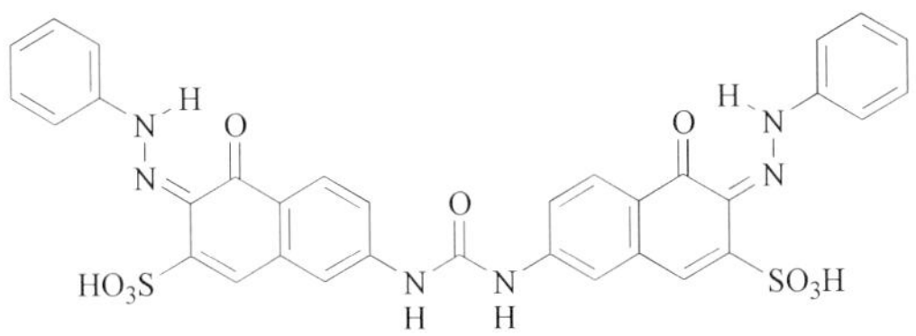

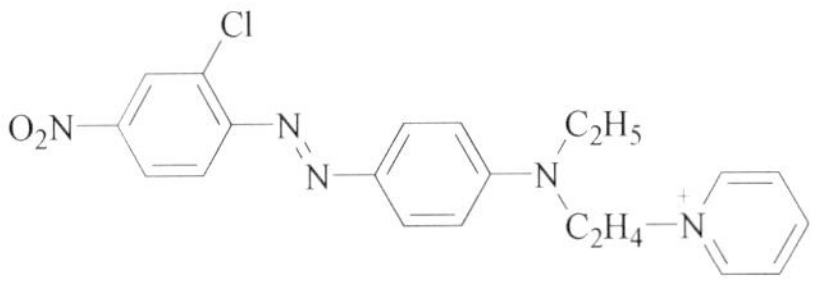

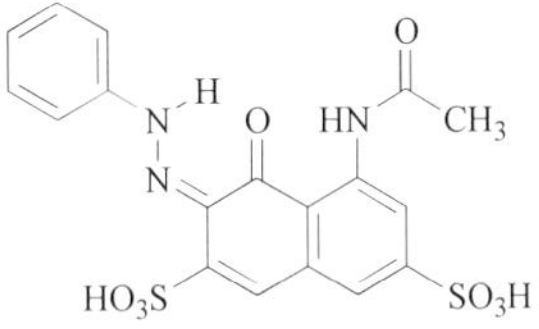

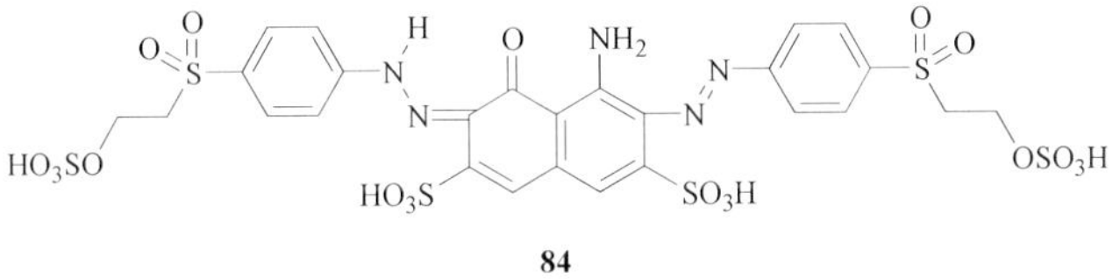

These dyes are the backbone of most commercial dye ranges. Based entirely on benzene and naphthalene derivatives, they provide yellow, red, blue, green, and black colors for all the major substrates such as polyester, cellulose, nylon, polyacrylonitrile, and leather. Typical structures are (**79**)–(**84**).

C.I. Disperse Yellow 3 (**79**) is used to dye polyester; *C.I. Reactive Orange 1* (**80**) is a cotton dye; *C.I. Direct Orange 26* (**81**) is a dye for paper; *C.I. Basic Red 33* (**82**) dyes acrylic fibres; *C.I. Acid Red 38* (**83**) dyes nylon and wool; *C.I. Reactive Black 5* (**84**) dyes cotton.

2.1 Azo Chromophore

Most azoic dyes belong to the carbocyclic azo class, but these dyes are formed in the fiber pores during the application process.

The carbocyclic azo dyes are highly cost-effective and have good all-around fastness properties. However, they generally lack brightness and consequently cannot compete with anthraquinone dyes in this respect. This shortcoming of carbocyclic azo dyes is overcome by heterocyclic azo dyes, as well as other dye classes such as triphenodioxazines and benzodifuranones.

2.1.3.4 Heterocyclic Azo Dyes

A long-term aim of dyes research has been to combine the brightness and fastness properties of anthraquinone dyes with the strength and economy of azo dyes. This aim is now being realized with heterocyclic azo dyes, which fall into two main groups: those derived from heterocyclic coupling components and those derived from heterocyclic diazo components.

All the heterocyclic coupling components that provide commercially important azo dyes contain only nitrogen as the heteroatom. They are indoles (**85**), pyrazolones (**86**), and especially pyridones (**87**); they provide yellow to orange dyes for various substrates.

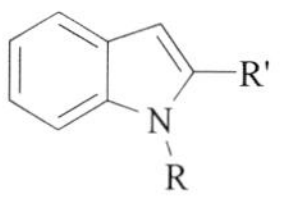

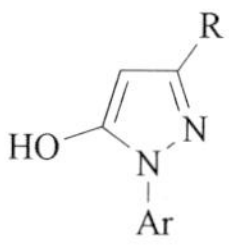

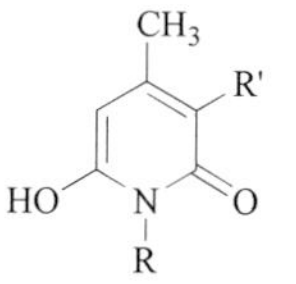

Many yellow dyes were of the azopyrazolone type, but these have now been largely superseded by azopyridone dyes. Azopyridone yellow dyes are brighter, stronger and generally have better fastness properties than azopyrazolone dyes. Both azopyrazolone and azopyridone dyes exist in the hydrazone tautomeric form. Typical dyes are the azopyrazolone *C.I. Acid Yellow 23* (**88**) and the azopyridone (**89**).

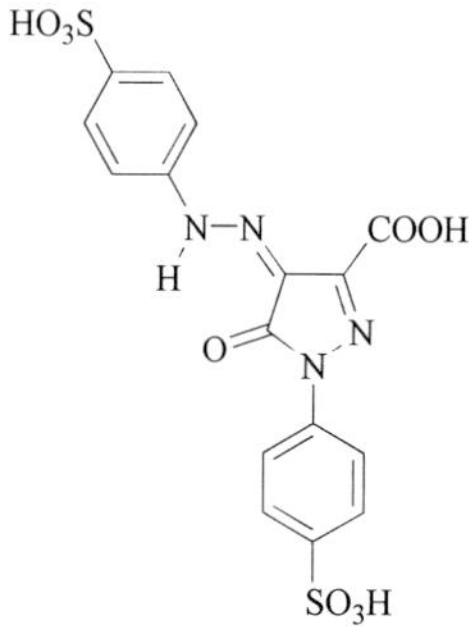

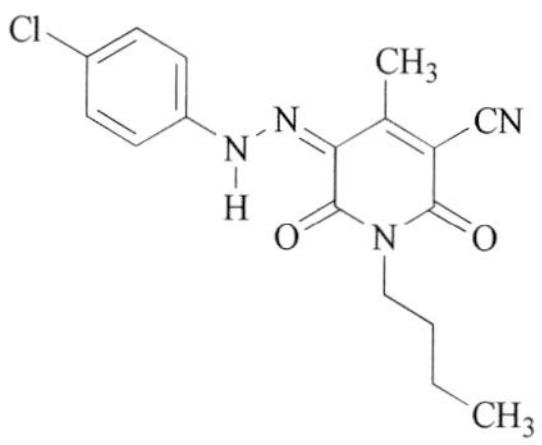

In contrast to the heterocyclic coupling components, virtually all the heterocyclic diazo components that provide commercially important azo dyes contain sulfur, either alone or in combination with nitrogen (the one notable exception is the triazole system). These S or S/N heterocyclic azo dyes provide bright, strong shades that range from red through blue to green and therefore complement the yellow-orange colors of the nitrogen heterocyclic azo dyes in providing a complete coverage of the entire shade range. Two representative dyes are the thiadiazole red (**90**) and a thiophene greenish blue (**91**). Both are disperse dyes for polyester, the second most important substrate after cellulose.

2.1.4 References

[1] *Colour Index*, Vol. 4, 3rd ed., The Society of Dyers and Colourists, Bradford, UK, 1971.
[2] P. F. Gordon, P. Gregory, *Organic Chemistry in Colour*, Springer-Verlag, Berlin, 1983, pp. 96–115.
[3] T. Zincke, H. Binderwald, *Chem Ber.* **17 (**1884) 3026.
[4] P. F. Gordon, P. Gregory, *Organic Chemistry in Colour*, Springer-Verlag, Berlin, 1983, pp. 104–108.

2.2 Anthraquinone Chromophore

2.2.1 Introduction

Once the second most important class of dyes, it also includes some of the oldest dyes; they have been found in the wrappings of mummies dating back over 4000 years. In contrast to the azo dyes, which have no natural counterparts, all the important natural red dyes were anthraquinones [1]. However, the importance of anthraquinone dyes has declined due to their low cost-effectiveness.

Anthraquinone dyes are based on 9,10-anthraquinone (**1**), which is essentially colorless. To produce commercially useful dyes, strongly electron donating groups such as amino or hydroxyl are introduced into one or more of the four α positions

(1, 4, 5, and 8). The most common substitution patterns are 1,4, 1,2,4, and 1,4,5,8-. To optimize the properties, primary and secondary amino groups (not tertiary) and hydroxyl groups are employed. These ensure the maximum degree of π-orbital overlap, enhanced by intramolecular hydrogen-bonding, with minimum steric hindrance. These features are illustrated in *C.I. Disperse Red 60* (**2**).

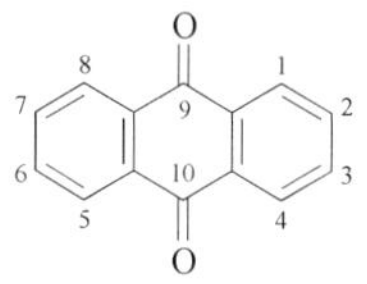

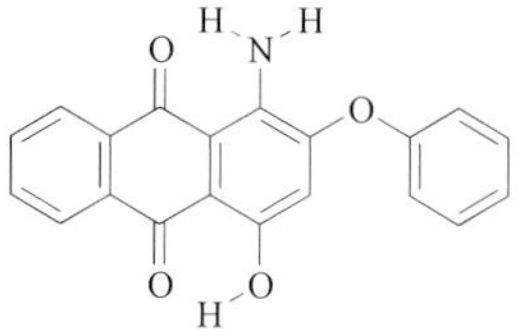

The strength of electron-donor groups increases in the order: OH<NH_2<NHR<HNAr. Tetrasubstituted anthraquinones (1,4,5,8-) are more bathochromic than di- (1,4-) or trisubstituted (1,2,4-) anthraquinones. Thus, by an appropriate selection of donor groups and substitution patterns, a wide variety of colors can be achieved.

2.2.2 General Synthesis

Anthraquinone dyes are prepared by the stepwise introduction of substituents onto the preformed anthraquinone skeleton (**1**) or ring closure of appropriately substituted precursors. The degree of freedom for producing a variety of different structures is restricted, and the availability of only eight substitution centers imposes a further restriction on synthetic flexibility. Therefore, there is significantly less synthetic versatility than in the case of azo dyes, and consequently less variety; this is a drawback of anthraquinone dyes.

Although *tautomerism* is theoretically possible in amino and hydroxy anthraquinone dyes, none has been observed. Studies by ^{13}C NMR spectroscopy have shown convincingly that amino and hydroxy dyes of 9,10-anthraquinone exist as such.

2.2.3 Principal Properties

The principal advantages of anthraquinone dyes are brightness and good fastness properties, including lightfastness, but they are both expensive and tinctorially weak. However, they are still used to some extent, particularly for red and blue shades, because other dyes cannot provide the combination of properties offered by anthraquinone dyes, albeit at a price. Metallization would deteriorate the brightness and fastness properties, and there is no need to improve the lightfastness. Consequently, metallized anthraquinone dyes are of very little importance.

2.2.3.1 Benzodifuranone Dyes

The benzodifuranone (BzDF) dyes are challenging anthraquinone dyes in the red shade area. The BzDF chromogen is one of the very few commercially useful novel chromogens to have been discovered in the twentieth century. As with many other discoveries [2] the BzDF chromogen was detected by accident. The pentacenequinone structure (**3**) assigned to the intensely red colored product obtained from the reaction of *p*-benzoquinone with cyanoacetic acid was questioned [3]. A compound such as (**3**) should not be intensely colored owing to the lack of conjugation. Instead, the red compound was correctly identified as the 3,7-diphenylbenzodifuranone (**4**) (λ_{max} = 466 nm; ε = 51 000 in chloroform).

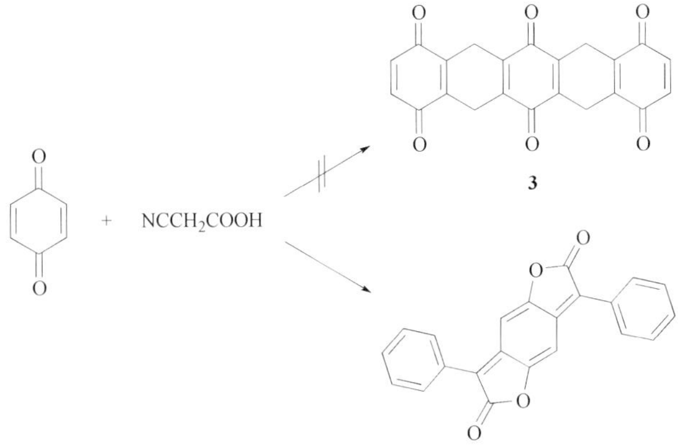

Improved syntheses from arylacetic acids and hydroquinone or substituted quinones have been devised for BzDF dyes [4] (Scheme 2.3).

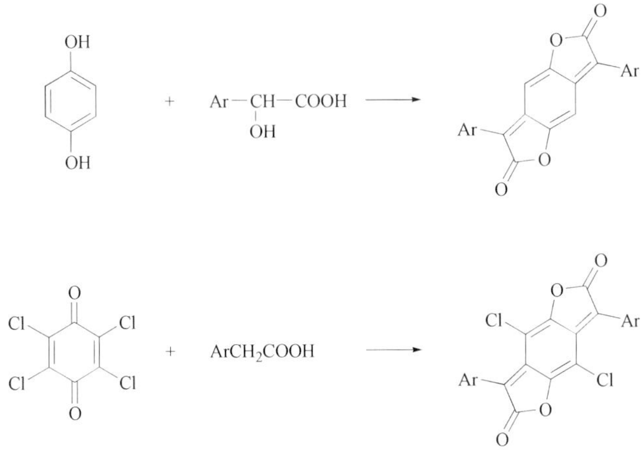

Scheme 2.3 Synthesis of benzodifuranone (BzDF) dyes

The BzDFs are unusual in that they span the whole color spectrum from yellow through red to blue, depending on the electron-donating power of the R group on the phenyl ring of the aryl acetic acid, i.e. Ar = C_6H_4R (R=H, yellow-orange; R = alkoxy, red; R = amino, blue). The first commercial BzDF, Dispersol Red C-BN, a red disperse dye for polyester, has made a significant impact. Its brightness even surpasses that of the anthraquinone reds, while its high tinctorial strength (ca. 3–4 times that of anthraquinones) makes it more cost-effective. In contrast, BzDF yellow and blue dyes have found it difficult to compete with other classes of dyes and have not found commercial use.

2.2.3.2 Polycyclic Aromatic Carbonyl Dyes

These dyes contain one or more carbonyl groups linked by a quinonoid system. They tend to be relatively large molecules built up from smaller units, typically anthraquinones. Since they are applied to the substrate (usually cellulose) by a vatting process, the polycyclic aromatic carbonyl dyes are often referred to as anthraquinonoid vat dyes.

In the vatting process the water-insoluble vat dye is reduced by sodium hydrogensulfite (hydros) to the water-soluble phenate salt. This is less colored than the vat dye and has affinity for cotton. The cloth "dyed" with the phenate salt is placed in water, normally containing soap, and air is bubbled through the solution to regenerate the vat dye, which is trapped in the pores of the cotton (Scheme 2.4). The phenate salt may be stabilized by forming the sulfate ester. These stabilized, water-soluble salts of vat dyes, such as *C.I. Solubilised Vat Blue 1* (**5**), are commercial products.

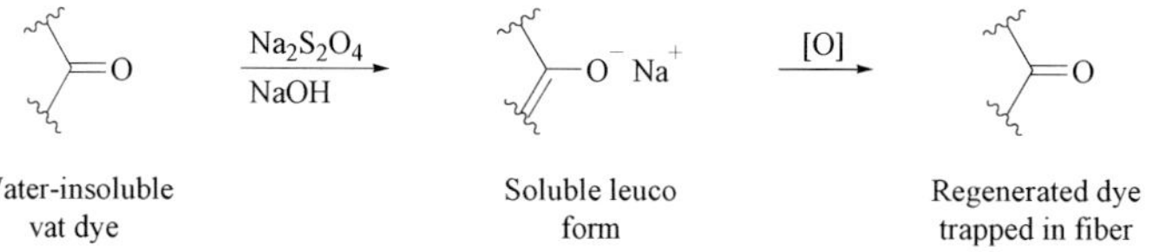

Scheme 2.4 The vatting process

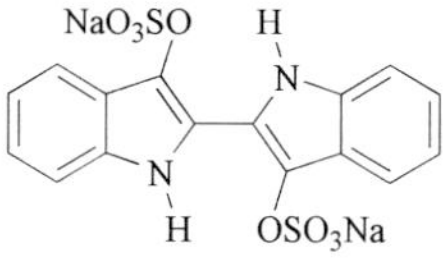

Although the polycyclic aromatic carbonyl dyes cover the entire color range, only the blues, browns, greens, and blacks are important commercially. Typical dyes are the blue indanthrone (**6**), *C.I. Vat Brown 3* (**7**), *C.I. Vat Black 27* (**8**), and the *C.I. Vat Green 1* (**9**), probably the best known of the polycyclic aromatic carbonyl dyes.

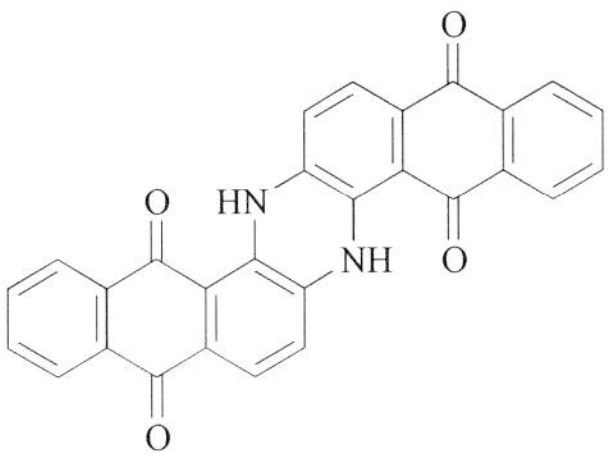

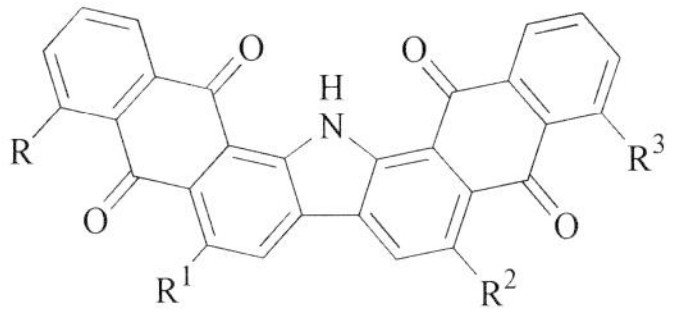

7: R = R² = C₆H₅CONH; R¹ = R³ = H
8: R = R³ = H; R¹ = R² = C₆H₅CONH

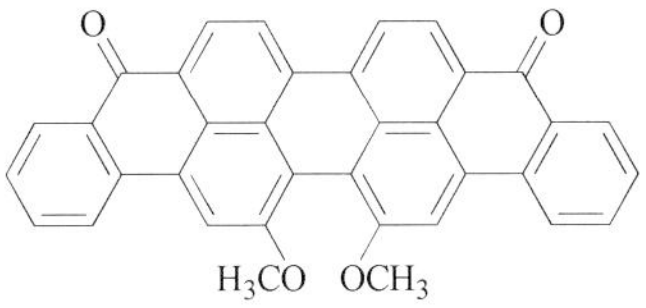

As a class, the polycyclic aromatic carbonyl dyes exhibit the highest degrees of lightfastness and wetfastness. The high lightfastness is undoubtedly associated with the absence of electron-donating and any electron-withdrawing groups other than carbonyl, so that the number of photochemically active sites in the molecule is restricted. The degree of aggregation is also a major contributory factor. The high wetfastness is a direct manifestation of the application process of the dyes.

2.2.4 References

[1] P. F. Gordon, P. Gregory, *Organic Chemistry in Color*, Springer-Verlag, Berlin, 1983.
[2] P. F. Gordon, P. Gregory, *Organic Chemistry in Color*, Springer-Verlag, Berlin, 1983, pp. 5–21.
[3] C. W. Greenhalgh, J. L. Carey, D. F. Newton, *Dyes and Pigments* **1** (1980) 103.
[4] British Patent 2,151,611 ICI 1985

2.3 Indigoid Chromophore

2.3.1 Introduction

Indigoid dyes represent one of the oldest known class of organic dyes. For 5000 years they have been used for dyeing textiles such as wool, linen, and cotton. For example, 6,6′-dibromoindigo is Tyrian Purple, the dye made famous by the Romans. Tyrian Purple was so expensive that only the very wealthy were able to afford garments dyed with it. Indeed, the phrase "born to the purple" is still used today to denote wealth. Although many indigoid dyes have been synthesized, only indigo itself (**1**) is of any major importance today. Indigo is the blue used almost exclusively for dyeing denim jeans and jackets and is held in high esteem because it fades to give progressively paler blue shades.

Adolf von Baeyer reported the chemical structure of indigo in 1883, having already prepared the first synthetic indigo from isatin (**2**) in 1870. Until then, isatin had only been obtainable by oxidation of indigo. It was not until 1878 that von Baeyer first synthesized it from phenylacetic acid, making a complete synthesis of indigo possible [1].

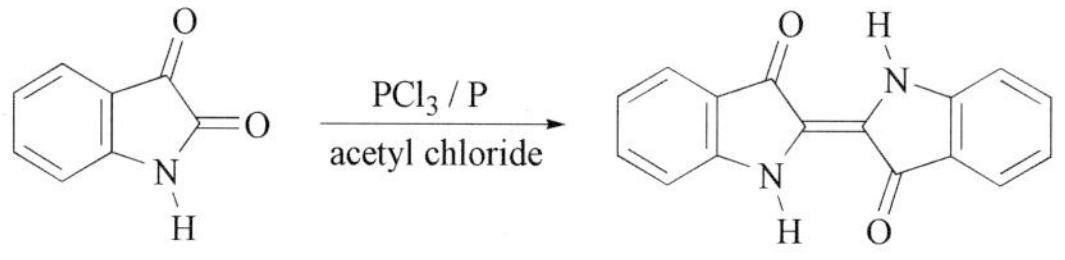

Isatin (**2**) →(PCl$_3$ / P, acetyl chloride)→ Indigo (**1**)

2.3.2 General Synthesis

In most indigo syntheses the indole structure is built up by ring closure of appropriate benzene derivatives by C–N or C–C bond formation. Examples of C–N bond formation include von Baeyer's 1878 synthesis from phenylacetic acid (**3**) via oxindole (**4**).

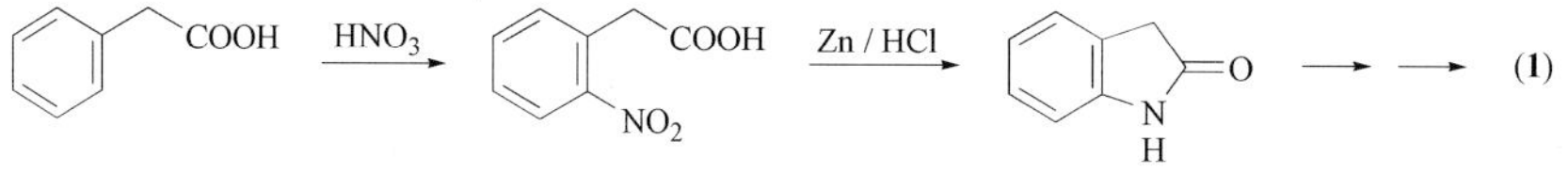

3 →(HNO$_3$)→ →(Zn / HCl)→ Oxindole (**4**) → → (**1**)

In the Heumann I synthesis, which is the basis of modern synthetic indigo production (see Section 3.5), ring closure takes place by C–C bond formation. Aniline is first treated with a chloroacetate salt to give phenylglycine salt **5**, which is converted under alkaline conditions to indoxyl salt **6**. Ring closure of the alkali metal salt of phenylglycine **5** gives particularly high yields in the presence of stoichiometric quantities of sodium amide. This suggests that in the first reaction step the phenylglycine salt is deprotonated to give the dianion, followed by a C–C bond formation and a 1,3-H shift to yield the indoxylate dianion **6**. The mechanism is similar to that of an intramolecular (aza)-Kolbe–Schmitt synthesis. Hydrolysis, oxidation, and dimerization of the indoxylate dianion (**6**) then result in indigo (**1**) (Scheme 2.5).

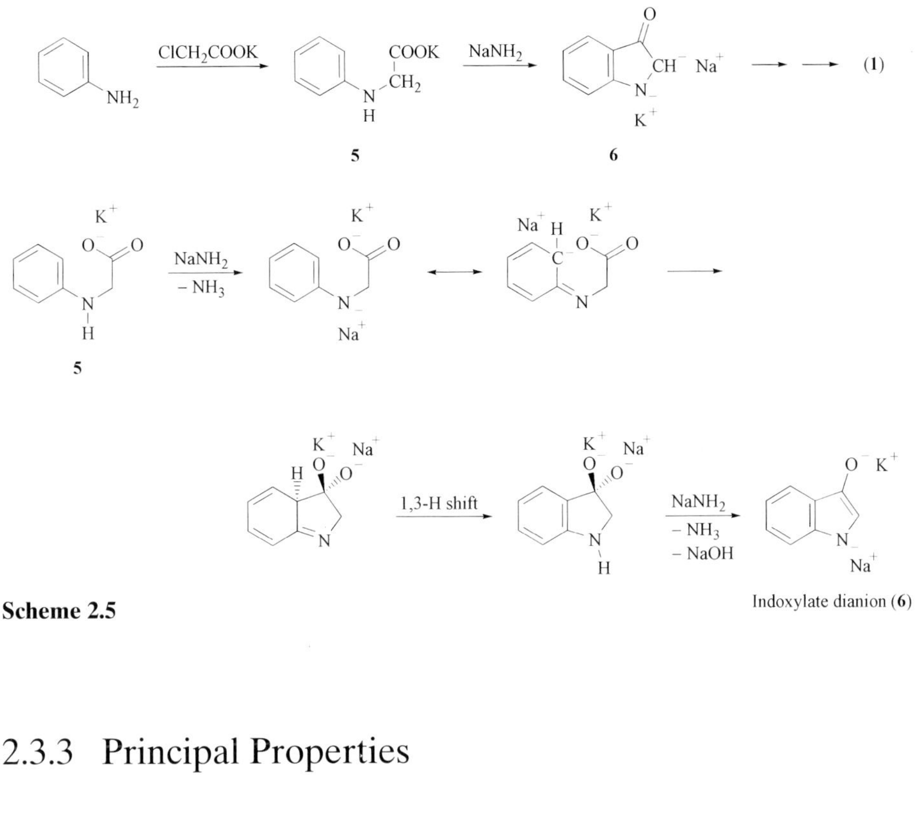

Scheme 2.5

2.3.3 Principal Properties

2.3.3.1 Color

One of the main fascinations of indigo is that such a small molecule should be blue. The reason for indigo's deep blue color was long unknown. Normally, extensive conjugation (e.g., phthalocyanines) and/or several powerful donor and acceptor groups (e.g., azo and anthraquinone dyes) are required to produce blue dyes. Only with the intensive studies of Lüttke et al. in the 1960s was any significant insight gained. After much controversy it was proven that the chromogen is the crossed conjugated system [6]. The deep indigo color is explicable in terms of the special arrangement of the atoms in the basic indigo chromophore and the high polarizability of the charge distribution, which is strongly influenced by the ability of the molecule to form hydrogen bonds [3]. The deep shade is determined by the doubly cross-conjugated system of two donor and two acceptor groups, the conformation of the two carbonyl groups with respect to the central C=C double bond, and the nature of the heteroatom [4, 6]. Thus the long-wave UV/Vis absorption band of indigo (610 nm in ethanol) is shifted hypsochromically as the heteroatom is replaced to give selenoindigo (562 nm), thioindigo (543 nm), and oxindigo (432 nm) [5].

2.3.3.2 Basic Chromophore

The question of the contribution of the benzene rings to the indigo chromogen led to a search for the basic chromophore. Consecutively removing parts of the mesomeric system leaves a cross-conjugated structure **7** that, as the basic chromophore, still exhibits the typical deep color and redox properties of indigo.

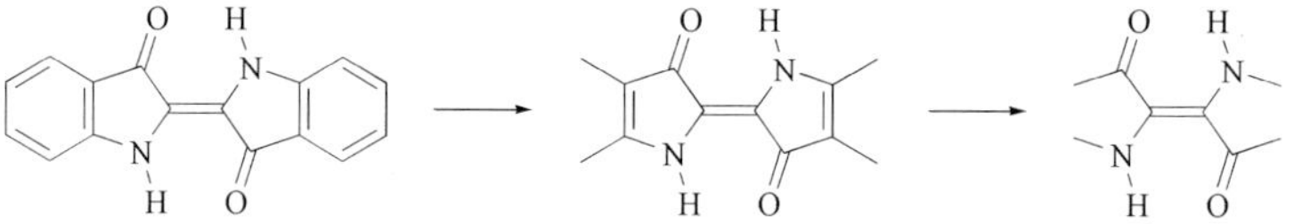

Basic chromophore (**7**)

A Pariser–Parr–Pople (PPP) calculation indicates that the long-wavelength absorption band is what would be expected for the *trans-s-cis-s-cis* configuration shown by crystal-structure analyses to occur in the indigo crystal [2].

It was possible to test the various theoretical structures proposed for the basic chromophore **7** by synthesizing urindigo [4,4′,4′-tetramethyl(2,2′-bipyrrolidinylidene)-3,3′dione] [7].

2.3.3.3 Solvatochromism

PPP calculations by Klessinger showed that the first excited state of indigo is more polar than the ground state, so that positive solvatochromism might be expected. This was in fact observed by Reichardt (see Table 2.1) [8].

Table 2.1 Absorption maximum of indigo in various media

Medium	Steam	CCl_4	Xylene	Ethanol	DMSO	Solid (KBr)
λ_{max} nm	540	588	591	610	620	660

2.3.3.4 Redox System

Indigo is readily reduced to leuco forms [cf., tetramethyl leucoindigo (**8**)] and oxidized to dehydroindigo (**9**). These reactions strongly influence the cross-conjugated system, and pronounced hypsochromic shifts of the long-wave absorption band in the UV/Vis spectrum have been measured [9]. The low frequency of the CO band in the IR spectrum of indigo at 1626 cm^{-1} points to the presence of hydrogen bonds (Table 2.2).

Table 2.2: IR and UV/VIS Spectra

	IR spectrum: $\tilde{\nu}$ / cm^{-1}	UV / VIS spectrum (ethanol): λ_{max} / nm
Tetramethyl leucoindigo (**8**)	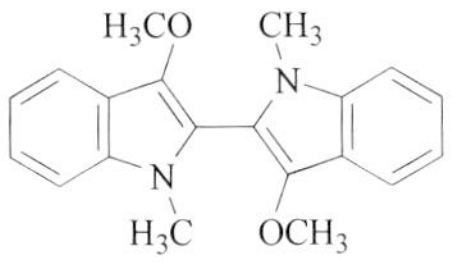 1239	306
Indigo (**1**)	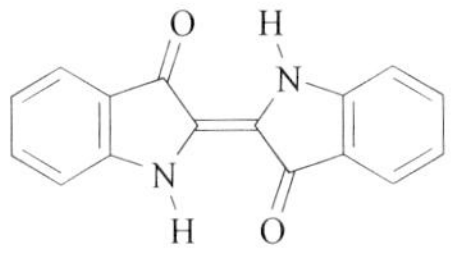 1626	610
Dehydro indigo (**9**)	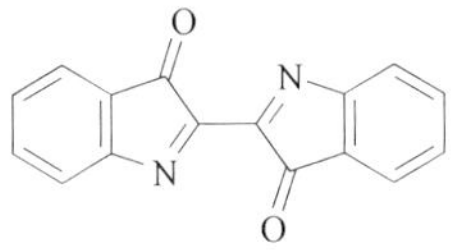 1724	447

2.3.4 References

[1] A. von Baeyer, "Zur Geschichte der Indigo-Synthese," *Ber. Dtsch. Chem. Ges.* **33** (1900) Anlage IV, LI–LXX.; H. Brunck "Die Geschichte der Indigo-Fabrikation", *Ber. Dtsch. Chem. Ges.* **33** (1900) Anlage V, LXXI–LXXXVI.
[2] M. Klessinger, *Tetrahedron* **22** (1966) 3355–3365.
[3] K. Klessinger, W. Lüttke, *Chem. Ber.* **99** (1966) 2136–2147.
[4] H. Klessinger, W. Lüttke, *Tetrahedron* **19** (1963) 315–335.
[5] W. Lohner, K. Praefke, "Telluroindigo", *J. Organomet. Chem.* **208** (1981) 35–37.
[6] W. Luttke, M. Klessinger, *Chem. Ber.* **97** (1946) 2342; W. Luttke, H. Hermann, M. Klessinger, *Angew. Chem. Int. Ed. Engl.* **5** (1966) 598.
[7] E. Wille, W. Lüttke, *Angew. Chemie* **83** (1971) 853–854; E. Wille, W. Lüttke, *Liebigs Ann.Chem.* **1980**, 2039–2054; H. Bauer, K. Kowski, H. Kuhn, W. Lüttke, P. Rademacher, *J. Mol. Struct.* **445** (1998) 277–286
[8] C.Reichardt, *Solvent Effects in Organic Chemistry*, Weinheim, VCH, 1979, p. 193.
[9] W. Lüttke, M. Klessinger, *Chem. Ber.* **97** (1964) 2342–2357.

2.4 Cationic Dyes as Chromophores

2.4.1 Introduction

Cationic dyes carry a positive charge in their molecule. The salt-forming counterion is in most cases the colorless anion of a low molecular mass inorganic or organic acid. Many of these dyes can be converted into water-insoluble dye bases by addition of alkali. For this reason, they were formerly called *basic dyes*; although still in use today, the term should be abandoned.

The positive charge of cationic dyes may be either localized or delocalized. In **1**, the positive charge is localized on an ammonium group:

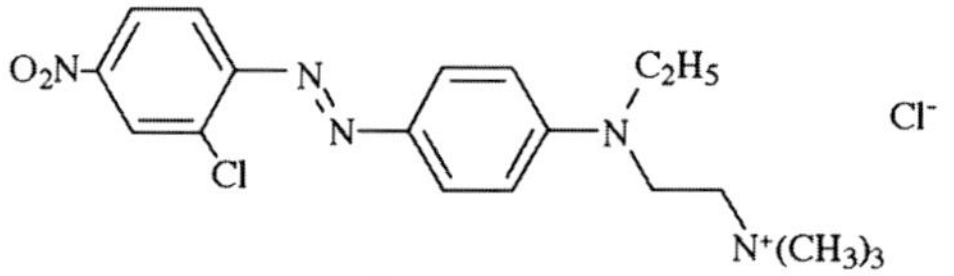

1

In the diazahemicyanine dye **2**, the positive charge is delocalized across the dye cation.

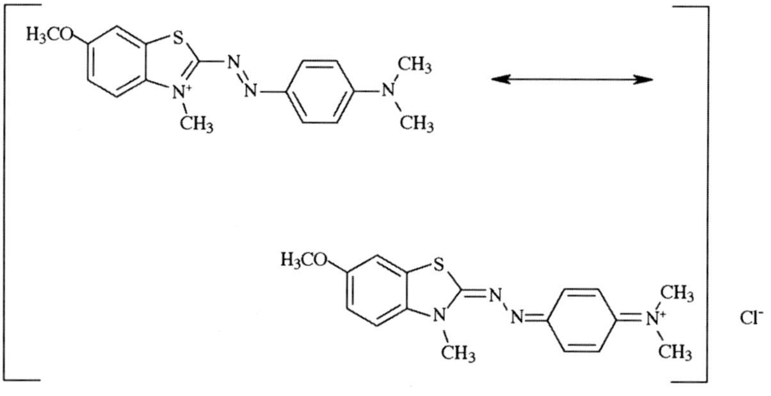

2

As in these examples, the charge-carrying atom is usually nitrogen, but in some dyes this function is adopted by an oxygen, sulfur, or phosphorus atom.

2.4.2 General Synthesis

Cationic functionality is found in various types of dyes, mainly in cationic azo dyes (Section 3.7) and methine dyes (Section 3.8), but also in anthraquinone (Section 3.4), di- and triarylcarbenium (Section 2.6), phthalocyanine dyes (Section 2.7), and in various polycarbocyclic and solvent dyes (Section 3.10).

2.4.3 Chemical Structure and Classification

This chapter gives a short survey on cationic dyes and shows typical representatives of each class of cationic dyes. In general, the dyes are classified according to their chemical structure.

2.4.3.1 Dyes with Delocalized Charge

Cationic dyes with delocalized charge are classified with the *methine dyes* (see Section 2.6). They may be viewed as vinylogous amidinium salts **3** [1]:

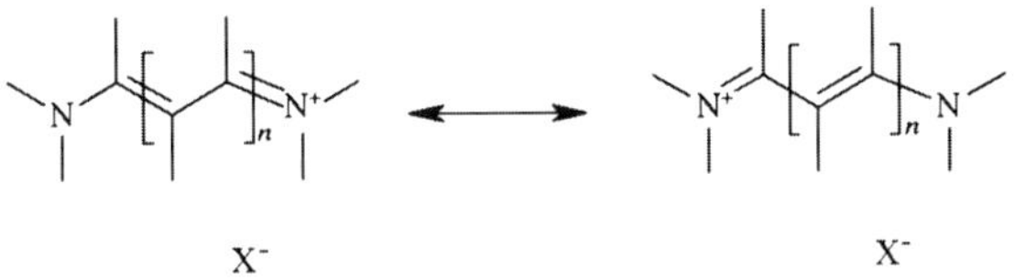

All of these dyes can be represented by various resonance formulas. They are characterized by high color strength.

Dyes in which both charged terminal atoms are not part of a heterocyclic ring are called *streptocyanine dyes*. An example is **4**, which dyes paper in a greenish yellow shade [2].

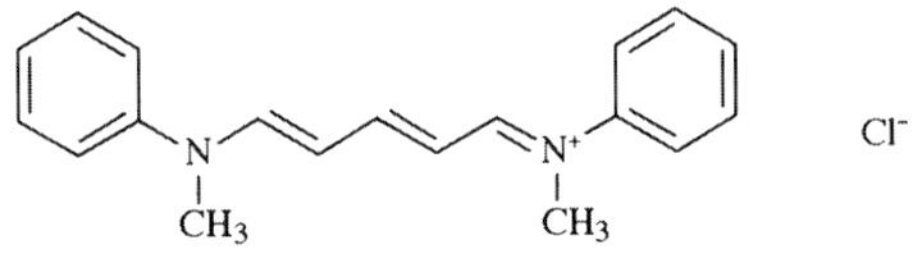

In the *hemicyanine dyes*, one of the charged terminal atoms is part of a heterocyclic ring, and the other is a nitrogen atom directly linked to the methine chain (**5**).

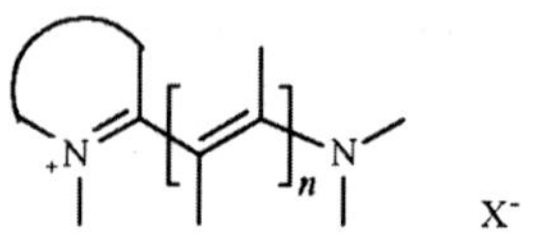

5

An example of this type is the *enamine dye* **6**, which confers a yellow color to polyacrylonitrile fibers [3].

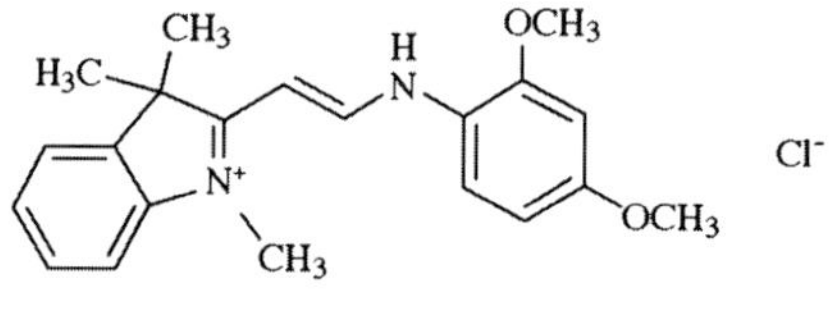

6

Insertion of an aryl residue between the second nitrogen atom and the methine chain leads to the *styryl dyes*, which dye polyacrylonitrile in red and purple shades, e.g., **7** [4].

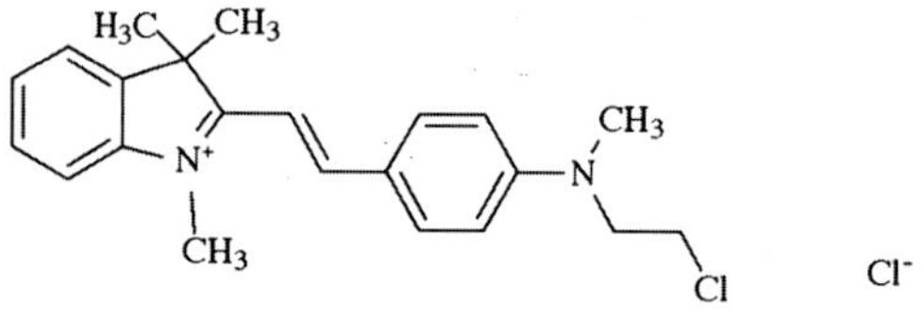

7

In a special form of this type, the chain may be completely missing. This leads to *zeromethine dyes* such as **8** [5], which dye polyacrylonitrile in very lightfast violet and blue shades.

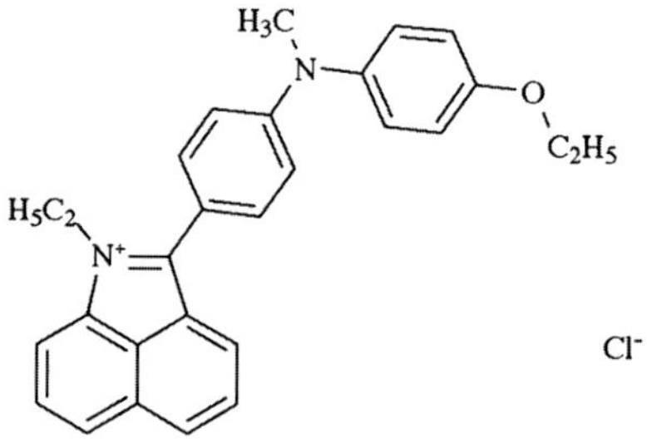

8

If the second charged terminal atom is also part of a heterocyclic ring, the *cyanine dyes* are obtained (**9**).

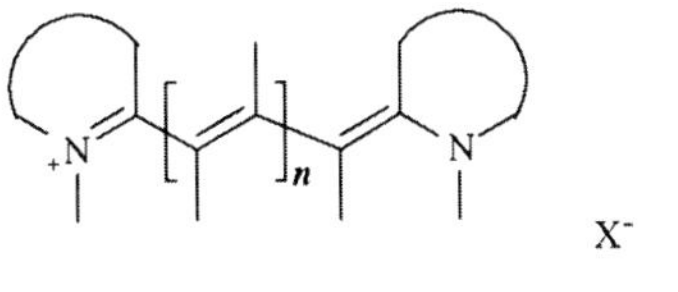

2.4.3 Chemical Structure and Classification

9

These dyes are used predominantly as photographic spectral sensitizers, but some years ago a new application in optical data storage media was found. Compound **10** dyes paper in brilliant pink shades [6]:

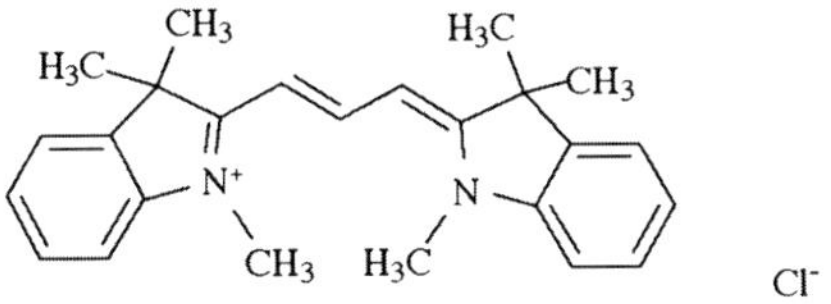

10

One or several of the carbon atoms in the chain may be replaced by nitrogen, as is indicated by the prefix "aza." Depending on the number of nitrogen atoms introduced into the chain, *mono-* or *diazadimethine dyes* are obtained. *Tri-* and *tetraazamethine* dyes are also known. An example is dye **11**, which gives polyacrylonitrile fibers a golden yellow color [7].

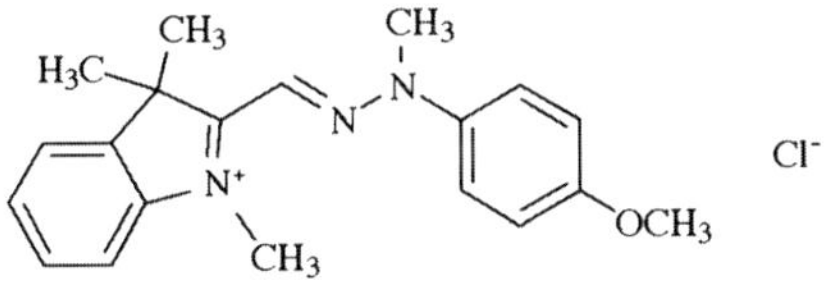

11

Another example is the above-mentioned dye **2**, which dyes polyacrylonitrile in lightfast blue shades [8].

Diphenylmethane and *triphenylmethane dyes* are monomethine dyes with two or three terminal aryl groups, of which at least one, but preferably two or three, are substituted by a donor group *para* to the methine carbon atom. The most important donor is the amino group. Important dyes belonging to this class include the well-known malachite green (**12**) [9] and crystal violet (**13**) [10], which are some of the oldest synthetic cationic dyes:

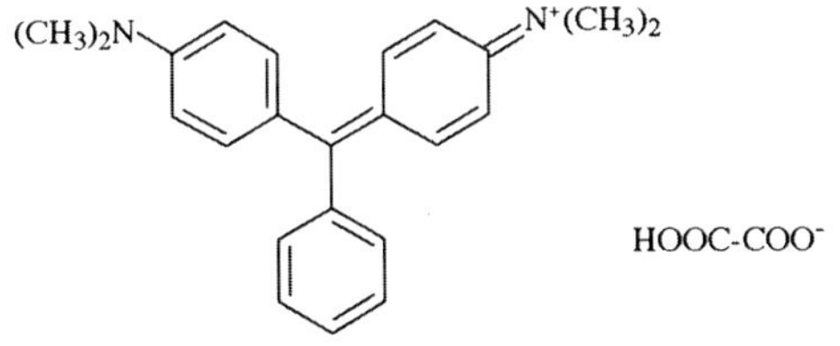

12

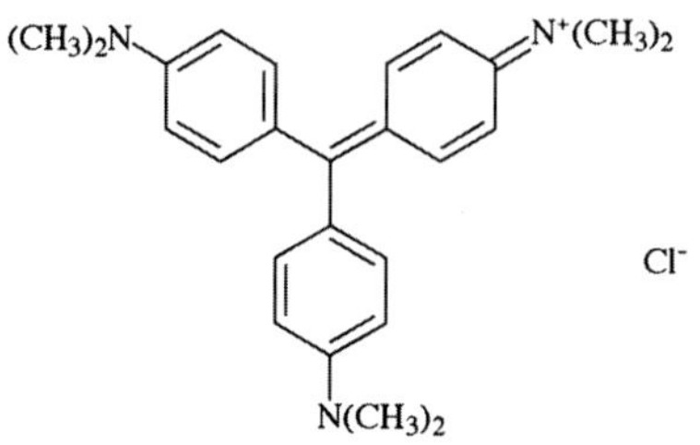

2.4 Cationic Dyes as Chromophores

13

Introduction of an oxygen bridge into the triphenylmethane dye molecule leads to the *xanthene dyes*. The color is shifted from blue to red. The restricted rotation of the phenyl groups inhibits radiationless de-excitation and gives rise to very strong fluorescence. Rhodamine B (**14**) is used for dyeing paper [11] and as a laser dye.

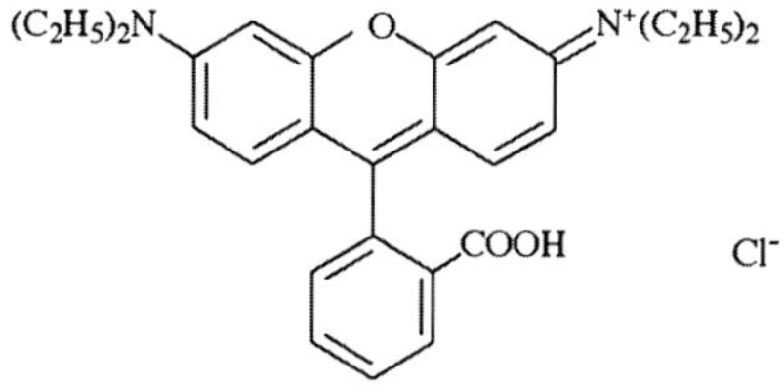

14

Azine dyes, which have the general formula **15**, are ring-closed azadiphenylmethane or indamine dyes. Examples are *phenazine dyes, oxazine dyes*, and *thiazine dyes*.

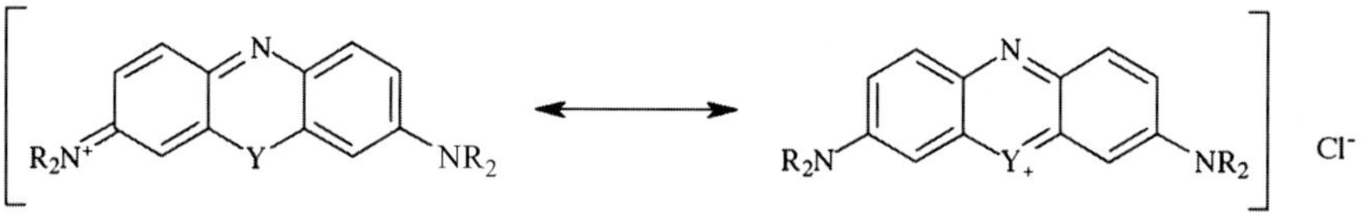

15

where Y = -O-, -S-, or -NR-
R = hydrogen, alkyl, aryl

The phenazine dye mauveine (**16**), discovered by PERKIN [12], was the first synthetic dye produced on an industrial scale.

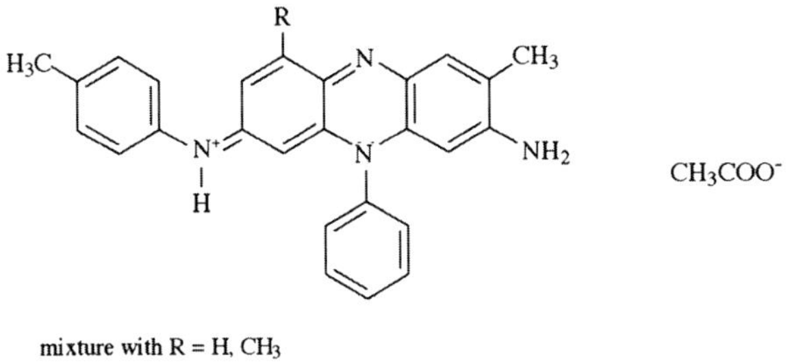

2.4.3 Chemical Structure and Classification

16 mixture with R = H, CH₃ ; CH$_3$COO$^-$

Of current importance are the oxazine dye **17** [13] and the thiazine dye **18** (methylene blue) [14].

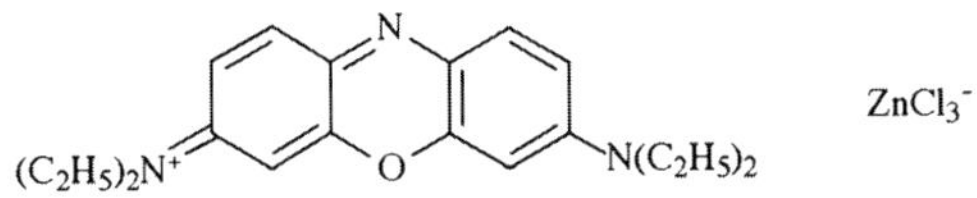

17 ZnCl$_3^-$

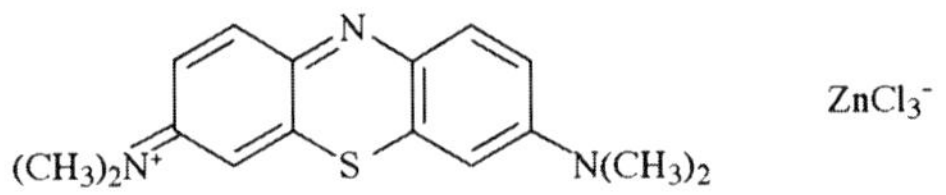

18 ZnCl$_3^-$

2.4.3.2 Dyes with Localized Charge

In contrast to dyes with delocalized charge, in which the charge is an essential part of the chromophore, in dyes with localized charge the charged atom in most cases is isolated from the chromophore by a nonconjugated group. Therefore, any chromophore can be used, and this chromophore is not or only slightly influenced by the charged atom. Representatives of cationic dyes with localized charge are found in the following classes of dyes:

1) *Azo Dyes.* The localized charge in cationic azo dyes may reside either in the diazo moiety (e.g., **19**) [15] or in the coupling component (e.g., **20**) [16]:

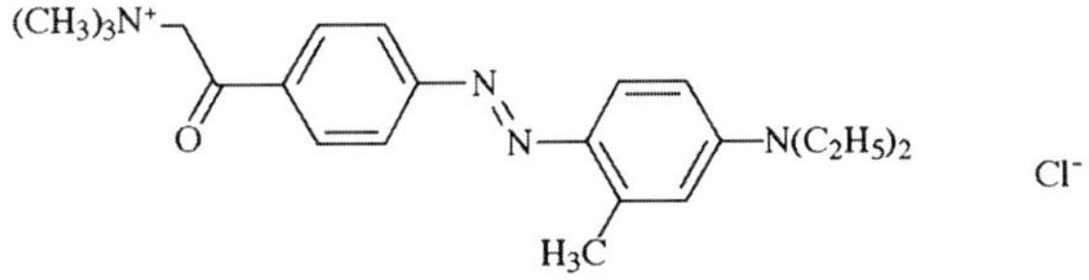

19 Cl$^-$

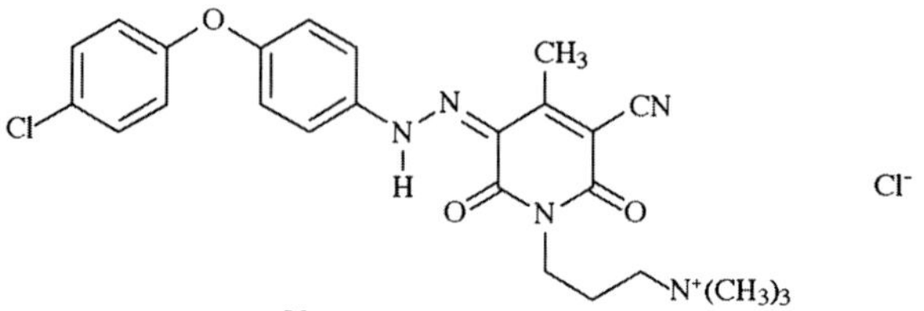

2.4 Cationic Dyes as Chromophores

20

2) *Anthraquinone dyes* (e.g., **21**) [17].

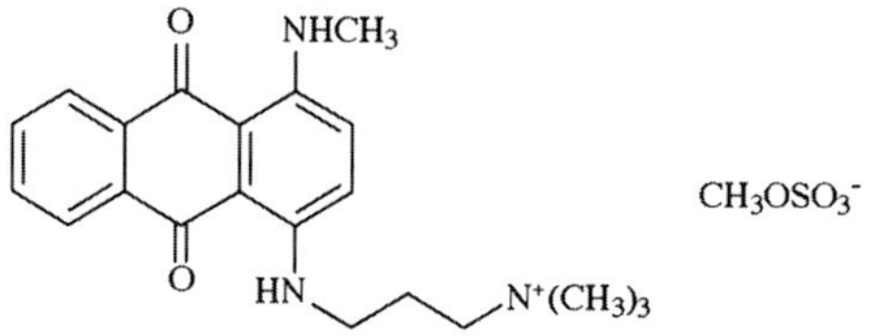

21

3) *Phthalocyanine dyes* (e.g., **22**) [18]:

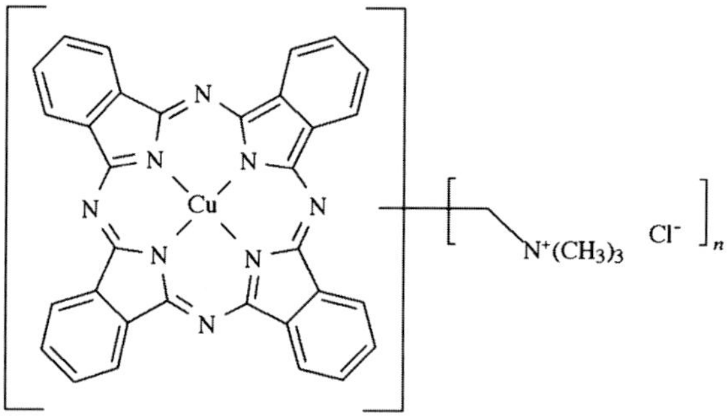

22

where *n* = 2-8

4) *Perinone dyes* (e.g., **23**) [19]:

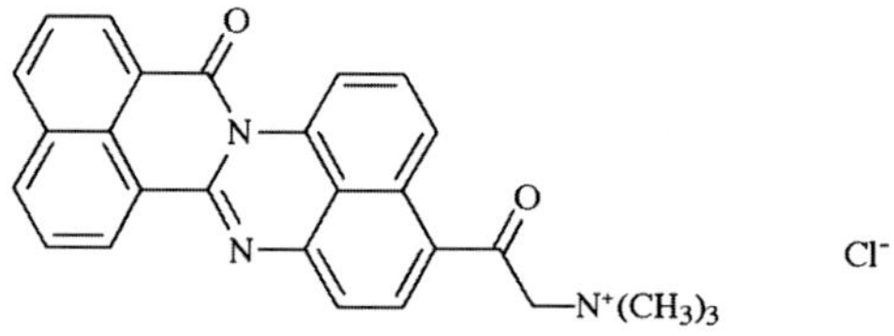

23

2.4.3 Chemical Structure and Classification

5) *Naphthalimide dyes* (e.g., **24**) [20]:

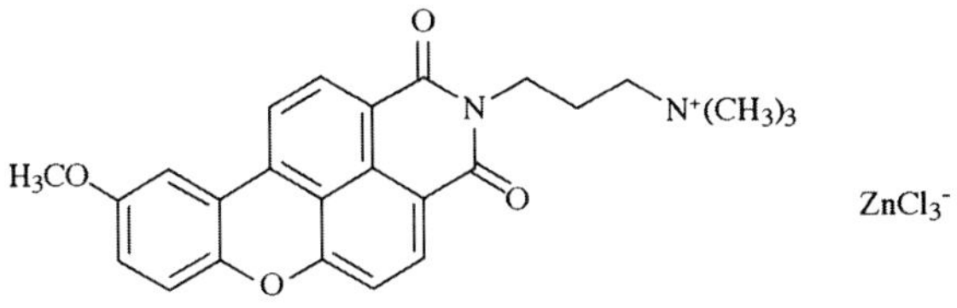

24, ZnCl$_3^-$

6) *Quinophthalone dyes* (e.g., **25**) [21]:

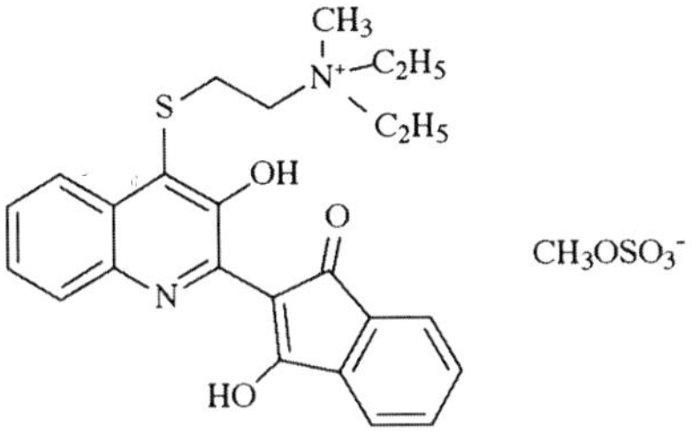

25, CH$_3$OSO$_3^-$

7) *Neutrocyanine dyes* (e.g., **26**) [22]:

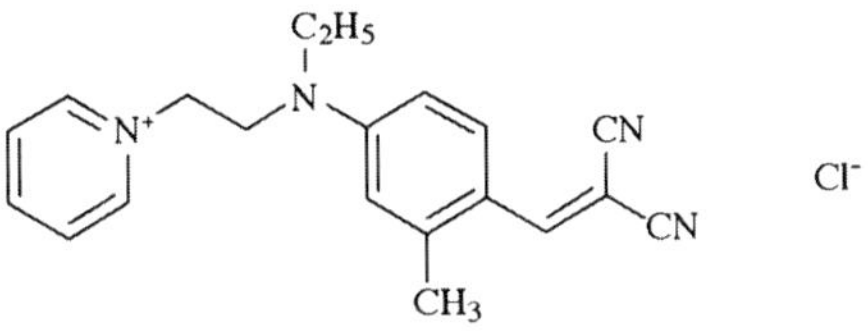

26, Cl$^-$

8) *Nitro dyes* (e.g., **27**) [23]:

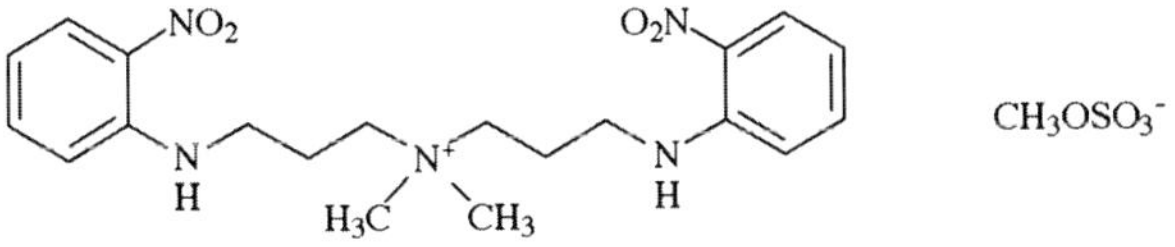

27, CH$_3$OSO$_3^-$

2.4.4 Principal Properties

Cationic dyes were used initially for dyeing of silk, leather, paper, and cotton mordanted with tannin, as well as for the production of ink and copying paper in the office supplies industry. Their high brilliance and color strength partly compensated for their poor lightfastness on these materials. With the development of synthetic fibers their most suitable substrates were found, and their importance for dyeing textiles increased greatly. Cationic dyes were employed initially for printing of cellulose acetate but gained much greater importance for dyeing of polyacrylonitrile, acid-modified polyester, and polyamide fibers. (see Chapter 4)

2.4.4.1 Cationic Dyes for Synthetic Fibers

Glenz and Beckmann [24] were the first to investigate color balance and surface potentials in polyacrylonitrile fibers. On the basis of their results, they proposed the following dyeing mechanism: The cationic chromophore is first absorbed by the negatively charged fiber surface and then diffuses, at elevated temperature, into the interior of the fiber; there it binds to active acid groups, the number of which is limited and the accessibility of which depends on temperature and fiber constitution. Therefore, the dyeing characteristics of a cationic dye are determined by affinity and diffusibility.

The *affinity* of cationic dyes is enhanced by increasing the size of the molecule and especially by introducing aromatic residues. The phenyl-substituted dye **28** is absorbed onto polyacrylonitrile materials at a higher rate than the dye **11** [25].

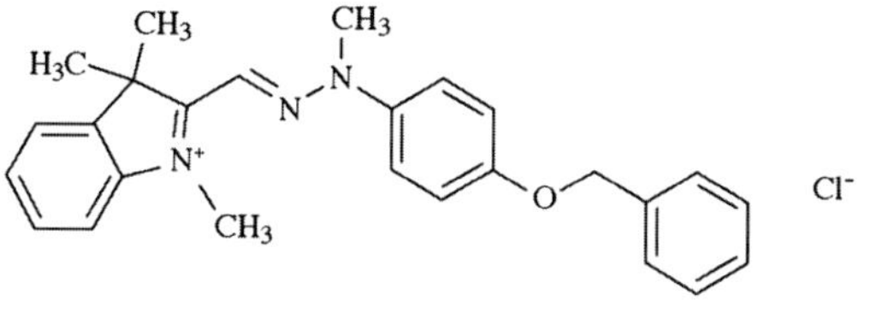

28

Residues which make the substance more hydrophilic, such as carboxyl and hydroxyethyl groups, lower the affinity of cationic dyes for polyacrylonitrile. Therefore, dye **29** is absorbed onto polyacrylonitrile materials much more slowly than dye **2** [26].

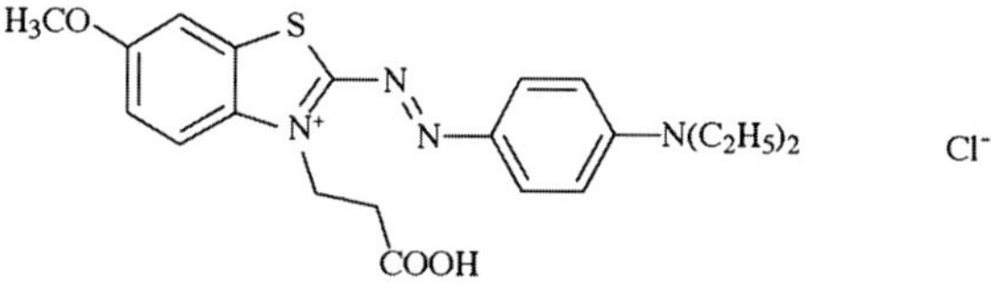

29

High affinity may, however, lead to problems in leveling, because these dyes are bound quickly and nearly irreversibly to the acid groups of the fiber.

Diffusibility increases with decreasing molecular mass. Dyes with a cation mass of less than 275 migrate well and are suitable for the production of level dyeings, especially in the lighter shades [27].

Cationic dyes are also used to dye polyacrylonitrile fibers during the spinning process either from dimethylformamide solution or in the gel stage subsequent to aqueous spinning processes.

In Western Europe, the bulk of acrylic fibers is targeted for the clothing industry. As much as two-thirds of acrylic fiber production is used in this sector, whereas only one-third is used for home furnishings and in technical applications.

2.4.4.2 Cationic Dyes for Paper, Leather, and Other Substrates

Cationic dyes exhibit good affinity for negatively charged wood pulp and unbleached pulp grades. Anionic fixing agents improve their wetfastness. Cationic dyes are chosen because of their brilliance and intensity, which makes them especially valuable for recycled paper grades. Their poor lightfastness is not a major problem because of the low light stability of the substrate itself. They suffer from poor affinity for bleached, lignin-free pulp grades, which are used, for instance, in the manufacture of table cloths and napkins. Therefore, dyes have been developed for this application that bear several cationic charges linked by substantive moieties, such as urea or triazine [28] (e.g., **30**).

30

Cationic dyes were the first synthetic organic dyes employed for dyeing leather, initially for dyeing of vegetable-tanned leather. Currently, brilliant cationic dyes are used for enhancing the brilliance of the shades obtained with 1:2 metal-complex dyes.

Because of their coloring strength, cationic dyes are used in the office supplies industry for the manufacture of ink, typewriter ribbon, and copying paper.

Colorless derivatives of the carbinol bases of cationic dyes have recently found application for the manufacture of carbonless copy paper and for special paper used for thermal printers. Examples of such color formers are the crystal violet derivative **31** [29], which develops a blue color, and the black-developing fluorane lactone **32** [30] and benzoxazine **33** [31].

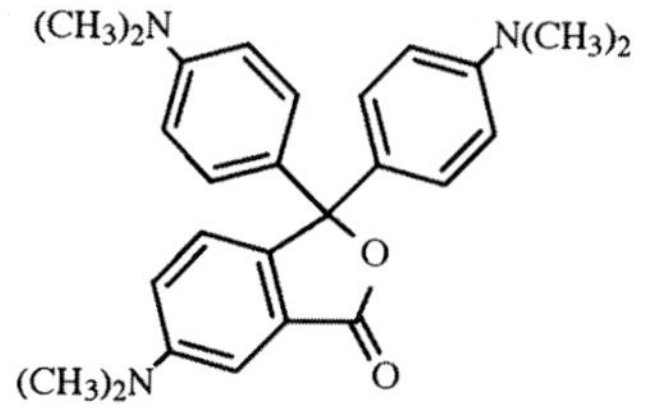

31

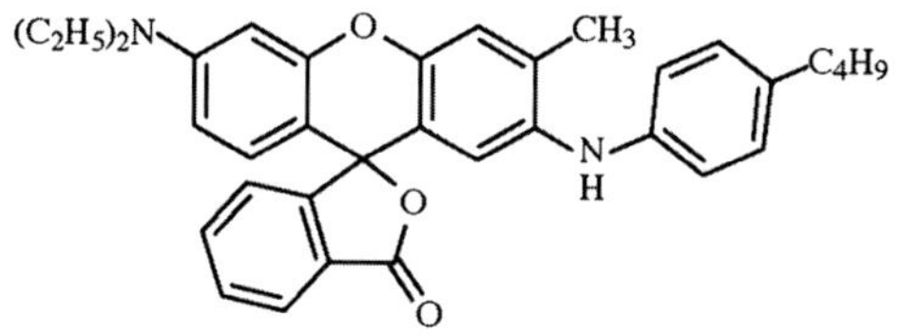

32

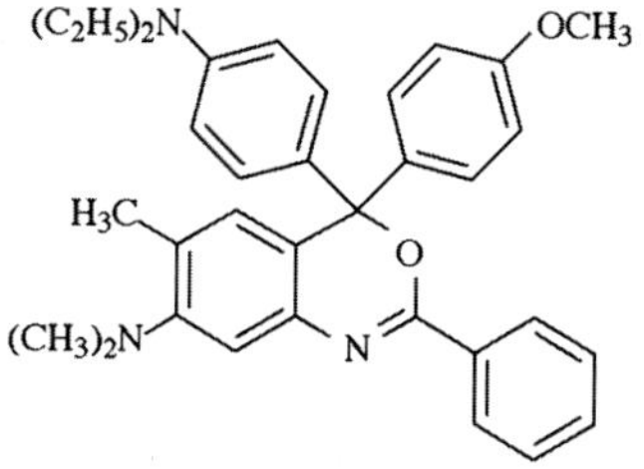

33

Cationic dyes which absorb in the near-infrared region of the spectrum are used in optical data storage disks, e.g., **34** [32]:

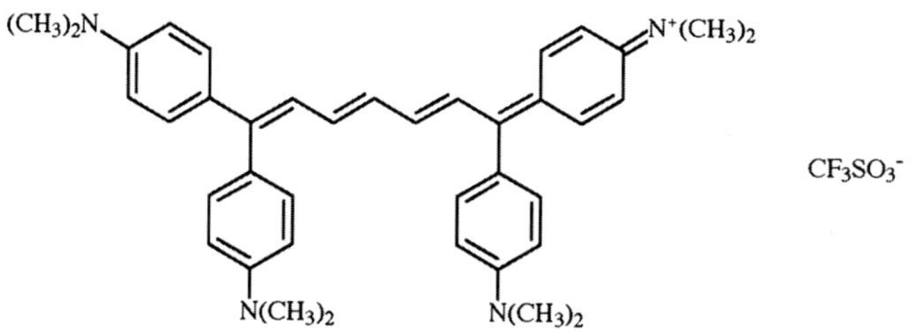

34

2.4.5 References

[1] H. A. Staab, *Einführung in die Theoretische Organische Chemie*, 4th ed., Verlag Chemie, Weinheim, 1964, pp. 236–336, and references cited therein; S. Dähne Z. *Chem.* **11** (1965) 441.
[2] Bayer, FR 1 490 350, 1965.
[3] IG-Farbenind., DE 686 198, 1935 (K. Winter, N. Roh, P. Wolff, G. Schäfer).
[4] IG-Farbenind./Bayer, DE 891 120, 1935 (W. Müller, C. Gerres).
[5] Bayer, DE 1 190 126, 1962 (A. Brack, D. Kutzschbach).
[6] Bayer, DE 410487, 1922 (W. König); DE 415534, 1923 (A. Blömer); IG Farbenind., DE 641799, 1928 (K. Schmidt).
[7] Bayer, DE 1 083 000, 1957 (R. Raue).
[8] Geigy, DE 1 044 023, 1954, (W. Bossard, J. Voltz, F. Favre); DE 1 050 940, 1954 (W. Bossard, J. Voltz, F. Favre).
[9] O. Fischer, *Ber. Dtsch. Chem. Ges.* **10** (1877) 1625; **11** (1878) 950; 1881, *14*, 2520.
[10] BASF, DE 26016, 27032, 27789, 1883 (Kern, H. Caro); Meister, Lucius und Brüning (today Hoechst AG), DE 34463, 1884.
[11] BASF, DE 44002, 1887 (M. Ceresole). Meister, Lucius und Brüning (today Hoechst AG), DE 48367, 1888.
[12] W. H. Perkin, *J. Chem. Soc.* **69** (1896) 596, and references cited therein; O. Meth-Cohn, M. Smith, *J. Chem. Soc. Perkin Trans. 1* **1994**, 5.
[13] Hoechst, DE 1 569604, 1966 (G. Schäfer, N. Ottawa).
[14] BASF, DE 1886, 1877 (H. Caro).
[15] Du Pont, DE 1 054 616, 1954 (S. N. Boyd, Jr.).
[16] Sandoz, DE 2 118 536, 1970 (R. Entschel, C. Müller, W. Steinemann); BE 756 820, 1970; CH 549 629, 1971 (R. Entschel, C. Müller, W. Steinemann).
[17] Bayer, DE 1 150 652, 1955 (G. Gehrke, L. Nüssler).
[18] Allied Chem. Corp., US 3 565 570, 1968 (C. K. Dien).
[19] Ciba, FR 1 489 639, 1965.
[20] Hoechst, DE 2 122 975, 1971 (H. Tröster, K. Löhe, E. Mundlos, R. Mohr).
[21] Sandoz, DE 2 055 918, 1969 (B. Gertisser, F. Müller, U. Zirngibl).
[22] Eastman Kodak, DE 1 250 947, 1963 (J. G. Fisher, D. J. Wallace, J. M. Straley).
[23] Hoechst, DE 1 941 376, 1969 (K. Lohe).
[24] O. Glenz, W. Beckmann, *Melliand Textilber.* **38** (1957) 296.
[25] Bayer, DE 2 040 872, 1970 (H.-P. Kühlthau, R. Raue).
[26] Sandoz, FR 2 013 732, 1969 (R. Entschel, V. Käppeli, C. Müller).
[27] Ciba-Geigy, DE 2 548 009, 1974 (J. Koller, M. Motter, P. Moser, U. Horn, P. Galafassi).
[28] Bayer, DE 2 933 031, 1979 (H. Nickel, F. Müller, P. Mummenhoff).
[29] BASF, DE 1 962 881, 1969 (D. Leuchs, V. Schabacker).
[30] Nisso Kako, JP 50 82 126, 1971; JP 48 62 506, 1971; DE 2 242 693, 1971 (M. Yagahi, T. Suzuki, T. Igaki, S. Kawagoe, S. Horiuchi). Yamamoto Kagaku, JP 51 15 445, 1971; JP 51 06561, 1974; DE 2 202 315, 1971 (Y. Hatano, K. Yamamoto). Nat. Cash Reg., DE 2 155 987, 1970 (Ch.-H. Lin). Fuji, DE 2 025 171, 1969 (S. Kimura, T. Kobayashi, S. Ishige, M. Kiritani).
[31] Bayer, EP 0 187 329, 1985 (H. Berneth); EP 0 254 858, 1986 (H. Berneth, G. Jabs).
[32] Minnesota Mining and Manufacturing, WO 83/2 027, 1982 (V. L. Bell, I. A. Ferguson, M. J. Weatherley).

2.5 Polymethine and Related Chromophores

2.5.1 Introduction

Methine dyes and polyene dyes are characterized by a chain of methine groups that forms a system of conjugated double bonds. The methine chain in *polyene dyes* is even-numbered and the terminal groups do not affect the excitation of electrons in the dye. The most important group of polyene dyes, the carotenoids, bear aliphatic or alicyclic groups at the end of the methine chain, and a very long conjugated chain is necessary to shift the excitation wavelength to the visible region. β-Carotene contains a chain of 22 methine groups and absorbs at 450 and 478 nm.

Methine dyes are characterized by a chain of conjugated double bonds with an odd number of carbon atoms between the two terminal groups X and Y, which may be considered as being capable of charge exchange.

$$\left[X=C-(C=C)_n-Y \longleftrightarrow X-C=(C-C)_n=Y \right]^q$$

One of the substituents X and Y is an electron acceptor and the other as an electron donor. The dyes can be cationic, anionic, or neutral.

Methine dyes have an extended system of conjugated double bonds. A large number of resonance structures may therefore be formulated, of which (**1**) and (**2**) are the most important, i.e., the electron density is lowest on the nitrogen atoms.

$$\left[N=C-(C=C)_n-N \longleftrightarrow N-C=(C-C)_n=N \right]^+$$
$$\quad\quad\quad\quad\quad\quad 1 \quad\quad\quad\quad\quad\quad\quad\quad 2$$

In all the following formulas only one resonance structure is shown. To indicate that a π-electron system with many possible resonance structures is involved, the total charge of the molecule is given outside a square bracket.

In the following classification of methine dyes only nitrogen and oxygen are considered as charge-carrying terminal atoms, although other heteroatoms are possible. The charge-carrying terminal atoms may be joined directly to the methine chain, may be linked via an aromatic group to the methine chain, or may form part of a heterocyclic ring.

$$\left[N=C-(CH=CH)_n-N \right]^+$$

Cyanine dyes (**3**) are the best known polymethine dyes. Nowadays, their commercial use is limited to sensitizing dyes for silver halide photography and as infrared absorbers for optical data storage and other (bio)imaging applications. However, derivatives of cyanine dyes provide important dyes for polyacrylonitrile.

n = 0, 1, 2,...
R groups are typically part of ring system

3

2.5.2 General Synthesis

The general synthesis of the most important dyes within this group, e.g. the cyanine dyes, is described in Section 5.7.

2.5.3 Principal Properties and Classification

2.5.3.1 Azacarbocyanines

A cyanine containing three carbon atoms between the heterocyclic rings is called a carbocyanine (**3**, *n* = 1). Replacing these carbon atoms by one, two, and three nitrogen atoms produces aza-, diaza-, and triazacarbocyanines. Dyes of these three classes are important yellow dyes for polyacrylonitrile, e.g., *C.I. Basic Yellow 28*, 48054 [54060-92-3] (**4**)

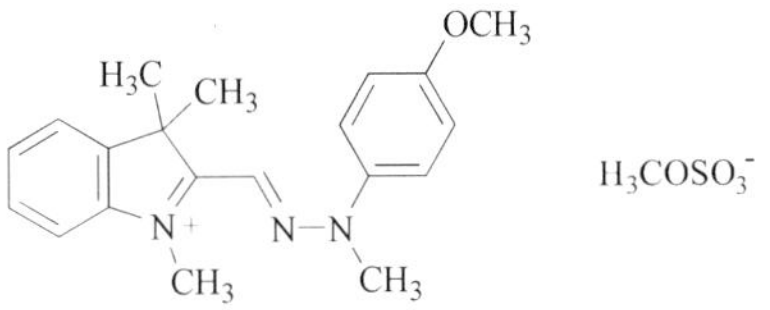

2.5.3.2 Hemicyanines

The hemicyanine dyes can be represented by structure (**5**). They may be considered as cyanines in which a benzene ring has been inserted into the conjugated chain. Hemicyanines provide some bright fluorescent red dyes for polyacrylonitrile.

2.5.3.3 Diazahemicyanines

Diazahemicyanine dyes are arguably the most important class of polymethine dyes. They have the general structure (**6**).

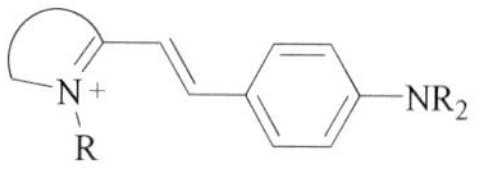

5 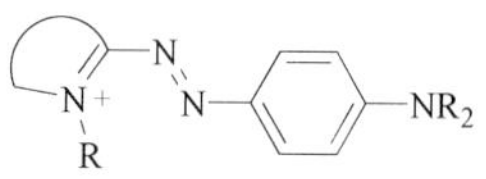 **6**

The heterocyclic ring normally contains one (e.g., pyridinium), two (e.g., pyrazolium and imidazolium), or three (e.g., triazolium) nitrogen atoms, or sulfur and nitrogen atoms, e.g., (benzo)thiazolium and thiadiazolium. Triazolium dyes (**7**) provide the market-leading red dyes for polyacrylonitrile, and the benzothiazolium dye (**8**) is the market-leading blue dye.

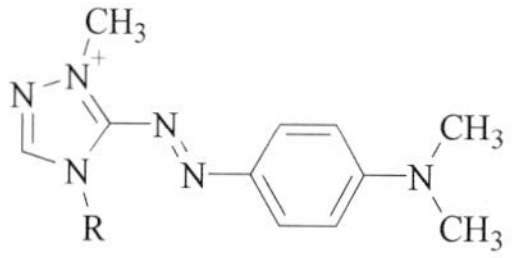

7 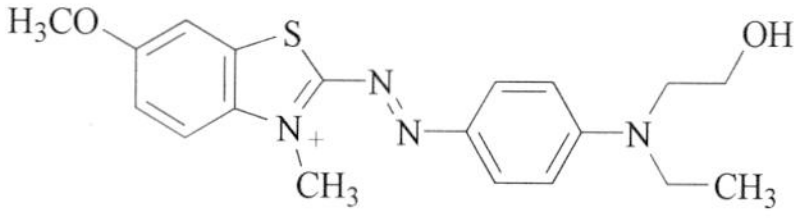 **8**

2.5.3.4 Styryl Dyes

The styryl dyes are neutral molecules containing a styryl group $C_6H_5CH=C$, usually in conjugation with an *N,N*-dialkylaminoaryl group. Styryl dyes were once a fairly important group of yellow dyes for a variety of substrates. They are synthesised by condensation of an active methylene compound, especially malononitrile (**9**, X = Y = CN), with a carbonyl compound, especially an aldehyde. Styryl dyes have small molecular structures and are ideal for dyeing densely packed hydrophobic substrates such as polyester. C.I. Disperse Yellow 31 (**57**; R = C_4H_9, R^1 = C_2H_4Cl, R^2 = H, X = CN, Y = $COOC_2H_5$) is a typical styryl dye.

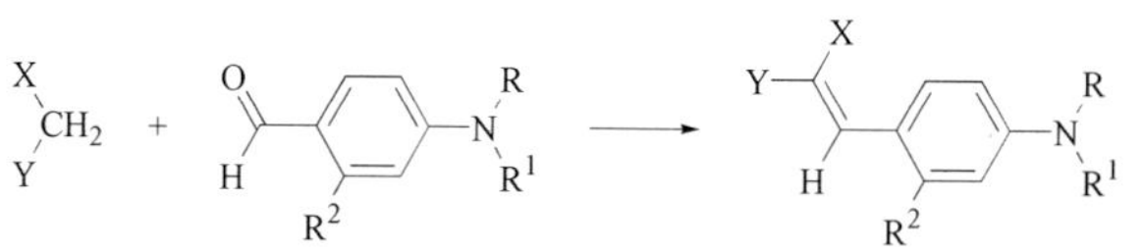

9 **10**

Yellow styryl dyes have now been largely superseded by superior dyes such as azopyridones, but there has been some interest in red and blue styryl dyes. The addition of a third cyano group to produce a tricyanovinyl group causes a large bathochromic shift: the resulting dyes (e.g., **11**) are bright red rather than the greenish yellow of the dicyanovinyl dyes. These tricyanovinyl dyes have been patented by Mitsubishi for the transfer printing of polyester substrates. Synthetic routes to the dyes are the replacement of a cyano group in tetracyanoethylene and the oxidative cyanation of a dicyanovinyl dye (**12**) with cyanide. The use of such toxic reagents is a hindrance to the commercialization of the tricyanovinyl dyes.

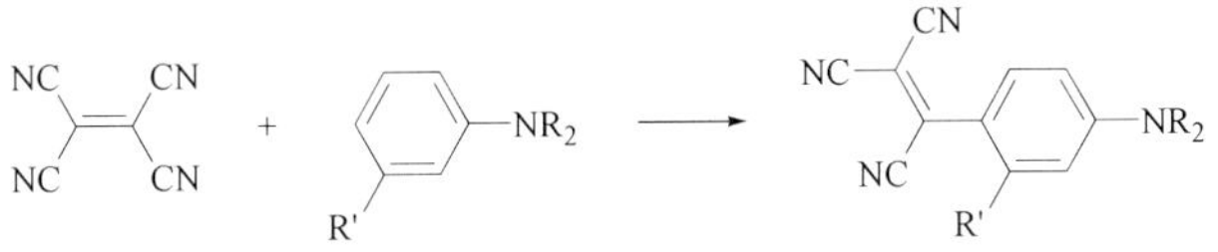

Blue styryl dyes are produced when an even more powerful electron-withdrawing group than tricyanovinyl is used. Thus, Sandoz discovered that the condensation of the sulfone (**13**) with an aldehyde gives the bright blue dye (**14**) for polyester. In addition to exceptional brightness, this dye also possesses high tinctorial strength (ε_{max} ca. 70 000). However, its lightfastness is only moderate.

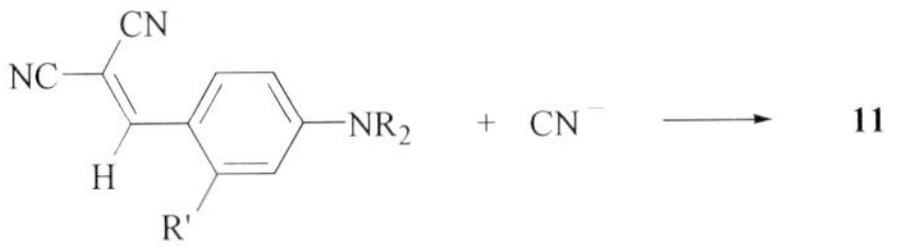

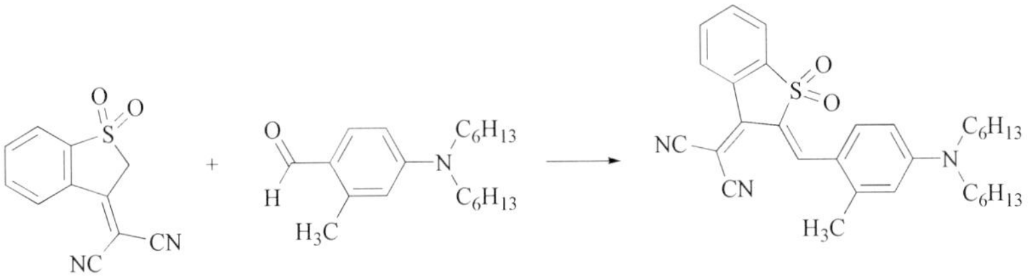

2.6 Di- and Triarylcarbenium and Related Chromophores

2.6.1 Introduction

Di- and triarylcarbenium dyes [1] belong to the class of polymethine dyes and can be considered as branched polymethines. The branches are created by two aryl rings, in which the polymethine chain is incorporated, and by another R group bonded to the central (*meso*) methine carbon atom (**1**).

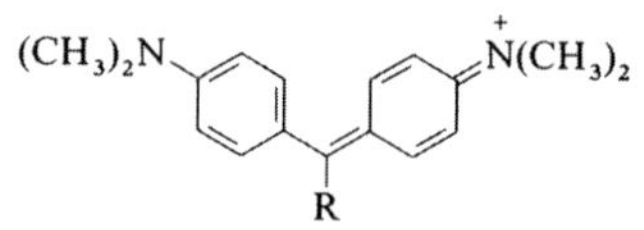

[Structure 1: (CH$_3$)$_2$N–C$_6$H$_4$–C(R)=C$_6$H$_4$=N$^+$(CH$_3$)$_2$]

The R group possesses π-electrons or lone pairs of electrons that can interact with the rest of the π-electron system. The most important electron donor is the amino group. *Triarylmethine dyes* are usually divided into mono-, di-, and triaminotriarylmethine dyes. In some di- and triarylmethine dyes, the ring carbon atoms *ortho* to the central methine carbon atom are bonded via a heteroatom to form a heterocyclic six-membered ring. These include the acridine, xanthene, and thioxanthene dyes.

2.6.2 Chromophores

The diphenylmethyl and triphenylmethyl cations can be considered the basic chromophores of di- and triarylmethane dyes [1–4]. However, the electronic state of the donor-substituted di- and triarylmethine dyes can be described better by breaking down the chromophore of these dyes into a straight- or branched-chain polymethine subchromophore, respectively, and two ethylene units (from the formal cleavage of the benzene rings) [2]. This model allows the similarities between the di- and triarylmethine dyes and the polymethines to be recognized.

The introduction of two dimethylamino groups into the diphenylmethyl cation gives Michler's hydrol blue (**1**, R = H). This compound shows a strong bathochromic shift and an increase in intensity of the absorption band at the longest wavelength relative to the diphenylmethyl cation [**1**, R = H; λ_{max} (log ε) = 607.5 nm (5.17)]. Donor substituents R in diarylmethane dye **1** cause a blue shift of the longest-wavelength absorption band, whereas acceptor groups generally cause a red shift [2].

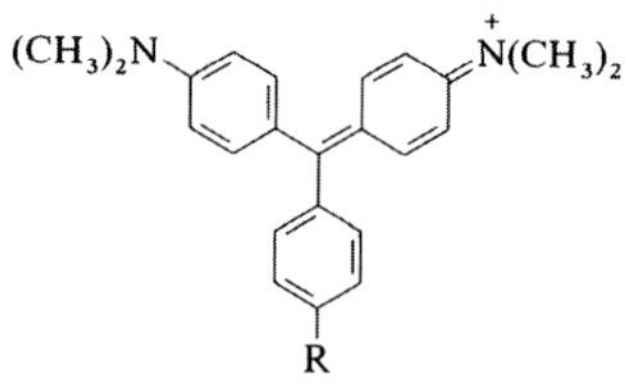

[Structure 2: Malachite green skeleton with two (CH$_3$)$_2$N– aryl groups and a third aryl group bearing R]

Malachite green (**2**, R = H) has two absorption bands in the visible spectrum at λ_{max} (log ε) = 621 (5.02) and 427.5 nm (4.30). The intense absorption band at longest wavelength (*x* band) is shifted to the red compared with Michler's hydrol blue. The second band (*y* band) is in the shorter-wavelength part of the spectrum and is significantly weaker [1–3].

2.6.2 Chromophores

Substituents in the *meta* and *para* positions of the *meso*-phenyl ring shift the position of the absorption band at longest wavelength. If a dimethylamino group is introduced into the *para* position of the *meso*-phenyl ring, crystal violet is obtained [**2**, R = N(CH$_3$)$_2$], instead of malachite green. The absorption spectrum of the former has a hypsochromically shifted x [λ_{max} (log ε) = 589 nm (5.06)] and no y band [9]. The *hydroxytriarylmethane dyes*, which are isoelectronic with cationic triarylmethane dyes, absorb at a shorter wavelength than the latter. The longest-wavelength absorption band of benzaurin (**3**) is at λ_{max}= 585 nm compared with λ_{max} = 621 nm in malachite green [2].

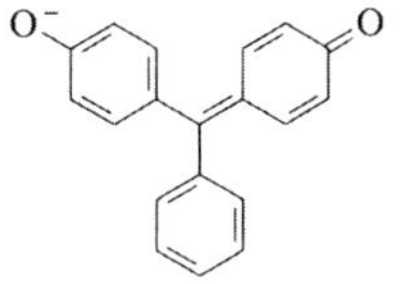

3

By bonding two phenyl rings via a *heteroatom bridge* in the 2- and 2'-positions, the yellow acridine dyes (**4a**), the red to violet xanthene dyes (**4b**), or similarly colored thioxanthene dyes (**4c**) in the case of amino-substituted di- or triarylmethane dyes are obtained. Because of their rigid molecular skeletons these compounds fluoresce. Hydroxyxanthenes behave similarly (e.g., fluorescein) [2].

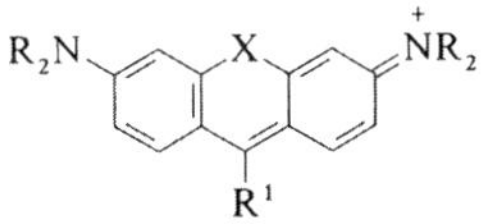

4a: X = NR^2 Acridines
4b: X = O   Xanthenes
4c: X = S   Thioxanthenes

On protonation of crystal violet, the color changes first to green and then to yellow. The green coloration is caused by protonation of one of the dimethylamino groups and can be explained by the appearance of the y band. Protonation of another amino group changes the color to yellow. Amino groups whose conjugation with the ring system is reduced for steric reasons are readily protonated so the intensity of the absorption bands is lowered in comparison with systems having unhindered conjugation because of the proportion of protonated species. In an alkaline medium, decoloration occurs because of the formation of triphenylmethanol (e.g., **5**) [2].

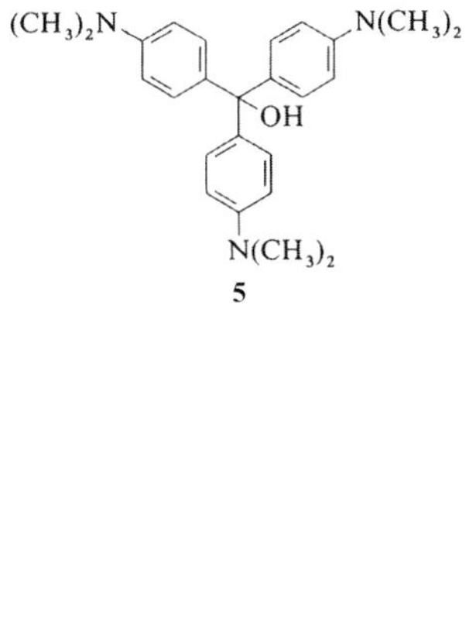

5

2.6 Di- and Triarylcarbenium and Related Chromophores

Color formers are very sensitive to the addition of acids and solvents. In acid, crystal violet lactone (**6**) is only partially converted to the colored carboxylic acid (**7**) with opening of the lactone ring [λ_{max} (log ε) = 603 nm (4.41) in methanol / acetic acid (1 / 1)].

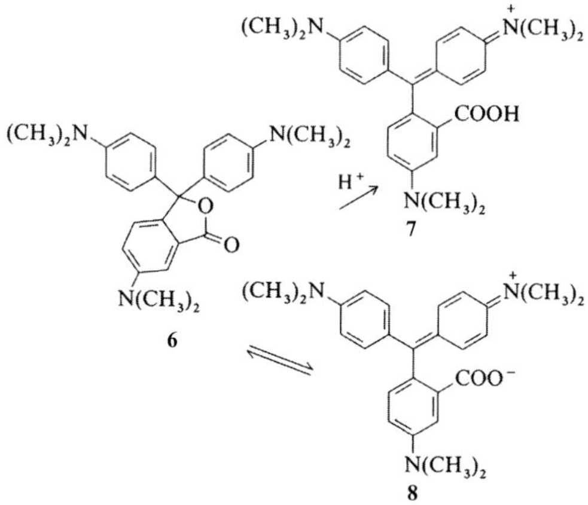

The acid is in equilibrium with colorless lactones protonated on the nitrogen atom. In highly polar solvents such as trifluoroethanol, hexafluoro-2-propanol, or phenol, the lactone ring opens to give the zwitterion (**8**). The longest-wavelength band of (**8**) [λ_{max} (log ε) = 593 nm (4.91) in hexafluoro-2-propanol] is shifted hypsochromically relative to that of (**7**). As in the case of rhodamine dyes [5], this reaction is a solvent-dependent intramolecular Lewis acid–base equilibrium. The high intensity of the band shows that the equilibrium is shifted strongly to the zwitterionic compound.

2.6.3 General Synthesis

Condensation Reactions. Many triaminotriarylmethane dyes can be produced by condensation of Michler's ketone (**9**) with aromatic amines. The ketone must generally be activated with phosgene or phosphorus oxychloride, whereby a reactive blue chloro compound (**10**) is formed from the intermediate geminal dichloride. With N-phenyl-1-naphthylamine, for example, (**13**) reacts to form Victoria blue B (**11**).

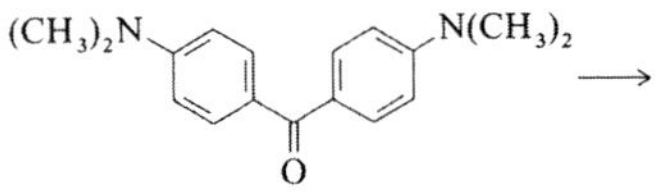

2.6.3 General Synthesis

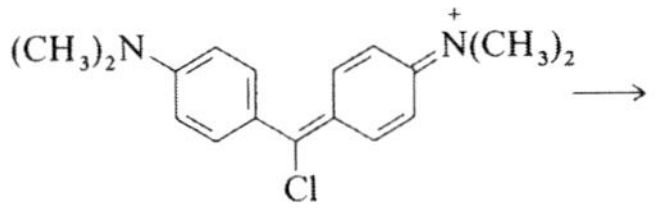

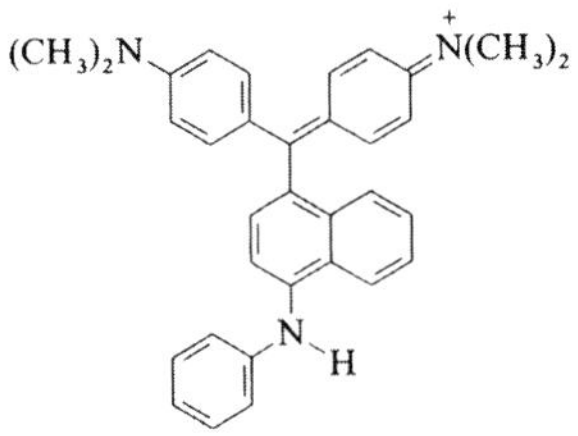

Condensation of Michler's hydrol (**12**) with 3-dimethylaminobenzoic acid (**13**) gives leuco crystal violet carboxylic acid (**14**), which is then oxidized to crystal violet lactone.

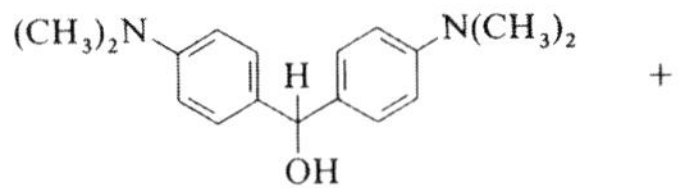

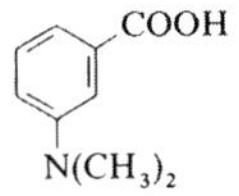

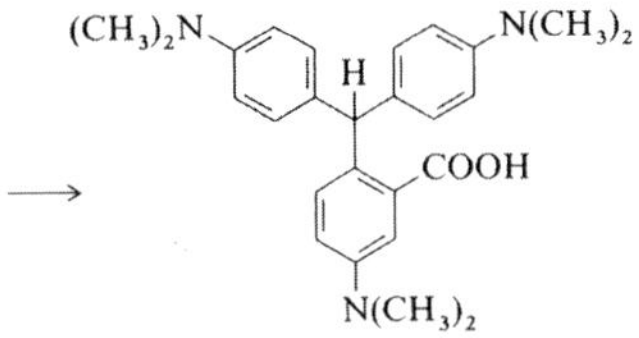

A leuco compound is also obtained when a benzaldehyde derivative is treated with 2 molar equivalents of an aniline derivative. Leuco malachite green (**17**) is thus formed from benzaldehyde (**15**) and dimethylaniline (**16**).

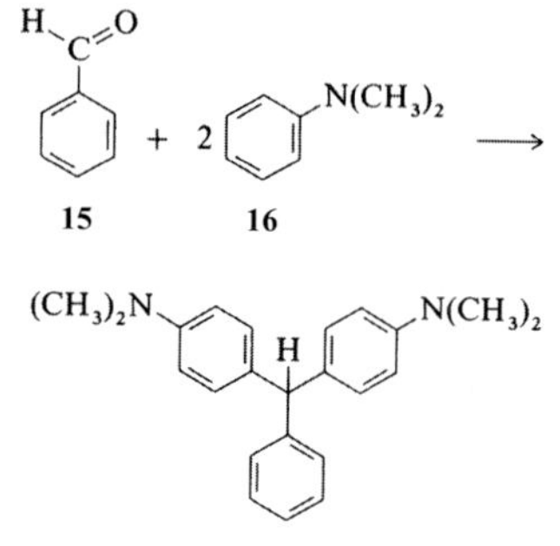

2.6 Di- and Triarylcarbenium and Related Chromophores

[Scheme showing benzaldehyde (15) + 2 equivalents of N,N-dimethylaniline (16) → compound 17: bis(4-dimethylaminophenyl)phenylmethane with central CH]

With benzaldehyde-2-sulfonic and benzaldehyde-2,4-disulfonic acids, important acid dyes are synthesized by this reaction. They are known as patent blue dyes.

Crystal violet (**20**) can be produced directly from formaldehyde and dimethylaniline. The methane base (**18**), N,N,N′,N′-tetramethyl-4,4′-methylenedianiline, is formed initially. This is then oxidized to Michler's hydrol (**12**), which condenses with another molecule of dimethylaniline to give leuco crystal violet (**19**). The latter is converted to the dye in a second oxidation step. Pararosaniline, methyl violet, and Victoria blue can also be obtained by this reaction sequence.

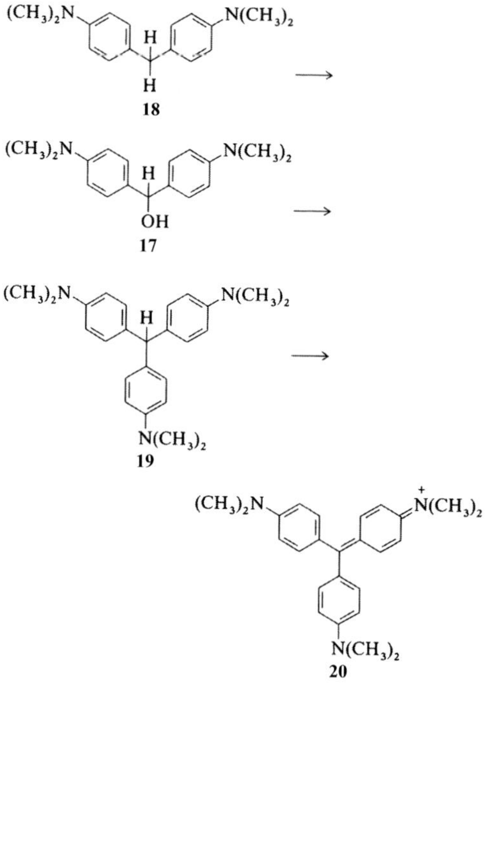

[Scheme: 18 (methane base, CH₂ bridge) → 17 (carbinol with OH) → 19 (leuco crystal violet, tris(4-dimethylaminophenyl)methane) → 20 (crystal violet cation)]

Nucleophilic substitution of chlorine, sulfonic acid, or amino groups by aromatic amines is a common process for producing triaminotriarylmethane dyes. Spirit blue (**23**) can be formed from the trichlorotriphenylmethyl cation (**21**) [6, 7] and from pararosaniline (**22**) [8, 9] by reaction with aniline.

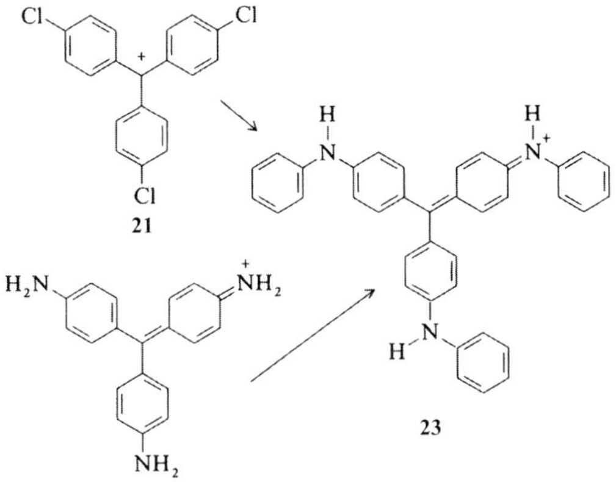

Synthesis of Xanthene Dyes. Phthalic anhydride reacts with 3-diethylaminophenol in the molten state to give rhodamine B, and with resorcinol to give fluorescein. For unsymmetrical xanthene dyes, dihydroxybenzoyl- (**24**) or an aminohydroxybenzoylbenzoic acid (**25**) is required.

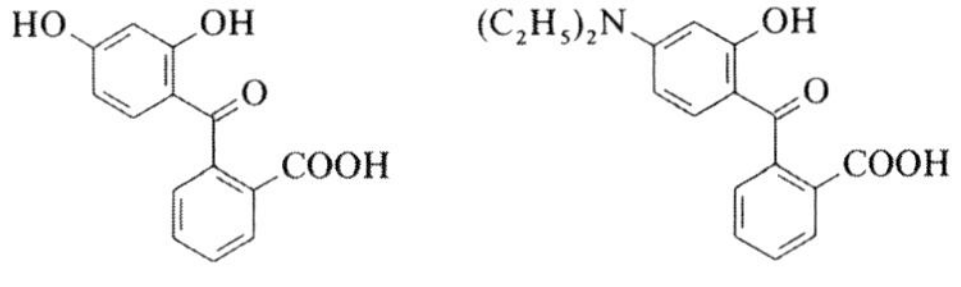

These compounds can condense and cyclize with resorcinol or substituted 3-aminophenols. Nucleophilic substitution of chlorine by amines leads to xanthene dyes which contain a sulfonic acid group [10] or hydrogen [11] instead of the carboxyl group.

2.6.4 Principal Properties

As a class, the dyes are bright and strong, but are generally deficient in lightfastness. Consequently, they are used in outlets where brightness and cost-effectiveness, rather than permanence, are paramount, for example, the coloration of paper. Many dyes of this class, especially derivatives of pyronines (xanthenes) are among the most fluorescent dyes known (see later).

2.6 Di- and Triarylcarbenium and Related Chromophores

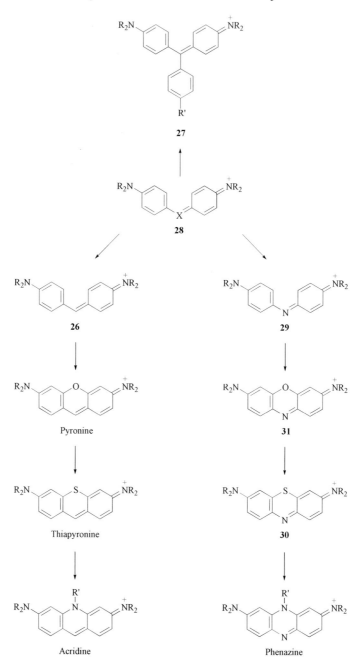

Figure 2.6 Structural interrelationships among diaryl- and triarylcarbenium dyes. Pyronine = xanthene; **29** (R = CH$_3$) is Bindschedler's Green.

Typical dyes are the diphenylmethane Auramine O (**28**; R = CH$_3$, X = C-NH$_2$; *C.I. Basic Yellow 2*); the triphenylmethane Malachite Green (**27**; R = CH$_3$; R' = H; *C.I. Basic Green 4*); the thiazine dye Methylene Blue (**30**; R = CH$_3$; *C.I. Basic Blue 9*) used extensively in the Gram staining test for bacteria; the oxazine dye *C.I. Basic Blue 3* (**31**; R = C$_2$H$_5$), and the xanthene dye *C.I. Acid Red 52* (**32**). The bright magenta dye (**32**) is used in several ink-jet magenta inks, normally to improve the chroma (vividness).

The structural interrelationships of the diarylcarbenium dyes (**26**), triarylcarbenium dyes (**27**) and their heterocyclic derivatives are shown in Figure 2.6.

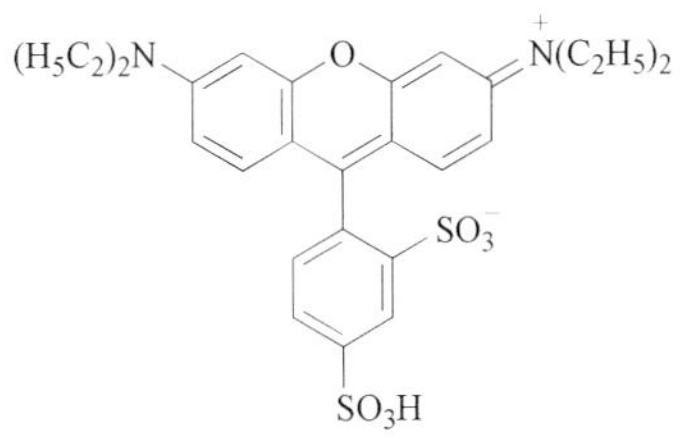

32

2.6.5 References

[1] J. Griffiths: *Colour and Constitution of Organic Molecules*, Academic Press, London, 1976, p. 250.
[2] J. Fabian, H. Hartmann: "Light Absorption of Organic Colorants", *Reactivity and Structure Concepts in Organic Chemistry*, vol. 12, Springer Verlag, Berlin, 1980, p. 137.
[3] H. Zollinger: *Color Chemistry*, 2nd ed., VCH Verlagsgesellschaft, Weinheim, 1991, p. 71.
[4] J. Griffiths, K. J. Pender, *Dyes Pigm.* **2** (1981) 37.
[5] C. Reichardt: *Solvents and Solvent Effects in Organic Chemistry*, 2nd ed., VCH Verlagsgesellschaft, Weinheim 1988, p. 107.
[6] Hoechst, DE 1 098 652, 1956.
[7] Hoechst, DE 1 644 619, 1967.
[8] Dahl, DE 36 900, 1886.
[9] ICI, DE 508 499, 1929.
[10] Hoechst, DE 2 460 491, 1974 (M. Haehnke, F. Kohlhaas, F. Meininger, T. Papenfuhs).
[11] Hoechst, DE 2 344 443, 1973 (R. Neeb, T. Papenfuhs).

2.7 Phthalocyanine Chromophore

2.7.1 Introduction

The term phthalocyanine was first used by R. P. Linstead in 1933 [1] to describe a class of organic dyes, whose colors range from reddish blue to yellowish green. The name phthalocyanine originates from the Greek terms *naphtha* for mineral oil and *cyanine* for dark blue. In 1930–1940, Linstead et al. elucidated the structure of phthalocyanine (H_2Pc) and its metal complexes [1–11]. The basic structure is represented by phthalocyanine (**1**) itself:

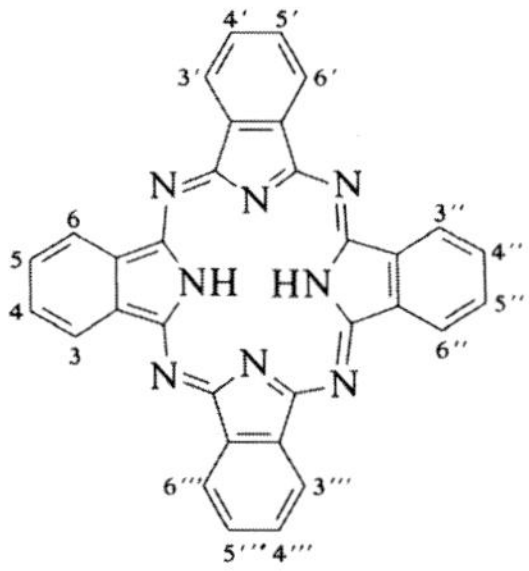

1

Phthalocyanine forms complexes with numerous metals of the Periodic Table. A large number of complexes with various elements are known [12–18]. Metal phthalocyanines MPc (**2**) and compounds with metalloids such as B, Si, Ge, and As or nonmetals such as P display a wide variety in their coordination chemistry.

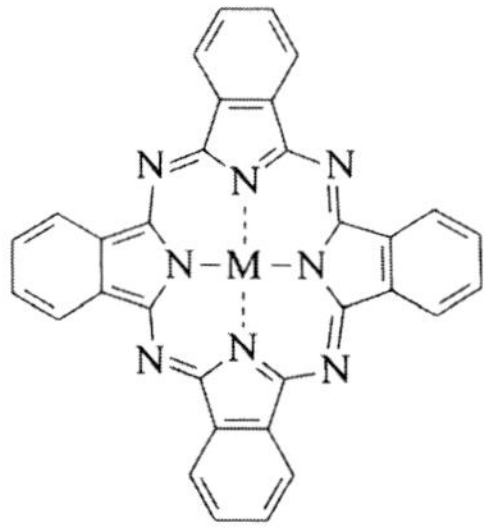

2

The coordination number of the square-planar complexes of Cu, Ni, or Pt is 4. Higher coordination numbers of 5 or 6 with one or two additional ligands such as water or ammonia result in square-based pyramidal, tetrahedral, or octahedral structures (**3**) [19–22].

2.7.1 Introduction

3

The phthalocyanines are structurally related to the macrocyclic ring system porphyrin (**4**). Formally, phthalocyanine can be regarded as tetrabenzotetraazaporphyrin and as the condensation product of four isoindole units.

4   5

The phthalocyanines are structurally similar to naturally occurring porphyrins such as hemoglobin (**5**), chlorophyll a, and vitamin B_{12}. Phthalocyanines themselves do not occur in nature.

History. Braun and Tschernak [23] obtained phthalocyanine for the first time in 1907 as a byproduct of the preparation of *o*-cyanobenzamide from phthalimide and acetic anhydride. However, this discovery was of no special interest at the time. In 1927, de Diesbach and von der Weid prepared CuPc in 23 % yield by treating *o*-dibromobenzene with copper cyanide in pyridine [24]. Instead of the colorless dinitriles, they obtained deep blue CuPc and observed the exceptional stability of their product to sulfuric acid, alkalis, and heat. The third observation of a phthalocyanine was made at Scottish Dyes, in 1929 [25]. During the preparation of phthalimide from phthalic anhydride and ammonia in an enamel vessel, a greenish blue impurity appeared. Dunsworth and Drescher carried out a preliminary examination of the compound, which was analyzed as an iron complex. It was formed in a chipped region of the enamel with iron from the vessel. Further experiments yielded FePc, CuPc, and NiPc. It was soon realized that these products could be used as pigments or textile colorants. Linstead et al. at the University of London discovered the structure of phthalocyanines and developed improved synthetic methods for several metal phthalocyanines from 1929 to 1934 [1–11]. The important CuPc could not be protected by a patent, because it had been described earlier in the literature [23]. Based on Linstead's work the structure of phthalocyanines was confirmed by several physicochemical measurements [26–32]. Methods such as X-ray diffraction or electron microscopy verified the planarity of this macrocyclic system. Properties such as polymorphism, absorption spectra, magnetic and catalytic characteristics, oxidation and reduction, photoconductivity and semiconductivity, solubility, and photochemical and dielectric properties were investigated from the 1930s to the 1950s.

Copper phthalocyanine was first manufactured by ICI in 1935, where its production from phthalic anhydride, urea, and metal salts was developed [33]. Use of catalysts such as ammonium molybdate improved the method substantially [34]. In 1936, I.G. Farbenindustrie began production of CuPc at Ludwigshafen, and in 1937 Du Pont followed in the United States. The most important of the phthalocyanines, CuPc, is now produced worldwide. The first phthalocyanine dye was a phthalocyanine polysulfonate [25]. Other derivatives, such as sulfonyl chlorides, ammonium salts of pyridyl phthalocyanine derivatives, sulfur and azo dyes, and chrome and triazine dyes, have been patented since 1930. At that time, the use of phthalocyanines as colorants for printing ink, paint, plastics, and textiles began. Of the industrial uses, the application of CuPc in printing inks is its most important use. The greenish blue CuPc shade is suitable for color printing. Other favorable properties such as light, heat, and solvent resistance led to the use of this blue pigment for paints and plastics. The chloro and bromo derivatives are important green organic pigments. Other derivatives are used in textile dyeing and printing or for the manufacture of high-quality inks (pastes for ballpoint pens, ink jets, etc.).

Of all the metal complexes evaluated, copper phthalocyanines give the best combination of color and properties and consequently the majority of phthalocyanine dyes (and pigments) are based on copper phthalocyanines; *C.I. Direct Blue 86* (**6**) is a typical case.

As well as being extremely stable, copper phthalocyanines are bright and tinctorially strong (ε_{max} ca. 100 000); this renders them cost-effective.

6

2.7.2 General Synthesis

Phthalocyanine complexes have been synthesized with nearly all the metals of the periodic table [18, 35, 36]. Despite the apparently complex structure of the Pc system, it is formed in a single-step reaction from readily available starting materials. The reaction is strongly exothermic. For example, the synthesis of CuPc from phthalodinitrile (4 $C_8H_4N_2$ + Cu <Pr> $C_{32}H_{16}N_8Cu$) has a reaction enthalpy of −829.9 kJ/mol. The low energy of the final product can be accounted for by resonance stabilization; this explains at least partially the relatively facile formation of the complex. The most important metal phthalocyanines are derived from phthalodinitrile, phthalic anhydride, Pc derivatives, or alkali metal Pc salts.

From o-Phthalonitrile.

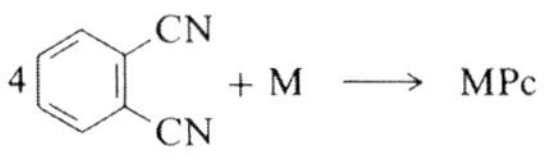

$$4 \, \text{C}_6\text{H}_4(\text{CN})_2 + \text{M} \longrightarrow \text{MPc}$$

where M is a metal, a metal halide (MX_2), or a metal alkoxide [$M(OR)_2$]. The reaction is carried out in a solvent at ca. 180 °C or by heating a mixture of solid reactants to ca. 300 °C.

From Phthalic Anhydride.

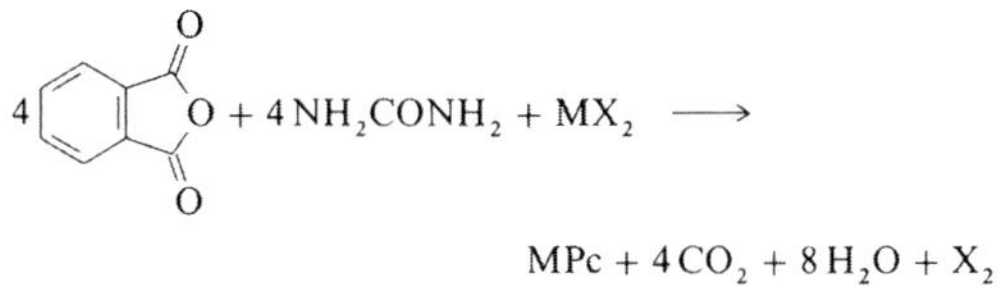

$$4 \, \text{C}_6\text{H}_4(\text{CO})_2\text{O} + 4\,NH_2CONH_2 + MX_2 \longrightarrow$$

$$MPc + 4\,CO_2 + 8\,H_2O + X_2$$

This synthesis is carried out either in a solvent at 200 °C or without solvent at 300 °C.

From phthalimide derivatives, e.g., diimidophthalimide:

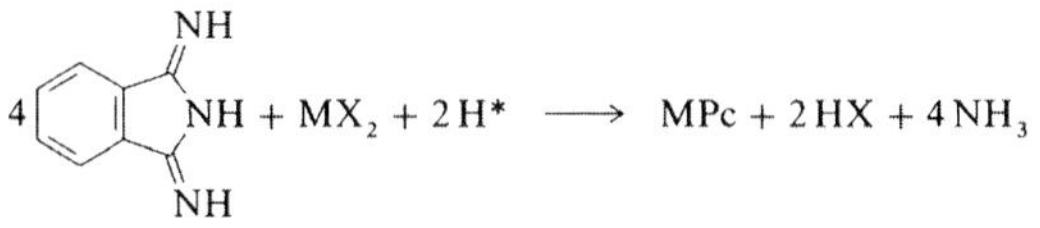

$$4 \, \text{C}_6\text{H}_4(\text{C=NH})_2\text{NH} + MX_2 + 2\,H^* \longrightarrow MPc + 2\,HX + 4\,NH_3$$

*from reducing agent

This synthesis is carried out in a solvent (e.g., formamide).

Metal-free phthalocyanine is obtained by the following procedures [15, 35–37].

1) Decomposition of an unstable MPc with alcohol or acid
   $PcNa_2 + 2\,H_3O^+ \rightarrow PcH_2 + 2\,Na^+ + 2\,H_2O$
2) Direct synthesis (e.g., from phthalodinitrile).

Syntheses of MPc from phthalodinitrile or phthalic anhydride in the presence of urea are the two most important laboratory and industrial methods. They were also used originally by Linstead et al. [8, 9]. This procedure allows the production of many phthalocyanine compounds [35–37]. Catalysts such as boric acid, molybdenum oxide, zirconium and titanium tetrachloride, or ammonium molybdate are used to accelerate the reaction and improve the yield [36, 37]. Ammonium molybdate is especially effective. Reaction is carried out either in a solvent or by heating the solid components. When metal chlorides and phthalodinitrile are used as starting materials, the reaction products are partially chlorinated (e.g., **7**).

$$\text{MCl}_2 + 4 \underset{\text{CN}}{\overset{\text{CN}}{\bigodot}} \longrightarrow \text{[phthalocyanine-M-Cl complex]} \mathbf{7} + \text{HCl}$$

Lowering the reaction temperature or adding urea or basic solvents decreases the extent of chlorination. Solvents such as nitrobenzene, trichlorobenzene, alcohols, glycols, pyridine, and aliphatic hydrocarbons are employed. By using substituted phthalic acids such as 4-chlorophthalic acid anhydride, 4-sulfophthalic acid anhydride, or 4-nitrophthalimide, phthalocyanines with inner substitution can be produced. The products can often be purified by sublimation in vacuo at 300–400 °C. Soluble Pc's can be purified by recrystallization.

2.7.3 Principal Properties

The color of most Pc's ranges from blue-black to metallic bronze, depending on the manufacturing process. Ground powders exhibit colors from green to blue. Most compounds do not melt but sublime above 200 °C, which can be used for purification.

H_2Pc, CuPc, and halogenated phthalocyanines have very poor solubility in organic solvents. Only in some high-boiling solvents such as quinoline, trichlorobenzene, and benzophenone is recrystallization possible at higher temperature. However, the solubilities have a maximum of several milligrams per liter. In common solvents such as alcohols, ethers, or ketones the solubility is considerable lower.

Phthalocyanine and its unsubstituted metal derivatives dissolve in highly acidic media such as concentrated sulfuric acid, chlorosulfuric acid, or anhydrous hydrofluoric acid, presumably due to protonation of the bridging nitrogen atoms [38–40]. In the presence of strong bases, reversible deprotonation of the central imino groups occurs [40]. The solubility in sulfuric acid depends on temperature and concentration [41]. The rate of decomposition of CuPc increases with increasing H_2SO_4 concentration, reaching a maximum at about 80 % H_2SO_4. The stability of metal phthalocyanines increases in the order: ZnPc<CuPc<CoPc< NiPc<CuPcCl$_{15}$ [42]. The color of phthalocyanine solutions in sulfuric acid depends on the degree of protonation (the N atoms of the ring systems are protonated by H_2SO_4; metals such as Cu influence this protonation): H_2Pc gives a brownish yellow color; CuPc, greenish yellow to olive. The phthalocyanines can be precipitated from these solutions by addition of water. Solubility can be

improved in some cases by reversible oxidation with organic peroxides or hypochlorites; the Pc's are oxidized to substances soluble in organic solvents, from which they can be regenerated by reduction [43].

Both H_2Pc and its derivatives exhibit high thermal stability. For example, CuPc can be sublimed without decomposition at 500–580 °C under inert gas and normal pressure [7]. In vacuum, stability up to 900 °C has been reported [44]. Polychloro CuPc is thermally stable up to 600 °C in vacuum. At higher temperature it decomposes without sublimation. CuPc decomposes vigorously at 405–420 °C in air. In nitrogen, sublimation and decomposition occur simultaneously at 460–630 °C [45, 46]. Generally all metal Pc's are more stable thermally in N_2 than in O_2. CuPc changes from the α- to the β-form at 250–430 °C [47].

2.7.4 Industrial Production

2.7.4.1 Copper Phthalocyanine

Two processes are commonly used for the production of copper phthalocyanine: the *phthalic anhydride–urea process* patented by ICI [33, 34] and the I.G. Farben *dinitrile process* [48]. Both can be carried out continuously or batchwise in a solvent or by melting the starting materials together (*bake process*). The type and amount of catalyst used are crucial for the yield. Especially effective as catalysts are molybdenum(IV) oxide and ammonium molybdate. Copper salts or copper powder is used as the copper source [35–37]; use of copper(I) chloride results in a very smooth synthesis. Use of copper(I) chloride as starting material leads to the formation of small amounts of chloro CuPc. In the absence of base, especially in the bake process, up to 0.5 mol of chlorine can be introduced per mole of CuPc with CuCl, and up to 1 mol with $CuCl_2$.

The patent literature gives details of modifications and refinements of the original processes. A review of older processes is given in [35, 36], and examples of more modern production methods are described in [37, 49].

As apparatus for the batch process, an enamel or steel reactor with an agitator and pressure steam or oil heating suffices. Apparatuses used in the continuous synthesis in the presence of solvents and in the bake process are described in [50] and [51, 52], respectively. The choice of process depends on the availability and cost of the starting materials phthalodinitrile or phthalic anhydride. Although the phthalodinitrile process has certain advantages over the phthalic anhydride process, the latter is preferred worldwide because of the ready accessibility of phthalic anhydride. In this process the molar ratio of phthalic anhydride, urea, and copper(I) chloride is 4:16:1, with ammonium molybdate as catalyst. The mixture is heated in a high-boiling solvent such as trichlorobenzene, nitrobenzene, or kerosene. The solvent is removed after the formation of copper phthalocyanine. Frequently a purification step follows. Carrying out the reaction under pressure gives a high-purity CuPc pigment [53].

Several *dry processes* have also been described [37, II, p. 32]. The solvent can be replaced by ammonium chloride [54], a fourfold excess of phthalic anhydride, sodium chloride [55], or a 1:1 NaCl–MgCl$_2$ mixture [37, 56]. In the dry reaction, the ammonium molybdate catalyst can be replaced by a molybdenum or molybdenum alloy agitator [36, 37].

Another dry process is run continuously [57]. The dry, powdered reaction mixture is fed into a rotary furnace kept at 180 °C, and the dry product is discharged into a drum at a yield of 96 % [36, 37]. A vacuum method for the preparation of relatively pure CuPc is described in [58].

One improvement of the process consists of grinding phthalodinitrile, anhydrous CuCl, and urea; mixing the powder thoroughly (or grinding in a ball mill); and heating it to 150 °C. The temperature increases to 310 °C due to the heat of reaction, thus completing the reaction within a few minutes. After purification the yield is 97 %, and the product contains 0.3 % Cl [37, II, p. 30; 59]. Carrying out the reaction in the presence of a salt that decomposes at 30–200 °C to form ammonia improves the yield [60].

The reaction has also been carried out in solvents such as trichlorobenzene in the presence of pyridine [61]. Pyridine converts the insoluble copper(I) chloride into a soluble complex, which reacts more quickly.

The reaction can be accelerated by the use of sodium hydroxide or sulfonic or carboxylic acids [62] instead of pyridine. Other high-boiling solvents such as nitrobenzene, benzophenone, or naphthalene can be used instead of trichlorobenzene [36, 37, 63–66].

Diiminoisoindolenine Process. An alternative route is the formation of the isoindolenines, which are then treated with copper(II) salts [67–69]. 1,3-Diiminoisoindolenine is prepared by reaction of phthalonitrile with ammonia. The isoindolenine is then treated with copper acetate in ethylene glycol and 2-chlorobenzonitrile at 60–70 °C for 1 h.

Copper phthalocyanine can also be made by milling diiminoisoindolenine, copper(I) chloride, anhydrous sodium sulfate, and ethylene glycol at 100–110 °C.

2.7.4.2 Phthalocyanine Derivatives

The copper phthalocyanine derivatives are of major industrial importance as green dyes and organic pigments (halogenated products). The first phthalocyanine dye was polysulfonated CuPc [25]. Since then, many patents describing various phthalocyanine compounds have been registered [35]. Substituted phthalocyanines are either accessible through synthesis from phthalocyanine derivatives (with the advantage of defined products) or by substitution of phthalocyanines [35, p. 255], [36, pp. 171, 192]. The latter method is favored in industry for economic reasons.

Synthesis. Usually a substituted phthalodinitrile or a substituted phthalic acid is used as starting material. A mixture of an unsubstituted and substituted starting material in approximate ratios, respectively, of 1:3, 2:2, or 3:1 can also be used. When reactivities of the two starting materials are approximately equal, Pc derivatives whose degree of substitution closely corresponds to the ratio of the starting materials are obtained. More often, however, a mixture of products results.

With the exception of tetrachlorophthalic acid, substituted phthalic acids, phthalimides, or phthalonitriles are industrially not readily accessible in pure form.

Substitution. Copper phthalocyanine is preferred as starting material. Very little is known about the position of substitution. With the exception of hexadecachloro CuPc, all commercial Pc substitution products, as well as the tetrasubstituted derivatives synthesized from monosubstituted phthalic acids, are mixtures of isomers. Despite the 16 hydrogen atoms that can be substituted, only two different monosubstituted Pc's are possible. The number of disubstituted isomers is higher. Mono- to heptasubstituted Pc derivatives have not yet been isolated in isomerically pure form. In addition, only a limited number of isomers are accessible in pure form by synthesis. Only symmetrically substituted phthalic acids, phthalimides, or phthalodinitriles (3,6-di-, 4,5-di-, or 3,4,5,6-tetrasubstituted derivatives) yield pure isomers of octa- or hexadecasubstituted phthalocyanine derivatives. All other substituted phthalic acids give mixtures of isomers.

2.7.4.3 Pthalocyanine Sulfonic Acids and Sulfonyl Chlorides

The sulfonic acids and sulfonyl chlorides, especially those of CuPc, are readily accessible. The sulfonic acids $CuPc(SO_3H)_n$ with $n=2$, 3, or 4 were significant direct cotton dyes (*C.I. Direct Blue 86*, 74186 and *87*, 74200). The sulfonyl chlorides are intermediates in the production of various copper phthalocyanine colorants [35, p. 259], [36, p. 192]. The water-soluble sulfonic acids are produced by heating copper phthalocyanines in oleum. By varying concentration, reaction temperature, and time, one to four sulfo groups can be introduced in the 4-position [70]. The products synthesized from 4-sulfophthalic acid exhibit slightly different properties. This is due to the different isomer distribution with respect to the 4- and 5-positions. Only one sulfo group is introduced in each benzene ring, as shown by the fact that only 4-sulfophthalimide is obtained on oxidative degradation. The most important dyes have two sulfonic acid groups per molecule.

2.7.5 References

[1] R. P. Linstead, *Br. Assoc. Adv. Sci. Rep.* 1933, 465.
[2] R. P. Linstead, *J. Chem. Soc.* 1934, 1016.
[3] G. T. Byrne, R. P. Linstead, A. R. Lowe, *J. Chem. Soc.* 1934, 1017.
[4] R. P. Linstead, A. R. Lowe, *J. Chem. Soc.* 1934, 1022.
[5] C. E. Dent, R. P. Linstead, *J. Chem. Soc.* 1934, 1027.
[6] R. P. Linstead, A. R. Lowe, *J. Chem. Soc.* 1934, 1031.
[7] C. E. Dent, R. P. Linstead, A. P. Lowe, *J. Chem. Soc.* 1934, 1033.
[8] P. A. Barret, C. E. Dent, R. P. Linstead, *J. Chem. Soc.* 1936, 1719.
[9] P. A. Barret, D. A. Frye, R. P. Linstead, *J. Chem. Soc.* 1938, 1157.
[10] J. S. Anderson, E. F. Bradbrook, A. H. Cook, R. P. Linstead, *J. Chem. Soc.* **1938**, 1151.
[11] R. P. Linstead, *Ber. Dtsch. Chem. Ges.* **A72** (1939) 93.
[12] A. B. P. Lever, *Adv. Inorg. Radiochem.* **7** (1965) 27.
[13] L. J. Boucher in G. A. Melson (ed.): *Coordination Chemistry of Macrocyclic Compounds*, Plenum Press, New York 1979, chap. 7.
[14] K. Kasuga, M. Tsutsui, *Coord. Chem. Rev.* **32** (1980) 67.
[15] P. Sayer, M. Goutermann, C. R. Connell, *Acc. Chem. Res.* **15** (1982) 73.
[16] C. Hamann et al. in P. Goerlich (ed.): *Organische Festkörper und dünne Schichten*, Akademische Verlagsgesellschaft, Leipzig 1978, chap. 2.
[17] D. Woehrle, G. Meyer: "Phthalocyanine – ein System ungewöhnlicher Struktur und Eigenschaften," *Kontakte* **3** (1985) 38; **1** (1986) 24.
[18] F. Lux in C. J. Kevane, T. Moeller, *Proc. Rare Earth Res. Conf. 10th* 1973, 871.
[19] M. S. Fischer, D. H. Templeton, A. Zalkin, M. Calvin, *J. Am. Chem. Soc.* **93** (1971) 2622.
[20] T. Kobayashi et al., *Bull. Chem. Soc. Jpn.* **44** (1971) 2095.
[21] F. Cariati, F. Morazzoni, M. Zocchi, *Inorg. Chim. Acta* **14** (1975) L 31.
[22] J. R. Mooney, C. K. Choy, K. Knox, M. Kenney, *J. Am. Chem. Soc.* **97** (1975) 3033.
[23] A. Braun, J. Tscherniak, *Ber. Dtsch. Chem. Ges.* **40** (1907) 2711.
[24] H. de Diesbach, E. von der Weid, *Helv. Chim. Acta* **10** (1927) 886.
[25] Scottish Dyes, GB 322 169, 1929; DE 586 906, 1929.
[26] R. P. Linstead, A. R. Lowe, *J. Chem. Soc.* 1934, 1031.
[27] I. M. Robertson, *J. Chem. Soc.* 1935, 615; 1936, 1195.
[28] I. M. Robertson, R. P. Linstead, C. E. Dent, *Nature (London)* **135** (1935) 506.
[29] I. M. Robertson, I. Woodword, *J. Chem. Soc.* 1937, 219; 1940, 36.
[30] R. P. Linstead, I. M. Robertson, *J. Chem. Soc.* 1936, 1195; 1936, 1736.
[31] E. W. Mueller, *Naturwissenschaften* **37** (1950) 933.
[32] A. J. Melmed, E. W. Mueller, *J. Chem. Phys.* **29** (1958) 1037.
[33] ICI, GB 464 126, 1935 (M. Wyler).
[34] ICI, GB 476 243, 1936 (A. Riley)
[35] G. Booth: "Phthalocyanines," in K. Venkataraman (ed.): *The Chemistry of Synthetic Dyes*, vol. V, Academic Press, New York 1971, p. 241.
[36] F. H. Moser, A. L. Thomas: *Phthalocyanine Compounds*. Reinhold Publ. Co., New York, Chapman and Hall, London 1963.
[37] F. H. Moser, A. L. Thomas: *The Phthalocyanines*, vol. I and II, CRC Press, Boca Raton, Fla. 33431, 1983.
[38] M. A. Dahlen, *Ind. Eng. Chem.* **31** (1939) 839.
[39] D. L. Ledson, M. V. Twigg, *Chem. Ind.* **3** (1975) 129.

[40] D. L. Ledson, M. V. Twigg, *Inorg. Chim. Acta* **13** (1975) no. 1, 43.
[41] B. D. Berezin, *Izv. Vyssh. Uchebn. Zaved. Khim. Khim. Technol.* **2** (1959) no. 10, 169.
[42] B. D. Berezin, *Izv. Vyssh. Uchebn. Zaved. Khim. Khim. Technol.* **6** (1963) no. 5, 841; **7** (1964) no. 1, 111.
[43] C. J. Petersen, *J. Org. Chem.* **22** (1957) 127.
[44] E. A. Lawton, *J. Phys. Chem.* **62** (1958) 384.
[45] Y. Taru, K. Takoaka, *Shikizai Kyokaishi* **55** (1982) no. 1, 2.
[46] M. Radulescu, R. Vîlceanu, *J. Therm. Analysis* **7** (1975) no. 1, 209.
[47] M. Zhan, B. Zheng, X. Gu, *Xiamen Daxue Xuebao Ziran Kexueban* **25** (1986) no. 2, 192.
[48] BIOS Final Report 960.
[49] Y. L. Meltzer: "Phthalocyanine Technology", *Chem. Process. Rev.* **42** (1970).
[50] BASF, US 3 412 102, 1968 (G. Schulz, R. Polster).
[51] Hoechst, DE 2 256 170, 1972 (W. Deucker, E. Spietschka, D. Steide).
[52] E. F. Klenke Jr., US 2 964 532, 1960.
[53] Dainichiseika Color Chem., JP 02 9851, 1981 (Y. Abe).
[54] Allied Chemical Corp., GB 991 419, 1965.
[55] Y. Abe, M. Muto, JP 70 07661, 1970.
[56] S. Ohira, M. Muto, JP 70 07662, 1970.
[57] D. E. Mack, US 3 188 318, 1965.
[58] T. P. Prassad, *Res. Ind.* **33** (1988) no. 2, 144.
[59] E. Spietschka, W. Deucker, DE-OS 1 644 679, 1970; GB 1 228 997, 1970.
[60] J. C. R. Nicaise, L. A. Cabut, US 3 985 767, 1977.
[61] FIAT 1313, vol. III, (1948).
[62] H. Hiller, W. Kirshenlohr, DE-OS 2 045 908, 1972; GB 1 353 049, 1972; FR 2 107 595, 1972.
[63] Y. Kuwabara, JP 65 27186, 1965.
[64] Y. Abe, S. Horiguchi, JP-Kokai 77 36130, 1977.
[65] R. Polster, R. Schroedel, D. von Pigenot, GB 1 073 348, 1967; FR 1 410 814, 1967.
[66] K. Hoelzle, CH 352 436, 1961.
[67] S. Susuki, Y. Bansho, Y. Sakashita, K. Ohara, *Nippon Kagaku Kaishi* **51** (1976) 1460.
[68] BASF, DE 2 250 938, 1972 (M. Gaeng).
[69] Dainichiseika Colour Chem., JP 48 022 117, 1973.
[70] R. P. Linstead, F. T. Weiss, *J. Chem. Soc.* 1950, 2977.

2.8 Sulfur Compounds as Chromophores

2.8.1 Introduction

Sulfur dyes are a special class of dyes with regard to both preparation and application, and knowledge of their chemical constitution [1]. They are made by heating aromatic or heterocyclic compounds with sulfur or species that release sulfur. Sulfur dyes are classified by method of preparation as sulfur bake, polysulfide bake, and polysulfide melt dyes. Sulfur dyes are not well-defined chemical compounds but mixtures of structurally similar compounds, most of which contain various amounts of both heterocyclic and thiophenolic sulfur.

On oxidation, the monomeric dye molecules cross-link into large molecules by forming disulfide bridges. When a sulfur dye is dissolved by heating with aqueous sodium sulfide solution, the disulfide groups are cleaved to mercapto groups. In this leuco form, the sulfur dye is applied substantively to cotton and other cellulosic fibers. The dyeing is oxidized after rinsing, and the dye molecules re-cross-link on the fiber through disulfide groups.

Pseudo sulfur dyes are dyes that can be applied in the same way as sulfur dyes but are not prepared by classical sulfurization.

History. Croissant and Bretonnière found in 1873 that heating sawdust with sulfur-containing substances yielded a dye that could be applied to cotton from an alkaline bath. Such products appeared on the market under the name "Cachou de Laval", but the true birth of sulfur dyes was in 1893, when Vidal first prepared them by heating well-defined aromatic compounds with sulfur or alkali metal polysulfides. Development then proceeded rapidly. Leopold Cassella & Co. brought Immedial Black V Extra, prepared by the sulfurization of 2,4-dinitro-4'-hydroxydiphenylamine, onto the market in 1897. Priebs and Kaltwasser, working for Agfa in 1899, obtained Sulfur Black T by sulfurizing 2,4-dinitrophenol; this appears to be the highest-tonnage textile dye even today. In 1900, Weinberg and Herz found that valuable blue to violet dyes result when indophenols are melted in aqueous or alcoholic media. The preparation of Hydron Blue from N-(4-carbazolyl)-p-aminophenol was discovered by Haas and Herz in 1909, and that of Indocarbon from 4-(2-naphthylamino)phenol by Hellmann in 1927. In the years 1934–1937, Hagge and Hagen found lightfast brown dyes based on decacyclene and its nitro derivatives; these were marketed by IG Farben. Working at ICI in 1941, Haddock prepared pseudo sulfur dyes by inserting mercapto or thiocyanate groups into phthalocyanine and perylene tetracarboxylic diimide pigments.

2.8.2 Chromophores

Sulfur dyes are synthesized by heating aromatic or heterocyclic compounds as amines, phenols, or nitro compounds with sulfur or, more usually, alkali metal polysulfides. Unlike most other dye types, it is not easy to define a chromogen for the sulfur dyes. It is likely that the chromophore of Sulfur Bake and Polysulfide Bake Dyes consists of macromolecular structures of the thiazole type 1, if the starting material contains amino and methyl groups. In those structures the sulfur is present as (sulfide) bridging links and thiazole groups (Scheme 2.6). Nothing is known about the structure of Sulfur Bake and Polysulfide Bake Dyes which are unable to form thiazole rings. These include, for instance, dyes, which are formed in the sulfur melt from decacyclene.

The characteristic properties of Polysulfide Melt (Chinonimin-) Dyes can be expressed to a large extent in a simplified model (see Scheme 2.8, p. 83).

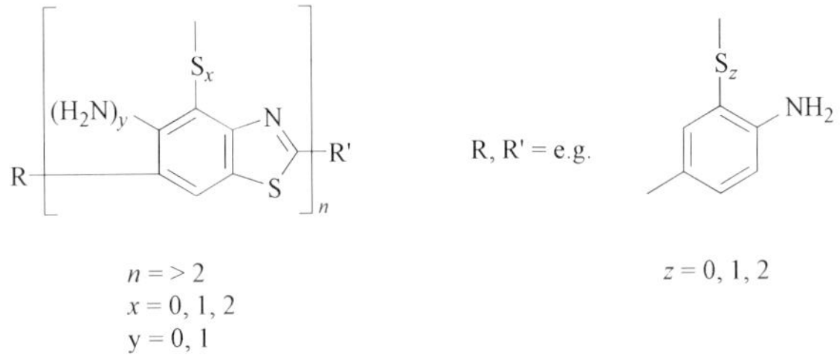

$n => 2$
$x = 0, 1, 2$
$y = 0, 1$

$z = 0, 1, 2$

Scheme 2.6

Even now, knowledge of the constitution of sulfur dyes is rather fragmentary. Two reasons are that the preparation process yields complicated mixtures of related compounds and that sulfur dyes could not be obtained in pure form because of their amorphous and colloidal structure and their insolubility in common solvents.

2.8.3 General Synthesis

2.8.3.1 Sulfur Bake and Polysulfide Bake Dyes

Research by Zerweck et al. [2] suggests that many bake dyes contain thiazole rings. It was already known that dehydrothio-*p*-toluidine, which can be made by sulfur baking of *p*-toluidine, and primulin bases, which are formed at higher temperatures, are both thiazole derivatives.

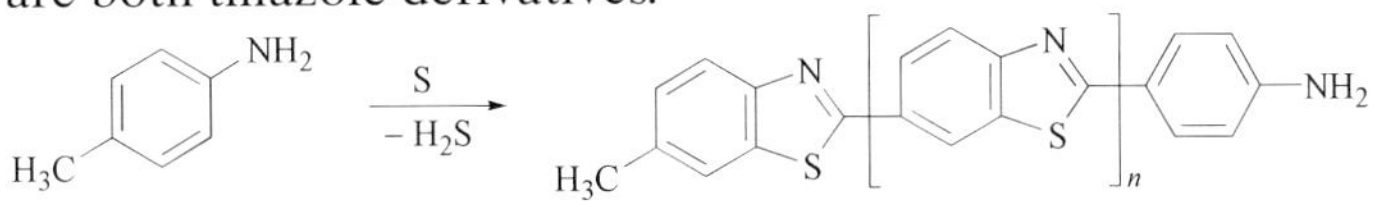

1

$n = 0$: Dehydrothio-*p*-toluidine
$n = 1\text{-}2$: Primulin bases

2.8 Sulfur Compounds as Chromophores

By melting *C.I. Sulphur Yellow 4*, 53160 [1326-75-6] (prepared by sulfur melting of dehydro-*p*-toluidine and benzidine) with KOH, Zerweck et al. obtained not only *p*-aminobenzoic acid but also three distinct *o*-aminothiophenols **2**, which they treated with chloroacetic acid and, after acidification, identified as lactams **3**.

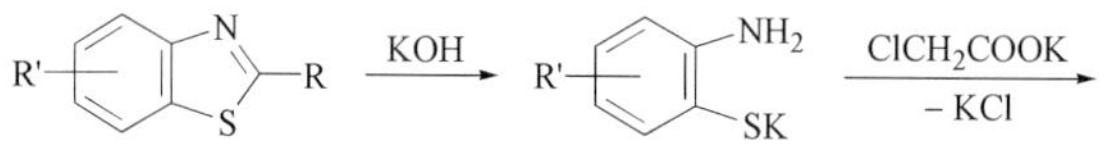

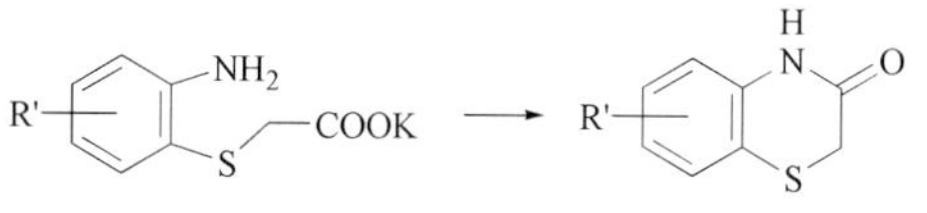

Hence, it could be inferred that *C.I. Sulphur Yellow 4* essentially consists of a mixture of compounds **4–7**, of which **4** is the principal constituent.

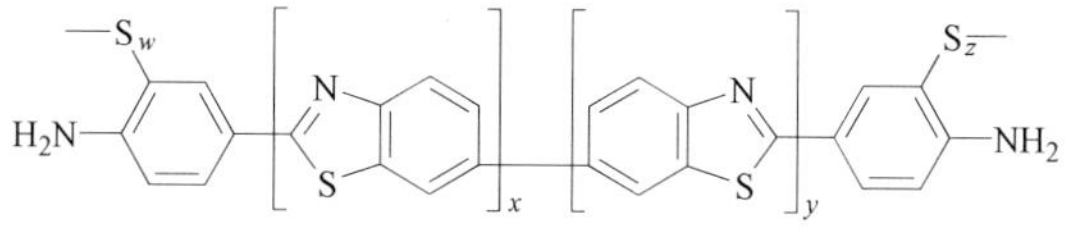

4: $x=2$ $y=0$ $w,z=1$
5: $x=2$ $y=0$ $w=0$ $z=1$
6: $x=2$ $y=0$ $w=1$ $z=0$
7: $x=2$ $y=2$ $w,z=1$

These compounds were synthesized and mixed in the calculated proportions; coloristically and in terms of fastness, the resulting mixture behaved like *C.I. Sulphur Yellow 4*.

Immedial Orange C, prepared by sulfur melting of 2,4-toluenediamine, was investigated in a similar way. The results led to the conjecture that some six to eight molecules of *m*-toluenediamine are linked by thiazole rings, possibly in a branched form (**8**).

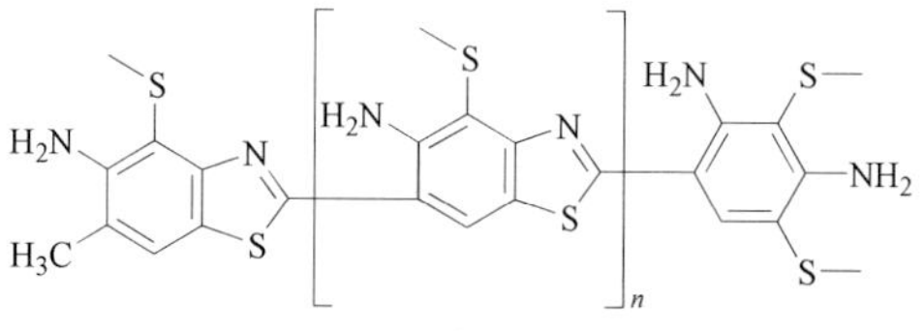

$n = 4 - 6$ **8**

The results of Zerweck et al. were confirmed by Marek and Marková [3], who investigated *C.I. Sulphur Yellow 5* [prepared (as is *C.I. Sulphur Yellow 4*) by sulfur melting of benzidine and primulin base (or dehydrothio-*p*-toluidine and *p*-toluidine), but at a higher temperature and with a longer baking time] and proposed the formula **9**.

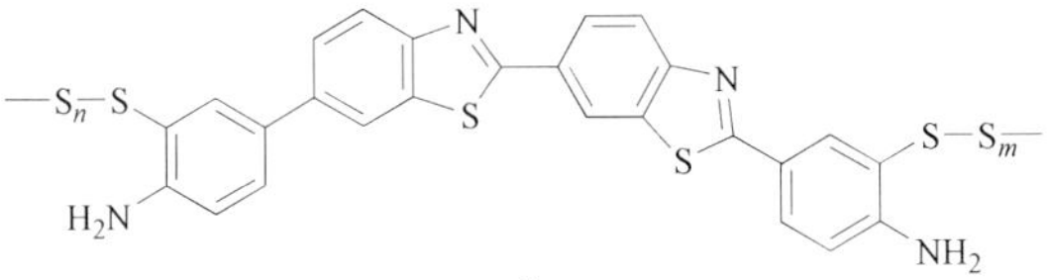

2.8.3 General Synthesis

9

$n + m = 3$

This work suggests that *C.I. Sulphur Yellow 4*, *C.I. Sulphur Orange 1*, 53050 [*1326-49-4*], and *C.I. Sulphur Yellow 5* contain thiazole (benzothiazole) rings as the chromophore.

Nothing is known of the structure of bake dyes made from nitrogen-free intermediates (i.e., from which thiazole rings cannot be formed). This group includes sulfur dyes prepared from decacyclene by sulfur baking (dry fusion process), such as *C.I. Sulphur Brown 52*, 53320 [*1327-18-0*] and also by the sulfurization of anthracene. The latter dye is thus exceptional because it is used only as a vat dye; it is largely insoluble in sodium sulfide.

The constitutional features of the sulfur-bake dyes can be summarized as follows:

1) The aromatic intermediates are combined, with the incorporation of sulfur, to form higher molecular weight chromophore systems. If the intermediate contains both CH_3 and NH_2 groups, thiazole rings are presumably the linking group as well as the chromophore.
2) The aromatic ring systems of the dyes in reduced form bear sodium thiolate groups.

The baking process yields chiefly yellow, orange, brown and olive dyes, depending on the intermediate.

2.8.3.2 Polysulfide Melt Dyes

The starting products for reflux thionation in aqueous or solvent media are chiefly derivatives of indophenol (**10**, Y = OH) and indoaniline (**10**, Y = NH_2). The starting compounds are often in the leuco form **11**.

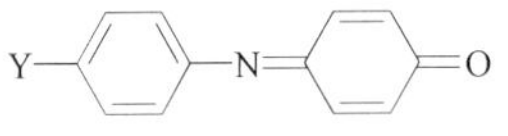

10 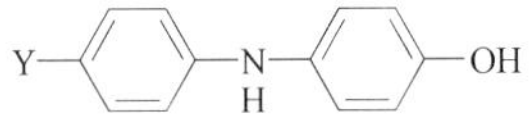 **11**

Y = OH   Indophenol
Y = NH_2 Indoaniline

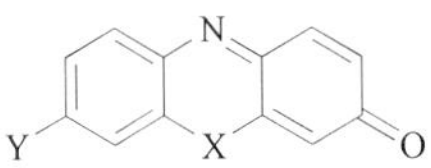

12

X = NH   Phenazone derivatives
X = S   Phenothiazone derivatives
X = O   Phenoxazone derivatives

2.8 Sulfur Compounds as Chromophores

This class of compounds also includes phenazone, phenothiazone, and phenoxazone derivatives (**12**), and so these polysulfide melt dyes are also referred to as quinoneimine or indophenol sulfur dyes.

Gnehm and Kaufler [4] showed that *C.I. Sulphur Blue 9*, 53430 [*1327-56-6*], which is prepared by sulfurization of the indophenol derivative **10** [Y = N(CH$_3$)$_2$], has the phenothiazone structure **12**.

Zerweck et al. [2], on the other hand, subjected well-defined phenothiazone derivatives to successive sulfurizations, obtaining sulfur dyes whose colors and other properties largely matched those of the corresponding unchlorinated indophenols or diphenylamine derivatives **14**. In the trichlorophenothiazones **13**, the chlorine atoms were replaced by sulfur in stepwise reactions (Scheme 2.7).

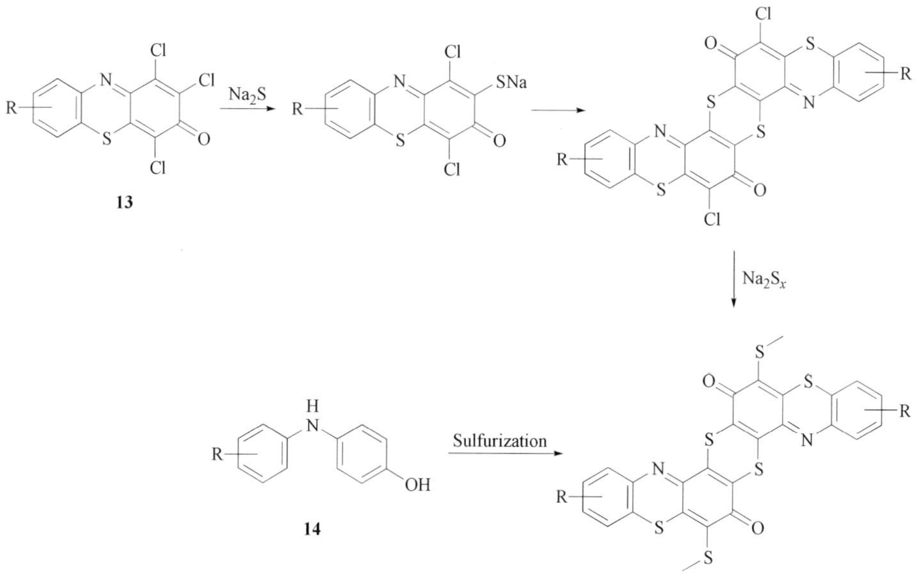

Scheme 2.7

In this way it was established that both thiazone and thianthrene structures are present in all the quinoneimine sulfur dyes studied (*C.I. Vat Blue 43*, 53630 [*1327-79-3*]; *C.I. Sulphur Black 6*, 53295 [*1327-16-8*], *C.I. Sulphur Black 7*, 53300 [*1327-17-9*]; *C.I. Sulphur Black 11*, 53290 [*1327-14-6*]; and *C.I. Sulphur Red 5*, 53820). The thianthrene structure was also found to contribute greatly to the substantivity of the dyes.

The structure **15** can thus be regarded as the prototype of the polysulfide-melt dyes (quinoneimine or indophenol sulfur dyes) (Scheme 2.8).

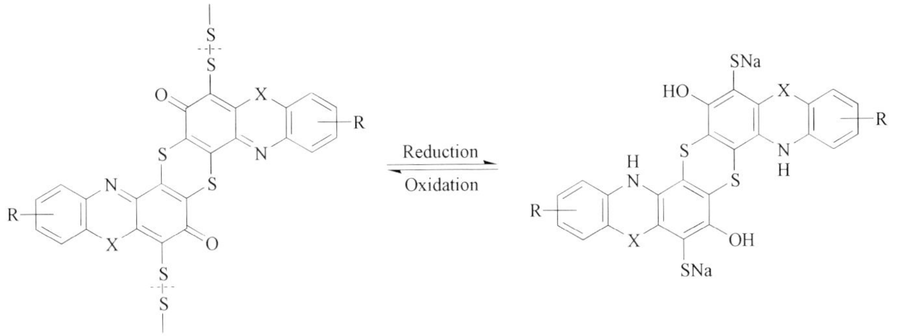

2.8.3 General Synthesis

Scheme 2.8

X = O, S, NH   15a   15b

The polysulfide melt dyes are similar to vat dyes by virtue of the quinoneimine structure. Reducing agents also convert them to a leuco form with a change of hue. The agent not only reduces the quinoneimine group but also transforms the disulfide groups of the dye molecule to sodium thiolate groups. The reduction and the associated cleavage of the disulfide groups render the molecule considerably smaller.

Polysulfide melt dyes, which are applied chiefly with dithionite and commonly show a higher degree of fastness, are called sulfur vat dyes. They include such dyes as *C.I. Vat Blue 43* and *C.I. Sulphur Black 11*.

The polysulfide melt process yields mainly reddish brown, violet, blue, green, and black sulfur dyes, depending on the intermediate.

2.8.3.3 Pseudo Sulfur Dyes

The term "pseudo (or synthetic) sulfur dyes" refers to dyes that have application properties similar to those of sulfur dyes but are obtained by the insertion of mercapto groups or their precursors into the dye precursors or pigments rather than by the classical sulfurization method. These dyes thus have some of the characteristics of sulfur dyes.

The principal chromophores in pseudo sulfur dyes are copper and cobalt phthalocyanines, e.g., in *C.I. Sulphur Green 25* (**16**), and the perylene tetracarboxylic diimide structure in *C.I. Sulphur Red 14* [81209-07-6] and *C.I. Solubilised Sulfur Red 11* [61969-41-3] (**17**). In contrast to the sulfur dye made from Cu phthalocyanine, the cobalt derivative can be applied with dithionite.

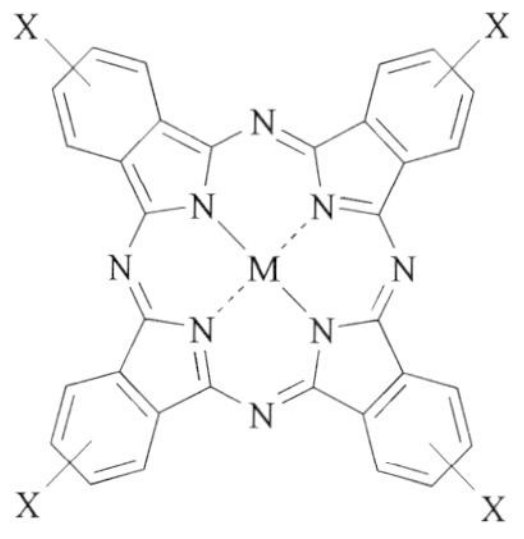

16

M = Cu, Co
X = SCN, SNa, SO$_3$Na

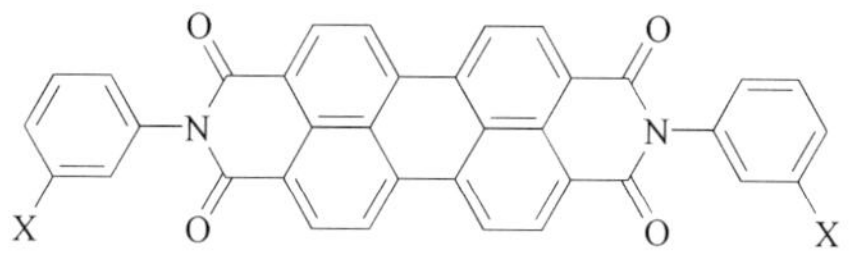

17

X = SCN, SNa

Some special reactive dyes that can be applied by sulfur-dye methods and can also be combined with sulfur dyes represent a special case. An example is *C.I. Solubilised Sulfur Red 11*. These dyes contain no thiolate groups at all.

2.8.4 Principal Properties

With a few exceptions, sulfur dyes are used for dyeing cellulosic fibres. They are insoluble in water and are reduced to the water-soluble leuco form for application to the substrate by using sodium sulfide solution. The sulfur dye proper is then formed within the fiber pores by atmospheric or chemical oxidation. Sulfur dyes constitute an important class of dye for producing cost-effective tertiary shades, especially black, on cellulosic fibers. One of the most important dyes is *C.I. Sulphur Black 1*, prepared by heating 2,4-dinitrophenol with sodium polysulfide.

2.8.5 References

[1] C. Heid, K. Holoubek, R. Klein, "100 Jahre Schwefelfarbstoffe" [100 Years of Sulfur Dyes], *Melliand Textilber. Int.* **54** (1973) 1314–1327.
[2] W. Zerweck, H. Ritter, M. Schubert, "Zur Konstitution der Schwefelfarbstoffe" [On the Constitution of Sulfur Dyes], *Angew. Chem.* **60** (1948) 141–147.
[3] J. Marek, D. Marková, "Über Schwefelfarbstoffe I, Zur Konstitution des lichtechten gelben Schwefelfarbstoffs Immedial-Lichtgelb GWL" [On Sulfur Dyes, I, On the Constitution of the Lightfast Yellow Sulfur Dye Immedial Supra Yellow GWL], *Collect. Czech. Chem. Commun.* **27**, 1533-1548 (1962).
[4] R. Gnehm and F. Kaufler, *Ber. dtsch. chem. Ges.* **37** (1904) 2617–2623, 3032–3033.

2.9 Metal Complexes as Chromophores

2.9.1 Introduction

Metal-complex dyes are coordination compounds in which a metal ion is linked to one or more ligands containing one or more electron-pair donors. Ligands with one and more donor groups are called mono-, di-, trifunctional ligands, etc. Coordination of two or more of the donor groups of such ligands to the same metal atom leads to di-, tri-, or tetradentate chelation, etc.; other names for these ligands are thus chelating agents or chelators. The metal complexes of these ligands are called chelates. The metals in metal-complex dyes are predominantly chromium and copper, and to a lesser extent cobalt, iron, and nickel.

The first use of metal-complex dyes was the process of mordant dyeing, a method that can be traced to the Middle Ages. Here the textile fabrics to be dyed were impregnated with a solution of salts of a metal such as aluminum, iron, chromium, or tin and then treated with a naturally occurring colorant containing a chelating system to achieve metallization within the fiber. The mordant dyeing leads to a bathochromic shift in color, albeit with duller hue, and to improved resistance to light and washing.

In the 1940s the so-called afterchrome method was developed and gained commercial importance, particularly in the dyeing of wool in dark shades with high fastness. In this process a chromium complex was formed on the fiber by first dyeing with a metallizable dye followed by aftertreating the dyed fabris with sodium or potassium dichromate ($M_2Cr_2O_7 \cdot 2\,H_2O$) or chromate (M_2CrO_4). Two other chrome methods, namely, the mordant chrome and the metachrome techniques, were employed. The former is a variant of the above-mentioned mordant dyeing process. It involves treating wool with sodium dichomate and an agent such oxalic acid to reduce chromium in the trivalent state and then dyeing with a chelating azo dye. In the metachrome method the dyeing and chroming are performed in a single bath containing the dye, sodium dichromate, and ammonium sulfate. This process has attained only limited application, since it requires dyes which are not oxidized by chromate in the dyebath.

Besides the afterchrome dyes premetallized dyes became commercially important and remained so until to the present day.

The milestones of the metal-complex dyes can be summarized as follows [1, 2]:

1887	Synthesis of the first metallizable azo dye, Alizarin Yellow (R. Nietzky)
1891	Formulation of the theory of coordination (A. Werner)
1893	Synthesis of the first metallizable 2,2'-dihydroxyazo dye (E. Bergmann and O. Borgmann)
1908	Elucidation of stereochemistry of metal complexes (A. Werner)

1912	Synthesis of the first premetallized 1:1 chromium complex (R. Bohn, BASF)
1920	Marketing of 1:1-chromium complex dyes as Palatin Fast serie (BASF) and Neolan serie (Ciba)
1927	First synthesis of copper phthalocyanine (H. de Diesbach and E. von der Weid)
1949	First water-soluble 1:2 chromium complexes lacking sulfo groups (G. Schetty, Geigy).
1962	Unsymmetrical 1:2 complex dyes containing only one sulfo group
1970	1:2 chromium complexes with two sulfo groups

2.9.2 Azo/Azomethine Complex Dyes

2.9.2.1 General Synthesis

In general, metallization is accompanied by the lost of protons, and therefore it can be monitored potentiometrically, similar to a neutralization reaction. With the exception of 1:1 chromium complex formation, all metallization reactions are promoted by acid-binding agents.

Copper Complexes. The preparation of copper and nickel complexes of tridentate metallizable azo and azomethine dyes is easily carried out in aqueous media with copper and nickel salts at pH 4–7 in the presence of buffering agents such as sodium acetate or amines. Sparingly water soluble precursors can be metallized in alkaline medium at up to pH 10 by using an alkali-soluble copper tetram(m)ine solution as coppering reagent, which is available by treating copper sulfate or chloride with an excess of ammonia or alkanolamines [3].

Three other approaches to copper complexes are also applicable, all of which do not start from *o,o'*-dihydroxyazo compounds. These are valuable and convenient methods in those cases where *o,o'*-dihydroxyazo compounds are difficult to prepare from diazotized *o*-aminophenols. The first method is the simultaneous dealkylation and coppering of *o*-alkoxy-*o'*-hydroxyazo dyes. The dealkylative reaction is assisted by the coordination of the alkoxyl oxygen atom to the copper ion and is carried out by heating above 80 °C with copper tetrammonium sulfate in presence of an alkanolamine or pyridine [4]. The 1:2 opper complex **1**, in which the the ether groups are uncoordinated, reacts only very slowly to give the copper chelate of the corresponding *o,o'*-dihydroxyazo dye [5].

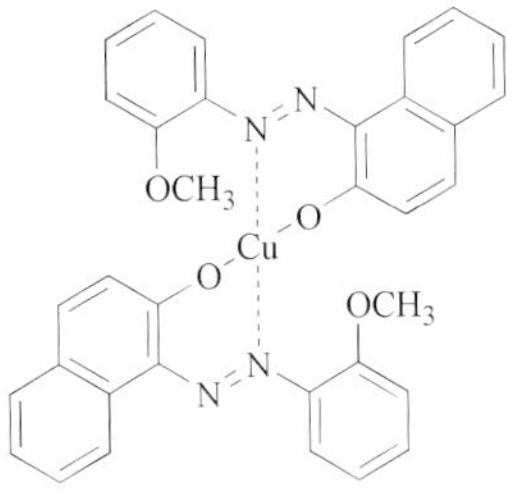

2.9.2 Azo/Azomethine Complex Dyes

1

A further coppering method is oxidative coppering [6], which involves treatment of an *o*-hydroxyazo dye with an oxidizing agent in the presence of copper(II) ions. The most widely used oxidizing agent is hydrogen peroxide, but Na_2O_2, peroxy acids, $Na_2S_2O_8$, and $NaBO_3$ have also been employed. The conditions must be mild, in general in aqueous medium at 40–70 °C and pH 4.5–7.0, because more drastic treatment degrades the original dye. The third method starts from *o*-chloro-*o*′-hydroxyazo compounds and gives the desired copper complexes on treatment with a copper salt in alkaline medium at 50 °C [7] (Scheme 2.9).

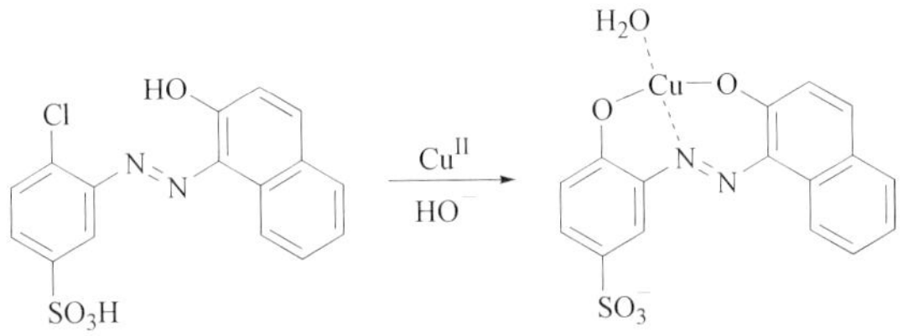

Scheme 2.9 Example for an oxidative coppering reaction

This nucleophilic replacement of halogen atoms proceeds under mild conditions due to the neighboring azo group and the presence of copper ions. The *o*-sulfonic acid group is also susceptible to the copper-mediated nucleophilic substitution, and other nucleophiles, such as alkoxy, alkylamino, cyano, and sulfinic acid can also replace the halogen atom in the position *ortho* to the azo group in the presence of copper ions.

Dealkylative and oxidative coppering and subsequent demetallation in aqueous mineral acids are technical processes to prepare *o,o*′-dihydroxyazo compounds, which are not available from *o*-aminophenols by diazotation and coupling reactions.

Chromium Complexes. Because of their high stability chromium complexes of tridentate azo dyes are the most important class of metal-complex dyes. This is due to the reluctance of hexacoordinated chromium(III) complexes to exchange ligands, which, however, complicates the preparation of chromium complex dyes from hexaaqua chromium(III) salts, and makes it possible to prepare triaqua 1:1 chromium complex dyes. Generally, 1:1 chromium complexes can be made in acid medium below pH 4, whereas 1:2 chromium complexes are prepared at higher pH in weakly acid to alkaline medium. The stability of the chrome dyes parallels the pH conditions for production. The 1:1 chromium complex dyes are only stable in the presence of mineral acids, and in slightly alkaline medium they revert to the metal-free dye and chromium hydroxide. The 1:1 chromium complexes based on o-carboxy-o'-hydroxyazo dyes are much less stable than those based on o,o'-dihydroxyazo dyes. For example, the 1:1 chromium complex of 1-phenyl-3-methyl-4-(2'-carboxy-5'-sulfophenylazo)pyrazol-5-one decomposes slowly already at room temperature and under neutral conditions. On the other hand, 1:2 chromium complex dyes are unstable at high acidity and disproportionate into the corresponding 1:1 chromium complex and the metal-free dye.

The 1:1 chromium complexes containing sulfonic acid groups are made in aqueous solution at 100 °C or at up to 140 °C under pressure for several hours by using an excess of a chromium salt of a strong acid such as HCl, H_2SO_4, or HF. In the case of azo compounds devoid of sulfonic acid groups it is valuable to add or to employ exclusively an organic solvent, such as alcohol, ethylene glycol, monoalkyl ethylene glycol, or monoalkyl diethylene glycol. The use of a solvent allows higher reaction temperature and shortened reaction time without the need to use pressure equipment. The resulting 1:1 chromium complexes having one sulfonic acid group can be conveniently isolated as insoluble betaines **2**, which become slightly water soluble above pH 4 in the form of salts of sulfonic acids **3** (Scheme 2.10).

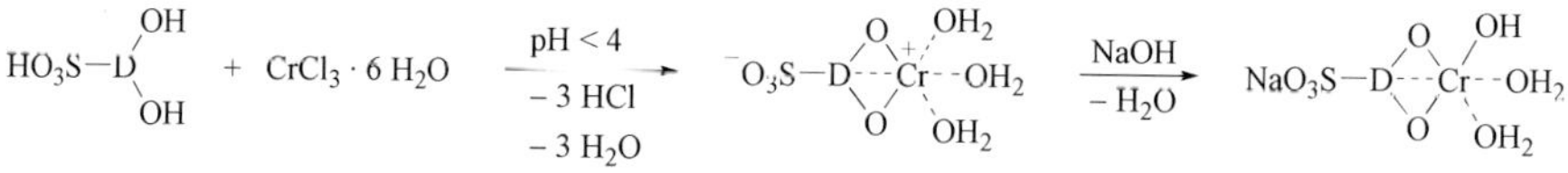

Scheme 2.10 Chromium 1:1 complexes as unsoluble betaines **2** and as slightly soluble sodium salts of sulfonic acids **3**.

Decreasing acidity of the reaction medium favors the formation of 1:2 chromium complexes, for example, by using strongly buffering formamide as solvent or by using chromium salts of buffering weak acids, in general acetic acid. Under conditions yielding 1:2 chromium complexes no 1:1 complex can be detected during the metallization, and this indicates that the second complex-formation step proceeds more rapidly than the first. This fact can be utilized in the synthesis of 1:1 chromium complexes by first making the 1:2 complex above pH 4 and then treating it with mineral acids to obtain the 1:1 complex [8].

Metallizable azo dyes having no water-solubilizing groups such as sulfonic acid, carboxylic acid, alkylsulfonyl, or sulfonamide are sparingly soluble in aqueous acidic medium and can not be chromed completely under these conditions. One method to overcome this difficulty is the addition of alcohols [9] and carbonamides [10]. Another method is the treatment of the azo dyes in alkaline solution to yield conventional 1:2 complexes. In this case a chromium complex of a chelating organic acid, such as oxalic acid [11], tartaric acid [12], or salicylic acid [13], or a mixture of chromium salt and chelating acid serves as chromium source.

2.9.2 Azo/Azomethine Complex Dyes

Using these agents, especially the particularly effective 1:2 complex **4** with salicylic acid, metallization can be carried out easily at high pH without precipitation of inert chromium hydroxide. In certain cases a 1:1 chromium complex that still contains chelating organic acid as ligand can also be obtained.

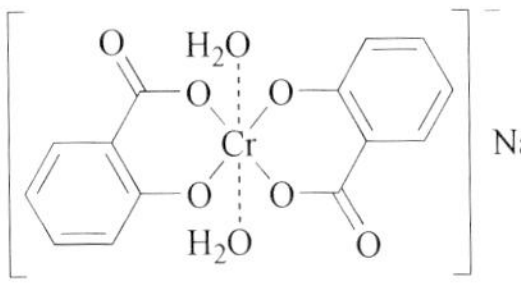

4 [68214-27-7]

An alternative method of preparing 1:2 chromium complexes under alkaline aqueous conditions is based on the use of an alkaline dichromate–glucose solution. In this mixture at boiling temperature, first Cr^{VI} is reduced to Cr^{II} with a reactive d^4 configuration followed by rapid formation of the 1:2 Cr^{II} complex and then the 1:2 Cr^{III} complex with partial reductive fission of the azo chromophore. The partial dye decomposition is apparent in the decrease of tinctorial yield compared to chromation with Cr^{III} reagents. Introduction of chromium by simultaneous demethylation requires the use of sodium dichromate and the high-boiling solvent glycol, which at 120–150 °C also serves as a reductant for Cr^{VI}.

The formation of 1:2 chromium complexes can also be performed in two steps, first by production of the 1:1 complex and then by treating it with a equimolecular quantity of a further metallizable dye to give a 1:2 complex [14]. This complex is symmetrical, for example, **5**, if the added dye is identical with the 1:1 chromium-based dye, and unsymmetrical, for example, **6**, if this is not the case (Scheme 2.11).

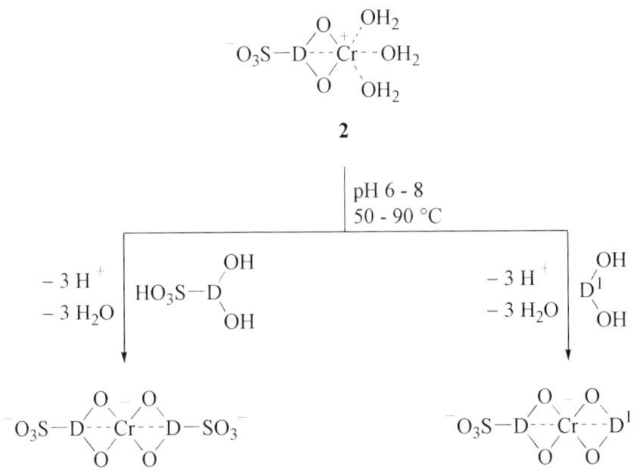

Scheme 2.11 Stepwise synthesis of symmetrical **5** and unsymmetrical 1:2 chromium complexes **6**

The latter stepwise method is the method of choice for the production of the commercially important 1:2 chromium complexes having only one sulfonic acid group (**6**). This class of metal-complex dyes is made by converting the sulfonic acid bearing dye to the 1:1 chromium complex, isolating the resulting betaine and addition of the sulfonic acid-free counterpart in aqueous weak alkaline solution under mild conditions at 50–90 °C.

Unsymmetrical 1:2 chromium complexes are also obtained in one step by so-called mixed metallization. This means 1:2 chromation of a mixture of at least two different metallizable dyes, leading in the case of two dyes to a mixture of two symmetrical 1:2 complexes and an unsymmetrical 1:2 complex as main products. The mixture is statistical if the affinity of the two metallizable dyes for chromium is comparable; otherwise, the two symmetrical 1:2 complexes prevail. The mixed method is technically exploited for producing the important class of 1:2 chromium complexes having a solubilizing group, such as sulfonamide or alkylsulfonyl, on each dye. Moreover, the method is very versatile. For extending the range of shades and matching a definite shade, one can increase the number of metallizable dyes or change their molecular proportions to another. For example, a mixture of 4 mol of dye A, 3 mol of dye B, and 3 mol of dye C with 5 mol of a chromation agent furnishes in statistical distribution the symmetrical and unsymmetrical 1:2 chromium complexes as outlined in Scheme 4, and 1:2-chromation of a mixture of 10 metallizable dyes yields 10 symmetrical and 45 unsymmetrical 1:2 chromium complexes.

$$5A + 3B + 2C + 5Cr$$
$$\downarrow$$
$$1.25\ ACrA + 0.45\ BCrB + 0.2\ CCrC + 1.5\ ACrB + 1\ ACrC + 0.6\ BCrC$$

Scheme 2.12 Products of the mixed 1:2-chromation of three dyes in the molecular proportions 5:3:2

Cobalt Complexes. Preparation of 1:2 cobalt complexes does not require such high reaction temperatures as the corresponding 1:2 chromium complexes, since the aqua cobalt complexes are less inert than those of chromium. The usual method is the reaction of Co^{II} salts in alkaline medium at about 60 °C, which leads rapidly to the diamagnetic 1:2 Co^{III} complexes. Atmospheric oxygen serves as oxidant. If the reaction is carried out under anaerobic conditions or if the cobaltization proceeds too slowly, partial reductive fission of the azo dye results. The usual remedy to minimize dye degradation is dropwise addition of dilute hydrogen peroxide simultaneously with the cobalt(II) salt solution. Direct use of Co^{III} is unfavorable, because it acts as strong oxidant and forms the 1:2 cobalt(III) complex more slowly than Co^{II}. Demethylative cobaltization is achieved in high-boiling glycols above 100 °C in the presence of a base such as alkanolamines or sodium hydroxide.

Co^{III} exhibits a greater tendency to form complexes with nitrogen-donor ligands than with oxygen-donor ligands. Thus the azo dye from diazotized 2-amino-1-hydroxybenzene-5-sulfonamide and acetonylbenzthiazole affords the 1:2 cobalt complex **7**, which is coordinated with the ring nitrogen of the thiazole nucleus and not with the enol oxygen atom of the acetyl group as described in [15]. An indication for the involvement of the ring nitrogen atoms is the brown shade of **7**.

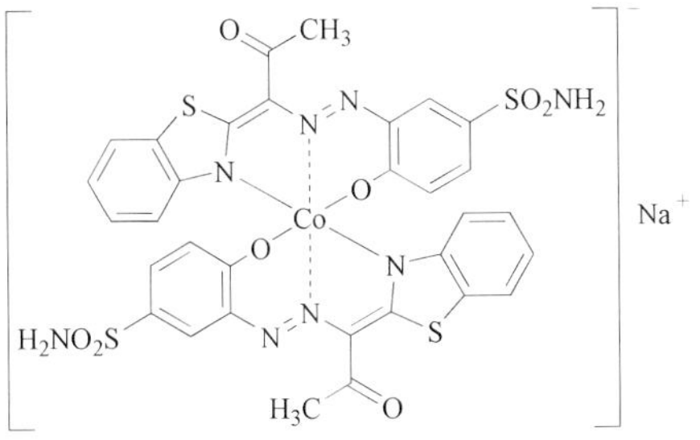

7

In contrast to chromium, triaqua 1:1 cobalt complexes cannot prepared in acidic medium. Owing to the high affinity of nitrogen for the Co^{III} ion, 1:1 cobalt complexes of tridentate dyes can be obtained when the coordination requirement is satisfied by nitrogen ligands [16]. The reaction is achieved, for example, in presence of a large excess of ammonia to give triammino 1:1 cobalt complexes. These complexes are insufficiently stable for technical application, but they can readily be converted with an equimolecular quantity of another metallizable azo dye to an unsymmetrical 1:2 cobalt complex [17]. Nitrite, as a strong nucleophile, also stabilizes intermediates for the production of unsymmetrical 1:2 cobalt complexes [18]. Bidentate nitrogen donor ligands, such as ethylenediamine, biguanide, and dipyridyl, are even more suitable for stabilizing 1:1-cobalt complexes [19], and tridentate nitrogen-donor ligands, such as diethylenetriamine and terpyridyl, are even better [20]. Complexes of this type are too stable to be converted into 1:2 cobalt complex dyes.

Iron Complexes. Iron complexes of tridentate *o,o'*-dihydroxyazo compounds are prepared under weakly acidic conditions at 40–80 °C. Both Fe^{II} and Fe^{III} salts can serve as iron source. The Fe^{III} complexes that result in both cases do not have sufficient stability to dye textile substrates, but the dyeings on leather have good fastness properties [21].

1-Nitroso-2-hydroxyaryls always form 1.3 Fe^{II} complexes with Fe^{II} and Fe^{III} salts. The 1:3 stoichiometry, that is, three bidentate nitrosohydroxy ligands around one Fe^{II} ion, is readily revealed by metallization of a mixture of two different compounds, for example 1-nitro-2-naphthol-6-sulfonic acid (A) and 1-nitroso-2-hydroxy-6-sulfoamide (B). In this case four products can be discriminated on the chromatogram in the following order: type AAA, type AAB, type ABB, and type BBB.

2.9.2.2 Principal Properties

The coordination number of the chelated metal atom determines the number of linkages to functional groups. It is typically greater than the valency of the metal ion; for example, the divalent ions of copper and nickel have coordination numbers of four, and the trivalent ions of chromium, cobalt, iron a coordination number of six. In the case of iron the coordination number six applies for the di- and trivalent forms.

The most important chelate types are illustrated in Figure 2.7. Compounds in column A display a metal ion with coordination number four. The coordination number six is represented by the chelates of column B and C. Rows a, b, and c cover bi-, tri- and tetrafunctional or -dentate ligands, respectively. The types Ab and the types of column B contain additional monofunctional ligands to complete the coordination numbers of four and six, respectively.

As far as dyes are concerned, chelates of type Cb are termed 1:2 or symmetrical 1:2 metal complexes if the two tridentate ligands are equal, and mixed or unsymmetrical 1:2 metal complexes in the other case. Chelates of type Bb represent 1:1 metal complexes. The types Cb and Bb include, in general, azo and azomethine metal complex dyes, whereas chelates of the quadridentate types Ac and Bc are derived predominantly from formazan and phthalocyanine chromophores.

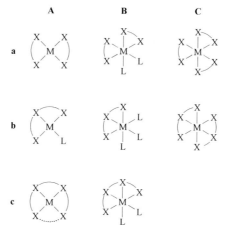

Figure 2.7 Types of chelates.
M = central metal atom, L = monofunctional ligand, X-X = bidentate ligand, X-X-X = tridentate ligand, X-X-X-X = tetradentate ligand.

Stability. Ligands with two or more donor groups afford monocyclic or polycyclic-cyclic metal chelates that result in enhancement of the complex stability relative to two or more monofunctional nucleophiles. The increasing tendency to form complexes with increasing number of donor groups per ligand, known as the chelate effect, was discovered by Schwarzenbach [22] and is discussed in [23].

2.9.2 Azo/Azomethine Complex Dyes

Thus, *o,o'*-dihydroxyazo (**8**), *o,o'*-hydroxyaminoazo (**9**), *o,o'*-carboxyhydroxyazo (**10**), *o,o'*-dihydroxyazomethine (**11**), and *o,o'*-carboxyhydroxyazomethine (**12**) residues yield with metals stable annelated 5,6- and 6,6-membered ring systems, respectively (Scheme 2.13). The residues in the azo compounds are in most cases benzene, naphthalene, pyrazolone or open-chain enol–keto systems, while in the azomethine compounds they are mostly benzene systems.

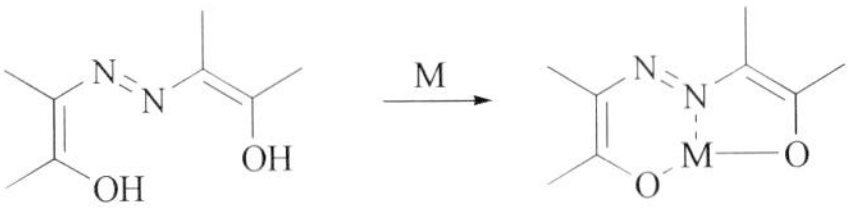

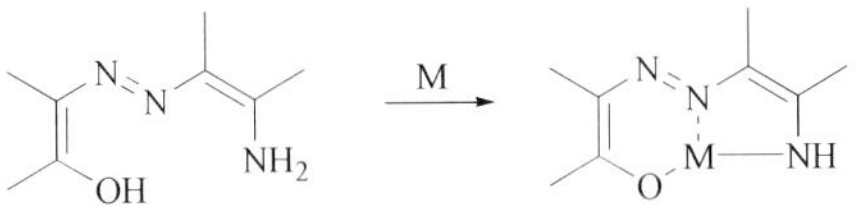

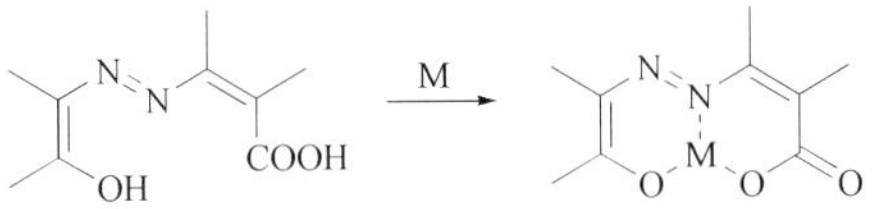

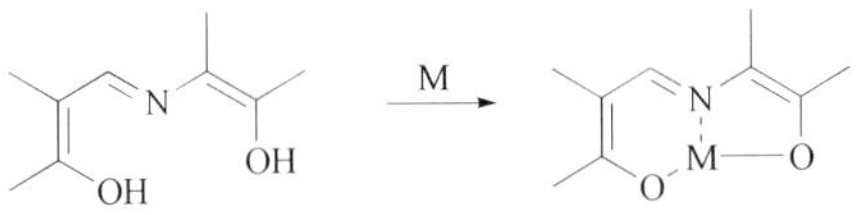

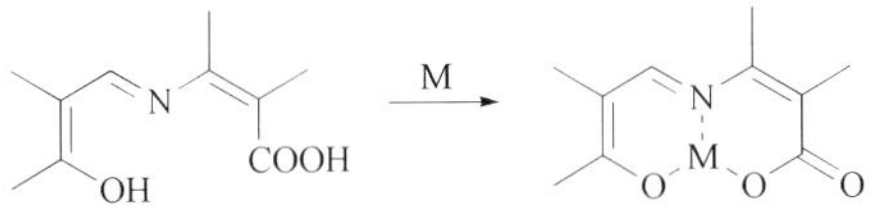

Scheme 2.13 Tridentate ligands and their metal chelates (M = Cu^{2+}, Cr^{3+}, Co^{3+}, etc)

Chelates in which one nitrogen atom of the azo group and that of the azomethine group form part of the coordination system can also be called medially metallized azo and azomethine dyes [24]. The bond joining metal atoms to the nitrogen, sometimes referred to as a secondary valence, may be represented by an arrow or dotted line. The latter is preferred here.

The stability of a metal complex is related to the formation of strain-free five- or six-membered rings, to the number of such chelate rings per ligand, to the basicity of the donor groups and the nature of the metal ion. Chromium(III) ions impart the greatest stability in the field of azo dye complexes. An in depth description of complex stability is given in [25].

Chelation of a chromphore can also result in a change in properties, including absorption spectra, brilliance, charge, solubility, and affinity for substrates. In general, metallization leads to a bathochromic shift and leads to a distinct improvement in lightfastness and in many cases an enhanced wetfastness compared to the metal-free compound.

Stereochemistry and Isomerism. Metals having a coordination number of four usually give square-planar complexes, whereas metals having a coordination number of six are octahedrally surrounded by the functional groups of ligands.

From the viewpoint of stereochemistry the most interesting metal complexes are the octahedrally coordinated 1:2 chromium and cobalt complex dyes, which are medially metallized azo and azomethine compounds with functional groups in the *o*- and *o'*-positions. Three types of isomerism can be discriminated: geometrical, N-α,β, and that arising from azo–hydrazone tautomerism.

Geometrical isomerism of 1:2 chromium and cobalt complexes is due to the arrangement of the two tridentate dye ligands in a meridial (*mer*) and a facial (*fac*) mode (Figure 2.8). The *mer* configuration is is also known as Drew–Pfitzner structure, after Drew [26] and Pfitzner [27] who independently proposed that the two tridendate ligands are arranged mutually perpendicular. This configuration permits only one pair of enantiomers, which were resolved by Pfeiffer and Saure into optically active forms [28]. The perpendicular arrangement has been verified by X-ray diffraction studies [29].

In the *fac* configuration, also called the Pfeiffer–Schetty structure, nine isomers are possible: four pairs of enantiomers and a centrosymmetric structure. Since the ligands are arranged parallel to one another, they sometimes are called sandwich complexes. The two *mer* and the nine *fac* orientations are depicted in [30].

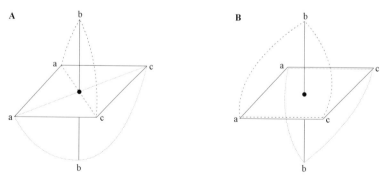

Figure 2.8 Geometrical isomers. A) Meridial arrangement (Drew–Pfitzner structure). B) Facial arrangement (Pfeiffer–Schetty structure).

The *fac* isomers were isolated by Schetty [31] by chromatographic separation and confirmed by Jaggi by X-ray crystal determination [32]. In solution, each separate isomer readily undergoes equilibration to generate the original mixture. The nine isomers of the *fac* configuration apply to symmetrical 1:2 metal complexes having two ligands A-N-B. In the case of unsymetrical 1:2-metal complexes having the ligands A-N-B and C-N-D, six pairs of enantiomers are possible.

In the *mer* configuration the three coordinating atoms A, N, and C form a isosceles triangle, while in the *fac* system they are situated at the corners of an equilateral triangle (Figure 2.9). The distance between the atoms A and B in the equilateral triangle is smaller than that in the isosceles triangle. Figure 2.9 illustrates that azo dyes forming annelated 6/6-membered ring chelates fit the isosceles triangle better and prefer to coordinate in the sandwich manner, while dye chelates having 5/6-membered ring systems prefer to adopt the isosceles triangle geometry. A broad discussion about distances, configuration, and coloristic behavior is given in [33].

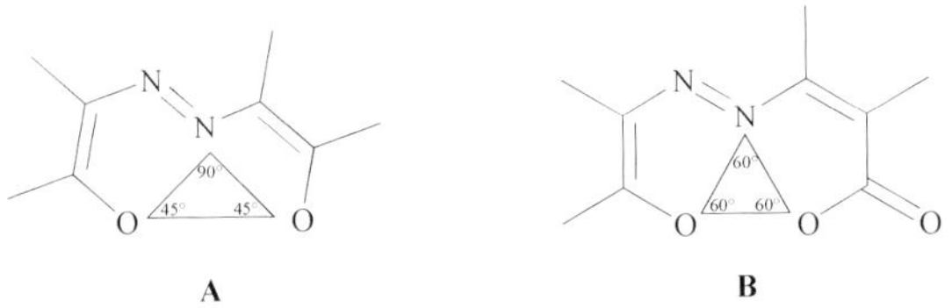

Figure 2.9 A) Isosceles triangle embedded in a 5/6-membered ring chelate of the Drew–Pfeiffer type. B) Isolateral triangle suiting the 6/6-memebred ring system of the sandwich type.

N-α,β isomerism arises from the fact that only one nitrogen atom of the azo group can act as third ligand. In the case of metallizable unsymmetrical azo compounds two isomers can be discriminated in a 1:1 metal complex [34] and three isomers in a symmetrical 1:2 metal complex [35] (Figure 2.10).

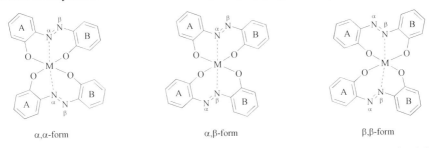

Figure 2.10 The three N-α,β isomers of symmetrical 1:2 metal complexes derived from unsymmetrical azo compounds, rings A and B being differently substituted.

With mixed 1:2 metal complexes derived from two distinct unsymmetrical azo compounds four N-α,β isomers become possible. The N-α,β isomerism has been affirmed by NMR spectroscopy on diamagnetic 1:2 cobalt(III) complexes [35] and X-ray investigations [36].

2.9 Metal Complexes as Chromophores

Steric factors strongly influences the relative proportions of N-α and N-β coordination. Thus, the azo compound **13** strongly favors the N-β coordinated configuration as a result of the interaction of the 6-methyl group with the electron lone pair of the N-β atom, and the azo compound **14** coordinates predominantly through the N-α atom, because this atom interacts with the hydrogen atom in the 8-position of the naphthalene nucleus.

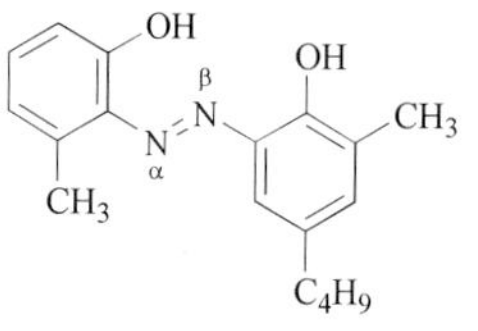

13

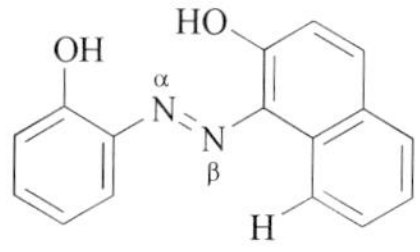

14

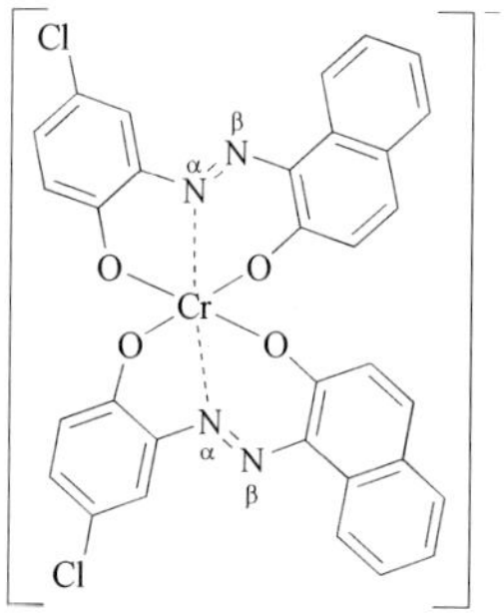

15 α,α-form

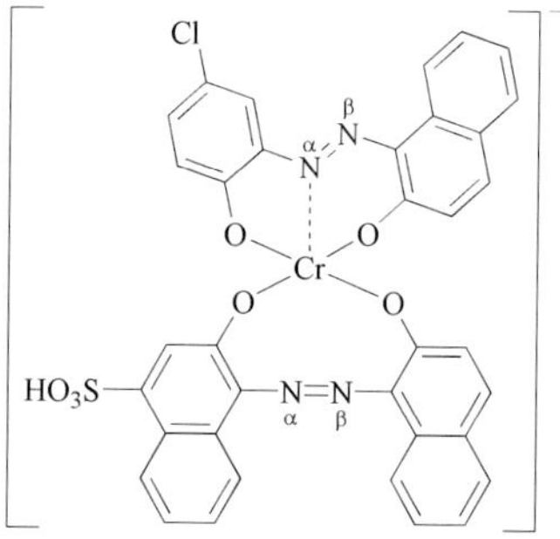

16 α,α-form
α,β-form

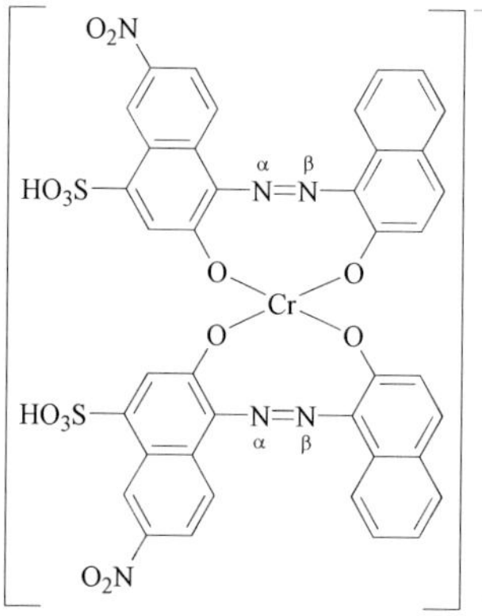

17 α,α-form
α,β-form
β,β-form

For example, by chromatographic techniques it can be shown that the 1:2 chromium complex **15** exhibits only one isomer (α,α form), the mixed 1:2 chromium complex **16** two (α,α and α,β forms), and the symmetric 1:2 chromium complex **17** three isomers (α,α, α,β and β,β forms), assuming that these dyes exist only in the *mer* configuration.

Figure 2.11 illustrates *azo-hydrazone tautomerism* in azo and azomethine compounds. Strongly polarized o,o'-dihydroxyazo dyes, such as 1-(2-hydroxyphenylazo)-2-naphthols or 4-(2-hydroxyphenylazo)-5-pyrazolones, exist predominantly in the hydrazone form. Metallizable hydrozone compounds coordinate only through the imino group due to its stronger donor property. They do not exhibit N-α,β isomerism, because only one hydrazone form is predominant. In contrast to the planar azo form, the donor nitrogen atom of the hydrazone group is sp^3 hybridized and therefore tetrahedral. In this situation the ligand can adopt two nonplanar configurations, leading to three enantiomeric conformers in 1:2 metal complexes if both ligands exist in the hydrazone form. If one ligand coordinates in the azo and the second in the hydrazone form, two *dl* conformers are possible. Only one pair of enantiomers arises if both ligands are in the azo form. Thus, six isomers and their mirror images can be considered if azo–hydrazo tautomerism is present [37]. As mentioned previously in the case of N-α,β isomerism, this type of isomerism is also based on the assumption that the o,o'-dihydroxyazo and -azomethine compounds form only *mer* complexes. The fact that these conformers coexist in equilibrium means that the colors tend to rather dull shades, whereas planar metal-complex dyes, such as copper phthalocyanines and copper formazans are brighter in hue.

Chromium, cobalt and copper chelates are applied mainly in textile dyeing. The use of the less stable iron chelates is restricted to leather dyeing, and the nickel chelates are employed mostly as pigments.

X = N, azo
X = CH, azomethine

X = N, hydrazone
X = CH, ketoamine

Figure 2.11 Tautomerism in azo and azomethine compounds.

2.9.3 Formazan Dyes

2.9.3.1 Introduction

Formazan dyes are closely related to azo dyes and are derived from the following basic structure:

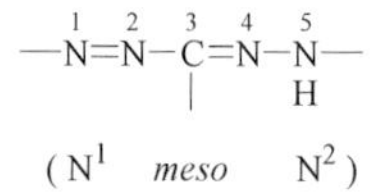

$$-\underset{1}{N}=\underset{2}{N}-\underset{3}{\underset{|}{C}}=\underset{4}{N}-\underset{5}{N}-$$
$$H$$

(N^1 *meso* N^2)

The position of the substituents is given according to IUPAC nomenclature using the prefixes 1-, 3-, and 5-. Formazans unsubstituted in the 1- and 5-positions and 1,5-dialkyl-substituted formazans are unknown. Aryl or heteroaryl groups are the most common 1,5-substituents. The 3- or *meso* position can be occupied by a variety of substituents (e.g., aryl, heteroaryl, H, OH, SR, halogen, NO_2, CN, and alkyl).

The 1,5-diphenyl and 1,3,5-triphenyl derivatives were discovered almost simultaneously by H. von Pechmann [38] and E. Bamberger [39] in 1892. Interest in those compounds was renewed in 1939 when G. Lakon [40] and R. Kuhn and D. Jerchel [41] found that colorless tetrazolium salts were very sensitive to biological reduction processes and were converted to deeply colored formazans:

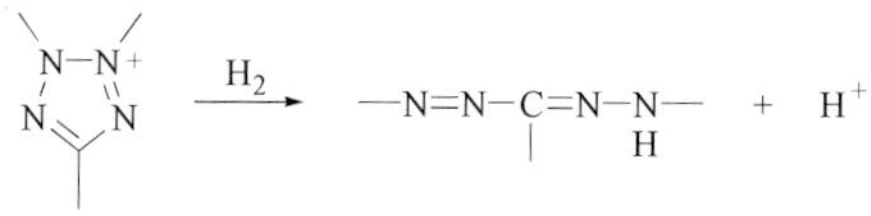

$$\text{tetrazolium} \xrightarrow{H_2} -N=N-\underset{|}{C}=N-N- + H^+$$
$$\phantom{tetrazolium \xrightarrow{H_2} -N=N-}H$$

In 1941 L. Hunter and L. B. Roberts produced metal complexes from formazan compounds [42].

R. Wizinger et al. introduced other complex-forming groups into formazans [43], which ultimately led to the discovery that metallized formazans could be used for textile dying.

Early reviews of formazans and their coordination compounds can be found in [44–50].

2.9.3.2 General Synthesis

Coupling of Diazonium Compounds with Hydrazones. The most common method for the production of formazans is the coupling of diazonium salts with aryl hydrazones in an alkaline medium, possibly in presence of an organic solvent.

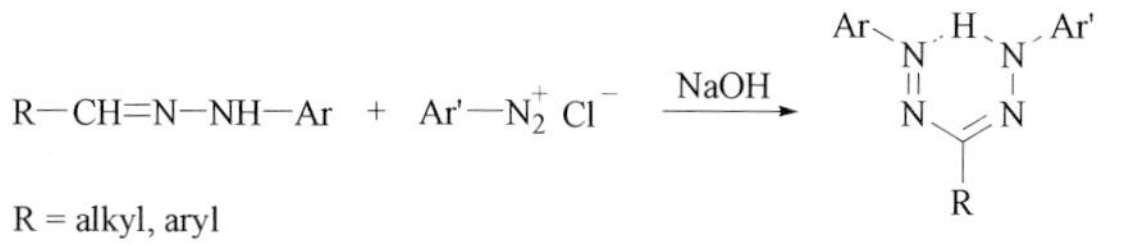

$$R-CH=N-NH-Ar + Ar'-N_2^+ Cl^- \xrightarrow{NaOH} \text{formazan}$$

R = alkyl, aryl

The industrially important 1-(2-hydroxyaryl)-3-aryl-5-(2-carboxyaryl)formazans are obtained by this process. The corresponding bis(2-hydroxy) compounds cannot be produced by this method because 2-hydroxyaryl hydrazines are very unstable compared to 2-carboxyaryl hydrazines.

2.9.3 Formazan Dyes

Double Coupling of Diazonium Compounds. Some compounds with an activated methyl group can form symmetric formazans with strong diazo compounds by double coupling:

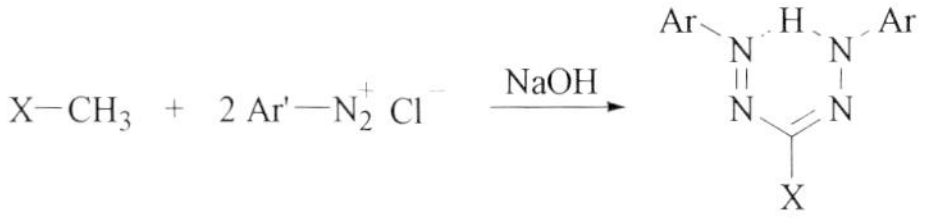

$$X-CH_3 + 2\ Ar'-N_2^+\ Cl^- \xrightarrow{NaOH} \underset{X}{\underset{|}{C}}\begin{smallmatrix}Ar-N=N\\ \\ Ar-NH-N\end{smallmatrix}$$

X = CH$_3$CO, C$_6$H$_5$CO, NO$_2$
some heterocycles, quaternized heterocycles

Examples of heterocycles with an activated methyl group are **18** [51], **19** [52], **20** [53], **21**, and **22** [54].

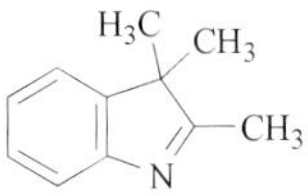

18, 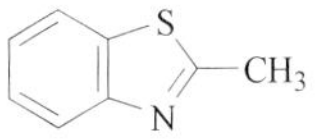 **19**, 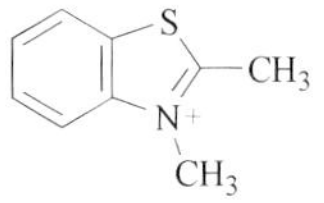 **20**, 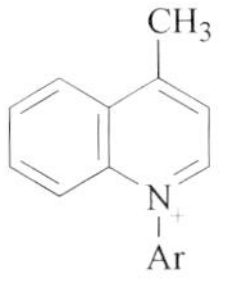 **21**, 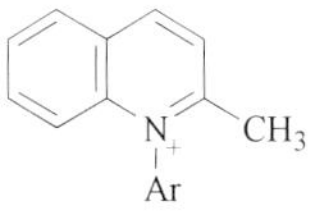 **22**

During coupling of many methylene groups linked to two activating groups X and Y, the activating group X is exchanged for an azo group in the second coupling step. If the first coupling step is easier than the removal of the activating group X, then unsymmetrical formazans can be synthesized by coupling the initially formed hydrazone with another diazonium compound (Scheme 2.14).

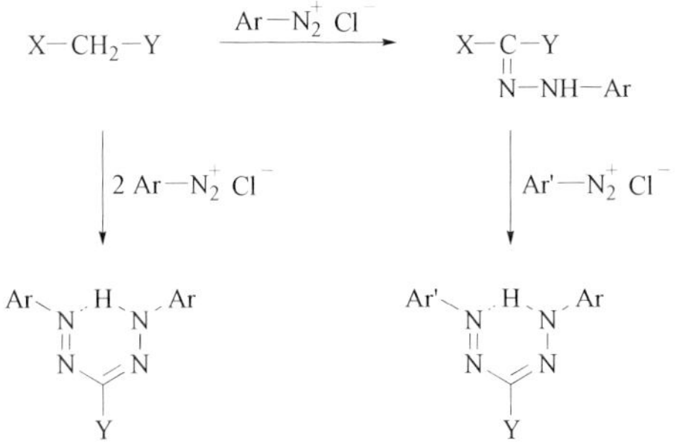

X = COOH, CONH$_2$, CHO, CO–alkyl, CO–aryl, COCOOH, SOCH$_3$
Y = X, COO–alkyl, CN, C=N–NH–aryl, SO$_2$–CH$_3$, SO$_2$–aryl

Scheme 2.14

Only symmetrical formazans can be obtained, for example, from **23** [55], **24** [56], and **25** [57].

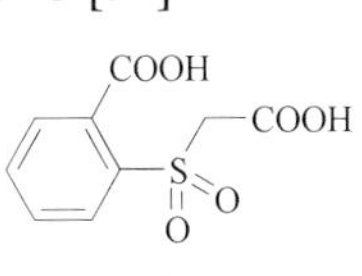

23, 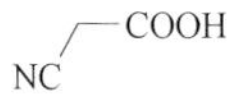 **24**, 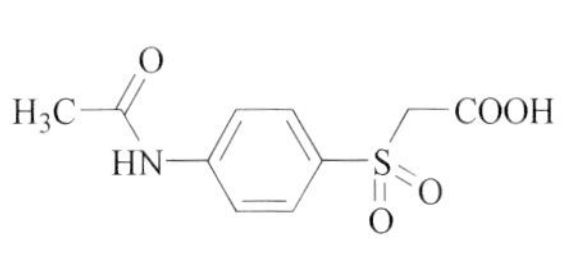 **25**

and from 4-oxopentanoic acid (levulinic acid), which couples to form the bis-formazan **26** [58]:

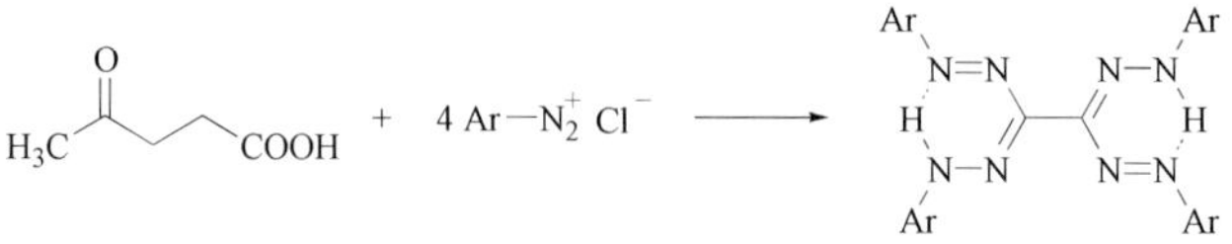

Unsymmetrical formazans can be prepared from **27** [56], **28** [59], **29** [60], and **30** [61] by stepwise coupling, preferably at different pH values.

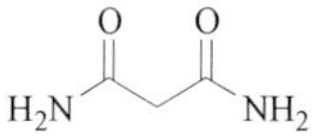

Doubly activated methane compounds with aryl substituents can also be converted to formazans **31** by elimination of both of the activating groups

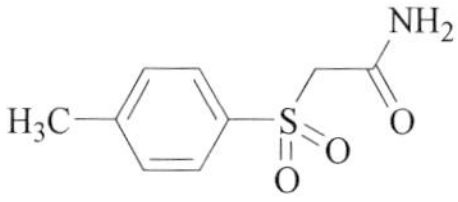

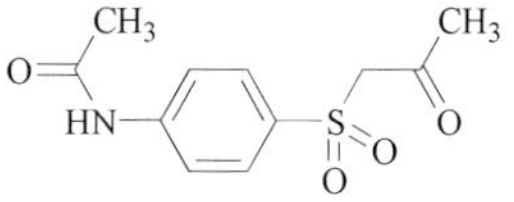

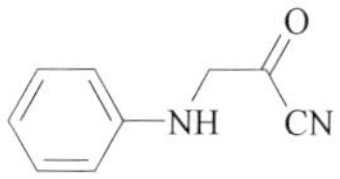

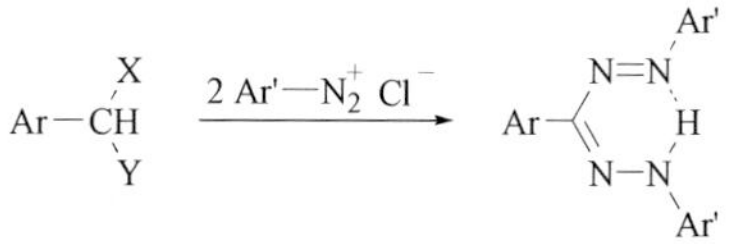

X = COOH, CONH$_2$, CHO, CO–alkyl, CO–aryl
Y = X, CN, COO–alkyl

Ester and cyano groups generally resist substitution by aryl azo groups and must first be hydrolyzed to produce carboxyl and carbonamide groups respectively, which can be more easily eliminated [62]. Unsymmetrical formazans can be produced according to Scheme 2.15.

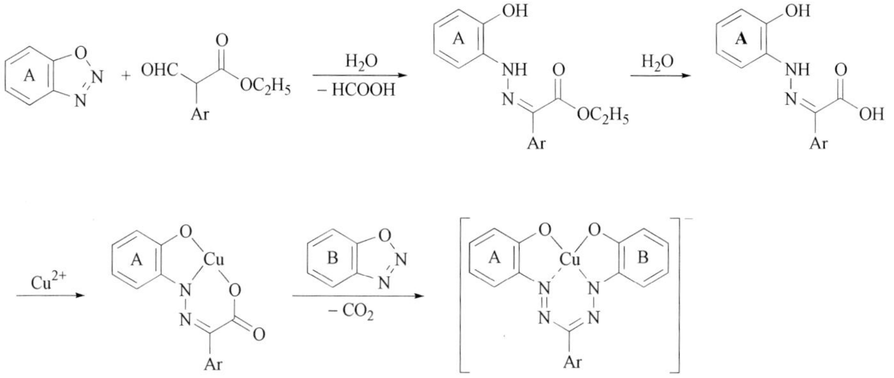

Scheme 2.15

2.9.3 Formazan Dyes

A special process for the production of coppered 1,5-bis(2-hydroxyphenyl)-formazans **32** is based on the demethylative copperization of 1,5-bis(2-methoxyphenyl)formazans [63]. The method involves heating for a short time in pyridine, formamide, or dimethylformamide.

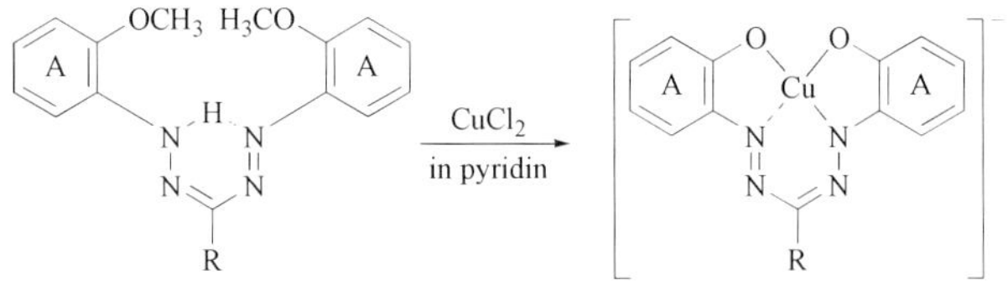

In an another process, the copper complexes are produced with alkylsulfonylamine as the complex-forming group [64].

Syntheses of formazans under phase-transfer conditions starting from hydrazones or from compounds with active methylene groups are also possible [65].

Oxidative Coupling of Azole Hydrazines with Hydrazones. Oxidative coupling of azole hydrazines with hydrazones takes place spontaneously in the presence of pyridine in air and is confined to the synthesis of heteroarylformazans, for example, **33** [66]:

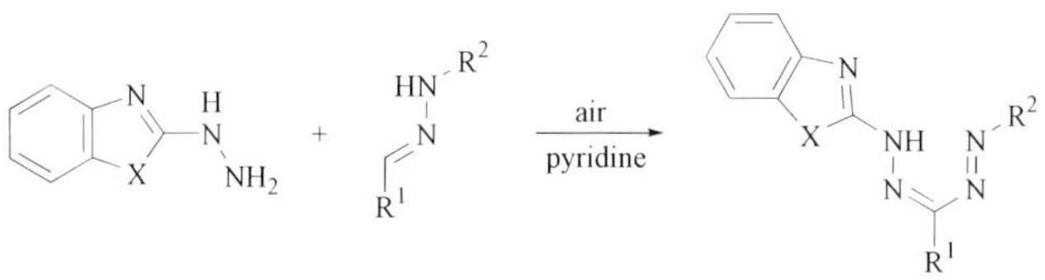

X = S, N-alkyl
R^1 = alkyl, aryl, heteroaryl
R^2 = aryl, heteroaryl

Metallization. Bidentate formazans that are insoluble in water can be warmed with cobalt, nickel, and copper salts (preferably acetates) to form metal chelates in solvents such as methanol, ethanol, acetone, and dimethylformamide. Metal complexes of tri- and tetradentate formazans are much more stable. Metallization with divalent salts occurs rapidly at room temperature. On reaction with diazotized 2-aminophenols or 2-aminonaphthols, coupling and metallization with divalent metal salts can take place concurrently under the same conditions. When coupling is complete, the dye is usually fully metallized.

For production of Co^{III} 1:2 complexes, a cobalt salt is generally added after coupling and warmed to 60–90 °C in a weakly alkaline medium. Reaction with chromium requires more severe reaction conditions, for example, refluxing for 20 h [67].

2.9.3.3 Principal Properties

Metal-Containing Bidentate Formazans. The most outstanding property of formazans is their ability to form coordination compounds with metals. They behave as monovalent, bidentate ligands and form neutral 1:2 metal chelates with divalent metals.

M = Cu, Ni, Co, Pd, Zn, Cd, Be, Mg

Opinions about the geometry of these compounds vary. A planar structure has been proposed on account of their magnetic properties [68]. However, with 1:2 Ni^{II} complexes their partial resolution into mirror-image isomers with (+)-quartz and their diamagnetism suggest a high nonplanar structure [69]. Reactions with Cu^{II} salts are dealt in [42, 70, 71].

Metal-Containing Tridentate Formazans. Formazans substituted with OH or COOH in the 2-position of the N^1- or N^5-aryl group have the same complexing properties as 2-hydroxy- (or carboxy)-2'-aminodiarylazo dyes. Similarly, they also form 1:1 complexes with four-coordinate metals, and 1:2 complexes with six-coordinate metals. Being N ligands formazans react more readily with cobalt salts than with chromium salts. The mostly blue 1:2 cobalt(III) complexes of type **34** [72] and the mostly gray-blue complexes of type **35** [73] are known.

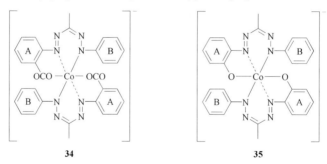

34   **35**

Rings A and B may contain other substituents

As with the tridendate azo compounds, ligands coordinated in 6/6-rings of type **34** have a *fac* arrangement, and those coordinated in 5/6-rings of type **35** a *mer* configuration. In the formazan system, unlike the azo system, the plane of the anellated 6/6-rings in the octahedron remains almost planar. Twisting of the aromatic groups does not occur during metallization and there is therefore no loss of color intensity. Compounds of type **34** have an even larger absorption area in the UV/Vis spectrum than their metal-free precursors [72].

Tridentate formazans form 1:1 complexes with divalent metal salts:

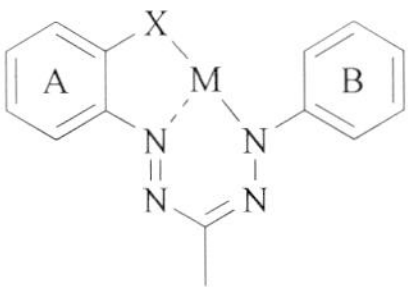

2.9.3 Formazan Dyes

X = COO, O

These coordinatively unsaturated complexes can undergo coordination reactions with compounds having a free electron-pair donor group, particularly pyridine, amines, ammonia, and water. Coordination is associated with a marked hypsochromic color change. For example, the unsaturated nickel complex **36** is green and its pyridine adduct **37** is violet [74].

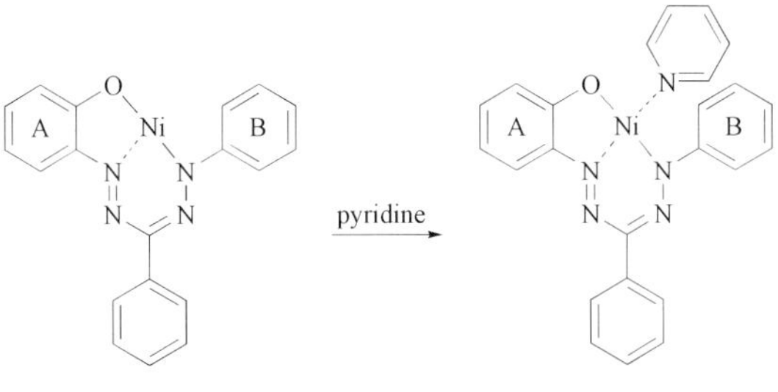

36   37

The coordinatively unsaturated complexes are even less soluble in organic solvents such as diethyl ether, benzene, and chloroform than their saturated equivalents. The magnetic moment of copper complexes of tridentate formazans with a 2-hydroxyaryl substituent at N^1 is very low (1.44 μ_B) compared to the expected value of copper(II) complexes with an unpaired electron and strong covalent bonding. The low solubility and magnetic moment suggest that the dimerization occurs with the formation of a saturated dinuclear complex **38**.

The saturated pyridine adduct **39**, which does not dimerize, has a magnetic moment of 1.7 μ_B. The copper complex **40** has an even higher magnetic moment of 1.81 μ_B [75].

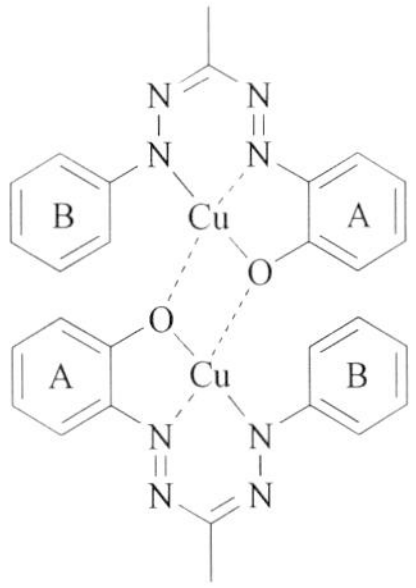

38 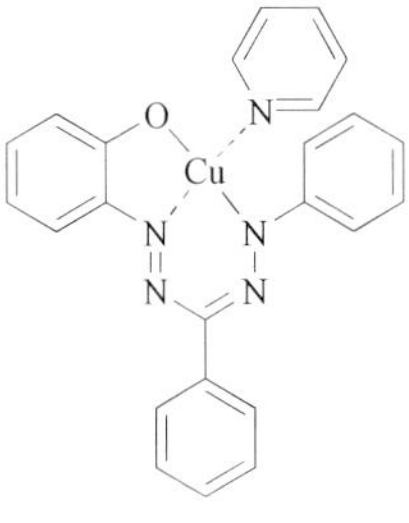   39 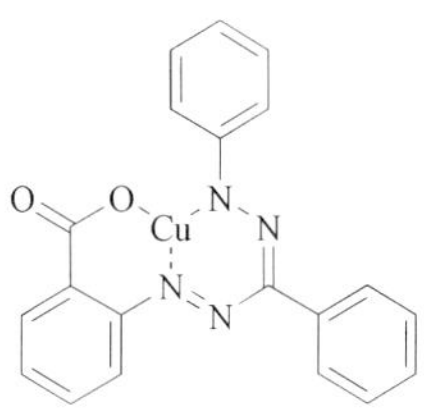   40
[69239-84-5]

2.9 Metal Complexes as Chromophores

Metal Containing Tetradentate Formazans. Formazans with two complex-forming groups in the 2- and 2′-positions of the N^1- and N^5-aryl substituents are potential quadridentate ligands. They form 1:1 coordination complexes with four- and six-coordinate metals:

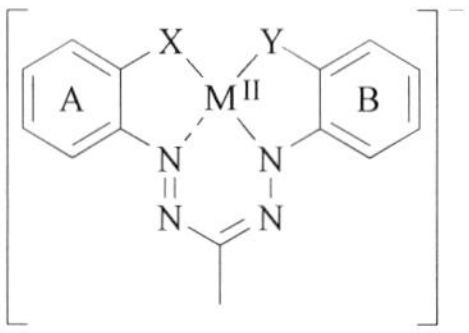

$$\left[\begin{array}{c} A \underset{N=N}{\overset{XY}{M^{II}}} B \\ NN \end{array}\right]$$

X, Y = COO, O

The complexes assume a planar tetragonal structure so that two monodentate ligands can occupy free coordination sites at the apices of an octahedron.

The copper complex of 1,5-bis(2-hydroxyphenyl)-3-cyanoformazan probably exists as tricyclic structure **41** because of the increased acidity of the phenolic hydroxyl group.

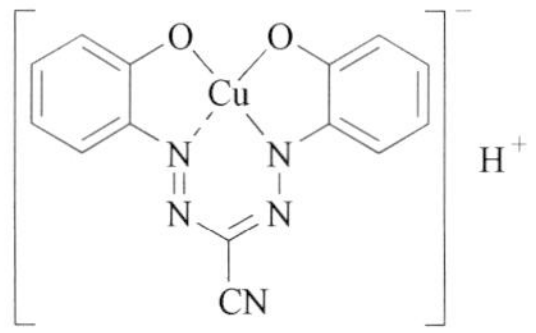

$$\left[\begin{array}{c} \overset{OO}{Cu} \\ N=N \\ NN \\ CN \end{array}\right] H^+$$

41 [36090-18-3]

Complexes of formazans having one or two carboxylic groups in the 2- or 2,2′-positions of the aryl substituents, however, appear to be bicyclic. Thus, a hydrate was obtained on reaction of 1,5-bis(2-carboxyphenyl)-3-cyanoformazan with copper acetate [76]. The coordinated water could not be removed by drying and produced a broad band in the IR spectrum at 3450 cm^{-1}. The copper complexes of formazans containing carboxyl groups were then assigned the structure **42**.

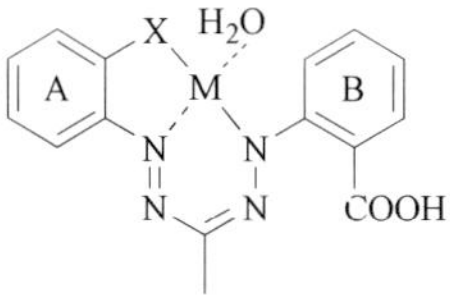

42 X = COO, O

In the patent literature there are indications that the structure of copper complexes of the 2-hydroxy-2′-carboxylic acid type can be modified. After coupling and metallization, lightfastness, color intensity, and color quality can be improved by stirring the reaction mixture at low pH [77]. The copper complex can be stabilized by refluxing for 15 h [78]. The coordinatively bound water may possibly be displaced by the carboxyl group in the 2-position during this aftertreatment, converting the dye to the tricyclic form **43** with a 5/6/6-membered ring system. A rearrangment may proceeds simultaneously, in the course of which the coordinative bond migrates from N^1 to N^2 of the formazan structure to give the less strained form with a 6/5/6-membered ring system **44**.

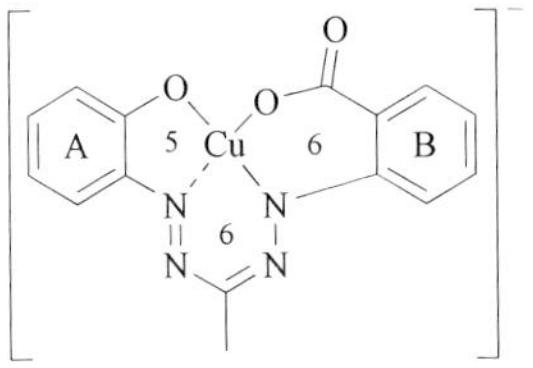

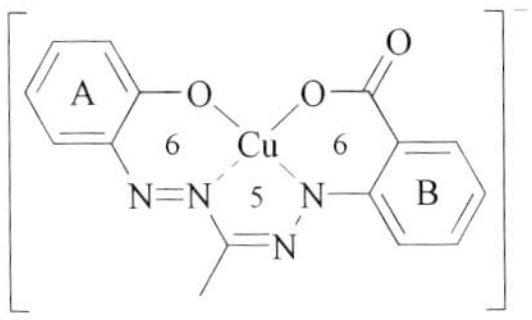

Formazan-like structures with a 6/5/6-ring system are exemplified by **45** [79] and **46** [80]:

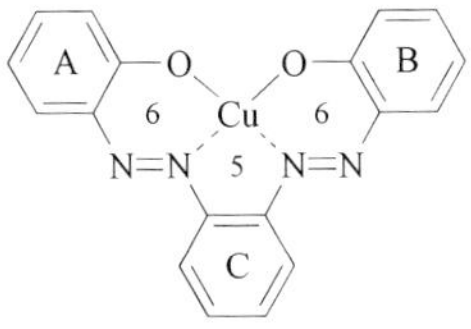

Formazan dyes have become important reactive dyes for cotton. Formazan dyes are also used in analytical chemistry because of the high color intensity of many of their metal complexes. The chemistry of formazans was first exploited in 1962 by MacDonald to produce color photographs [81].

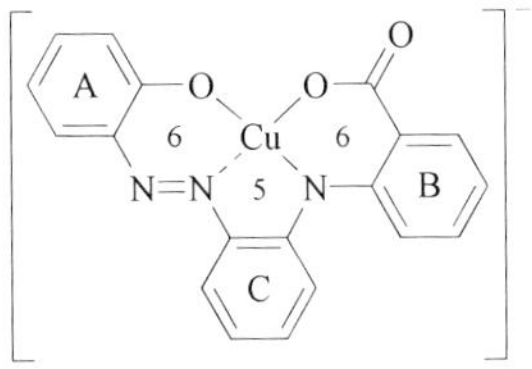

2.9.4 References

[1] H. Baumann, R. H. Hensel: "Neue Metallkomplexfarbstoffe, Struktur und färberische Eigenschaften", *Fortschr. Chem. Forsch.* **7** (1967) 643–783.
[2] F. Beffa, G. Back, *Rev. Prog. Color.* **14** (1984) 33.
[3] Houben-Weyl, **10/3** pp. 452–463.
[4] Ciba, BP 644 883, 1946.
[5] G. P. Grippa, *Gazz. Chim. Ital.* **57** (1927) 20, 497, 593; **58** (1928) 716; V. I. Mur, *Zh. Obschch. Khim.* **24** (1954) 572; *Chem. Abstr.* **49** (1955) 6198.
[6] H. Pfitzner, H. Baumann, Angew. Chem. **70** (1958) 232.
[7] I.G. Farbenindustrie, DE 571 859, 1931 (D. Delfs, R. Knoche); DE 658 841, 1934 (D. Delfs); DE 738 900, 1936 (D. Delfs).
[8] R. Price in: K. Venkataraman: *The Chemistry of Synthetic Dyes*, Academic Press, New York, 1970, vol. III, p. 336.
[9] BASF, DE 479 373, 1926 (H. Krzikalla, W. Müller).
[10] D. Brown, H. R. Hitz, L . Schäfer, *Chemosphere* **10** (1981) 245.
[11] ICI, BP 740 272, 1953 (A. H. Knight, C. H. Reece).
[12] Sandoz, BP 753 550, 1953; Francolor, BP 810 207, 1954 (R. F. M. Sureau, G. R. H. Mingasson).
[13] Geigy, BP 756 874, 1952; Ciba, BP 637 404, 1946.

[14] Ciba, BP 765 355.
[15] Bayer, EP-A 73 998, 1982 (W. Mennicke, H. Fürstenwerth).
[16] ICI, BP 1 094 746, 1965 (A. Johnson, P. A. Mack, R. Price, A. Warwick).
[17] ICI, BP 1 089 826, 1965 (A. Johnson, P. A. Mack, R. Price).
[18] Sandoz DE-AS 2 153 548, 1971 (J. Dore).
[19] ICI, FR 1 466 877, 1966 (P. A. Mack, R. Price).
[20] Geigy, BE 671 099; Ciba FR 1 486 661.
[21] VEB Farbenfabrik Wolfen, BP 1 051 219, 1964 (G. Essbach. H. Modrow, E. Schick, H. Baumann).
[22] G. Schwarzenbach, *Helv. Chim. Acta* **35** (1952) 2344.
[23] H. Zollinger: *Color Chemistry*, VCH Verlagsgesellschaft, Weinheim, 1987, pp. 121–122.
[24] R. Price in: K. Venkataraman: *The Chemistry of Synthetic Dyes*, Academic Press, New York, 1970, vol. III, p. 312.
[25] R. Price: "Some Aspects of the Chemistry of Metal Complex Dyes", *Chimia* **28** (1974) 221.
[26] H. D. K. Drew, R. E. Fairbairn, *J. Chem. Soc.* **1939**, 823.
[27] H. Pfitzner, *Angew. Chem.* **62** (1950) 242.
[28] P. Pfeiffer, S. Saure, *Ber. Dtsch. Chem. Ges.* **74** (1941) 935.
[29] R. Grieb, A. Niggli, *Helv. Chim. Acta* **48** (1965) 317.
[30] R. Price in: K. Venkataraman: *The Chemistry of Synthetic Dyes*, Academic Press, New York, 1970, vol. III, p. 328.
[31] G. Schetty, W. Kuster, *Helv. Chim. Acta* **44** (1961) 2193; G. Schetty, *Helv. Chim. Acta* **45** (1962) 1095, 1473; **47** (1964) 921; **49** (1966) 49; **50** (1967) 15; *Chimia* **18** (1964) 244.
[32] H. Jaggi, *Helv. Chim. Acta* **51** (1968) 580.
[33] G. Schetty, *Am. Dye Reporter*, August 2, **1965** 37 (P 589).
[34] H. Pfitzner, *Angew. Chem.* **84** (1972) 351; Angew. Chem Int. Ed. Engl. **11** (1972) 312.
[35] G. Schetty, E. Steiner, *Helv. Chim. Acta* **57** (1974) 2149.
[36] U. Lehmann, G. Rihs in: *Chemistry of Functional Dyes*, Z. Yoshida, T. Kitao (eds.), Mita Press Tokyo, **1989**, p. 215.
[37] G. Schetty, *Helv. Chim. Acta* **53** (1970) 1437.
[38] H. von Pechmann, *Ber. Dtsch. Chem. Ges.* **25** (1892) 3175.
[39] E. Bamberger, E. Wheelwright, *Ber. Dtsch. Chem. Ges.* **25** (1892) 3201.
[40] G. Lakon, *Ber. Dtsch. Bot. Ges.* **57** (1939) 191; **60** (1942) 299, 434.
[41] R. Kuhn, D. Jerchel, *Ber. Dtsch. Chem. Ges.* **74** (1941) 949.
[42] L. Hunter, L. B. Roberts, *J. Chem. Soc.* **1941**, 820, 823.
[43] R. Wizinger, , V. Biro, *Helv. Chim. Acta*, **32** (1949) 901.
[44] R. Wizinger, *Chimia Suppl.* **1968**, 82–94.
[45] A. W. Nineham, *Chem. Rev.* **55** (1955) 355–483.
[46] R. Pütter, *Houben-Weyl*, **10/3** (1965) 627–694.
[47] H. Baumann, R. H. Hensel: "Neue Metallkomplexfarbstoffe, Struktur und färberische Eigenschaften," *Fortschr. Chem. Forsch.* **7** (1967) 714–727.
[48] H. Iida, *Senryo to Yakuhin* **14** (1969) 39–45.
[49] R. Price: *Comprehensive Coordination Chemistry*, vol. 6, Pergamon Press, Oxford, pp. 77–83.
[50] N. Bednyagina, I. Postovskii, A. Garnovskii, O. Osipov: "Hetarylformazans", *Russ. Chem. Rev. (Engl. Transl.)* **44** (1975) 493–509.
[51] H. Wahl, F. Gault, *C. R. Serie A* **251** (1960) 1523.
[52] H. Wahl, M.-T. le Bris, F. Gault, F. Pierrot, *Teintex* **21** (1956) 697, 786.
[53] H. Wahl, M.-T. le Bris, *Bull. Soc. Chim. Fr. Ser. 5* **1954**, 248, 587; Francolor, FR 1 115 086, 1954 (H. Wahl, M.-T. le Bris).

[54] O. Stashkevich, G. Pilyugin, V. Stashkevich, *Khim. Geterotsikl. Soedin.* **1970**, 1104.
[55] M. Lozinskij, S. Sanova, P. S. Pel'kis, *Z. Org. Chim.* **1** (1965) 314.
[56] Sandoz, FR 1 338 068, 1962 (W. Steinemann).
[57] Sandoz, FR 1 284 174, 1962 (W. Steinemann).
[58] E. Bamberger, F. Kuhlemann, *Ber. Dtsch. Chem. Ges.* **26** (1893) 2978; Y. Gok, B. Senturk, *Org. Prep. Proced. Int.* **27** (1995) 87.
[59] Sandoz, GB 984 451, 1960 (W. Steinemann).
[60] Sandoz, GB 984 452, 1960 (W. Steinemann).
[61] F. M. Abdel Galil, F. A. Khalifa, T. S. Abdin, *Dyes Pigm.* **12** (1990) 1.
[62] Soc. Carbochimique, DE-AS 1 062 851, 1952, 153 (H. Ziegler).
[63] Du Pont, US 3 663 525, 1972 (V. Witterholt, J. Dalgarn).
[64] Geigy, DE-OS 1 419 838, 1959 (B. Beffa, G. Schetty).
[65] A. R. Katritzky, S. A. Belyakov, D. Cheng, H. D. Durst, *Synthesis* **1995**, 577.
[66] N. Bednyagina, Y. Sedov, G. Petrova, I. Postovskii, *Khim. Geterotsikl. Soedin.* **1972**, 390.
[67] J. Poskocil, Z. J. Allan, GB 927 128, 1963.
[68] M. Ermakova, E. Krylov, I. Postovskii, *Zh. Obshch. Khim.* **30** (1960) 849.
[69] H. Irving, J. Gill, W. Cross., *J. Chem. Soc.*, **1960**, 2087.
[70] B. Hirsch, *Justus Liebigs Ann. Chem.* **637** (1960) 169; B. Hirsch, E. Jauer, *Justus Liebigs Ann. Chem.* **682** (1965) 99.
[71] R. Price, *J. Chem. Soc.* A **1971**, 3379.
[72] F. Beffa, P. Lienhard, E. Staner, G. Schetty, *Helv. Chim. Acta* **46** (1963) 1369.
[73] Instytut Przemyslu Organicznego, Warschau, PL 55 361, 55 362, 55 363, and 56 636, 1966 (E. Kubicky); PL 66 469, 1968 (K. Alchimowicz, E. Kubicky); Pabianickie Zaklady Farmaceutyczne "POLFA", Polytechnika Lodzka PL 66 469 (Z. Jankowski, R. Stolarski, W. Cieslak, R. Salagacki); Geigy, DE-AS 1 252 829 and DE-OS 1 444 688, 1962 (F. Beffa, P. Lienhard, E. Steiner).
[74] R. Wizinger, *Angew. Chem.* **61** (1949) 33.
[75] M. Kishita, Y. Muto, M. Kubo, *Aust. J. Chem.* **11** (1958) 309.
[76] R. Price, *J. Chem. Soc.* A **1971**, 3385.
[77] Hoechst, DE-OS 2 945 493, 2 945 537, 1979 (E. Schwaiger, E. Hoyer).
[78] Bayer, EP-A 321 784, 1987 (E. Köcher, H. Jäger).
[79] Bayer, EP-A 240 839, 1986 (H. Henk).
[80] Bayer, EP-A 201 026, 1985 (H. Henk).
[81] Ilford, GB 908 299, 1962 (E. MacDonald).

2.10 Fluorescent Dyes

For fluorescence to occur, it is necessary to increase the lifetime of the first excited singlet state of the molecule so that radiative processes can compete with nonradiative processes (vibrational heat loss). This is achieved by making the molecule more rigid, thus restricting the vibrational (and rotational) degrees of freedom. In addition, a nonplanar group must also be present, otherwise the planar molecule would be pigmentary, and the excited state energy dissipated by rapid vibrational relaxation via the crystal lattice of the pigment. Indeed, this is the major reason why pigments have high lightfastness.

2.10 Fluorescent Dyes

Fluorescent molecules span the entire region from the ultraviolet, through the visible, to the near infrared. Many dyes exhibit fluorescence, but to be of practical use, fluorescent dyes must satisfy certain requirements: they must produce a pure color dictated by their absorption and emission spectra, they must have a high molar extinction, and most important, they must have a high quantum yield. These requirements are met by very few dyes. A disadvantage shared by many fluorescent dyes is their poor lightfastness, but there are some exceptions.

The main types which find industrial use are:

Colourless molecules with blue fluorescence are optical brightening agents that are used as blue whiteners in paper production and washing powders (see Chapter 7). The two main classes are heterocylics such as pyrazolines and particularly stilbenes (**1**).

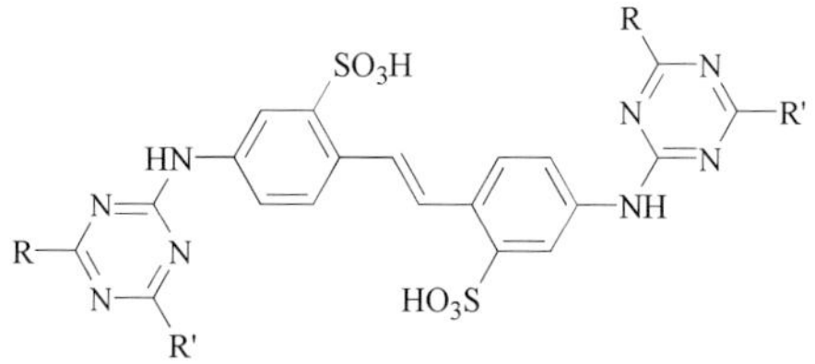

1

Yellow Dyes with Blue/Green Fluorescence. Fluorescein, *C.I. Solvent Yellow 94*, 45350:1 [*518-45-6*] (**2**), is perhaps the best known example, but equally important are coumarins, such as *C.I. Basic Yellow 40* [*12221-86-2*], and especially naphthalimides such as *C.I. Solvent Yellow 44*, 56200 [*2478-20-8*] (**3**).

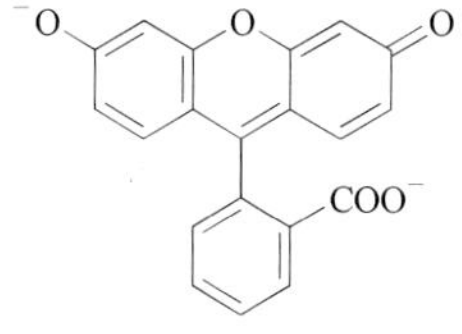

2 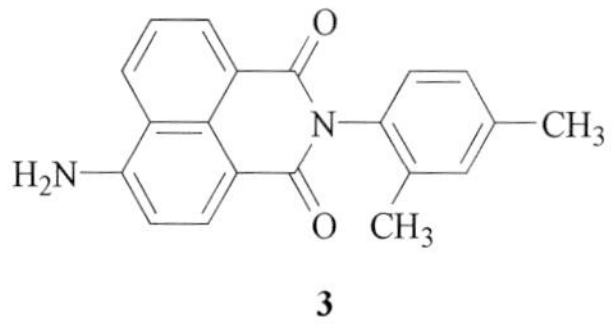   **3**

Red Dyes with Yellow/Orange/Red Fluorescence. The best known dyes of this type are the rhodamines (xanthenes), such as *C.I. Solvent Red 49* [*509-34-2*] (**4**). Another type are hemicyanines, e.g. *C.I. Basic Red 12*, 48070 [*6320-14-5*] (Astraphloxine).

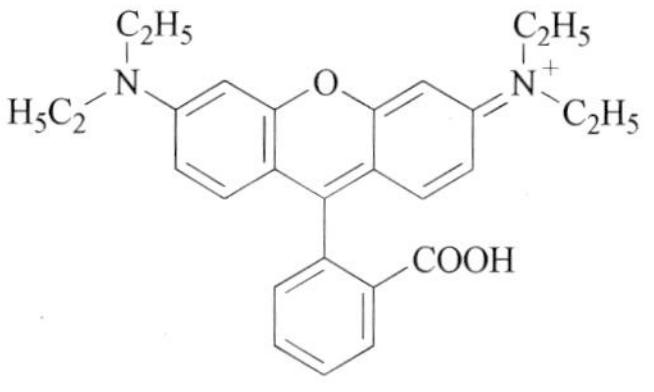

4

Blue Dyes with Red or Near-Infrared Fluorescence. These are less important than the above types and are of little importance industrially. Indeed, whilst the blue-green fluorescence of a yellow dye and the red fluorescence of a red dye increase the visual attractiveness of the dyes by making them brighter, a blue dye with red fluorescence is visually weird.

Near-Infrared Dyes with Near-Infrared Fluorescence. This type is becoming more important, particularly in biomedical applications (see Chapter 6). Phthalocyanines and cyanines provide this type of fluorescence.

2.11 Other Chromophores

2.11.1 Quinophthalone Dyes

Like the hydroxy azo dyes, quinophthalone dyes can, in theory, exhibit tautomerism. Because the dyes are synthesised by the condensation of quinaldine derivatives (**1**) with phthalic anhydride, they are often depicted as structure (**2**), but this is incorrect, since the two single bonds prevent any conjugation between the two halves of the molecule.

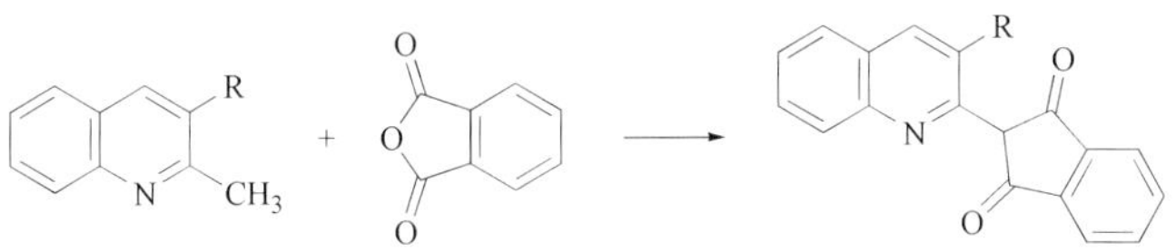

The dyes exist as structure (**3**), in which the donor pyrrole-type nitrogen atom is conjugated to the two acceptor carbonyl groups through an ethylenic bridge. In addition to the increased conjugation, structure (**3**) is stabilized further by the six-membered intramolecular hydrogen bond between the imino hydrogen atom and the carbonyl oxygen atom.

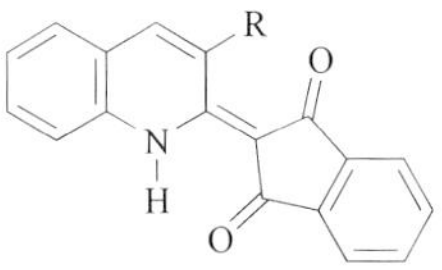

Quinophthalones provide important dyes for the coloration of plastics, e.g., C.I. Solvent Yellow 33 (**3**, R = H), and for the coloration of polyester. C.I. Disperse Yellow 54 (**3**, R = OH) is the leading yellow dye for the transfer printing of polyester.

2.11.2 Nitro and Nitroso Dyes

These dyes are now of only minor commercial importance, but are of interest for their small molecular structures. The early nitro dyes were acid dyes used for dyeing natural animal fibers such as wool and silk. They are nitro derivatives of phenols, e.g., picric acid (**4**) or naphthols, e.g., C.I. Acid Yellow 1, 10316 [846-70-8] (**5**).

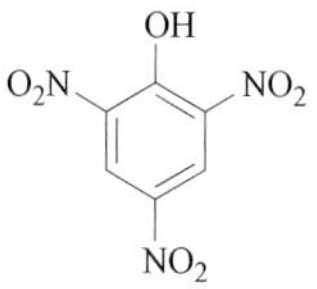

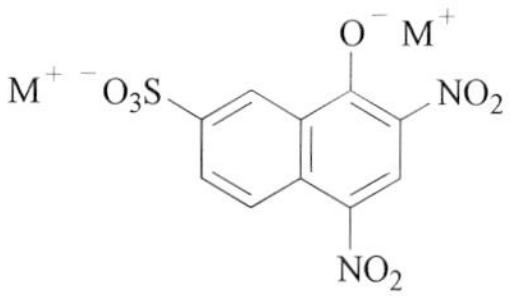

The most important nitro dyes are the nitrodiphenylamines of general structure (**6**).

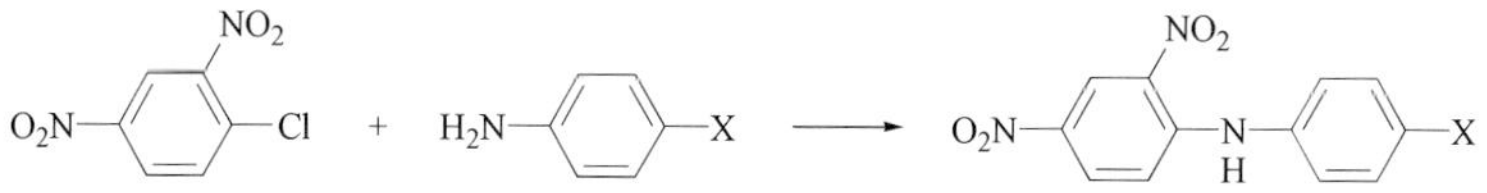

These small molecules are ideal for penetrating dense fibers such as polyester and are therefore used as disperse dyes for polyester. All the important dyes are yellow: (**6**) (X = H) is C.I. Disperse Yellow 14, 10340 [961-68-2] and (**6**) (X = OH) is C.I. Disperse Yellow 9, 10375 [6373-73-5]. Although the dyes are not terribly strong (ε_{max} ca. 20 000), they are cost-effective because of their easy synthesis from inexpensive intermediates. C.I. Disperse Yellow 42 and C.I. Disperse Yellow 86 are important lightfast dyes for automotive-grade polyester.

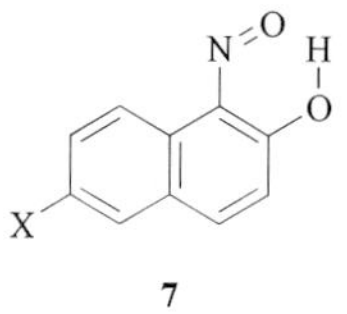

Nitroso dyes are metal-complex derivatives of *o*-nitrosophenols or -naphthols. Tautomerism is possible in the metal-free precursor between the nitrosohydroxy tautomer (**7**) and the quinoneoxime tautomer (**8**).

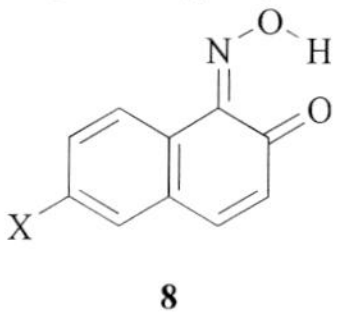

The only nitroso dyes important commercially are the iron complexes of sulfonated 1-nitroso-2-naphthol, e.g., C.I. Acid Green 1, 10020 [19381-50-1] (**9**); these inexpensive colorants are used mainly for coloring paper.

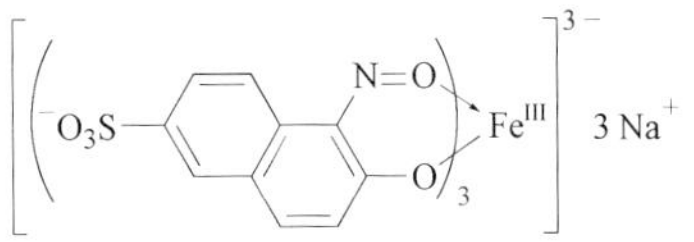

2.11.3 Stilbene Dyes

Stilbene dyes are in most cases mixtures of dyes of indeterminate constitution that are formed from the condensation of sulfonated nitroaromatic compounds in aqueous caustic alkali, either alone or with other aromatic compounds, typically arylamines [1]. The sulfonated nitrostilbene (**10**) is the most important nitroaromatic, and the aminoazobenzenes are the most important arylamines. *C.I. Direct Orange 34*, 40215 [*1325-54-8*], the condensation product(s) of (**10**) and the aminoazobenzene (**11**), is a typical stilbene dye.

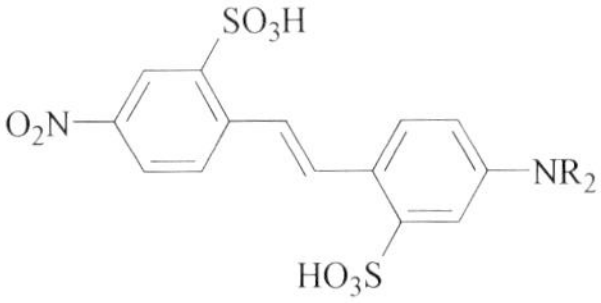

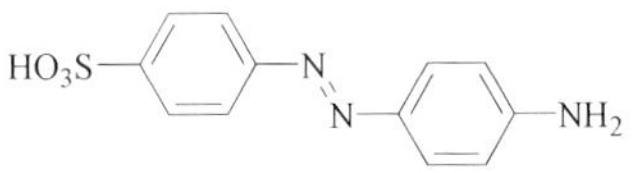

2.11.4 Formazan Dyes

Formazan dyes bear a formal resemblance to azo dyes, since they contain an azo group, but have sufficient structural dissimilarities to be considered as a separate class. The most important formazan dyes are the metal complexes, particularly copper complexes, of tetradentate formazans (see Section 2.10.3). They are used as reactive dyes for cotton; (**12**) is a representative example.

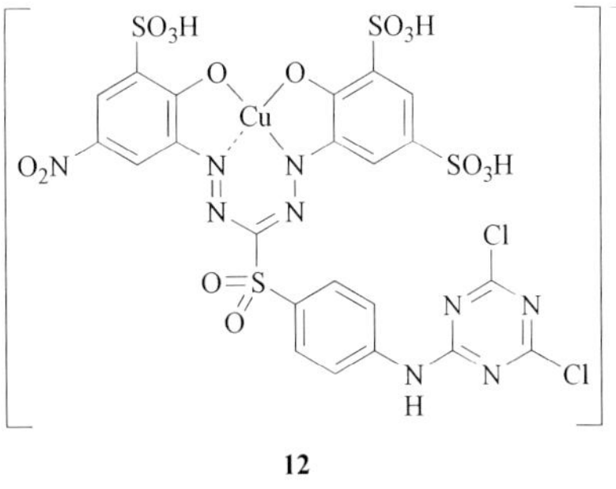

2.11.5 Triphenodioxazine Dyes

Notable advances have been made in recent years in triphenodioxazine dyes. Triphendioxazine direct dyes have been known for many years; *C.I. Direct Blue 106*, 51300 [*6527-70-4*] (**13**) is a typical dye. Resurgence of interest in triphendioxazine dyes arose through the successful modification of the intrinsically strong and bright triphendioxazine chromogen to produce blue reactive dyes for cotton [2]. These blue reactive dyes combine the advantages of azo dyes and anthraquinone dyes. Thus they are bright, strong dyes with good fastness properties. Structure (**14**) is typical of these reactive dyes, where R represents a reactive group.

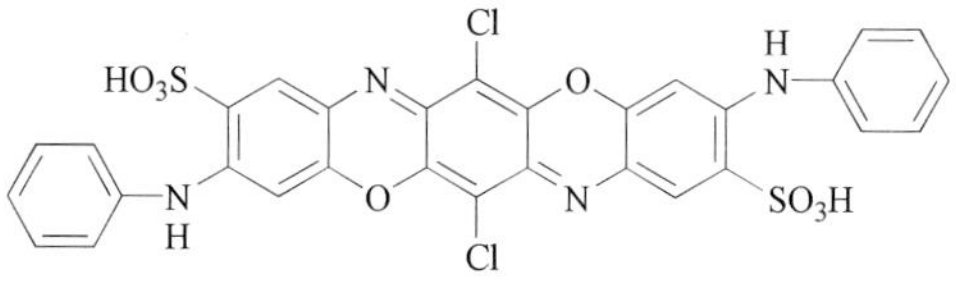

13

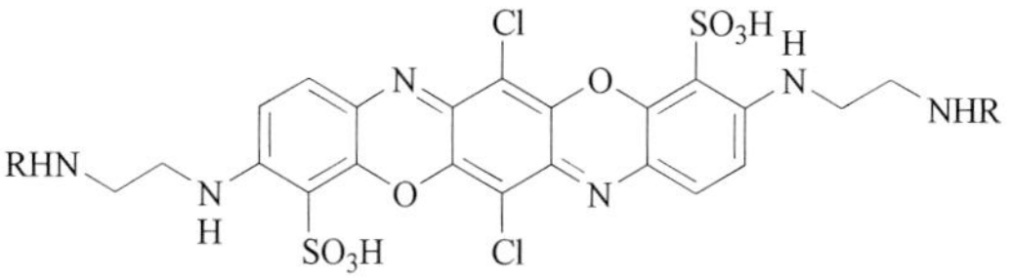

14

Like phthalocyanine dyes, triphenodioxazine dyes are large molecules, and therefore their use is restricted to coloring the more open-structured substrates such as paper and cotton.

2.11.6 References

[1] *Colour Index*, Vol. 4, 3rd ed., The Society of Dyers and Colourists, Bradford, UK, 1971.
[2] ICI, GB 1 349 513, 1970.

3 Dye Classes For Principal Applications

3.1 Reactive Dyes

3.1.1 Introduction

The most important distinguishing characteristic of reactive dyes is that they form covalent bonds with the substrate that is to be colored during the application process. Thus, the dye molecule contains specific functional groups that can undergo addition or substitution reactions with the OH, SH, and NH_2 groups present in textile fibers.

Cross and Bevan first succeeded in fixing dyes covalently onto cellulose fibers (in 1895) [1], but their multistep process was too complicated for practical application. Early work by Schröter with sulfonyl chloride-based dyes was unsuccessful [2], but Günther later did succeed in fixing derivatives of isatoic anhydride onto cellulose fibers [3].

An alternative approach is to modify the fibers themselves and then introduce the coloration. For example, cellulose can be reacted with 4-nitrobenzyldimethylphenylammonium hydroxide or 3-nitrobenzyloxymethylpyridinium chloride, followed by reduction, diazotization, and coupling to a dye [4]. Another process, invented by Haller and Heckendorn, involves treating the cellulose with cyanuric chloride in an organic solvent and then treating the product with an amino group containing dye [5].

The first industrially important reactive dye systems were developed for wool, and took advantage of the chloroacetylamino [6] and the chloroethanesulfonyl groups [7]. Vinylsulfonyl- and 2-sulfooxyethanesulfonyl groups were found to be applicable to both wool and cellulose. Heyna and Schumacher patented some of the first dyes of this type in the 1940s [8, 9], and vinylsulfone dyes continue to be of great importance.

The dyes described in [8] entered the market for wool in 1952 under the trade name Remalan, followed a year later by the Cibalan dyes from Ciba-Geigy, which contained monochlorotriazinyl reactive groups [10]. In 1953 Rattee and Stephen were able to fix chlorotriazine-containing dyes onto cellulose in basic, aqueous solvents, representing an economic breakthrough for the reactive dye concept [11].

Introduction of the Procion-MX dyes by ICI in 1956 was followed in 1957 by the appearance of the Remazol dyes from Hoechst and the Cibacron dyes from Ciba-Geigy. The year 1960 brought the Levafix dyes from Bayer and the Drimaren range from Sandoz. Other significant developments included the introduction of multiple-anchor dyes (Solidazol, Cassella, 1975; Sumifix-Supra, Sumitomo, 1981), fluorotriazine-containing reactive dyes (Ciba-Geigy, 1978), and fluorotriazine-containing multiple-anchor dyes (Ciba-Geigy, 1988).

The development of reactive dyes has continued to be rapid.

3.1.2 Chemical Constitution of Reactive Systems

Contrary to other dye groups reactive dyes are characterized by known chromophore systems, as described in this book, and by bearing various reactive moieties in their molecules. Thus, dye systems which are used to form reactive dyes comprise all industrial important groups of dyes. Consequently, the classification in this section is primarily according to the various reactive groups.

3.1.2.1 Mono-Anchor Dyes

Most mono-anchor dyes are derivatives of cyanuric chloride (2,4,6-trichloro-1,3,5-triazine [*108-77-0*]), a molecule of wide synthetic potential because the three chlorine atoms on the triazine ring differ in their reactivities [12]. The first chlorine atom exchanges with nucleophiles in water at $0-5\,°C$, the second at $35-40\,°C$, and the third at $80-85\,°C$. A wide variety of triazinyl dyes can thus be prepared by careful selection of the reaction conditions. Condensation of cyanuric chloride with a chromophore ("Chrom.") containing an amino group yields the highly reactive dichlorotriazinyl dyes **1** [11]. These very reactive dyes are sensitive to hydrolysis, and a suitable buffer is usually added to the powdered dye to increase its stability [13].

$$\text{Chrom.}-\text{NH}-\underset{\underset{\mathbf{1}}{}}{\overset{}{\underset{N=}{\overset{N=}{\bigg\langle}}}}\!\!\!\!\!\!\!\!\!\!\!\!\overset{Cl}{\underset{Cl}{\overset{N}{\bigg\rangle}}}$$

When two of the chlorine atoms are substituted, for example with amino or alkoxyl groups, monochlorotriazinyl dyes **2** are obtained. These are considerably less reactive, and hence react with cellulose in the exhaust dyeing process only at relatively high temperature (80 °C). Such dyes are especially advantageous for printing [14].

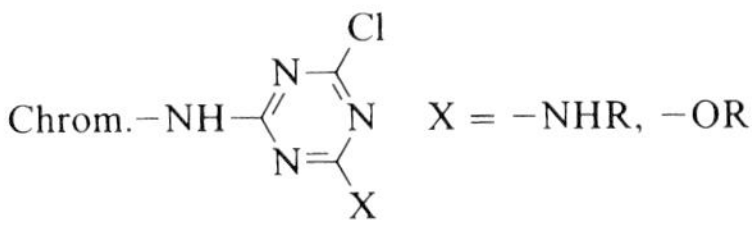

3.1.2 Chemical Constitution of Reactive Systems 115

$$\text{Chrom.}-NH-\underset{\mathbf{2}}{\underset{X}{\left\langle\begin{array}{c}N=\\N=\end{array}\right\rangle}}N \quad X = -NHR, -OR$$

The reactivity of monochlorotriazinyl dyes can be increased by replacing chlorine with fluorine (**3**), which allows the exhaust process to be carried out at 40 °C [15].

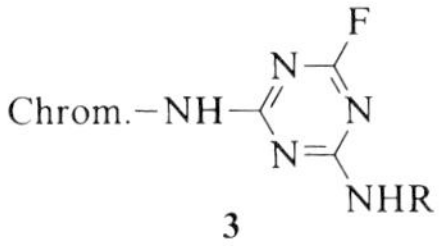

$$\text{Chrom.}-NH-\underset{\mathbf{3}}{\underset{NHR}{\left\langle\begin{array}{c}N=\\N=\end{array}\right\rangle}}N$$

Reaction of monochlorotriazinyl dyes with tertiary amines also yields highly reactive dyes **4**. One advantage conferred by the ammonium groups in these dyes is high water solubility [16].

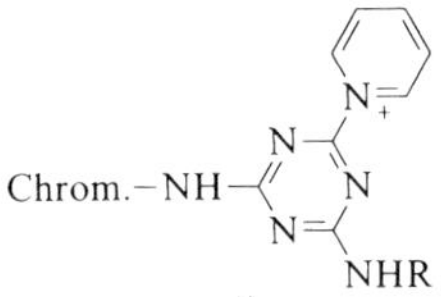

$$\text{Chrom.}-NH-\underset{\mathbf{4}}{\underset{NHR}{\left\langle\begin{array}{c}N=\\N=\end{array}\right\rangle}}N-\overset{+}{N}\text{C}_5H_5$$

The reactivity of dyes of this type is strongly influenced by the choice of tertiary amine. Nicotinic acid is preferred because it acts as a good leaving group and introduces comparatively little odor into the dyeing bath. The ammonium residue is also a better leaving group than, for example, chlorine, and allows the exhaust process to be conducted at 40 – 60 °C. In contrast to halogen-containing triazinyl dyes, these dyes do not require alkali in the application process, so dyeing can be conducted under neutral conditions.

In addition to the 1,3,5-triazines, other classes of heterocycles are important for the dyeing of cotton. Especially noteworthy are the halopyrimidine dyes **5** [17].

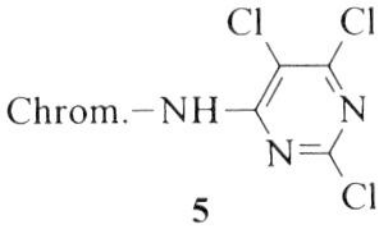

$$\text{Chrom.}-NH-\underset{\mathbf{5}}{\underset{Cl}{\left\langle\begin{array}{c}Cl\quad Cl\\N=\end{array}\right\rangle}}N$$

These dyes are less reactive than the triazines because the extra carbon atom reduces the ability of the ring to stabilize a negative charge. The reactivity of the system can be increased by introducing strongly electron withdrawing groups, including cyano [18], fluoro [19], or methylsulfonyl groups [20]. The pyrimidine ring can also be activated by inserting a carbonyl group between the chromophore and the heterocycle (**6**) [21].

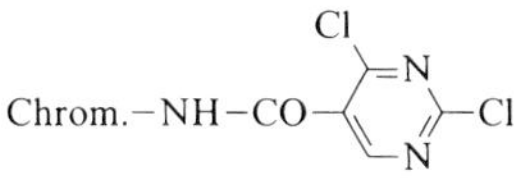

$$\text{Chrom.}-NH-CO-\underset{\mathbf{6}}{\underset{N}{\left\langle\begin{array}{c}Cl\\=N\end{array}\right\rangle}}-Cl$$

Other diazine moieties are also often incorporated into reactive dyes, e.g., **7** and **8** [22, 23]:

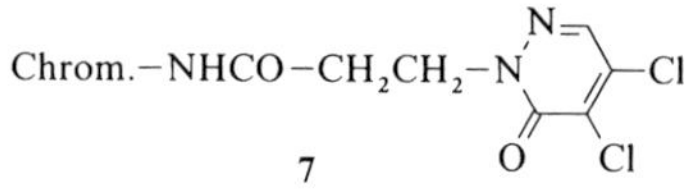

Chrom.–NHCO–CH$_2$CH$_2$–N⟨pyrimidinone with Cl, Cl⟩

7

Chrom.–NH–CO–⟨pyridazine with Cl, Cl⟩

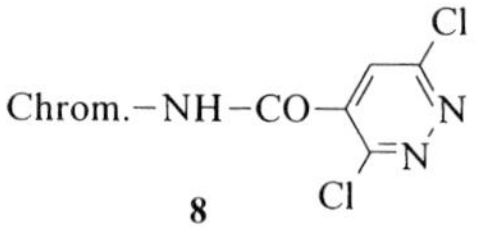

8

The list of economically attractive reactive heterocyclic intermediates also includes dichloroquinoxalines **9** and dichlorophthalazines **10**. Anchor systems in both cases are bound to the chromophore via a carbonylamino group, and the reactivities correspond roughly to those of dichlorotriazinyl dyes.

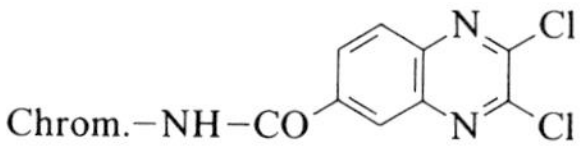

Chrom.–NH–CO–⟨dichloroquinoxaline⟩

9

Chrom.–NH–CO–⟨dichlorophthalazine⟩

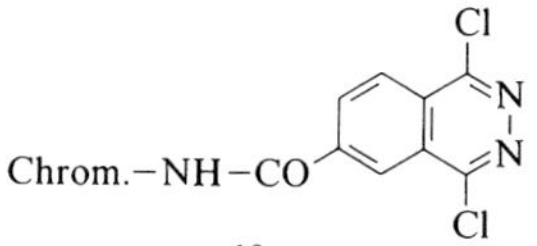

10

Benzothiazole derivatives with good leaving groups at the 2-position of the heterocycle can also be used as reactive anchors. An example with chlorine as the leaving group is **11**:

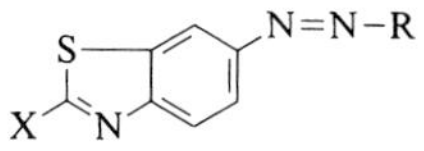

⟨benzothiazole with X at 2-position, N=N–R⟩   X=Cl

11

The reactive group that has had the greatest impact on the market is the 2-sulfooxyethylsulfonyl group [9, 24, 25]. Treatment with alkali in this case causes the elimination of sulfuric acid to form a vinylsulfonyl moiety that reacts with cotton to give a dye – fiber bond [26]. Describing this as an elimination – addition sequence is not meant to rule out the possibility that the nucleophile attacks the β-carbon atom directly, without intervention of a vinyl intermediate [27]. Reactive vinylsulfones are also prepared from 2-chloroethylsulfonyl derivatives, which lead to the desired intermediates by elimination of hydrogen chloride:

Chrom.–SO$_2$–CH$_2$–CH$_2$–X

↓ + Alkali

Chrom.–SO$_2$–CH=CH$_2$ + HX
X = OSO$_3$H, Cl

Numerous derivatives of ethylsulfonyl and vinylsulfonyl groups have been prepared in recent years, though none have approached the economic importance of the sulfate esters [28, 29].

3.1.2.2 Double-Anchor Dyes

Since the mid-1980s considerable interest has been shown in strongly fixing reactive dyes, and the required high fixation values have increasingly been achieved with the aid of double anchors. Double-anchor dyes can be divided into two categories: those containing two equivalent reactive groups, and those with mixed-anchor systems.

The first group includes structures in which two monochlorotriazine units are connected by a suitable bridge (e.g., **12**). The synthetic approach to these compounds makes it possible to combine two different chromophores in a single system, opening the way to certain color shades that are not easily accessible via a single chromophore [30].

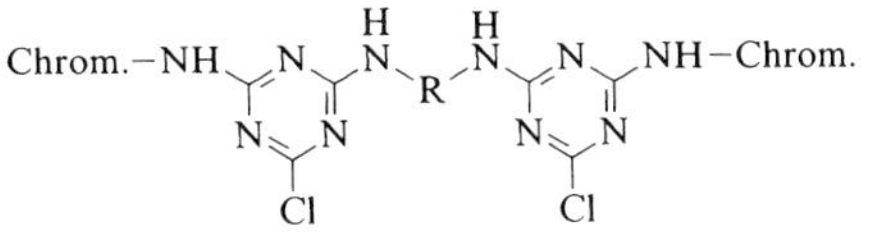

$$\text{Chrom.}-NH-\underset{\underset{Cl}{\|}}{\underset{N\diagdown N}{\bigtriangleup}}-\underset{H}{N}-R-\underset{H}{N}-\underset{\underset{Cl}{\|}}{\underset{N\diagdown N}{\bigtriangleup}}-NH-\text{Chrom.}$$

12

The category of reactive dyes with two equivalent anchors also encompasses products containing two vinylsulfonyl or 2-sulfooxyethylsulfonyl groups. An example is *C.I. Reactive Black 5*, 20505 [*17095-24-8*] (**13**), which not only displays a high degree of fixation but is also accessible by a synthetic route that is both simple and economical:

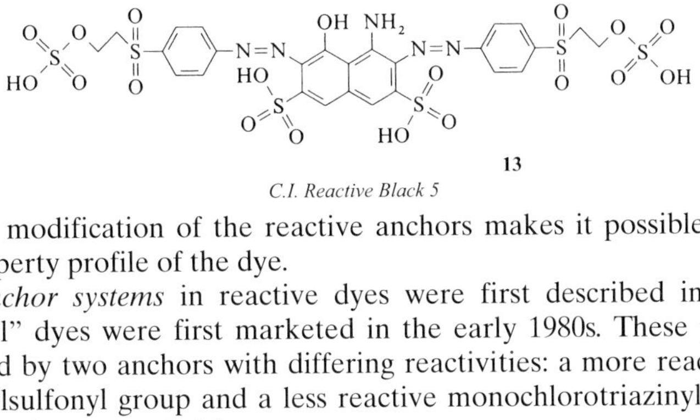

13

C.I. Reactive Black 5

Selective modification of the reactive anchors makes it possible to vary the physical property profile of the dye.

Mixed-anchor systems in reactive dyes were first described in 1959. Such "bifunctional" dyes were first marketed in the early 1980s. These products are characterized by two anchors with differing reactivities: a more reactive 2-sulfohydroxyethylsulfonyl group and a less reactive monochlorotriazinyl residue. An example of such a system is **14**:

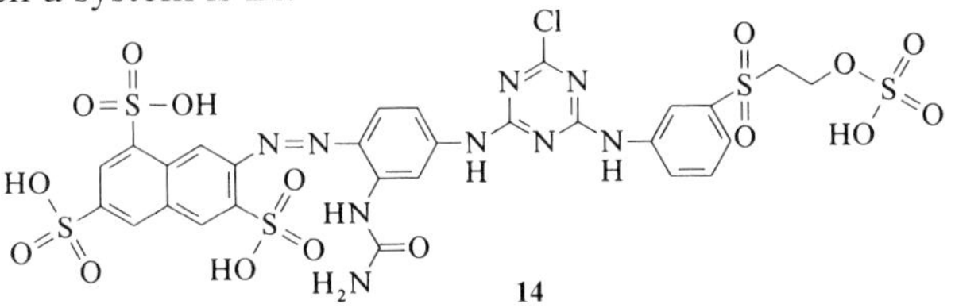

14

Warm- and cold-dyeing double-anchor dyes are prepared by incorporating halotriazinyl and vinylsulfonyl reactive anchors. The bond between the triazine ring and a fiber is stable under basic conditions, whereas that to the vinylsulfonyl group is stable to acid. A combination of the two anchor systems therefore produces a dye with good fastness over a wide pH range. Relative to dyes based on dichlorotriazinyl, dichloroquinoxalinyl, or difluorochloropyrimidyl groups, mixed double-anchor dyes show greater acid fastness and more satisfactory behavior with respect to laundry water containing peroxide detergents. The dyes **15**, introduced in 1988, include products in which a fluorotriazine unit bridges a vinylsulfonyl anchor and a chromophore [31]. Here the two reactive systems exhibit similar reactivities.

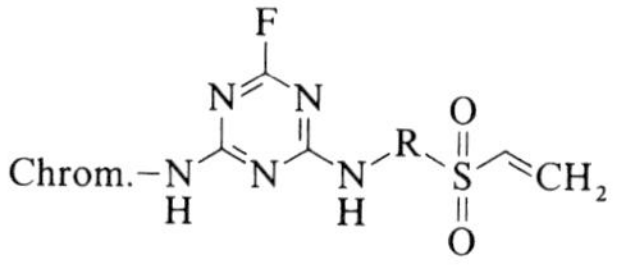

$$\text{Chrom.}-\underset{H}{N}-\underset{}{\overset{N\diagup\diagdown N}{\underset{N}{\bigcirc}}}\kern-2pt-\underset{H}{N}-R-\underset{O}{\overset{O}{\underset{\|}{S}}}-CH_2$$
(F on top of triazine)

15

Other mixed reactive systems have also been patented in addition to the double-anchor dyes so far described, including combinations of monofluorotriazine with difluorochloropyrimidine [32], alkoxychlorotriazine [33], or 2-(sulfothio)ethylsulfonyl groups [34]. Other examples incorporate a difluorochloropyrimidine ring in conjunction with a vinylsulfonyl group [35].

Despite the growing importance of mixed reactive systems, only the monochloro-triazinyl/vinylsulfonyl and monofluorotriazinyl/vinylsulfonyl dyes described above have become commercially established.

3.1.2.3 Multiple-Anchor Dyes

The introduction of more than two anchor groups has only a minor influence on dye fixation characteristics, so multiple-anchor dyes play only a subordinate role in the world market. One example is a multiple-anchor system, in which cyanuric chloride is bound to an amine with two aliphatic 2-chloroethylsulfonyl chains (**16**) [36]:

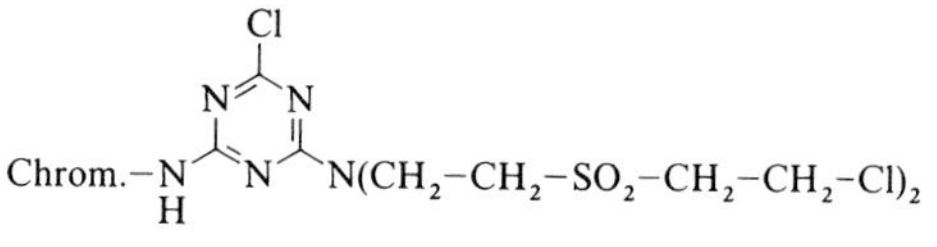

$$\text{Chrom.}-\underset{H}{N}-\text{(triazine with Cl)}-N(CH_2-CH_2-SO_2-CH_2-CH_2-Cl)_2$$

16

3.1.3 Dye Classes (Chromogens) for Reactive Dyes

Virtually every conceivable chromophore has been used in the synthesis of reactive dyes, including monoazo and disazo species, metal complexes of azo dyes, formazan dyes, anthraquinones, triphenodioxazines, and phthalocyanines. The product lines offered by the major dye producers in most cases feature comparable chromophores, differing primarily in the nature of the reactive systems and the particular substitution patterns adopted.

3.1.3.1 Azo Dyes [12], [37]

Most reactive dyes fall in the category of azo dyes. Virtually every hue in the dye spectrum can be achieved by appropriate structural modifications (mono- and disazo dyes, combinations involving either single or multiple aromatic and heterocyclic ring systems).

Monoazo dyes with heterocyclic coupling components, such as pyrazolone or pyridone, are yellow to greenish yellow. Other suitable coupling components include aniline and naphthylamine derivatives. Substituted anilines or aminonaphthalenesulfonic acids are employed as diazo components. Examples are *C.I. Reactive Yellow 4*, *C.I. Reactive Yellow 17* (see Section 3.1.5).

A particularly important class of coupling components are the aminohydroxynaphthalenesulfonic acids. Appropriate variation of the disazo component permits the development of shades ranging from orange to black. Orange and scarlet are achieved with I-acid (6-amino-1-hydroxy-naphthalene-3-sulfonic acid) and γ-acid (7-amino-1-hydroxynaphthalene-3-sulfonic acid), whereas H-acid (8-amino-1-hydroxynaphthalene-3,6-disulfonic acid) and K-acid (8-amino-1-hydroxynaphthalene-3,5-disulfonic acid) derivatives are useful for red to bluish-red hues. Extremely lightfast red shades are also accessible with disazo dyes ("brown dyes"). For chemical structures see Section 3.1.5).

When H-acid is used as a double coupling component, as in **13**, in which reactive groups are present in both diazo functions, the resulting dye color ranges from blue to black.

3.1.3.2 Metal-Complex Azo Dyes

Exceptionally lightfast colors are obtained with metal-complexed azo dyes. Copper complexes of *o,o'*-disubstituted azo compounds produce a wide range of colors (yellow, ruby, violet, blue, brown, olive, black).

3.1.3.3 Anthraquinone Dyes [38]

Anthraquinone-based dyes are significant because of their brilliance, good lightfastness, and chromophore stability under both acidic and basic conditions. Until quite recently they dominated the market for brilliant blue reactive dyes in spite of their relatively low color strength (ε ca. 15 000) and comparatively high cost. The shades of commercial reactive anthraquinone dyes range from violet to blue, e.g., *C.I. Reactive Blue* 19 (**17**). Almost all the important dyes are descendants of "bromaminic acid". This is treated with a reactive amine component, whereby the reactive group may be linked to the amino group by way of various aliphatic or, more often, aromatic bridging units.

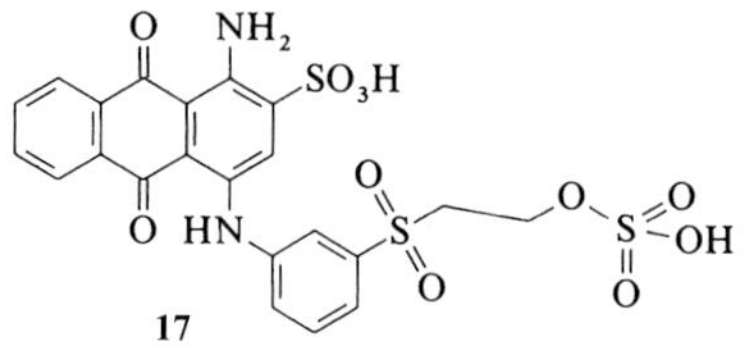

17

C.I. Reactive Blue 19, 61200 [2580-17-1]

3.1.3.4 Triphenodioxazine Dyes

Dyes derived from the triphenodioxazine ring system (**18**) have been commercially available since 1928 when Kränzlein and coworkers discovered dyes with this basic structure augmented by sulfonic acid groups. The unsubstituted triphenodioxazine (which is of no importance as a colorant) was first obtained by G. Fischer in 1879 [39], and its structure **18** was elucidated in 1890 [40].

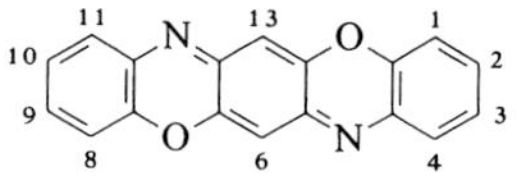

18

By varying the substituents on the orange parent substance **19**, particularly in the positions *para* to the imino groups, its color can be modified. Red to red-violet shades result when X=OH or OR, and blue colors when X=NH$_2$ or NHR'. A blue shade also results if the external phenyl groups of the dioxazine system are part of an annulated and highly condensed aromatic ring system, for example, pyrene (*C.I. Direct Blue 109*, [*33700-25-3*]).

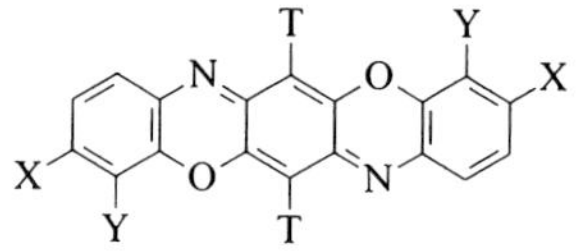

19

Table 3.1 Absorption maxima of triphenodioxazines (**19**) as a function of substituent groups

T	Y	X	λ_{max}	ε
Cl	SO$_2$R	OR	530 nm	45 000
Cl	SO$_2$R	NH-allyl	570 nm	80 000
Cl	SO$_2$R	NHR	612 nm	80 000
Br	SO$_2$R	NHR	613 nm	80 000
Cl	SO$_2$NHR	NHR	620 nm	60 000
Cl	SO$_2$R	NHAr	623 nm	75 000

3.1.3 Dye Classes (Chromogens) for Reactive Dyes

Table 3.1 summarizes observed absorption maxima as a function of various substituents. In particular, blue triphenodioxazines have very high molar extinctions ε, comparable to those of bisazo and phthalocyanine dyes. Until recently, anthraquinone dyes (ε ca. 15 000) were predominant in most applications requiring brilliant blue dyes, but the much stronger triphenodioxazine dyes now represent a less expensive alternative in many applications.

Dioxazine chromophores show a sensitivity to acid and base that is more pronounced in dye solutions than in the fixed dyes. Electron-withdrawing substituents enhance this effect, a major reason why red reactive dyes based on triphenodioxazine have not yet been commercialized [41] even though they are described in most dioxazine patents. Much research and development effort has succeeded in suppressing this sensitivity in the blue dye products currently on the market. Generally, triphenodioxazine chromophores show a high degree of agglomeration and thus increased substantivity because of their planar arrangement. A higher substantivity causes problems during dyeing, such as unsatisfactory leveling in the exhaust process, tailing from dark to light in the padding process, or limited removal of unfixed dye in the rinsing process. Studies have been directed toward overcoming these problems either with new products displaying less critical substantivity or by control of the substantivity through more suitable dyeing conditions.

The first commercial products of this type were sold as blue direct dyes with unmatched tinctorial strength, brilliance, and high lightfastness. An example is *C.I. Direct Blue 106*, 51300 [6527-70-4] (**20**):

20

The triphenodioxazine chromophore was tested in almost every class of dyes, but only recently has it been introduced into reactive dyes. Dioxazines were investigated in some of the earliest work on reactive dyes in the late 1950s, and they are included in some of the first patents covering reactive anchors and reactive dyes [42]. In particular, compound **21** was invoked as a chromophore, with reactive anchors attached to the free amino groups of the molecule.

Z = reactive anchor

21

The first mixed azo – triphenodioxazine dye containing a monochloro-*s*-triazine group emerged in 1953 [43], with the triazine residue serving as a link between the two chromophores. In the years that followed, the dioxazine chromophore was also mentioned in most of the patents related to new reactive systems, although none led to a commercial product. At the beginning of the 1970s, the major activity seemed to be concentrated, leading to several exclusive patents for triphenodioxazine dyes [44]. Within a few years, the first commercial triphenodioxazine reactive dye was introduced: *C.I. Reactive Blue 163*, [72847-56-4]. Other products followed, all designed to achieve brilliant blue shades that could compete with the then-dominant anthraquinone chromophores in certain market segments [45].

The launching of these new products initiated research by other dye manufacturers. This led to numerous new dioxazine reactive dyes (Examples see Table 2.3).

3.1.3.5 Formazan Dyes [12]

Copper complexes of the formazan dye series (see Sections 2.10 and 3.11) are another alternative to reactive anthraquinone dyes; they produce red to greenish-blue shades. Like triphenodioxazine dyes, copper formazans exhibit high molar extinctions (ε = 25 000 –30 000). These materials are derived from 1-(2-hydroxyphenyl)-3-phenyl-5-(2-carboxyphenyl)formazan (**22**), in which all three rings are capable of supporting groups that increase the compound's reactivity and solubility.

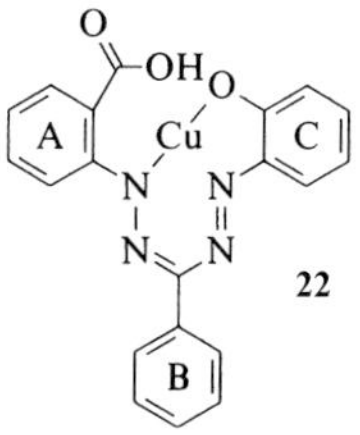

3.1.3.6 Phthalocyanine Dyes [12], [37]

The water-soluble reactive phthalocyanine dyes (see Section 2.8) yield brilliant turquoise and green shades not available from any other dye category. The most important reactive phthalocyanine dyes contain copper or nickel as their central atom; they are substituted with sulfonic acid groups and also with reactive groups joined via sulfonamide bridges. An example is *C.I. Reactive Blue 15*, 74459 [*12225-39-7*] (**23**):

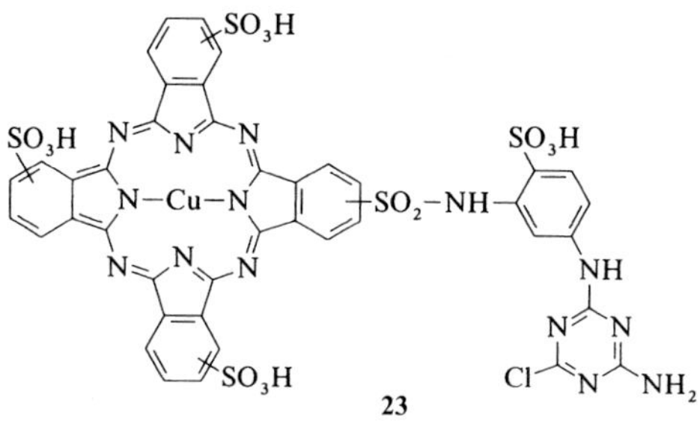

C.I. Reactive Blue 15

3.1.4 Synthesis

The synthesis of selected reactive dyes is illustrated by the following examples taken from the patent literature.

Azo Dyes. The dye obtained by coupling diazotized 4-(2-sulfooxyethylsulfonyl)aniline with 1-amino-8-naphthol-3,6-disulfonic acid at pH 1 – 2 is coupled at pH 6 – 7 with a diazonium compound synthesized from the condensation product derived from 1,3-diaminobenzene-6-sulfonic acid and 2,4,6-trifluoro-5-chloropyrimidine. The crude material is salted out of the solution and isolated. This product (**24**) gives a blue solution in water and dyes cotton in a marine blue to black color:

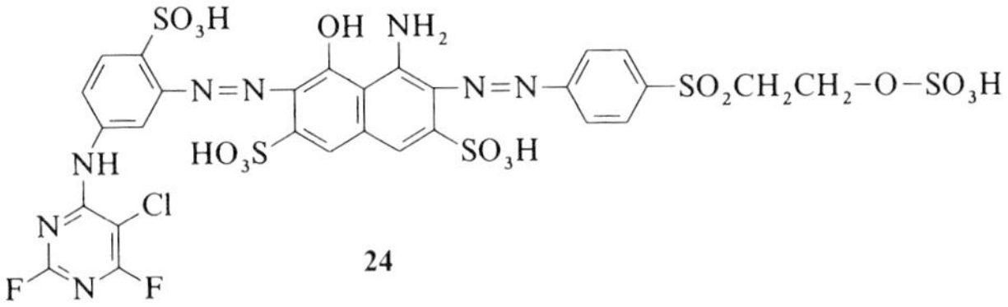

Metal-Complex (Formazan) Dyes. The hydrazone from 2-carboxyphenylhydrazine-4-sulfonic acid and benzaldehyde is suspended in water and then dissolved by adding aqueous sodium hydroxide to obtain pH 6.5 –7.0. This solution is added to the aqueous diazonium salt solution obtained from a typical aqueous diazotization of 4-(2-sulfooxyethylsulfonyl)-2-aminophenyl-6-sulfonic acid. The mixture is then dripped into an aqueous solution of copper sulfate, while the pH is maintained with soda at 5.5 – 6.5. After complete coupling the pH is adjusted to 1 with concentrated hydrochloric acid. The strongly acidic solution is then neutralized with alkali to pH 5.5. The copper – formazan complex is salted out along with sodium chloride, filtered, washed with dilute aqueous sodium chloride solution, and dried. A dark powder results which gives a dark blue solution in water. It consists of an electrolyte-containing powdered sodium salt of the acid **25**:

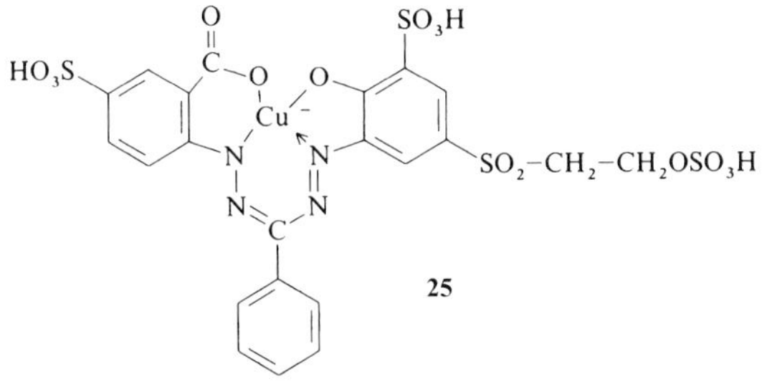

This compound is a very effective dye that renders cotton and regenerated cellulose fibers a clear blue shade on prolonged treatment in the presence of an acid-binding agent. The resulting color prove to be very light- and waterfast.

Dioxazine Dyes. The synthetic routes (see below) to almost all dioxazines yield products with a symmetrical structure, resulting in at least two reactive anchors. The highly reactive double-anchor dyes are suitable for both the exhaust and the padding processes. Ecological limits for wastewater are satisfied as a result of a high degree of fixation and low salt requirements in the exhaust process, whereby the latter also has a positive influence on leveling.

The most important industrial process for synthesizing triphenodioxazines for cellulose fibers involves reaction of two equivalents of an aromatic amine with 2,3,5,6-tetrachloro-1,4-benzoquinone (*p*-chloranil) [*118-75-2*] and subsequent oxidative closure of the ring. All commercially available dioxazines thus contain chlorine atoms at the 6- and 13-positions. The process was discovered in 1911 [46], and its simplicity, coupled with excellent availability of raw materials, was the basis for the commercial success of early dioxazines. In the first reaction step the starting materials are condensed in the presence of an acid-binding agent such as soda ash or magnesium oxide to form the 2,5-diarylamino-3,6-dichloro-1,4-benzoquinone (**26**). Ring closure is then induced by heating either with a metal halide or an acid halide catalyst in a high-boiling solvent (e.g., nitrobenzene) or, in the case of sulfonated dyes, by heating with an oxidant in concentrated sulfuric acid.

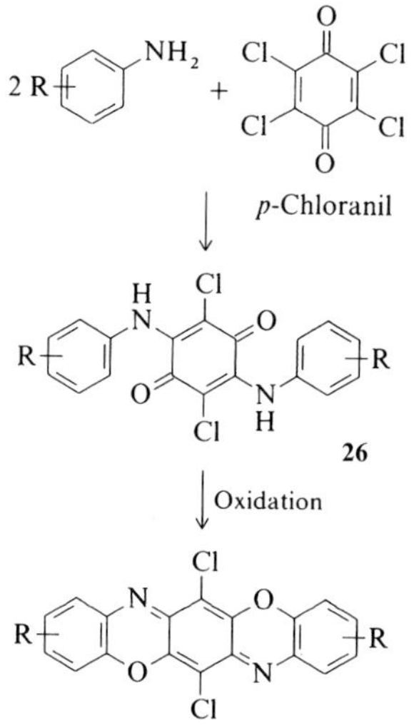

The cyclization process needs a very careful choice of starting materials and reaction parameters to obtain a product with the desired brilliance and tinctorial strength. The use of oleum in the second step is a significant improvement, because oleum serves not only as a solvent but also as an oxidant, permitting cyclization to proceed at or below room temperature. Ring closure continued to prove difficult with starting materials containing sulfonic acid groups, which are of great importance in reactive dyes, but this problem was overcome by resorting to a mixture of oleum and peroxosulfate compounds [47]. Later patents describe the use of halogens [48], manganese dioxide [49], and aqueous hydrogen peroxide [50] as alternatives to peroxosulfate.

3.1.4 Synthesis

The triphenodioxazine chromophore can be converted to a reactive dye by incorporating anchors that are attached to the dioxazine core either directly or via bridging groups. The first patents, like the first commercial products, were based almost exclusively on the use of bridging groups. The usual starting material is 2-chloro-5-nitrobenzenesulfonic acid (**27**), which is treated with a diamine (**28**).

Reduction produces the *p*-phenylenediamine derivative **29**, which is condensed with chloranil, and the resulting diarylamino-3,6-dichloro-1,4-benzoquinone (**30**) is isolated, dried, and then cyclized in oleum containing peroxosulfate to yield the desired dye base. The base is converted to a reactive dye by reaction of the two primary amino groups with two equivalents of an acylating agent Z–Cl. Typical diamine bridges include 1,4-phenylenediamine-2-sulfonic acid, ethylenediamine, and 1,3-diaminopropane. The reactive anchors are almost always derivatives of *s*-triazines [51–58] (Scheme 3.1).

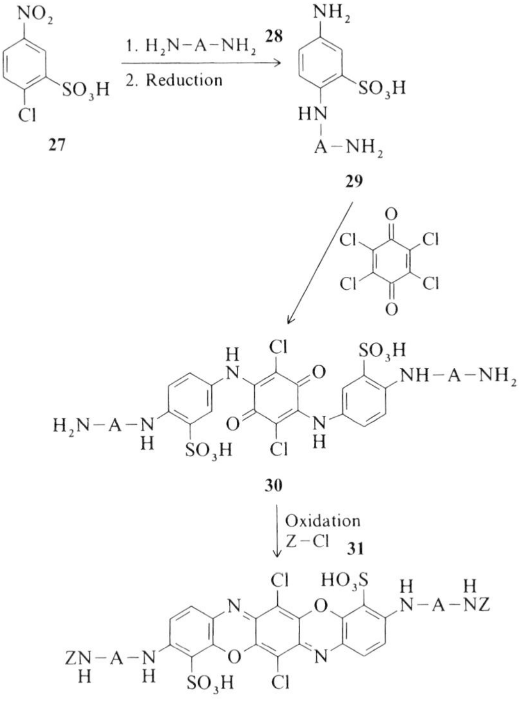

A = bridging group
Z = reactive anchor

Scheme 3.1

Because of their symmetry, all such triphenodioxazine reactive dyes feature double or even fourfold anchor systems. However, dyes with only one anchor group can be prepared by adding a single equivalent of acylating agent. Most asymmetric products of this type are based on 1,4-phenylenediamine-2-sulfonic acid [59] and characterized by somewhat reddish-blue hues (e.g., **32**, Z = monochloromonomethoxy-*s*-triazine, λ_{max}= 560 nm).

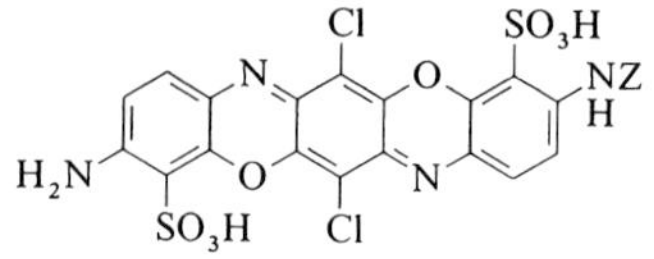

32

Red dioxazines can be synthesized by substituting an amino alcohol for the diamine derivative [60].

Replacing the sulfonic acid **27** with the appropriate carboxy-, carboxamido-, sulfonyl-, or sulfon-amido-4-nitrochlorobenzene leads to analogues described in [61], [62].

Another alternative is attaching the reactive anchor to the dioxazine ring via sulfonamide or carboxamide groups instead of via amine groups, as in **33** [62], [63]:

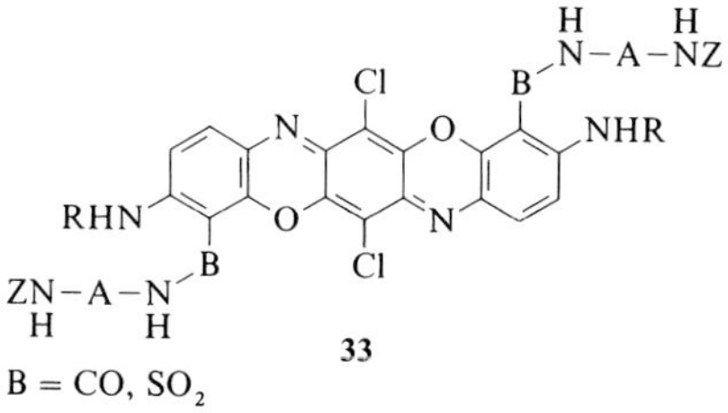

33

B = CO, SO$_2$

Compound **27** or one of the above-mentioned derivatives can also be reacted with a 2-hydroxyethylsulfonylaniline instead of with a diamine. After reduction of the 4-nitroaniline product **34** followed by condensation with chloranil, the cyclization takes place in oleum where, besides ring closure, the 2-hydroxyethylsulfonyl groups are also esterified to β-sulfooxyethylsulfonyl groups [vinylsulfone (VS) groups]. Thus, the reactive anchor is formed concurrently during the cyclization reaction [64, 65] (Scheme 3.2):

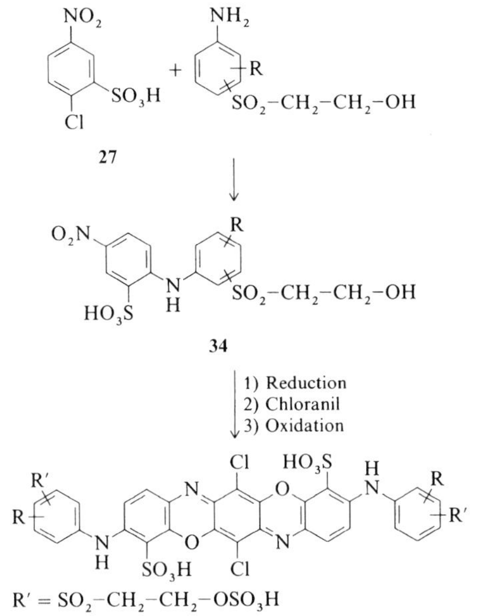

3.1.4 Synthesis 127

Scheme 3.2

Other similar fiber-reactive triphenodioxazines are prepared by replacing the sulfonic acid group in **26** with a 2-hydroxyethylsulfonyl group. The starting material here is 4-chloro-3-(2-hydroxyethylsulfonyl) nitrobenzene (**35**), which leads to products containing a VS group bound directly to the dioxazine ring (**36**) [64, 66] (Scheme 3.3).

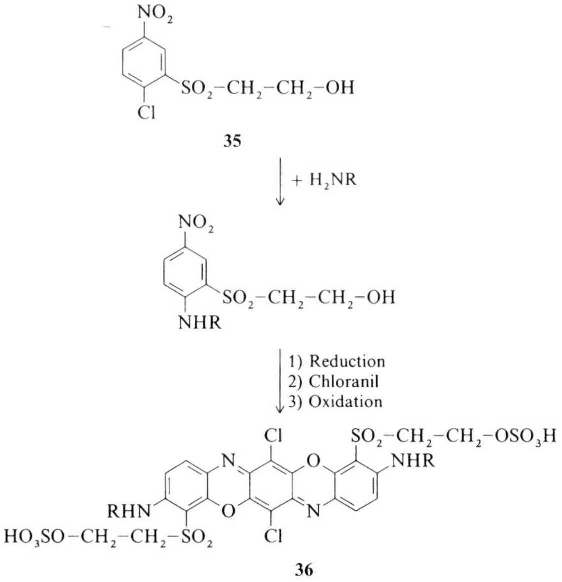

Scheme 3.3

3.1 Reactive Dyes

In an analogous reaction, use of a diamine affords dyes with both VS and amino groups. Subsequent acylation of the free amino groups produces the four-anchor system **37** [67].

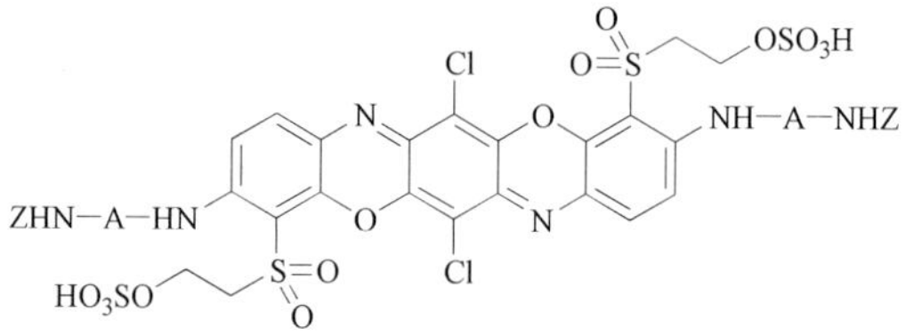

37

Z = reactive anchor

Because of the symmetrical nature of the chloranil condensation product **7**, all the syntheses described so far afford exclusively symmetrical triphenodioxazines; i.e., the two phenyl rings fused onto the dioxazine skeleton bear the same substituents. An elegant synthesis for an asymmetric chloranil condensation product has not yet been developed.

3.1.5 Examples of Commercially Available Dyes

3.1.5.1 Azo Dyes

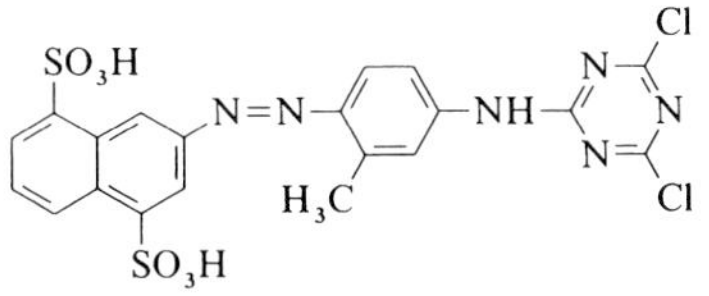

C.I. Reactive Yellow 4 [12226-45-8]

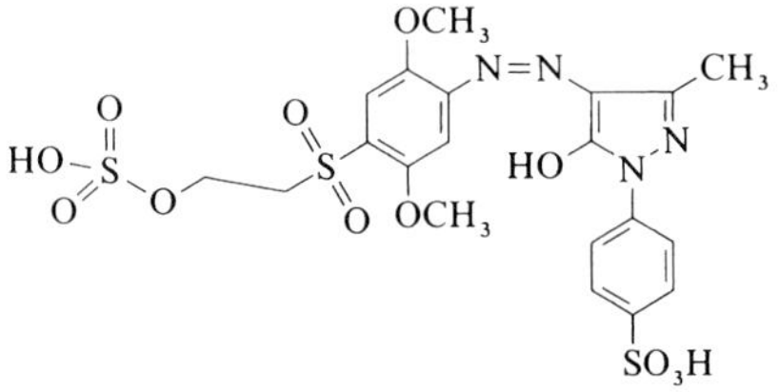

C.I. Reactive Yellow 17, 18852 [20317-19-5]

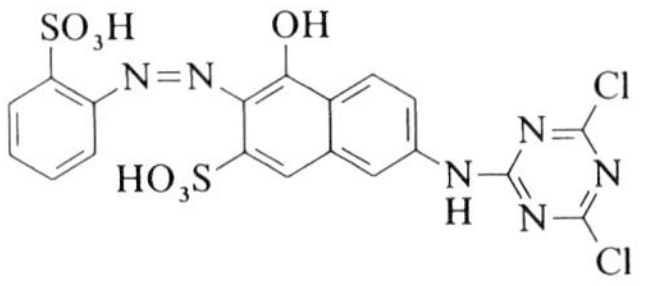

C.I. Reactive Orange 1 [18886-16-3]

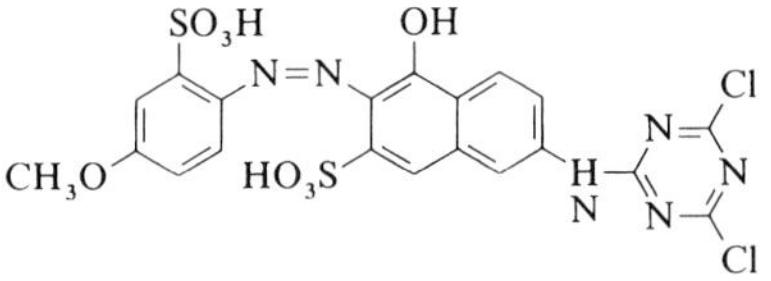

C.I. Reactive Red 8 [25489-36-5]

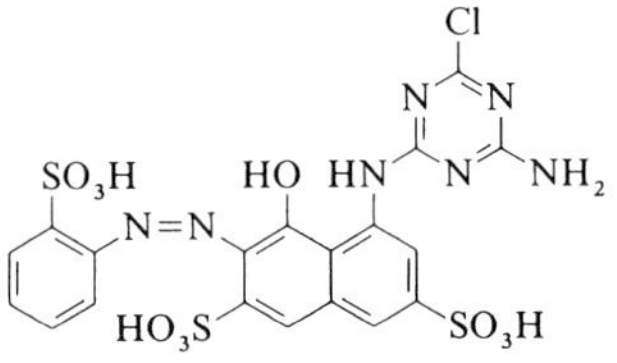

C.I. Reactive Red 12 [41423-92-1]

3.1.5.2 Metal Complex Azo Dyes

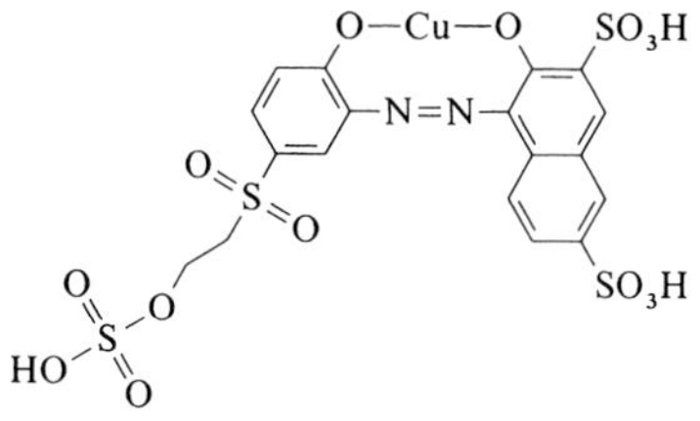

C.I. Reactive Red 23, 16202 [12769-07-2]

3.1.5.3 Formazan Dyes

C.I. Reactive Green 15 [61969-07-1]
C.I. Reactive Blue 70 [61968-92-1]
C.I. Reactive Blue 83 [12731-65-8]
C.I. Reactive Blue 84 [12731-66-7]
C.I. Reactive Blue 104 [61951-74-4]
C.I. Reactive Blue 160, 137160 [71872-76-9]
C.I. Reactive Blue 182 [68912-12-9]
C.I. Reactive Blue 209 [110493-61-3]
C.I. Reactive Blue 212 [86457-82-1]
C.I. Reactive Blue 216, 137155 [89797-01-3]
C.I. Reactive Blue 220 [128416-19-3]
C.I. Reactive Blue 221 [93051-41-3]
C.I. Reactive Blue 235 [106404-06-2]
C.I. Reactive Blue 157, 202, 218, 226, 260, 263

3.1.5.4 Anthraquinone Dyes

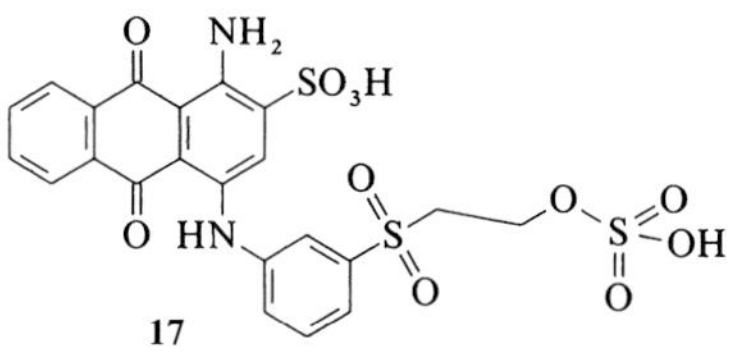

C.I. Reactive Blue 19, 61200 [2580-17-1]

3.1.5.5 Diphenodioxazine Dyes

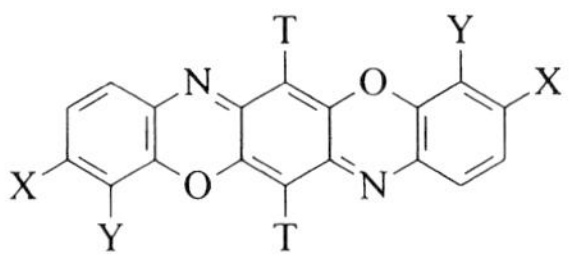

Table 3.2 Examples of triphenodioxazine dyes 19 with different reactive anchors

C.I. name	CAS no.	Anchor	Year of introduction
Reactive Blue 163	[72847-56-4]	dichlorotriazine	1975
ReactiveBlue 172	[85782-76-9]	phosphonic acid	1978
Reactive Blue 187	[79771-28-1]	triazylammonium	1979
Reactive Blue 198	[12448-55-1]	monochlorotriazine	1982
Reactive Blue 204	[85153-92-0]	monofluorotriazine	1983
Reactive Blue 224	[122390-99-2]	sulfatoethylsulfone	1986

At present, only blue dioxazines are used commercially to color cellulose, where R' is varied to give reddish to greenish-blue tints. Variation of T has only a minor effect on the shade.

3.1.6 Forms of Supply

In the past, reactive dyes were marketed mainly as powders, but recently other physical forms of these products have gained importance. Dyes are now expected to be dedusted to minimize hygiene-related problems in the workplace. Moreover, the growing trend toward automated dye works has encouraged the development of more readily meterable products. Reactive dyes should also display good cold-water solubility. These demands are all met by liquid dyes (highly concentrated, aqueous solutions) and by solids supplied in the form of pourable granules and finished powders [68]–[72].

3.1.7 References

[1] C. F. Cross, E. J. Bevan, *J. Res. Cellulose* **34 – 39** (1895 – 1900).
[2] G. Schröter, *Ber. Dtsch. Chem. Ges.* **39** (1906) 1559.
[3] BASF, GB 259 634, 1924.
[4] D. H. Peacock, *J. Soc. Dyers Colour.* **42** (1926) 53.
[5] Ciba, DE 560 035, 1930;583 398, 1932 (R. Haller,A. Heckendorn).
[6] I. G. Farbenind., GB 341 461, 1929.
[7] H. Schweizer, O. Bayer, *Angew. Chem.* **73** (1961) 343.
[8] Hoechst AG, DE 965 902, 1949.
[9] J. Heyna, *Angew. Chem.* **74** (1962) 966; *Angew. Chem. Int. Ed. Engl.* **2** (1963) 20.
[10] *Ullmann*, 3rd ed., **14**, 615.
[11] ICI, DE-AS 1 019 025, 1 062 367, 1955
[12] D. Waring: *The Chemistry and Application of Dyes*, Plenum Press, New York 1990.
[13] ICI, GB 838 337, 1957 (R. N. Heslop, W. E. Stephen).
[14] ICI, GB 805 562, 1956;Ciba-Geigy, DE-AS 1 105 541, 1956.
[15] Bayer, DE-OS 1 644 208, 1967.
[16] ICI, GB 937 182, 1963 (G. A. Gamlen, C. Morris, D. F. Scott, J. Twichett). M. Miujumoto, R. Parham, *AATCC Int. Conf. Exhib.* (1986) 153.
[17] M. Capponi, E. Metzger, A. Giamara, *Am. Dyestuff Rep.* **50** (1961) 23.
[18] ICI, GB 917 780, 1963 (H. F. Andrew, V. D. Poole).
[19] Bayer, GB 1 169 254, 1969 (H. S. Bien, E. Klauke).
[20] Bayer, US 3 853 840, 1974 (K.-H. Schündehütte, K. Trautner).
[21] Geigy, BE 644 495, 1964.
[22] BASF, DE-AS 1 193 623, 1960.
[23] Cassella, FR 1 403 233 (1963).
[24] *Ullmann*, 4th ed., **20**, 113.
[25] E. P. Sommer, *Am. Dyestuff Rep.* **47** (1958) 895.
[26] H. Zimmermann, *Melliand Textilber.* **39** (1958) 1026.
[27] Hermann Rath: *Lehrbuch der Textilchemie*, 3rd ed., Springer Verlag, Berlin 1972.
[28] Hoechst, DE 966 651, 1950, US 3 387 914, 1963, DE-AS 1 103 886, 1958.
[29] F. Wolf, S. Stefaniak, *Melliand Textilber.* **47** (1966) 767.
[30] ICI, DE-OS 2 244 537, 1973 (D. Waring).
[31] Ciba-Geigy, EP 70 806, 1981.
[32] Ciba-Geigy, EP 97 119, 1982.
[33] Ciba-Geigy, EP 89 923, 1982.
[34] Ciba-Geigy, EP 85 654, 1982.
[35] Ciba-Geigy, EP 133 843, 134 193, 1983.
[36] Cassella, DE-AS 2 614 550, 1976.
[37] K. Venkataraman: *The Chemistry of Synthetic Dyes*, vol. VI: *Reactive Dyes*, Academic Press, New York –London 1972.
[38] A. H. M. Renfrew, *Rev. Prog. Color. Relat. Top.* **15** (1985) 15.
[39] G. Fischer, *J. Prak. Chem.* **19** (**2**) (1879) 317.
[40] P. Seidel, *Ber. Dtsch. Chem. Ges.* **23** (1890) 182.
[41] J. Wolff, H. Henk, *Textilveredlung* **25** (**6**) , 1990, 213 –218.
[42] Ciba, CH 358 188, 1957; FR 1 205 384, 1958; DE-AS 1 058 467, 1957; BE 567 435, 1958. Hoechst, BE 566 648, 1958; BE 570 437, 1958. Bayer, BE 569 439, 1958. Sandoz, BE 573 299, 1958;FR 1 233 530, 1958.
[43] Ciba, FR 1 116 564, 1954.

- [44] ICI, GB 1 384 749, 1972;GB 1 368 158, 1972.
- [45] A. H. M. Renfrew, DE 2 503 611, 1975 *Rev. Progr. Colouration* **15** (1985) 15 – 20.
- [46] Meister, Lucius und Brüning, GE 253 091, 1911.
- [47] ICI, GB 1 589 915, 1976.
- [48] Bayer, DE-OS 3 444 888, 1984; EP-A 296 411, 1987; EP-A 141 359, 1984.
- [49] Bayer, DE-OS 3 510 613, 1985.
- [50] Hoechst, EP-A 292 906, 1988.
- [51] ICI, GB 1 368 158, 1972; GB 2 059 985, 1981.
- [52] ICI, DE 2 503 611, 1975.
- [53] Ciba, EP 070 807, 1982. Bayer, DE-OS 3 410 236, 1984.
- [54] Ciba, EP-A 101 665, 1983.
- [55] Sumitomo, JP-Kokai 62 048 768, 1987; JP 57 14 654, 1980.
- [56] Sumitomo, EP-A 325 246, 1988. Mitsubishi, JP 12 89 867, 1988.
- [57] Sumitomo, EP-A 281 799, 1987.
- [58] ICI, EP 135 381, 1984.
- [59] Bayer, DE-OS 3 423 581, 1984. A. H. M. Renfrew, *J. Soc. Dyers Colour* **105** (1989) 262 – 4. Sumitomo, EP-A 385 120, 1990.
- [60] Sumitomo, JP 22 38 064, 1989. Bayer, DE-OS 3 635 312, 1988.
- [61] ICI, EP-A 279 122, 1987. Bayer, DE-OS 3 336 362, 1985; EP-A 199 036, 1986; EP-A 158 857, 1985. Hoechst, DE-OS 3 544 982, 1985.
- [62] Bayer, DE-OS 3 409 439, 1984.
- [63] Bayer, EP-A 205 080, 1986;EP-A 204 245, 1986;EP-A 190 603, 1986.
- [64] Bayer, EP-A 141 996, 1984.
- [65] Hoechst, DE-OS 3 625 386, 1986;DE-OS 3 625 347, 1986.
- [66] Hoechst, DE-OS 3 628 084, 1986;DE-OS 3 627 458, 1986;EP-A 355 735, 1988;DE-OS 3 426 727, 1984;Bayer, EP-A 334 172, 1988;DE-OS 3 439 756, 1984.
- [67] Bayer, DE-OS 3 439 755, 1984.
- [68] H.-U. von der Eltz, *Melliand Textilber.* **68** (1987) 132.
- [69] W. Beckmann, F. Hoffmann, J. Wolff, *Textilveredlung* **24** (1989) 81.
- [70] D. Link, J. Moreau, *Textilveredlung* **24** (1989) 87.
- [71] W. Zysset, *Textilveredlung* **24** (1989) 91.
- [72] B.-T. Gröbel, M. Kunze, U. Mrotzeck, *TPI. Text. Prax. Int.* **10** (1991) 1111.

3.2 Disperse Dyes

3.2.1 Introduction

Disperse dyes are colorants with low water solubility that, in their disperse colloidal form, are suitable for dyeing and printing hydrophobic fibers and fabrics. Forerunners of the disperse dyes were the ionamine dyes of British Dyestuffs Corp.; these were *N*-methanesulfonic acids of aminoazo or aminoanthraquinone dyes that released the *N*-methanesulfonic acid group in the dyeing process and, thereby, precipitated as disperse dyes on the acetate fibers. The understanding of this mechanism in 1923 initiated the development of genuine disperse dyes. British Celanese and British Dyestuffs Corp. were the first companies to introduce these dyes into the market for coloring acetate fibers. The dyes were dispersed with sulforicinoleic acid, soap, or Turkey red oil [1].

From 1924 to 1930, products of other companies appeared on the market, initially as pastes; later, when the materials could be dried successfully without interfering with their dispersibility, they were also marketed as powders. Since 1950, the production of disperse dyes has increased sharply, closely following the growth in worldwide production of synthetic fibers, especially polyester [poly(ethylene terephthalate)] fibers, production of which has grown steadily from ca. 2×10^6 t/a in 1970 to about 16×10^6 t/a in 1999 [2]. Furthermore, new dyeing processes necessitated the development of special disperse dyes. For instance, dyes characterized by special ease of sublimation are preferred for transfer printing [3]. The demand for new fastness properties such as thermomigration fastness and automotive light fastness [4–7] also led to new dyes, as has the ongoing pressure on market prices.

Models for the dyeing of polyester fibers with disperse dyes have been developed [8]. When the dye is applied from aqueous medium, it is adsorbed from the molecularly dispersed aqueous solution onto the fiber surface and then diffuses into the interior of the fiber. The following parameters determine the rate of dyeing and, to some extent, the leveling properties: (1) the dissolution rate during the transition from the dispersed crystalline state of the dye into the molecularly dispersed phase, and (2) the diffusion rate at the fiber surface and, especially, in the interior of the fiber. The rates of both processes vary with temperature.

Differences in geometry and polarity of the dye molecules can lead to wide variations in these finishing or dye-specific properties and can have a marked effect on the absorption characteristics of all dyes, irrespective of whether single-component or combination dyeing processes are used. For instance, uneven dyeing may occur when an unequal distribution of particle size results in insufficient dispersion stability and hence crystal growth and precipitation at the substrate surface.

3.2.2 Chemical Constitution

Industrially applied disperse dyes are based on numerous chromophore systems. Approximately 60 % of all products are azo dyes, and ca. 25 % are anthraquinone dyes, with the remainder distributed among quinophthalone, methine, naphthalimide, naphthoquinone, and nitro dyes [9].

Azo dyes are currently employed to create almost the entire range of shades; anthraquinone derivatives are used for red, violet, blue, and turquoise. The remaining dye classes are used mainly to produce yellow shades. Fastness properties and often also the structures are disclosed in the Colour Index [10].

3.2.2.1 Azo Dyes

Azo dyes represent the largest group of disperse dyes for two reasons: (1) the ease with which an extraordinary number of molecular combinations can be generated by varying the diazo and coupling components and (2) the relatively simple process by which the dyes can be produced. With this class of dyes, manufacturers can respond much more easily to customers' and end users' requests for special shades and fastness characteristics.

In the following survey, the most important disperse azo dyes are divided into mono- and disazo types; then each of these classes is subdivided according to the diazo and the coupling components. The diazo component is further subdivided where appropriate into aromatic and heteroaromatic amines.

Monoazo Dyes. About 50 % of all disperse dyes are monoazo dyes, which thus represent the largest single group [9]. Relatively simple syntheses enable a range of shades from greenish yellow to cyan to be produced with this chromophore system.

With *aromatic amine coupling components* and *carbocyclic aromatic amine diazo components*, dyes of structure **1**, in which aminobenzenes are the coupling components, predominate (D denotes an aromatic or heterocyclic group of the diazo component).

Of all the disperse azo dyes, this class has the greatest economic importance. Commercial products are most often represented by structure **2**, in which 4-nitroaniline [*100-01-6*] and its substituted derivatives constitute the diazo component.

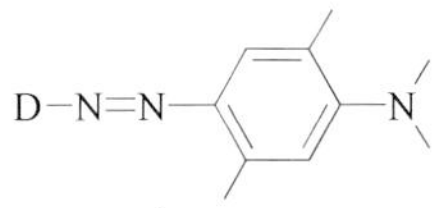

$$D-N=N-\underset{\mathbf{1}}{\underset{}{\text{C}_6\text{H}_2(\text{CH}_3)_2}}-N(\text{CH}_3)_2$$

3.2 Disperse Dyes

$$O_2N \text{—} \underset{Y}{\overset{X}{\bigcirc}} \text{—} N=N \text{—} \underset{A}{\overset{B}{\bigcirc}} \text{—} N \overset{R^1}{\underset{R^2}{\diagdown}}$$

2

X, Y = H, Cl, Br, CN, NO_2, CH_3SO_2
A = H, CH_3, Cl, NHCOR
B = H, CH_3O, C_2H_5O, Cl
R^1, R^2 = H, CN, OH, OCOR, COOR, aryl (R = alkyl)

The most important diazo components are 4-nitroaniline [100-01-6], 2-chloro-4-nitroaniline [121-87-9], 2-cyano-4-nitroaniline [17420-30-3], 2,4-dinitroaniline [97-02-9], 2,6-dichloro-4-nitroaniline [99-30-9], 2-bromo- [1817-73-8] and 2-chloro-4,6-dinitroaniline [3531-19-9], and 2-bromo-6-cyano-4-nitroaniline [17601-94-4]. The coupling components are derived from aniline, 3-aminotoluene [108-44-1], 3-chloroaniline [108-42-9], 3-aminoacetanilide [102-28-3], and 3-amino-4-alkoxyacetanilide by N-alkylation.

This type of dye quickly became important for dyeing acetate fibers. It has been adapted to the requirements of polyester fibers and of different dyeing processes, mainly by varying the substituents R^1 and R^2 [11–14].

For example, the β-hydroxyalkyl groups (R^1 and/or R^2 = OH), typical of many acetate dyes, were replaced by less hydrophilic groups such as CN, OCOR, and COOR, which enhanced the affinity for polyester fibers and, in many instances, the fastness to light and sublimation [9, 15].

The shade of dyes with formula **2** depends largely on the substituents X, Y, A, B, R^1, and R^2. If the coupling component is kept constant and X = H, variation of Y results in a bathochromic shift, which increases in the following order: H, Cl, NO_2, CH_3SO_2, CN.

If X and Y are other than H, the situation is more complicated because of possible steric effects on the azobenzene system [16]. For example, if X = Y = Cl, steric hindrance prevents a planar alignment of the azobenzene molecule, which leads to a hypsochromic shift with simultaneous loss of clarity. This effect is exploited in the industry to produce brown shades. Cyano groups exert no steric effects, and the strongest bathochromic shifts can be obtained by using X = Y = CN; X = CN, Y = NO_2; and X = CN, Y = CH_3SO_2. For a given diazo component, a bathochromic shift occurs which increases in the order A = Cl, H, CH_3, NHCOR. The most extreme bathochromic effect is obtained when A = NHCOR and B = OR. In combination with 2-bromo- [1817-73-8] or 2-chloro-4,6-dinitroaniline [3531-19-9] as the diazo component, commercially important navy blue dyes are obtained; of all the monoazo dyes, these are produced in the largest quantity.

The significant influence of R^1 and R^2 on color is quite surprising because these substituents are not components of the chromophore system. Ester and, especially, cyano groups in these positions cause a distinct hypsochromic shift [9].

3.2.2 Chemical Constitution

Developments in the field of *heteroaromatic amines as diazo components* started with the finding that 2-amino-5-nitrothiazole [121-66-4] [17] gave bright blue shades of reasonable fastness in azo dyes for cellulose acetate (e.g., **3**) [18], [19]. Because of their good dischargeability but limited fastness properties, these dyes were also used on polyester. The importance of this class of dyes has diminished since some of them have been found to have sensitizing properties (see Chapter 8).

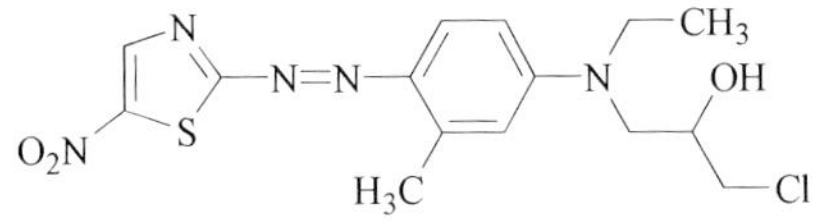

3

Blue [70865-21-3]

2-Amino-6-nitrobenzthiazole [6285-57-0] and 2-amino-5,6- or 6,7-dichlorobenzthiazole [24072-75-1], [25150-27-0] are widely employed to give scarlet to ruby shades because of their facile accessibility by ring closure of 4-nitrophenylthiourea (e.g., **4**) [20] or 3,4-dichlorophenylthiourea, respectively. In the latter case the resulting mixture of isomers is directly utilized for diazotization and transformed into the mixture of dye isomers **5**.

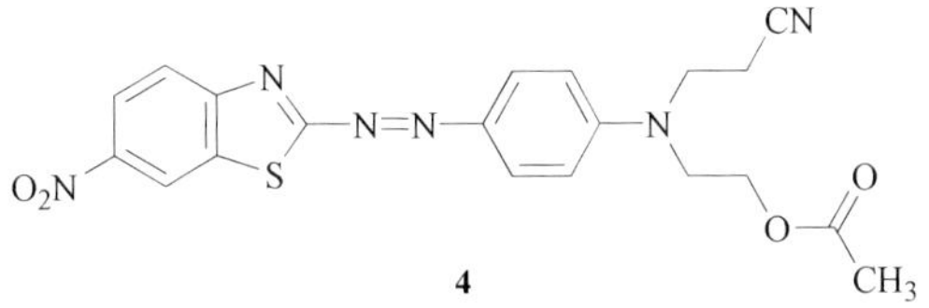

4

C.I. Disperse Red 177, 11122 [68133-69-7] [21]

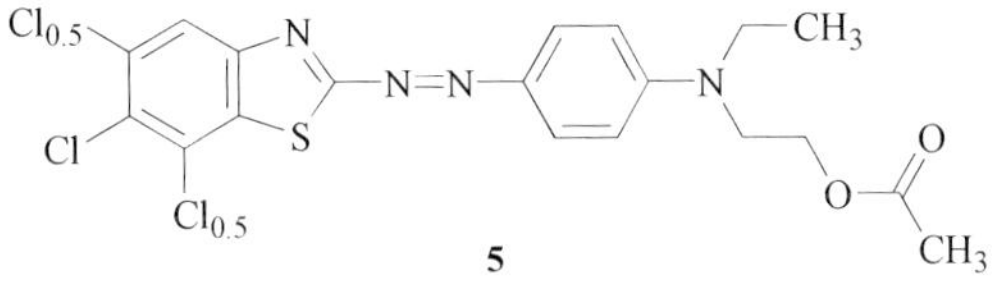

5

[117541-97-6], [117541-98-7] [22–24]

Dyes with *aromatic hydroxy compounds as coupling components*, of which C.I. Disperse Yellow 3, 11855 [2832-40-8] is a representative example, have long been used to dye acetate fibers. Its use has been questioned for ecological reasons (see Chapter 8). For application to polyesters, sublimation fastness has been enhanced by increasing the size of the molecule.

With regard to *heterocyclic compounds as coupling components*, the importance of the formerly widespread pyrazolone-, aminopyrazole-, and 4-hydroxyquinolone-based yellow azo dyes has greatly diminished with the advent of the tinctorially superior and therefore more economical pyridone azo dyes. Only a few examples have survived and then only for special applications such as for dyeing of acetate fibers. Examples are *C.I. Disperse Orange 56* and *C.I. Disperse Yellow 5*. (for structure, see Section 3.2.5).

Disazo Dyes. About 10 % of all disperse dyes are disazo compounds [9]. Even the simplest hydroxy disazo dyes, such as 4-aminoazobenzene coupled to phenol and 4-aminoazobenzene coupled to *o*-cresol, have a good affinity for polyester fibers and yield lightfast reddish yellow hues. However, these shades are frequently less bright than those obtained with monoazo dyes.

The introduction of an alkoxy group into the central benzene ring causes a distinct shift toward orange. A similar bathochromic effect is obtained by replacing this benzene with naphthalene.

Substitution of the first benzene nucleus by electron acceptors also causes bathochromic shifts. Disazo dyes of this type such as **6** are frequently incorporated as components of black mixtures.

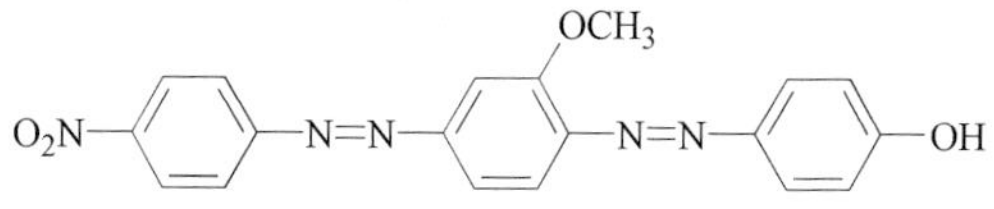

$$O_2N-\langle\bigcirc\rangle-N=N-\langle\bigcirc\rangle\text{(OCH}_3\text{)}-N=N-\langle\bigcirc\rangle-OH$$

6

Dispers Orange 29, C.I. 26077 [*19800-42-1*] [25]

3.2.2.2 Anthraquinone Dyes

The brilliant red, blue, and turquoise anthraquinone dyes have major industrial significance. The most important red shades are produced by alkyl or aryl ethers of 4-amino-1,3-dihydroxyanthraquinone [*81-51-6*] (**7**).

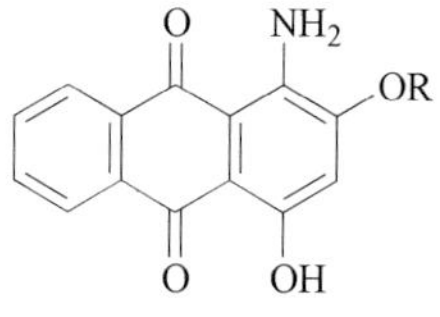

7

R = alkyl, aryl

Blue and turquoise dyes also play an important role. The most important blue dyes come from ring- or *N*-substituted derivatives of the two isomers of aminodihydroxy-anthraquinone **8** and **9**.

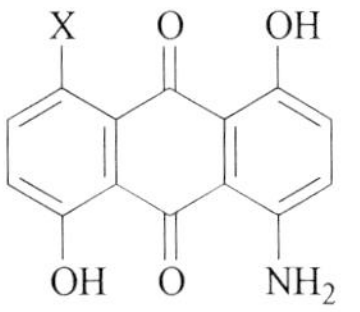

X = NH₂, NO₂

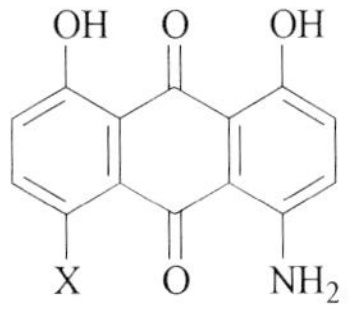

In comparison to azo dyes of similar shades, the anthraquinone derivatives are frequently characterized by greater clarity and better stability against hydrolysis and reduction (see Sections 2.3 and 3.4).

A serious handicap of anthraquinone dyes is their poor dyeing strength, which is indicated by a low molar extinction coefficient ε. The ε value usually has a range of 12 000–34 000 $L\,mol^{-1}\,cm^{-1}$ for anthraquinone dyes, depending on the shade, whereas azo dyes of similar shade attain ε values of 30 000–84 000 $L\,mol^{-1}\,cm^{-1}$. This is the reason why disperse anthraquinone dyes have been increasingly displaced by azo dyes. Reviews of this ongoing process are given in [26, 27]. Research on new dyes has therefore been concentrated on non-anthraquinoid dyestuffs. The only exception to this is the brilliant blue dye (**49**) [28].

3.2.2.3 Other Chromophores

Quinophthalone Dyes. Currently, 3′-hydroxyquinophthalone [*7576-65-0*] is employed extensively in a number of product lines used to dye synthetic fibers in greenish yellow hues with good lightfastness and generally sufficient sublimation-fastness. Suitable substitution in the phthalic acid residue or the quinoline nucleus may improve thermosetting fastness [29]. An example is *C.I. Disperse Yellow 54*, 47020 [*7576-65-0*] (**10**):

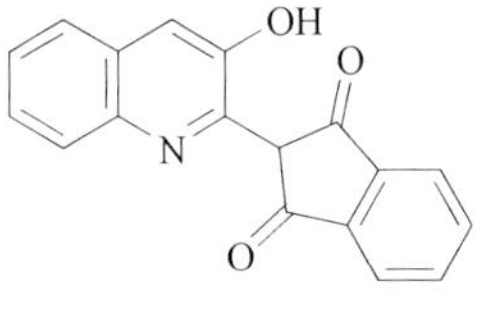

Methine Dyes. The condensation products of 4-dialkylaminobenzaldehydes with cyanoacetic esters have long been used to dye acetate fibers. Brilliant greenish yellow dyes with excellent lightfastness are obtained on polyester fibers with the corresponding condensation products of malonodinitrile. The sublimation fastness of this dye type can be improved by introducing suitable substituents into the alkyl residue of the amino group or by doubling the molecular size, e.g., *C.I. Disperse Yellow 99*, [*25857-05-0*] (**11**) [12, 30, 31].

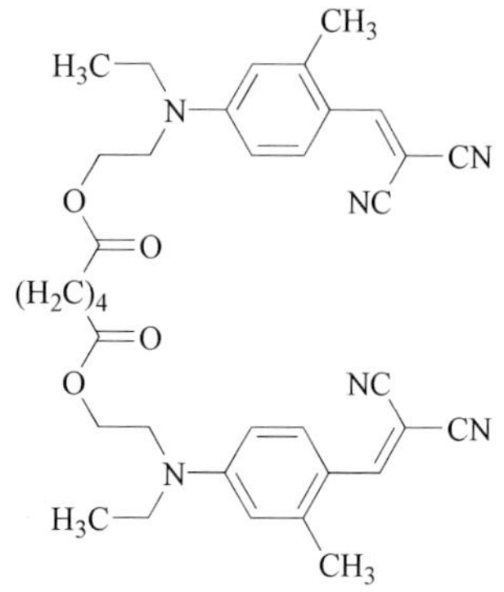

11

Brilliant blue dyes such as *C.I. Disperse Blue 354*, 48480 [*74239-96-6*] (**12**) [32] are produced from aminobenzaldehydes and benzothiophene derivatives.

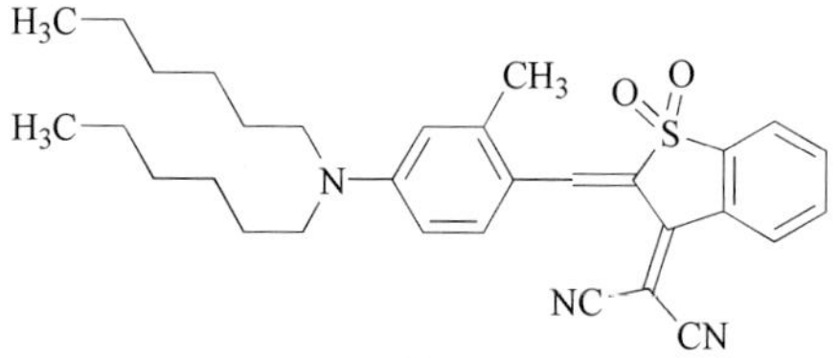

12

By condensation of 2-hydroxy-4,4′-diethylaminobenzaldehydes with hetarylacetonitriles, 3-heteroaryl-7-diethylaminocoumarins such as **13** and **14** are obtained, which dye polyester fibers fluorescent yellow with a strong green tint.

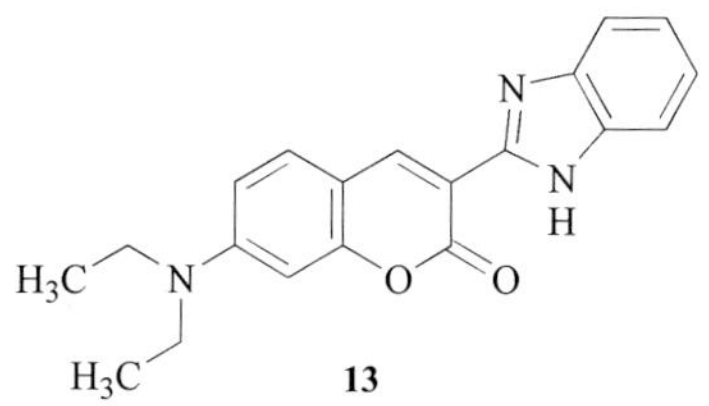

13

C.I. Disperse Yellow 82, [*27425-55-4*] [33]

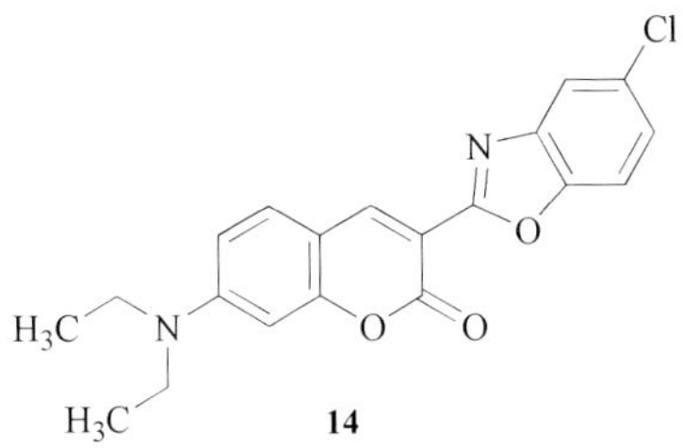

3.2.2 Chemical Constitution 141

C.I. Disperse Yellow 332, 55165 [*35773-43-4*] [34]

Further condensation with malonodinitrile results in fluorescent red dyes such as (**15**).

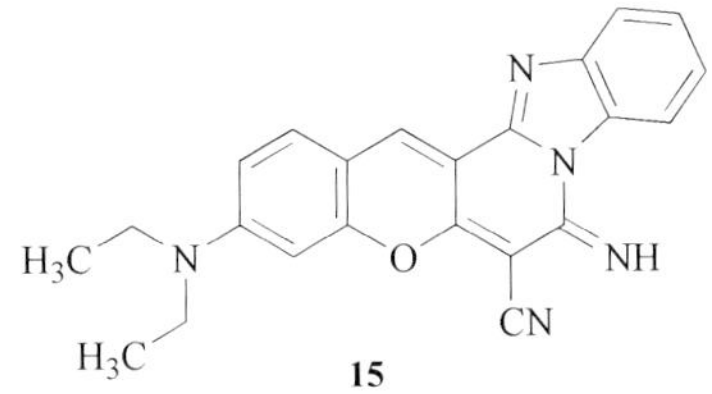

Red [*53272-39-1*] [35]

Oxidative cyanation of yellow 3-heteroaryl-4*H*-7-diethylaminocoumarin dyes such as **13** and **14** introduces a 4-cyano group and yields fluorescent red dyes [36]. For example, treatment with NaCN followed by Br$_2$ gives **16**.

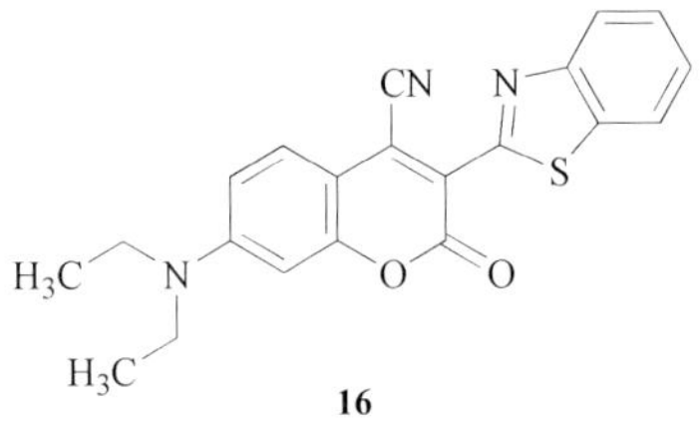

By elongation of the methine chromophore even brilliant blues such as **17** can be produced. Because of limited fastness properties, this interesting new dye could not yet pose a threat to anthraquinone dyes.

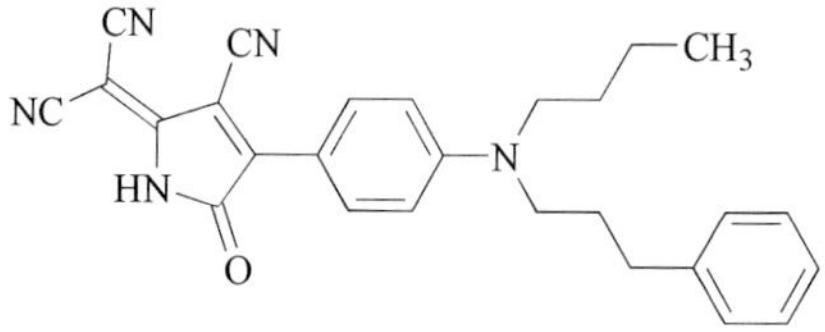

17

C.I. Disperse Blue 365 [108948-36-3] [37]

Condensation of mandelic acids and hydroquinone, followed by oxidation, results in yellow to red polyester dyes [38]. This new chromophore shows excellent thermo-migration fastness and has been frequently varied to cover all red shades (e.g., 18) and to improve build up.

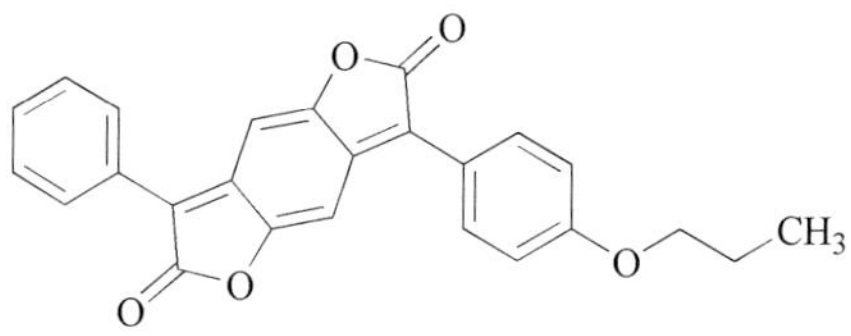

18

Red [*79694-17-0*] [38]

Naphthalimide Dyes. Derivatives of 4-aminonaphthalimide were used initially to dye acetate fibers. For polyesters, condensation products of 1,8-naphthalenedicarboxylic acid (e.g., **19**) or 1,4,5,8-naphthalenetetracarboxylic acid [*128-97-2*] with 1,2-diaminobenzenes are used.

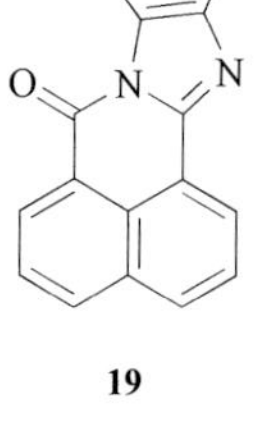

19

Yellow [*15220-29-8*]

Indigo Dyes. Whereas indigo itself is not suitable for exhaustion dyeing of polyester, thioindigo gives fast brilliant red shades in pale to medium depth [39].

Nitro Dyes. 2-Nitrodiphenylamines are readily obtained by condensation of derivatives of 2-nitrochlorobenzene [*88-73-3*] with suitable aromatic amines. Because of their accessibility and good lightfastness, these dyes became very important for dyeing cellulose acetate and, more recently, have gained a solid position as disperse dyes for polyester fibers. This is especially true for the reaction product of 1 mol of 3-nitro-4-chlorobenzenesulfonyl chloride [*97-08-5*] and 2 mol of aniline. An exhaustive review of the constitution and color of nitro dyes is given by Merian [40]. The yellow nitroacridones may also be classified in this group.

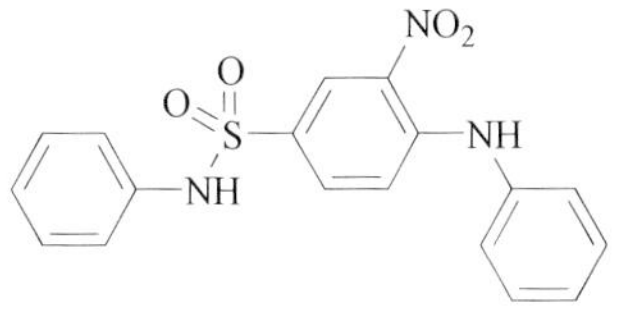

20

C.I. Disperse Yellow 42, 10338 [*5124-25-4*]

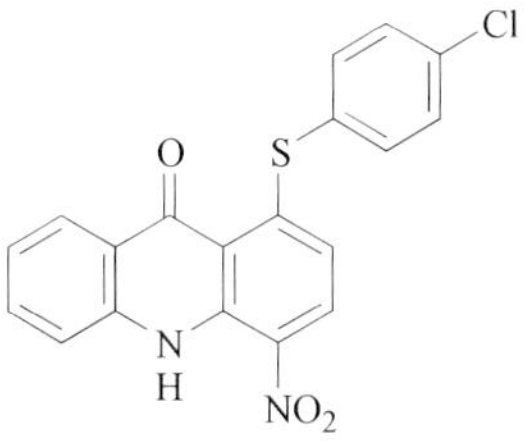

21

Yellow [*19309-55-8*] [41]

3.2.3 Synthesis

3.2.3.1 Monoazo Dyes

The synthesis of monoazo disperse dyes by diazotization and coupling is described under Azo Dyes (see Section 2.2). Additional chemical treatment subsequent to synthesis of the azo compounds is performed only rarely.

Frequently, weakly basic amines, such as negatively substituted 4-nitroanilines or aminoheterocycles, must be diazotized. These reactions require concentrated sulfuric acid, phosphoric acid (85 wt %), glacial acetic acid, propionic acid, or mixtures of thereof as the reaction medium, with nitrosylsulfuric acid as the reagent. On an industrial scale, this reaction is conducted preferably in enameled, stirred vessels. Whenever the stability of the amine or the diazonium salt permits, diazotization is carried out in concentrated sulfuric acid at slightly elevated temperature (10–40 °C). With regard to safety, factors such as temperature and concentration must be controlled carefully to avoid explosion [42].

The sulfuric acid diazotization mixture is generally added directly to the ice-cold solution or suspension of the coupling component. If the diazonium salt is stable in dilute sulfuric acid, the diazotization mixture may first be quenched with ice, possibly clarified, and only then combined with the coupling component.

Strongly negatively substituted 4-nitroanilines, such as 2,4,6-trinitroaniline [*489-98-5*], 2,4-dinitro-6-cyanoaniline [*22603-53-8*], 2-methanesulfonyl-4-nitro-6-cyanoaniline, and 2,6-dicyano-4-nitroaniline [*20033-48-1*], are difficult to diazotize, and their diazonium salts tend to decompose extensively during the coupling process [15]. This results in diminished yields and loss of clarity of the dyes.

These difficulties can be avoided if 2-halogo azo dyes, preferably 2-bromo azo dyes, are generated first, and the halogen is then exchanged for the NO_2 or CN group. These copper-catalyzed reactions proceed especially smoothly in dipolar aprotic solvents such as ethylene glycol monoalkyl ethers, dimethylformamide, dimethyl sulfoxide, or *N*-methylpyrrolidone, and they occur under considerably milder conditions than those in the starting amine [43–45]. In this process, the bromo–cyano exchange yields clear blue polyester dyes (e.g., **22**) of a shade which, until recently, could only be obtained by using anthraquinones.

22

X = CN, NO_2

By modification of the substituent in the 4-position of the diazo component, this process yields further 2,6-dicyanoazo dyes, for example, 4-methyl- or 4-bromo-2,6-dicyano derivatives, for which the basic diazo component is otherwise not readily accessible. In this way clear red shades on polyester can be obtained.

3.2.3.2 Disazo Dyes

The simplest route to disazo dyes is by using *p*-aminoazobenzene [*60-9-3*]. As *p*-aminoazobenzene has been classified in the EU as carcinogenic, the use of derived disazo dyes like *Disperse Red 151*, *Disperse Yellow 7*, *Disperse Yellow 23*, and *Disperse Yellow 56* has diminished.

3.2.3.3 Anthraquinone Dyes

For synthesis of anthraquinone dyes, see Sections 2.3 and 3.4.

3.2.3.4 Other Chromophores

For synthesis of methin dyes see Section 3.8, and for quinaphtholone dyes see Section 2.12.1.

3.2.4 Aftertreatment

Chemical synthesis produces dyes of varying particle size. When the dyes are applied in this form, uneven and spotty dyeing results, and the dyeing process may be slow and frequently accompanied by incomplete absorption [46]. To assure high yield, good reproducibility, and faultless dyeing and printing in commercial use, especially when densely woven fabric or wound material is involved, the dye must be applied as a fine dispersion that is stable under the process conditions.

Therefore, the first requirement is that the dye itself be present in a stable crystal modification that does not change during application. If such a modification cannot be obtained by synthesis, attempts must be made to generate it by thermal posttreatment, for instance, by heating the dye in an aqueous suspension in the presence of nonionic detergents.

The eventual dispersion process consists of grinding the moist dye, generally in a ball, bead, or sand mill, in the presence of dispersing agents. Since the dyeing rate depends on particle size, the aim is to reduce the particle size to <1 μm. A narrow particle-size distribution is important to minimize recrystallization during storage and application.

The dispersion agents used are primarily lignin sulfonates (sulfite cellulose liquors); condensation products of naphthalene, sulfuric acid, and formaldehyde; condensation products of *m*- and *o*-cresol, formaldehyde, and 2-hydroxynaphthalene-6-sulfonic acid [*93-01-6*]; or mixtures of these products [46]. Literature on dispersing agents is relatively scarce [47–49].

Difficulties in the production of disperse dyes may be encountered, for example, in grinding or drying dyes with low melting points. Therefore, only dyes that have a melting point > 120 °C, and preferably > 140 °C, should be used. Disperse dyes are sold either as aqueous pastes obtained by grinding or as powders obtained by gentle drying of the paste. Powders are generally produced by spray drying. Important requirements for industrial products are the storage stability of the paste and the ease with which the powder can be redispersed in water. The actual dye content of commercial products is generally 20–40 % for pastes and 20–50 % for powders.

3.2.5 Examples of Commercially Available Dyes

Disperse dyes are currently used to dye cellulose 2.5-acetate, cellulose triacetate, synthetic polyamides, and to a lesser degree, polyacrylonitrile and polypropylene. Their major application is clearly for dyeing polyesters.

Since only a few individual dyes have satisfactory properties for all synthetic fibers, many dye producers offer separate lines for individual fibers as well as for various application methods.

3.2.5.1 Monoazo Dyes

Monoazo dyes derived from *aromatic amines as coupling components* and *carbocyclic aromatic amines as diazo components* are the class of disperse azo dyes with the greatest economic importance. Commercial products are most often represented by structure **2**, in which 4-nitroaniline [*100-01-6*] and its substituted derivatives constitute the diazo component.

A few examples of the many dyes from this series that are used to dye polyesters are listed below:

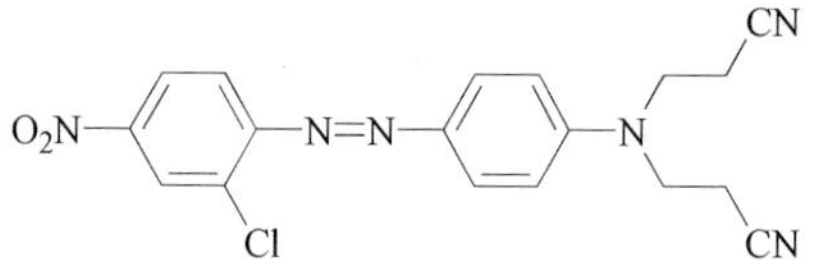

C.I. Disperse Orange 44, [*4058-30-4*] [50]

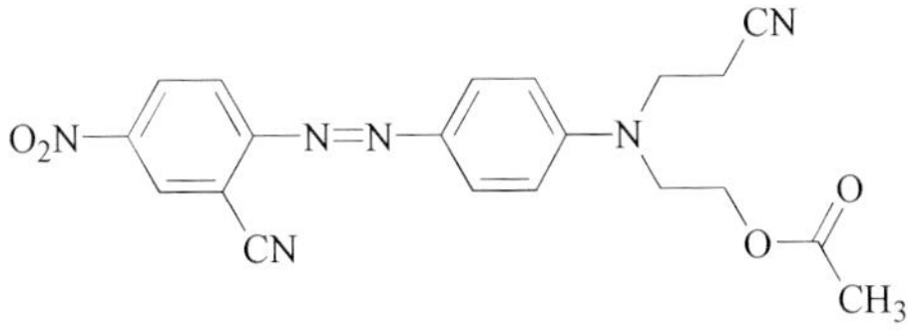

3.2.5 Examples of Commercially Available Dyes

C.I. Disperse Red 72, 11114, [*12223-39-1*] [51]

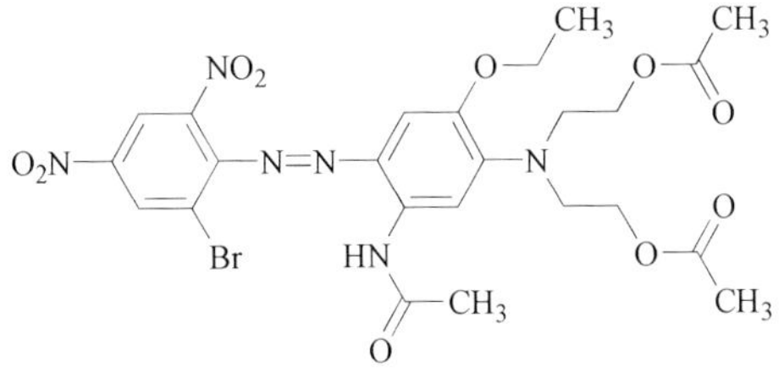

C.I. Disperse Blue 79, 11345, [*3956-55-6*] [52]

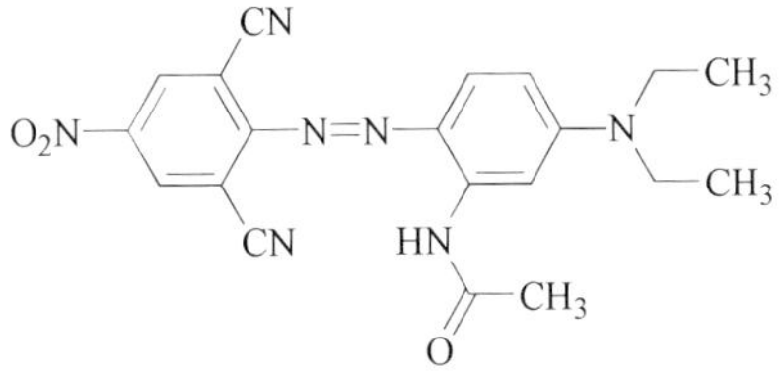

C.I. Disperse Blue 165 [*41642-51-7*] [53]

Representatives of this class are also suitable for dyeing cellulose triacetate. However, sufficient lightfastness is not obtained on synthetic polyamides [10].

Other industrially important 2,6-dicyanoazo dyes are obtained by using diethyl-*m*-toluidine and 3-diethylaminophenylmethansulfonamide [*52603-47-1*] as coupling components (see **23** and **24**, respectively).

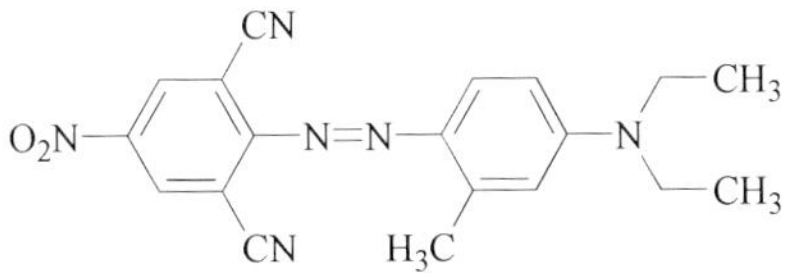

23

C.I. Disperse Blue 366 [*84870-65-5*] [54]

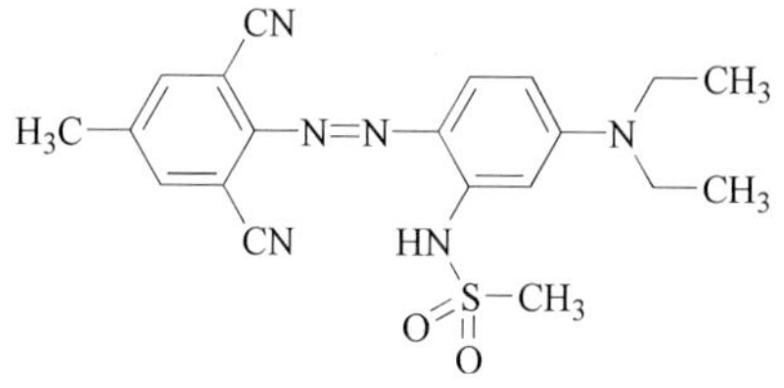

24

Red [*68385-96-6*] [55]

By mixing the reddish blue *C.I. Disperse Blue 366* with *C.I. Disperse Blue 165* [56] or heteroarylazo dyes [57], technically equivalent replacement products [13] for the anthraquinoid Disperse Blue 56 were obtained. The red dye **24** has been similarly used as a substitute [26] for the anthraquinoid *C.I. Disperse Red 60*, 60765 [*1741858-5*].

With *heteroaromatic amines as diazo component*, e.g., the red dye **25**, which is an example for the use of the methanesulfonyl group as an acceptor in azo dyes.

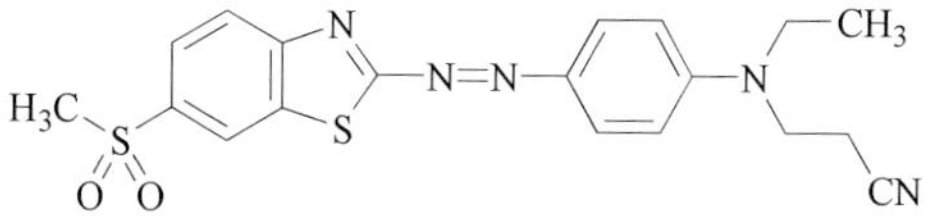

25

Red [*63467-01-6*] [58]

2-Amino-5-ethylsulfanyl-1,3,4-thiadiazole [*25660-70-2*], accessible by alkylation of bithiourea with diethyl sulfate and ring closure [59], is suitable for the manufacture of bright reds, e.g., **26**.

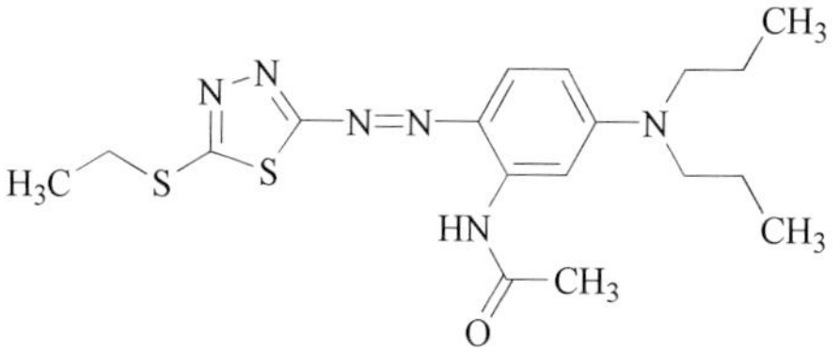

26

C.I. Disperse Red 338, 111430 [*63134-15-6*] [60]

2-Amino-4,5-dicyanoimidazole [*40953-34-2*] can be made by reaction of diaminomaleonitrile with cyanogen chloride. By coupling with aniline couplers and alkylating the resultant azo dye precursor bright reds such as **27** are obtained.

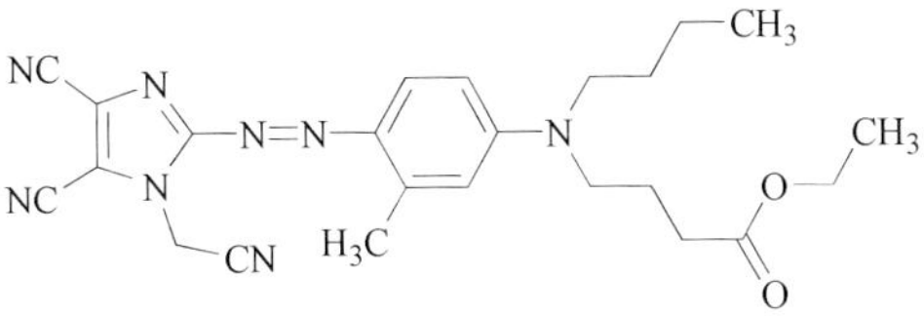

3.2.5 Examples of Commercially Available Dyes

27

Red [*86772-44-3*] [61]

3-Amino-5-nitro-2,1-benzisothiazole is manufactured by thiolysis of 2-amino-5-nitrobenzonitrile and oxidative ring closure. With aniline coupling components, shades from navy to greenish blue are obtainable (e.g., **28** and **29**).

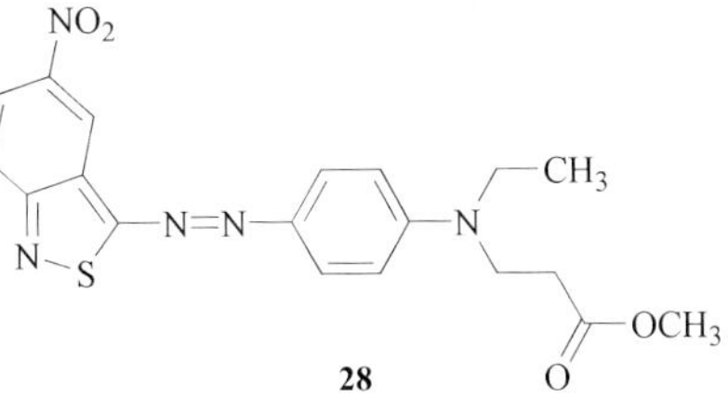

28

C.I. Disperse Blue 148, 11124 [*52239-04-0*] [62]

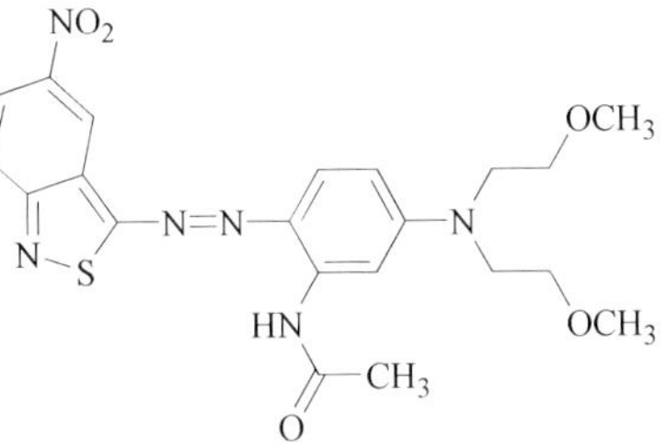

29

Blue [*105076-77-5*] [63]

Since the first appearance of thienylazo dyes like *Disperse Green 9* in the early 1970s, this field has been intensely exploited. The interest originates from the fact that thienylazo dyes are nearly 100 nm more bathochromic than their carbocyclic counterparts (cf. *C.I. Disperse Green 9* and **30**) and exceed these in the molar extinction coefficient and hence also in dyeing strength.

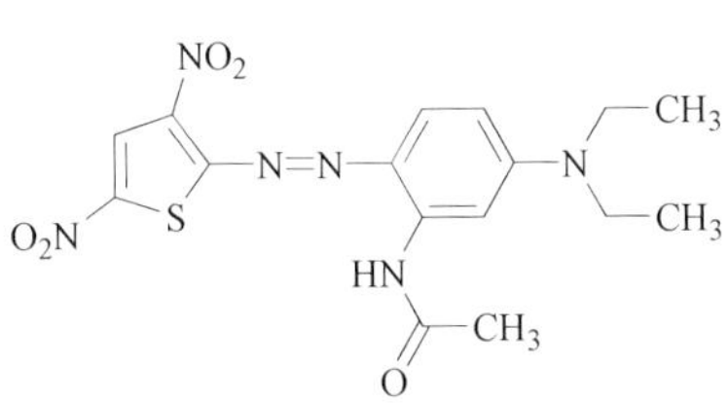

C.I. Disperse Green 9, 110795 [*58979-46-7*] [64]
$\lambda = 644$ nm $\varepsilon = 87\,400$ L mol^{-1} cm^{-1} [26]

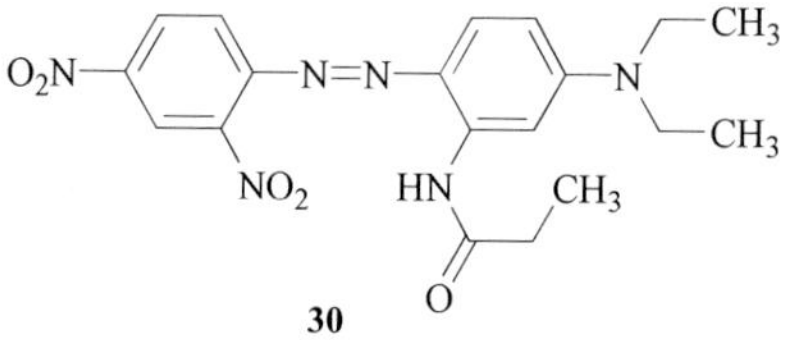

30

$\lambda = 545$ nm $\varepsilon = 46\,000$ Lmol^{-1} cm^{-1} [65]

Furthermore, the Gewald reaction [66] opened a simple and economic route from aliphatic reactants to aminothiophenes that were formerly only accessible from multistep reactions [67] (see, e.g., the dye **31**). For example, 2-amino-3-carbethoxy-5-nitrothiophene, a precursor for the dye **32**, can be prepared by reaction of 1,4-dithiane with ethyl cyanoacetate and subsequent nitration.

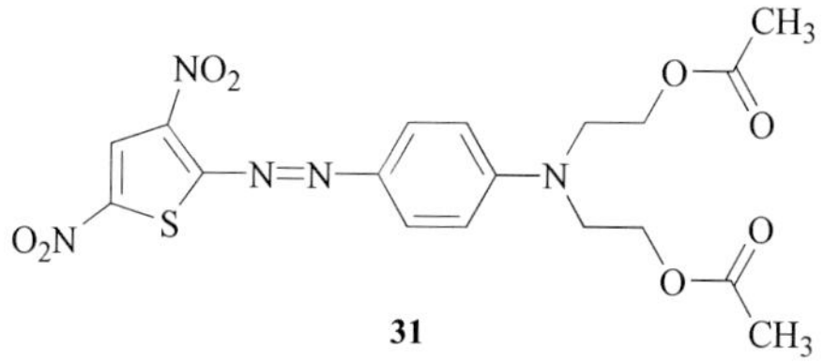

31

Blue [*42783-06-2*] [68]

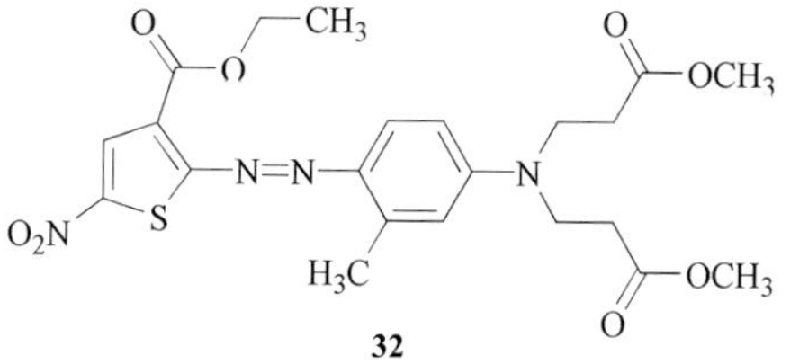

32

Blue [*210758-04-6*] [69]

Whereas the blue dyes **31** and **32** are claimed to have good thermomigration fastness, **33**, **34**, and **35** have superior pH stability and are proposed as substitutes for bright anthraquinone blues.

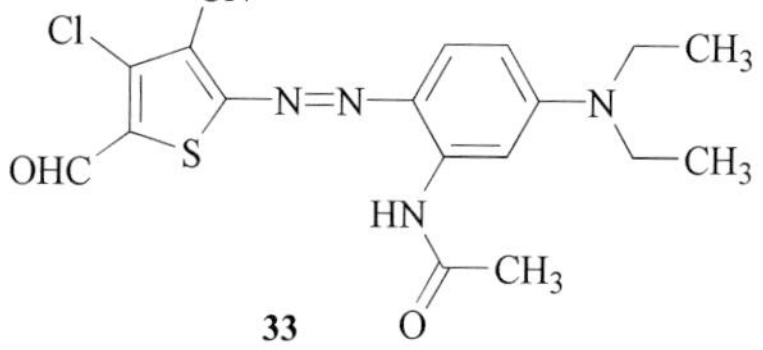

33

Blue [*104366-25-8*] [70]

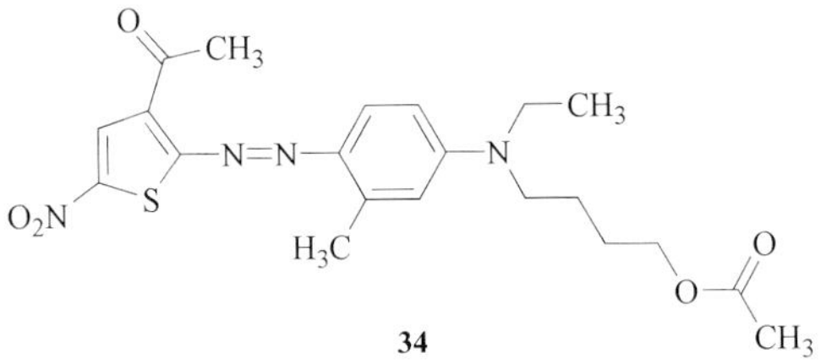

3.2.5 Examples of Commercially Available Dyes

34

Blue [*122063-39-2*] [71]

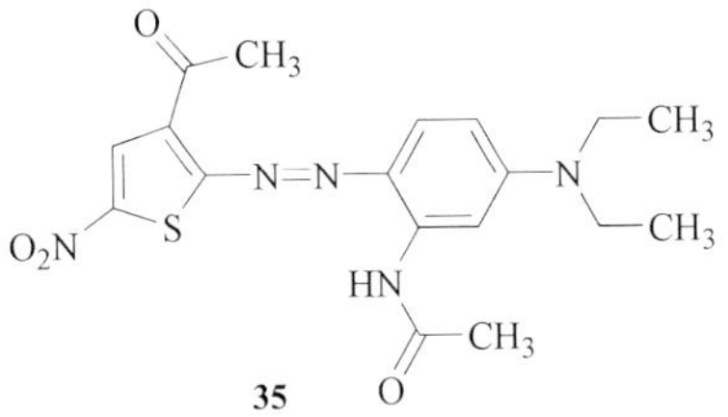

35

Blue [*167940-11-6*]

By combination of the thiophene diazo moiety with a diaminopyridine coupler, the hue is shifted to a bright red with good lightfastness (**36**).

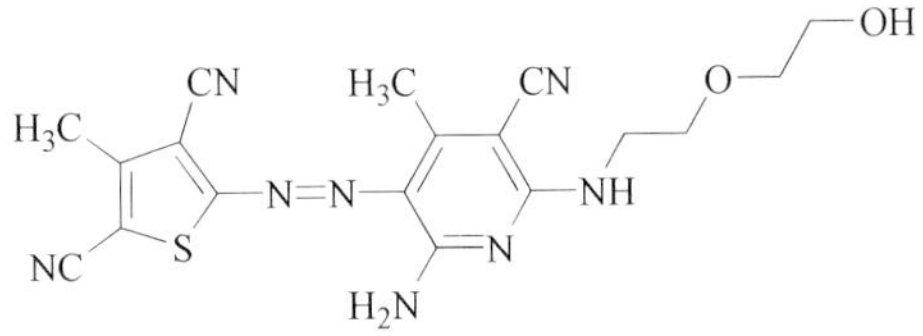

36

Red [*107815-88-3*] [72]

The chemistry and physics of azo dyes based on heterocyclic amines has been surveyed in many reviews [73–75].

Monoazo dyes with *aromatic hydroxy compounds as coupling components*:

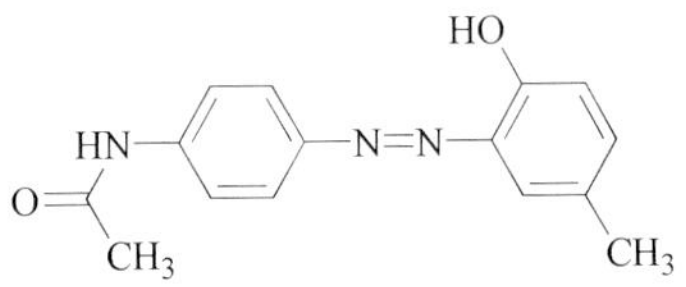

Disperse Yellow 3, C.I. 11855 [*2832-40-8*]

When 2-hydroxynaphthalene [*135-19-3*] and its derivatives are used as coupling components, clear orange (e.g., **37**) and red shades result. Their commercial importance is minor.

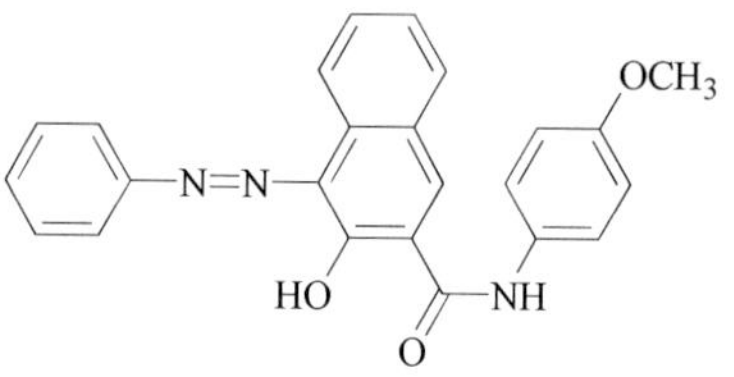

37

Orange [*17947-32-9*] [76]

The chemistry of 4-hydroxynaphthalimides as coupling components has mainly been investigated in Japan. Brilliant reds with good fastness properties are obtainable (e.g., **38**).

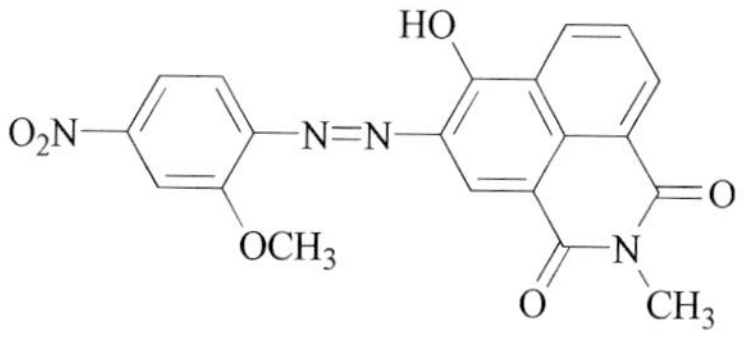

38

Red [*42357-98-2*] [77]

Monoazo dyes with *heterocyclic compounds as coupling components*:

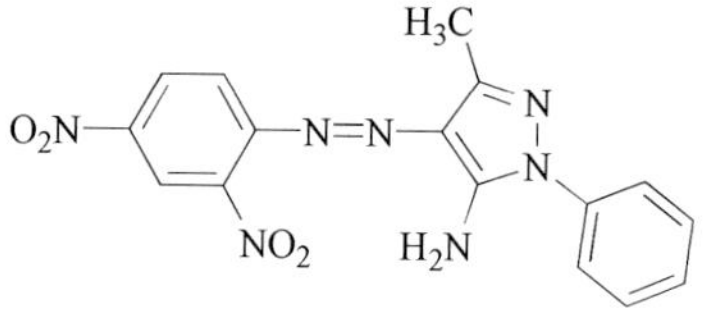

C.I. Disperse Orange 56 [*67162-11-2*] [78]

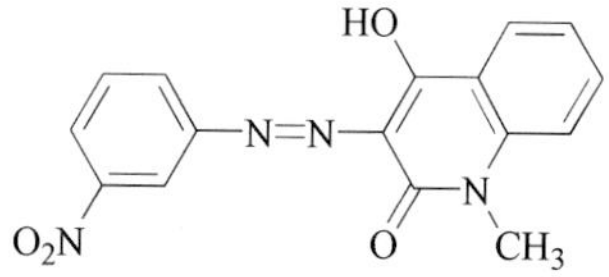

C.I. Disperse Yellow 5, 12790 [*6439-53-8*]

By reaction of ethyl acetoacetate with cyanoacetamides, pyridones are easily accessible. By selection of suitable substituents in the diazo components, shade and fastness properties and build up can be controlled (e.g., **39–42** and *Disperse Yellow 211*).

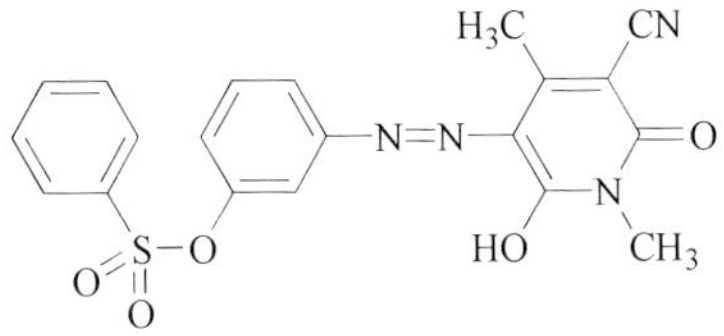

3.2.5 Examples of Commercially Available Dyes

39

Greenish Yellow [*59312-61-7*] [79]

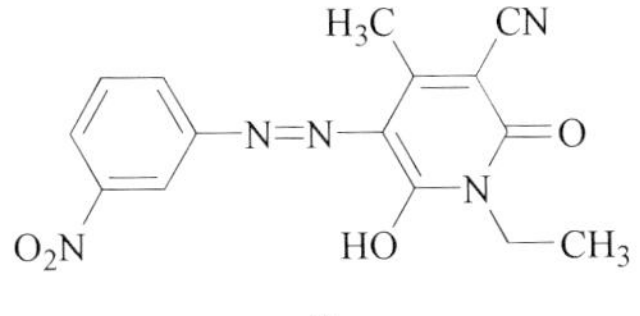

40

Greenish Yellow [*37781-00-3*] [80]

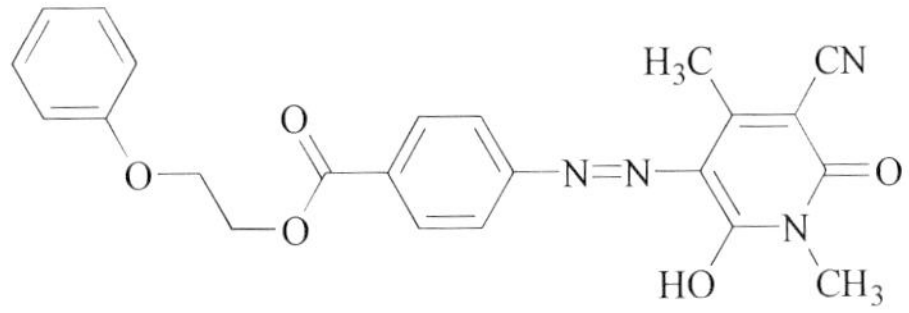

41

Greenish Yellow [*88938-37-8*] [81]

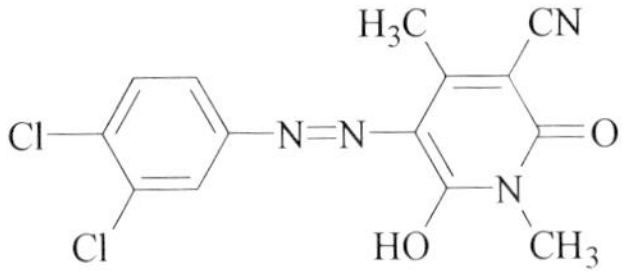

42

C.I. Disperse Yellow 241, 128450 [*83249-52-9*] [82]

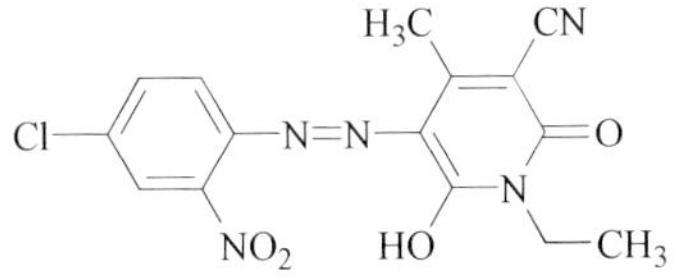

C.I. Disperse Yellow 211, 12755 [*70528-90-4*] [83]

Chlorination of 2,6-dihydroxy-3-cyano-4-methylpyridine with phosphorus oxychloride and successive reaction with amines yields 2,6 diaminopyridines. With these coupling components brilliant orange and red dyes with excellent lightfastness are obtained (e.g., **43** [4]).

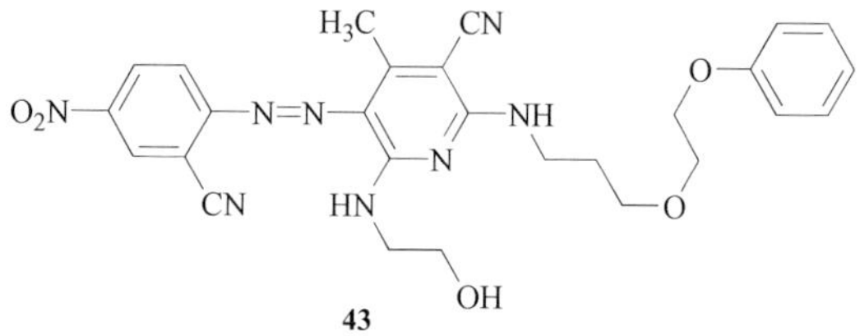

43

Red [63833-78-3] [84]

Many disazo dyes with derivatives of aniline as the terminal component (e.g., **44**, **45**) have been described in the literature, but only a few have become industrial products. Replacement of the central benzene ring by heterocycles such as thiazole or thiophene causes marked bathochromic shifts.

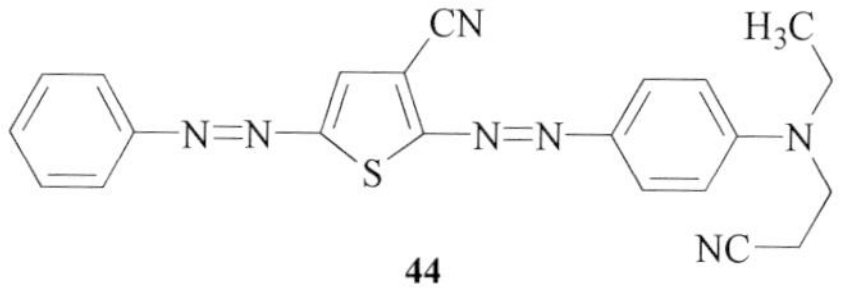

44

Blue, [84425-43-4] [85]

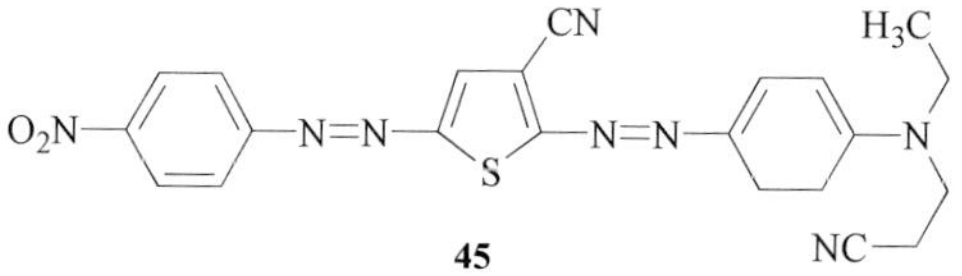

45

Blue [87606-56-2]

3.2.5.2 Disazo Dyes

About 10 % of all disperse dyes are disazo compounds [9]. Even the simplest hydroxy disazo dyes, such as 4-aminoazobenzene coupled to phenol and 4-aminoazobenzene coupled to o-cresol, have a good affinity for polyester fibers and yield lightfast reddish yellow hues. However, these shades are frequently less bright than those obtained with monoazo dyes.

The introduction of an alkoxy group into the central benzene ring causes a distinct shift toward orange. A similar bathochromic effect is obtained by replacing this benzene ring with naphthalene.

Substitution of the first benzene nucleus by electron acceptors also causes bathochromic shifts. Disazo dyes of this type, such as **46**, are frequently incorporated as components of black mixtures.

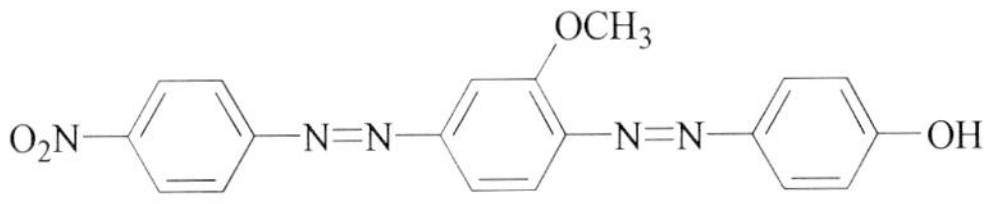

3.2.5 Examples of Commercially Available Dyes

C.I. Disperse Orange 29, 26077 [*19800-42-1*] [25]

3.2.5.3 Anthraquinone Dyes

Turquoise shades are obtained from derivatives of 1,4-diaminoanthraquinone-2,3-dicarboximide [*128-81-4*] (**47**).

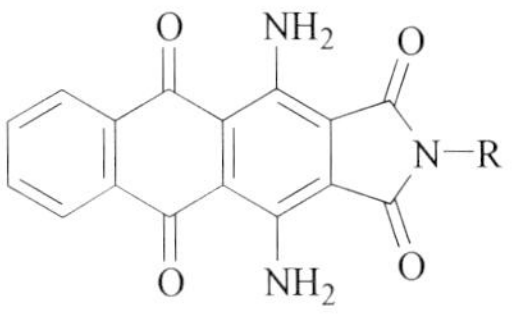

R = alkyl

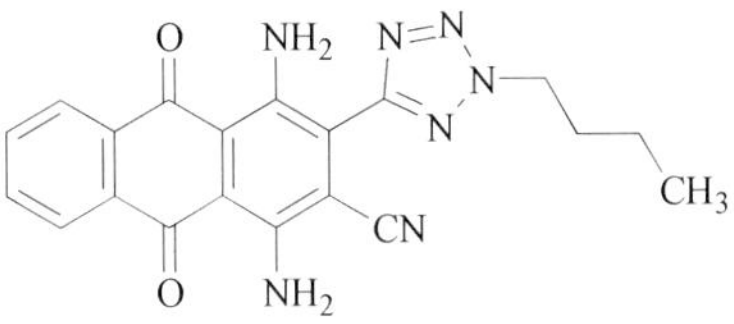

[*93686-63-6*] [28]

3.2.5.4 Other Chromophores

Quinapththolones:

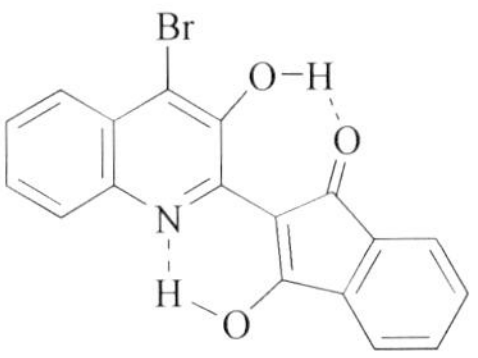

C.I. Disperse Yellow 64, 42023 [*10319-14-9*] [86]

C.I. Disperse Yellow 160 [75216-43-2] [87]

3.2.6 References

[1] R. K. Fourness, *J. Soc. Dyers Colourists* **72** (1956) 513.
[2] Acordis, The Man-made Fibre Year 1998.
[3] N. L. Moore, *J. Soc. Dyers Colourists* **90** (1974) 318.
[4] T. Hihara, *Textile Art* **3** (1997) 23.
[5] T. Lampe, P. Neumayer, W. Roggenbach, *Melliand Textilberichte* **1992**, 402–406.
[6] K. H. Ulbrich, *Melliand Textilberichte* **1992**, 406–410.
[7] P. Vonhöne, J. Stuck, *Melliand Textilberichte* **1922**, 411–416.
[8] W. McDowell, *Melliand Textilber.* **61** (1980) 946. E. Schollmeyer, G. Heidemann, A. Bossmann, S. Dugal, C. Heinrichs, *Textilveredlung* **20** (1985) 190.
[9] C. Müller, *Chimia (Aarau)* Suppl. 1968, 69.
[10] *Colour Index International*, 3rd ed., The Society of Dyers and Colourists, American Association of Textile Chemists and Colorists, Bradford/England, 1992.
[11] C. V. Stead, *Rev. Prog. Color. Relat. Top.* **1** (1970) 26.
[12] J. F. Dawson, *Rev. Prog. Color. Relat. Top.* **3** (1972) 18.
[13] *Ullmann*, 3rd ed., Suppl., p. 222.
[14] J. F. Dawson, *Rev. Prog. Coloration* **9** (1978) 25.
[15] *Ullmann*, 3rd ed., 4, 128.
[16] A. T. Peters, *J. Soc. Dyers Colourists* **101** (1985) 361.
[17] Monsanto, US 2 617 809, 1950 (H. L. Hubbard, G. W. Steahly).
[18] Eastmann Kodak, US 2 683 709, 1951 (J. B. Dickey, E. B. Towne).
[19] ICI, GB 840 903, 1957 (J. R. Atkinson, G. Booth, E. L. Johnson).
[20] A. T. Peters, S. S. Yang, *Dyes Pigments* **28** (1995) 151.
[21] Saul and Co, US 2 891 942, 1959.
[22] A. T. Peters, E. Tsatsaroni, M. Xisai, *Dyes Pigments* **20** (1992) 41.
[23] Mitsubishi, JP 631 355 79, 1988 (K. Himeno, J. Yoshihara).
[24] A. T. Peters, N. M. A. Gbadamosi, *Dyes Pigments* **18** (1992) 115.
[25] Koppers, BE 604 935, 1960 (D. A. Zanella, C. R. Conway).
[26] O. Annen, R. Egli, R. Hasler, B. Henzi, H. Jakob, P. Matzinger, *Rev. Progr. Coloration* **17** (1987) 2772.
[27] A. T. Leaver, B. Glover, P. W. Leadbetter, *Textile Chemist and Colorist* **24** (1992) 18
[28] Ciba-Geigy, EP 79 862, 1981 (J. M. Adam).
[29] B. K. Manukian, A. Mangini, *Chimia (Aarau)* **24** (1970) 328.
[30] J. M. Strayley in: *The Chemistry of Synthetic Dyes*, vol. III, Academic Press, New York 1970, p. 385.
[31] Ciba, GB 1 201 925, 1966 (R. Peter, H.-J. Anliker).

3.2.6 References

[32] Sandoz, DE-OS 2 929 001, 1978 (W. Baumann).
[33] Geigy, GB 1 085 456, 1964 (J. Voltz, H. Heusermann).
[34] Bayer, DE-OS 2 065 076, 1972 (H. Harnisch).
[35] BASF, DE-OS 2 253 538, 1972 (W. Mach, D. Augart, H. Scheuermann).
[36] Ciba-Geigy, US 4 547 579, 1985 (P. Möckli).
[37] Nippon Kayaku, DE-OS 3 716 840, 1987 (H. Matsumoto, H. Imai, S. Tada).
[38] ICI, EP 33 583, 1980 (C. W. Greenhalgh, N. Hall, D. F. Newton).
[39] Hoechst, EP 26 463, 1981.
[40] E. Merian, *Angew. Chem.* **72** (1960) 766.
[41] Geigy, GB 1 127 721, 1966 (K. Burdeska, H. Bosshard, A. Pugin).
[42] P. Bersier, L. Valpina, H. Zubler, *Chem. Ing. Tech.* **43** (1971) 1311.
[43] Bayer, DE-OS 1 544 563, 1966.
[44] ICI, DE-OS 1 807 642, 1967.
[45] Sandoz, DE-OS 1 966 124, 1968.
[46] J. F. Dawson, *Rev. Prog. Coloration* **14** (1984) 90.
[47] A. Murray, K. Mortimer, *J. Soc. Dyers Colourists* **87** (1971), no. 6.
[48] N. J. Sardesai, *Colourage* **36** (1989), Moscow Seminar Special Issue.
[49] "Dispersing Agents, Review of Progress in Colouration and Related Topics", *Soc. Dyers Colourists* **23** (1993) 48.
[50] DuPont, US 2 782 187, 1954 (M. F. Sartori).
[51] Sandoz, DE-AS 1 065 112, 1955 (E. Merian).
[52] Sandoz, DE-AS 1 069 313, 1956 (E. Merian).
[53] Bayer, DE-OS 1 544 563, 1966 (A. Gottschlich, K. Leverenz).
[54] Synthetic Dye Technology Research Association, JP 03 006 275, 1991 (M. Kurosawa et al.).
[55] Bayer, DE-OS 2 711 130, 1977 (R. Hamprecht).
[56] Cassella, EP 277 529, 1987 (U. Bühler, K. Hofmann, M. Hähnke).
[57] Bayer, EP 347 685, 1988 (M. Hoppe, D. Wiegner, K.-P. Sagner, H. D. Jordan, H. Brandt).
[58] Eastman, DE-AS 1 065 542, 1952 (J. M. Straley, D. J. Wallace).
[59] Eastmann Kodak, US 4 374 993, 1981.
[60] Eastman, US 3 657 215, 1969 (M. A. Weaver, D. J. Wallace).
[61] R. W. Sabnis, D. W. Rangnekar, N. D. Sonawane, *J. Heterocycl. Chem.* **36** (1999) 333. Mitsubishi, JP 5 829 859, 1981.
[62] BASF, DE-AS 1 544 375, 1965 (M. Seefelder, H. G. Wippel, H. Armbrust).
[63] Bayer, EP 167 913, 1984 (W. Kruckenberg, K. Leverenz, H.-G. Otten).
[64] ICI, DE-OS 2 304 218, 1972 (D. B. Baird, A. T. Costello, B. R. Fishwick, R. D. McClelland, P. Smith).
[65] A. T. Peters, *Dyes Pigments* **22** (1993) 223.
[66] K. Gewald, *Chimica* **34** (1980) 101.
[67] J. B. Dickey, E. B. Towne, M. S. Bloom, W. H. Moore, B. H. Schmith, Jr., D. G. Hedberg, *J. Soc. Dyers Colour.* **74** (1958) 123.
[68] ICI, GB 1 394 365, 1972 (B. D. Baird, A. T. Costello, R. B. Fishwick, R. McClelland).
[69] ICI, DE-OS 2 612 792, 1976 (D. B. Baird, A. T. Costello, B. R. Ribbons R. D. McClelland, P. Smith).
[70] Clariant, DE 3 529 831, 1984 (R. Egli, B. Henzi).
[71] ICI, EP 297 710, 1988 (D. Brierley, R. W. Kenyon, D. R. A. Ridyard).
[72] BASF, EP 214 445, 1986 (J. Dehnert, G. Lamm, H. Loeffler).
[73] E. Hahn, *J. Heterocycl. Chem.* **24** (1987) Suppl. Vol. 9, 9–13.
[74] M. A. Weaver, L. Shuttleworth, *Dyes Pigments* **7** 93 (1982) 81.

[75] A. D. Towns, *Dyes Pigments* **42** (1999) 3.
[76] Hoechst, DE-OS 1 810 305, 1968.
[77] Mitsubishi, GB 1 384 457, 1972 (S. Imahori, M. Kaneko, Y. Kato).
[78] Geigy, BE 594 867, 1959 (W. Bossard, F. Farre).
[79] Casella, DE-OS 1 813 385, 1968 (H. von Brachel, E. Heinrich, O. Gräwinger, K. Hintermeier, H. Kindler).
[80] ICI, DE-AS 1 932 806, 1968 (A. H. Berrie, N. Hughes).
[81] Mitsubishi, JP 58 136 656, 1982
[82] Bayer, EP 74 562, 1981 (H. Brandt, K. Leverenz, H.-G. Otten).
[83] ICI, DE-OS 1 932 806, 1970 (H. H. Berrie, N. Hughes).
[84] BASF, DE-OS 2 525 505, 1975 (N. Zimmermann, E. Daubach).
[85] Mitsubishi, DE-OS 3 151 114, 1980 (S. Imahor, K. Himeno, S. Maeda).
[86] BASF, GB 1 036 389, 1965 (A. Schöllig).
[87] Mitsubishi, JP 7 308 319, 1973.

3.3 Direct Dyes

3.3.1 Introduction

Direct or substantive dyes are colored compounds that are mainly used to dye materials made from natural or regenerated cellulose (e.g., cotton, jute, viscose, or paper) without employing mordants as auxiliaries. The essential requirement for classification of a dye in this group is its substantivity, i.e., its absorption from an aqueous salt-containing solution onto cellulosic materials. Absorption onto cotton takes place in a neutral to soda alkaline medium, and onto paper in a weakly acid to neutral medium.

Substantivity was initially attributed to secondary valence bonding between fiber and dye. The fact that coplanar molecules are always more substantive than nonplanar ones later led to the coplanarity theory with its assumption that coplanar dyes are in contact with the cellulose molecule along their entire length.

The presence of hydrogen bonds has also been expounded as a possible explanation for high affinity between fiber and dye [1]; however, such bonds are probably prevented by a water layer between fiber and dye [2].

The nature of substantivity has been convincingly explained [3]. According to this explanation, single dye molecules are adsorbed by the intermicellary cavities of the cellulosic fibers and, unlike nonsubstantive dyes, they form aggregates in these cavities. Because of their size, these aggregates can no longer be directly washed out with water, but only after further solvation has taken place.

Because direct dyes become aggregated in aqueous solutions at normal temperatures, substantivity often cannot take effect until the temperature has risen.

Only then is diffusion into the fiber possible. The tendency toward aggregation is therefore characteristic of substantive dyes, which also explains why coplanar dyes possess greater substantivity than nonplanar ones.

There is no exact delineation between substantive and nonsubstantive dyes; the boundaries between them are fluid.

Because of their ease of application and moderate price, direct dyes still represent one of the largest groups of azo dyes, although their wetfastness, in particular, often only satisfies moderate requirements.

3.3.2 Chemical Constitution

3.3.2.1 Structural Characteristics

Azo dyes make up the major proportion of the direct dyes; apart from these, only a few azine, phthalocyanine, and nonazo metal-complex dyes possess a certain significance. The following structural characteristics are necessary for high substantivity:

a) *Coplanarity of the Dye Molecules.* The importance of coplanarity is shown by the example of 4,4′-diaminodiphenyl derivatives. Whereas 3,3′-disubstituted diaminodiphenyls (e.g., *o*-tolidine or *o*-dianisidine) as diazo components produce high absorptive power in dyes, this substantivity is neutralized in 2,2′-disubstituted derivatives and coplanarity of these molecules is no longer possible.
b) *Long Chain of Conjugated Double Bonds.* Dye molecules of this kind (see also Schirm's rule [4]) can contain bridges, such as $-CH=CH-$, in addition to the aromatic rings, thus ensuring the development of resonance forms across the entire molecule. Exceptions to this rule are the bridges $-NHCONH-$, $-CONH-$, and $-NH-$, which increase substantivity in spite of the interruption in the conjugated system. In these groups, the ability to form hydrogen bonds is the decisive factor. Other groups causing an interruption in the conjugated system, e.g., $-CH_2-$, $-CH_2CH_2-$, $-CO-$, $-S-$, $-SO_2-$, and cyclohexylene, decisively lower substantivity. Only the substantivity of the linked segments remains.

3.3.2.2 Precursors

Based on the above requirements, the following compounds are especially suited for the synthesis of direct dyes:

1) Compounds of the type

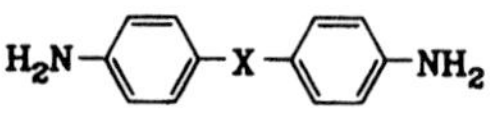

$$H_2N-\langle\rangle-X-\langle\rangle-NH_2$$

where X = NH, SO_2, $-N=N-$, $-CONH-$, $-CSNH-$, $-SO_2NH-$, $-CH=CH-$, $-HNCONH-$, $-HNCSNH-$, $-O_2SNHSO_2-$.

2) 1,4-Diaminonaphthalene and 1,5-diaminonaphthalene
3) p- and m-Phenylenediamine and other 1,3-disubstituted diamines, e.g., 2,6-diaminotoluene-4-sulfonic acid or bis(3-aminophenyl)urea.
4) Heterocyclic diamines

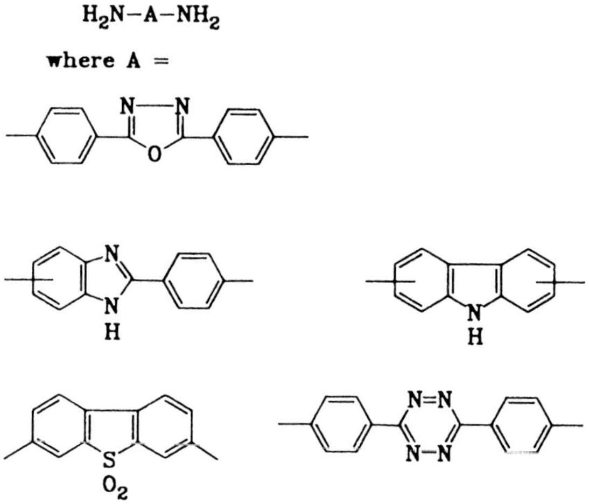

$$H_2N-A-NH_2$$

where A =

In the past 3,3′-disubstituted 4,4′-diaminodiphenyl derivatives with the structure

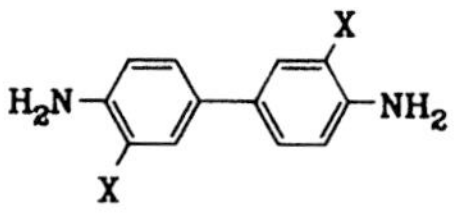

$$H_2N-\langle\rangle-\langle\rangle-NH_2$$

where X = H, CH_3, OCH_3 were considered to be the best bridging groups for direct dyes. Because of the carcinogenic properties of unsubstituted benzidine in humans and animal carcinogenicity of compounds with X = CH_3 and CH_3O, almost all major dye manufacturers worldwide stopped production of dyes based on these substances.

3.3.2.3 Classification According to the Method of Application

Conventional direct dyes include monoazo, disazo, trisazo, and tetrakisazo dyes. It is advantageous to subdivide them according to the nature of their chemical structure. Disazo dyes, for example, can be divided according to chemical synthesis principles into primary and secondary. Conventional azo direct dyes further include symmetric urea derivatives, dyes obtained by oxidation of amines, and triazinyl and copper-containing dyes.

The group of *direct dyes with aftertreatment* includes direct dyes that, after being applied to the fiber by the usual method, are subjected one of the following aftertreatments: (1) Aftertreatment with cationic auxiliaries, (2) aftertreatment with formaldehyde, 3) diazotization of the dye on the fiber and coupling with suitable components (diazotization dyes), and (4) aftertreatment with metal salts.

3.3.3 Dye Classes

3.3.3.1 Monoazo Dyes

Only a few monoazo dyes possess the necessary substantivity to render them suitable for direct dyeing of cellulosic fibers.

Of practical importance as diazo components are only the derivatives of 2-(4'-aminophenyl)-6-methylbenzthiazole, the so-called dehydrothio-*p*-toluidine (**1**).

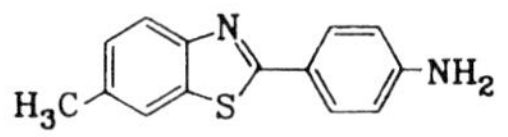

Compound **1** is obtained by melting *p*-toluidine with sulfur at 130 – 230 °C. The derivatives **2–5** are used to produce monoazo direct dyes.

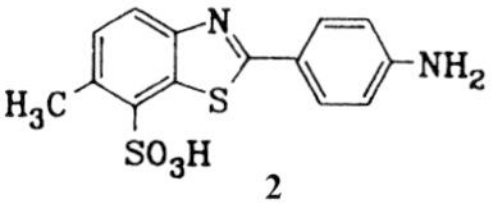

2 Dehydrothio-*p*-toluidine-sulfonic acid

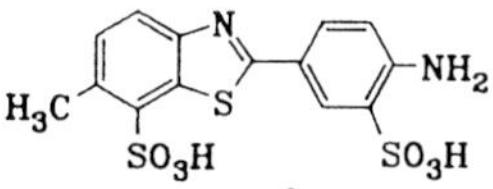

3 Dehydrothio-*p*-toluidine-disulfonic acid

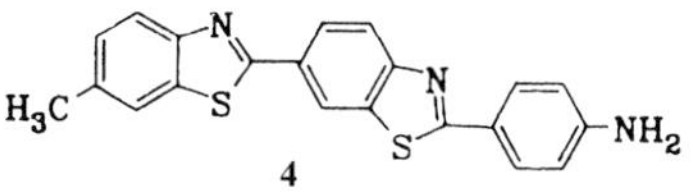

4 Primulin base

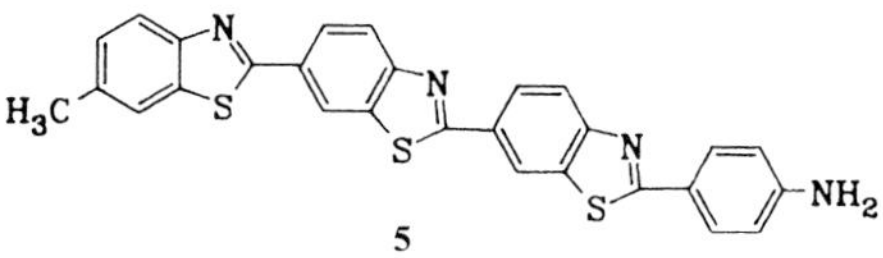

5

Whereas **2** and **3** are obtained by sulfonation of dehydrothio-*p*-toluidine (**1**) with oleum, **4** and **5** are formed in a mixture with **1** during the sulfur melting process and are separated by vacuum distillation.

Acetoacetic acid arylides and pyrazolone derivatives are the coupling components mainly used. Naphtholsulfonic acids are of only minor importance here. Dehydrothio-*p*-toluidinesulfonic acid (**2**) and the sulfonated primulin base are also used as diazo components for certain orange disazo direct dyes.

3.3.3.2 Disazo Dyes

Primary Disazo Dyes. Primary disazo dyes are obtained in accordance with the pattern $K_1 \leftarrow D \rightarrow K_2$ by bis-diazotization of a bifunctional diazo component D and coupling to two equivalents of coupling component K_1 and K_2, or according to the pattern $D_1 \rightarrow K \leftarrow D_2$ by coupling of two equivalents of diazo component (D_1 and D_2) to one equivalent of a bifunctional coupling component. Depending on whether K_1 and K_2 or D_1 and D_2 are the same or different, symmetrical or unsymmetrical dyes are obtained.

Primary symmetrical disazo dyes of the $K_1 \leftarrow D \rightarrow K_1$ type with diaminodiphenyl derivatives as bis-diazo component were long considered to be the prototypical direct dyes. Today, no individual chemical group enjoys such significance. Instead, other groups, such as 4,4'-diaminostilbene-2,2'-disulfonic acid, 4,4'-diaminodiphenylurea-3,3'-dicarboxylic or -3,3'-disulfonic acid, and heterocyclic groups, have taken the place of the diaminodiphenyl compounds (see Section 3.3.4).

3.3.3 Dye Classes

The most important coupling component for the $D_1 \rightarrow K \leftarrow D_1$ and $D_1 \rightarrow K \leftarrow D_2$ types is **6**, which is produced by treatment of 6-amino-1-hydroxynaphthalene-3-sulfonic acid (I acid) with phosgene.

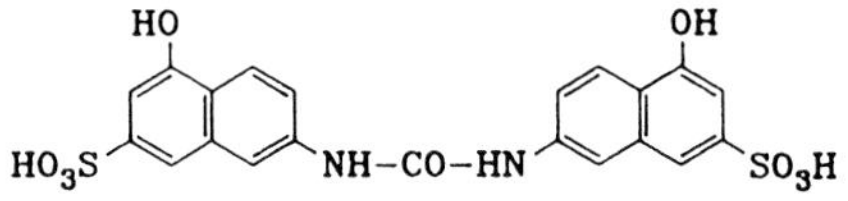

6

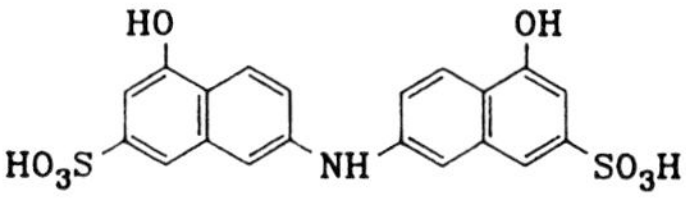

7

Dyes with a urea bridge, $-HN-CO-NH-$, are also obtained by phosgene treatment of aminoazo compounds. The so-called I-acid imide (**7**), resorcinol, and *m*-phenylenediamine are also of importance.

Examples of dyes are *C. I. Direct Orange 26*, 29150 [3626-36-6] (**8**)

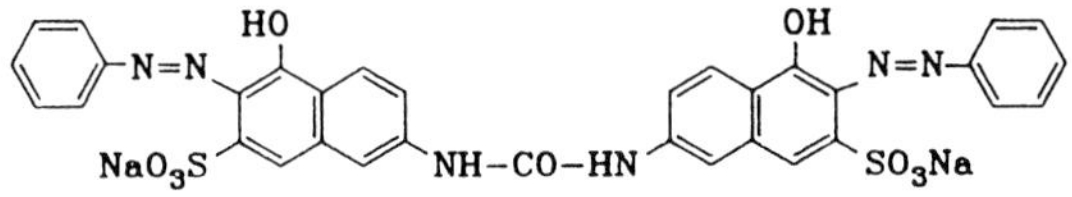

8

C. I. Direct Red 250, 29168 (**9**),

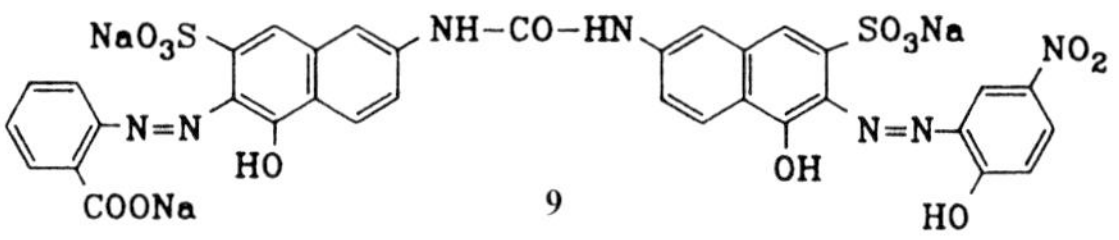

9

and *C. I. Direct Red 23*, 29160 [3441-14-3], (**10**).

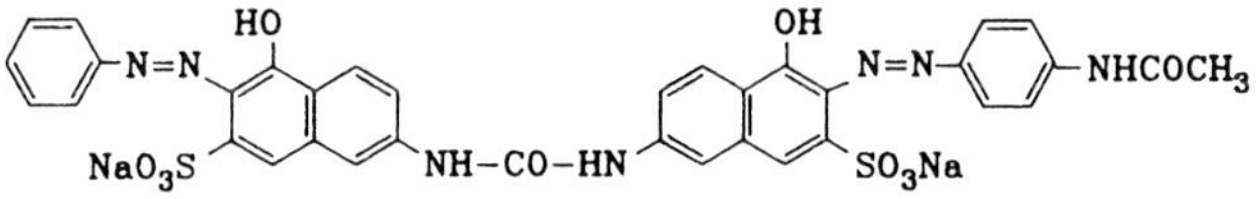

10

The dye **10** is obtained by simultaneous coupling of one equivalent each of diazotized aniline and 4-aminoacetanilide to I-acid urea (**6**).

Coupling diazotized dehydrothio-*p*-toluidinesulfonic acid to resorcinol and subsequent coupling of diazotized aniline to the monoazo dye formed gives *C. I. Direct Orange 18*, 20215 [5915-59-3] (**11**), an important substantive dye.

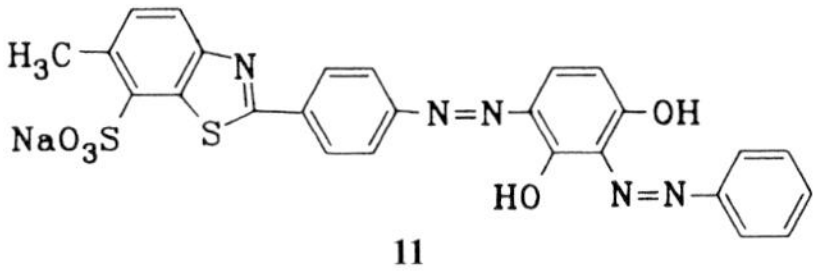

11

Secondary Disazo Dyes. For the manufacture of secondary disazo dyes of the D→M→K type, a diazotized amine D is coupled to an amine middle component M with a free *para* position, which in turn is diazotized and coupled to a coupling component K (series coupling).

Dyes of this type mostly have a straight-chain structure and often contain groups that increase substantivity, such as I-acid and I-acid derivatives, in particular with acylated amino groups, urea bridges, or benzoylamino groups. The shades in this range extend from orange through red, violet, and blue to black.

Examples: From 4-aminoazobenzene-4′-sulfonic acid and benzoyl I-acid, the important red dye *C. I. Direct Red 81*, 28160 [*2610-11-9*] (**12**) is obtained.

12

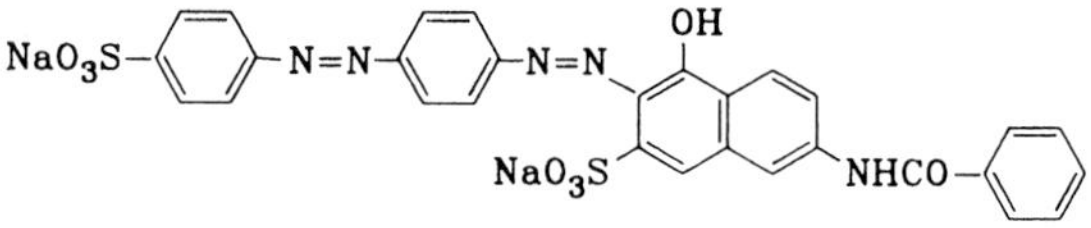

C. I. Direct Black 51, 27720 [*3442-21-5*] (**13**) is produced by coupling a mixture of diazotized 3- and 5-aminosalicylic acid to 1-naphthylamine and subsequent alkaline coupling to γ acid.

13

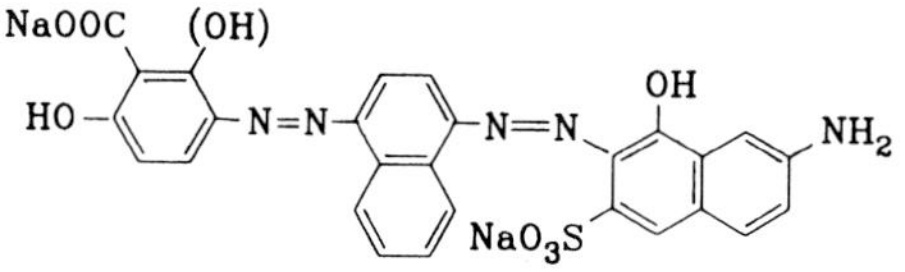

3.3.3.3 Trisazo Dyes

The trisazo dyes include in particular blue, green, and black grades. Diamines used are 4,4′-diaminobenzanilide (DABA) and 4,4′-diaminodiphenylamine-2-sulfonic acid (DADPS). The most important synthetic principle in this range is the type $K_1 \leftarrow D_1 \rightarrow K_2 \leftarrow D_2$. The sequence of the three couplings depends on the nature of the components.

An example is *C. I. Direct Black 166*, 30026 [*57131-19-8*] (**14**). The first step is the acid coupling of 3,3′-diaminobenzanilide to 1-amino-8-hydroxynaphthalene-3,6-disulfonic acid (H-acid), followed by alkaline coupling of aniline to the monoazo dye and finally coupling of the disazo dye to *m*-phenylenediamine.

14

$$m\text{-phenylenediamine} \xleftarrow{3} \text{DABA} \xrightarrow[\text{a}]{1} \text{H acid} \xleftarrow[\text{alk}]{2} \text{aniline}$$

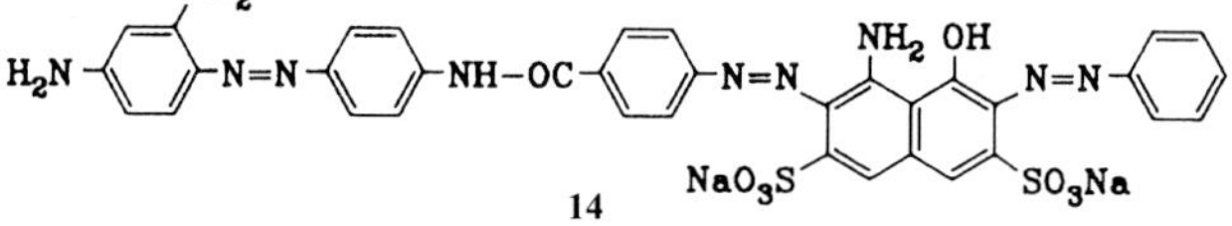

3.3.3 Dye Classes

C. I. Direct Black 150, 32010 [6897-38-7] (**15**), is produced by double (alkaline) coupling of bis-diazotized 4,4′-diaminodiphenylamine-2-sulfonic acid to γ acid and subsequent single coupling to *m*-phenylenediamine.

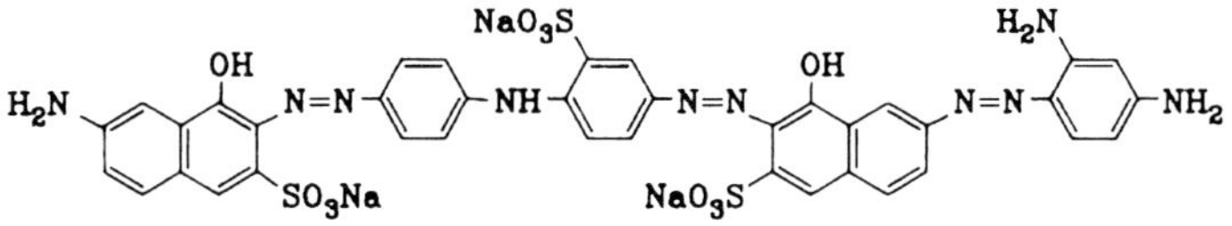

15

$$\gamma \text{ acid} \xleftarrow{\frac{1}{\text{alk}}} \text{DADPS} \xrightarrow{\frac{1}{\text{alk}}} \gamma \text{ acid} \xrightarrow{2} m\text{-phenylenediamine}$$

Among the trisazo dyes, series coupling of the type D→M$_1$→M$_2$→K is also of importance. This group contains a number of dyes with good to very good general fastness properties.

Anilinesulfonic acids are chiefly used as starting component D, naphthylamine and Cleve acids as middle components M$_1$ and M$_2$, and I acid, its *N*-phenyl derivatives, H acid, 2-amino-8-hydroxynaphthalene-6-sulfonic acid (γ acid), and their derivatives as final component K. This type mainly possesses blue and green shades. Example: *C. I. Direct Blue 78*, 34200 (**16**) [2503-73-3].

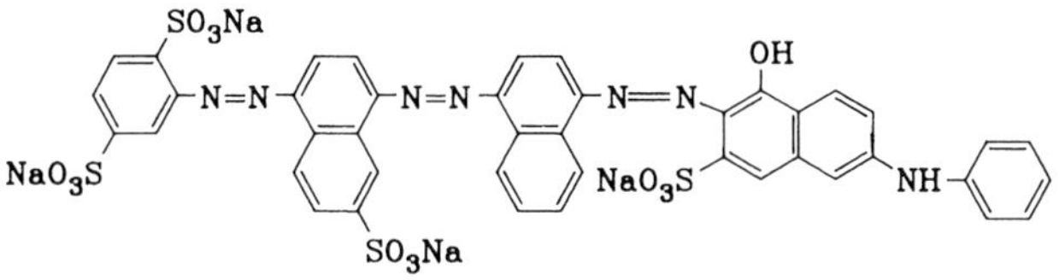

16

3.3.3.4 Tetrakisazo Dyes

Numerous syntheses are possible for these dyes, which all yield only dark shades. Here, too, only a few types are of industrial importance.

A further extension to the trisazo dyes on the principle of series coupling D→M$_1$→M$_2$→M$_3$→K offers no advantages, because the intermediate isolation that is frequently necessary leads to yield losses, and a chain extension is therefore ruled out on economic grounds. This is also reflected in the number of polyazo dyes listed in the Colour Index [5]. Although 78 tetrakisazo dyes with eleven different synthesis principles are listed, only 14 dyes with five and more azo groups are mentioned, two of which are specified with eight azo groups.

An example of type D→M$_1$→M$_2$→M$_3$→K is the brown *C. I. Direct Dye*, 35850 [8002-98-0] (sulfanilic acid→*m*-toluidine→Cleve acid 6→*m*-toluidine→1-nitrophenylene-2,4-diamine).

A number of important direct dyes are derived from the type K$_1$←M$_1$←D→M$_1$→K$_1$: *C. I. Direct Black 22*, 35435, (**17**) [6473-13-8] is produced

3.3 Direct Dyes

via the primary disazo dye from 4,4′-diaminodiphenylamine-2-sulfonic acid and two equivalents of γ acid. This dye is bis-diazotized and coupled to two equivalents of *m*-phenylenediamine.

Pattern:

4,4′-diaminodiphenylamine-2-sulfonic acid → γ acid ⟶ *m*-phenylenediamine
↘ γ acid ⟶ *m*-phenylenediamine

Another example is C. I. Direct Black 151, 35436 (**18**)

$$m\text{-phenylenediamine} \xleftarrow{3} \gamma\text{ acid} \xleftarrow[\text{alk}]{1} \text{DADPS} \xrightarrow[\text{alk}]{1} \gamma\text{ acid} \xrightarrow{2} \gamma\text{ acid}$$

Here, too, the disazo dye is initially prepared from bis-diazotized 4,4′-diaminodiphenylamine-2-sulfonic acid with 2 equivalents of γ acid. Coupling is then again carried out stepwise to γ acid and finally to *m*-phenylenediamine.

C. I. Direct Black 19, 35255 [6428-31-5] (**19**) is also an important dye. It is obtained by coupling two equivalents of *p*-nitraniline to H acid, sulfhydrate reduction of the two nitro groups, and coupling of the bis-diazotized intermediate compound with two equivalents of *m*-phenylenediamine.

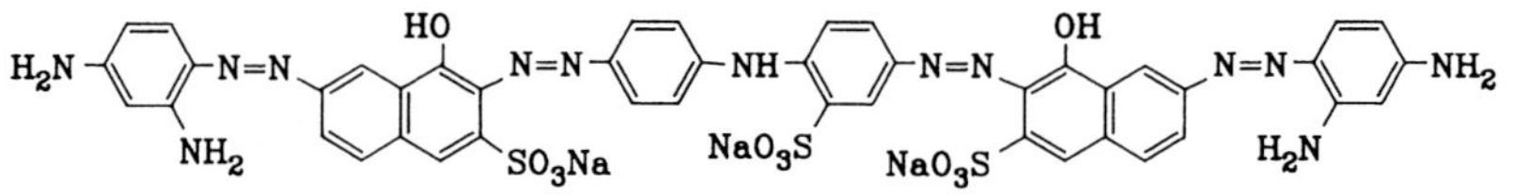

17

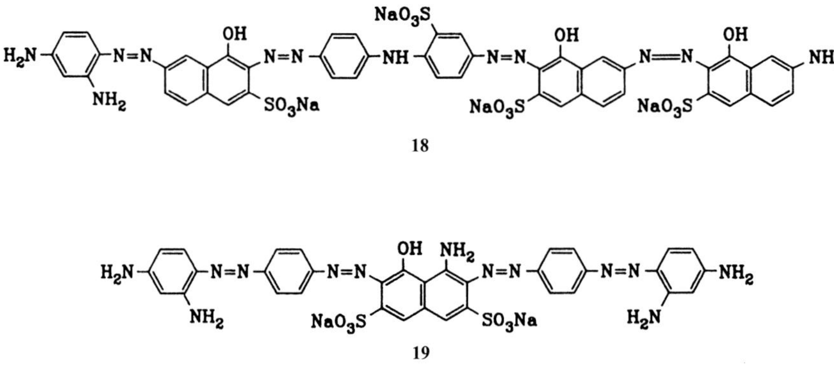

18

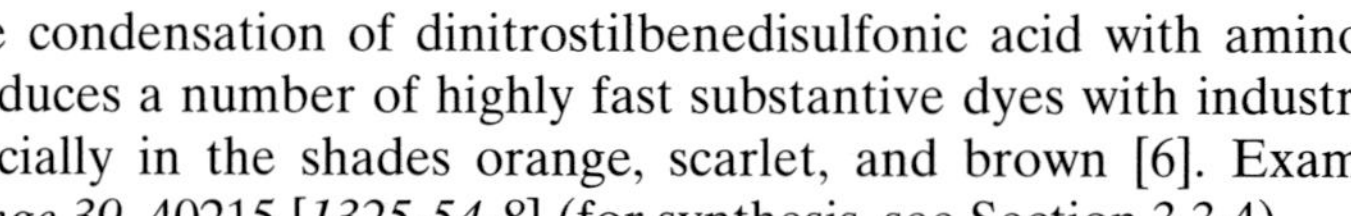

19

Condensation dyes are produced by condensation of nitro compounds with amines.

Alkaline condensation of dinitrostilbenedisulfonic acid with aminoazo compounds produces a number of highly fast substantive dyes with industrial importance, especially in the shades orange, scarlet, and brown [6]. Example: C. I. Direct Orange 39, 40215 [1325-54-8] (for synthesis, see Section 3.3.4).

Table 3.1 (3.3.6.5) shows further examples of condensation dyes obtained from the corresponding aminoazo compounds and 4,4′-dinitrostilbene- 2,2′-disulfonic acid.

Other examples of symmetrical direct dyes with a urea bridge are shown in Table 3.2 (3.3.6.6).

Direct Dyes with a Urea Bridge (see Section 3.3.4).

Triazinyl Dyes. The principle of linking two or three azo dyes by means of a triazine ring was established by Ciba patents [7]. The triazine bridge increases the substantivity of the dyes, in a similar manner to the $-NH-CO-NH-$ group (for synthesis see Section 3.3.4).

Copper Complexes of Substantive Azo Dyes. Under certain structural conditions, azo dyes are capable of forming metal complexes (see Sections 2.9, 3.11). A process for the complexing of o,o'- dihydroxyazo dyes with copper by reaction of these dyes with copper sulfate in a weakly acid solution was described for the first time in 1915 [8].

The introduction of copper into conventional substantive dyes often increases their light and wetfastness considerably, since complexing blocks the hydrophilic groups. Copper complexes of o,o'-disubstituted azo dyes are chiefly of interest for direct dyes. Substituents especially capable of complexing are hydroxyl, methoxy, carboxy, and carbomethoxy groups. BASF developed the oxidative copper treatment process in which on the basis of o-hydroxyazo dyes, copper complexes of the corresponding o,o'-dihydroxyazo dyes are obtained [9].

The possibility of first converting suitable azo compounds into copper complexes, and subsequently using them as coupling components for the synthesis of azo direct dyes, was also described.

As in the case of the metal-free direct dyes, a large diversity of structures is encountered; here, too, there are direct dyes with pyrazolones, substituted 4,4′-diphenylamines, or the 1,3,5-triazine ring, urea and stilbene derivatives, and condensation dyes.

For each dye it is necessary to establish the most favorable conditions under which copper-containing products with improved properties are obtained. Because none of the known complexing processes provides optimum results in all instances, the production of certain dyes or complexing variations is protected by a large number of patents.

Especially for ecological reasons, however, the importance of copper complex dyes is decreasing. This applies in particular to copper aftertreatment.

3.3.4 Synthesis

3.3.4.1 Monoazo Dyes

Diazotization of dehydrothio-*p*-toluidinesulfonic acid and coupling to acetoacetic acid *o*-anisidide results in the lightfast yellow C. I. Direct Yellow 27, 13950 [*10190-68-8*] (**20**).

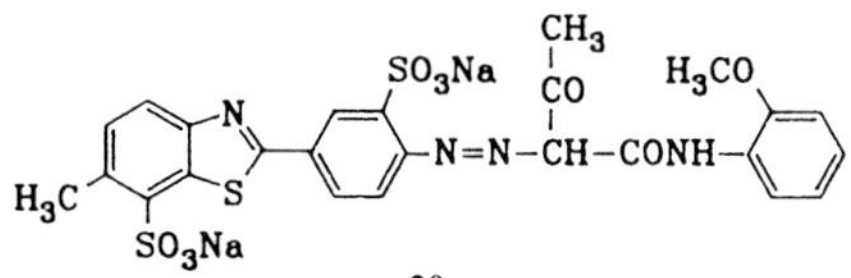

[Structure 20]

Coupling the diazotized monosulfonic acid of the primulin base to acetoacetic acid anilide produces the clear, greenish yellow C. I. Direct Yellow 22, 13925 [*10190-69-9*] (**21**).

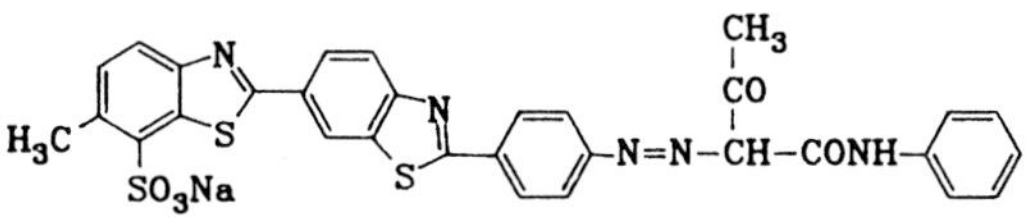

[Structure 21]

Substantive yellow monoazo dyes with very good light fastness are also obtained by oxidation of dehydrothio-*p*-toluidine and its derivatives. Oxidation of the sodium salt of the monosulfonic acid of the primulin base in the alkaline pH range with sodium hypochlorite results, for example, in C. I. Direct Yellow 29, 19556 [*6537-66-2*] (**22**).

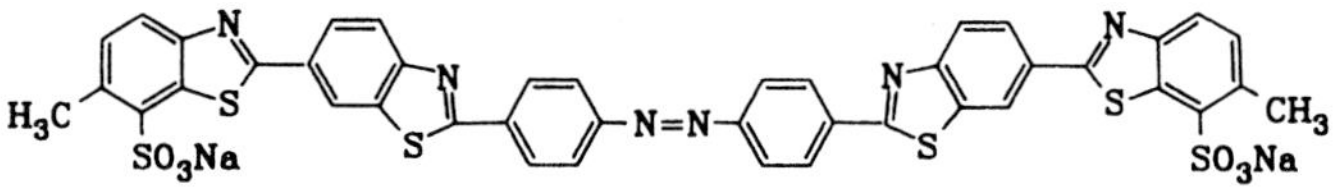

[Structure 22]

3.3.4.2 Disazo Dyes

Diamines used for the synthesis of symmetrical substantive disazo dyes are **23** and **24**.

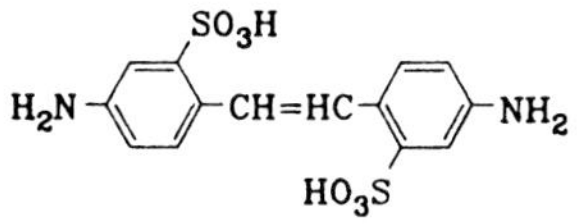

$$H_2N-\underset{}{\bigcirc}-CH=HC-\underset{HO_3S}{\overset{SO_3H}{\bigcirc}}-NH_2$$
23

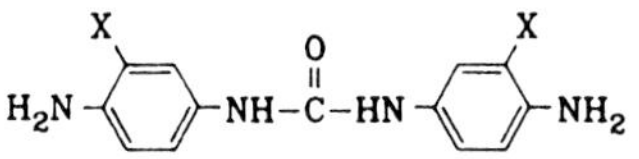

$$H_2N-\overset{X}{\underset{}{\bigcirc}}-NH-\overset{O}{\underset{}{C}}-HN-\overset{X}{\underset{}{\bigcirc}}-NH_2$$
24

X = COOH or SO₃H

The sodium salt of 4,4′-dinitrostilbene-2,2′-disulfonic acid itself belongs to the group of stilbene dyes. It is mainly used as *C. I. Direct Yellow 11*, 40000 [1325-37-7] for the coloration of paper. The product is obtained by coupling bis-diazotized 4,4′-diaminostilbene-2,2′-disulfonic acid with two equivalents of phenol and subsequent ethylation. It is a very high strength reddish yellow with very good leveling power and average fastness properties, which is excellently suited for the dyeing of fabric blends, e.g., cotton– polyamide.

The following abridged specification illustrates the synthesis of a disazo direct dye.

C. I. Direct Yellow 12, 24895 [2870-32-8]: 4,4′-Diaminostilbene-2,2′-disulfonic acid is bis-diazotized with an aqueous solution of sodium nitrite at 5 °C. Next, phenol is dissolved in water, and sodium hydroxide solution and soda are added. To this solution is added the above bis-diazotized solution. Then, more 30 % sodium hydroxide solution is added. On the following morning, the solution is heated to 70 °C, and after addition of 30 % hydrochloric acid, the dye is salted out with rock salt. The damp press cake, ethanol, soda, 30 % sodium hydroxide solution, and ethyl chloride are kept in a closed stirrer vessel for 24 h at 100 °C (5 – 5.5 bar). The mixture is then cooled to 70 °C and transferred at its own pressure to a distilling vessel, from which the ethanol is distilled. After being cooled to 80 °C, the mixture is suction filtered and dried at 100 °C.

An example of the use of **24** as diamine is *C. I. Direct Red 75*, 25380 [2829-43-8] (**25**).

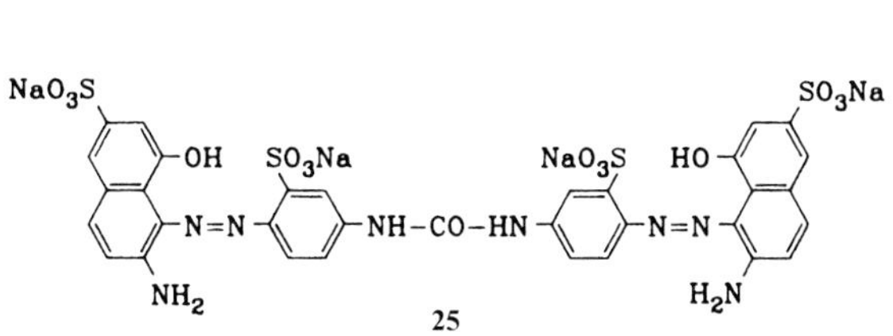

25

Other direct dyes with the diphenylurea group can be produced from aminomonoazo dyes by subsequent treatment with phosgene.

3.3.4.3 Trisazo Dyes

In the manufacture of trisazo dyes, good yield and purity during final coupling are often obtained only in the presence of pyridine or other bases as coupling accelerators [10]. Intermediate isolation and separation of impurities prior to continuation of coupling are also frequently necessary.

The use of pyridine during final coupling (pyridine coupling) is illustrated in the following abridged manufacturing specification for *C. I. Direct Green 33*, 34270: 2-aminonaphthalene-8-sulfonic acid→3,5-xylidine→2-ethoxy Cleve acid-6→*N*-acetyl H acid

Final Diazotization: The disazo dye is produced by the usual method aminocroceic acid→sym. *m*-xylidene→2-ethoxy-Cleve acid 6. Aminocroceic acid is stirred in water with 50 % acetic acid. Then sodium nitrite solution is added at 25 °C. After stirring overnight, a reddish brown solution is obtained. *Final coupling:* To H acid, dissolved in water, are added 40 % sodium hydroxide solution and soda; the mixture is heated to 70 °C and acetic acid anhydride is added. After acetylation is completed, the mixture is transferred to pure pyridine and ice. The diazotizing solution is added at 0 – 12 °C over 30 min. After a further hour, sodium hydrogencarbonate is added, and after 1 h the mixture is heated to 60 °C. Following the addition of rock salt, and maintaining 60 °C for 1 h, the dye is filtered and dried at 110 °C.

Condensation Dyes. *C. I. Direct Orange 39*, 40215 [*1325-54-8*] is obtained by condensation of 4-aminoazobenzene-4′-sulfonic acid with dinitrostilbenedisulfonic acid in aqueous sodium hydroxide solution. The main product formed is the tetrakisazo dye of the stilbene-2,2′-disulfonic acid, as well as the corresponding azoxy compounds. No uniform product is obtained, since part of the aminoazo compound is consumed as reducing agent. Condensation dyes are often purified by aftertreatment with reducing agents such as glucose or sodium sulfide.

Direct Dyes with a Urea Bridge. By the end of the 19th century, it was discovered at BASF that the reaction of aminoazo dyes with phosgene in aqueous solution in the presence of soda resulted in valuable symmetrical urea azo dyes [11]:

$$2 \ A-N=N-\!\!\!\left\langle\!\!\!\bigcirc\!\!\!\right\rangle\!\!\!-NH_2 + COCl_2 \longrightarrow$$

$$A-N=N-\!\!\!\left\langle\!\!\!\bigcirc\!\!\!\right\rangle\!\!\!-NH-CO-HN-\!\!\!\left\langle\!\!\!\bigcirc\!\!\!\right\rangle\!\!\!-N=N-A$$

A = azo dye radicals

The urea bridge can also be introduced into direct dyes with 4,4′-diaminodiphenylurea derivatives by means of bis-diazotization and coupling (see **25**).

C. I. Direct Yellow 50, 29025 [*3214-47-9*] (**26**) is obtained, for example, by phosgene treatment of two equivalents of the aminoazo dye prepared from diazotized 2-aminonaphthalene-4,8-disulfonic acid coupled to *m*-toluidine.

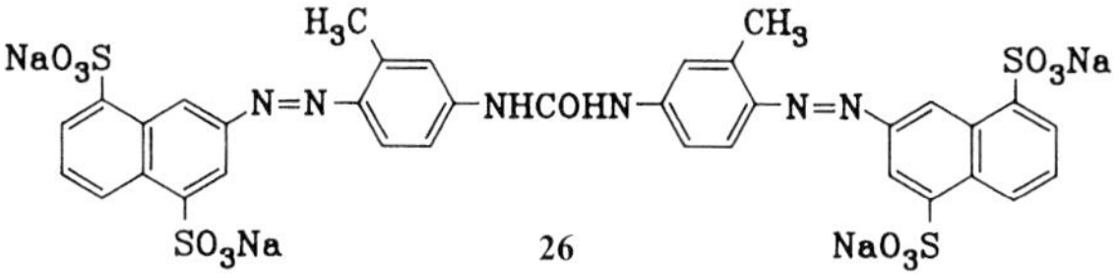

[Structure 26: symmetrical bisazo dye with urea bridge]

Other examples of symmetrical direct dyes with a urea bridge are shown in Table 3.2.

Aminodisazo and aminotrisazo dyes can also be treated with phosgene, largely red and brown shades being obtained.

Example: *C. I. Direct Red 80*, 35780 [*2610-10-8*], is rendered accessible by phosgene treatment of the aminodisazo dye **27**.

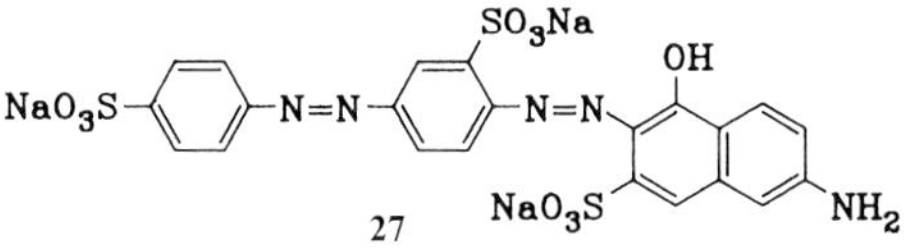

[Structure 27]

By mixing two different aminoazo compounds ANH_2 and BNH_2 and subsequent phosgene treatment, the asymmetrical diphenylurea direct dye **28** is obtained in addition to the two symmetrical members.

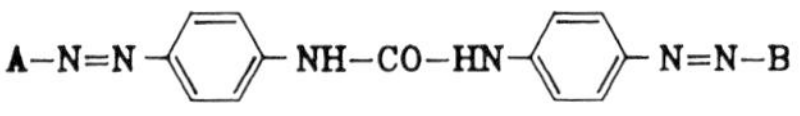

A—N=N—⟨ ⟩—NH—CO—HN—⟨ ⟩—N=N—B

28

There are a number of yellow substantive dyes synthesized by this principle. Example: *C. I. Direct Yellow 41*, 29005 [*8005-53-6*] (**29**) is obtained by phosgene treatment of an equimolar mixture of bases A (sulfanilic acid→cresidine) and B (*p*-nitroaniline→salicylic acid and reduction of the NO_2 group). The mixture additionally contains the two corresponding symmetrical dyes.

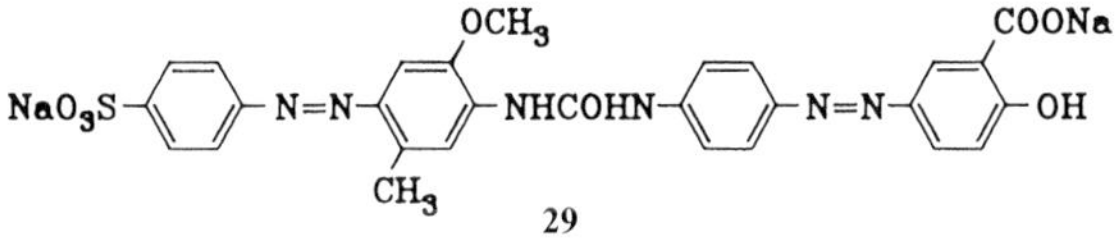

[Structure 29]

Triazinyl Dyes. The synthesis is based on cyanuric chloride, the three chlorine atoms of which can be replaced in stages by nucleophilic radicals under different conditions.

The procedure that can be adopted is for intermediate products with amino groups (e.g., H acid) to be condensed with cyanuric chloride and the reaction product subsequently attached to the diazo component, or aminoazo dyes are treated directly with cyanuric chloride. In the relevant commercial azo dyes, the third chlorine atom in the cyanuric chloride is usually treated with aniline or ammonia, and less frequently left unchanged.

Of interest here are dyes that yield green shades through combination of a blue and yellow component.

Example: *C. I. Direct Green 26*, 34045 [*6388-26-7*] (**30**).

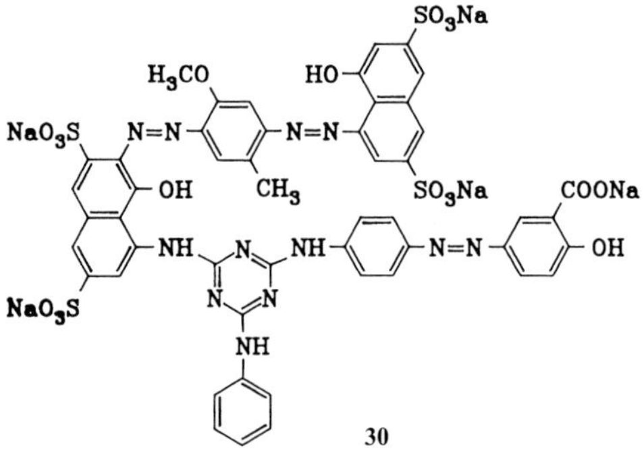

30

The blue component (H acid→cresidine→H acid) is linked with the triazine ring by means of the amino group of the H acid; after reduction of the nitro group, the yellow component (*p*-nitroaniline→salicylic acid) replaces the second chlorine atom of cyanuric chloride, and aniline the third chlorine atom. The dye constitutes a very pure, clear, bluish green with very good fastness properties.

3.3.4.4 Anthraquinone Direct Dyes

Compared to direct azo dyes, the direct anthraquinone dyes have lower tinctorial strengths and are therefore far less economical to use. They have lost most of their importance. Only a few special green dyes have retained their importance. Direct green cotton dyes can be produced by coupling a blue bromamine acid dye and a yellow azo dye via ureido or diaminotriazine bridges.

An example is compound **31**, *C.I. Direct Green 28*, 14155 [*6471-09-6*].

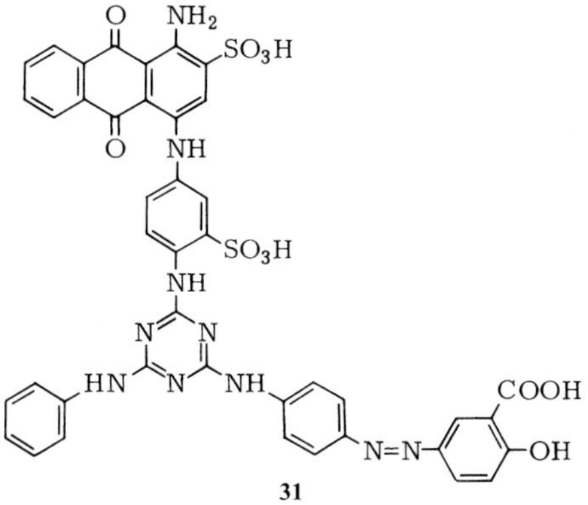

31

3.3.5 Direct Dyes with Aftertreatment

In spite of the reactive dyes with their outstanding wet-fastness, the direct dyes continue to hold a large share of the market for inexpensive cellulose and paper dyes. Apart from the new developments to replace the old established benzidine dyes, considerable efforts are being made to improve the fastness properties of direct dyes by aftertreatment. The various aftertreatment methods are described below in the order of their present importance. For supplementary literature see [12, 13]

3.3.5.1 Aftertreatment with Cationic Auxiliaries

Aftertreatment of substantive dyeings with cationic organic substances has lately begun to gain in importance. Improvements are obtained in particular in wetfastness properties, especially fastness to water, washing, and wet pressing, as well as fastness to perspiration and cross-dyeing.

The cationic compounds precipitate the (anionic) dyes from their aqueous solutions and on the fiber form higher molecular mass, sparingly soluble, saltlike compounds with the dyes. Removal of the latter from the fiber is thus made more difficult. A number of companies have compiled their own ranges of this type of aftertreatment agent.

Cationic textile auxiliaries can be divided into two groups: quaternary ammonium compounds and cationic formaldehyde condensation resins.

Quaternary Ammonium Compounds. These contain aliphatic and/or cycloaliphatic radicals (R^1–R^4 in Eq. 1), of which at least one is a long-chain alkyl radical (with more than five C atoms). The compounds react with the dyes in accordance with the following pattern:

$$R^1 - \overset{\overset{R^2}{|}}{\underset{\underset{R^4}{|}}{N^+}} - R^3 \; Cl^- + Na^+ \; \bar{O}_3S-\text{\textcircled{F}} \rightarrow R^1 - \overset{\overset{R^2}{|}}{\underset{\underset{R^4}{|}}{N^+}} - R^3 \; \bar{O}_3S-\text{\textcircled{F}} + NaCl \qquad (1)$$

$\text{\textcircled{F}}$ = dye radical

Use is made, for example, of cetyldimethylbenzylammonium chloride and pyridinium compounds of the type

R = long–chain alkyl radical
X = Cl, SO_4H, etc.

In addition, the quaternary ammonium compounds mostly possess strong affinity for the cellulosic fiber, whereby the insoluble reaction product is anchored even more firmly with the dye on the fiber. This high affinity can, however, also result in nonuniform improvements in fastness properties. This behavior also prevents the use of circulating liquor dyeing machines, because even with addition of aftertreatment agent in portions (e.g., in the case of wound packages) completely uniform aftertreatment can not be guaranteed.

The action of quaternary ammonium compounds on the dyed fiber in many cases produces a change of shade. A further disadvantage is the possible elimination of aliphatic amines, e.g., in an alkaline medium in the presence of anionic substances or at excessive drying temperatures. This can cause an unpleasant (fishy) odor to arise. In such cases, an acid aftertreatment is necessary.

Cationic Formaldehyde Condensation Resins. An appreciable improvement in wet-fastness properties is obtained by aftertreating the dyeings with polymeric condensation resins containing strongly basic groups. This leads to the formation of insoluble salts from resin cation and dye anion.

The most important resins of this type are obtained from dicyanodiamide (**32**) by condensation with formaldehyde and hydrochloric acid.

$$\underset{\substack{\| \\ H_2N-C-NH-CN}}{NH} \xrightarrow[H_2O,\ HCl]{HCHO} \underset{\substack{\| \\ HOH_2C-HN-C-NH-CO-NH_2}}{{}^+NH_2}\ Cl^-$$
$$\mathbf{32}$$

Suitable blending of ammonium salts of fairly strongly basic amines (e.g., melamine or guanidine) and acid condensation with formaldehyde enables favorable conditions to be obtained with regard to solubility and cross-linking of the cationic condensation resins.

The water-soluble resins are inexpensive, and compared with quaternary ammonium compounds relatively small quantities often suffice to achieve a marked improvement in wetfastness and fastness to perspiration. A disadvantage is a certain influence on lightfastness. This may be lowered by one to two steps on the eight-step blue scale when these resins are used on materials dyed by substantive dyes. Formaldehyde condensation resins are therefore used especially for articles in which wet fastness and fastness to perspiration are important, but light fastness is less crucial, e.g., lining fabrics.

3.3.5.2 Aftertreatment with Formaldehyde

The wet-fastness properties of substantive azo dyes containing free amino or hydroxyl groups as terminal groups (final component of coupling, e.g., resorcinol or *m*-phenylenediamine) can be improved by aftertreating the dyed fiber material with aqueous formaldehyde solution. The free amino groups can also be generated by subsequent reduction of nitro groups or hydrolysis of acylamino groups. During reaction with formaldehyde, methylene bridges are formed between two dye molecules; i.e., an enlargement of the molecule occurs and hence an improvement in wetfastness, whereas lightfastness is not influenced.

Because of a lack of storage stability in dyeings treated by this method, formaldehyde aftertreatment is mainly confined to dark shades (brown and black). Although the basic, simple process has remained virtually unchanged, its importance has declined sharply for ecological reasons.

3.3.5.3 Diazotization Dyes

Direct dyes containing one or more diazotizable amino groups in the dye molecule permit further diazotization on the fiber and subsequent coupling with a "developer." β-Naphthol can be used as a developer for orange, red, brown, blue, and black shades, and 1,3-phenylenediamine and 2,4-diaminotoluene for brown, gray, and black shades.

The shade of the dye applied to the material usually changes considerably as a result of this aftertreatment. Wetfastness and fastness to perspiration are greatly improved, but lightfastness remains unchanged. The amino group needed for diazotization is introduced by final coupling with an amine, an aminonaphtholsulfonic acid, or an aminoacylaminonaphtholsulfonic acid, or by reduction of a nitro or hydrolysis of an acylamino group. "Development" results in enlargement of the dye molecule without further solubilizing groups being added. The time-consuming process necessitates careful handling. Exposure to sunlight or partial drying of the diazotized dyeing should thus be avoided. Diazotization dyes now enjoy only minor importance.

Example: *C. I. Direct Red 145*, 17805 [*6771-94-4*] (**33**), yields a clear yellowish red when developed with β-naphthol.

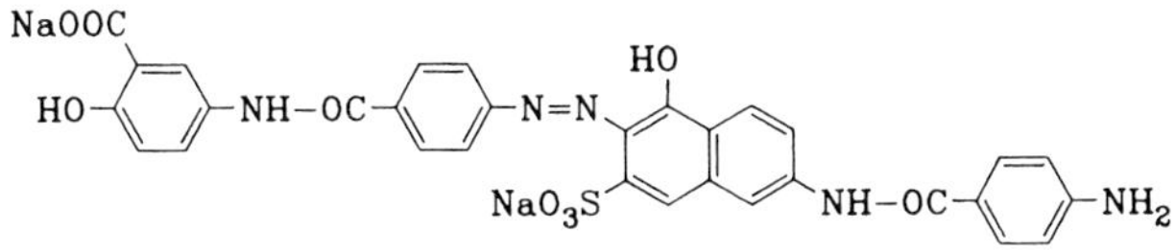

3.3.5.4 Aftertreatment with Metal Salts

Metal complexation results in blocking of solubilizing groups; aggregation increases, the dye becomes less soluble, and an improvement in wetfastness occurs. Although the lightfastness is generally also improved by this aftertreatment, fastness of the dyeings to perspiration is frequently reduced. The shades generally become duller and flatter. Catalytic action of traces of heavy-metal ions that are present can cause fiber tendering to occur in the presence of detergents containing oxidizing agents (peroxides).

3.3.6 Examples of Commercially Available Dyes

3.3.6.1 Monoazo Dyes

C.I. Direct Yellow 22, 13925 [*10190-69-9*] (**21**)
C.I. Direct Yellow 27, 13950 [*10190-68-8*] (**20**)
C.I. Direct Yellow 28, 19555 [*8005-72-9*]
C.I. Direct Yellow 29, 19556 [*6537-66-2*] (**22**)
C.I, Direct Yellow 137 [*71838-47-6*]
C.I. Direct Yellow 147 [*35294-62-3*]

3.3.6.2 Disazo Dyes

Primary Disazo Dyes
  C.I. Direct Yellow 11, 40000 [1325-37-7]
  C.I. Direct Yellow 12, 24895, [2870-32-8] (Chrysophenine, **34**)

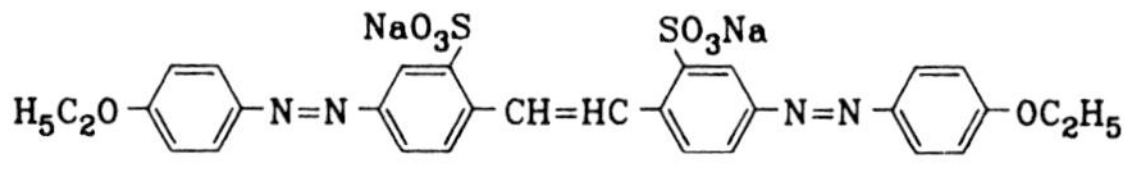

$$H_5C_2O-\langle\rangle-N{=}N-\underset{\underset{\textbf{34}}{}}{\langle\rangle}\overset{NaO_3S}{}-CH{=}HC-\overset{SO_3Na}{\langle\rangle}-N{=}N-\langle\rangle-OC_2H_5$$

  C.I. Direct Yellow 51, 29030 [6420-29-7]
  C.I. Direct Red 75, 25380 [2829-43-8] (**25**)
  C.I. Direct Red 239 [60202-36-6]
  C.I. Direct Orange 26, 29150 [3626-36-6] (**8**)
  C.I. Direct Orange 102, 29156 [6598-63-6]
  C.I. Direct Red 23, 29160 [3441-14-3] (**10**)

Although **10** only possesses average fastness properties, it enjoys great importance and, because of its good leveling power, is used for dyeing union fabrics, polyamide, and chrome leather.
  C.I. Direct Red 250, 29168 (**9**)

Secondary Disazo Dyes
  C.I. Direct Red 81, 28160 [2610-11-9] (**12**)
  C.I. Direct Red 253 [142985-51-1]
  C.I. Direct Blue 71, 34140 [4399-55-7]
  C.I. Violet 51, 27905 [5489-77-0]
  C.I. Direct Black 51, 27720 (**13**) [3442-21-5]

3.3.6.3 Trisazo Dyes

  C.I. Direct Black 166, 30026 [75131-19-8] (**14**)
  C.I. Direct Black 150, 32010 [6897-38-7] (**15**)
  C.I. Direct Blue 78, 34200 [2503-73-3] (**16**)

3.3.6.4 Tetrakisazo Dyes

  C.I. Direct Black 22, 35435, [6473-13-8] (**17**)
  C.I. Direct Black 151, 35436 (**18**)
  C.I. Direct Black 19, 35255 [6428-31-5] (**19**)

3.3.6.5 Condensation Dyes

Table 3.3 Examples of condensation dyes made from aminoazo compounds and 4,4'-dinitro-stilbene-2,2'-disulfonic acid

Aminoazo compound	Condensation dye	Aftertreatment with
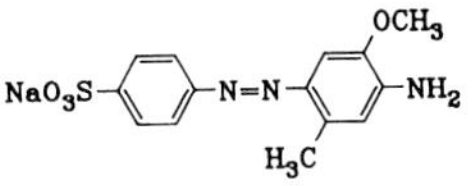 NaO₃S–⟨⟩–N=N–⟨OCH₃, H₃C⟩–NH₂	C.I. Direct Orange 40, 40265 [1325-62-8]	Na₂S
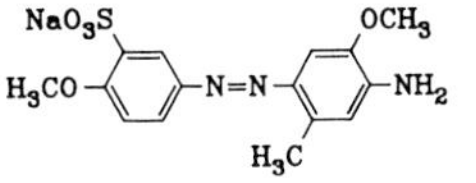 NaO₃S, H₃CO–⟨⟩–N=N–⟨OCH₃, H₃C⟩–NH₂	C.I. Direct Red 76, 40270 [1325-63-9]	glucose
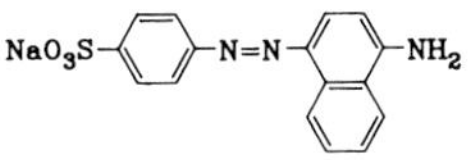 NaO₃S–⟨⟩–N=N–⟨naphthyl⟩–NH₂	C.I. Direct Brown 78, 40290 [1325-65-1]	–

3.3.6.6 Direct Dyes with a Urea Bridge

C. I. Direct Yellow 50, 29025 [*3214-47-9*] (**26**)
C. I. Direct Red 80, 35780 [*2610-10-8*]
C. I. Direct Yellow 41, 29005 [*8005-53-6*](**29**)

Table 3.4 Aminoazo compounds for symmetrical direct dyes with a urea bridge

Aminoazo dye	Direct dye
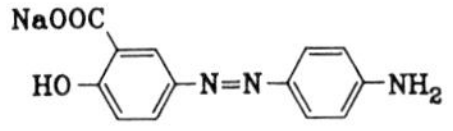 NaOOC, HO–⟨⟩–N=N–⟨⟩–NH₂	C.I. Direct Yellow 26, 25300 [*2829-42-7*]
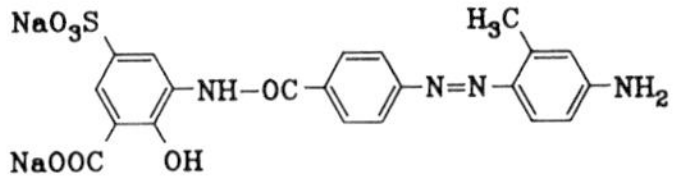 NaO₃S, NaOOC, OH–⟨⟩–NH–OC–⟨⟩–N=N–⟨H₃C⟩–NH₂	C.I. Direct Yellow 33, 29020 [*6420-28-6*]
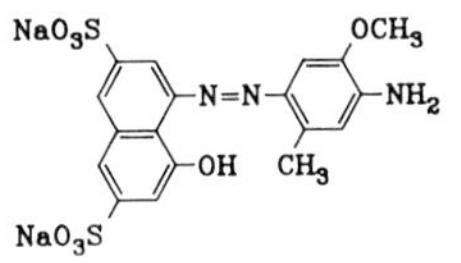 NaO₃S, OH, NaO₃S–⟨naphthyl⟩–N=N–⟨OCH₃, CH₃⟩–NH₂	C.I. Direct Red 79, 29065 [*1937-34-4*]

3.3.6.7 Triazinyl Dyes

C. I. Direct Green 26, 34045 [*6388-26-7*] (**30**).

3.3.7 References

[1] J. Boulton, *J. Soc. Dyers Colour.* **67** (1951) 522.
[2] I. D. Rattee, M. M. Breuer: *The Physical Chemistry of Dye Absorption*, Academic Press, London – New York 1974.
[3] H. Bach, E. Pfeil, W. Phillippar, M. Reich, *Angew. Chem.* **75** (1963) 407.
[4] E. Schirm, *J. Prakt. Chem.* **144** (1935) no. 2, 69.
[5] *Colour Index*, Society of Dyers and Colourists and American Association of Textile Chemists and Colorists, 3rd ed., Bradford 1971, First Revision 1975, Second Revision 1982, vol. 4, p. 4325, 4th ed. Online, 2002: www.colour-index.org.
[6] Cassella, DE 204212, 1907 (A. Greßly).
[7] Ciba, DE 436179, 1923.
[8] Ciba, DE 335809, 1915.
[9] BASF, DE 807289, 1948 (H. Pfitzner, H. Merkel); H. Pfitzner, H. Baumann, *Angew. Chem.* **70** (1958) 232.
[10] H. Zollinger, *Helv. Chim. Acta* **38** (1955) 1597, 1623; R. Pütter, *Angew. Chem.* **63** (1951) 186.
[11] BASF, DE 46737, 1888; DE 47902, 1889 (C. L. Müller).
[12] J. Offenbach: *Deutscher Färbekalender*, Eder Verlag, Stuttgart 1972, pp. 185 – 194.
[13] L. Diserens: *Die neuesten Fortschritte in der Anwendung der Farbstoffe*, Birkhäuser, Basel, vol. 1, 3rd ed., 1951; vol. 2, 2nd ed., 1949.

3.4 Anthraquinone Dyes

3.4.1 Introduction

The introduction of auxochromes into the almost colorless anthraquinone permits the tailoring of compounds to cover nearly all shades of dye colors: types and positions of the substituents in the molecule determine the hue. As a rule the bathochromic shift in simple anthraquinones increases with increasing basicity of the substituents. This generalization is clearly shown by the wavelengths of the longest wavelength absorption maximum of anthraquinones monosubstituted in the 1-position [1] (Table 3.5):

Table 3.5 Absorption maxima of 1-substituted anthraquinones

Substituent	λ, nm	Substituent	λ, nm	Substituent	λ, nm
H	327	$NHCOCH_3$	410	$N(CH_3)_2$	504
Cl	337	$NHCOC_6H_5$	415	$NHCH_3$	508
OCH_3	380	SCH_3	438	NHC_6H_5	508
OH	405	NH_2	465		

The wavelengths of the absorption maxima of hydroxy- and aminoanthraquinones are a function of position and number of OH and NH_2 groups [1] (Table 3.6):

Table 3.6 Absorption maxima of OH- and NH_2-substituted anthraquinones

Position	2	1	1,2	1,5	1,8	1,4
OH; λ, nm	365	405	416	428	430	476
NH_2; λ, nm	410	465	480	480	492	550

The large shift from the 1,8- to the 1,4-disubstituted anthraquinones, which even exceeds that between the α- and the β-substituted compounds, is remarkable. For α-substituents, additional changes in shade and color fastness occur on formation of hydrogen bonds to the neighboring carbonyl group.

Alteration of color in isomeric compounds can be demonstrated by comparing various diaminohydroxyanthraquinones:

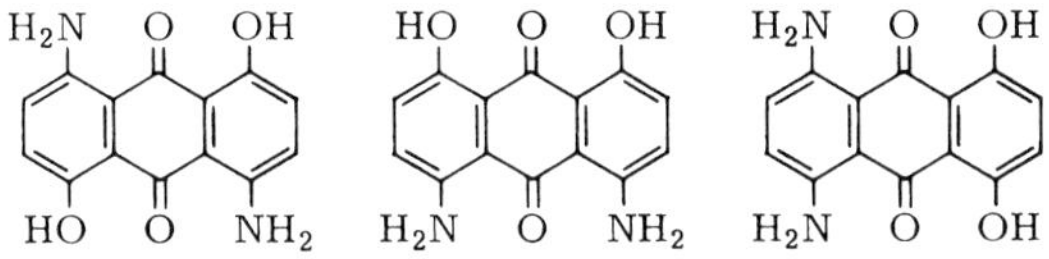

Neutral Blue — Slightly Greenish Blue — Greenish Blue

The effect of β-substituents on the hue is best demonstrated by the 1,4-diaminoanthraquinones:

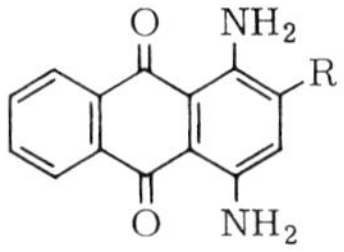

R = OCH_3, Red Violet
= OC_6H_5, Slightly Red Violet
= H, Violet
= $SO_2OC_6H_5$, Red Blue
= $COOCH_3$, Slightly Red Blue
= NO_2, Blue Green

Furthermore, the shade may be modified strongly by ring closure and condensation reactions:

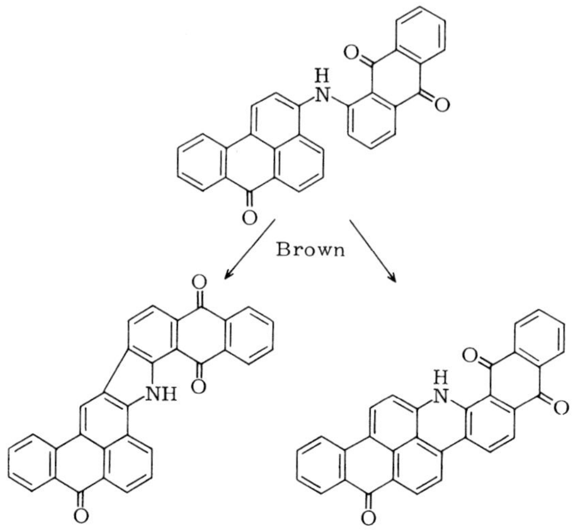

Brown

Yellow Brown   Olive Green

3.4.2 Chemical Constitution and Properties

Not every colored anthraquinone derivative is a dye. In addition to tinctorial strength and depending on the intended use, such other properties as affinity to the fiber and resistance to atmospheric conditions must be present also. The choice of a particular textile dye is determined also by its application properties and its fastness.

The three major dye types, i.e., neutral, anionic, and cationic, are subdivided by use: the *neutral dyes* comprise disperse, vat, and solvent dyes; the *anionic dyes*, acid, direct, and reactive dyes; and the *cationic dyes*, dyes for polyacrylonitrile fibers and other fibers modified to contain acid groups.

3.4.2.1 Disperse Dyes

The disperse dyes are water-insoluble, colloidally dispersed materials that are used to dye synthetics, including polyester, polyamide, and acetate fibers having 2.5 or 3 acetate groups per glucose subunit (see Section 3.2). Disperse dyes are used for dyeing polyacrylonitrile fibers when good leveling of light shades is required. Although disperse dyes were developed many decades ago for coloring cellulose acetate fibers, the most important current application is to dye polyester fibers.

The simple anthraquinone dyes, containing hydroxy or amino functions as their major auxochromes, are used for brilliant red to blue shades. Yellow and orange shades are obtained by other dye classes. Choosing appropriate substituents allows the best dyes for various fibers and colors to be obtained.

Dyes for Polyester Fibers. Anthraquinone dyes for polyester fibers can be classified into five basic types:

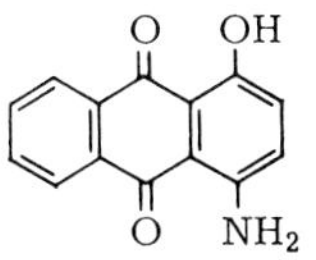

3.4.2 Chemical Constitution and Properties

Yellow Red to Rubine

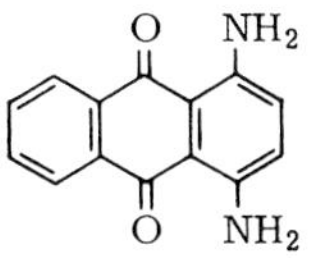

Violet to Blue Green

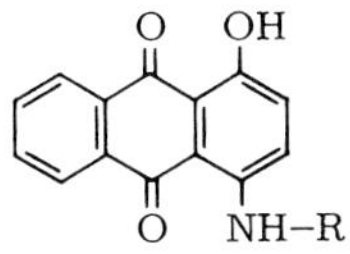

Red Blue to Blue

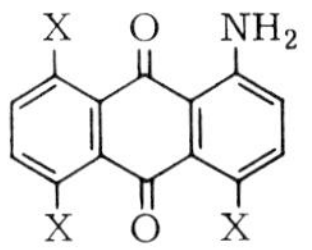

1 X = NH$_2$, 2 X = OH
Blue to Green Blue

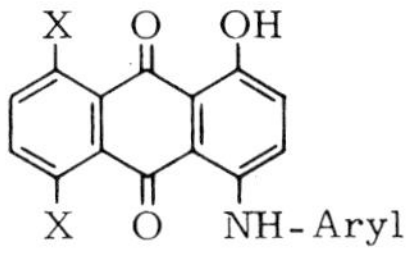

1 X = OH, 1 X = NO$_2$
Green Blue

The most important factors in the selection of anthraquinone dyes are affinity, lightfastness, and resistance to sublimation. Large substituents in the side chain tend to improve the sublimation fastness considerably. On the other hand, increased sublimation resistance generally decreases the affinity characteristics of the dye. The affinity is greatly affected by the position and the hydrophilic characteristics of the substituents. Mixtures of appropriate dyes or the presence of contaminants formed during synthesis may increase affinity by synergistic action. Introduction of negatively charged substituents, for instance, carboxylic esters, halogen, or sulfone groups, may improve lightfastness.

1-Amino-4-hydroxyanthraquinones possess good lightfastness and affinity for polyester fibers. They are bright, red dyes, whose brilliance may be improved significantly by introduction of ether groups ortho to the amino groups. The aliphatic ethers surpass the aromatic ethers with respect to lightfastness and are much more yellow and somewhat brighter. Additional substituents in the side chains may improve sublimation resistance. Isomeric compounds with alkyl or aryl ether groups in addition to hydroxyl entities are of little commercial interest because their shades are more blue and duller.

Examples are **1** and **2**, both bright yellow reds [2, 3]; **3**, bright blue red (R = H: *C.I. Disperse Red 60*, C.I. 60756 [*17418-58-5*]); for **3b** see [4]; for **3c** see [5].

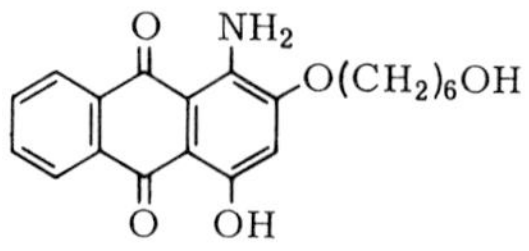

3.4 Anthraquinone Dyes

1 Bright Yellow Red

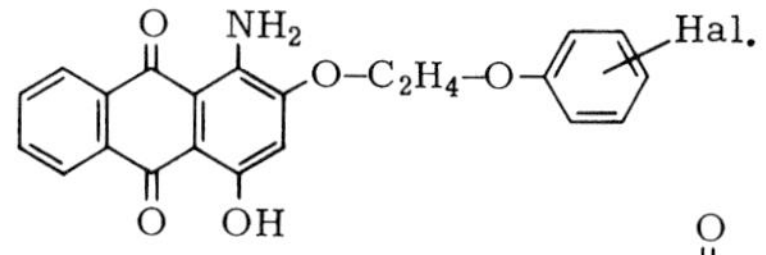

2 Bright Yellow Red

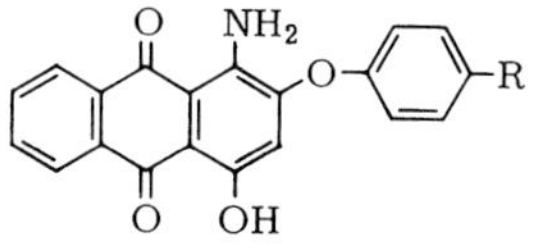

3 Bright Blue Red

a R = H

b R = CH$_2$-N(cycloheptanone)

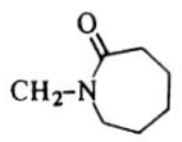

c R = O-SO$_2$-C$_6$H$_4$-CH$_3$

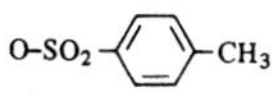

The poor lightfastness of *1,4-diaminoanthraquinone* can be improved by appropriate negatively charged substitutents. The 2-sulfophenyl ester is a sublimation-resistant, brilliant blue dye with a reddish cast and good lightfastness. The introduction of a nitro group into the 2-position shifts the shade into bluish green and improves resistance to sublimation and fading. Introduction of chlorine atoms into the β position considerably improves lightfastness with little affect on the basic sublimation characteristics.

β-Phenoxy groups in the 2,3-positions shift the shade to a bright, somewhat reddish violet with good stability to fading and sublimation. A very bright turquoise color with excellent lightfastness and good resistance to sublimation are properties of the 2,3-dicarboximides.

Examples are **4a**, *C.I. Disperse Violet 28*, 61102 [*81-42-5*]; **4b**, a red [6, p. 48]; **5a**, blue-green [7]; **5b**, brilliant red-blue [8]. Compounds **6** and **7** are both turquoise blue dyes [9], [10].

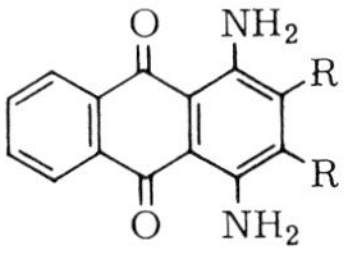

3.4.2 Chemical Constitution and Properties

4
- **a** R = Cl Disperse Violet
- **b** R = O—⟨Ph⟩ Red Violet

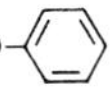

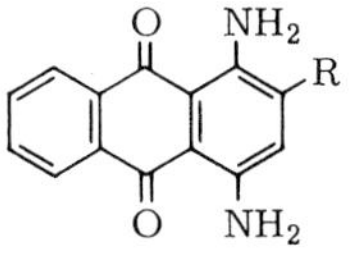

5
- **a** R = NO₂ Blue Green
- **b** R = SO₂—O—⟨Ph⟩ Brilliant Red Blue

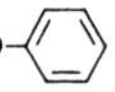

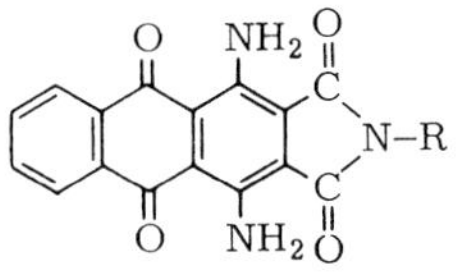

6 Turquoise Blue
R = H, Alkyl, Hydroxyalkyl

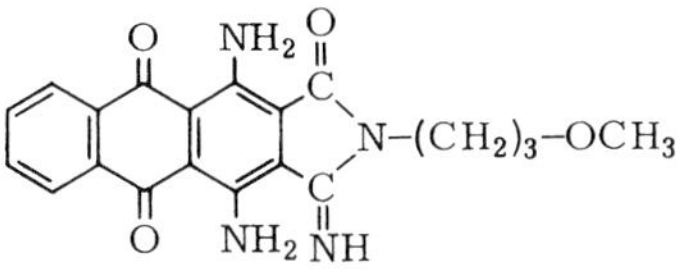

7 Turquoise Blue

N-Alkyl- or N-arylaminohydroxyanthraquinones, which show good affinity and lightfastness and give violet to blue shades, generally do not satisfy the requirements with respect to sublimation fastness. Compared to the dyes of the tetrasubstituted series, they are of lower tinctorial strength. Such substituents as carboxylic esters, arylsulfonic esters, amides, hydroxyethylether, and methoxy groups in arylamino compounds improve sublimation resistance, whereas optimized blends prevent lowering of affinity.

Examples are **8a**, C.I. Disperse Violet 27, 60724 [19286-75-0]; **8b**, C.I. Disperse Blue 72, 60725 [81-48-1], and **8c**, a violet dye [11].

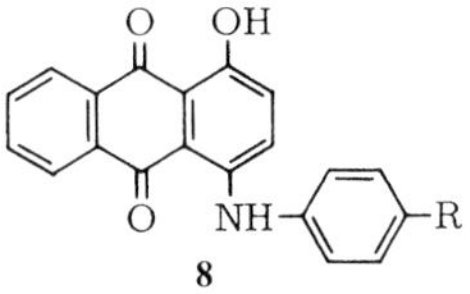

8
- **a** R = H C.I. Disperse Blue 27
- **b** R = CH₃ C.I. Disperse Blue 72
- **c** R = O—SO₂CH₃ Violet

Derivatives of α-*diaminodihydroxyanthraquinones* are the most important disperse dyes with respect to shade and affinity. The dye properties may be optimized by introducing suitable substituents, selecting the positions of the isomers, and blending. Some of these properties can be illustrated with the three basic structures **9–11**.

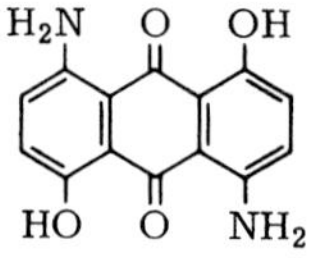

3.4 Anthraquinone Dyes

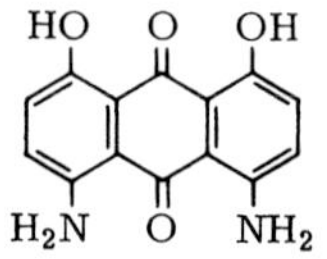

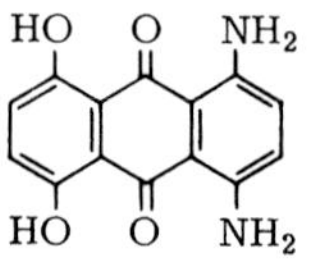

[Structures 9, 10, 11: diaminodihydroxyanthraquinone isomers]

The bathochromic shift from **9** to **11** affects the color; the lightfastness increases in the order of **10** to **11** to **9**; and the affinity increases from **9** to **11** to **10**. The sublimation fastness of all three is moderate.

Halo, alkoxy, hydroxyaryl, and phenylmercapto derivatives have attained commercial importance. Compared to substitution next to the hydroxy group, substitution next to the amino group leads to brighter dyes and improved affinity.

Examples are **12**, C.I. Disperse Blue 56, 63285 [12217-79-7] and **13**, C.I. Disperse Blue 73, 63265 [12222-75-2].

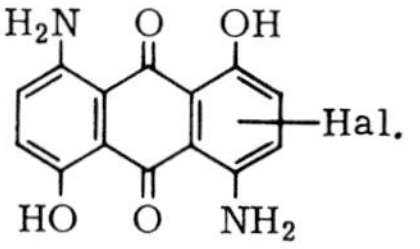

12 C.I. Disperse Blue 56

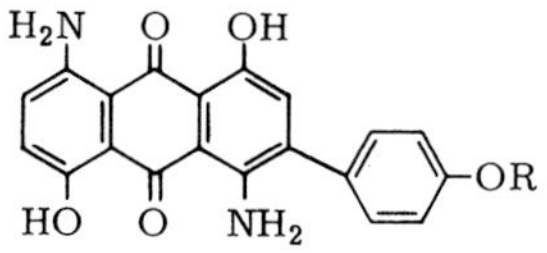

13 C.I. Disperse Blue 73

R = H, CH$_3$

Nitroarylaminodihydroxyanthraquinones are valued because of their greenish-blue shade and their good fastness to light and sublimation. Reduction improves their affinity somewhat but decreases lightfastness.

Example: **14** is a blue dye of this kind [12].

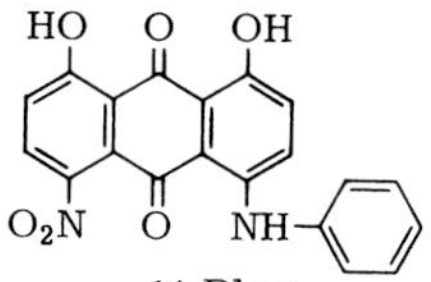

14 Blue

Dyes for Cellulose Ester and Synthetic Polyamide Fibers. The first disperse dyes were developed for dyeing cellulose fibers, but the importance of these diminished considerably when other synthetic fibers appeared on the market. Synthetic polyamide fibers could be dyed with dyes used for acetate fibers; very few new dyes had to be developed specifically for polyamide fibers.

The basic type of dye closely resembles that used for dyeing polyester fibers, but the selection of compounds is based on other criteria. Requirements regarding sublimation resistance are not as stringent, whereas fastness to ozone, exhaust gases, and washing are important. Substitution by amino, and especially by alkylamino, groups tends to decrease lightfastness of the dyes in polyester fibers. This is not the case for acetate and polyamide fibers.

The orange derivatives of 1-aminoanthraquinone are of little importance in this context because of their low tinctorial strength. This contrasts with the

3.4.2 Chemical Constitution and Properties 185

1-amino-4-hydroxyanthraquinone derivatives, which provide brilliant red dyes. The most important dyes are derived from 1,4-diaminoanthraquinone, with shades ranging from violet to greenish blue. Affinity may be enhanced vastly by blending similar compounds.

Examples of this class of dyes are **15**, a brilliant red dye [13]; **16**, *C.I. Disperse Blue 14*, 61500 [*2475-44-7*] [6, p. 54]; **17**, a brillant blue dye [14]; **18**, *C.I. Disperse Blue 31*, 64505 [*1328-23-0*] [6, p. 53]; **19**, *C.I. Disperse Blue 7*, 62500 [*3179-90-6*] [15, p. 201].

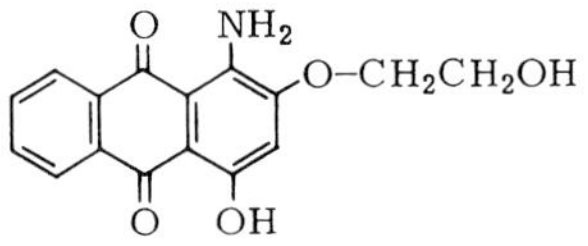

15 Brilliant Red

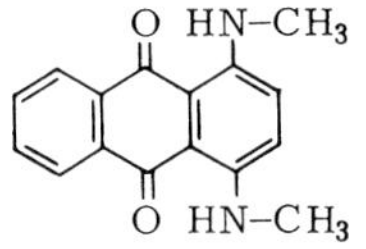

16 C.I. Disperse Blue 14 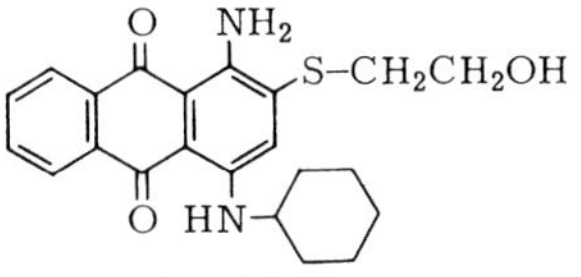   **17 Brilliant Blue**

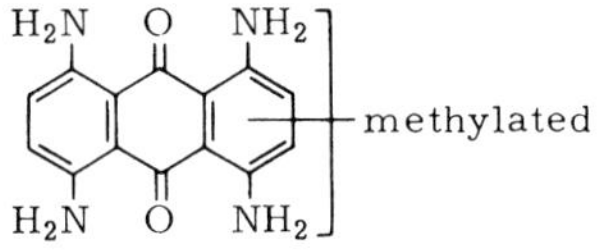

18 C.I. Disperse Blue 31

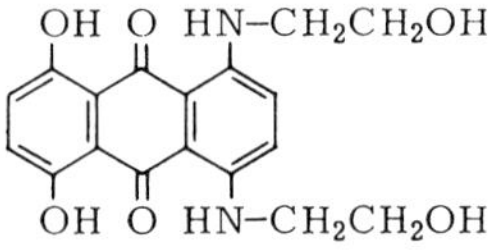

19 C.I. Disperse Blue 7

Transfer Dyes. In transfer printing the dye is supplied in the form of a coating on transfer paper. The fabric is pressed closely against the paper, and the dye is sublimed at ca. 200 °C and diffuses into fibers. This process is used primarily for printing on polyester fabrics. Originally it relied on available disperse dyes with good sublimation characteristics. New dyes, specifically developed for this process, have appeared on the market recently (see also [16, vol. VIII, pp. 191 ff.]).

Examples are **20**, *C.I. Disperse Red 60*, 60756 [*17418-58-5*], and **21**, *C.I. Disperse Blue 26*, 63305 [*3860-63-7*].

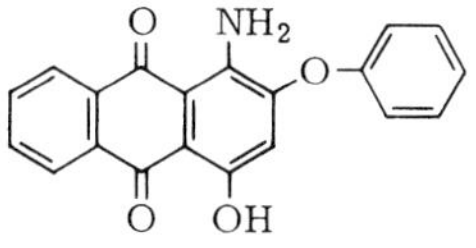

20 Disperse Red 60 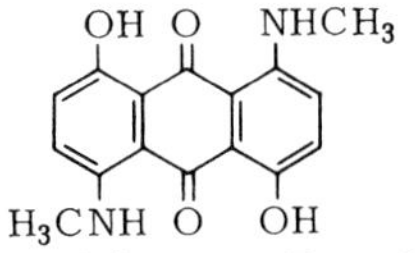   **21 Disperse Blue 26**

Dyes for Cotton–Polyester Fabrics. Anthraquinone dyes of medium molecular mass are suitable for direct printing and dyeing of cellulose fibers, especially cot-

3.4 Anthraquinone Dyes

ton – polyester fabrics pretreated with water. Most of these dyes are classified as disperse dyes having excellent resistance to sublimation, but vat dyes of low molecular mass are included also. The dyes are applied generally together with higher boiling, water-miscible solvents (glycol and glycol derivatives [17] or boric acid esters of species with one to six hydroxy groups [18]) to fabrics preswollen with water. Heat treatment at ≈ 200 °C evaporates the water, and the dye enters the fiber via its solution phase. The polyester component of the fabric is dyed simultaneously.

Examples: Compounds **22** [17] and **23** [19] are blue dyes; **24** [20] is green.

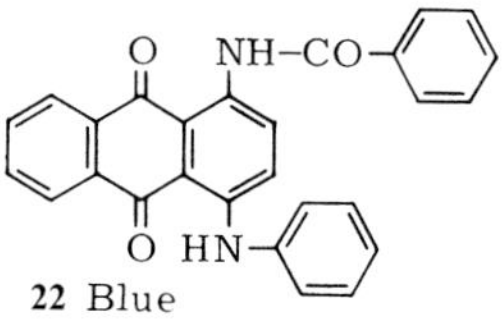

22 Blue

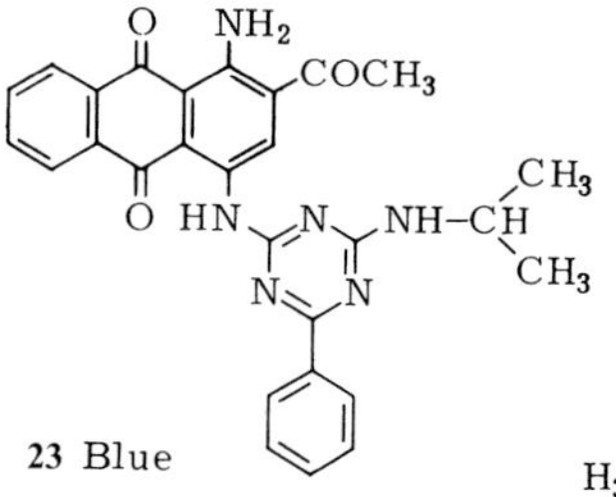

23 Blue

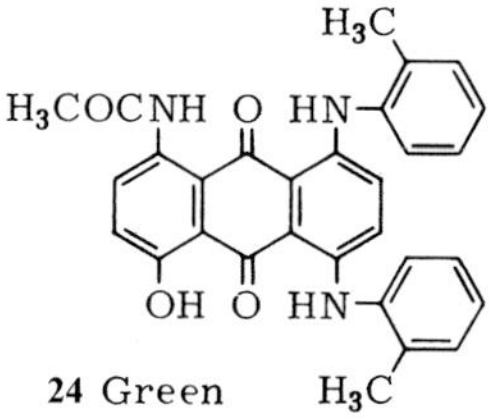

24 Green

Reactive disperse dyes also make it possible to dye cotton–polyester fabrics in level shades. Recently fluorotriazine reactive dyes have been proposed for this application [21–23].

Examples: Compound **25** [21] is a blue, and **26** [22] a red dye.

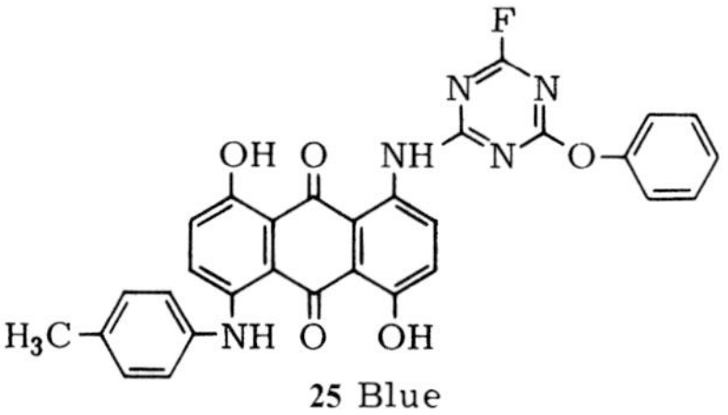

25 Blue

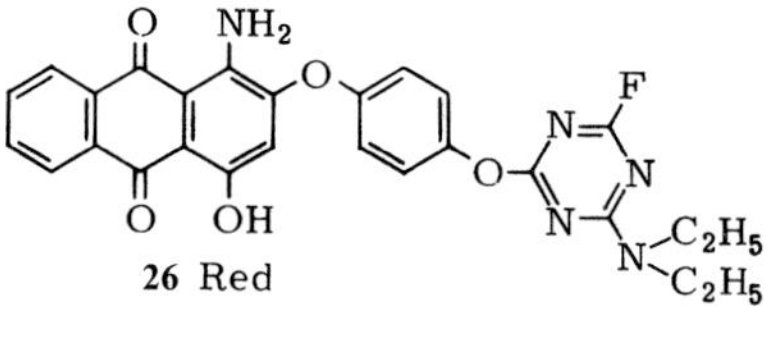

26 Red

Dyes Soluble in Organic Solvents. Anthraquinone dyes soluble in organic solvents are of relatively simple structure. They are used for coloring gasoline, oil, and plastics. For example, the highly soluble bis(alkylamino)anthraquinones are suitable additives for gasoline, whereas quinizarin and its derivatives are used for marking heating oils (see Section 3.10).

3.4.2.2 Vat Dyes

Vat dyes have been used for many decades to color cotton and other cellulose fibers. Despite their high cost and their muted colors, these dyes are extremely important for certain textiles because of their superior fastness. Very few new vat dyes have been developed over the past few years.

Water-insoluble vat dyes are converted to soluble "anthrahydroquinones" by reducing agents, such as sodium dithionite (hydrosulfite) in the presence of sodium hydroxide. The sodium salts of these mostly deep colored leuco compounds penetrate cellulose fibers. The insoluble dye is attached firmly to the fiber after reoxidation (see Section 4.5). Representative of a special form are the water-soluble sulfuric acid esters of the "anthrahydroquinone" compounds, namely, the leuco esters of vat dyes).

On the basis of their chemical constitutions the anthraquinoid vat dyes may be classified in the following major groups: acylaminoanthraquinones, anthraquinoneazoles, anthrimides and other linked anthraquinones, anthrimidocarbazoles, phthaloylacridones, benzanthrone dyes, indanthrones, and other polycondensed ring systems.

Acylaminoanthraquinones. Acylation of aminoanthraquinones with benzoic acid or benzoyl chloride, for example, affords vat dyes with satisfactory affinity to cellulose fibers. The simplest dyes of this type, 1,4- and 1,5-dibenzoylaminoanthraquinones, are no longer important. Bridging linkages, such as the dicarboxylic acids oxalic or phthalic acid, permit coupling of two anthraquinone units; three aminoanthraquinones may be combined by use of a triazine, such as cyanuric chloride.

Acylation of 1-amino- or 1,5-diaminoanthraquinones yields yellow vat dyes and affords red to ruby colored dyes when 1,4-diaminoanthraquinones are used. Use of 4,8-diamino-1,5-dihydroxy-anthraquinones gives violet to blue dyes. The relatively low lightfastness of the yellow acylaminoanthraquinones may be improved greatly by using azodiphenyl-4,4'-dicarboxylic acid. Acylaminoanthraquinones are relatively sensitive to atmospheric conditions.

Examples of this class of dyes are **27**, *C.I. Vat Violet 15*, 63355 *[6370-58-7]* [24, p. 8]; **28**, *C.I. Vat Yellow 12*, 65405 *[6370-75-8]* [24, p. 56]; **29**, *C.I. Vat Orange 17*, 65415 *[6370-77-0]* [25]; **30**, *C.I. Vat Yellow 10*, 65430 *[2379-76-2]* [15, p. 178]; **31**, *C.I. Vat Yellow 20*, 68420 *[4216-01-7]* [26]; **32**, *C.I. Vat Red 28*, 65710 *[6370-82-7]* [27, p. 353].

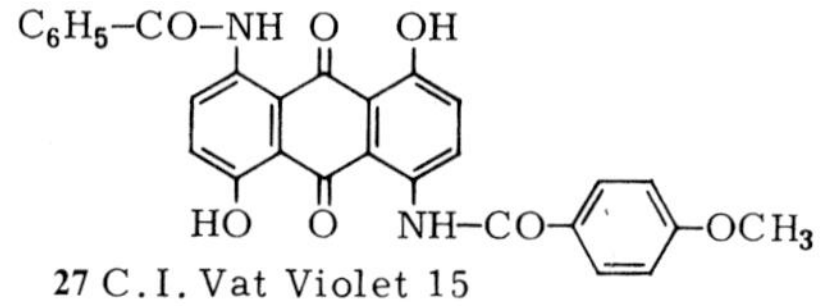

3.4 Anthraquinone Dyes

27 C.I. Vat Violet 15

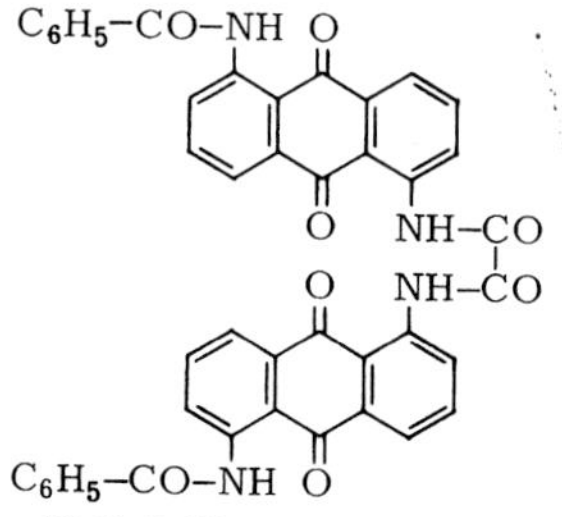

28 C.I. Vat Yellow 12

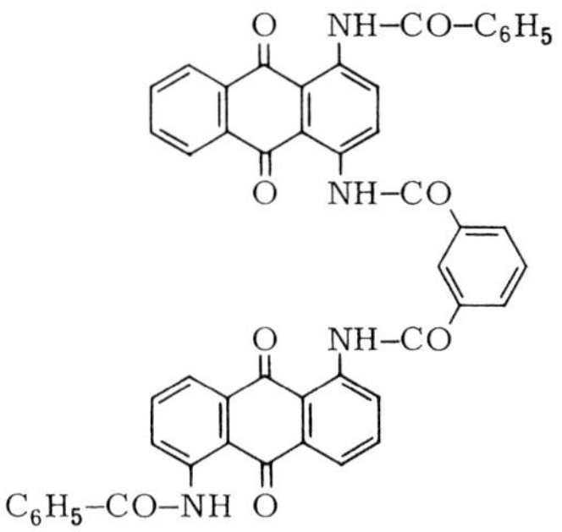

3.4.2 Chemical Constitution and Properties

29 C.I. Vat Orange 17

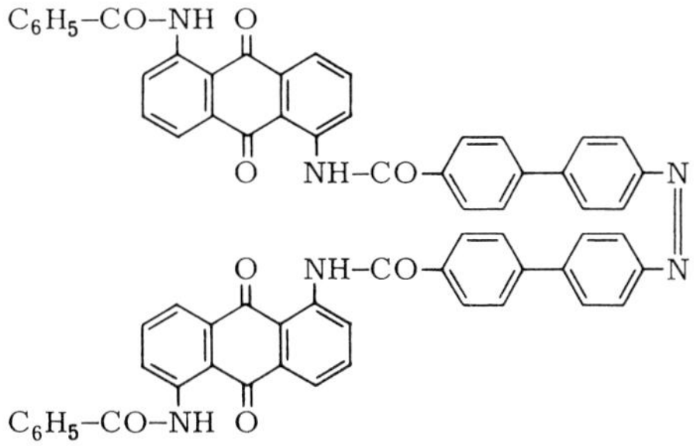

30 C.I. Vat Yellow 10

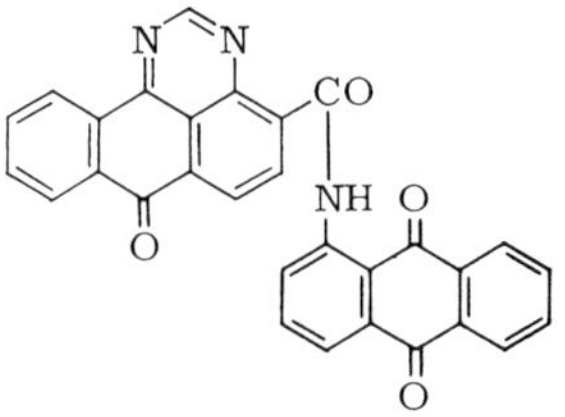

31 C.I. Vat Yellow 20

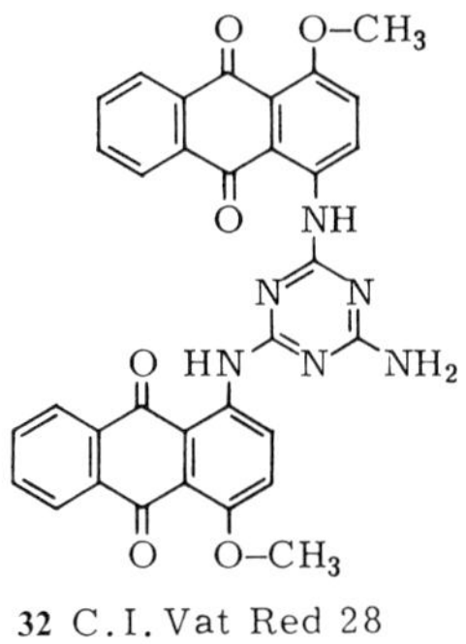

32 C.I. Vat Red 28

Anthraquinoneazoles. In contrast to the older yellow Algol dyes, which contain two thiazole rings (e.g., 2,2′-bisanthra [2,1d] thiazole-6,11-quinonyl), the red to blue oxazoles and thiazoles derived from 1-aminoanthraquinone-2-carboxylic acid and 3-amino-2-hydroxy- or -mercaptoanthraquinones exhibit good lightfastness. The good fastness to atmospheric conditions and chlorine of the blue deriv-

3.4 Anthraquinone Dyes

atives of 1-amino-4-aroylaminoanthraquinone-2-carboxylic acid deserves special mention.

Examples are **33**, *C.I. Vat Red 10*, 67000 [*2379-79-5*] [15, p. 152]; **34**, *C.I. Vat Blue 30*, 67110 [*6492-78-0*] [28, p. 17], [15, p. 72].

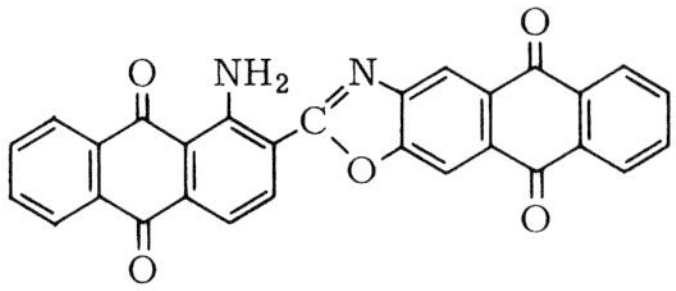

33 C.I. Vat Red 10

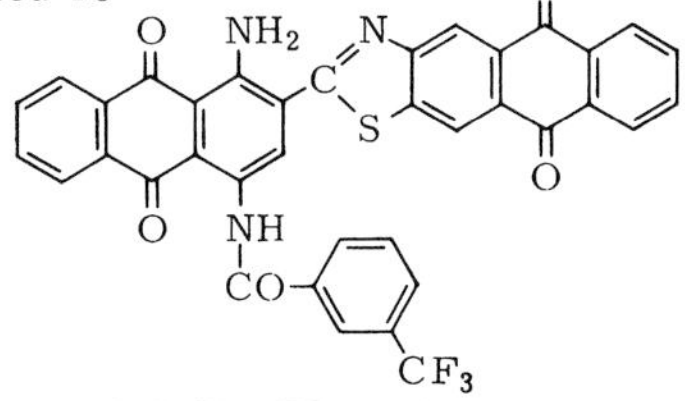

34 C.I. Vat Blue 30

Anthrimides and Other Linked Anthraquinones. Among the anthrimides (dianthraquinonyl-amines), only the α,β derivatives have achieved limited importance as vat dyes. Coupling two anthraquinone molecules via functional derivatives of the 2-aldehyde (or 2-carboxy) group offers another type of building block for vat dyes. Such compounds, e.g., 1-aminoanthraquinones, are linked in the 2-position via an azine or oxadiazole group, and all have good fastness.

Examples are **35**, *C.I. Vat Violet 16*, 65020 [*4003-36-5*] [24, p. 14]; **36**, *C.I. Vat Red 18*, 60705 [*6409-68-3*] [28, p. 64]; **37**, *C.I. Vat Blue 64*, 66730 [*15935-52-1*] [29]; **38**, *C.I. Vat Red 13*, 70320 [*4203-77-4*] [15, p. 159]

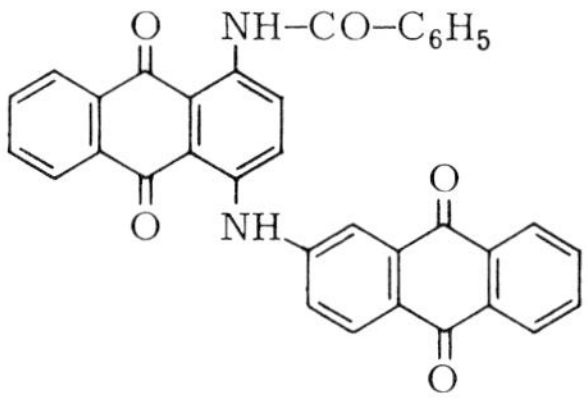

3.4.2 Chemical Constitution and Properties 191

35 C.I. Vat Violet 16

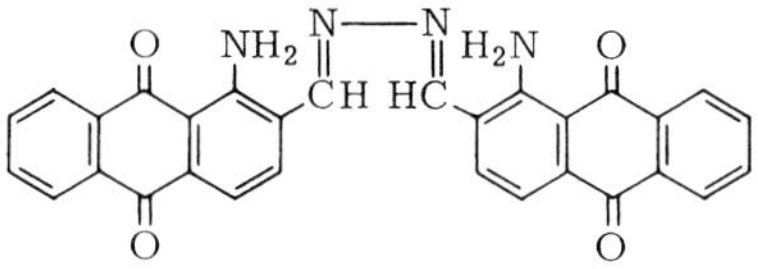

36 C.I. Vat Red 18

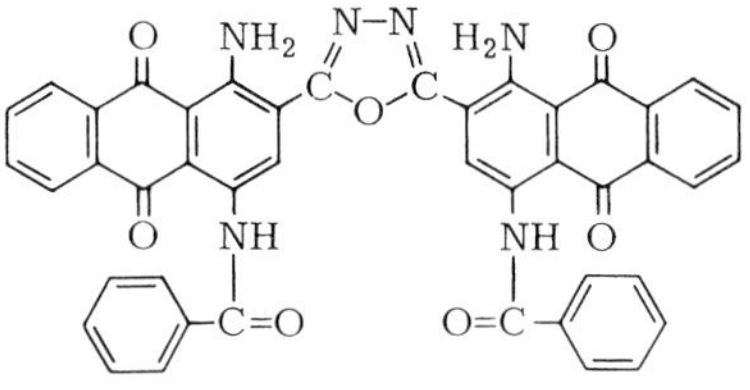

37 C.I. Vat Blue 64

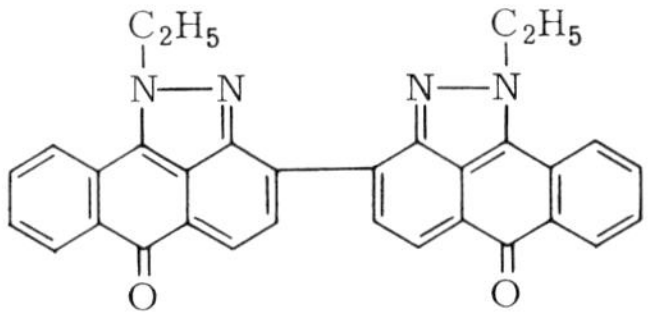

38 C.I. Vat Red 13

Anthrimide Carbazoles. Fast vat dyes are produced by carbazole ring closure from α,α-dianthraquinonylamines (anthrimides). The shade is determined by the number and position of the carbazole systems and by additional substituents, especially acylamino or alkoxyl groups. Anthraquinone carbazoles make it possible to dye cellulose fibers in level, very fast yellow, orange, brown, gray, and olive shades. However, this series lacks dyes with bright shades.

Examples include **39**, *C.I. Vat Yellow 28*, 69000 [*4229-15-6*] [24, p. 53]; **40**, *C.I. Vat Orange 15*, 69025 [*2379-78-4*] [24, p. 15], [15, p. 119]; **41**, *C.I. Vat Brown 3*, 69015 [*131-92-0*] [24, p. 10]; **42**, *C.I. Vat Black 27*, 69005 [*2379-81-9*] [24, p. 29], [15, p. 135]; **43**, *C.I. Vat Brown 1*, 70800 [*2475-33-4*] [15, p. 101]; **44**, *C.I. Vat Green 8*, 71050 [*14999-97-4*] [24, p. 23] [15, p. 129]

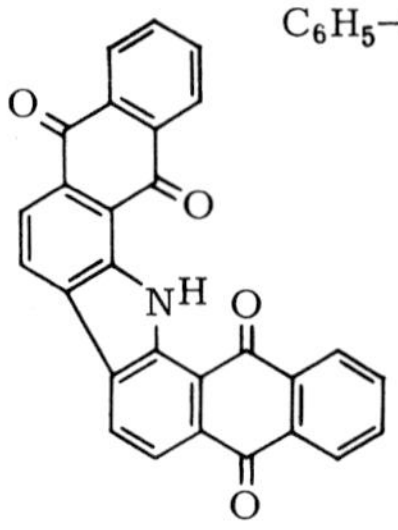

3.4 Anthraquinone Dyes

39 C.I. Vat Yellow 28 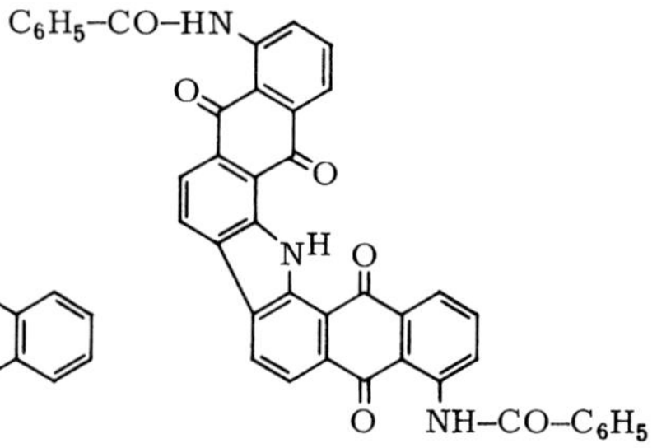   **40** C.I. Vat Orange 15

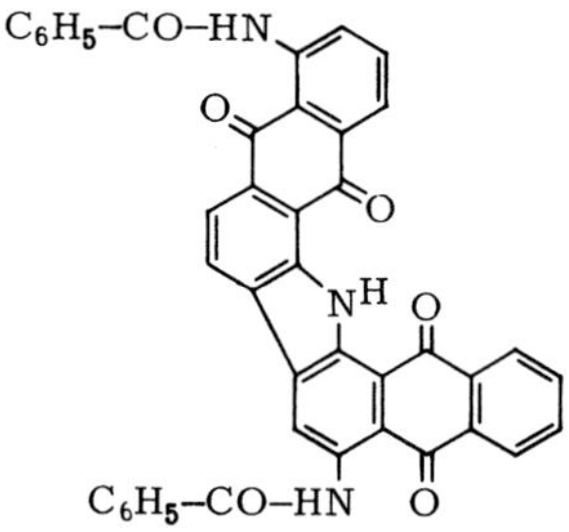

41 C.I. Vat Brown 3

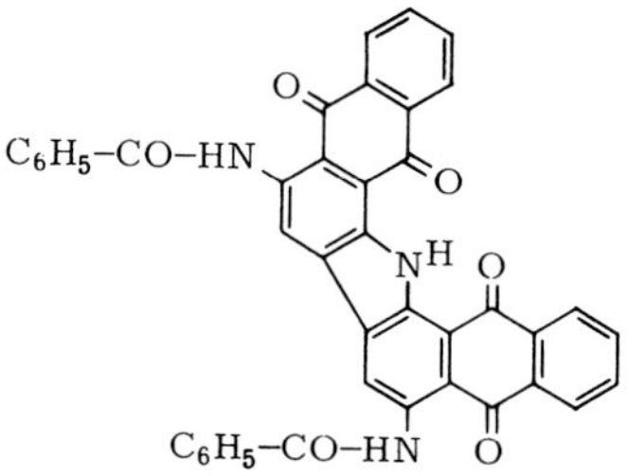

42 C.I. Vat Black 27

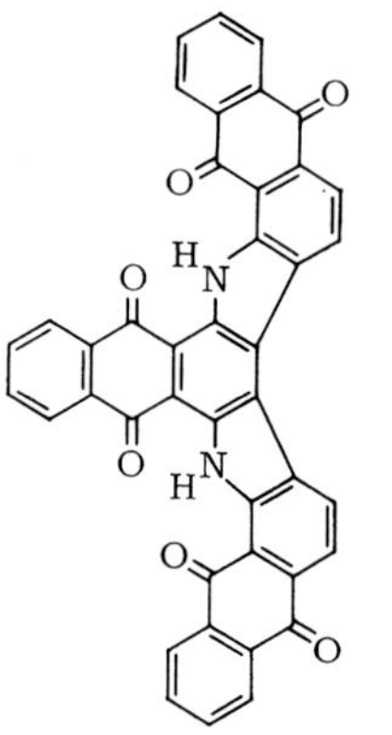

43 C.I. Vat Brown 1 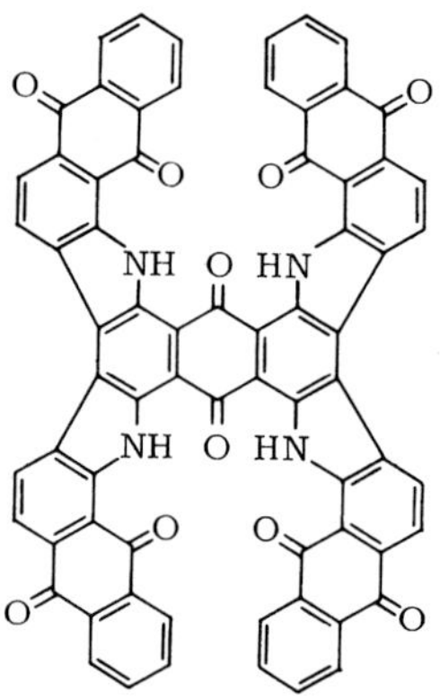   **44** C.I. Vat Green 8

3.4.2 Chemical Constitution and Properties 193

Phthaloylacridone. Depending on the substituent in the 2-position, the 3,4-phthaloylacridones supply shades varying between red and green. The acridone vat dyes combine good lightfastness with somewhat poorer washfastness. Dyes of this series are especially suited for printing purposes because of the ease of vatting.

Examples are **45**, *C.I. Vat Blue 21*, 67920 [*6219-97-2*] [24, p. 34], [15, p. 143]; **46**, *C.I. Vat Green 12*, 70700 [*6661-46-7*] [24, p. 19].

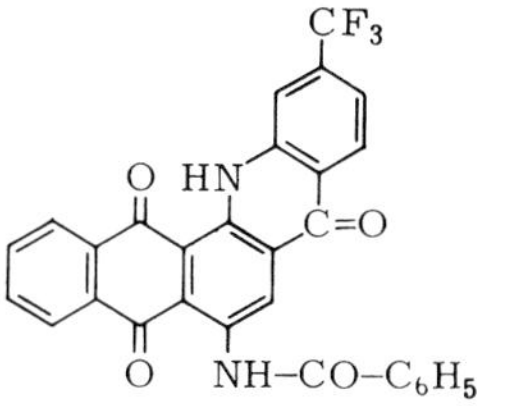

45 C. I. Vat Blue 21

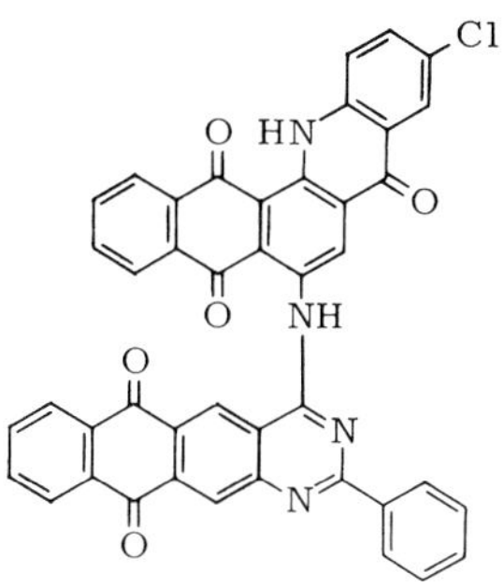

46 C.I. Vat Green 12

Benzanthrone Dyes. Vat dyes derived from benzanthrone can be subdivided into two major groups: the *peri* ring-closure products of 3-anthraquinonylaminobenzanthrone, referred to as "imide-green" dyes, and the dyes of the violanthrone and the isoviolanthrone series.

The first class gives the muted colors olive green, olive, khaki, and gray and excels in its resistance to light and atmospheric conditions. The blue color of violanthrone is shifted to a brilliant green by introducing two alkoxyl groups into the 16- and 17-positions. An additional shift may be achieved by halogenation.

A redder and brighter shade is obtained with isoviolanthrone. Its halogenation products are marketed as brilliant violet dyes. Related to violanthrone is the alkaline ring-closure product of 3-pyrazolanthronylbenzanthrone, which is used as a navy blue vat dye.

Examples are **47**, *C.I. Vat Green 3*, 69500 [*3271-76-9*] [28, p. 71]; **48**, *C.I. Vat Black 25*, 69525 [*4395-53-3*] [28, p. 76]; **49**, *C.I. Vat Blue 25*, 70500 [*6247-39-8*] [24, p. 26]; **50**, *C.I. Vat Blue 20*, 59800 [*116-71-2*] [15, p. 108]; **51**, *C.I. Vat Green 1*, 59825 [*128-58-5*] [28, p. 69], [15, p. 83]; **52**, *C.I. Vat Violet 1*, 60010 [*1324-55-6*] [30]; **53**, *C.I. Vat Blue 26*, 60015 [*4430-55-1*] [31].

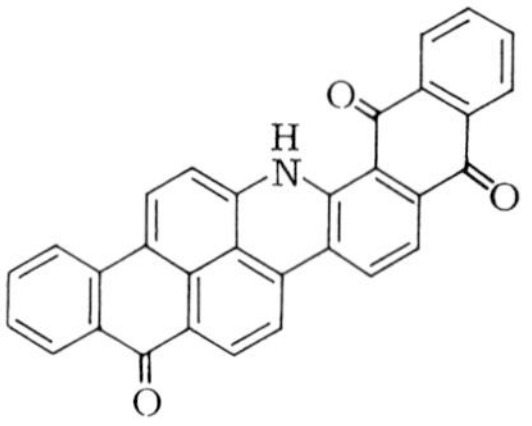

3.4 Anthraquinone Dyes

47 C.I. Vat Green 3

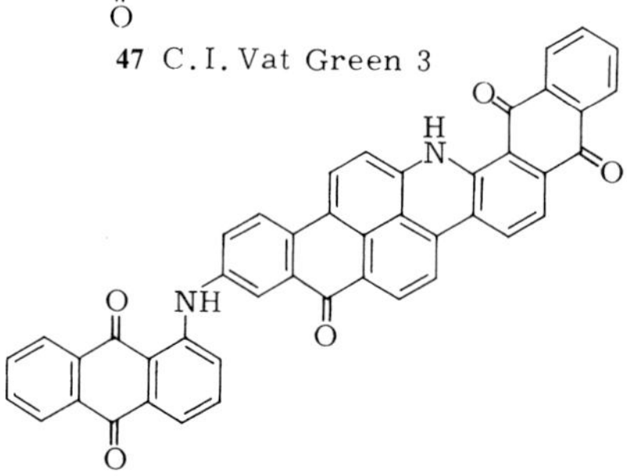

48 C.I. Vat Black 25

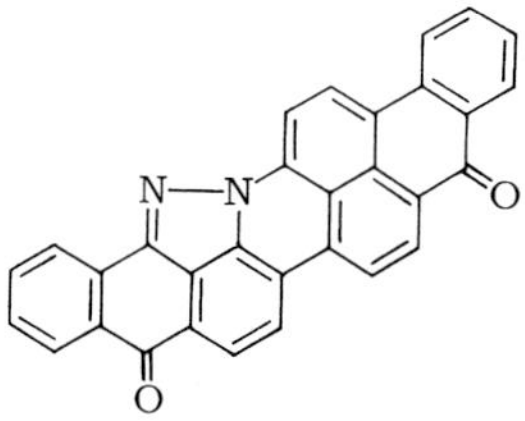

49 C.I. Vat Blue 25

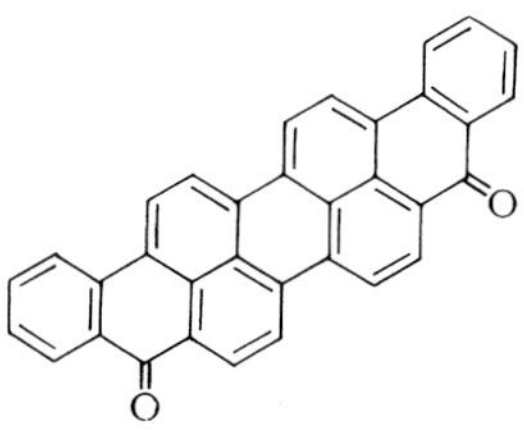

50 C.I. Vat Blue 20

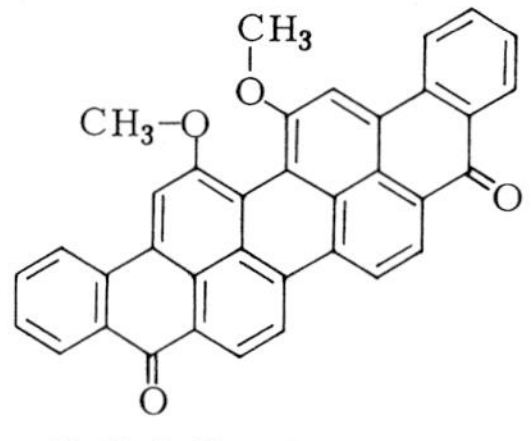

51 C.I. Vat Green 1

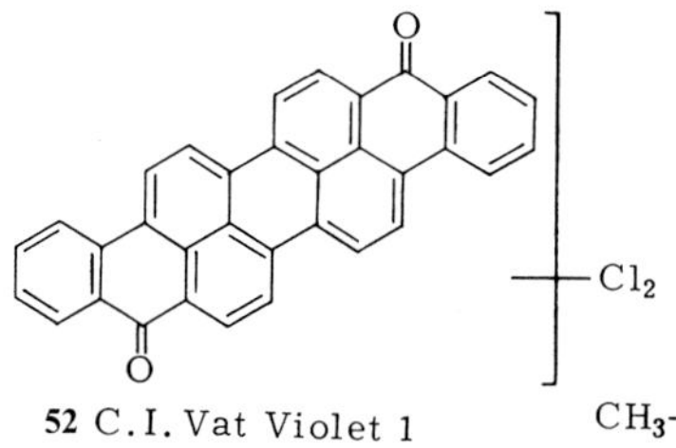

52 C.I. Vat Violet 1

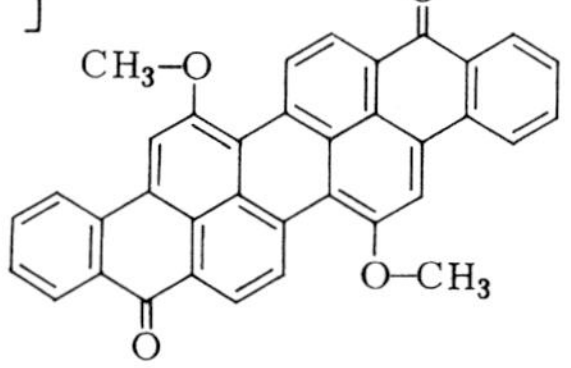

53 C.I. Vat Blue 26

3.4.2 Chemical Constitution and Properties

Indanthrones. The blue indanthrone was the first synthetic vat dye of the anthraquinone series. Because of its excellent fastness and bright colors it has remained the most important vat dye for a long time despite its low resistance to chlorine. Its chlorine resistance can be improved somewhat by post-halogenation. Introduction of hydroxy or amino groups shifts the shade to green.

Examples are **54**, *C.I. Vat Blue 4*, 69800 [*81-77-6*] [28, p. 52], [15, p. 73]; **55**, *C.I. Vat Green 11*, 69850 [*1328-41-2*] [24, p. 18].

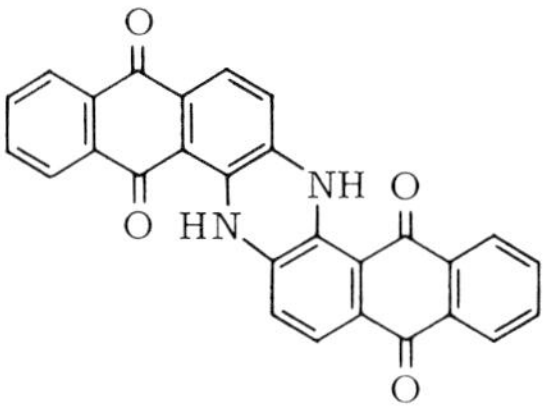

54 C.I. Vat Blue 4

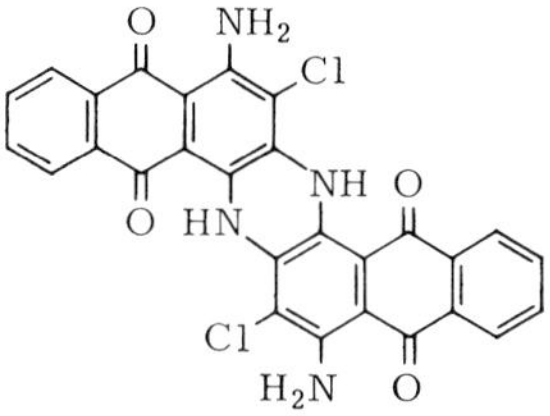

55 C.I. Vat Green 11

Highly Condensed Ring Systems. A valuable supplement to the anthraquinone vat dyes is found in the more highly condensed carbocyclic quinones dibenzpyrenequinone, anthanthrone, and pyranthrone. These substances are yellow to red vat dyes without additional auxochromic substituents. Halogenation may improve their substantivity and change their shades. The halo derivatives can be converted to anthrimide-like compounds by reaction with aminoanthraquinones. These are in themselves useful as vat dyes and dye cotton in brown to gray shades but can be subjected to further carbazolyation.

Examples are *C.I. Vat Orange 1*, 59105 [*1324-11-4*] [28, pp. 117, 119]; *C.I. Vat Orange 3*, 59300 [*4378-61-4*] [28, p. 129], [15, p. 90]; *C.I. Vat Black 29*, 65225 [*6049-19-0*] [28, p. 23]; *C.I. Vat Orange 2*, 59705 [*1324-35-2*] [28, p. 61], [15, p. 116]); *C.I. Vat Brown 45*, 59500 [*6424-51-7*] [24, p. 47], [15, p. 149]; *C.I. Vat Orange 9*, 59700 [*128-70-1*] [28, p. 60], [15, p. 114]

3.4.2.3 Acid Dyes

Acid dyes are used for dyeing wool, synthetic polyamides, and silk in aqueous media. They may be subdivided into the following basic types:

The majority of the acid anthraquinone dyes available commercially give bright blue shades not obtainable with azo dyes. The red and yellow anthraquinone dyes are of little importance. Dyes of green shades obtained by combining yellow and blue dyes possess mostly slight washfastness. Here, the uniformly dyeing green dyes of the anthraquinone series have proved their special value. The acid anthraquinone dyes are classified for particular applications according to their leveling characteristics, lightfastness, and washfastness. When synthetic polyamide fibers were introduced to the market appropriate types were selected from the existing stocks. Recently, special acid dyes have been developed for polyamides.

1-Aminoanthraquinone-2-sulfonic Acids. Condensation of bromamine acid (1-amino-4-bromoanthraquinone-2-sulfonic acid) with aromatic or cycloaliphatic amines is used to produce a large number of blue acid dyes. The shade, leveling characteristics, washfastness, and lightfastness may be varied over a wide range by choosing particular amines. Cycloaliphatic amines provide the same brightness as aliphatic dyes but impart greater lightfastness. Arylamines substituted with alkyl, halogen, aryl, aryloxy, or sulfonic ester groups yield dyes with better washfastness and affinity in neutral media but less uniform leveling. Substituents at the *o*-position cause a hypochromic shift and increase brilliance.

The arylamino residue may be altered subsequently by sulfonation, halogenation, acylation, or by the Einhorn reaction. Substitution of anthraquinone in the 5-, 6-, 7-, or 8-positions offers an additional possibility to change the characteristics of the dye. For instance, halogen atoms and sulfonic acid groups cause bathochromic effects that are most pronounced when the substituents are introduced in the β-position. The solubility of the 2,6- (or 2,7-) disulfonic acids is higher than that of the 2,5- (or 2,8-) series.

Examples are **56**, *C.I. Acid Blue 25*, 62055 [*6408-78-2*] [6, p. 41]; **57**, *C.I. Acid Blue 62*, 62045 [*4368-56-3*] [6, p. 30]; **58**, *C.I. Acid Blue 129*, 62058 [*6397-02-0*] [32]; **59**, *C.I. Acid Blue 40*, 62125 [*6424-85-7*] [28, p. 135]; **60** [33] and **61** [34] are brilliant blue dyes.

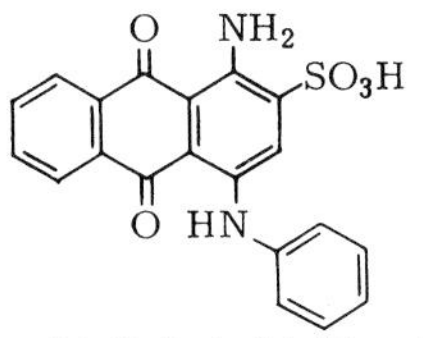

3.4.2 Chemical Constitution and Properties

56 C.I. Acid Blue 25

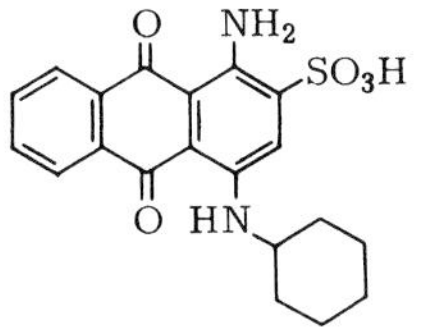

57 C.I. Acid Blue 62

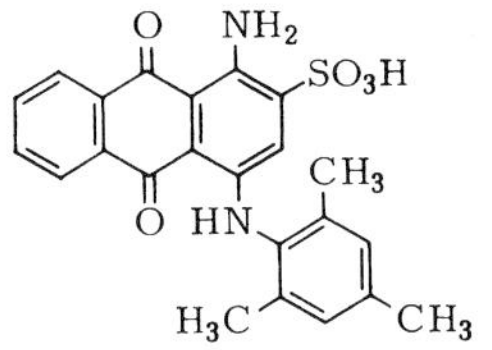

58 C.I. Acid Blue 129

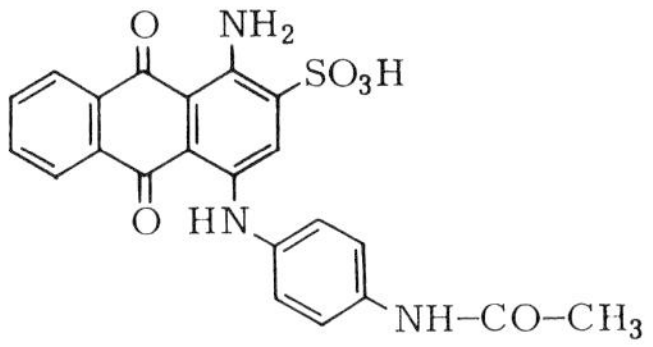

59 C.I. Acid Blue 40

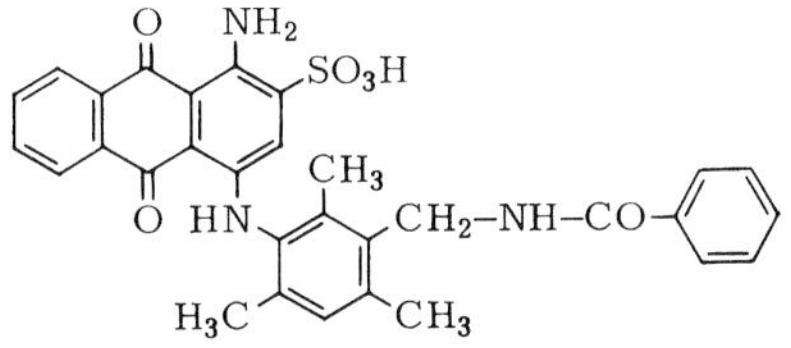

60 Brilliant Blue

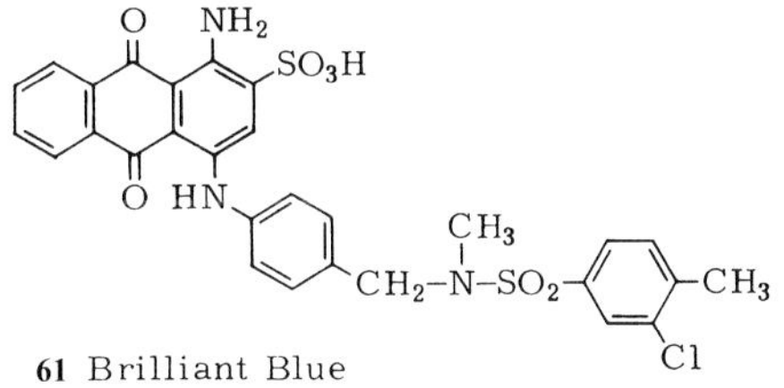

61 Brilliant Blue

Diaminodihydroxyanthraquinonesulfonic Acids. These dyes belong to the oldest synthetic acid wool dyes, but their importance has decreased considerably. An example for this class is **62**, *C.I. Acid Blue 43*, 63000 [*2150-60-9*] [6, p. 42].

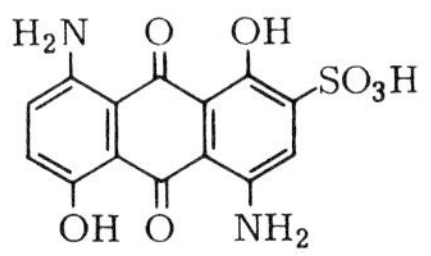

62 C.I. Acid Blue 43

1,4-Diaminoanthraquinones with External Sulfonic Acid Groups. A common feature of dyes of this group is their manufacture by sulfonation of the corresponding dye base derived from quinizarin or haloaminoanthraquinones. The reaction products of quinizarin with aromatic or araliphatic amines predominate in number and importance. The introduction of hydroxy groups into the 5- or the 5,8-positions brings about the expected bathochromic shift. Washfastness and leveling properties may be altered by substitution. The shade can be varied from brilliant blue to green by appropriate amines. Araliphatic and cycloaliphatic amines lead to brilliant blue shades. A similar effect is exhibited by sterically hindered aromatic amines such as mesidine. These products are more lightfast than the derivatives of bromamine acid. Among the unsymmetrically substituted dyes derived from 1-amino- or 1-alkylamino-4-haloanthraquinones, the 1-alkylaminoanthraquinones have the lower lightfastness. Exceptions are the derivatives of the 1-*sec*-alkylamino-4-haloanthraquinones. Introduction of alkoxy- (or aryloxy) groups into the 2-position shifts the shade of the 4-arylamino-1-aminoanthraquinones to a bright violet.

Examples are **63**, *C.I. Acid Green 25*, 61570 [4403-90-1] [15, p. 215]; **64**, *C.I. Acid Green 41*, 62560 [4430-16-4] [6, p. 35]; **65** [35] and **66** [36] are greenish-blue dyes; **67**, *C.I. Acid Violet 42*, 62026 [6408-73-7] [6, p. 48].

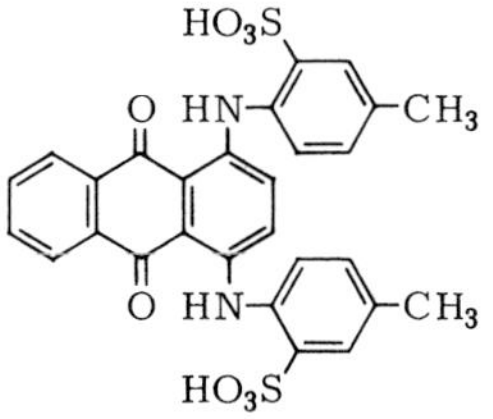

63 C.I. Acid Green 25

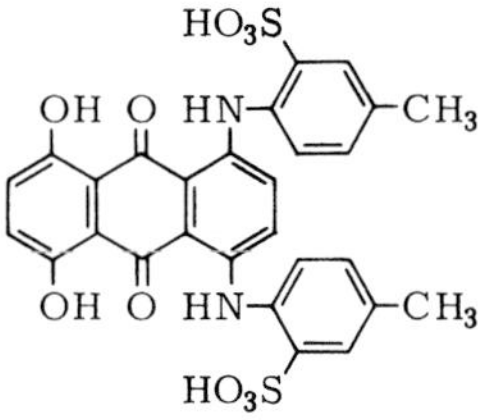

64 C.I. Acid Green 41

65 Greenish Blue

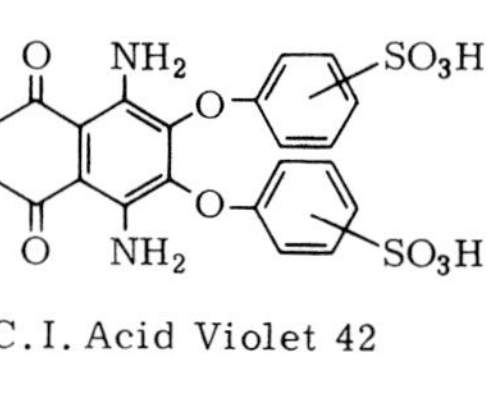

66 Greenish Blue

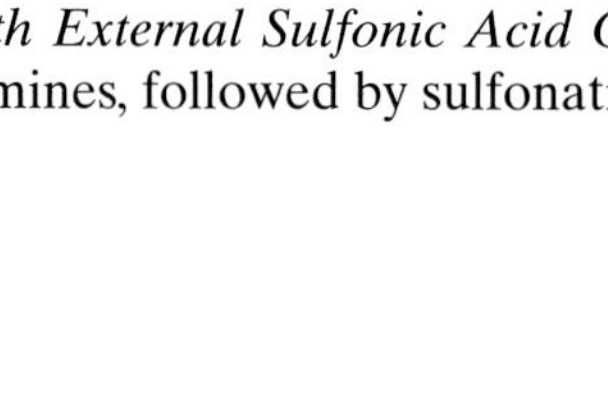

67 C.I. Acid Violet 42

1-Amino-4-hydroxyanthraquinones with External Sulfonic Acid Groups. By one-sided reaction of quinizarin with arylamines, followed by sulfonation, violet level-

3.4.2 Chemical Constitution and Properties

ing dyes are obtained. Derivatives of 1-amino-4-hydroxy-2-phenoxyanthraquinones were developed specially for polyamide fibers.

Examples are **68**, *C.I. Acid Violet 43*, 60730 [4430-18-6] [6, p. 48]; **69** is a bluish brilliant red dye [37].

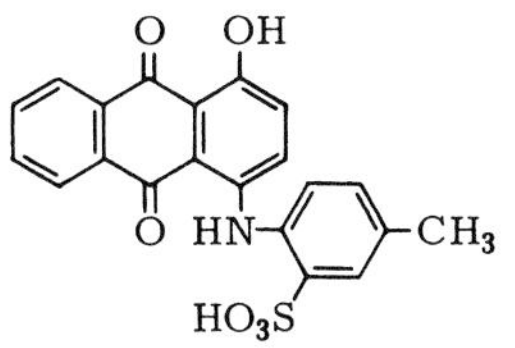

68 C.I. Acid Violet 43

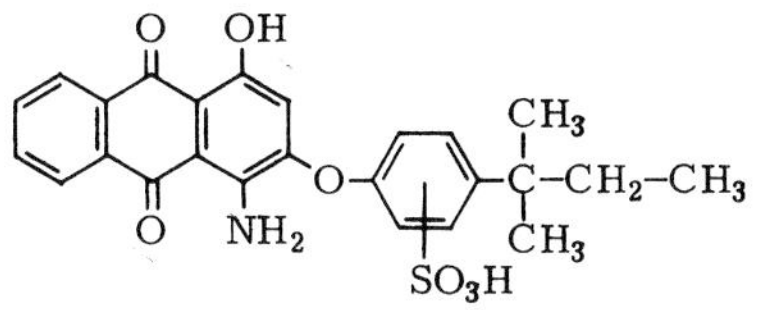

69 Bluish Brilliant Red

Other Acid Anthraquinone Dyes. In addition to the dyes in the preceding classes, a whole series of specially developed products is available. For instance, derivatives of the anthrimide or carbazole series are known to be very light-fast gray and brown wool dyes. The post-sulfonation products of 1,5- and 1,8-diarylaminoanthraquinones are violet dyes commonly applied as mixtures.

An example is **70**, *C.I. Acid Black 48*, 65005 [*1328-24-1*] [15, p. 216].

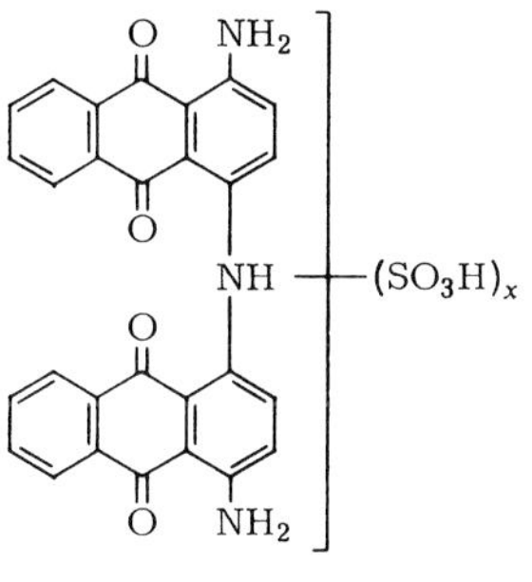

70 C.I. Acid Black 48

3.4.3 Synthesis

3.4.3.1 Introduction

Due to the wide range of reactions for the synthesis of anthraquinone dyes and to remain within the scope of this book, this section concentrates on the most important basic reactions and refers to the literature for more details. For further information, see [38].

The production of anthraquinone dyes generally proceeds from a few key products generated by electrophilic substitution of unsubstituted anthraquinone or by synthesis of the nucleus. The major methods employed to prepare anthraquinone derivatives substituted in the α-position are sulfonation and nitration. Preparation of β-substituted anthraquinones and of quinizarin (1,4-dihydroxyanthraquinone) generally is accomplished by synthesis of the nucleus starting from phthalic anhydride and a benzene derivative.

Until fairly recently, preparation of almost 80 % of all important anthraquinones was based on anthraquinonesulfonic acids. However, nitration of anthraquinone is gaining in importance presently, a development triggered mainly by environmental considerations (e.g., production of large volumes of waste dilute acids during anthraquinone sulfonation). However, great progress has been made recently in solving the problems associated with sulfonation.

Preparation of nearly all important anthraquinones starts from the following key intermediates: anthraquinonesulfonic acids, nitroanthraquinones, and the products of nucleus synthesis, 1,4-dihydroxy-, 2-methyl-, and 2-chloroanthraquinone. The only exceptions are derivatives with condensed rings, e.g., benzanthrone and derived products, which are prepared directly from anthraquinone via anthrone.

3.4.3.2 Anthraquinonesulfonic Acids

are generally obtained by sulfonation. The sulfonic acid group in anthraquinone enters readily into nucleophilic exchange reactions, whereby the α-position is much more reactive than the β-position. Sulfonic acid groups are exchanged readily against amino, alkylamino, hydroxyl, and alkoxyl groups. Exchange against chlorine atoms also proceeds smoothly (Fischer synthesis).

Reduction also may be used to remove the sulfonic acid group from certain substituted anthraquinones, for example, from 1-amino-4-arylaminoanthraquinone-2-sulfonic acid.

Desulfonation of the α-position may be accomplished by heating in 70 – 90 % sulfuric acid in the presence of mercury. This method is used to convert undesired α,α′-disulfonic acids back to anthraquinones [39, 40].

3.4.3.3 Haloanthraquinones

are important intermediates for all classes of dyes, especially for vat and disperse dyes.

Unsubstituted halogenated anthraquinones are prepared mainly by substitution reactions or by synthesis of the nucleus. Only the chloro derivatives are of practical importance. Aminoanthraquinones, hydroxyanthraquinones, phenoxyanthraquinones, and anthrimides (dianthraquinonylamines) may be obtained by replacing the chlorine atom. The β-chloroanthraquinones are much more suitable for preparing β-aminoanthraquinones than the corresponding sulfonic acids. In contrast to the 1,5- and 1,8-dichloroanthraquinones, a single chlorine atom of 2,3-dichloroanthraquinone can be replaced by ammonia to give 2-amino-3-chloroanthraquinone in good yield [15, p. 30].

Haloanthraquinones with additional substituents commonly are prepared by direct halogenation of the corresponding anthraquinones in water, hydrochloric acid, sulfuric acid, or organic solvents. Electron-withdrawing substituents, such as sulfonic acid or nitro groups, direct the halogenation to the other nucleus.

Anthraquinones with electron-donating substituents, such as amino or hydroxyl groups, permit selective halogenation of one or the other nucleus by appropriate choice of reaction conditions. Only the chloro and bromo derivatives are commercial products.

3.4.3.4 Nitroanthraquinones

Nitration of anthraquinones has considerable industrial importance. Examples of compounds that may be nitrated are anthraquinone and the halo-, hydroxy-, amino-, and acylaminoanthraquinones. As a rule the nitro group attacks at the α-position; choice of reaction conditions frequently permits mono- or dinitration.

Prior to the discovery of α-sulfonation of anthraquinone, nitration was the only useful method for preparing α-substituted anthraquinones. The nitro group of α-nitroanthraquinones can be replaced in a manner similar to the sulfonic acid moiety, e.g., by chlorine atoms and amino, hydroxy, alkoxy, or mercapto groups. Reduction readily yields aminoanthraquinones. Nitration of anthraquinone has gained increasing importance because of environmental considerations, this method offering an economical alternative to α-sulfonation

3.4.3.5 Aminoanthraquinones

especially 1-amino- and 1,5-diaminoanthraquinone, are key products for essentially all classes of anthraquinone dyes. Important production methods are the replacement of sulfonic acid and nitro groups or of halogen atoms by ammonia or primary or secondary amines. With 1,4-dihydroxy-, 1,4-aminohydroxy-, and 1,4-diaminoanthraquinones, the replacement of hydroxyl and amino groups is also successful. Primary aminoanthraquinones are also prepared by reduction of nitroanthraquinones. Modifications of the amino groups also have industrial importance. These include alkylation, arylation, acylation, and hydrolysis of acyl-aminoanthraquinones. The choice of production method for a desired aminoan-thraquinone depends on the position and type of amino group, as well as on the availability of starting materials.

In most cases the presence of the amino group facilitates the introduction of further substituents into the same ring. Reaction media and conditions as well as type and position of the amino group determine the position of substitution.

In addition to conversion of existing amino groups to other amino groups, the amino groups also can be replaced by halogen or hydroxyl substituents by diazotization followed by the Sandmeyer reaction.

3.4.3.6 Hydroxyanthraquinones

are prepared primarily by synthesis of the nucleus, by exchange reactions, and by ether cleavage of alkoxy-anthraquinones. Examples for exchange reactions are the conversion of anthraquinone-α,α'-disulfonic acids into α,α'-dihydroxyanthraquinones and the direct hydrolysis of a halogen in the 4-position in 1-amino-2,4-dihaloanthraquinone using boric–sulfuric acid. As the direct replacement of the nitro group by a hydroxy group frequently leads to side reactions, nitroanthraquinones are converted to hydroxyanthraquinones most readily via the corresponding alkyl or aryl ethers. In some cases, under rather special reaction conditions, direct replacement of one or more nitro groups is also possible. Replacement of halogen by hydroxy groups in otherwise unsubstituted haloanthraquinones is best carried out via the corresponding anthraquinone ethers. In individual cases the hydroxy groups also may be inserted at defined positions by oxidation.

Amino groups in 1,4-diamino- or 1,4-aminohydroxyanthraquinones are converted readily into hydroxyl groups by reduction in alkaline medium via the 2,3-dihydro compounds or in an acid medium by oxidation via the quinoneimines.

Halogenation, nitration, and sulfonation of hydroxyanthraquinones present no special difficulties. Modification of the hydroxyl group (boric esters, ethers) alters the mode of substitution. Derivatives of the hydroxyl groups frequently enable a different or more selective substitution than the free hydroxy compounds.

3.4.4 Examples of Commercially Available Dyes

Disperse Dyes
    C.I. Disperse Red 60, 60756 [*17418-58-5*]
    C.I. Disperse Violet 28, 61102 [*81-42-5*]
    C.I. Disperse Violet 27, 60724 [*19286-75-0*]
    C.I. Disperse Blue 72, 60725 [*81-48-1*]

C.I. Disperse Blue 56, 63285 [*12217-79-7*]
C.I. Disperse Blue 73, 63265 [*12222-75-2*]
C.I. Disperse Blue 14, 61500 [*2475-44-7*]
C.I. Disperse Blue 31, 64505 [*1328-23-0*]
C.I. Disperse Blue 7, 62500 [*3179-90-6*]
C.I. Disperse Blue 26, 63305 [*3860-63-7*]

Vat Dyes
C.I. Vat Red 13, 70320 [4203-77-4]
C.I. Vat Orange 15, 69025 [*2379-78-4*]
C.I. Vat Brown 3, 69015 [*131-92-0*]
C.I. Vat Black 27, 69005 [2379-81-9]
C.I. Vat Brown 1, 70800 [*2475-33-4*]
C.I. Vat Green 8, 71050 [*14999-97-4*]
C.I. Vat Green 3, 69500 [*3271-76-9*]
C.I. Vat Black 25, 69525 [*4395-53-3*]
C.I. Vat Blue 20, 59800 [*116-71-2*]
C.I. Vat Green 1, 59825 [*128-58-5*]
C.I. Vat Violet 1, 60010 [*1324-55-6*]
C.I. Vat Blue 4, 69800 [*81-77-6*]
C.I. Vat Orange 1, 59105 [*1324-11-4*]
C.I. Vat Black 29, 65225 [*6049-19-0*]
C.I. Vat Orange 9, 59700 [*128-70-1*]

Acid Dyes
C.I. Acid Blue 25, 62055 [*6408-78-2*]
C.I. Acid Blue 62, 62045 [*4368-56-3*]
C.I. Acid Blue 129, 62058 [*6397-02-0*]
C.I. Acid Blue 40, 62125 [*6424-85-7*]
C.I. Acid Green 25, 61570 [*4403-90-1*]

3.4.5 References

[1] H. Labhart, *Helv. Chim. Acta* **40** (1957) 1410.
[2] BASF, DE-AS 1209680, 1962 (K. Maier).
[3] Bayer, DE-OS 2531557, 1975 (V. Hederich, H.-S. Bien, G. Gehrke).
[4] Bayer, FR 1503492, 1965.
[5] Sumitomo Chem., FR 1497689, 1965.
[6] BIOS Final Report 1484.
[7] Bayer, DE-AS 1105837, 1958 (G. Gehrke).
[8] Bayer, DE-AS 1644587, 1965 (M. Groll, K. Wunderlich, H.-S. Bien).
[9] Du Pont, US 2628963, 1951 (J. F. Laucius, S. B. Speck).
[10] BASF, DE-AS 1918696, 1969 (E. Hartwig).
[11] Geigy, BE 650734, 1963.

[12] Ciba, DE-AS 1065959, 1954 (P. Grossmann, W. Jenny, W. Kern).
[13] Ciba, FR 1594324, 1968.
[14] Sandoz, FR 1490805, 1965 (M. F. Müller).
[15] FIAT Final Report 1313 II.
[16] K. Venkataraman: *The Chemistry of Synthetic Dyes*, Vol. VIII, Academic Press, New York, 1978, pp. 191 ff.
[17] Du Pont, DE-AS 1811796, 1968 (J. Blackwell, R. E. Starn, W. H. Gumprecht).
[18] BASF, DE-OS 2524243, 1975 (H. Schwab, K. Oppenlaender, A. Blum).
[19] BASF, DE-OS 2654434, 1976 (W. Elser, H. Eilingsfeld, G. Meyer).
[20] BASF, DE-OS 2651975, 1976 (G. Epple, W. Elser).
[21] Bayer, DE-OS 2918881, 1979 (W. Harms et al.).
[22] Mitsubishi Chem. Ind., DE-OS 3218957, 1981 (T. Niwa, T. Hihara).
[23] Mitsubishi Chem. Ind., JP-Kokai 55-164250, 1979.
[24] Bios Final Report 1493.
[25] IG-Farbenind., DE 653386, 1935 (W. Mieg, F. Wieners); *Friedländer*, **vol. 24**, p. 862.
[26] IG-Farbenind., DE 696423, 1935 (K. Köberle); *Friedländer*, Suppl. vol. I/2, p. 239.
[27] H. R. Schweizer: *Künstliche organische Farbstoffe und ihre Zwischenprodukte*, Springer, Berlin, 1964.
[28] BIOS Final Report 987.
[29] Bayer, DE 911493, 1952 (H. W. Schwechten).
[30] BASF, DE 217570, 1909; *Friedländer*, vol. 9, p. 827.
[31] Scottish Dyes, FR 543910, 1921.
[32] Sandoz, US 2121928, 1934 (A. Peter).
[33] Ciba-Geigy, DE-OS 2850996, 1977 (R. Lacroix, J.-M. Adam, J. Vincze).
[34] Bayer, GB 1099756, 1966 (J. Singer).
[35] Sandoz, FR 1455722, 1964 (M. J. Günthard).
[36] Geigy, FR 1461074, 1964.
[37] Crompton & Knowles, FR 1478769, 1965 (M. R. W. Eltonhead).
[38] Ullmann's Encyclopedia of Industrial Chemistry, Sixth Edition, 2002 Electronic Release.
[39] Bayer, DE 160104, 1903; *Friedländer*, vol. 8, p. 236.
[40] Bayer, DE-OS 3106933, 1981 (K. Ebke, J. Ohm, J. Schroeder).

3.5 Indigoid Dyes

3.5.1 Introduction

Thousands of years ago it was discovered how by suitable treatment of the plant *Indigofera tinctoria*, a stable blue dye could be extracted that could be used both in painting and for dyeing textiles such as wool and linen. Excavations in what is now Pakistan show that dyeings were already being carried out with indigo in the Indus valley around 3000 B.C. The name "indigo" is derived from the Greek for "of India" and refers to the Indian subcontinent [1].

In the period 1700 to 1900, the production of what came to be known as natural indigo was carried out on a very large scale. The indigo plant was cultivated in enormous plantations, chiefly in India and the countries of South-East Asia and America. As much as one square kilometer was needed to obtain a tonne of indigo a year [2]. The reason for such a high land requirement was simply that the plant material contains barely 1 % of the dye's precursor indican, the rest being biomass.

Blue dyeings were also produced in Europe. Here the process involved using the woad plant (*Isatis tinctoria*) without isolation of the dye. The cities of Erfurt and Toulouse were important centres of cultivation [3]. In the 1600s, Indian indigo captured the market in Europe because of its lower price and more brilliant shade.

In the flourishing chemical industry of the 1800, novel dye syntheses played a very significant role. The synthesis of indigo, the "king of dyes", was a particular challenge. The motivation was partly academic curiosity as to the nature of the unknown coloration principle involved. But the economic interest of being able to produce this 5000-t/a dye more cheaply and in higher quality also provided a powerful impetus.

3.5.2 Chemical Constitution and Properties

The first laboratory synthesis of indigo was achieved by Adolf von Baeyer in 1878, and in 1883 he revealed the structural formula **1** to Heinrich Caro, at that time head of the research laboratories at BASF [4].

3.5.2.1 Physical Properties

The systematic name of indigo, also known as indigotin, is 2-(1,3-dihydro-3-oxo-2*H*-indol-2-ylidene)-1,2-dihydro-3*H*-indol-3-one or 2,2′-biindolinylidene-3,3′-dione, *C.I. Vat Blue 1*, 73000 *[482-89-3]*. It exists as blue-violet needles or prisms with a pronounced coppery luster. It sublimes above 170 °C as a red-violet vapor that condenses on cooling to form dark violet needles. The melting point is 390–392 °C.

Indigo is practically insoluble in water, dilute acids, and dilute alkalis, but slightly soluble in polar, high-boiling solvents such as aniline, nitrobenzene, phenol, phthalic anhydride, and dimethyl sulfoxide. Some polar solvents destroy indigo when it is dissolved in them at the boil.

The dye is positively solvatochromic, the absorption maximum in a polar solvent such as dimethyl sulfoxide being 620 nm, that is, 12 nm higher than in a nonpolar solvent such as carbon tetrachloride [5]. Characteristic of indigo is the unusually deep shade compared with other conjugated systems of similar size [6]. This is explicable in terms of the special arrangement of the atoms in the basic chromophore and the high polarizability of the charge distribution, which is strongly influenced by the ability of the molecule to form hydrogen bonds (see also Section 2.4).

Intra- and intermolecular hydrogen bonding are the explanation for indigo's extremely low solubility and high melting point. X-ray analysis and IR studies demonstrate the existence of intermolecular hydrogen bonds in the solid state. The very long-wave IR absorption of the carbonyl band at 1626 cm^{-1} can be regarded as characteristic of the indigo structure. By comparison, the carbonyl band of dehydroindigo is situated at 1724 cm^{-1}.

Bond orders and charge densities in the indigo molecule have been calculated [7] and compared with the results of X-ray analysis [8]. These studies confirm the structural formula **1** and answer questions about the basic chromophore of the dye (see Section 2.4).

3.5.2.2 Chemical Properties

Indigo is very stable to light and heat. The molecule does not readily undergo electrophilic or nucleophilic substitution. However, it can be successfully sulfonated in concentrated sulfuric acid to give the tetrasulfonic acid, and halogenated in nitrobenzene to introduce up to six halogen atoms.

Indigo is readily reduced by various reducing agents such as zinc dust, sodium dithionite, hydroxyacetone, and hydrogen, or by electrochemical means. In an alkaline medium, a salt (for example the sodium salt) of leuco indigo is produced (**3**), which can be converted by acids to so-called indigo white (**2**).

Indigo (**1**)

R = H: Leuco indigo, indigo white (**2**)
R = Na: Leuco indigo disodium salt, indigo vat (**3**)

The yellow-brown soluble vat form **3** of indigo has affinity for animal and vegetable fibers, making it possible for indigo to be used in dyeing. Upon oxidation with air, it forms blue indigo again, fixed on the fiber. The dye tends to adhere mainly to the surface of cotton fibers, whereas on the polypeptide fibers of wool or silk the bonding is more saltlike. This explains the fact that the fastness to light, rubbing, and washing are poorer on cotton than on wool. The difference in shade between dyeings on the two fiber types is also due to the different bonding mechanisms.

Oxidation of indigo results in dehydroindigo (**4**). Oxidation with permanganate or chromate splits the molecule, forming isatin (**5**). Oxidation and reduction of the indigo system are accompanied by corresponding changes in the spectroscopic properties (see Section 2.4).

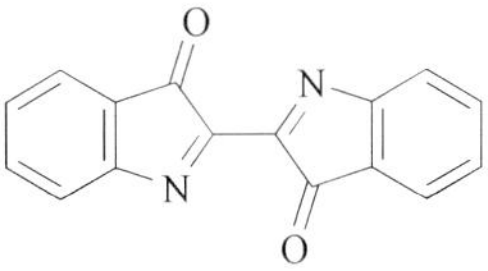

Dehydro indigo (**4**)

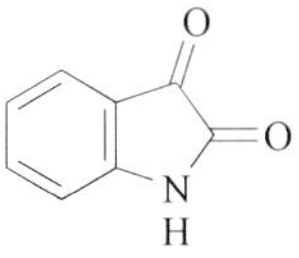

Isatin (**5**)

3.5.3 Synthesis

For thousands of years, indigo was produced from plant material containing low concentrations of indican, a precursor of the dye. Indican is split by enzymes and converted to indigo by oxidation. Indigo can be synthesized from D-glucose by genetically modified strains of coli bacteria, presumably by a process that resembles biosynthesis in plants.

3.5.3.1 Chemical Synthesis

Synthetic indigo was made by A. von Baeyer in 1870 by treating isatin (**5**) with phosphorus trichloride and phosphorus in acetyl chloride [9]. He obtained isatin by oxidizing indigo. The first complete synthesis of indigo (Scheme 1) was achieved in 1878, when von Baeyer succeeded in deriving isatin from phenylacetic acid (**6**) [10].

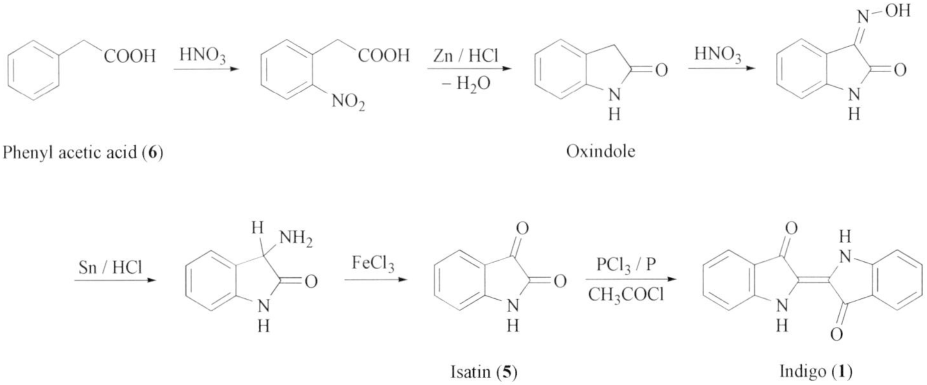

Scheme 3.4 First total synthesis of indigo

Another synthetic route proposed by von Baeyer began with *o*-nitrocinnamic acid (**7**) and led to *o*-nitrophenylpropiolic acid (**8**), which could be converted to indigo directly on the textile fiber with mild reducing agents under alkaline conditions [11]. For a few years o-nitrophenylpropiolic acid was sold commercially as "little indigo".

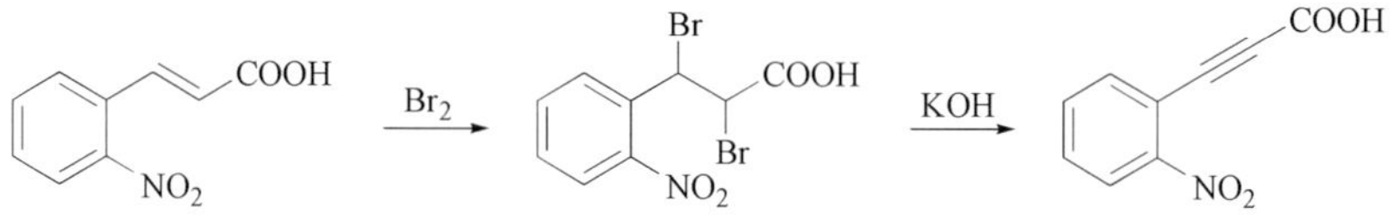

$$\underset{\underset{NO_2}{}}{\text{Ph-CH=CH-COOH}} \xrightarrow{Br_2} \underset{\underset{NO_2}{}}{\text{Ph-CHBr-CHBr-COOH}} \xrightarrow{KOH} \underset{\underset{NO_2}{}}{\text{Ph-C≡C-COOH}}$$

7   8

A. von Baeyer also synthesized indigo by a fascinatingly simple reaction between o-nitrobenzaldehyde (**9**) and acetone in alkaline solution. The product, o-nitrophenyllactic acid ketone (**10**), splits off acetic acid and water and dimerizes to form indigo [12]:

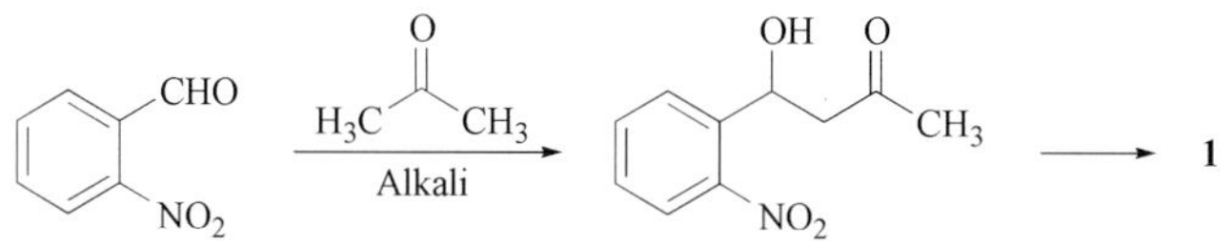

o-Nitrobenzaldehyde (**9**)   o-Nitrophenyllactic acid ketone (**10**)

Hoechst and BASF tried to develop the von Baeyer processes industrially. However, a breakthrough to a cost-effective industrial synthesis was not possible. The nitration step was in each case insufficiently selective and therefore expensive [4].

Heumann I Process. In 1890 Karl Heumann published a synthetic method based on aniline (Heumann I synthesis) [13]. It involved treating aniline with chloroacetic acid to form N-phenylglycine salt **11**, fusing this in potassium hydroxide to convert it to indoxylate **12** (di-salt), and finally hydrolyzing and oxidizing the indoxylate to indigo (**1**). Since the high reaction temperature (300 °C) needed caused partial decomposition, the yield (10 % of theoretical) was too low for large-scale production. Only the stoichiometric addition of sodium amide, discovered by J. Pfleger of Degussa in 1901, as a highly effective condensing agent for the indoxylate melt, produced yields of up to 90 % at reaction temperatures of around 220 °C. Hoechst and BASF launched this process on an industrial scale [14].

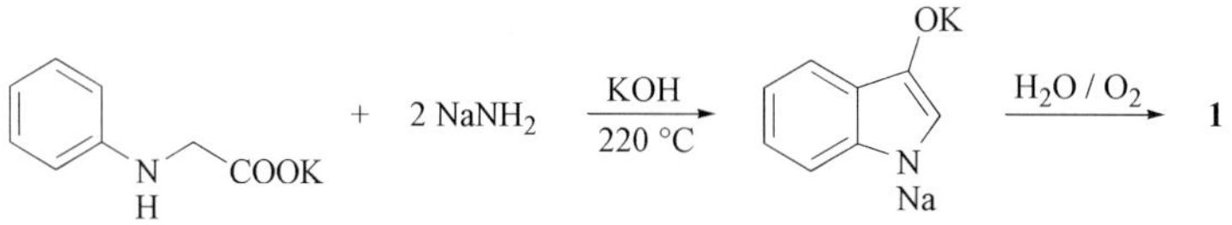

11   **12**
N-Phenylglycine salt   Indoxylate

After the ring-closure reaction, the yellow indoxylate present as the di-Na/K salt in the alkaline melt is hydrolyzed with water. Oxidation of the monosalt of indoxylate takes place in air at 80–90 °C. A suspension of blue indigo in an aqueous alkaline medium results. Small amounts of byproducts, such as aniline, N-methylaniline, and anthranilic acid, are also produced. Details of the manufacturing process are described in the literature [2, 14–18].

Indigo is isolated from the indigo suspension by cake filtration, washed with water, and further processed into the various commercial forms of indigo or vat indigo. The mother liquor from the filtration step can be regenerated and reused in the manufacturing process as an anhydrous alkaline melt.

N-Phenylglycine in the form of an alkali metal salt (**11**) is the starting material for the Heumann I synthesis. It can be prepared in various ways [15] (Scheme 2).

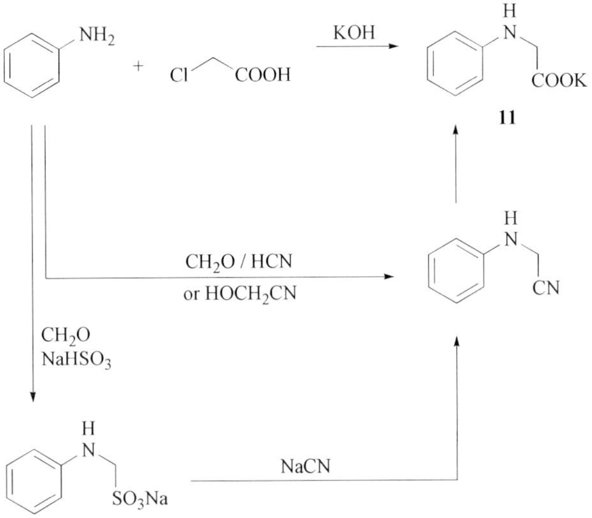

Scheme 3.5 Heumann PG salt synthesis

Heumann II Process. In the Heumann II process, starting from anthranilic acid, N-phenylglycine-o-carboxylic acid, prepared from anthranilic acid and chloroacetic acid, is added in the form of the alkali metal salt **13** to a KOH/NaOH melt at 200 °C to produce indoxylcarboxylic acid salt **14**. After hydrolysis and decarboxylation, the product is oxidized in air to yield indigo [19].

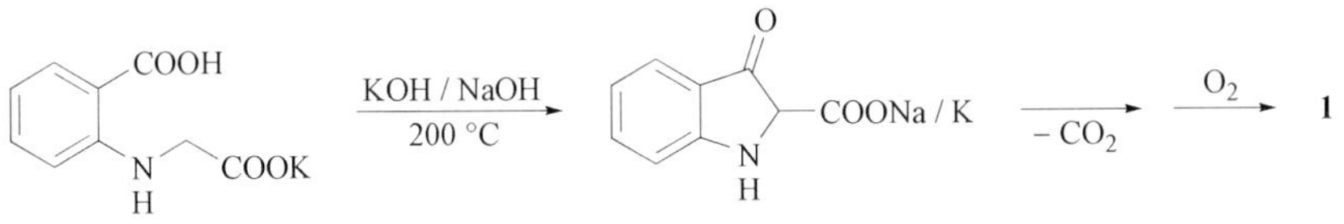

On account of the carboxyl group in the *ortho* position, ring closure occurs more readily than in the case of N-phenylglycine salt **11**. This makes it possible to attain yields of between 70 and 90 %, even without the use of sodium amide. Isolation of indigo from the suspension is carried out as in the Heumann I synthesis by filtration, washing, and drying. This method was employed by BASF from 1897 onwards to produce and market the first synthetic indigo on an industrial scale [4].

Indigo is insoluble in aqueous alkaline media and to be useful in dyeing it must be converted into soluble *leuco indigo* (**3**). In dyehouses this is done mostly by reducing it with sodium dithionite in the presence of alkali.

Indigo (**1**) + $Na_2S_2O_4$ + 4 NaOH ⟶ **3** Leuco indigo disodium salt, vat indigo + 2 Na_2SO_3 + 2 H_2O

(Sodium dithionite; Sodium sulfite)

The idea of liberating the dyer from the burden of vatting the dye and of integrating this reduction step into the synthesis process is not new. For example, indigo can be vatted with reducing agents that were commonly used in the past, such as zinc dust and iron(II) sulfate. However, reduction with hydrogen in the presence of a catalyst such as nickel or palladium is preferable, also from an ecological viewpoint.

Various forms of leuco indigo are available today. BASF Indigo Vat 60 % Grains is produced by evaporating to dryness an aqueous leuco indigo solution in the presence of molasses. The molasses stabilizes the leuco indigo against oxidation. The resulting dye, with an indigo content of 60 %, is a specialty product that, because it is so easy to use, mainly finds application in the production of indigo dyeings for commercial art purposes.

Liquid commercial forms of leuco indigo (**3**) are becoming increasingly significant. Thus, a 20 % and, as new market standard, a 40 % solution of an alkali metal salt of leuco indigo are available. In addition to simple metering, the liquid dye allows dyehouses to dispense with much of the sodium dithionite and alkali needed for dyeing with indigo granules. Furthermore, the resinlike residues produced by reduction with hydrosulfite no longer occur [20, 21].

Environmental Aspects. On account of its low solubility, indigo is degraded to only a very small extent in biological wastewater clarification plants. However, over 90 % is adsorbed onto activated sludge and is thus eliminated from the wastewater. Indigo's classification in Germany's official list of water-polluting substances is WGK 1, corresponding to the lowest polluting potential. A strain of bacteria capable of degrading indigo was discovered by the Hong Kong Institute of Biotechnology [22].

3.5.3.2 Biotechnological Synthesis

General. Suitable plants for producing indigo are the indigo plant (*Indigofera tinctoria*), woad (*Isatis tinctoria*), and Chinese indigo (*Polygonium tinctorium*). The latter is still grown on the island of Shikoku in Japan and is used for blue dyeing [4]. These plants all contain up to 0.8 % of the glucoside indican (**15**). Enzymatic splitting of indican into indoxyl (**16**) and glucose is carried out in a fermentation vat containing a carbohydrate-based material, such as bran or starch, and alkaline additives (potash, lime, or ammonia).

Indican (**15**) →[Fermentation, − glucose]→ Indoxyl (**16**) →[O_2]→ Indigo (**1**)

The enzyme responsible for cleaving indican is indoxyl-β-D-glucosidase, which is formed on the plant by microorganisms. The resulting indoxyl (**16**), which under alkaline conditions is soluble in water, is discharged from the fermentation vat and oxidized to indigo (**1**) by aeration. The insoluble dye is isolated by precipitation and air drying.

The fermentation mix can also be used directly for dyeing. The textile is impregnated with the fermenting mash and then "blued" by oxidation in the air. This type of dyeing was carried out with woad (*Isatis tinctoria*). Pure indigo dye was not extracted from woad because of its low content of indican (**15**).

Microbiological Synthesis [23]. In the 1980s, indigo was successfully produced by microbiological techniques in the USA. The development of a fermentation process involving genetically modified strains of coli bacteria capable of forming usable amounts of indigo from D-glucose represented a new, biosynthetic route to indigo.

Biotransformation of Indoles. It was reported in the technical literature as early as 1956 that bacteria are able to convert indole to indigo [24]. The bacteria use the enzyme naphthalene dioxygenase to oxidize indole to indoxyl. Apparently the substrate specificity of this enzyme is not very pronounced, so that besides naphthalene, which is oxidized to 1,2-dihydroxynaphthalene, it also accepts indole as a substrate [25]. Two process variants for carrying out this biotransformation are described. In the first, the biotransformation is carried out in a homogeneous aqueous system. The critical quantity is the indole concentration, since too high an indole concentration causes the biomass to die off [26]. In the second, the biotransformation is carried out in a two-phase system. The biomass, containing the catalyst, is in the aqueous phase, while the substrate is supplied in an organic solvent [27]. In both process variants, the indoxyl formed is spontaneously oxidized to indigo by atmospheric oxygen.

3.5 Indigoid Dyes

Bacterial De-Novo Synthesis. The basic idea behind this variant is to use the synthetic potential of bacteria to produce the indole precursor (Scheme 3). Although indole (**23**) does not occur as an intermediate in bacterial metabolism, it appears as an enzyme-linked intermediate in the biocatalytic transformation of D-glucose (**17**) to L-tryptophan (**22**). The crucial biosynthetic step is the conversion of indole-3-glycerine phosphate (**21**) to L-tryptophan (**22**) by the enzyme tryptophan synthase.

The essential stages of the multistep route used by nature to synthesize aromatic amino acids were elucidated in the 1950s by studies on mutant bacteria (e.g. *Aerobacter* and *Escherichia coli*): the cyclization of D-glucose (**17**) to 5-dehydroquinic acid (**18**) and the formation of shikimic acid (**19**) [28]. The first aromatic compound in the reaction chain is anthranilic acid (**20**):

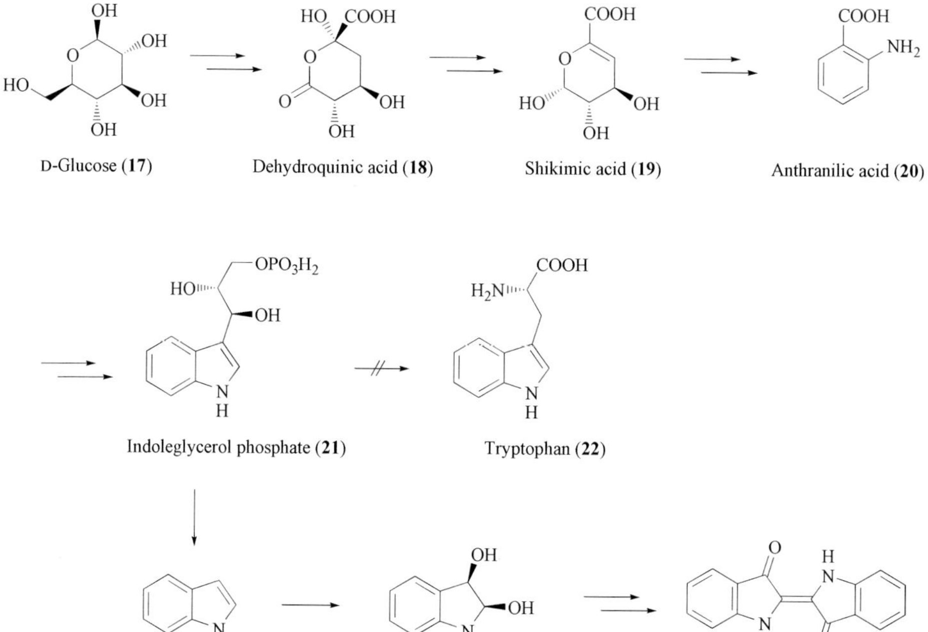

Scheme 3.6 Bioindigo–Shikimic acid path (simplified). Preparation of indigo by bacterial de novo synthesis.

Biotechnological techniques allow the genetic coding for the enzyme tryptophan synthase in *E. coli* to be modified in such a way as to liberate the originally enzyme-linked intermediate indole [29]. If *E. coli* is simultaneously implanted with the coding for naphthalene dioxygenase from the bacterium *Pseudomonas putida* [30], the modified microorganism can convert the liberated indole (**23**) directly to *cis*-indole-2,3-dihydrodiol (**24**), which splits off water and is oxidized to indigo.

A problem in the biotechnological synthesis of indigo is the disposal of the large amounts of biomass produced. Application as a fertilizer is not yet a ready option, because of the possible liberation of genetically modified microorganisms. Alternative disposal methods, such as an efficient clarification plant or incineration, are associated with additional costs.

Recent literature [31] treats the economic aspects and gives theoretical target production costs for biologically produced indigo, but these are scarcely likely to be realized in the foreseeable future.

3.5.4 Commercially Available Dyes

3.5.4.1 Indigo, *C.I. Vat Blue 1*, 73000, [482-89-3]

Commercial grades:
The following commercial forms of indigo are available worldwide:
In nonreduced form as
– indigo granules
– indigo powder
– indigo paste (alkaline)
In reduced form as
– vat indigo 60 %
– vat indigo 40 % (20 %) solution

3.5.4.2 Halogen Derivatives

5,7,5′,7′-Tetrabromindigo, *C.I. Vat Blue 5*, 73 065 [7475-31-2]

Direct bromination of indigo results in bromine derivatives, which have a certain value as colorants [32].

5,7,5′,7′-Tetrabromindigo is prepared by brominating indigo in glacial acetic acid. It has a brilliant, bathochromically shifted shade.

Brominated indigo with a lower degree of bromination is also commercially available.

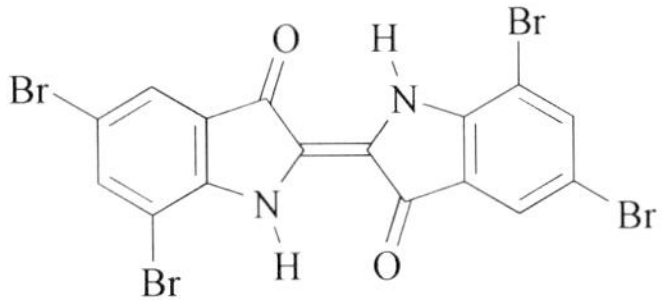

25

3.5.4.3 Other Indigo Derivatives

Indigo-5,5'-disulfonic acid, disodium salt, *C.I. Acid Blue 74*, 73015 [*860-22-0*], Indigotin I, indigo carmine, is approved almost worldwide as a food colorant, and has gained some significance as a dye for foodstuffs, cosmetics, and pharmaceuticals.

3.5.5 References

[1] M. Seefelder, *Indigo: Kultur, Wissenschaft und Technik*, Ecomed Verl.-Ges., Landsberg, 1994; CD Römpp Chemie-Lexikon, 9. Aufl., Thieme, Stuttgart, 1995.
[2] *Ullmanns Enzyklopädie der technischen Chemie*, 4. Aufl., Bd. 13, Verlag Chemie, Weinheim, 1977, pp. 177–181.
[3] R. Neu, *Naturwissenschaft im Unterricht Chemie* 9 (1998) Nr. 48, 44–45.
[4] H. Schmidt, *Chem. Unserer Zeit* 31 (1997) 121–128.
[5] C. Reichardt, *Solvent Effects in Organic Chemistry*, VCH, Weinheim1979, p. 193.
[6] W. Lüttke, M. Klessinger, *Chem. Ber.* 97 (1964) 2342–2357.
[7] M. Klessinger, W. Lüttke, *Tetrahedron* 19 (1963) Suppl. 2, 315–335; E. Wille, W. Lüttke, *Angew. Chem.* 83 (1971) 803.
[8] H. von Eller, *Bull. Soc. Chim. Fr.* **1955**, 1426–1444.
[9] A. Baeyer, A. Emmerling, *Ber. Dtsch. Chem. Ges.* 3 (1870) 514–517.
[10] A. von Baeyer, "Zur Geschichte der Indigo-Synthese", *Ber. Dtsch. Chem. Ges.* 33 (1900) Anl. IV, LI–LXX; H. Brunck, "Die Entwicklungsgeschichte der Indigo-Fabrication", *Ber. Dtsch. Chem. Ges.* 33 (1900) Anl. V, LXXI–LXXXVI.
[11] A. Baeyer, *Ber.Dtsch.chem.Ges.* 13 (1880) 2254.
[12] A. Baeyer, V. Drewsen, *Ber. Dtsch. Chem. Ges.* 15 (1882) 2856; *Ber. Dtsch. Chem. Ges.* 16 (1883) 2205; DRP 19768, 1882.
[13] K. Heumann, *Ber. Dtsch. Chem. Ges.* 1890, 23, 3043.
[14] *Ullmanns Enzyklopädie der technischen Chemie*, 3. Aufl., Bd. 8, Urban u. Schwarzenberg, Berlin, 1957, pp. 748–763.
[15] *Ullmanns Encyclopedia of Industrial Chemistry*, 5th ed., Vol. A14, VCH, Weinheim, 1989, pp. 149–156.
[16] F. Henesy, *J. Soc. Dyers Colour.* 54 (1938) 105–115; J. G. Kern, H. Stenerson, *Chem. Eng. News* 24 (1946) 3164–3181; *Ullmanns Enzyklopädie der technischen Chemie*, 2. Aufl., Bd. 6, Urban u. Schwarzenberg, Berlin, 1930, pp. 233–247; K. Holzach, *Angew. Chem.* 60 (1948) 200–204; *Kirk-Othmer*, 2nd ed., Vol. 7, Interscience, New York, 1965, pp. 621–625; Vol. 11, 1966, pp. 62–580; *Kirk-Othmer*, 1979, 3rd ed., Vol. 8, Wiley, New York, 1979, pp. 363–368.
[17] *Chemistry of Synthetic Dyes*, K. Venkataraman (Ed.), Vol. 2, Academic Press, New York, 1952, pp. 1003–1022.
[18] *Ullmanns Enzyklopädie der technischen Chemie*, Bd. 6, Urban u. Schwarzenberg, Berlin, 1919, pp. 494–508.
[19] K. Heumann, *Ber. Dtsch. Chem. Ges.* 23 (1890) 3431.
[20] H. Schmidt, *Melliand Textilber.* 6 (1997) 418–421.

[21] P. Miederer, Moderne Farbstoffproduktion am Beispiel der BASF-Indigosynthese, in BASF Ökokompendium (Produkte für die Textilveredlung), Ausg. Aug. 1994, Kap. 7, pp. 37–41.
[22] *Am. Dyest. Rep.* 1991, July 6.
[23] H. Schmidt, internal report, BASF Aktiengesellschaft, 1997.
[24] P. G. Miles, H. Lund, J. R. Raper, *Arch. Biochem. Biophys.* **62** (1956) 1.
[25] "Expression of Naphthalene Oxidation Genes in Escherichia coli Results in the Biosynthesis of Indigo", *Science* 1983, *222*, 167–169.
[26] JP 04287691, Mitsubishi, M. Terasawa, M. Goto, K. Uchida.
[27] EP 0315949, Idemitsu Kosan Company, Doi, SeiJi.
[28] I. I. Salamon, B. D. Davis, *J. Am. Chem. Soc.* **75** (1953) 5567, 5572; B. D. Davis, *J. Biol. Chem.* **191** (1951) 315; B. D. Davis, *Science* **118** (1953) 251; U. Weiss, C. Gilvarg, E. S. Mingioli, B. D. Davis, *Science* **119** (1954) 774.
[29] WIPO (=World Intellectual Property Organization) 94/08035.
[30] US 437035 EP 0109583, US 4520103, Amgen Inc., B. D. Ensley; A. Crossway, H. Hauptli, C. M. Houk, J. M. Irvine, J. V. Oakes, L. A. Perani, *BioTechniques* **4** (1986) 320–334.
[31] J.W. Frost, J. Livense, *New J. Chem.* **18** (1994) 341–348.
[32] DRP 128575, 1900, A. Rahtjen; DRP 209078, 1907, Gesellschaft für chemische Industrie in Basel; DRP 226611, 1908, Farbenfabriken vorm. Friedr. Bayer; DRP 144249, 1902, Farbwerke vorm. Meister Lucius & Brüning.

3.6 Sulfur Dyes

3.6.1 Introduction

With a few exceptions, sulfur dyes [1] are employed to dye cellulosic fibers. The main applications are the dyeing of corduroy, velveteen, denim articles, twill work clothes, tarpaulins, and backpack fabrics. The dyeing of the cellulosic fiber component of blended fabrics (polyester/cotton blends above all) is increasingly important.

Sulfur dyes of the Indocarbon type (*C.I. Sulphur Black 6*, 53295 [*1327-16-8*], *C.I. Sulphur Black 7*, 53300 [*1327-17-9*] and *C.I. Sulphur Black 11*, 53290 [*1327-14-6*]) are also used for printing on these fibers. Sulfur dyes find limited use with polyamide fibers, silk, leather, paper, and wood.

3.6.2 Chemical Constitution

Sulfur dyes with a constitution number are defined in the Colour Index by the starting materials, the type of sulfurization, and in some cases the type of work-up. These definitions naturally cannot be precise and were therefore only chosen because little is known about the structures of sulfur dyes.

Even in sulfurization of a defined compound the sulfur dye formed consists of a mixture of products, because sulfurization involves various reactions, including substitution, ring formation, reduction, and oxidation. The mixture contains positional isomers as well as compounds with different sulfur contents and degrees of condensation. The composition and properties of a sulfur dye depend on the nature and the composition of the sulfurization agent, on the ratio of sulfurization agent to starting material, on the ratio of the starting materials to one another (for two or more starting materials), on the reaction medium, on the reaction conditions, on the method of work-up, and sometimes on the subsequent treatment. Commercial products of higher value are the result of extensive optimization.

3.6.3 Synthesis

For further information, see [2].

3.6.3.1 Sulfur and Polysulfide Bake or Dry Fusion

The intermediates for sulfur-bake dyes are heated to between 200 and 350 °C with sulfur alone, and those for polysulfide bake dyes, with sodium polysulfide. The process is carried out primarily in gas-fired rotating steel drums.

In the polysulfide bake process, the starting materials are first steam-treated with aqueous sodium polysulfide solution; the thickened mass is then heated to above 180 °C. In a first reaction step, the nitro groups, if present, are reduced to amino groups.

Hydrogen sulfide generated by dehydrogenation reactions in the fusion process is absorbed by caustic soda solution. The crude melt is ground and dissolved. Insoluble crude melts are digested with caustic soda or sodium sulfide. From the resulting solution, after clarification, the dye is precipitated by air blowing or acidification, then filtered and dried.

Tables 3.7 and 3.8 present some examples in schematic form.

Table 3.7 Sulfur bake dyes [3]

C.I. Name	C.I. no.	Color	Intermediates	Method
Sulphur Brown 52	53320	reddish brown	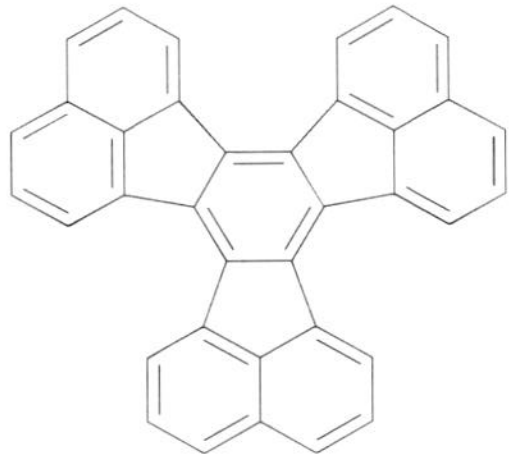 **1**	8 h at 350 °C, digest with NaOH, precipitate dye with H_2SO_4
Sulphur Yellow 4	53160	yellow	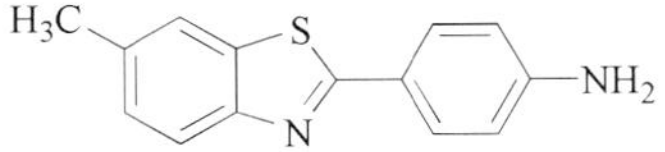 **2**	27 h at 190–225 °C, digest with NaOH, blow with air to precipitate dye
Vat Green 7	58825	olive	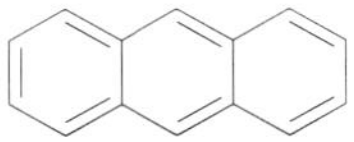 **3**	24 h at 270–300 °C
Sulphur Orange 1	53050	orange	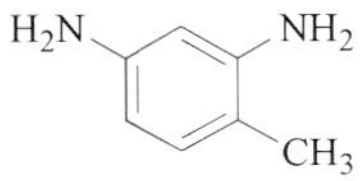 **4**	15–24 h at 190–250 °C, then treat with NaOH or Na_2S; blow with air

3.6.3 Synthesis 217

Table 3.8 Polysulfide bake dyes [3]

C.I. name	C.I. no.	Color	Intermediates	Method
Sulphur Yellow 9	53010	yellow	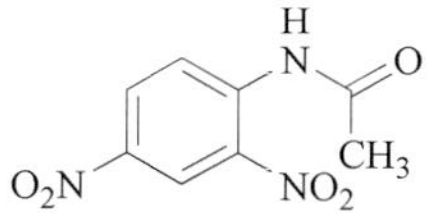 **5** 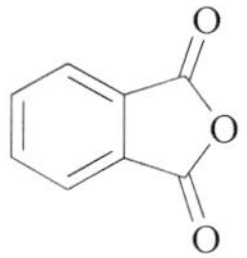	95–130 °C; 10–11 h at 130 °C after addition of salt, then 3–4 h at 180 °C; grind
Sulphur Brown 1	53000	black-brown	sulfite waste liquor from cellulose production	8 h at 270 °C; grind
Sulphur Brown 16	53285	black-brown	**6**	43–45 h at 180–280 °C; precipitate dye with acid after dissolving in H_2O
Sulphur Brown 60	53325	yellow-brown	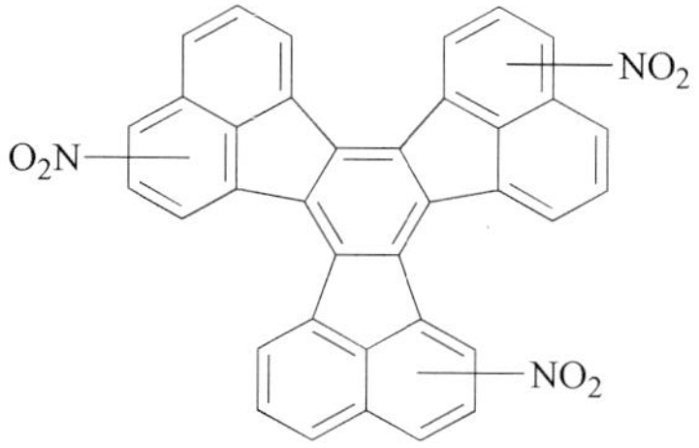 **7**	24 h at 114–116 °C, then 8 h at 240 °C; digest with Na_2S, precipitate dye with acid
Sulphur Black 3	53180	black	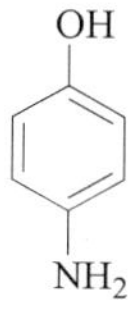 **9** or  **10**	180–210 °C

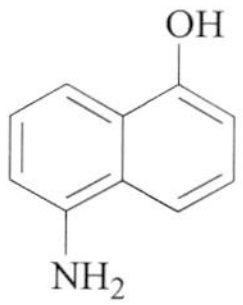

3.6.3.2 Polysulfide Melt, Solvent Reflux, or Reflux Thionation Process

In this process, the reaction of the intermediates — essentially indophenols, chiefly in leuco form, or indophenol-like substances — with sulfur is effected by heating with alkali metal polysulfide in an aqueous or alcoholic medium under reflux. When monoethers of ethylene glycol or diethylene glycol (such as Carbitol) are used, the dye can be obtained in solution form after melting, without distillation and elimination of the solvent. Accordingly, these solvents are used chiefly in the preparation of ready-to-dye sulfur dye solutions. The addition of hydrotropic substances such as sodium xylenesulfonate improves the homogeneity of the melt and hinders sedimentation of the dye on storage.

The process is commonly carried out in jacketed stainless steel vessels with stirrer and reflux condenser. As in dry fusion, dehydrogenation reactions give off hydrogen sulfide, which is absorbed in caustic soda solution.

The crude melt for powdered grades is then processed in a similar way to the dry-fusion product.

Tables 3.9 and 3.10 present some examples of the preparation of polysulfide-melt dyes.

3.6.3.3 Indophenols

All quinoneimine derivatives are commonly called "indophenols" by dye manufacturers, even though they are often indoanilines or indamines. The following methods can be used to prepare indophenols [4]:

1) Oxidation of the starting materials with hypochlorite at 0 °C, with MnO_2, or with dichromate.
   a) A 1,4-phenylenediamine with at least one primary amino group and a phenol with free *para* position [Eq. (1)].

$$R_2N-\langle\!\!\!\bigcirc\!\!\!\rangle-NH_2 \;+\; \langle\!\!\!\bigcirc\!\!\!\rangle-OH \xrightarrow{[O]} R_2N-\langle\!\!\!\bigcirc\!\!\!\rangle-N=\langle\!\!\!\bigcirc\!\!\!\rangle=O \quad (1)$$

   b) An amine with free *para* position and *p*-aminophenol [Eq. (2)].

$$R_2N-\langle\!\!\!\bigcirc\!\!\!\rangle \;+\; H_2N-\langle\!\!\!\bigcirc\!\!\!\rangle-OH \xrightarrow{[O]} R_2N-\langle\!\!\!\bigcirc\!\!\!\rangle-N=\langle\!\!\!\bigcirc\!\!\!\rangle=O \quad (2)$$

2) Condensation of the starting materials in concentrated H_2SO_4, in some cases at –25 to –35 °C.
   a) Secondary or tertiary amine with free *para* position (carbazole, diphenylamine, dimethylamine) and *p*-nitrosophenol [Eq. (3)].

$$R_2N-\langle\!\!\!\bigcirc\!\!\!\rangle \;+\; ON-\langle\!\!\!\bigcirc\!\!\!\rangle-OH \xrightarrow{H_2SO_4} R_2N-\langle\!\!\!\bigcirc\!\!\!\rangle-N=\langle\!\!\!\bigcirc\!\!\!\rangle=O \quad (3)$$

Table 3.9 Sulfur dyes made by polysulfide reflux melting in aqueous medium [3]

C.I. name	C.I. no.	Color	Intermediate	Method
Sulphur Red 6	53720	red-brown	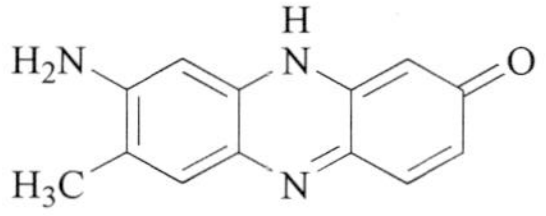 **11** (aminophenazinone with H₂N, H₃C, N–H, N, =O substituents)	30 h at 115–116 °C, add NaCl, precipitate dye by air blowing
Sulphur Red 7	53810	Bordeaux	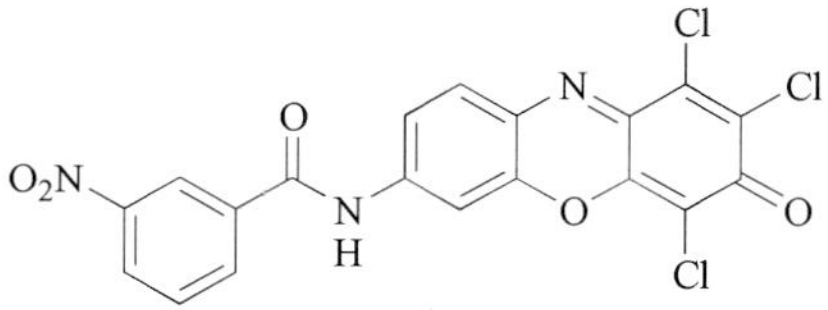 **12** (trichloro-phenoxazinone with O₂N-benzamide group)	30 h at 70–100 °C, then precipitate dye with NaCl
Sulphur Black 1	53185	black	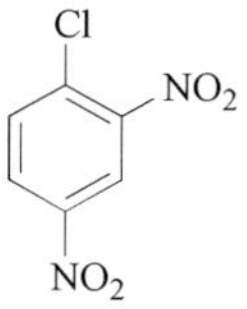 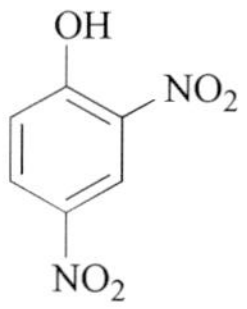 **13** (1-chloro-2,4-dinitrobenzene) or **14** (2,4-dinitrophenol)	saponify DNCB* with NaOH; heat intermediate under reflux for 48–72 h at 110–120 °C or for shorter time at 130–140 °C under pressure; precipitate dye by air blowing

Sulphur Blue 1, 3, 4	53235	navy blue	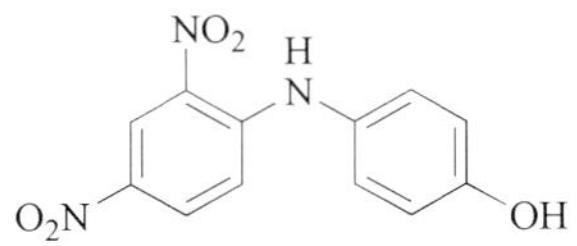 **15** (structure: 2,4-dinitro-N-(4-hydroxyphenyl)aniline)	3 h at 106 °C; add NaCl, then 10 h at 109–110 °C; filter; oxidize slurry in NaOH solution with air, 14–16 h at 40–45 °C; add NaCl and filter
Sulphur Black 9	53230	black	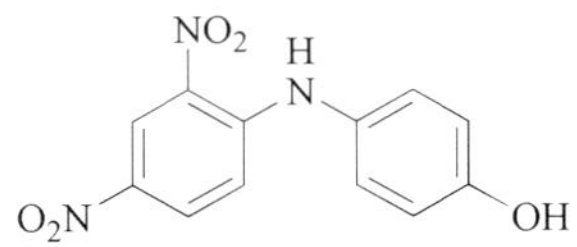 **16** (structure: 2,4-dinitro-N-(4-hydroxyphenyl)aniline)	114–115 °C; dilute, pour into $FeCl_2$ solution, dry
Sulphur Brown 51	53327	brown	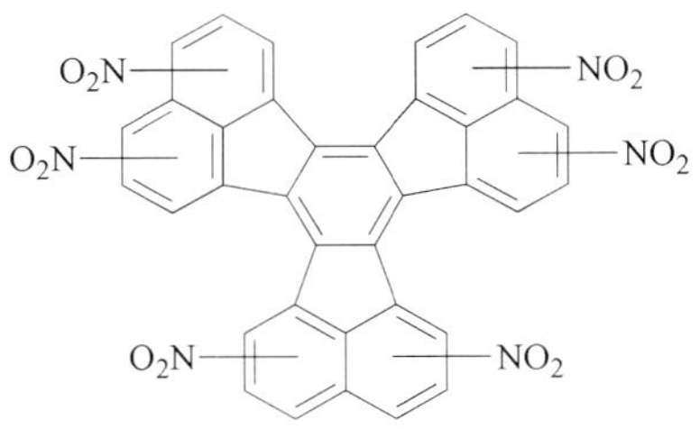 **17** (polycyclic hexanitro structure)	24 h at 114–116 °C, then 8 h at 240 °C; digest with Na_2S; precipitate dye with acid
Sulphur Black 8	53520	gray	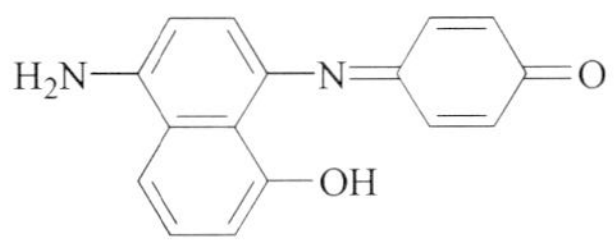 **18** (structure: quinone-imine naphthol with H_2N)	50 h at 111 °C; precipitate dye by air blowing after dissolving in Na_2S/NaOH

* Dinitrochlorobenzene

Table 3.10 Dyes made by polysulfide reflux melting in organic solvent medium [3]

C.I. name	C.I. no	Color	Intermediate	Method
Sulphur Black 11	53290	black	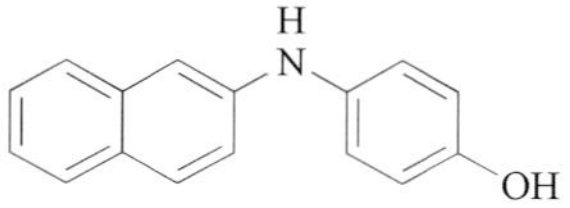 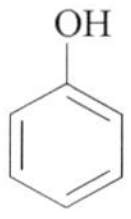 **19**: 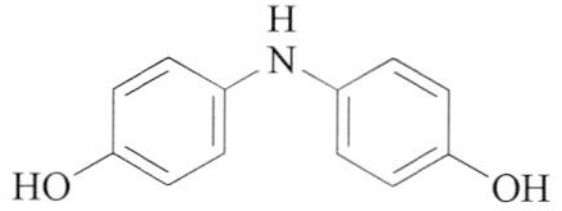 2-naphthyl-NH-C₆H₄-OH; **20**: HO-C₆H₄-NH-C₆H₄-OH; **21**: C₆H₅-OH	30 h at 108 °C in 1-butanol; add NaNO₂, distill off 1-butanol; filter dye and acidify with HCl
Sulphur Blue 10	53470	clear blue	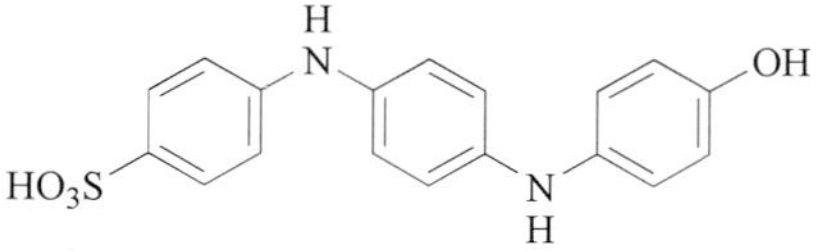 **22**: HO₃S-C₆H₄-NH-C₆H₄-NH-C₆H₄-OH	165–170 h at 84 °C in dilute ethanol; distill off ethanol and blow with air
Vat Blue 43	53630	navy blue	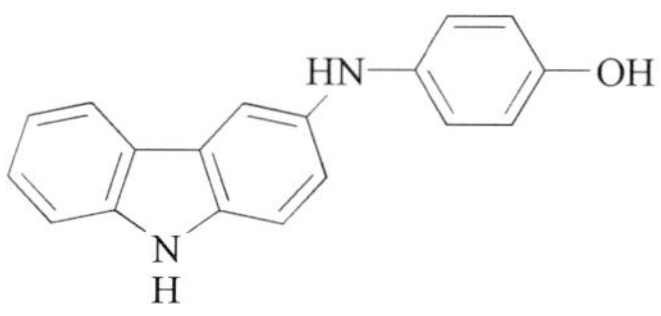 **23**: 3-(4-hydroxyphenylamino)carbazole	24 h at 107 °C in 1-butanol; distill off 1-butanol; precipitate dye by air blowing

Preparation of N-(4-carbazolyl)-p-aminophenol (intermediate for *C.I. Vat Blue 43*) [5]: Dissolve carbazole in 95 % H_2SO_4, cool with stirring to –28 °C; with intensive cooling add to solution of *p*-nitrosophenol in 95 % H_2SO_4 cooled to –27 °C; hold temperature at –23 °C. Add water, ice and powdered iron, using ice to hold temperature below 0 °C. Stir first at –5 °C; filter at 20 °C.

b) *p*-Nitroso derivatives of tertiary amines and phenols with free *para* position [Eq. (4)].

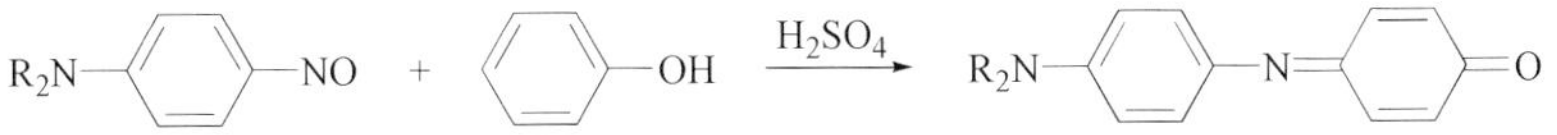

$$R_2N{-}C_6H_4{-}NO + C_6H_5{-}OH \xrightarrow{H_2SO_4} R_2N{-}C_6H_4{-}N{=}C_6H_4{=}O \quad (4)$$

3) Condensation of compounds containing activated chlorine with amines or aminophenols [Eq. (5)].

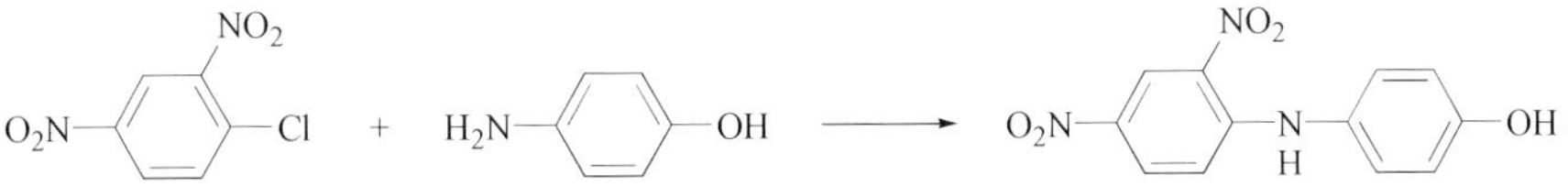

$$O_2N{-}C_6H_3(NO_2){-}Cl + H_2N{-}C_6H_4{-}OH \longrightarrow O_2N{-}C_6H_3(NO_2){-}NH{-}C_6H_4{-}OH \quad (5)$$

4) Condensation of naphthols with *p*-aminophenol in $NaHSO_3$ solution or by dry heating [Eq. (6)].

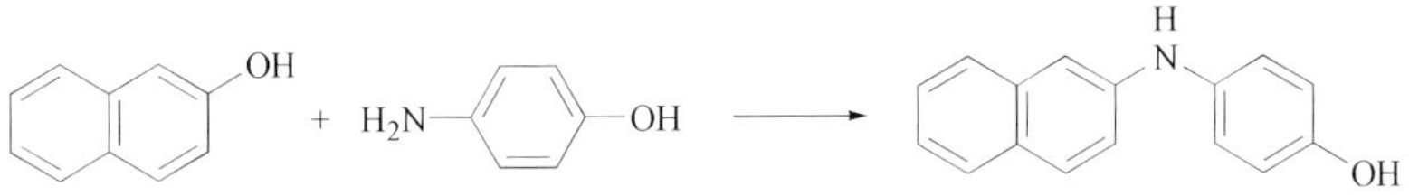

$$\text{2-naphthol} + H_2N{-}C_6H_4{-}OH \longrightarrow \text{naphthyl-NH-}C_6H_4{-}OH \quad (6)$$

5) Condensation of amines with hydroquinone [Eq. (7)].

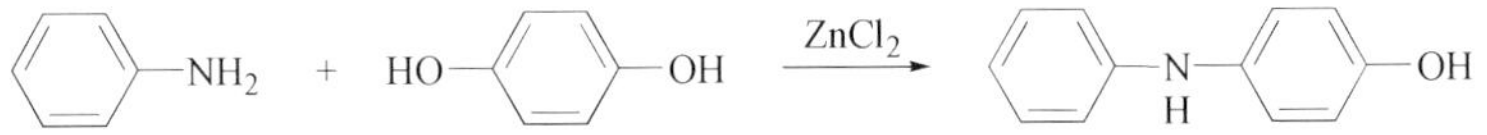

$$C_6H_5{-}NH_2 + HO{-}C_6H_4{-}OH \xrightarrow{ZnCl_2} C_6H_5{-}NH{-}C_6H_4{-}OH \quad (7)$$

Because indophenols are unstable and reactive (they are hydrolyzed by dilute acids), iron, sodium sulfide or other reducing agents are used to reduce them to the more stable leuco form, either in situ or before application.

3.6.4 Modifications; Commercial Forms; Types of Sulfur Dyes

For practical reasons, sulfur dyes are marketed in several forms, which are classified under the following General Generic Names in the Colour Index:

3.6.4.1 C.I. Sulphur Dyes

The first group consists of amorphous powders, insoluble or partly soluble in water, which must be solubilized by heating with sodium sulfide and water according to the Equation (8):

$$\text{DS-SD} + 2\,\text{S}^{2-} \rightleftharpoons 2\,\text{DS}^- + \text{S}_2^{2-} \qquad (8)$$

$$\underset{\text{sulfur dye}}{\text{water-insoluble}} \qquad\qquad \underset{\text{(leuco) dye}}{\text{reduced}}$$

For the sake of simplicity, each dye molecule D is assumed to have only one mercapto group. In strongly caustic alkaline baths, the reaction mode of Equation (9) is also possible:

$$4\,\text{DS-SD} + 2\,\text{S}^{2-} + 6\,\text{OH}^- \longrightarrow 8\,\text{DS}^- + \text{S}_2\text{O}_3^{2-} + 3\,\text{H}_2\text{O} \qquad (9)$$

This formula requires only one-fourth of the reducing agent usually needed. In the case of polysulfide melt dyes, the reducing agent can act not just on the disulfide groups but also on the quinoneimine group (as in vat dyes).

The second group comprises dispersible pigments similar to the Colloisol grades of indanthrene dyes (Paste Fine Caledon, Stabilosol Hydron). These are used particularly in pad dyeing. The dispersible reduced pigments with "Sol" in their names are a special case, containing a portion of reducing agent along with a dispersing agent.

3.6.4.2 C.I. Leuco Sulphur Dyes

These ready-to-dye liquid dyes are in reduced form and contain additional reducing agents for stabilization. Low-sulfide grades are prepared by reduction of the base dyes with caustic/glucose.

3.6.4.3 C.I. Solubilised Sulphur Dyes (Bunte Salts, S-Aryl Thiosulfate Dyes)

These dyes are thiosulfate derivatives of sulfur dyes, also known as Bunte salts. They are prepared by reacting water-insoluble sulfur dyes with sulfite and bisulfite in the presence of atmospheric oxygen [Eq. (10)]:

$$\text{D-S-S-D} + 2\,\text{HSO}_3^- + \tfrac{1}{2}\text{O}_2 \longrightarrow 2\,\text{D-SSO}_3^- + \text{H}_2\text{O} \qquad (10)$$

The dyes dissolve in hot water. They develop fiber affinity only when alkali and reducing agent are added [Eq. (11)]:

$$\text{D-SSO}_3^- + \text{S}^{2-} \longrightarrow \text{DS}^- + \text{S}_2\text{O}_3^{2-} \qquad (11)$$

3.6.5 Pseudo (Synthetic) Sulfur Dyes

If NaSH is used in place of Na$_2$S, alkali must be added to capture the H$^+$ ions produced in reduction, because these would lead to evolution of H$_2$S [Eq. (12)]:

$$\text{D-SSO}_3^- + \text{HS}^- + \text{CO}_3^{2-} \longrightarrow \text{DS}^- + \text{S}_2\text{O}_3^{2-} + \text{HCO}_3^- \quad (12)$$

3.6.5 Pseudo (Synthetic) Sulfur Dyes

Pseudo sulfur dyes are not prepared by the classical sulfurization method. Instead, mercapto groups or their precursors are introduced into dye precursors or pigments by preparative means.

In the case of the dye from copper phthalocyanine (*C.I. Sulphur Green 25*), the synthesis can follow one of two paths:

1) Preparation of a tetranitro copper phthalocyanine by condensation of 4-nitrophthalimide with Cu^{2+} salts in, e.g., nitrobenzene, followed by reduction to the tetramino derivative with, e.g., NaSH, diazotization, and reaction with NaSCN (Sandmeyer reaction) to tetrathiocyanato copper phthalocyanine [6].

2) Sulfochlorination of copper phthalocyanine with chlorosulfonic acid/thionyl chloride to yield the tetrasulfochloride, followed by reduction to the tetramercapto compound (e.g., with Fe/HCl or thiourea) [7] [Eq. (13)].

$$\text{CuPc} \xrightarrow{\text{ClSO}_3\text{H / SOCl}_2} \text{CuPc(SO}_2\text{Cl)}_4 \xrightarrow{\text{Fe / HCl}} \text{CuPc(SNa)}_4 \quad (13)$$

Some of the sulfochloride groups may be saponified to sulfo groups, depending on the reaction conditions; a bluer hue is then obtained.

In the case of the dye from perylenetetracarboxylic diimide [*C.I. Sulphur Red 14* [67923-45-9]], the dianhydride of perylenetetracarboxylic acid is treated with 1,3-diaminobenzene, the diamino compound is then diazotized, and the SCN groups are inserted by the Sandmeyer reaction with NaSCN (**24**)] [8].

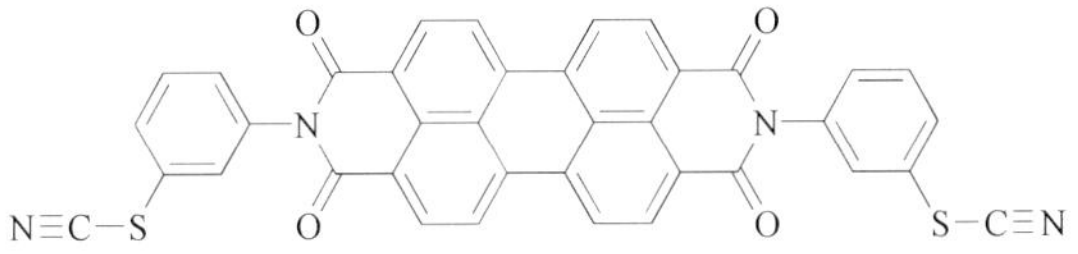

24

The dye *C.I. Solubilised Sulphur Red 11* [90218-69-2] (**25**) contains neither thiolate groups nor their precursors. It is prepared by treating perylenetetracarboxylic dianhydride with 1,4-diaminobenzene-2-sulfonic acid, then with cyanuric chloride and with NH$_3$. It is not known whether the chlorine of the monochlorotriazine groups of the precursor is replaced by a mercapto group in dyeing.

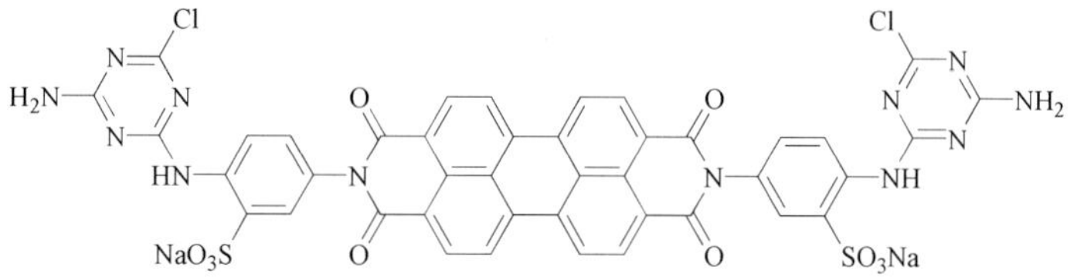

25

3.6.6 Commercially Available Dyes

Exact data on world output or world consumption of sulfur dyes are not available. The quantities cited in statistics refer to commercial dyes with no account taken of their concentration. Thus, highly concentrated powdered types are lumped together with relatively weak liquid grades.

The world production of sulfur dyes may still be more than about 100 000 t/a. The most important sulfur dyes, today as in the past, are *C.I. Sulphur Black 1*, 53185 [*1326-82-5*] and dyes of the Hydron Blue type such as *C.I. Vat Blue 43*, 53630 [*1327-79-3*].

3.6.7 References

[1] C. Heid, K. Holoubek, R. Klein, "100 Jahre Schwefelfarbstoffe" [100 Years of Sulfur Dyes], *Melliand Textilber. Int.* **54** (1973) 1314–1327.
[2] O. Lange, *Die Schwefelfarbstoffe, ihre Herstellung und Verwendung* [The Sulfur Dyes, Their Preparation and Use], 2nd Ed., Spamer, Leipzig, 1925.
[3] *Colour Index*, 3rd Ed., Vol. 3, pp. 3649–3704 and Vol. 4, pp. 4475–4501 (1971); Revised 3rd Ed., Vol. 6, pp. 6375–6385 (1975); The Society of Dyers and Colourists, Bradford.
[4] P. Karrer, *Lehrbuch der organischen Chemie* [Textbook of Organic Chemistry], 9th Ed., Thieme, Leipzig, 1943 pp. 610–612.;. Houben-Weyl, *Methoden der organischen Chemie* [Methods of Organic Chemistry], 4th Ed., Vol. VII/3b, *Chinone* [Quinones], Part II (1979).
[5] FIAT Final Report No. 1313, Vol. II, pp. 309–310.
[6] ICI, GB 541 146, 1941 (N. H. Haddock).
[7] ICI, GB 544 943, 1941 (N. H. Haddock); FR 1 526 096, 1968 (U. Kuhlmann).
[8] ICI, GB 547 853, 1942 (N. H. Haddock).

3.7 Cationic Azo Dyes

3.7.1 Introduction

Cationic functionality is found in various types of dyes, mainly in cationic azo dyes and methine dyes, but also in di- and triarylcarbonium dyes (Section 2.6) and solvent dyes (Section 3.10). Due to their importance, in this book the group of cationic dyes is divided into azo dyes and methine dyes.

Basic dyes of the azo class were among the earliest known synthetic dyes. They were used originally for dyeing cotton mordanted with tannin and potassium antimonyl tartrate and wool from neutral solution. They continue to have minor importance for dyeing leather, paper, plastics, and waxes, and as constituents of graphic arts colors.

The preparation of the first cationic azo dye, Vesuvin, was described by C. Martius in 1863. It is obtained by coupling diazotized m-phenylenediamine to an excess of the same amine. An analogous dye from toluylenediamine was reported by P. Gries in 1878. Chrysoidines, coupling products of aniline or toluidines to m-phenylenediamine or toluylenediamine, were reported by H. Caro in 1875 and O. N. Witt in 1876.

When polyacrylonitrile fibers appeared on the market, an intense research effort in the area of cationic azo dyes was stimulated worldwide at all the leading dye plants. These materials now occupy a place of importance in all significant product lines.

Azo dyes with several cationic charges, which are substantive dyes for cellulose, are increasingly being used for coloring bleached sulfite cellulose.

3.7.2 Chemical Constitution and Synthesis

Due to the great variability of different types of cationic azo dyes chemical constitution and synthesis are outlined together in this section. The general synthesis of azo dyes is described in Section 2.1.

The cationic charge may be introduced into the dye molecule via either the coupling component or the diazo component.

3.7.2.1 Cationic Charge in the Coupling Component

Amino groups, amidine residues, trialkylammonium, and cyclic ammonium groups can serve as carriers of the cationic charge in the coupling components.

3.7 Cationic Azo Dyes

Polyamines as Coupling Components. **Chrysoidines** are obtained by coupling diazonium salts of aniline, toluidines, or mixtures thereof to polyamine coupling components such as 1,3-phenylenediamine, 2,4- or 2,6- toluylenediamine, or mixtures of these diamines. These dyes confer muted yellow to orange shades to paper, leather, and polyacrylonitrile fibers. When they are mixed with malachite green and fuchsin, medium-fast black shades are obtained on polyacrylonitriles. A mixture of chrysoidine with Crystal Violet or Victoria Pure Blue is used to adjust the color of nigrosine hectograph inks [1].

Chrysoidine dye salts of dodecylbenzenesulfonic acid are soluble in glycols and glycol ethers and are used in the production of inks, printing inks, and varnishes [2].

Vesuvin and *Bismarck Brown* products are obtained when *m*-phenylenediamine, 2,4- and 2,6-diaminotoluene, or mixtures thereof are treated with nitrite in acidic solution, or when a mixture of the amines and nitrite is acidified. The final products consist of mixtures of monoazo, disazo, and polyazo dyes, with the disazo species probably predominating in commercial products. The dye is salted out immediately after its formation, and thereby the formation of polyazo dyes is prevented.

Liquid Vesuvin dye solutions can be obtained by diazotization and coupling of aromatic diamines [3] or, if desired, also in mixture with aromatic monoamines [4] in carboxylic acid solutions.

Vesuvines are used chiefly for dyeing paper containing wood pulp.

Heterocycles as Coupling Components. If heterocyclic polyamine coupling components, e.g., 2,4-diamino-6-hydroxypyrimidine, are used as coupling components with aromatic diazonium compounds, dyes are obtained that color polyacrylonitrile in lightfast yellow shades and are characterized by excellent leveling properties [5]. An example is **1** [6979-64-2].

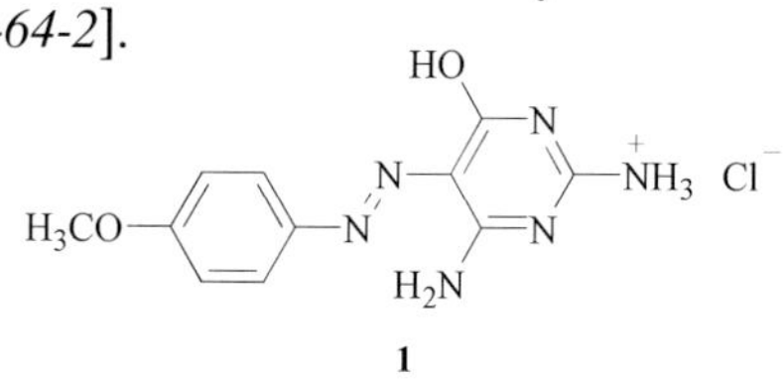

Well leveling azo dyes are also obtained when pyrazolones substituted in the 1-position by an amidine residue are used as coupling components. The diazo components are predominantly substituted by methyl and methoxyl groups [6]. An example is **2** [68936-08-3].

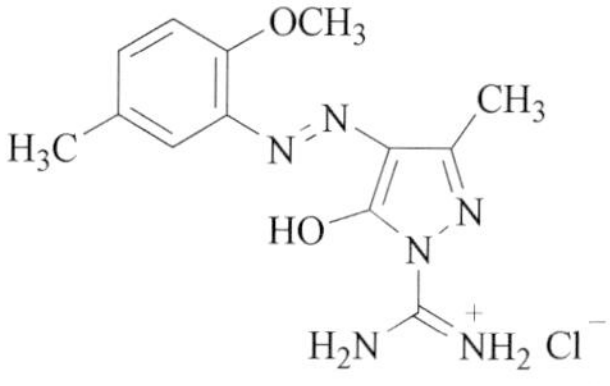

3.7.2 Chemical Constitution and Synthesis

By reaction of these dyes with formaldehyde and acetone, condensation products suitable for wetfast dyeing of cotton are obtained [7].

Coupling Components With Alkylammonium Groups. Coupling components with alkylammonium groups give rise to numerous cationic azo dyes containing a dialkylaminoalkyl or a trialkylammonium group as the charge carrier.

Dyes bearing one or more dialkylaminoalkyl substituents are use for coloring paper because they behave as substantive dyes which, when properly substituted, result in a practically colorless waste liquor. A suitable starting material for the preparation of such dyes is the reaction product of 2 mol dialkylaminopropylamine with 1 mol cyanuric chloride. This can be converted to an acetoacetarylide, which when coupled with diazotized 2-(4'-aminophenyl)-5-methylbenzothiazole, yields the yellow dye **3** [*91458-38-7*] with greenish cast suitable for coloring paper [8].

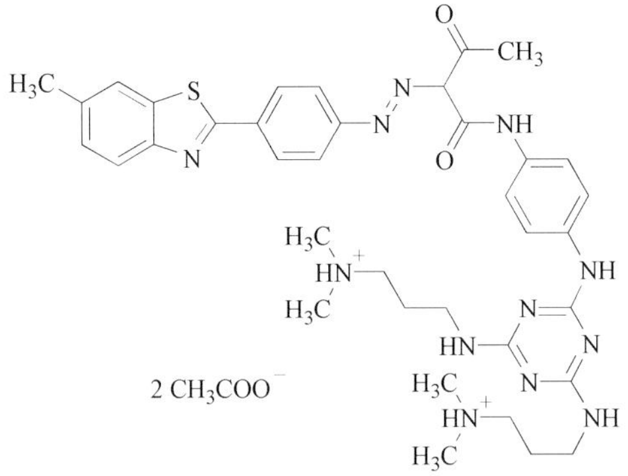

3

Reaction of a similar triazine component with I acid (1-hydroxy-6-aminonaphthalene-3-sulfonic acid), followed by coupling with diazotized 4-aminoazobenzene gives the cationic substantive azo dye **4** [*71032-95-6*], which colors paper in red hues.

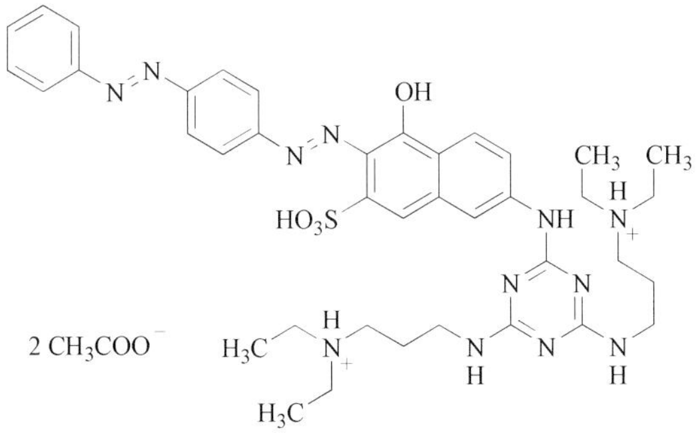

4

Copper complexes of these materials or of analogous dyes derived from γ acid (1-hydroxy-7-aminonaphthalene-3-sulfonic acid) confer gray-violet [9] and blue hues [10, 11] to paper, respectively, or dye particles of glass in navy blue tones [12]. These kinds of triazine moieties may also be part of pyrazolone and pyridone coupling components [13, 14].

The dialkylaminoalkyl residue may also be connected directly to the phenyl residue of the acetoacetanilide. By coupling with diazotized 2-(4-aminophenyl)-benzimidazole, a yellow dye suitable for paper is formed [15]. The dialkylaminoalkyl group may be connected to the N1 atom of 3-methylpyrazolone [16], to the amide nitrogen atom of N-phenylpyrazolone-3-carbonamide [17], to the N2 atom of 3-cyano-2,4,6-trisaminopyridine [18], or to the amino group of 2-amino-4,6-dihydroxypyridine [19].

Amides of 3- or 6-hydroxy-2-naphthoic acid bearing dialkylaminoalkyl residues on the nitrogen atom lead to orange dyes for polyacrylonitrile or paper when coupled with diazotized aniline [20, 21] or to red dyes, soluble in glycol ethers and useful for ink-jet printing, when coupled with diazotized 3-amino-4-methoxybenzenesulfonic acid N,N-diethylamide [22], and to substantive red dyes when the diazo component is a 2-(4'-aminophenyl)benzotriazole [23].

Dialkylaminoalkyl amides of I acid have been used to prepare red dyes for polymers and especially paper, with 4-acetylaminoaniline as diazo component [24, 25].

Exhaustive alkylation of aminonaphthols, such as 2-amino-7-naphthol, and coupling with diazotized aromatic amines generates yellow, orange, red, and brown cationic azo dyes. Compound **5** dyes human hair and polyacrylonitrile in brown shades [26].

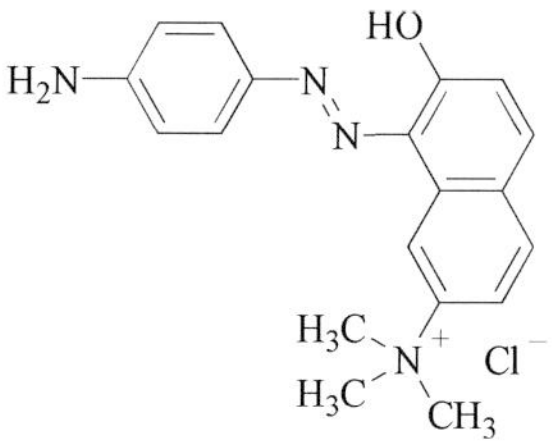

Exhaustively alkylated 4-aminodiphenylamine is also suitable as a coupling component. Dyes such as **6** [*41025-69-8*], obtained by reaction with diazotized nitroanilines, confer fast yellow, orange, and red shades to polyacrylonitrile [27].

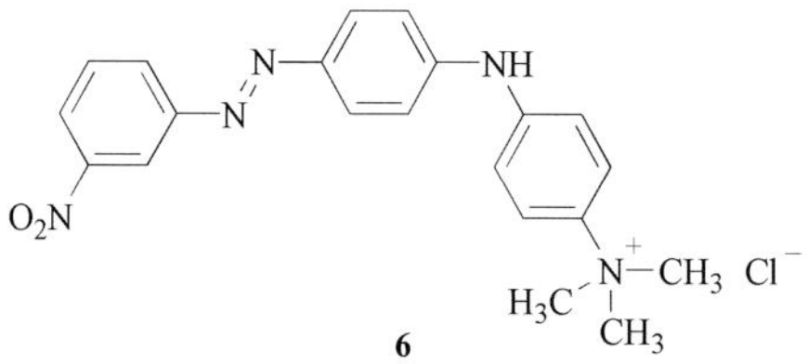

By exhaustive alkylation of 3-aminophenylpyrazolone compounds, a trimethylammonium moiety can be introduced into this coupling component. The disazo dye **7** [*86565-98-2*] colors paper in orange shades [28].

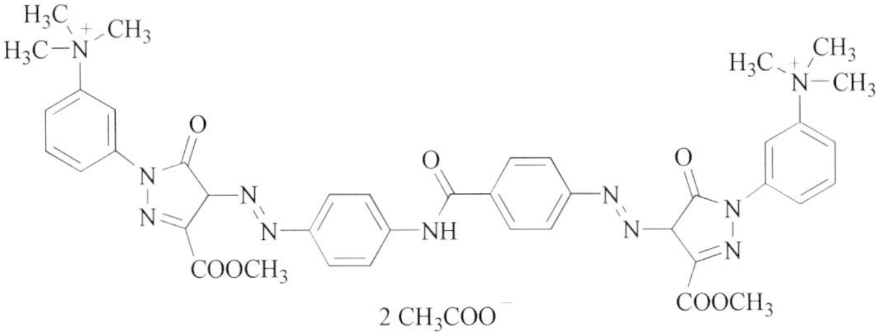

3.7.2 Chemical Constitution and Synthesis

[Structure 7]

The trialkylammoniumphenyl group may also be connected to the coupling component via an ether link, e.g., **8** [*84041-74-7*] [29].

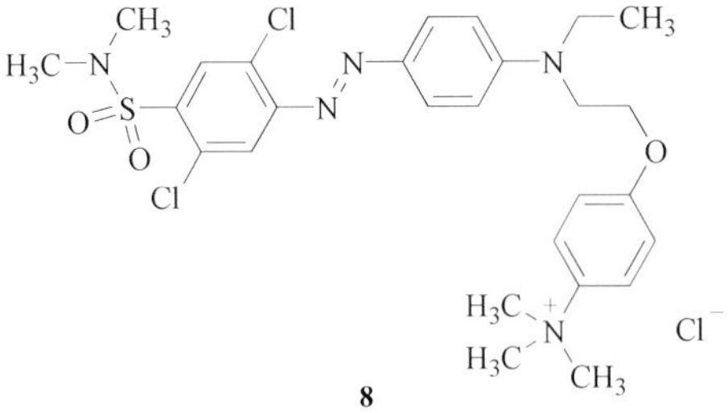

[Structure 8]

Dialkylanilines that bear a cationic substituent at one of the alkyl groups are among the most important coupling components of cationic azo dyes. By reaction of *N*-ethyl-*N*-(2-chloroethyl)aniline or of *N*-ethyl-*N*-(2-chloroethyl)-*m*-toluidine with trimethylamine, ammonium salts are obtained that, upon coupling with diazotized aromatic amines, yield a large number of valuable dyes for coloring polyacrylonitrile. Red shades with a blue cast are obtained with 2-cyano-4-nitroaniline as the diazo component.

Other suitable diazo components are 4-nitroaniline (orange), 2,6-dichloro-4-(*N,N*-dimethylsulfamoyl)aniline (orange), 4-aminobenzophenone (yellow), 2,4,6-trichloroaniline (yellow), 2,4,6-tribromoaniline (yellow), 2,6-dibromo-4-nitroaniline (yellow-brown) [30], and 2-cyano-5-chloroaniline (orange).

Among heterocyclic diazo components an especially favored starting material is 3-phenyl-5-amino-1,2,4-thiadiazole, which is also used in the production of disperse dyes. It yields very bright dyes for coloring polyacrylonitrile [31]. An example is **9** [*85283-77-8*].

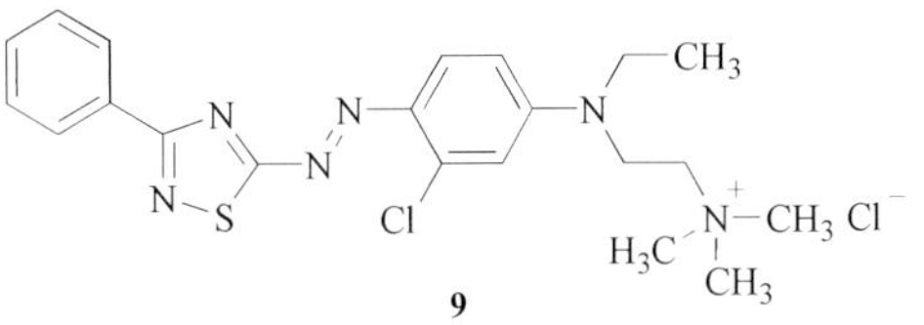

[Structure 9]

By coupling aminobenzothiazoles to the quaternary ammonium bases, red cationic dyes (e.g., **10**) are obtained [32].

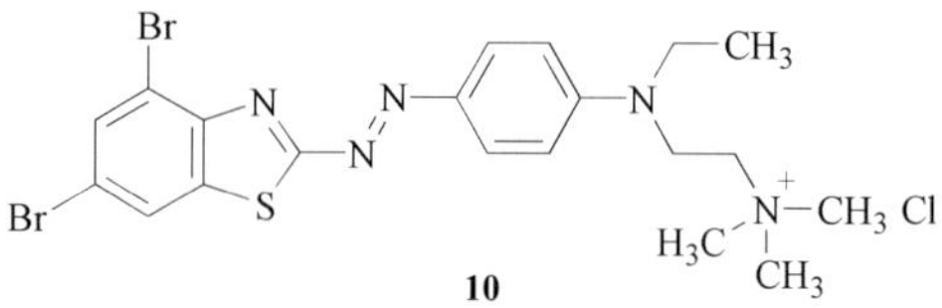

3.7 Cationic Azo Dyes

10

Dye affinity to polyacrylonitrile and hence the dyeing rate can be enhanced by introducing an aralkyl residue [33] or an aryloxyalkyl residue [34] in place of an alkyl group at the trialkylammonium group.

The solubility of these dyes is significantly enhanced by introduction of alkenyl [35], hydroxypropyl [36], or polyetheralkyl [37] moieties into the trialkylammonium group. The trialkylammonium group may also be a component of a heterocyclic ring [38]. When the trialkylammonium residue contains an alkyl chain of more than 11 carbon atoms, the dyes become soluble in aliphatic and alicyclic hydrocarbons and can be used for the production of printing inks [39]. With aromatic diamines as the diazo components, such as 4,4'-diaminodiphenyl sulfone, basic disazo dyes are obtained which are suitable for dyeing acid-modified polyamide materials [40].

The coupling component 2,6-dihydroxypyridine, already an acknowledged component of high-intensity chromophores in disperse dyes, has also been applied successfully to the synthesis of cationic azo dyes. The trialkylammoniumalkyl group can be linked either to the pyridine nitrogen atom [41] (**11**) [*71873-54-6*] or to C3 of the pyridine ring [42] (**12**).

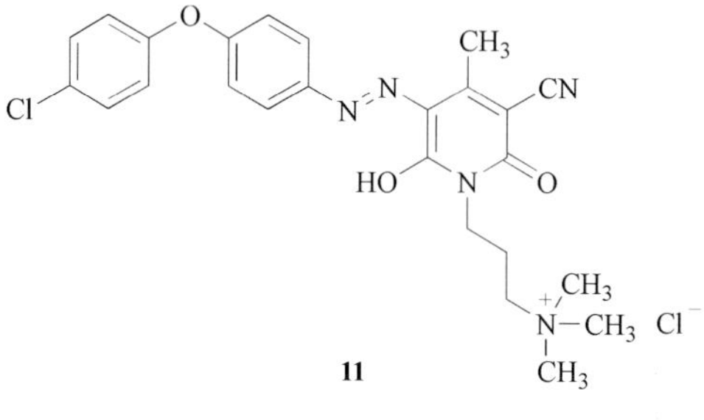

11

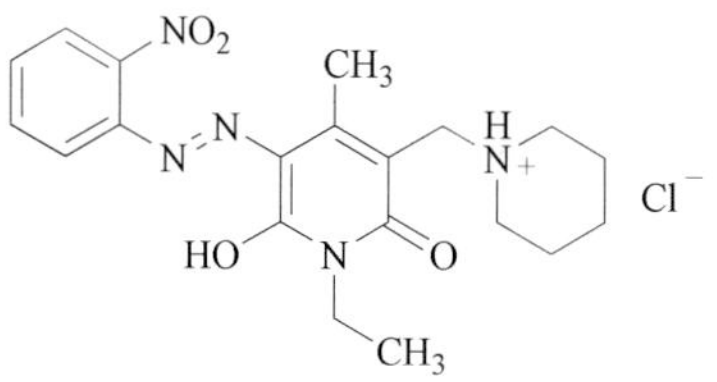

12

Aromatic coupling components can have the trialkylammoniumalkyl linked via a carboxamide group [43], e.g., **13**.

The red dye **14** [*4531-45-7*] is formed with 2-chloro-4-nitroaniline:

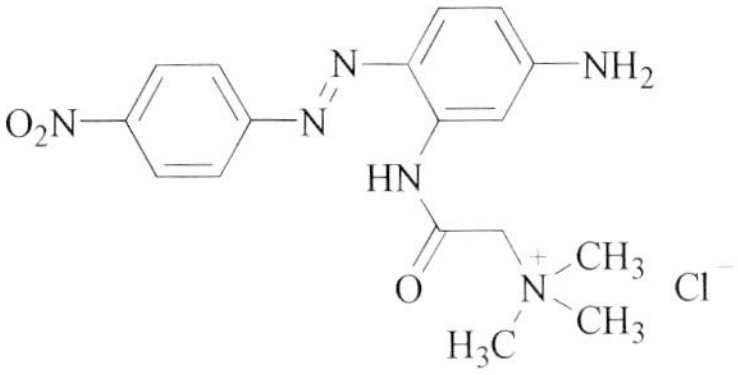

3.7.2 Chemical Constitution and Synthesis 233

13

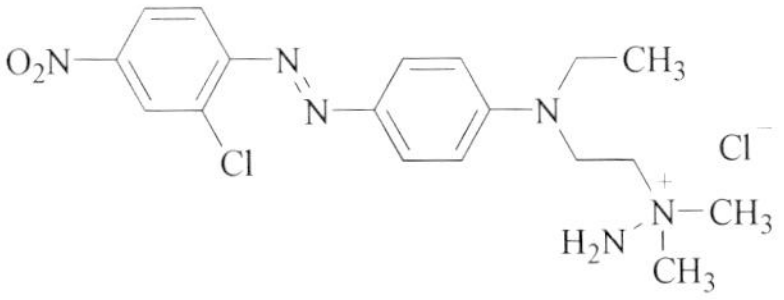

14

Various colors are obtained with different diazo components:

1-amino-2,6- dichloro-4-(*N,N*-dimethylsulfamoyl)benzene:	yellow
1-amino-2,5-dichloro- 4-(*N,N*-dimethylsulfamoyl)benzene:	orange
1-amino-2,6-dichloro-4-nitrobenzene:	brownish yellow
3-phenyl-5-amino-1,2,4-thiadiazole [45]:	red
1-amino-2-cyano-4-nitrobenzene:	ruby
1-amino-2,4-dinitro-6-bromobenzene [44]:	violet

Heterocyclically Linked Cyclic Ammonium Groups. Reaction of chloroacetamide with pyridine, followed by condensation with acetoacetic esters in alcoholic sodium hydroxide, leads to a 2,6-dihydroxy-4-methylpyridine that is substituted in the 3-position by a pyridinium moiety. If this is coupled with diazotized aromatic or heterocyclic amines, yellow to red cationic azo dyes (e.g., **15** [*92691-25-3*]) are obtained [46].

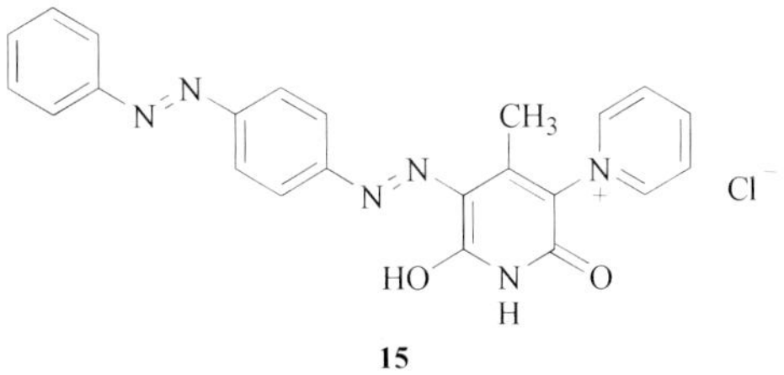

15

With aromatic diamines and triamines as the diazo components, cationic substantive disazo and trisazo dyes (e.g., **16** [*62073-65-8*]) are obtained, which are suitable for bulk dyeing of paper [47].

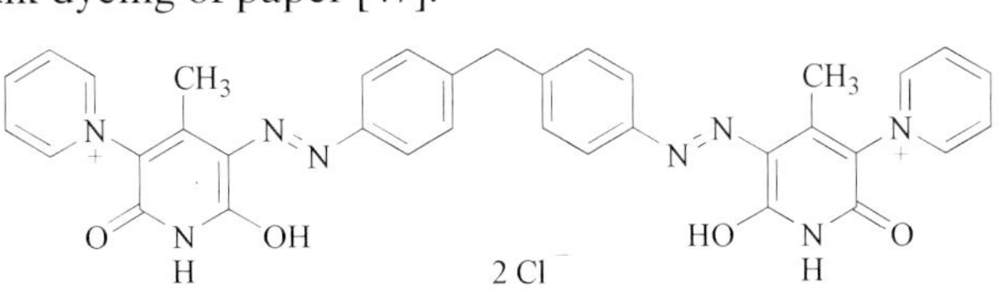

16

Aliphatically Linked Cyclic Ammonium Groups are obtained by reaction of *N*-ethyl-*N*-chloroethylaniline, *N*,*N*-bis(chloroethyl)aniline, or *N*-ethyl-*N*-chloroethyl-*m*-toluidine with pyridine. The red cationic azo dye **17** [*36986-04-6*], obtained with 2-chloro-4-nitroaniline [48], is represented in numerous cationic dye product lines. Introduction of a second pyridiniumalkyl residue (**18**) [*24447-84-5*] leads to a brightening of the hue to orange.

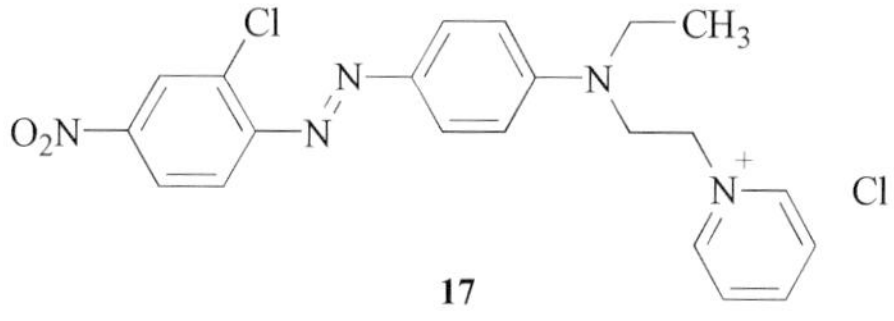

In addition, the following have been described as diazo components: 2,5-dichloro-4-nitroaniline (red) [49], 2-chloro-5-trifluoromethylaniline (yellow) [50], 4-nitro-2-trifluoromethylaniline (red) [51], 2-amino-5-trifluoromethyl-1,3,4-thiadiazole (red) [52], 3-methylmercapto-5-amino-1,2,4-thiadiazole (red) [53], 3-methyl-4-nitro-5-aminoisothiazole (blue-violet) [54], 2-amino-6-chlorobenzothiazole (red) [55], and 3-amino-5-nitro-7-bromobenzisothiazole (blue with a red cast) [56].

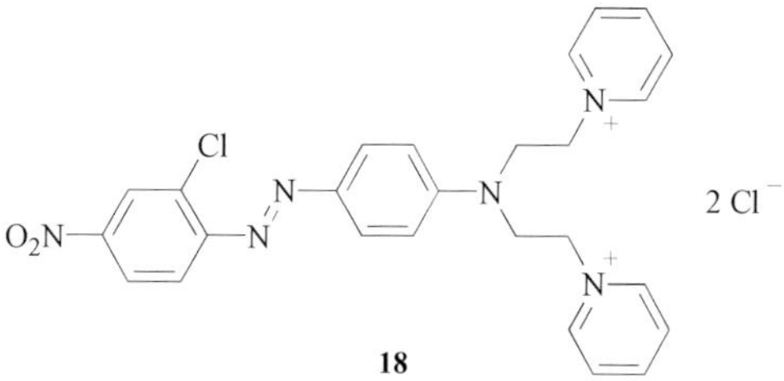

The solubility of these dyes is enhanced when picolines are used for quaternization in place of pyridine [57].

If monoalkylanilines are treated with vinyl heterocycles, especially vinylpyridine, and if the heterocyclic nitrogen is then quaternized, coupling components are obtained that carry the pyridinium component on an alkylidene linkage [58], e.g., **19**.

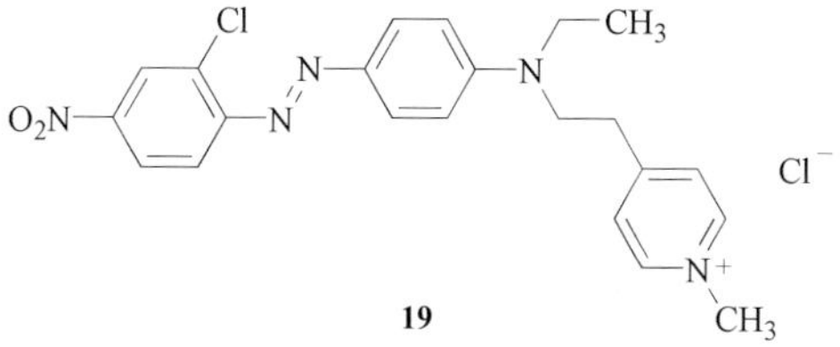

The place of the pyridinium residue may also be taken by the 1,2,4-triazolium [59], the benzimidazolium [60], or the imidazolium [61] moieties. An example is **20** [*47660-05-9*].

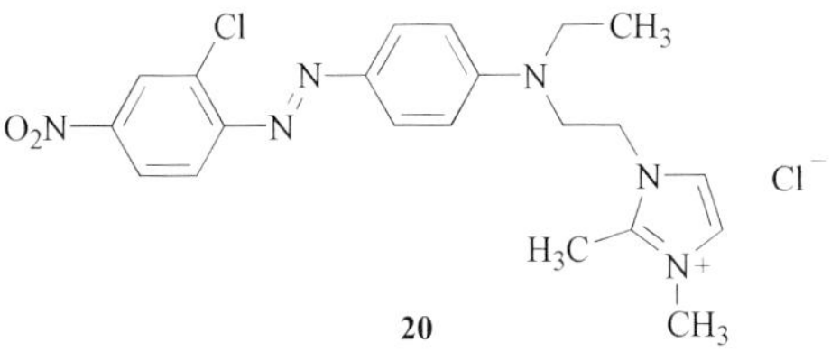

3.7.2 Chemical Constitution and Synthesis

20

The cyclic ammonium residue may also be linked to the coupling component via a sulfonamide [62] or an ester [63] group.

Coupling Components With Condensed Cyclic Ammonium Residues. Heterocyclic compounds in which the condensed benzene ring is substituted by a hydroxyl or an amino group can be coupled with diazonium compounds and may also be quaternized, either prior or subsequent to the coupling reaction, to yield cationic azo dyes. 1,2-Dialkyl-6-nitroindazolium salts are reduced to the 6-amino compounds and then coupled with diazonium salts of aromatic amines. These dyes (e.g., **21**) color polyacrylonitrile in bright yellow to orange shades [64].

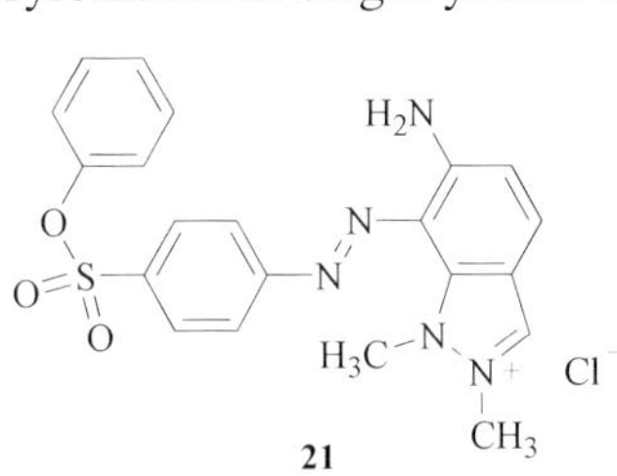

21

6-Amino-1,3-dimethylbenzotriazolium chloride may also be used as the coupling component [65]. By hydrolysis of 6-amino-1,2-dialkylindazolium chloride in dilute sulfuric acid at 180–190 °C, a hydroxyl group is introduced into the 6-position. The 6-hydroxy-1,2-dialkylindazolium salt formed is suitable for the generation of mono- [66] (**22**) and disazo dyes [67] (**23**) [*92888-19-2*].

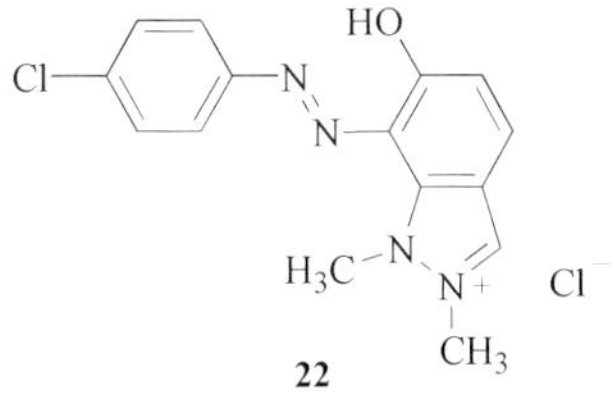

22

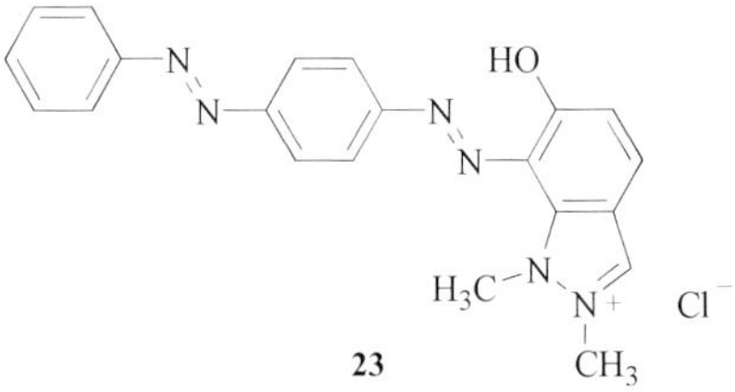

23

In a similar manner 6-amino-1,3-dimethylbenzotriazolium chloride can be converted into the 6-hydroxy compound for use as a coupling component [68].

Coupling Components With Two Different Cationic Groups. Mono- and polyazo dyes have been prepared in which the coupling component carries at least two different basic and/or cationic groups. Of these, mostly yellow to orange dyes are derived from pyridones with a cyclic ammonium group and an amino- or trialkylammoniumalkyl group. To obtain sufficient affinity to cellulose fibers bis-diazo components are preferentially used, either with the same coupling component [69–73], e.g., **24**, or with different ones [74, 75], e.g., **25**.

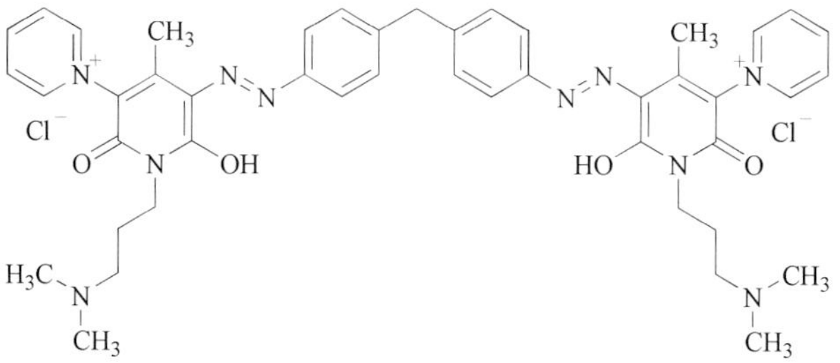

24

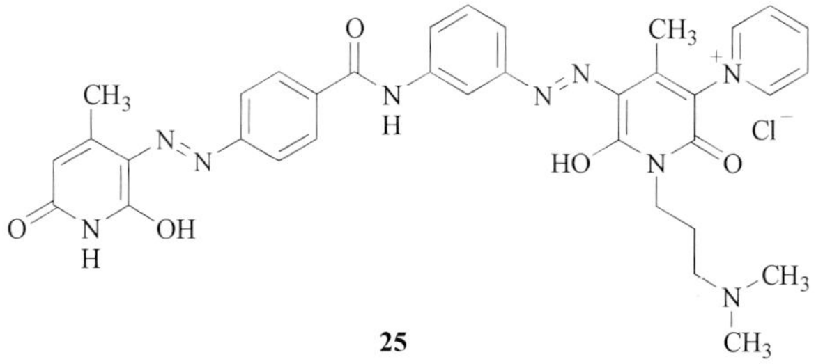

25

A step further is the connection of two suitable molecules by using a dihaloalkane [76, 77], e.g., **26**.

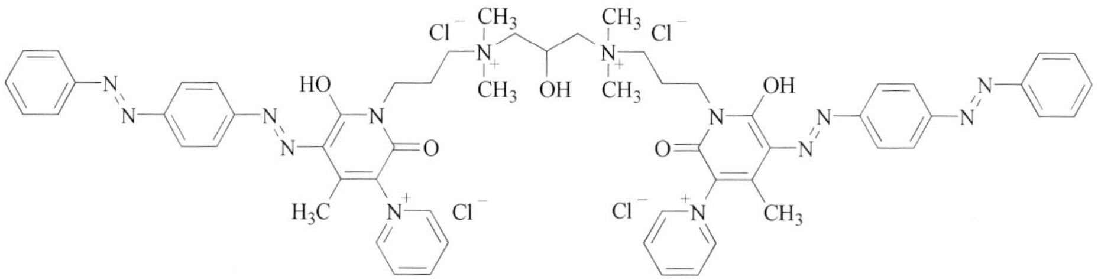

26

Dyes of this kind can be used to color anodically generated oxide layers on aluminum [78].

Orange, red, and violet dyes can be obtained by using as coupling components aminonaphtholsulfonic acids that are connected via the N atom to a 1,3,5-triazine with different aminoalkylamino, trialkylammonium alkylamino, or cyclic ammonium substituents in the 4- and 6-positions [79], e.g., **27**. Dye **28** dyes paper in a scarlet shade [80].

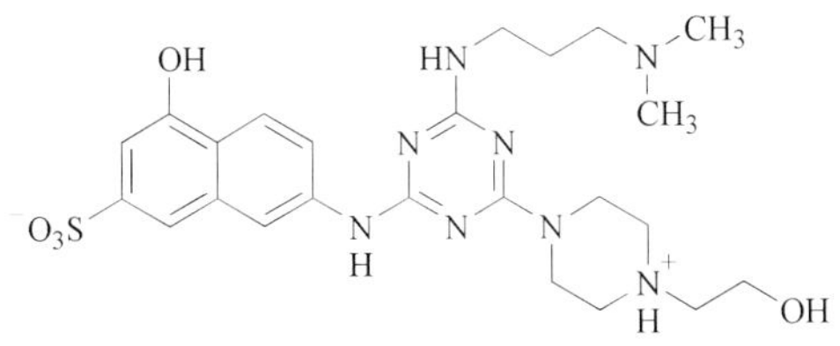

3.7.2 Chemical Constitution and Synthesis 237

27

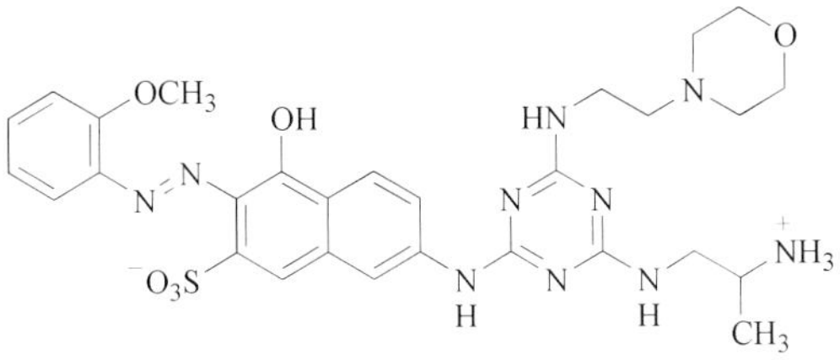

28

Between the N atom of the aminonaphtholsulfonic acid and the triazine ring, spacer groups can be introduced, e.g., aminoaroyl or aminophenylcarbamoylgroups [81, 82].

A different way to synthesize a coupling component with two different basic groups is to treat *N*-dibromopropionylaminonaphtholsulfonic acids with dialkylaminoalkylamines to give aziridine derivatives; red dyes are obtained by coupling these with a diazonium salt of *p*-aminoazobenzene [83], e.g., **29**.

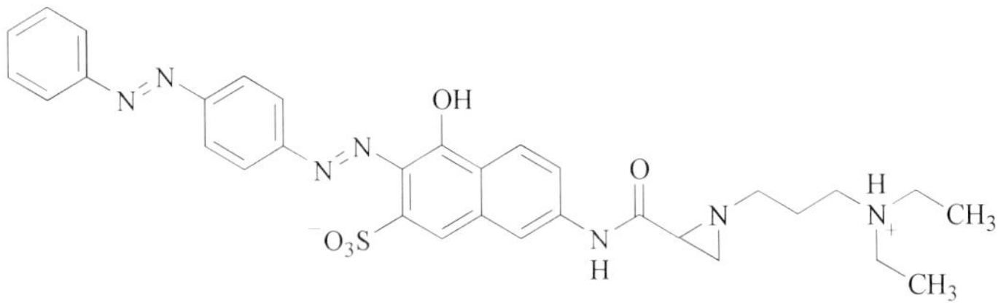

29

3.7.2.2 Cationic Charge in the Diazo Component

Diazo components may contain the cationic charge at amino, dialkylamino, trialkylammonium, or cycloammonium groups.

Diazo Components With Aminoalkyl Groups. Yellow dyes are prepared by aminomethylation of 2-(4′-aminophenyl)-5-methylbenzothiazole followed by diazotization and coupling with acetoacetarylides. Red cationic dyes are obtained with naphthols or hydroxynaphthoic acid arylides. All are suitable for dyeing paper. The bleachability of these dyes is important for the recycling of waste paper [84]. The aminomethylation of aromatic amines is carried out by reaction with formaldehyde and phthalimide, followed by hydrolytic scission of the phthalic acid residue. By diazotization of 4-methoxy-3,5-bis(aminomethyl)aniline trihydrochloride and coupling with 3-hydroxynaphthoic acid arylides, red cationic substantive paper dyes are obtained [85]. Of importance in the manufacture of cationic substantive paper dyes are reaction products of cyanuric chloride with 2 mol dialkylaminoalkylamine and 1 mol 1,3- or 1,4-diaminobenzene.

Yellow paper dyes (e.g., **30**) are obtained by diazotization of these aromatic amines and coupling with acetoacetarylides [86].

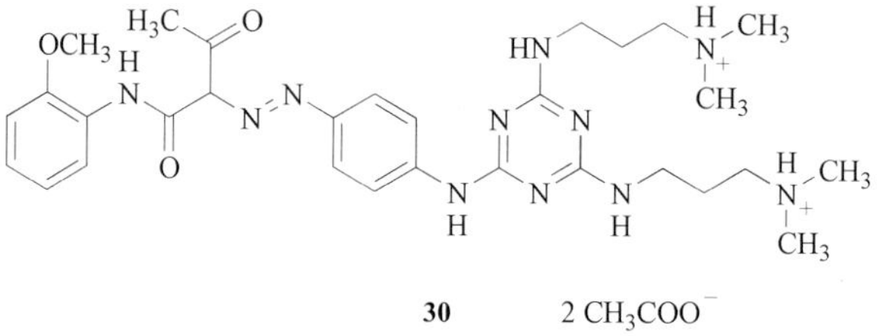

By coupling to derivatives of I acid, substantive dyes (e.g., **31**) are obtained in which the charge on the sulfonic acid group is more than compensated for by the multiple cationic charges. These dye paper in shades of red [87].

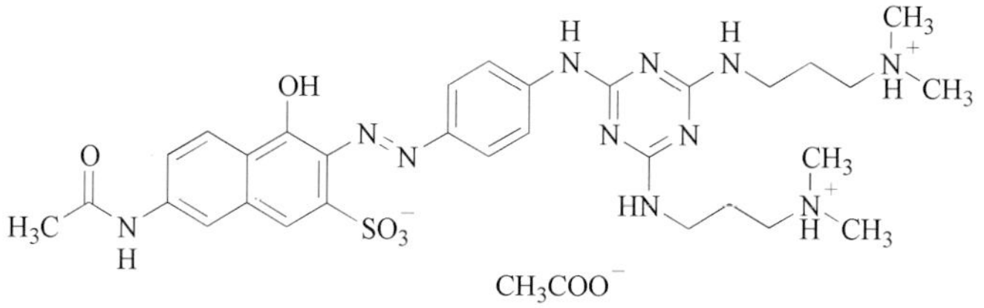

Dyes in which the dialkylaminoalkyl group is connected to the diazo component via a carboxamide [88] or a sulfonamide [89] residue have also been described.

Diazo Components With Trialkylammonium Residues. The trialkylammonium group can be linked to the diazo component either via aromatic substitution or by way of an aliphatic residue.

Aminophenyltrimethylammonium chloride was already an important diazo component of the Janus line of dyes. The fastness of these dyes, e.g., **32**, on polyacrylonitrile is enhanced by introduction of halogens. Mono- and dialkylanilines are suitable coupling components [90].

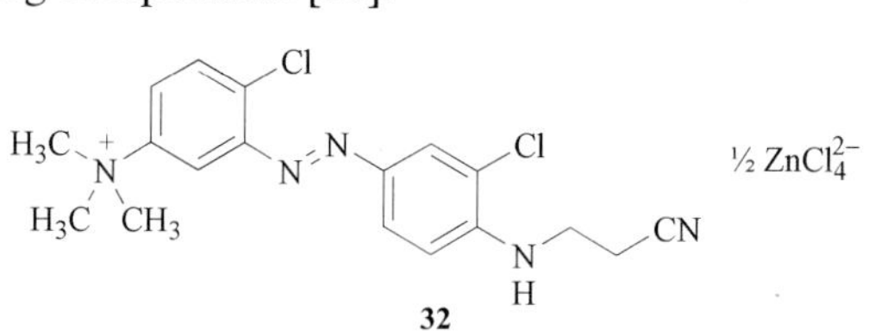

3.7.2 Chemical Constitution and Synthesis

Disazo dyes of this series (e.g., **33**) are suitable for dyeing polyacrylonitrile and paper [91].

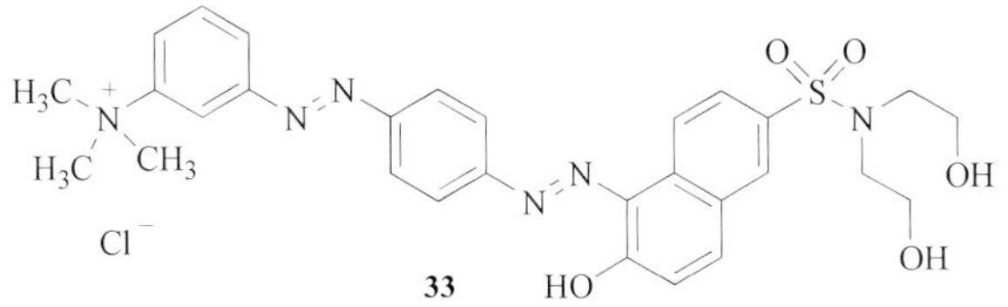

4-Hydroxynaphthalimide as a coupling component yields cationic dyes that confer an orange-red shade to polyacrylonitrile [92].

Heterocyclic coupling components that have been coupled with diazotized aminophenyltrimethylammonium chloride are 1-alkyl-6-hydroxy-2-pyridone [93], 1-amino-3-hydroxy-isoquinoline [94], and 2,4-diamino-6-hydroxypyrimidine [95]. The trialkylammoniumaryl residue may also be connected to the aromatic diazo component via a sulfone or a sulfonamido function [96]. Disazo dyes in this series (e.g., **34**) [77901-21-4] may also be generated from monoazo dyes that still contain a primary amino group by dimerization using phosgene [97] or cyanuric chloride [98].

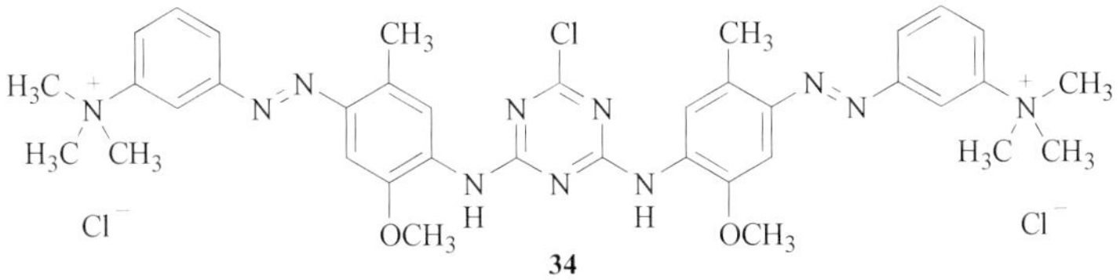

The Friedel–Crafts acylation of acetanilide with chloroacetyl chloride yields 1-acetamido-4-chloroacetylbenzene. The trimethylammonium group is introduced by reaction with trimethylamine, followed by hydrolysis of the acetamide group. This diazo component is a constituent of numerous yellow, orange, and red cationic azo dyes. Using diethyl-*m*-toluidine as the coupling component, the lightfast red dye **35** [67905-12-8] is obtained [99].

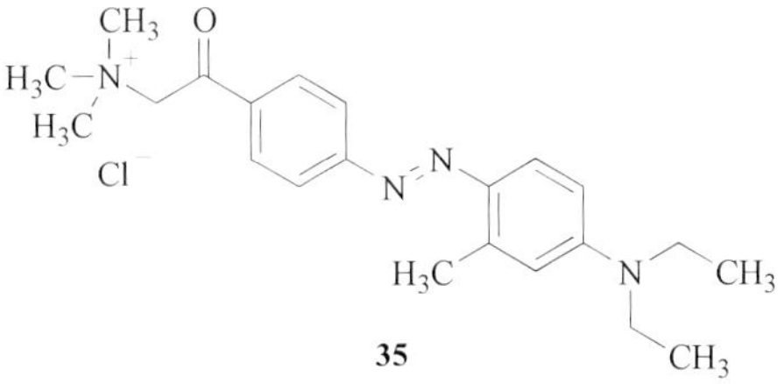

Introduction of an unsaturated alkyl residue improves the solubility of the dye [100]. *N*-Methyl-*N*-cyanoethylaniline yields an orange dye. The shade is shifted to yellow with a red cast by using *N*-cyanoethyl-2-chloroaniline.

Use of 2-methylindole as the heterocyclic coupling component has led to commercially important dyes such as **36** [25784-16-1] [101].

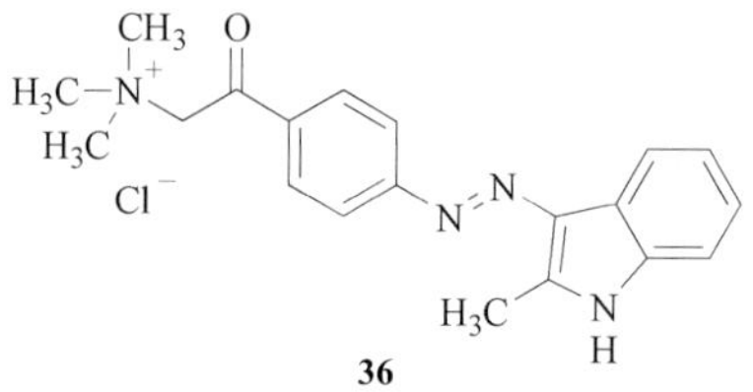

3.7 Cationic Azo Dyes

36

However, dyes have also been generated from 1-phenyl-3-methyl-5-aminopyrazole [101], 2,4-dihydroxyquinoline [102], 2,6-dihydroxypyridine [103], 2,6-diaminopyridine [104], 2,4,6-triaminopyridine [105], and 3-phenyl-5-hydroxyisoxazole [106]. The solubility of such dyes may be enhanced by introduction of polyether moieties [107].

Cationic dyes with reactive functional groups, e.g., dichlorotriazine or vinylsulfone, suitable for dyeing wool and silk, have also been generated in this series [108]. Bis-cationic dyes such as **37** are suitable for dyeing mixed-fiber fabrics containing acid-modified and unmodified polyamide, whereby the latter does not take up the dye [109].

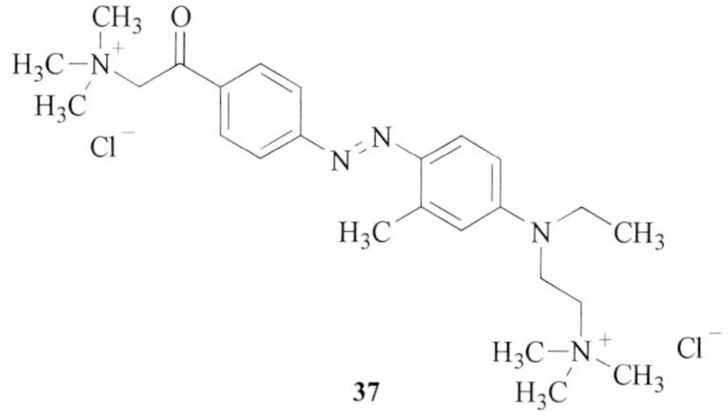

37

The trialkylammonium alkyl residue may also be attached directly to the aryl residue of the diazo component [110]. Dyes that derive from aminomethylated 2-aminonaphthalene and contain acetoacetarylides, naphthols, or pyrazolones as the coupling components are suitable for dyeing paper. An example is the yellow monoazo dye **38**.

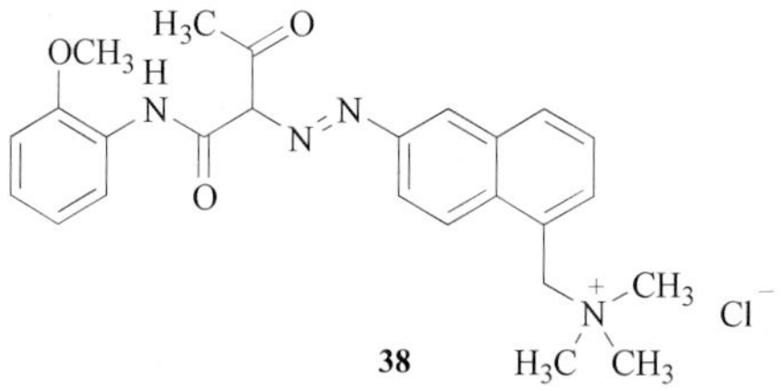

38

Disazo dyes are obtained by using coupling components with a primary amino group and dimerization with phosgene [111]. The diazo components are prepared by exhaustive alkylation of aminomethylated 2-aminonaphthalene-1-sulfonic acid, followed by scission of the sulfonic acid group in the presence of mineral acids [112].

3.7.2 Chemical Constitution and Synthesis 241

The trialkylammonium residue may also be connected to the diazo component by way of an ether [113], a sulfone [114], or a sulfonamide group [115].

Bis-cationic metal-complex dyes are suitable for dyeing leather. Cationic monoazo dyes in which a trialkylammonium group is linked via a carboxamide group to the diazo component [116] and disazo dyes are known. The latter are prepared from dialkylamino compounds by quaternization with alkylene dibromides. They are suitable for dyeing paper [117].

A *cyclic ammonium residue* can either be linked directly to the aromatic ring or attached to the diazo component via an aliphatic residue. Cyclic ammonium residues linked via carbon bonds to the aromatic ring of the diazo component include 1,3-dialkylbenzimidazolium [118], benzothiazolium [119], and pyridinium [120] moieties, linked via nitrogen, as well as the benzotriazolium moiety [121].

Upon reaction with pyridine, azo dyes that carry a chloroethylsulfonamido group in the diazo component form cationic azo dyes, e.g., **39** [*33869-97-5*], which color polyacrylonitrile in brilliant shades [122].

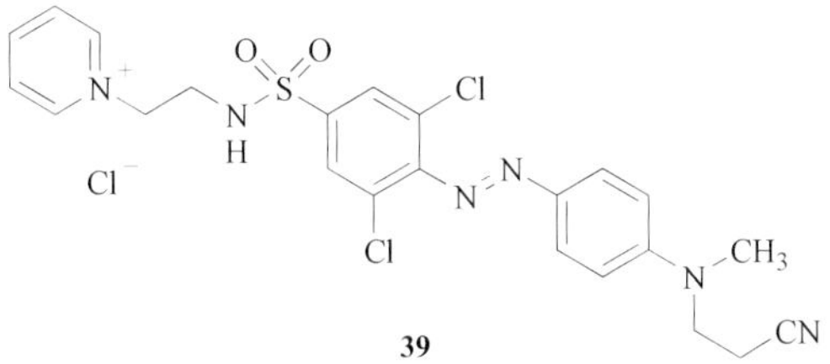

The chloroethylsulfonamide may also be treated with 1,2,4-triazole [123].

Dyes containing the pyridinium group linked to the diazo component via a carboxylic ester function (e.g., **40**) [*32017-47-3*] are obtained by condensation of 4-nitrobenzoyl chloride with chloroethanol, reaction of the ester with pyridine, reduction of the nitro to an amino group, diazotization, and coupling with aromatic amines [124].

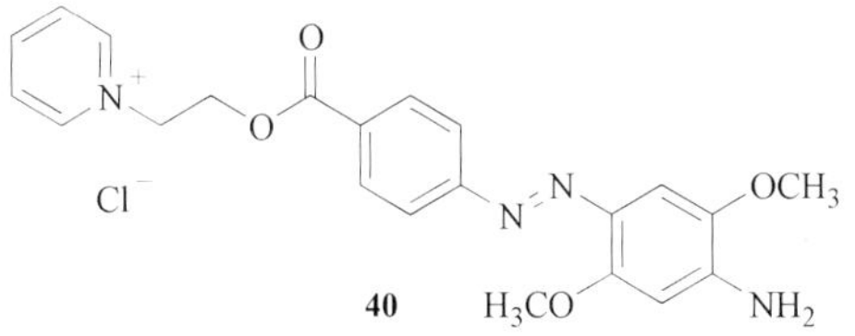

This dye imparts an orange shade to polyacrylonitrile. The cyclic ammonium residues may also be linked to the diazo component via a carboxamide [125] or an ether group [126] or directly by means of an alkylidene residue [127].

Diazo Components With Two Different Cationic Residues. Mono- and disazo dyes have been prepared in which the diazo compound carries at least two different basic and/or cationic groups. Reaction of 2,4- or 2,5-diaminobenzenesulfonic acids with 2-chloro-1,3,5-triazines that are substituted in positions 4 and 6 by aminoalkylamines and cyclic ammonium alkylamines, respectively, leads to diazo compo-

3.7 Cationic Azo Dyes

nents which, when diazotized and coupled with derivatives of aminonaphtholsulfonic acids, give rise to dyes for cotton and leather, and especially for paper [128]. An example is **41**, which colors paper in a violet hue.

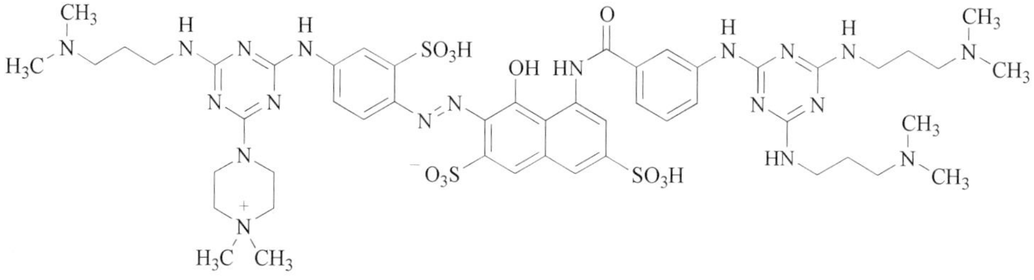

41

Reaction of chloromethyl derivatives of 4′-aminophenyl-2-benzothiazoles with, e.g., *N,N′,N″*-permethyldiethylenetriamine leads to compounds with alkylaminoalkylammonium residues that can be diazotized and coupled with, e.g., acetoacetarylides [129]. Dye **42** is an example that colors paper in a greenish shade of yellow.

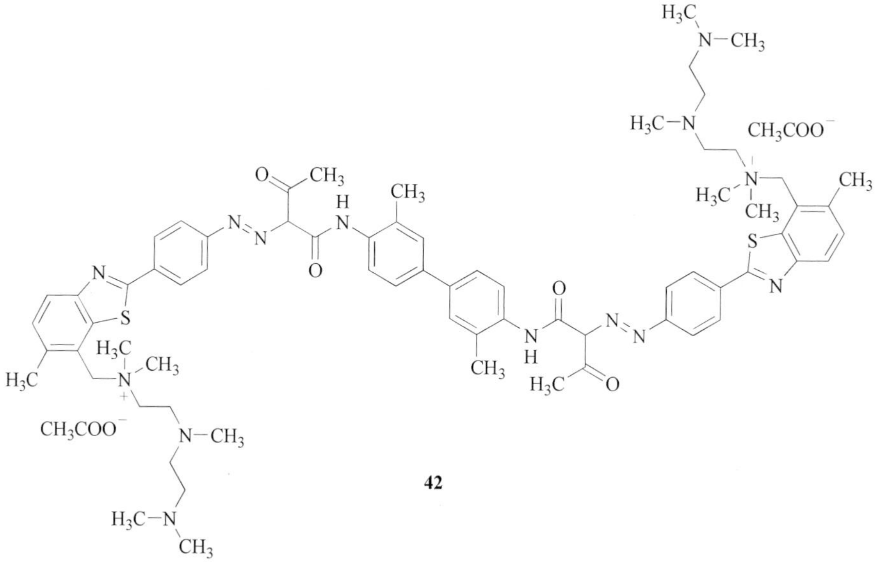

42

Diazo Components With Cycloammonium Groups. Polynuclear N heterocycles that carry an amino group in the carbocyclic aromatic ring can be diazotized and then joined to azo dyes by using aromatic or heterocyclic coupling components. Quaternization at the heterocyclic nitrogen atom may occur before or after coupling. The azo dye **43**, from 2-methyl-5-aminobenzimidazole and 1-phenyl-3-methyl-5-pyrazolone, used in the form of its hydrochloride, dyes paper and leather in clear yellow shades [130].

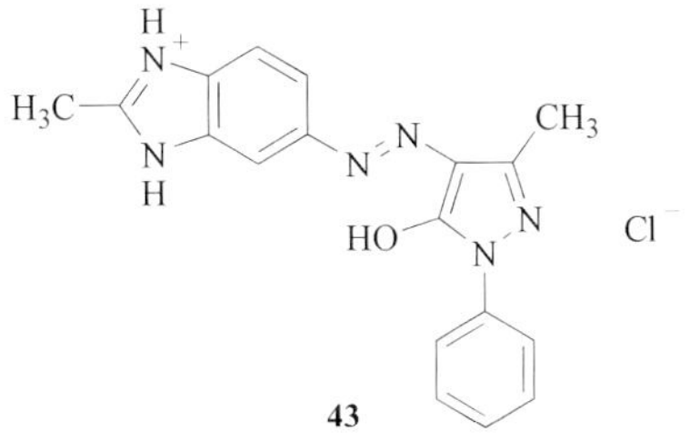

3.7.2 Chemical Constitution and Synthesis 243

43

By exhaustive alkylation, dyes are obtained that are suitable for polyacrylonitrile [131].

Introduction of halogen yields dyes with especially valuable properties. A suitable diazo component is 4-amino-6-chlorobenzimidazole. The sequence of diazotization, coupling with aromatic amines, and quaternization gives colorants (e.g., **44**) [*36116-31-1*] that confer fast orange and red shades to polyacrylonitrile [132].

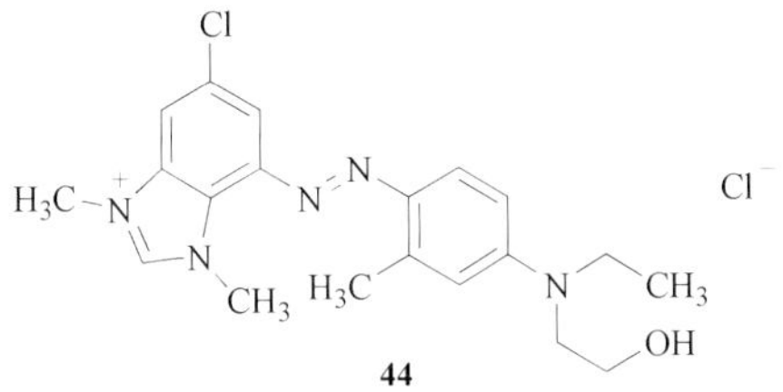

44

A bathochromic shift is obtained by using 4-amino-6-chlorobenzotriazole in place of the benzimidazole. The quaternized coupling products with dialkylanilines (e.g., **45**)

[*36116-25-3*] dye polyacrylonitrile in red shades with a blue cast [133].

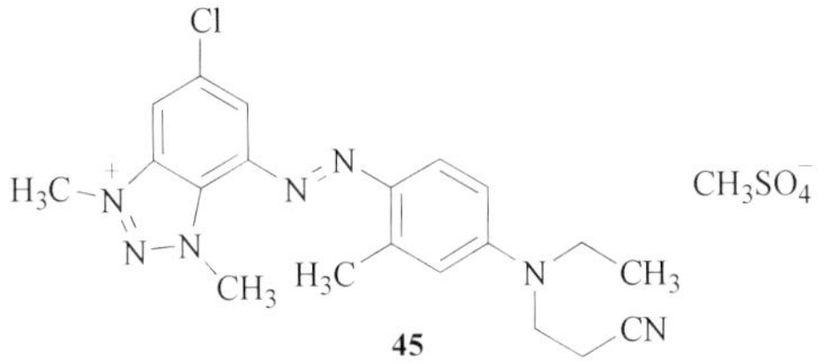

45

Coupling of diazotized N-heterocyclic amino compounds that carry the amino group in the 2- or 4-position to the heterocyclic nitrogen to aromatic or heterocyclic coupling components and subsequent quaternization of the heterocyclic nitrogen atom leads to diazacyanine dyes. If the starting materials are heterocyclic amino compounds with the amino group in a *meta* position, a mesomeric charge exchange is not possible. Therefore, dyes of this type are grouped with the cationic azo dyes.

By coupling of diazotized 3-aminopyridine to 2-naphthol, followed by quaternization, dyes (e.g., **46** [*41313-61-5*]) are obtained that color polyacrylonitrile in fast yellow shades and exhibit excellent leveling characteristics [134].

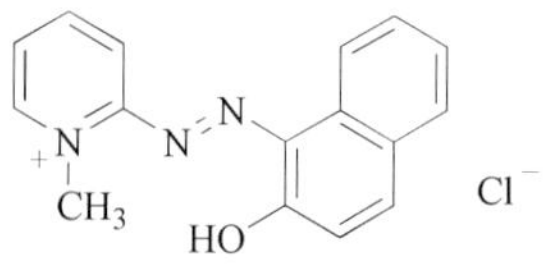

3.7.2.3 Different Cationic Charges in Both the Coupling and the Diazo Component

A further means to introduce different basic and/or cationic residues into a dye has been developed by combining diazo and coupling components that each carry different basic or cationic groups. They are synthesized according to the procedures described above. Dyes have been prepared that color paper in hues ranging from orange [135], e.g., **47**, to red [136–138], e.g., **48**, to black [136], e.g., **49**.

The yellow dye **50** can be used in ink-jet systems to flocculate anionic pigment dispersions, thus improving the bleedfastness [139, 140].

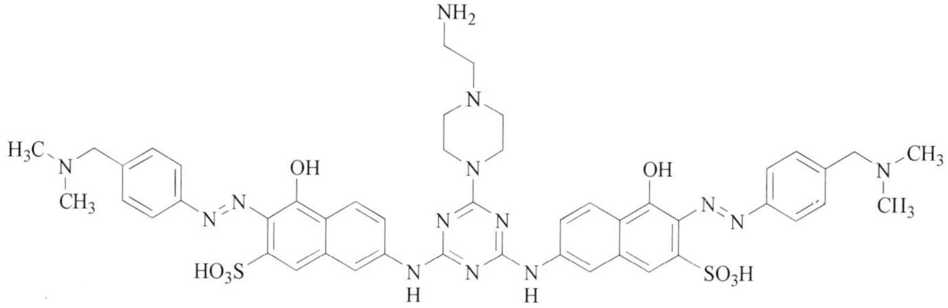

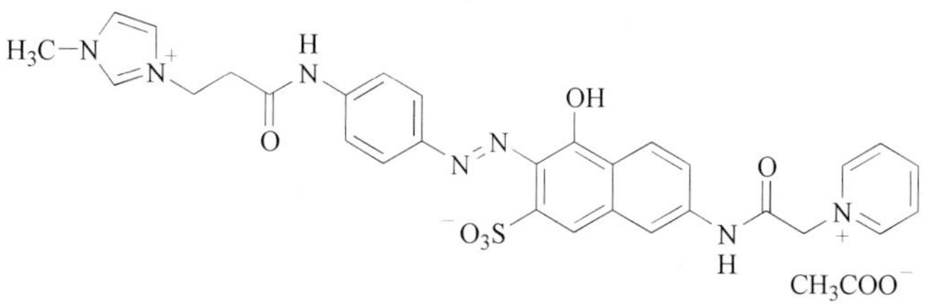

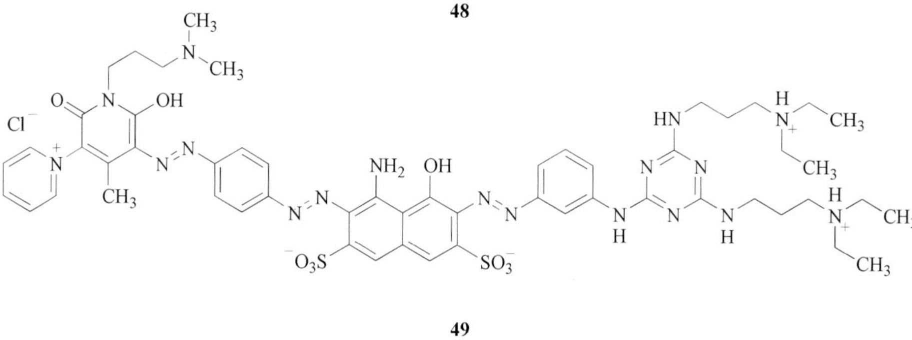

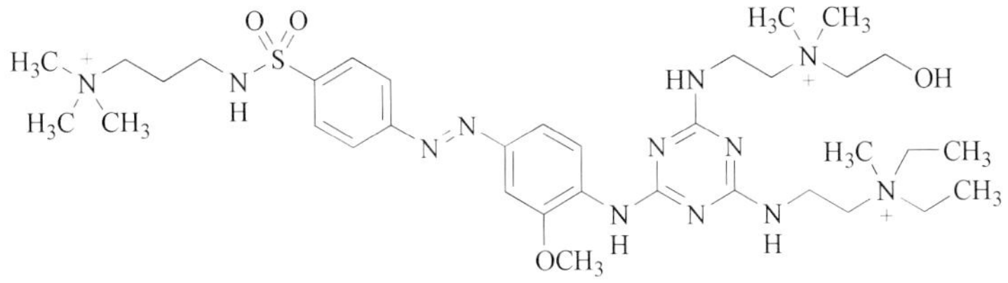

50

3.7.2.4 Introduction of Cationic Substituents into Preformed Azo Dyes

Cationic substituents can also be introduced into preformed azo dyes. In this manner, mixtures of isomers are obtained which may carry one to four cationic groups in various positions. The multiplicity of isomers is a cause for good water solubility. These dyes are predominantly used for coloring paper.

Reaction of chlorosulfonic acid with mono- [141] or disazo dyes [142] yields sulfonyl chlorides, which can then be transformed into cationically substituted sulfonamides by reaction with dialkylaminoalkylamines. The coupling product of diazotized 2-anisidine with 2-hydroxynaphthoic acid arylide upon such treatment dyes paper red, and the azo dye from tetrazotized dianisidine and 1-phenyl-3-methyl-5-pyrazolone gives yellowish orange shades.

Numerous cationic azo dyes are prepared by the action of N-hydroxymethyl-chloroacetamide on azo dyes in sulfuric acid medium, followed by displacement of the reactive chloro substituent by pyridine or trialkylamine. Of special significance for dyeing paper are dyes that are prepared by coupling of diazotized 2-(4′-aminophenyl)–5-methylbenzothiazole to acetoacetarylides, pyrazolones, naphthols [143], or barbituric acid derivatives [144], followed by reaction with N-hydroxymethylchloro-acetamide and pyridine. The azo dye obtained by oxidative dimerization of 2-(4′-aminophenyl)-5-methylbenzothiazole may also be subjected to this conversion [145]. Dye **51** colors paper yellow.

Disazo dyes with β-naphthol as the coupling component give red shades [146].

By removal of the chloroacetyl residue aminomethylated dyes are obtained which are used for coloring paper either as such or upon further alkylation [147]. An example is the yellow dye **52**.

Reaction of azo dyes with formaldehyde and 4-methylimidazole, followed by alkylation, produces cationic azo dyes that dye paper with excellent wetfastness [148]. An example is
the disazo dye **53**, which confers a brilliant red shade to paper.

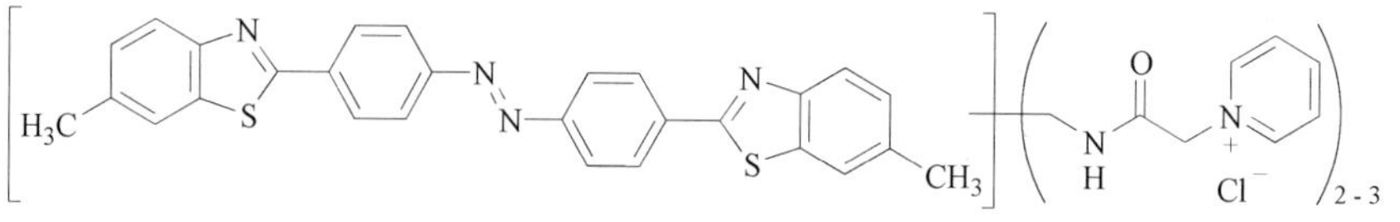

3.7.2.5 Cationic Dyes with Sulfur or Phosphorus as Charge-Carrying Atoms

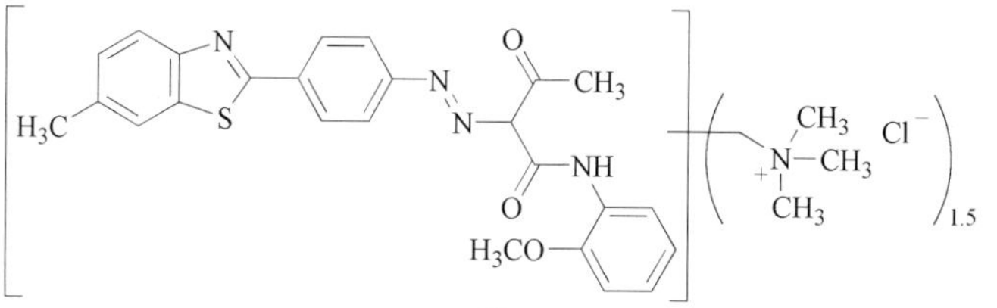

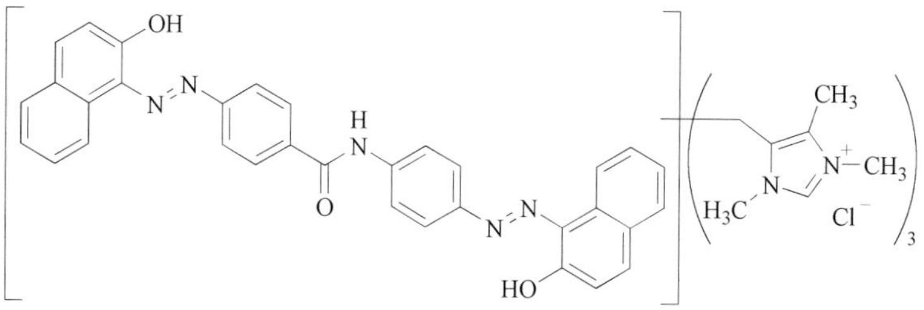

When dyes containing anionically displaceable groups are reacted with thioethers, cationic dyes (e.g., **54**) are obtained in which the sulfonium moiety carries the charge [149].

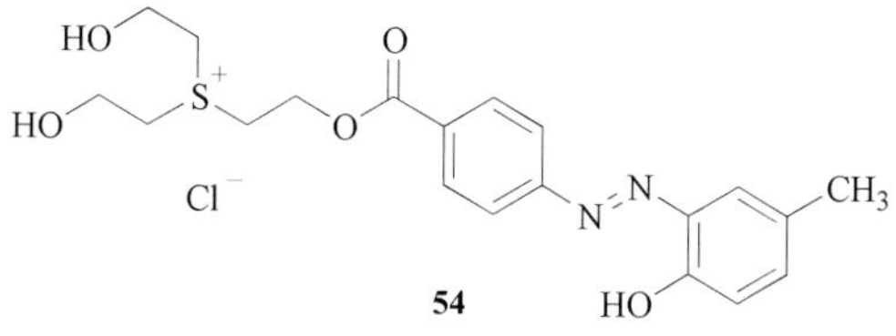

By treating dyes that contain a chloroethyl group with thiourea, the cationic charge may be introduced in the form of an isothiouronium residue (e.g., **55**) [150].

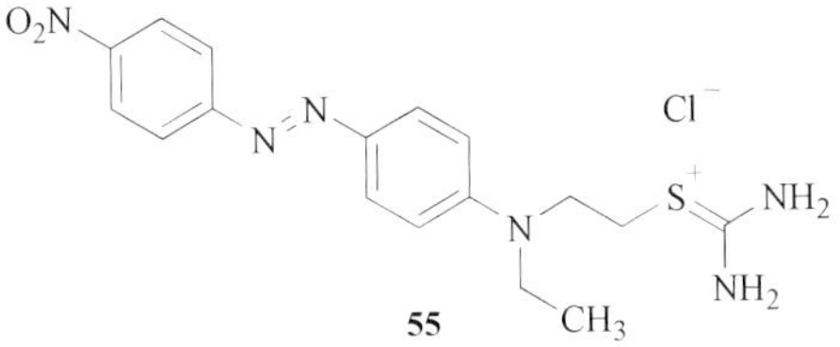

[Structure 55]

To introduce phosphorus as the charge-carrying moiety, dyes that contain haloalkyl groups are treated with phosphines such as dimethylphenylphosphine (e.g., **56**) [151].

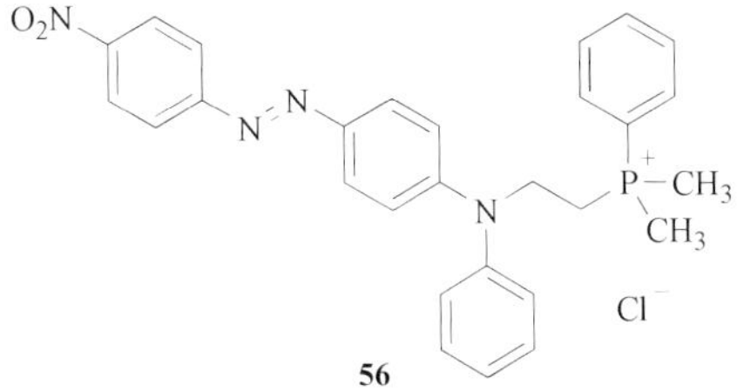

[Structure 56]

3.7.2.6 Dyes with Releasable Cationic Groups

Dyes with releasable cationic groups are converted to water-insoluble dyes when heated in the dye bath. They can then enter textile materials that are amenable to dyeing with disperse dyes. Using this method, for instance, it is possible to dye polyester materials without having first to generate a finely dispersed form of the colorant with the aid of surfactants. An example of such a releasable group is the isothiouronium moiety [152]. Dyes with isothiouronium groups are also suitable for wool [153].

Reaction of dyes containing a primary amino group with dimethylformamide and an inorganic acid chloride, e.g., phosphoryl chloride, permits introduction of the formamidinium group (e.g., **57**), which is also scissioned off upon heating in the dye bath [154]. A similar reaction occurs with the trialkylhydrazinium moiety obtained by reacting formyl-substituted azo dyes with dialkylhydrazines and subsequent quaternization [155].

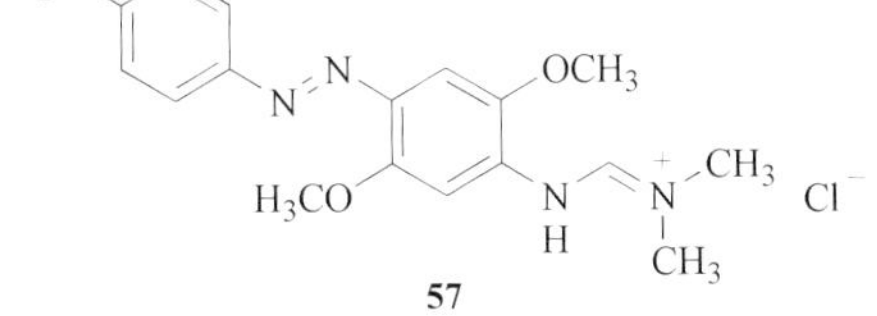

[Structure 57]

3.7.3 Synthesis

The syntheses of the numerous groups of cationic azo dyes are described under each specific group in Section 3.7.2.

3.7.4 Examples of Commercially Available Dyes

C.I. Basic Yellow 15, 11087 (**58**) [72208-25-4].

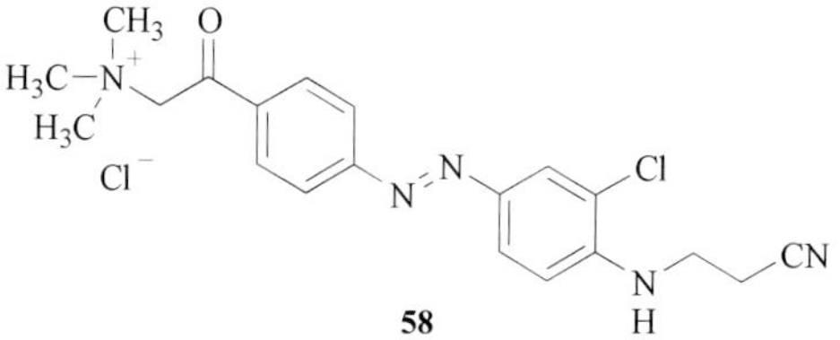

C.I. Basic Orange 1, 11320 (**59**) [4438-16-8] is the azo dye from diazotized aniline and 2,4-diaminotoluene.

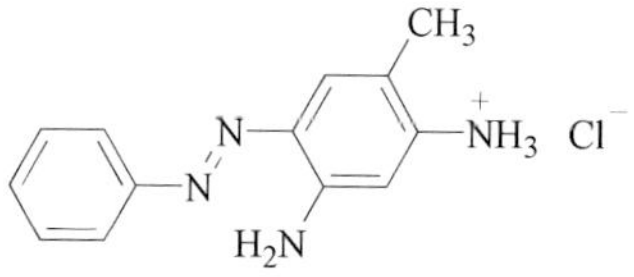

C.I. Basic Orange 2, 11270 [532-82-1] (**60**) is formed from aniline and *m*-phenylenediamine:

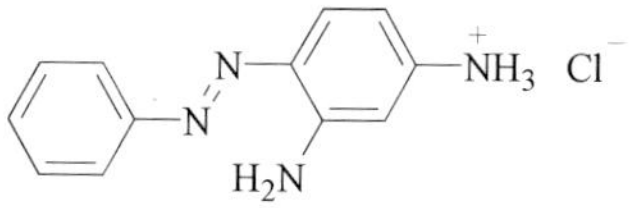

C.I. Basic Red 18, 11085 [25198-22-5] (**61**) is derived from the diazo component 2-chloro-4-nitroaniline [30].

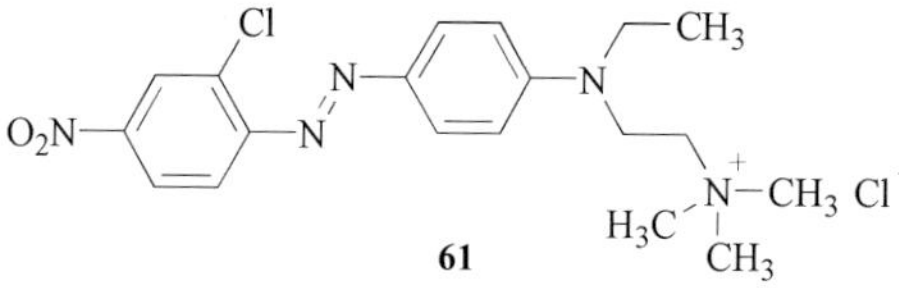

3.7.4 Examples of Commercially Available Dyes

It is a component of all major product lines used for dyeing polyacrylonitrile. The corresponding dye from 2,6-dichloro-4- nitroaniline as the diazo component is also of commercial importance.

C.I. Basic Red 29, [42373-04-6].

C.I. Basic Red 46, 110825 [12221-69-1] (**62**).

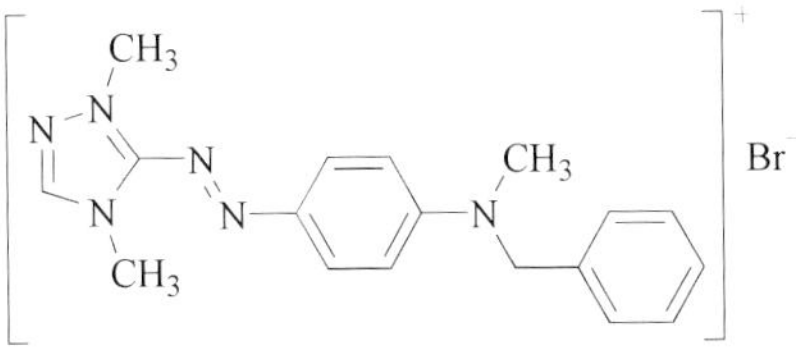

62

C.I.Basic Blue 41, 11105 [12270-13-2] (**63**).

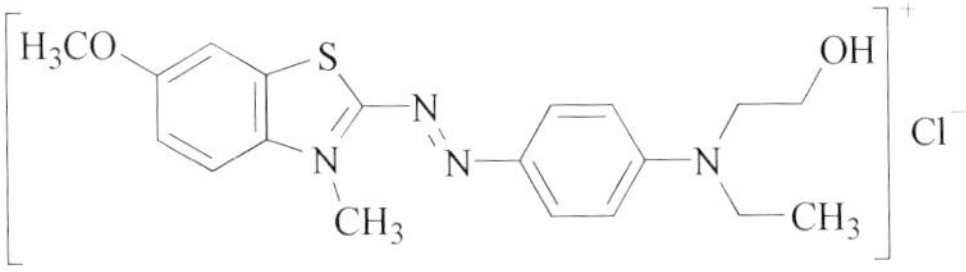

63

C.I.Basic Blue 54, 11052 [15000-59-6] (**64**).

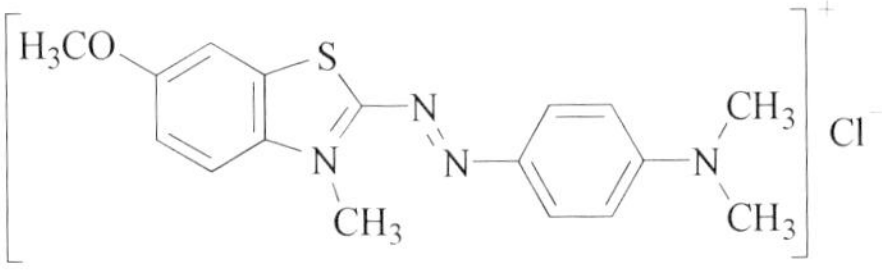

64

C. I. Basic Brown 1, 21010.
A commercial product in this series is [3068-75-5] (**65**).

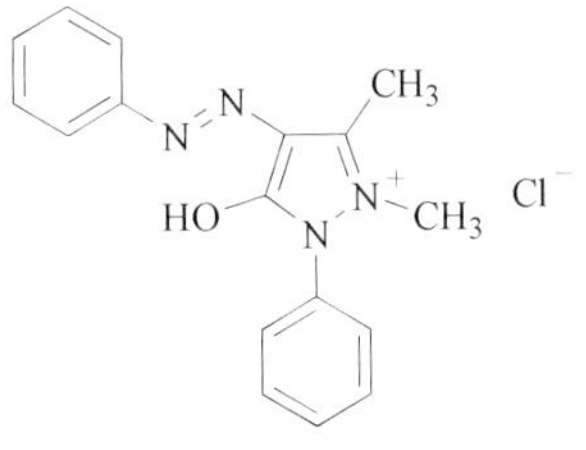

65

3.7.5 References

[1] GAF, DE 901419, 1950; DE 901540, 1950.
[2] BASF, DE-OS 2557561, 1975 (H. Kast, G. Riedel).
[3] BASF, DE-OS 3011235, 1980 (K. Schmeidl).
[4] Bayer, DE-OS 3025557, 1980 (K. Linhart, H. Gleinig, R. Raue, H. -P. Kühlthau).
[5] Amer. Cyanamid, DE-AS 1072222, 1958.
[6] Bayer, DE-AS 1044310, 1956.
[7] VEB Bitterfeld, DE-OS 2038411, 1970.
[8] Bayer, DE-OS 3048998, 1980 (K. Kunde).
[9] Sandoz, DE-OS 2915323, 1979; 3030197, 1980 (R. Pedrazzi).
[10] Sandoz, DE-OS 33 11 091, 1982 (J. Troesch, W. Portmann).
[11] Sandoz, DE-OS 33 29 817, 1982 (J. Troesch, W. Portmann).
[12] Sandoz, DE-OS 34 12 762, 1983 (J. Troesch).
[13] Sandoz, DE-OS 35 03 844, 1984 (R. Pedrazzi).
[14] Sandoz, EP 341 214, 1988 (H. A. Moser, R. Wald).
[15] Bayer, DE-OS 3133360, 1981 (K. Kunde).
[16] Nippon Chem. Works, JA 04 93 365, 1990.
[17] BASF EP 281 920, 1987 (J. Dehnert).
[18] Bayer EP 632 104, 1993 (K. Hassenrueck, P. Wild).
[19] Bayer EP 312 838, 1987 (A. Engel).
[20] Intreprenderea "Sintofarm" RO 88 681, 1984.
[21] BASF DE-OS 42 02 566, 1992 (E. Hahn, H. Hengelsberg, U. Mayer).
[22] Orient Chem. Works JA 08 048 924, 1994.
[23] Nippon Chcm. Works JA 04 183 754 .
[24] BASF DE-OS 42 35 154, 1992 (U. Schloesser, U. Mayer).
[25] BASF EP 666 287, 1994 (U. Schloesser, U. Mayer).
[26] Sandoz, DE-OS 1644138, 1966.
[27] Ciba, FR 1322766, 1961.
[28] Bayer, DE-OS 3138182, 1981.
[29] ICI, GB 951667, 1961.
[30] Bayer, DE-OS 2850706,1978.
[31] Bayer, DE-AS 1011396, 1955.
[32] Sumitomo, JP 7313753, 1969.
[33] Bayer, DE-OS 2024184, 1970; DE-OS 2036952, 1970; DE-OS 2036997, 1970; DE-OS 2041690, 1970; DE-OS 2065685, 1970 (W. Kruckenberg). ICI, DE-OS 2646967, 1975 (P. Gregory).
[34] Sandoz, DE-OS 2338729, 1972 (B. Henzi).
[35] Bayer, DE-OS 2059947, 1970.
[36] Bayer, DE-OS 2309528, 1973; DE-OS 3030918, 1980 (K. Linhart, H. Gleinig, G. Boehmke, K. Breig); DE-OS 3101140, 1981 (K. Linhart, H. Gleinig, G. Boehmke).
[37] Bayer, DE-OS 2508884, 1975; DE-OS 2631030, 1976 (W. Kruckenberg).
[38] Ciba, DE-OS 2011428, 1970 (G. Hegar). Sandoz, CH 546269, 1964; CH 552032, 1964 (R. Entschel). Du Pont, US 3890257, 1973; US 3987022, 1973 (D. S. James).
[39] Pentel K. K., JP 77101125, 1976.
[40] Sumitomo, JP 7237675, 1969.
[41] Sandoz, DE-OS 2118536, 1970; DE-OS 2712265, 1976; DE-OS 2741010, 1976; DE-OS 2805264, 1978 (M. Greve); BE 756820, 1970; CH 549629, 1971. Montedison, DE-OS 2851373, 1977 (F. Merlo).

3.7.5 References

[42] ICI, DE-OS 2200270, 1971; DE-OS 2364592, 1973; DE-OS 2364593, 1972.
[43] Ciba, BE 587048, 1959. GAF, FR 1295862, 1960.
[44] Sandoz, FR 1325176, 1961.
[45] Sandoz, CH 508709, 1969. Bayer, DE-OS 2042662, 1970.
[46] Sandoz, DE-OS 2054697, 1969 (W. Steinemann); DE-OS 2752282, 1976 (M. Greve). ICI, DE-OS 2638051, 1975; DE-OS 2656879, 1975; DE-OS 2657147, 1975; DE-OS 2657149, 1975; DE-OS 2702777, 1976; DE-OS 2702778, 1976 (B. Parton).
[47] Sandoz, DE-OS 2627680, 1975 (M. Greve, H. Moser); CH 601433, 1976; FR 2360632,.
[48] Ciba, BE 698730, 1966; BE 698917, 1966; EP 55221, 1980 (P. Loew). Yorckshire Dyeware Corp., GB 1047293, 1964; GB 1157867, 1967; GB 1349511, 1970 (J. F. Dawson, J. Schofield). ICI, GB 2009208, 1977 (M. G. Hutchings).
[49] ICI, GB 1211078, 1967.
[50] Yorckshire Dyeware Corp., GB 1233729, 1968.
[51] GAF, DE-OS 2244459, 1971.
[52] Yorckshire Dyeware Corp., GB 1380571, 1971.
[53] ICI, GB 1247683, 1968.
[54] ICI, DE-OS 2331953, 1972.
[55] Yorckshire Dyeware Corp., GB 1361542, 1972.
[56] Ciba-Geigy, DE-OS 2125911, 1970. ICI, DE-OS 3008899, 1979 (P. Gregory, D. Thorp).
[57] ICI, DE-OS 2548879, 1974; DE-OS 2703383, 1976;DE-OS 2817638, 1977 (P. Gregory). Sandoz, CH 627487, 1977 (U. Blass).
[58] Sandoz, BE 668688, 1964. ICI, DE-OS 2817638, 1977;DE-OS 3008899, 1979; GB 2074598, 1980 (P. Gregory).
[59] ICI, GB 1374801, 1971.
[60] Ciba-Geigy, DE-OS 2154477, 1970.
[61] BASF, DE-OS 2046785, 1970.
[62] ICI, GB 1282281, 1967; GB 1290321, 1967.
[63] Ciba-Geigy, CH 745871, 1974; CH 745872, 1974.
[64] Kuhlmann, BE 626882, 1962; DE-OS 2314406, 1972 (J. P. Stiot, C. Brouard).
[65] Kuhlmann, FR 1364560, 1962.
[66] Kuhlmann, FR 1364647, 1963.
[67] Kuhlmann, FR 1389432, 1963; FR 1414876, 1964; DE-OS 2140511, 1970; DE-OS 2314406, 1972 (J. P. Stiot, C. Brouard).
[68] Kuhlmann, FR 1467822, 1965.
[69] Sandoz, DE-OS 3609590, 1985.
[70] Sandoz, DE-OS 3516809, 1984.
[71] Sandoz, DE-OS 3715066, 1986.
[72] Sandoz, DE-OS 3340483, 1982.
[73] Sandoz, DE-OS 3311294, 1983.
[74] Sandoz, DE-OS 3538517, 1984.
[75] Sandoz, DE-OS 19629238, 1984.
[76] Sandoz, DE-OS 3932566,1988 (H. A. Moser, R. Wald).
[77] Sandoz DE-OS 19500203, 1994 (H. A. Moser).
[78] Sandoz DE-OS 44 39 004, 1994 (J. -P-Chavannes, G. Schöfberger, G. Scheulin).
[79] Bayer DE-OS 31 14 088, 1981 (F. -M. Stöhr, H. Nickel).
[80] Sandoz DE-OS 3625576, 1985 (J. Doré, R. Pedrazzi).
[81] Bayer DE-OS 3133568, 1981 (F. -M. Stöhr, H. Nickel).
[82] Clariant WO 97/35925, 1996 (R. Pedrazzi).
[83] Ciba-Geigy EP 122 458, 1983 (J. -M. Adam, H. Schwander).

[84] Sandoz, DE-OS 2250676, 1971 (H. Moser).
[85] Sterling Drug, DE-OS 2604699, 1975 (N. C. Crounse).
[86] Bayer, EP 54616, 1980; EP 74589, 1981; DE-OS 3222965, 1982 (K. Kunde).
[87] Sandoz, DE-OS 2915323, 1978 (R. Pedrazzi).
[88] Sandoz, DE-OS 3030196, 1980 (R. Pedrazzi).
[89] Sterling Drug, US 4376729, 1980 (N. C. Crounse).
[90] Hoechst, DE-OS 2028217, 1970; DE-OS 2057977, 1970.
[91] Mitsubishi, JP 2181/66, 1963. ICI, GB 1516978, 1974 (P. Gregory). Bayer, DE-OS 3110223, 1981 (P. Wild, H. Nickel).
[92] Mitsubishi, JP 4986424, 1972; JP 49116384, 1973.
[93] ICI, NL 6918341, 1968; DE-OS 2216206, 1971.
[94] Hoechst, DE-OS 2347756, 1973.
[95] Hoechst, FR 1601787, 1967.
[96] Sandoz, CH 555876, 1971 (R. Entschel).
[97] Bayer, EP 24322, 1979 (H. Nickel, F. Müller, P. Mummenhoff).
[98] Bayer, EP 24321, 1979 (H. Nickel, F. Müller, P. Mummenhoff).
[99] Du Pont, DE-AS 1054616, 1954.
[100] Bayer, DE-OS 2135152, 1971.
[101] Ciba, DE-AS 1156188, 1959.
[102] Du Pont, US 2965631, 1958.
[103] Ciba, DE-OS 1956142, 1968. Sandoz, CH 553241, 1969 (R. Entschel, C. Müller, W. Steinemann).
[104] Hoechst, DE-OS 2222099, 1972 (E. Fleckenstein, R. Mohr, E. Heinrich).
[105] Ciba-Geigy, DE-OS 2263109, 1971.
[106] ICI, GB 1414503, 1972 (D. B. Baird, J. L. Lang, D. F. Newton).
[107] Bayer, DE-OS 2559736, 1975 (W. Kruckenberg).
[108] Hodogaya, JP 7215473, 1969; JP 7215476, 1969.
[109] Du Pont, DE-OS 2137548, 1970.
[110] Sterling Drug, GB 1299080, 1968; JP 4702127,1970; US 3935182, 1973; US 3996282, 1974. Scott Paper, GB 2020703, 1978.
[111] Bayer, DE-OS 3114075, 1981; DE-OS 3114087, 1981 (H. Nickel, P. Wild, F. M. Stöhr).
[112] Bayer, DE-OS 3048694, 1980 (H. Nickel, P. Wild).
[113] Mitsubishi, JP 49118721, 1973.
[114] Eastman Kodak, US 3804823, 1972. Hoechst, DE-OS 2339713, 1973 (E. Fleckenstein).
[115] Bayer, DE-OS 2315637, 1973 (K. L. Moritz, K. -H. Schündehütte). Sandoz, DE-OS 2120878, 1970; EP 41040, 1980; EP 93828, 1980; EP 93829, 1980 (P. Doswald, E. Moriconi, H. Moser, H. Schmid). Kuhlmann, DE-OS 2119745, 1970. Eastman Kodak, US 3836518, 1972 (G. T. Clark).
[116] Bayer, DE-OS 2101999, 1971 (M. Wiesel,G. Wolfrum). Mitsubishi, JP 5043284, 1973.
[117] Sandoz, DE-OS 3313965, 1982.
[118] Hoechst, FR 1484503, 1965. Amer. Cyanamid, DE-AS 1155547, 1957; DE-AS 1162014, 1957.
[119] IG Farbenind., DE 477913, 1927. ICI, DE-OS 2319090, 1972.
[120] ICI, DE-OS 2421822, 1973 (J. L. Leng, D. F. Newton); GB 2011454, 1977 (M. G. Hutchings). Ciba-Geigy, GB 1257346, 1968.
[121] Kuhlmann, FR 1391676, 1963.
[122] ICI, GB 1246632, 1967; GB 1365625, 1972 (J. S. Hunter, J. L. Leng, C. Morris).
[123] Bayer, DE-OS 2141987, 1971 (M. Wiesel, G. Wolfrum).

3.7.5 References

[124] GAF, DE-OS 2050246, 1969 (N. A. Doss).
[125] Sandoz, CH 554917, 1969 (R. Entschel, C. Müller, W. Steinemann). Hodogaya, JP 7130110, 1968.
[126] Ciba, DE-OS 1644110, 1966.
[127] MLB Hoechst, DE 105202, 1898. Bayer, DE 272975, 1911. Geigy, DE 473 526, 1927.
[128] Sandoz DE-OS 37 17 869, 1986 (R. Pedrazzi)
[129] Sandoz DE-OS 33 40 486, 1982 (M. Greve, H. Moser).
[130] Bayer, DE 181783, 1906.
[131] Kuhlmann, FR 1380628, 1963. VEB Bitterfeld, DE-OS 2031561, 1970.
[132] Hoechst, DE-AS 1242773, 1963; FR 2019355, 1968; DE-OS 2516687, 1975; DE-OS 2522174, 1975 (R. Mohr, E. Mundlos, K. Hohmann).
[133] Hoechst, BE 677795, 1965; DE-OS 2221989, 1972; DE-OS 2222042, 1972 (E. Mundlos, R. Mohr, K. Hohmann); DE-OS 2357448, 1973 (K. Hohmann); FR 2016982, 1968.
[134] Amer. Cyanamid, FR 1340397, 1961; DE-OS 2741395, 1976 (R. B. Balsley). Hodogaya, JP 56134280, 1980.
[135] Ciba-Geigy, EP 116 513, 1983 (V. Ramanathan, P. Möckli).
[136] BASF, DE-OS 35 29 968, 1985 (M. Ruske, H. -J. Degen).
[137] Sandoz, DE-OS 34 16 117, 1983 (M. Greve, H. Moser, R. Pedrazzi, R. Wald).
[138] Sandoz, DE-OS 33 13 965, 1982 (J. Doré, R. Pedrazzi).
[139] Lexmark, GB 23 15 493, 1996 (J. F. Feeman, J. X. Sun).
[140] Lexmark, B 23 15 759, 1996 (B. L. Beach, J. F. Feeman, T. E. Franey, A. P. Holloway J. M Mrvos, J. X. Sun, A. K. Zimmer.
[141] Sterling Drug, US 4379088, 1980.
[142] Sterling Drug, US 4379089, 1980.
[143] Sandoz, DE-OS 2403736, 1973 (H. Moser).
[144] Ciba-Geigy, EP 84372, 1982 (J. M. Adam).
[145] Ciba-Geigy, EP 14677, 1979 (J. M. Adam, V. Ramanathan, P. Galafassi).
[146] Ciba-Geigy, EP 57159, 1981 (J. M. Adam).
[147] Sandoz, DE-OS 1903058, 1969; DE-OS 1965994, 1968. Sterling Drug, DE-OS 2810246, 1977.
[148] BASF, EP 34725, 1980 (M. Patsch, M. Ruske).
[149] ICI, BE 601144, 1960.
[150] ICI, BE 601145, 1960; BE 601146, 1960; DE-OS 2436138, 1973 (C. W. Greenhalgh, D. S. Phillips).
[151] ICI, BE 603028, 1960. V. V. Kormachev, G. P. Pavlov, V. A. Kukhtin, R. S. Tsekhawski. *Zh. Obshch. Khim.* **49** (1979) no. 111, 112479.
[152] Bayer, BE 661025, 1964.
[153] G. B. Guise, I. W. Stapleton, *J. Soc. Dyers Colour.* **91** (1975) 259.
[154] Bayer, BE 657179, 1963; BE 657302, 1963.
[155] Bayer, FR 1459140, 1964.

3.8 Cationic Methine Dyes

3.8.1 Introduction

Cationic methine dyes carry a positive charge in the molecule. The salt-forming counterion is, in most cases, the colorless anion of a low molecular mass inorganic or organic acid. Many of these dyes can be converted into water-insoluble dye bases by addition of alkali. For this reason, they were formerly called *basic dyes*; although this term still in use today, and they are listed as such in the Colour Index), it should be abandoned. The positive charge of cationic dyes can be either localized or delocalized. Of the methine dyes, the cationic compounds are the most important group, especially those in which nitrogen is the cationic charge carrier.

3.8.2 Chemical Constitution and Synthesis

3.8.2.1 Streptocyanine Dyes

In streptocyanine dyes both ends of the methine chain are joined directly to nitrogen atoms, and a double enamine structure is thus present. The dyes are extremely susceptible to hydrolysis, particularly if they contain secondary nitrogen atoms. Stable dyes are obtained if the nitrogen is part of a heterocyclic ring system. Streptocyanine dyes are brilliant yellow dyes that dye polyacrylonitrile and acid-modified polyamide fibers with outstanding lightfastness [1].

Trimethinestreptocyanine Dyes have a coloring strength sufficient for practical use if the nitrogen atom is part of a ring and if the aromatic ring bears an alkoxy group in the 4-position relative to the nitrogen atom (e.g., **1**).

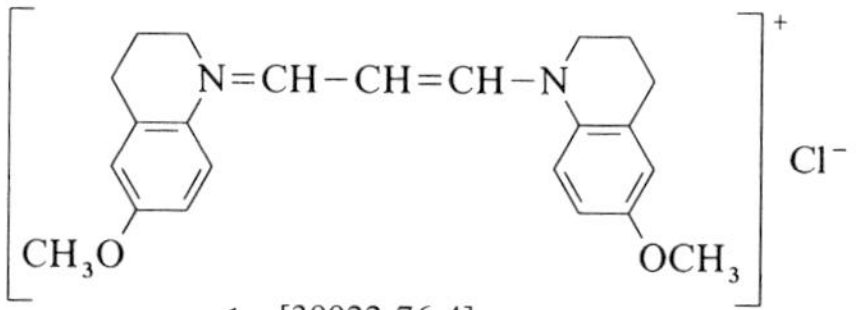

1 [39922-76-4]

Pentamethinestreptocyanine Dyes are formed by reaction of bromine cyanide, chlorine cyanide, or 2,4-dinitrochlorobenzene with pyridine and cleavage of the resultant compounds with primary or secondary amines. The tetraacetal of glutaconaldehyde is also used to form the methine chain [2]. The dyes can be used to dye paper.

Streptocyanine dyes are used to synthesize long-chain cyanine dyes.

Aminoarylenamine Dyes are phenylogous streptocyanines, and are formed by condensing aromatic aminoaldehydes with aromatic amines. If the aromatic amine bears a substituent with electron-donor properties, the enamine dyes are sufficiently stable to be adsorbed on polyacrylonitrile without hydrolytic decomposition. The yellow dye **2** is very lightfast [3].

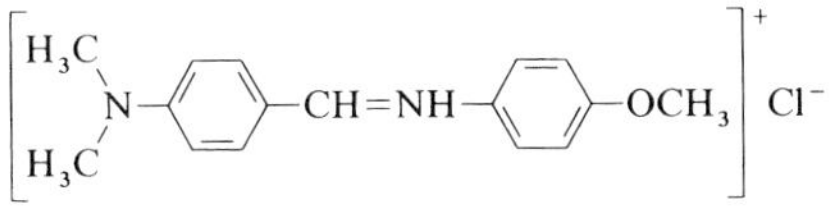

$$\left[(H_3C)_2N\text{−}C_6H_4\text{−}CH\text{=}NH\text{−}C_6H_4\text{−}OCH_3 \right]^+ Cl^-$$

2 [85291-90-3]

The dye bases of these compounds are colorless or only slightly colored, and are used in carbonless copying paper. Solutions of the dye bases are encapsulated in microcapsules made of gelatin or polycondensation products and coated onto paper. If this is brought into contact with paper bearing a coating of a compound that acts as an electron acceptor (e.g., attapulgite) and the capsules are destroyed by pressure, the color bases are protonated to the enamine dyes, thereby producing yellow or orange dyeings of high intensity. Compound **3** is particularly suitable in this respect and is also used for the preparation of black copying mixtures [4].

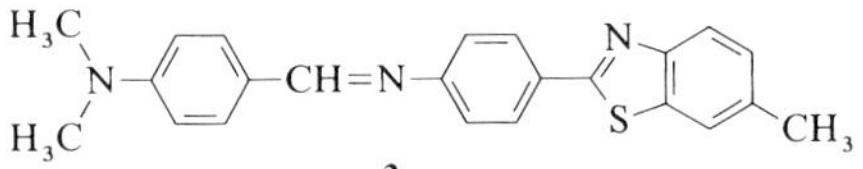

3

The color bases can also be sublimed image-wise from a transfer ribbon onto an acid-developing receiver sheet. They then form the yellow component in color-transfer prints [5].

3.8.2.2 Hemicyanine Dyes

In hemicyanine dyes one of the charge-carrying terminal atoms of the methine chain forms part of a heterocyclic ring. The lowest vinylogous hemicyanine dyes (cyclomethine dyes) do not contain a methine chain, and the donor and acceptor groups are in the same ring system.

Cyclomethine Dyes can be prepared by condensing N-heterocyclic carbonyl compounds with secondary amines. Intensely colored products such as **4** are obtained by condensing 3-phenylpyrazoline with compounds such as N-ethyl-naphtholactam. This compound dyes polyacrylonitrile a brilliant, lightfast yellow. Orange-red dyes are obtained by condensing acridone with secondary amines [6].

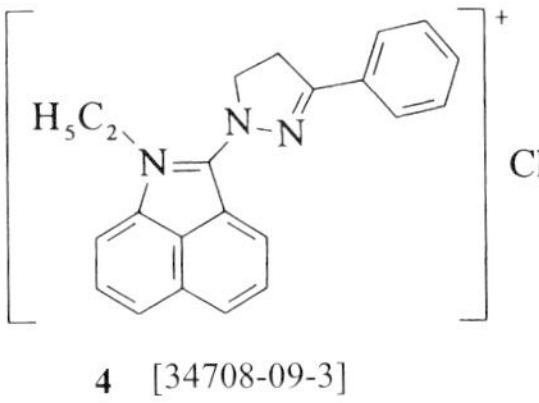

4 [34708-09-3]

Dimethinehemicyanine Dyes and their aza analogues are important for dyeing polyacrylonitrile. Most of the dyes that color polyacrylonitrile in greenish to reddish-yellow shades belong to this class.

Enamine Dyes are obtained by condensation of heterocyclic methylene-ω-aldehydes with aromatic amines in an acid medium. Technically important dyes contain 1,3,3-trimethyl-2-methyleneindoline-ω-aldehyde as aldehyde component [7]. *C.I. Basic Yellow 11*, the condensation product formed with 2,4-dimethoxyaniline, is of particular importance (see 3.8.4). This compound dyes polyacrylonitrile a lightfast, brilliant, greenish-yellow shade.

The following amine components have also been proposed for condensation with methyleneindoline-ω-aldehydes: 4-aminophenyl glycol ether [8]; 4-aminoacetanilide [9]; and 2-(4-aminophenyl)benzoxazole, -benzothiazole, -benzimidazole, and -benzotriazole [10]. Diamines react with methyleneindoline-ω-aldehydes in a molar ratio of 2:1 to form yellow dyes suitable for dyeing paper. Suitable components are bis(4-aminophenyl) ether, 4,4'-diaminodiphenylsulfone [11], and 4-aminophthalic anhydride dimerized by aliphatic diamines [12]. Cyanuric chloride can be treated twice or three times with 4-aminoacetanilide, and the acetyl group can then be split off by acid hydrolysis. Reaction of the product with an equivalent amount of 1,3,3-trimethyl-2-methyleneindoline-ω-aldehyde yields yellow dyes with a high affinity for paper [13].

The low molecular mass bases of enamine dyes can be sublimed by means of a thermohead from a transfer ribbon onto a receiving material and serve as yellow components for producing color copies in sublimation transfer printing [14].

Vinylogous enamine dyes are obtained by reaction of aromatic amines with mucochloric acid (2,3-dichloro-4-oxo-2-butanoic acid), followed by condensation with quaternary quinaldine. After methylation of the enamine nitrogen atom with dimethyl sulfate, the dye **5** is obtained, which dyes bleached sulfite pulp red [15].

5

Enhydrazine dyes such as **6**, synthesized from methyleneindoline-ω-aldehydes and the monoalkyl hydrazones of aromatic aldehydes, fluoresce in solution and on the fiber [16].

6 [56190-67-1]

Dyes prepared by condensation of methyleneindoline-ω-aldehydes with cyclic amino compounds have high lightfastness and resistance to hydrolysis [17].

3.8.2 Chemical Constitution and Synthesis

C.I. Basic Yellow 21 is synthesized from 1,3,3-trimethyl-2-methyleneindoline-ω-aldehyde and 2-methyl-2,3-dihydroindoline. It dyes polyacrylonitrile in brilliant, greenish yellow shades and is very lightfast (see 3.8.4):

The shade is changed to golden-yellow by introducing a phenylazo group into *C.I. Basic Yellow 21*. Alkoxy substitution of the cyclic amine component in the 4-position to the nitrogen atom increases the coloring strength considerably (**7**) [18].

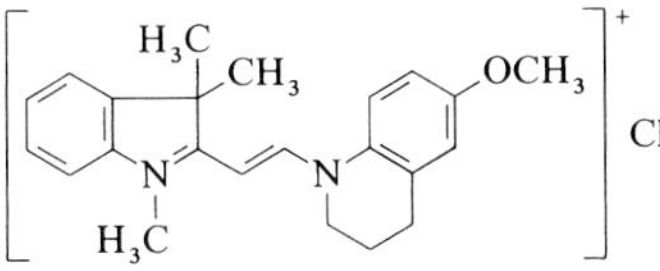

7 [27326-17-6]

Suitable N heterocycles also include tetrahydrobenzoxazine [19] and tetrahydroquinoxaline. The latter produces a marked bathochromic shift and dyes polyacrylonitrile a neutral yellow shade (**8**) [20]:

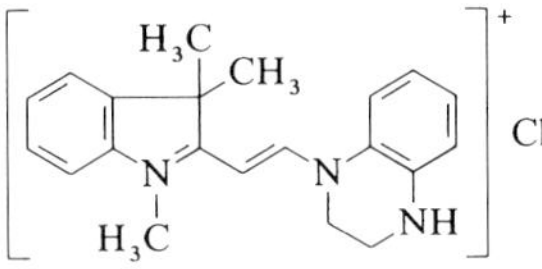

8 [28766-22-5]

The brilliant greenish-yellow dye **9** is obtained by condensing methyleneindoline-ω-aldehydes with 3-phenylpyrazoline [21].

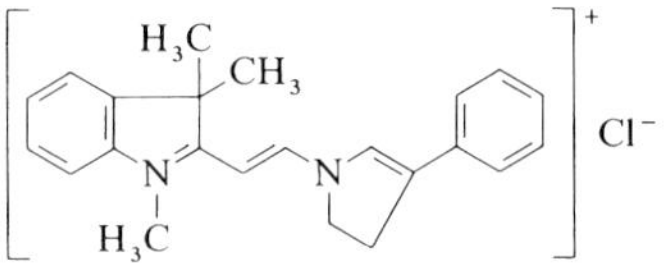

9 [31612-53-0]

If methyleneindoline-ω-aldehydes are condensed with cyclic amines substituted by an amino group in the aromatic ring, dyes are obtained that can be dimerized either by reaction with cyanuric chloride in a molar ratio of 2:1 or with phosgene. On account of their high substantivity these dyes are suitable for dyeing bleached sulfite pulp [22].

Hydrazone Dyes. If the diazonium salts of aromatic amines are coupled with the methylene derivatives of N heterocycles, hydrazone dyes (i.e., the monoaza derivatives of the enamine dyes) are obtained. Paper and leather can be dyed in yellow to red shades with these dyes [23]. They are moderately lightfast on polyacrylonitrile. If the diazo components contain substituents in the 2-position relative to the amino group that can form a hydrogen bond (e.g., 2-nitroaniline [24] or 1-aminoanthraquinone [25]), lightfastness is improved considerably (**10**).

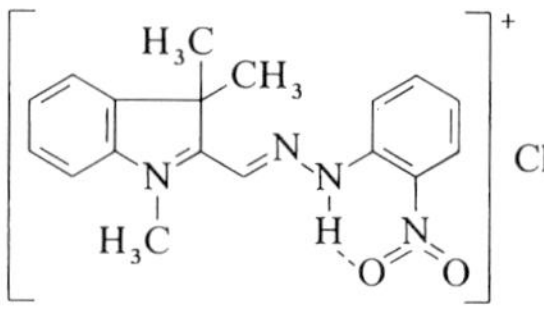

3.8 Cationic Methine Dyes

10 [27564-03-0]

Such dye salts are converted to azo bases by alkali. Alkylating agents attack the nitrogen atom of the azo bond, forming alkylarylhydrazone dyes which dye polyacrylonitrile in lightfast, yellow to orange shades [26].

Diazotization, coupling, and methylation may be combined in a one-pot process by dissolving the aromatic amine and the heterocyclic methylene base in an organic acid and effecting the simultaneous diazotization and coupling by addition of sodium nitrite. The dye base is released by addition of alkali and is methylated with dimethyl sulfate to form the alkylarylhydrazone dye (e.g., **11**) [27].

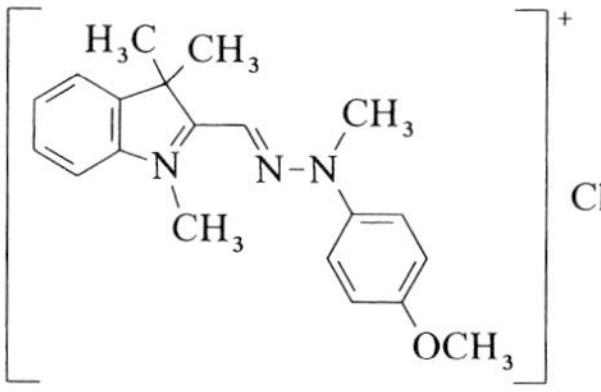

11 [36528-10-6]

Hydrazone dyes with an additional external cationic charge are suitable for use in writing inks, stamp pad inks, and ballpoint pen pastes [28]. Hydrazone dyes with acylamino groups can be hydrolyzed to compounds with a free amino group which dye polyacrylonitrile in red shades [29]. These compounds can be dimerized by reaction with phosgene, dicarboxylic acid chlorides, or cyanuric chloride in a molar ratio of 2:1. The dimerized compounds have a high affinity for cellulose and are suitable for dyeing bleached sulfite pulp [30]. The compounds with a free amino group can also be diazotized and treated with aromatic or heterocyclic coupling components to form azohydrazone dyes [31].

In addition to the methyleneindolines, 4-methylene derivatives of 1,3-dialkyl-pyrimidin-2-one have also been used to prepare hydrazone dyes (e.g., **12**) [26, 32]. A dye of this series is *C.I. Basic Yellow 49*.

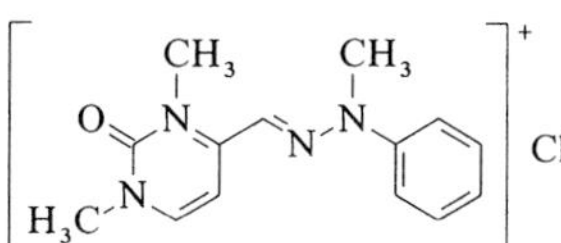

12 [73297-08-2]

Violet and blue hydrazone dyes are obtained with a methylene base derived from naphtholactam (**13**) [33].

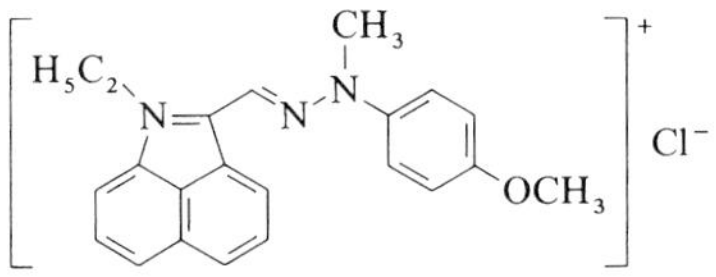

3.8.2 Chemical Constitution and Synthesis 259

13

Heterocyclically substituted acetic acids (e.g., benzothiazolyl-, benzimidazolyl- [34], and pyrimidylacetic acids [35]) are also suitable as coupling components for preparing cationic hydrazone dyes. In these compounds the methyl group in the 2-position is additionally activated. The carboxyl group is split off after coupling and the azo base is methylated.

Alkylarylhydrazone dyes can also be produced by treating suitable heterocyclic aldehydes with arylhydrazines. Subsequent alkylation of the heterocyclic nitrogen atom produces dyes that dye polyacrylonitrile extremely lightfast (e.g., **14**) [36]:

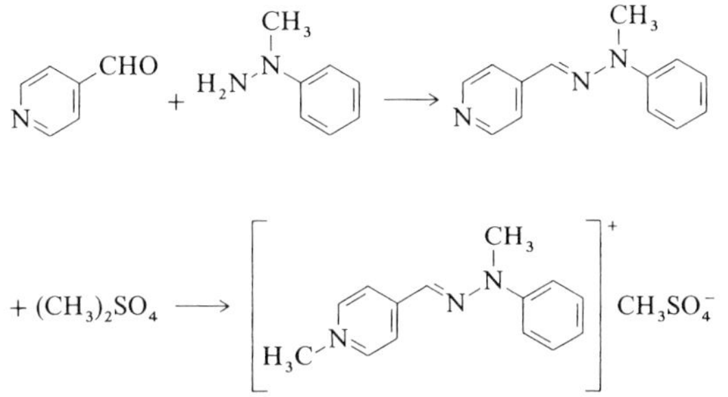

14 [42476-20-0]

The hydrazone dyes of the pyridine, quinoline, and acridine series can be prepared by this route. Hydrazone dyes that are derived from pyridine exhibit a particularly good leveling ability when dyeing polyacrylonitrile [37].

If N-amino nitrogen heterocycles are used as the hydrazine component, cyclic hydrazone dyes are obtained whose absorption maximum is shifted bathochromically with respect to the open-chain hydrazone dyes [38]. The dye **15** dyes polyacrylonitrile fibers a clear bluish red shade:

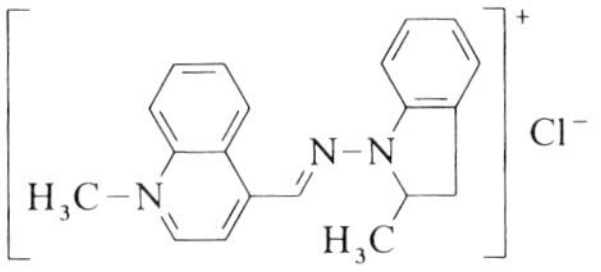

15 [72828-74-1]

Triazene Dyes. If diazonium salts of N-heterocyclic amines are coupled to N-alkylanilines, e.g., diazotized 2-aminoinomethoxybenzothiazole to N-methyl-4-nitroaniline, a triazene is obtained. A cationic triazene dye (**16**, diazadimethinehemicyanine dye) is formed by subsequent N-alkylation, and dyes polyacrylonitrile a very lightfast yellow shade [39].

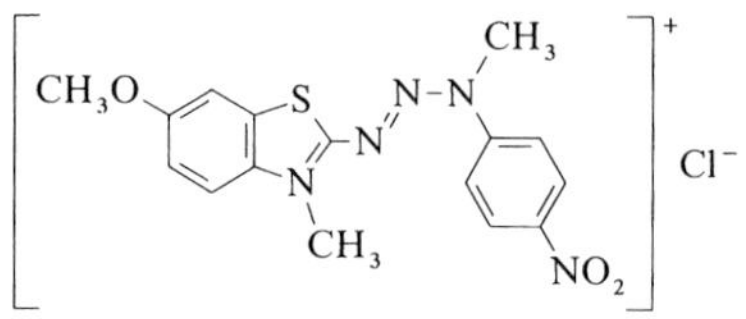

16 [14970-38-8]

Of the enamine and hydrazone dyes, those with donor groups in the aryl system are particularly lightfast, whereas the lightfastness is reduced on introduction of nitro groups. In contrast, triazene dyes with a nitro group attached to the aryl group are substantially more lightfast than those with a methoxyl group attached to the aryl group.

Triazene dyes in which the terminal nitrogen atom is cyclically bound are obtained by coupling heterocyclic diazonium salts with partially hydrogenated N heterocycles. In **17** alkylation was subsequently performed with ethylene oxide [40].

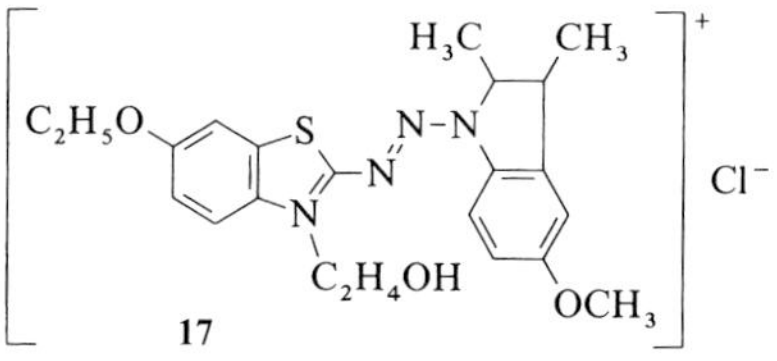

3.8.2.3 Higher Vinylogues of Hemicyanine Dyes

Streptocyanine dyes are semihydrolyzed under mild conditions to give ω-aminopolyenealdehydes. If these aldehydes are condensed with heterocyclic quaternary salts containing a reactive methyl group, the higher vinylogous hemicyanine dyes are obtained. Trimethinestreptocyanines yield tetramethinehemicyanine dyes, while pentamethinestreptocyanines yield hexamethinehemicyanine dyes such as **18** [41]. These dyes are used to sensitize photographic emulsions.

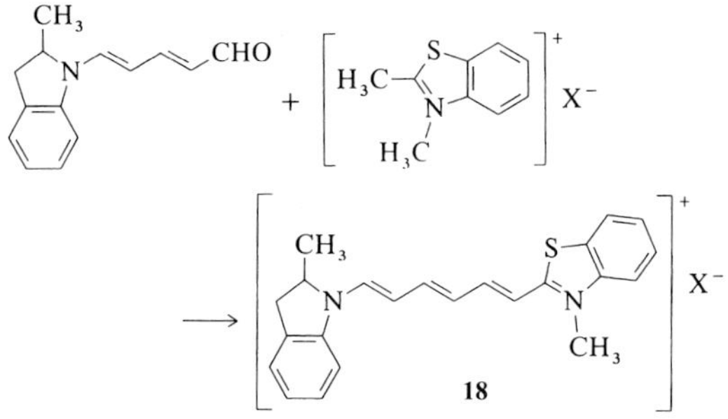

3.8.2.4 Phenylogous Hemicyanine Dyes

Phenylogous hemicyanine dyes are obtained by introducing a phenyl group between the charge-carrying nitrogen atom and the remaining dye system. The zeromethine (apocyanine) dyes are thus derived from the cyclomethine dyes. The styryl dyes are derived from the dimethinehemicyanine dyes by introducing a phenyl group between the charge-carrying nitrogen atom and the methine chain.

Zeromethine (Apocyanine) Dyes. Some representatives of the zeromethine dyes (apocyanine dyes) have been known for a long time, e.g., *C.I. Basic Yellow 1*, 49005 [2390-54-7] (**19**, Thioflavine T).

$$\left[H_3C \underset{\underset{CH_3}{N}}{\overset{S}{\bigcirc}} \bigcirc -N(CH_3)_2 \right]^+ Cl^-$$

$$\mathbf{19}$$

This dye is obtained by heating *p*-toluidine with sulfur and subsequent alkylation with methyl or ethyl sulfate. The dye colors silk and tanned cotton a greenish yellow shade. Thioflavine S [1326-12-1] is obtained by sulfonation [42].

Apocyanine dyes of the acridine series are obtained by treating acridinium iodide [43] or acridinium chloride [44] with tertiary aromatic amines. The dyes can also be obtained by *ipso* substitution of acridinium iodide with 4-haloarylamines in dimethyl sulfoxide at room temperature [45].

The most important method for preparing apocyanine dyes is the reaction of N heterocycles containing 2- or 4-carbonyl groups with secondary or tertiary aromatic amines in the presence of dehydrating agents (e.g., phosphorus oxychloride or tin tetrachloride).

Naphtholactam is a particularly important condensation component for preparing apocyanine dyes. At first only dyes unsubstituted at the lactam nitrogen atom were prepared and recommended for dyeing tanned cotton and wool [46]. However, it was later found that the *N*-alkyl derivatives dye polyacrylonitrile in very lightfast violet and blue shades (e.g., **20**).

$$\left[H_5C_2 \underset{N}{\diagdown} \text{...} \underset{\underset{H_3C}{N}}{\diagdown} \bigcirc -OC_2H_5 \right]^+ Cl^-$$

20 [35114-09-1]

The dyes are synthesized either from *N*-alkylnaphtholactams, obtained by intramolecular Friedel–Crafts reaction of the 1-alkylnaphthylamine carboxylic acid chlorides, or by subsequent alkylation of the dye bases obtained by condensation of naphtholactam with aromatic amines [47].

Secondary aromatic amines, in particular those that also bear a substituent in the 2-position to the amino group, condense with *N*-alkylnaphtholactams to form dyes that are particularly lightfast on polyacrylonitrile [48]. A marked bathochromic shift occurs if the aminoaryl nitrogen atom is part of a heterocyclic ring. By condensation of *N*-ethylnaphtholactam with 1-phenyl-3,5,5-trimethylpyrazoline, **21** is obtained, which dyes polyacrylonitrile a lightfast blue shade [49]:

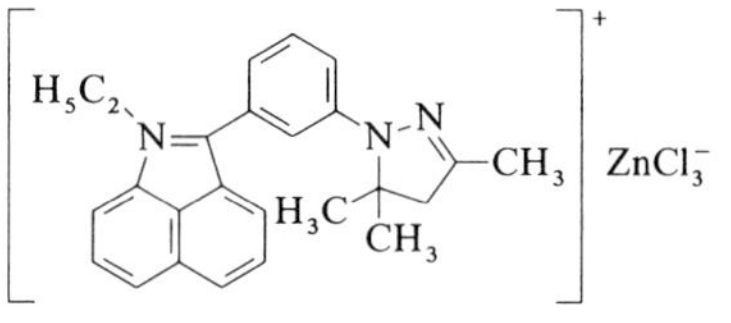

21 [40496-35-3]

Particularly lightfast dyes such as **22** are obtained if 1-alkyl derivatives of 2,2,4-trimethyltetrahydroquinoline are used as amine components [50]. This dye colors polyacrylonitrile a clear blue shade.

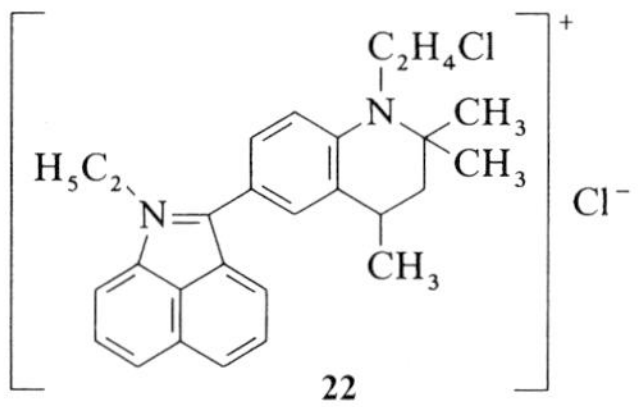

22

N-Alkylnaphtholactam can be condensed with formaldehyde to form a dinaphthylmethane derivative and then treated with aromatic amines or indole derivatives. These dyes are suitable for spin dyeing polyacrylonitrile and for dyeing paper [51]. The dyes with an unsubstituted lactam nitrogen atom can be converted into dye bases and used in this form for sublimation transfer printing on polyacrylonitrile fabrics [52]. Use of the carbinols of naphtholactam dyes in heat- or pressure-sensitive copying or recording materials has been proposed [53]. If the lactam nitrogen atom bears a carbonamidoethyl group as substituent, intramolecular lactam ring closure occurs under the action of alkali. These compounds can also be used as color formers [54].

Dye bases of apocyanine dyes derived from pyridine which is unsubstituted on the nitrogen atom are color formers for heat- or pressure-sensitive copying paper and develop into yellow to orange shades (e.g., **23**) [55].

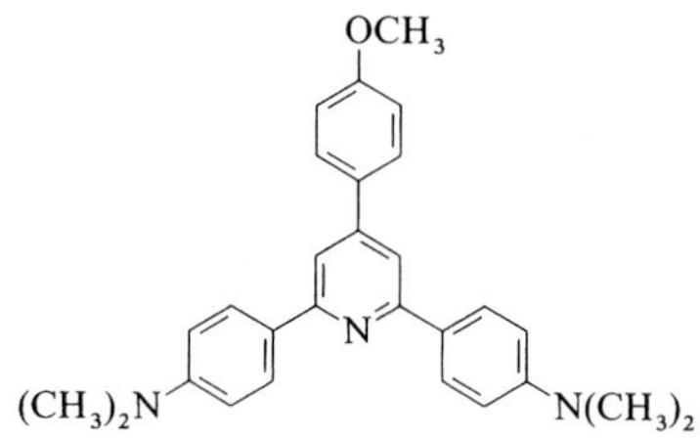

23 [84754-34-7]

Dye bases of apocyanine dyes are also used as organic photoconductors in electrophotography [56].

3.8.2 Chemical Constitution and Synthesis

Phenylogous Dimethinehemicyanine Dyes. Phenylogous dimethinehemicyanine dyes are used widely in photography and textile dyeing. The diazadimethinehemicyanine dyes are particularly important in the textile sector.

Styryl Dyes. They are produced by condensation of 4-aminobenzaldehydes with N heterocycles bearing an activated methyl group in the 2- or 4-position. The starting components are usually briefly heated in glacial acetic acid. The aminoaldehydes can also be condensed with quaternary heterocyclic salts in the presence of basic catalysts such as piperidine, diethylamine, or ammonia. The styryl dyes include pinaflavol (**24**) [57], which is used as a sensitizer in silver-based photography. Dyes containing nitro groups act as desensitizers (see Section 5.7).

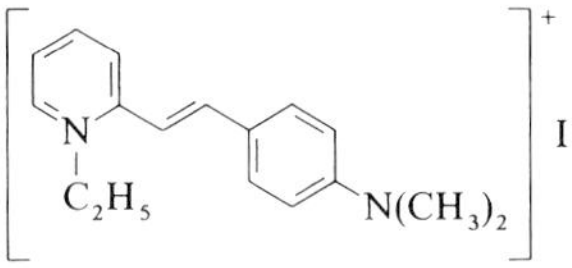

24 [3785-01-1]

Styryl dyes for dyeing paper in brilliant shades are derived from quaternized γ-picoline. The dye **25** obtained by reaction with 4-(*N*-ethyl-*N*-benzylamino)benzaldehyde has a brilliant orange color [58].

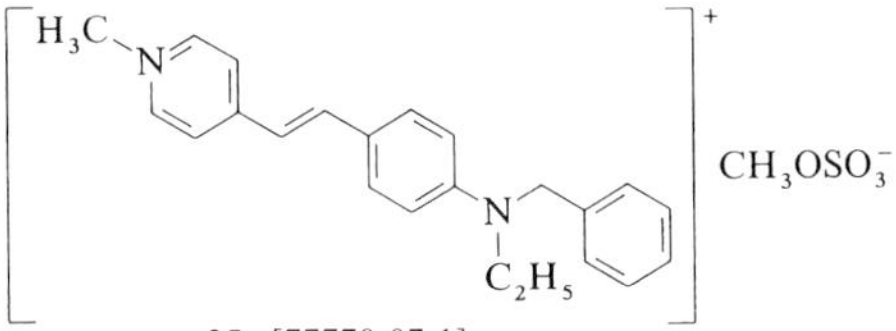

25 [77770-07-1]

The shade is shifted bathochromically by ring closure at the nitrogen atom of the amino group. Using 1-ethyl-2,2,4-trimethyl-6-formyltetrahydroquinoline gives a dye that colors paper a brilliant bluish red [59]. The solubility and affinity of the dyes are increased if amino and hydroxyl groups are introduced into the quaternized pyridine substituents.

The most important styryl dyes for textile dyeing are obtained by condensation of 4-aminobenzaldehydes with 1,3,3-trimethyl-2-methyleneindoline and its derivatives. The dye **26**, synthesized from 4-dimethylaminobenzaldehyde and 1,3,3-trimethyl-2-methyleneindoline dyes polyacrylonitrile, a brilliant red shade.

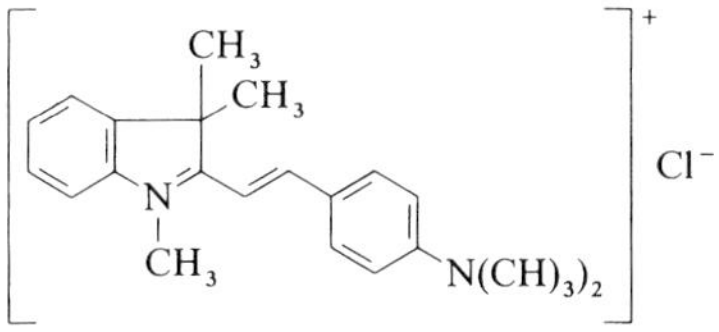

26 [54268-66-5]

3.8 Cationic Methine Dyes

The corresponding *C.I. Basic Violet 16*, 48013 [6359-45-1], obtained from 4-diethylaminobenzaldehyde, dyes polyacrylonitrile a more intense bluish red shade.

In the Vilsmeier formylation of N-(β-hydroxyethyl)anilines the hydroxyl group is simultaneously replaced by chlorine [60]. The resulting aldehydes condense with 1,3,3-trimethyl-2-methyleneindoline to form very brilliant dyes [61]. With 4-(N-methyl-N-β-chloroethylamino)benzaldehyde a commercial dye is obtained. A methyl group in the 2-position to the methine chain produces a bluish-red color (**27**).

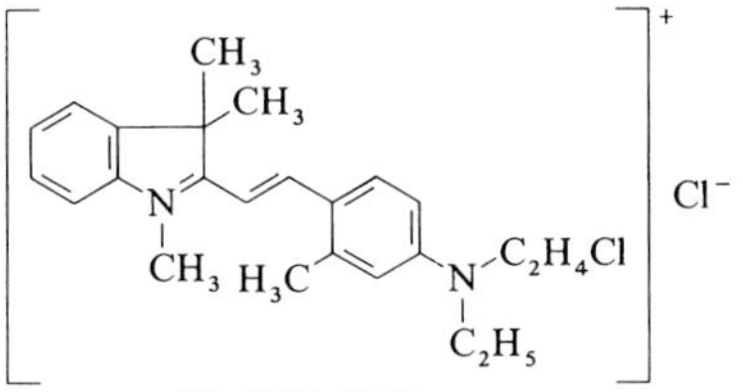

27 [6441-82-3]

Replacement of the chlorine by a cyano substituent produces a marked hypsochromic shift [62]. The resulting dye (*C.I. Basic Red 14* [12217-48-0]) is widely used to produce brilliant red dyeings on polyacrylonitrile. For this dye, as well as the aforementioned styryl dyes from 1,3,3-trimethyl-2-methyleneindoline, a new and particularly environmentally friendly process has been developed in which no wastewater is generated [63–65].

The cyanoethyl group can be hydrolyzed to a carbamoyl group by a strong acid. The resulting dyes are exhausted from the dyebath more slowly than compounds with an unsubstituted alkyl group [66]. The adsorption rate of the styryl dyes can also be reduced by introducing alkoxyalkyl groups at the aminoaryl nitrogen atom. Lightfastness can be improved by introducing benzyl groups (e.g., **28**) [67].

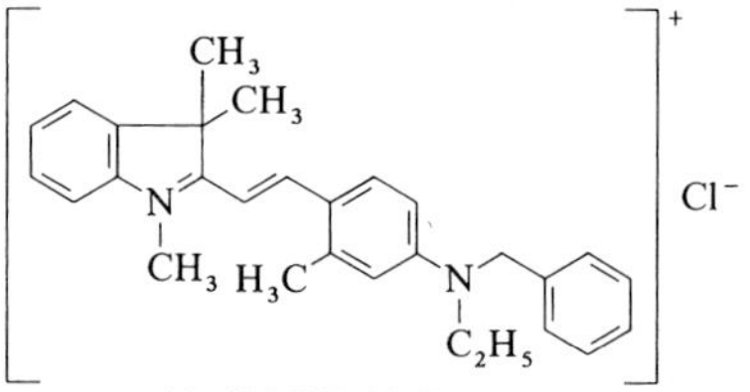

28 [106303-28-0]

This group of dyes is used in electrophotography to make organic photoconductors [68]. In addition to the unsubstituted 1,3,3-trimethyl-2-methyleneindoline, the 5-chloro derivative is often used as the heterocyclic condensation component [69].

Styryl dyes with external amino groups (e.g., **29**) are particularly suitable for dyeing polyacrylo-nitrile from a strongly acid medium [70]:

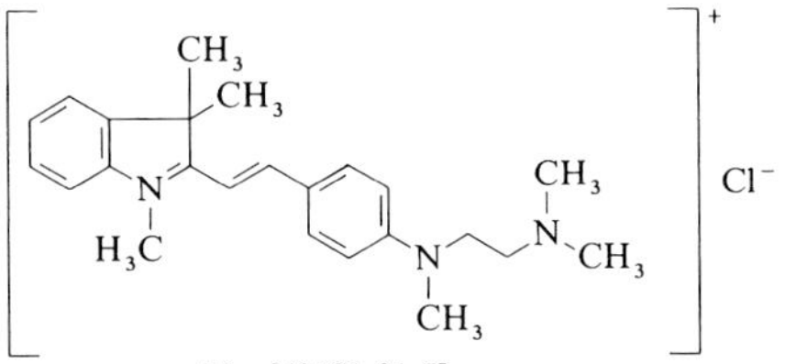

29 [43181-01-7]

The styryl dyes also include compounds in which the aminoaryl nitrogen atom is part of a carbazole (as in the commercial dye **30**) [71], phenoxazine [72], or tetrahydrobenzoxazine system [73].

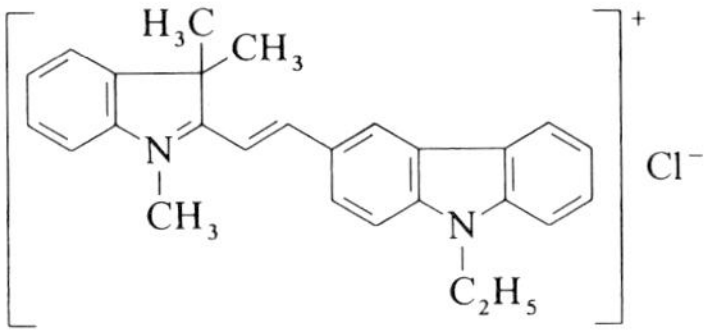

30 [72828-91-2]

Styryl dyes that contain the carbazole ring system are used in the production of photoconducting layers in electrophotography [74].

If 1,2-diaminobenzene is condensed with butyrolactone and the product quaternized, a benzimidazole derivative with a reactive methylene group is obtained. Condensation of this compound with 4-aminobenzaldehydes (which can also be dimerized via the nitrogen atom) produces styryl dyes (e.g., **31**) which, on account of ring closure, can be classified as monomethinehemicyanine dyes. These dyes are used to dye paper [75]:

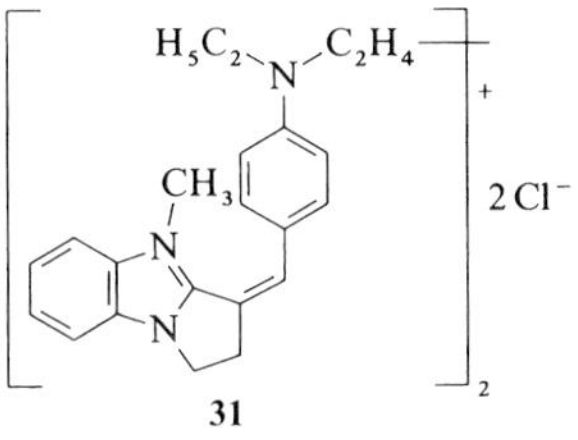

31

Phenylogous Diazadimethinehemicyanine Dyes. If N-heterocyclic compounds with an amino group in the 2- or 4-position are diazotized and then coupled to aromatic amines, yellow to red azo dyes are obtained. If these azo dyes are treated with alkylating agents, alkylation occurs at the heterocyclic nitrogen atom. The absorption maximum of the compounds thereby undergoes a considerable bathochromic shift and becomes more intense; this indicates delocalization of the cationic charge. The resultant dyes are *phenylogous diazadimethinehemicyanine dyes*.

Dyes with 2-amino-6-methoxybenzothiazole as diazo component are of industrial importance because their alkylated products dye polyacrylonitrile in clear, very lightfast, blue shades. With *N,N*-dimethylaniline as coupling component, *C.I. Basic Blue 54*, 11052 [*1500-59-6*] is obtained.

3.8 Cationic Methine Dyes

With butylcyanoethylaniline a blue dye is obtained that produces very steamfast dyeings on polyacrylonitrile [76]. The dyes with N-methyldiphenylamine as coupling component dye polyacrylonitrile in greenish-blue shades. C.I. Basic Blue 41, 11105 [12270-13-2], for instance,
is exhausted somewhat more slowly and can therefore be combined particularly well with other dyes. It is contained in most ranges for dyeing polyacrylonitrile.

Many diazadimethine dyes decompose during dyeing in a weakly acid to neutral medium, whereby accompanying fibers can be stained. This hydrolysis is suppressed by complexing anions such as benzoate, benzenesulfonate, or phenolate ions [77]. Dyes that are more stable to hydrolysis are obtained if derivatives of indoline [78] or tetrahydroquinoline [79] are used as coupling components.

The quaternization of heterocyclic azo dyes to form diazadimethine dyes can also be carried out in an acid medium if compounds containing vinyl groups are added to the protonated azo dyes. Acrylic acid [80], acrylates [81], and, in particular, acrylamide [82] are suitable for this purpose. The dye **32** colors polyacrylonitrile in greenish-blue shades.

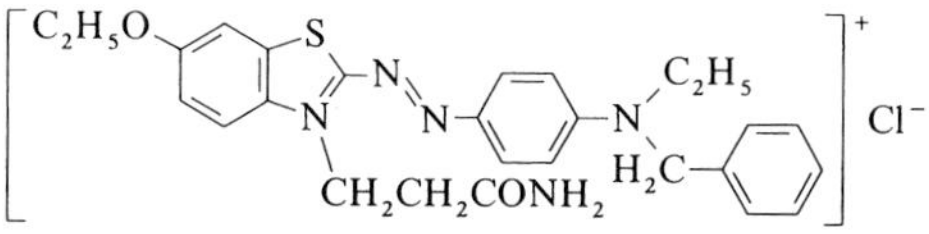

32 [85959-18-8]

Bis-cationic dyes, which are usually used for spin dyeing polyacrylonitrile, are derived from thiazole [83]. Compound **33** dyes acrylic fibers blue with good steamfastness [84].

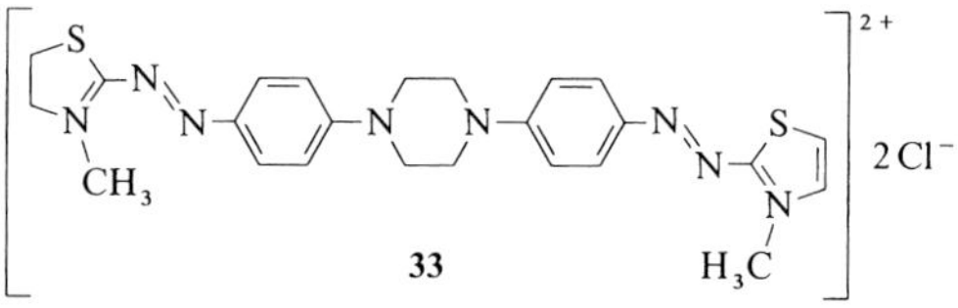

33

The coupling of diazotized 3-amino-1,2,4-triazole to aromatic amines affords azo dyes that require two equivalents of an alkylating agent to convert them to diazadimethine dyes. A mixture of two isomeric dyes is obtained, e.g., **34a** and **34b**.

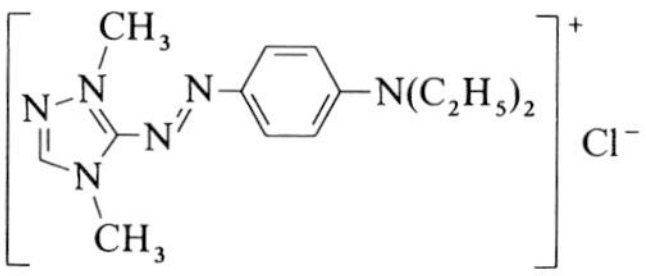

34a Main component [74109-48-1]

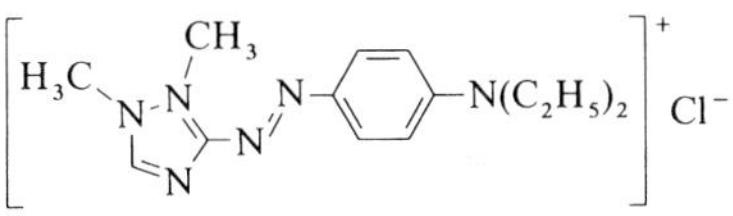

3.8.2 Chemical Constitution and Synthesis

$$\left[\begin{array}{c} H_3C-N\underset{N}{\overset{CH_3}{\underset{|}{N}}}\overset{}{\underset{N}{C}}-N=N-\underset{}{\underbrace{}}-N(C_2H_5)_2 \end{array} \right]^{+} Cl^{-}$$

34b Byproduct [74089-11-2]

Dyes that are obtained with *N,N*-dimethyl- and *N,N*-diethylaniline as coupling component and with dimethyl or diethyl sulfate as alkylating agent dye polyacrylonitrile in lightfast bluish-red shades [85], [86]. They have a lower affinity for polyacrylonitrile, but on account of their low molecular mass they are exhausted with a high diffusion rate by the fibers. They are therefore particularly suitable for producing level dyeings in light shades.

The affinity of the dyes for polyacrylonitrile and accordingly their exhaustion rate is increased by introducing aromatic groups. *C.I. Basic Red 46*, 110825 [*12221-69-1*] is of particular industrial importance.

N,N-Dibenzylaniline and *N*-methyldiphenylamine, as well as the arylcarboxylates and aryl carbonates of *N*-(hydroxyethyl)aniline, are also used to increase the exhaustion rate. Arylalkyl groups in the 1-position of the triazole ring likewise increase the exhaustion rate [87].

Two series of dyes are derived from 3,5-diamino-1,2,4-triazole [*1455-77-2*], also known as guanazole. The monoazo dyes form red diazadimethine dyes (e.g., **35**) after alkylation [88].

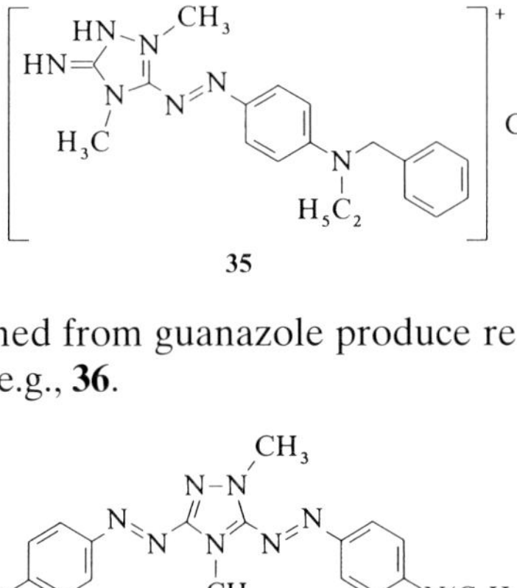

35

The disazo dyes formed from guanazole produce reddish blue diazadimethine dyes on alkylation [89], e.g., **36**.

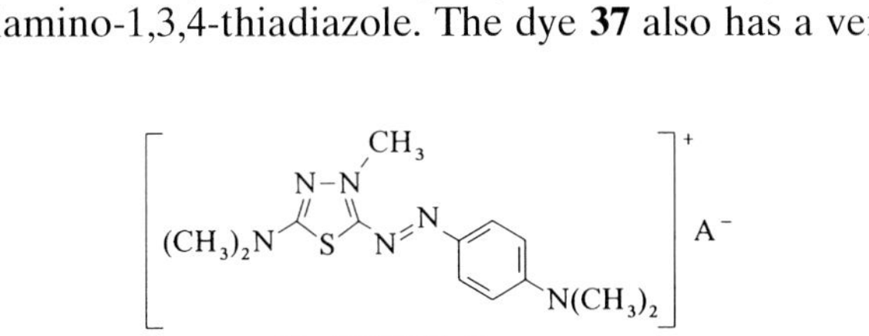

36

Blue diazadimethine dyes with a high dye bath stability are derived from 2-amino-5-dialkylamino-1,3,4-thiadiazole. The dye **37** also has a very good leveling ability [90].

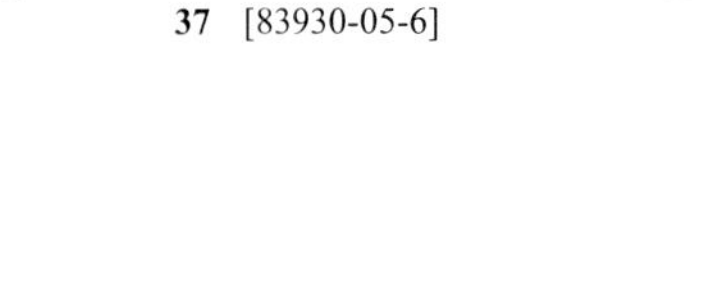

37 [83930-05-6]

Orange to red diazadimethine dyes are obtained in a similar manner with 3-aminopyrazole as diazo component [91]. The nitrogen atom in the 2-position of pyrazole can be substituted by a cyclohexyl, benzyl (**38**) [92], or phenyl group [93].

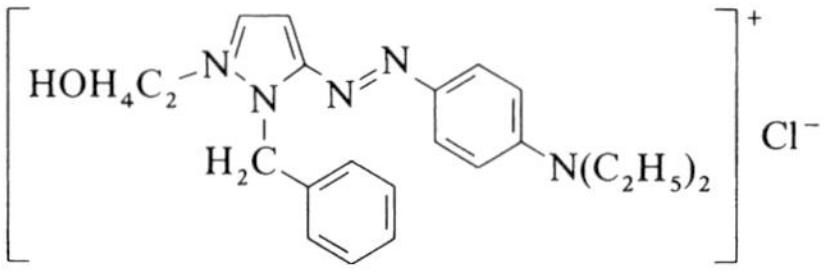

$$\left[\text{HOH}_4\text{C}_2-\overset{\text{N}}{\underset{\underset{\text{H}_2\text{C}}{|}}{\text{N}}}\diagdown\text{N}=\text{N}-\text{C}_6\text{H}_4-\text{N}(\text{C}_2\text{H}_5)_2 \right]^+ \text{Cl}^-$$

38 [68123-01-3]

In contrast to the aforementioned N-heterocyclic compounds, imidazole may be used as coupling component for diazotized aromatic amines. Dyes with primary amino groups are obtained by coupling diazotized 2- or 3-chloro-4-aminoacetanilide with imidazole, methylating the resultant azo dye to form the diazacyanine, and finally hydrolyzing the acetyl group in acid medium [94]. The dye **39** has an outstanding migration ability and accordingly dyes polyacrylonitrile in a lightfast orange shade with high levelness.

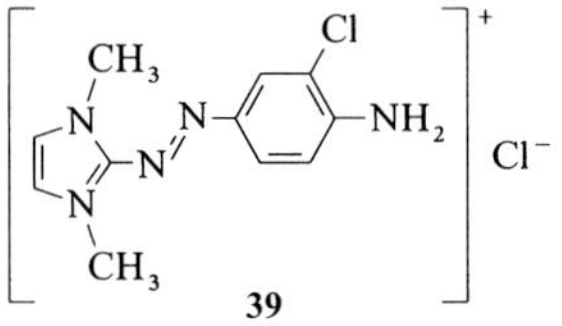

$$\left[\text{imidazole(CH}_3\text{)} - \text{N} = \text{N} - \text{C}_6\text{H}_3(\text{Cl}) - \text{NH}_2 \right]^+ \text{Cl}^-$$

39

Reaction of diazo components containing chlorine or a methoxy group in the 4-position to the amino group with imidazole, and subsequent quaternization yields compounds in which the halogen or alkoxy substituents (in the 4-position to the azo group) can be replaced by amino groups [95]. The dyes obtained with lower aliphatic amines [e.g., *C.I. Basic Red 51* [*12270-25-6*] (**40**)] color polyacrylonitrile in lightfast red shades [96]. On account of their low molecular mass they have a high diffusion rate and are therefore particularly suitable for producing level combination dyeings.

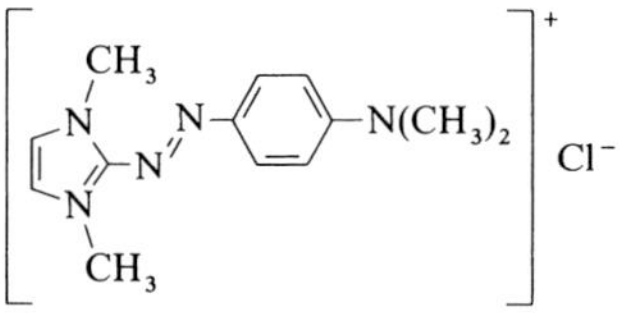

$$\left[\text{imidazole(CH}_3\text{)} - \text{N} = \text{N} - \text{C}_6\text{H}_4 - \text{N}(\text{CH}_3)_2 \right]^+ \text{Cl}^-$$

40 [12270-25-6]

3.8.2.5 Cyanine Dyes

In cyanine dyes both charge-carrying terminal atoms of the methine chain are constituents of heterocyclic rings. Depending on the nature of the heterocycle, the methine chain can be coupled in the 2-, 3-, or 4-position to the charge-carry-

3.8.2 Chemical Constitution and Synthesis

ing terminal atom. Accordingly, either one, two, or three ring atoms participate directly in the mesomeric methine system.

According to the general definition of methine dyes, there must always be an odd number of atoms in the conjugated system. In symmetrical dyes the sum of the charge-carrying atoms in both heterocyclic rings is always even. The methine chain must therefore contain an odd number of atoms. In cyanine dyes with different heterocyclic groups, however, the sum of the atoms involved in the charge-conducting system in the two rings can be an odd number. In this case the methine chain must contain an even number of atoms. Cyanine dyes with a substituted methine chain or in which parts of the chain are constituents of a ring are particularly important in photography (see Section 5.7).

Zeromethinecyanine Dyes. Naphtholactam and its N-alkyl derivatives can be condensed with indoles in the presence of acid condensation agents such as phosphorus oxychloride to form red zeromethinecyanine dyes that dye polyacrylonitrile very lightfast shades [97]. The indole may contain a methyl or phenyl group in the 2-position. If the 2-position is not substituted, the coloring strength increases because there is less steric hindrance [98], e.g., **41**.

41

Condensation of indoles with N-alkylquinolones or N-alkylpyridones yields yellow and orange zeromethine dyes that dye polyacrylonitrile in lightfast shades [99]. If the indole nitrogen atom is unsubstituted, intensely colored anhydro bases are formed from the zeromethinecyanine dyes. Dealkylation of the heterocyclic quaternary nitrogen atom also occurs [100].

Monomethinecaynine Dyes. The oldest dye of the this series is cyanine, a blue dye after which the whole class of dyes was named. Cyanine (**42**) was synthesized by Williams in 1856 from crude tar quinoline by quaternization with pentyl iodide, followed by treatment with alkali. The unsymmetrical monomethine dye coupled in the 2- and 4-positions of quinoline is termed isocyanine.

42 [862-57-7]

Dimethinecyanine Dyes. In the most important dimethinecyanine dyes the methine chain links two indole moieties in the 2- and 3-positions. The dyes were initially used for dyeing and printing cellulose acetate [101], but became increasingly important when it was discovered that they can dye polyacrylonitrile in brilliant and lightfast shades. *C.I. Basic Orange 21* is prepared either from 2-methylindoline-3-aldehyde and 1,3,3-trimethyl-2-methyleneindoline or from 1,3,3-trimethyl-2-methyleneindoline-ω-aldehyde and 2-methylindole. It is used in many ranges for dyeing polyacrylonitrile. A red-orange dye is prepared in a similar manner from 1-methyl-2-phenylindole.

Diazadimethinecyanine dyes are formed by coupling diazotized N-heterocyclic amines in the β-position to the nitrogen atom of other heterocycles followed by alkylation. The most important coupling agent is 1-methyl-2-phenylindole. Reaction of this compound with diazotized 2-aminothiazole and subsequent methylation with dimethyl sulfate, yields **43**, a dye that draws up very quickly on polyacrylonitrile and dyes it a lightfast red shade [102].

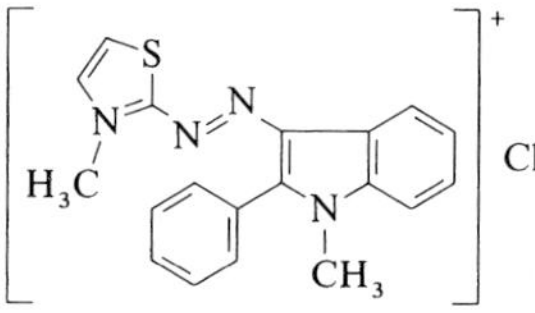

43 [42373-04-6]

Coupling diazotized 3-amino-1,2,4-triazole with 1-methyl-2-phenylindole and double alkylation yields dyes such as **44** that dye polyacrylonitrile in lightfast, reddish-yellow shades [103].

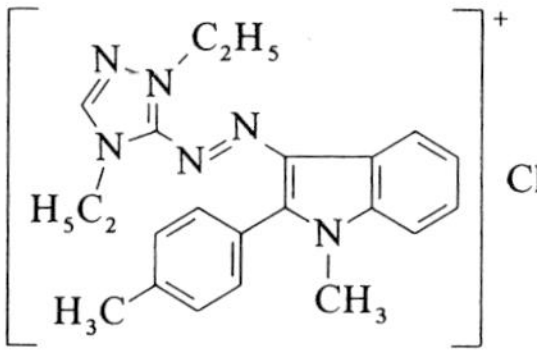

44

Diazacyanine dyes with two different alkyl substituents on the triazole nitrogen atoms are obtained by coupling diazotized 3-amino-1-benzyl-1,2,4-triazole to 1-ethyl-2-phenylindole and quaternizing the azo dye with dimethyl sulfate or, in the acid pH range, with acrylamide to form the diazacyanine dye [104].

Mono- and bis-diazacyanine dyes are derived from guanazole. Whereas the monodiazacyanine dyes dye polyacrylonitrile reddish yellow shades [105], the bis-diazacyanines [106] produce red dyeings, e.g. **45**.

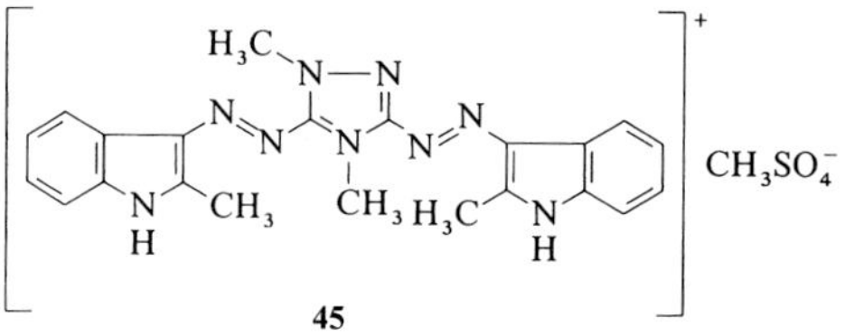

45

Dyes with 3-aminopyrazole or its derivatives [107] as diazo component dye polyacrylonitrile yellow, e.g. **46**. The tetrafluoroborate of this dye is used for spin dyeing polyacrylonitrile [108].

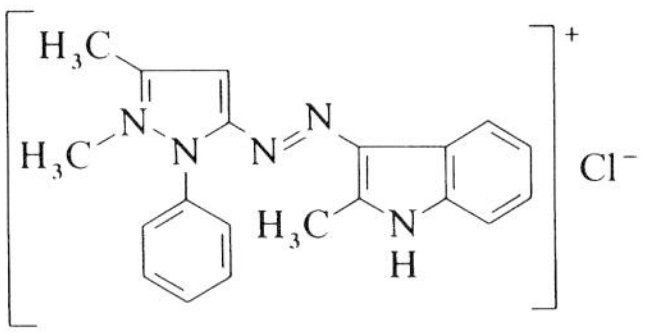

46 [33639-59-7]

Trimethinecyanine Dyes.. They are important sensitizers in photography (see Section 5.7). The first sensitizer for the long-wavelength red region was cryptocyanine [4727-50-8] (**47**), obtained from quaternized lepidine.

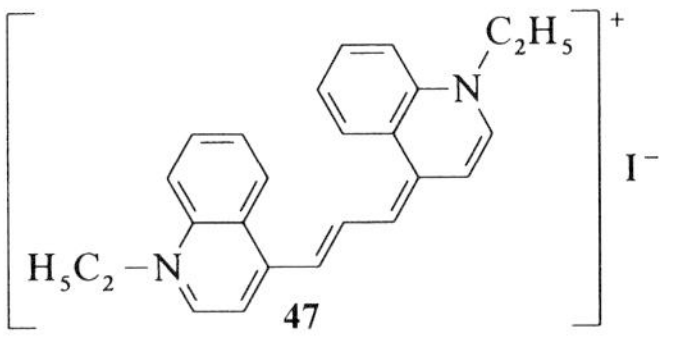

Sensitization maximum: 735 nm

Even more important was pinacyanol [605-91-4] (**48**), which was obtained from quaternary quinaldine by reaction with formaldehyde to form the leuco compound and atmospheric oxidation in an alkaline medium [109].

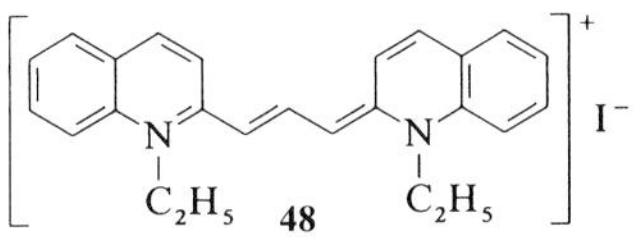

Sensitization maximum: 637 nm

Higher Vinylogs of Cyanine Dyes. Pentamethinecyanine dyes can be obtained analogously to trimethinecyanine dyes from the methyl-substituted heterocyclic quaternary salts by condensation. Heptamethinecyanine dyes are obtained by treating the quaternary salts of 2- or 4-methyl-*N*-heterocycles or the corresponding methylene compounds with pentamethinestreptocyanine dyes.

3.8.3 Synthesis

Due to the great variety of different types of cationic methine dyes the syntheses of the numerous groups of these dyes are described under each specific group in Section 3.8.2. Any further division would have led to an unreasonably large expansion of this book.

3.8.4 Examples of Commercially Available Dyes

C.I. Basic Yellow 11 [4208-80-4] (**49**) is included in practically all important dye ranges for dyeing polyacrylonitrile fibers.

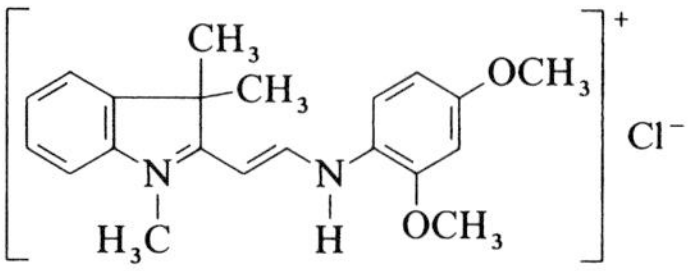

49

C.I. Basic Yellow 13 [12217-50-4] is obtained in a similar manner to Basic Yellow 11 by condensation of 4-anisidine with 1,3,3-trimethyl-2-methyleneindoline-ω-aldehyde. It dyes polyacrylonitrile a greener shade.

C.I. Basic Yellow 21, 48060 [6359-50-8] (**50**).

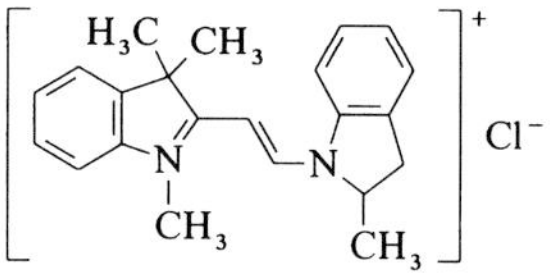

50

C.I. Basic Yellow 24 [52435-14-0] (**51**).

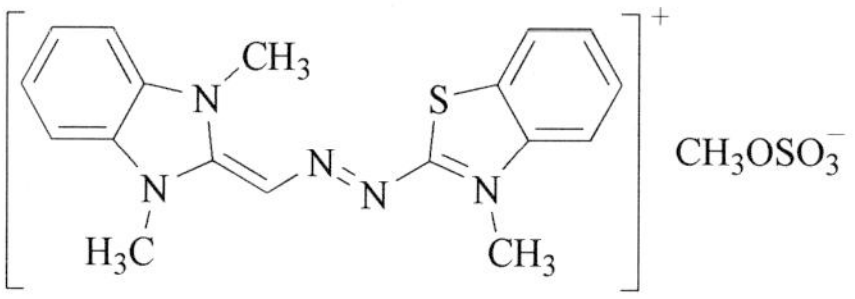

51

C.I. Basic Yellow 28, 48054 [54060-92-3] (**52**).

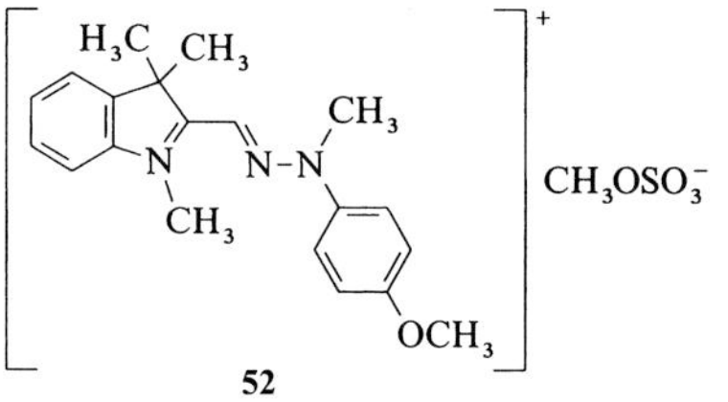

52

C.I. Basic Yellow 29 [38151-74-5] (**53**).

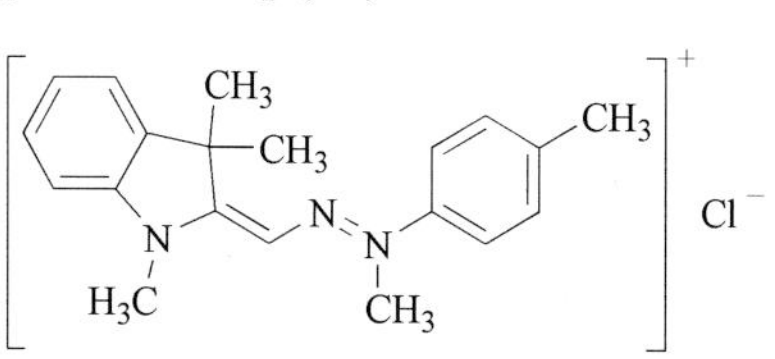

53

C.I. Basic Yellow 51 480538 [*83949-75-1*].
C.I. Basic Yellow 66 [*71343-19-6*].
C.I. Basic Yellow 70 [*71872-36-1*].
C.I. Basic Orange 21 [*47346-66-7*] (**54**).

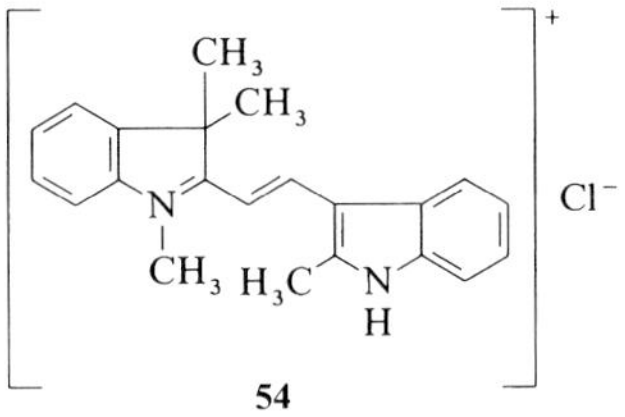

C.I. Basic Red 14 [*12217-48-0*] (**55**).

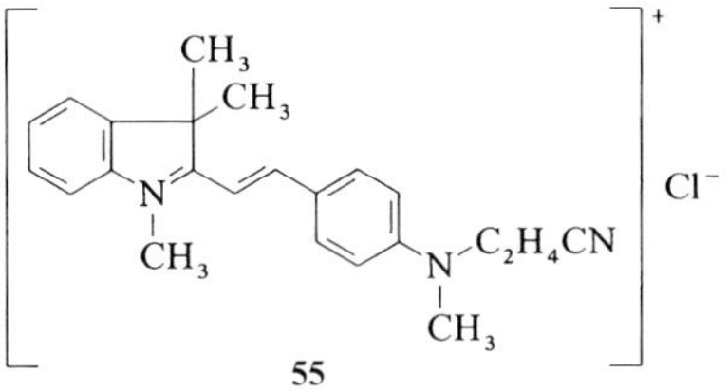

C.I. Basic Red 22, 11055 [*12221-52-2*].
C.I. Basic Violet 16, 48013 [*6359-45-1*] (**56**).

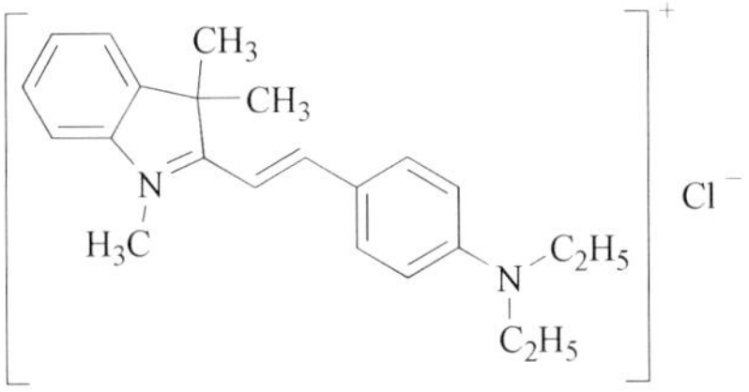

3.8.5 References

[1] Bayer, DE-OS 2 135 156, 1971 (A. Brack, H. Psaar).
[2] Mobil Finishes, SU 732 246, 1977.
[3] Bayer, DE-AS 1 108 659, 1956 (R. Raue, W. Müller).
[4] Ciba-Geigy, DE-OS 2 718 225, 1976 (R. Garner, B. Greenmount).
[5] Matsushita Electric, EP 97 493, 1982.
[6] Bayer, DE-OS 2 036 505, 1970 (A. Brack).
[7] IG Farben, DE 686 198, 1935 (K. Winter et al.)
[8] BASF, DE-OS 2 735 263, 1977 (H. Kast).
[9] Hoechst, FR 2 011 334, 1968.

[10] Nihon Kasei Kogyo, JP 47 32 183, 1971.
[11] ICI, DE-OS 2 658 638, 1975 (J. Hodgkinson et al.).
[12] VEB Chemiekombinat Bitterfeld, DD 244 348, 1985 (H. Noack et al.).
[13] Ciba-Geigy, EP 38 299, 1980 (W. Stingelin, P. Loew); EP 94 642, 1982 (W. Stingelin et al.); EP 145 656, 1983 (W. Stingelin et al.).
[14] BASF, EP 273 307, 1986 (J. P. Dix et al.).
[15] BASF, DE-OS 2 839 379, 1978 (K. Grychtol).
[16] Bayer, DE-OS 2 352 247, 1973 (A. Brack, R. Raue).
[17] IG Farben, DE 710 750, 1933 (P. Wolff).
[18] Du Pont, DE-OS 2 002 609, 1969 (F. R. Hunter). Bayer, DE-OS 2 130 790, 1971 (H. Psaar).
[19] Sandoz, DE-OS 2 917 996, 1978 (B. Gertisser). Ciba-Geigy, DE-OS 2 926 183, 1978 (P. Loew, R. Zink).
[20] Du Pont, DE-OS 1 956 158, 1968 (F. R. Hunter).
[21] Bayer, DE-OS 1 945 053, 1969 (A. Brack).
[22] Bayer, DE-OS 3 322 318, 1983 (K. Meisel, R. Raue).
[23] Bayer, DE-AS 1 044 588, 1956 (R. Raue).
[24] Ciba, DE 1 004 748, 1954 (R. Ruegg).
[25] Ciba, DE 1 061 931, 1954 (R. Ruegg).
[26] Bayer, DE 1 083 000, 1957 (R. Raue).
[27] Bayer, EP 42 556, 1980 (R. Raue, H.-P. Kühlthau).
[28] Hoechst, DE-OS 2 945 028, 1979 (R. Mohr, M. Hähnke).
[29] Bayer, DE 1 133 053, 1959 (R. Raue).
[30] Bayer, EP 87 706, 1982 (R. Raue).
[31] BASF, DE 1 150 474, 1960 (H. Baumann et al.).
[32] BASF, GB 1 186 745, 1966.
[33] Bayer, DE 1 233 520, 1963 (R. Raue, A. Brack, E.-H. Rohe).
[34] Hodogaya Chem., JP 55 102 654, 1979.
[35] Bayer, DE-OS 3 235 613, 1982 (H. Fürstenwerth).
[36] Bayer, DE 1 150 475, 1959 (R. Raue); DE 1 133 054, 1959 (R. Raue).
[37] Ciba-Geigy, DE-OS 2 548 009, 1974 (J. Koller et al.).
[38] Bayer, DE 1 209 679, 1961 (R. Raue, E.-H. Rohe).
[39] Geigy, BE 579 303, 1958.
[40] Hodogaya Chem., JP 71 35 630, 1968.
[41] IG Farben, DE 714 882, 1939.
[42] Cassella, DE 51 738, 1888 (J. Rosenheck).
[43] O. N. Chupakhin et al., *Zh. Org. Khim.* **12** (1976) no. 7, 1553–1557.
[44] V. N. Charushin et al., *Zh. Org. Khim.* **14** (1978) no. 1, 140–146.
[45] O. N. Chupakhin et al., *Dokl. Akad. Nauk. SSSR* 239 (1978) no. 3, 614–616.
[46] IG Farben, DE 483 234, 1926 (G. Kränzlein, M. Heyse, P. Ochwat).
[47] Bayer, DE 1 190 126, 1961 (A. Brack, D. Kutzschbach).
[48] Bayer, DE-OS 2 138 029, 1971.
[49] BASF, DE-AS 1 569 645, 1966 (H. Baumann, G. Hansen).
[50] Bayer, DE-OS 2 351 296, 1973 (H.-P. Kühlthau). DE-OS 2 557 503, 1975 (H.-P. Kühlthau).
[51] Bayer, DE-OS 3 148 104, 1981 (H.-P. Kühlthau, H. Harnisch).
[52] Bayer, DE-OS 2 747 110, 1977 (A. Brack, H.-P. Kühlthau).
[53] Ciba-Geigy, DE-OS 2 233 261, 1972.
[54] Ciba-Geigy, DE-OS 3 321 130, 1982 (R. Zink).
[55] Sterling Drug, EP 61 128, 1981 (P. J. Schmidt et al.). Matsushita Electric, JP 59 71 898, 1982.

[56] Ricoh, JP 59 75 257, 1982. Canon, JP 60 195 550, 1984.
[57] MLB Hoechst, DE 394 744, 1922 (R. Schuloff).
[58] Bayer, DE-OS 2 932 092, 1979 (H. Beecken).
[59] Bayer, DE-OS 3 012 599, 1980 (H. Beecken).
[60] IG Farben/Bayer, DE 711 665, 1935.
[61] IG Farben/Bayer, DE 891 120, 1935 (W. Müller, C. Berres).
[62] IG Farben, DE 721 020, 1935 (K. Winter, N. Roh).
[63] Bayer, EP 74 569, 1981 (R. Raue, V. Hühne, H.-P.Kühlthau).
[64] Bayer, DE-OS 3 213 966, 1982 (R. Raue, H.-P.Kühlthau).
[65] Bayer, EP 89 566, 1982 (R. Raue, H.-P. Kühlthau, K.-F. Lehment).
[66] Bayer, DE-OS 2 243 627, 1972 (H. Psaar).
[67] IG Farben, DE 721 020, 1935 (K. Winter, N. Roh).
[68] Canon, JP 63 74 069, 1986.
[69] IG Farben, DE 742 039, 1938 (G. Kochendörfer et al.). Sandoz, DE-OS 2 101 223, 1970 (R. Entschel, W. Käppeli).
[70] Nippon Kayaku, JP 74 04 530, 1970.
[71] Bayer, DE 835 172, 1950.
[72] Bayer, DE 1 569 606, 1967 (A. Brack).
[73] Ciba-Geigy, EP 37 374, 1980 (R. Zink).
[74] Vickers PLC, EP 112 169, 1982 (A. J. Chalk et al.).
[75] BASF, DE-OS 2 902 763, 1979 (H.-J. Degen, K. Grychtol); DE-OS 2 733 468, 1977 (H.-J. Degen, K. Grychtol).
[76] BASF, DE 1 619 328, 1966 (H. R. Hensel, M. Rosenkranz).
[77] A. P. D'Rozario et al., *J. Chem. Soc. Perkin Trans. 2* **1987**, 1785–1788.
[78] Bayer, DE-OS 2 255 058, 1972 (H.-P. Kühlthau); DE-OS 2 344 901, 1973 (H. P. Kühlthau).
[79] Bayer, DE-OS 2 255 059, 1972 (H. P. Kühlthau); DE-OS 2 344 672, 1973 (H. P. Kühlthau).
[80] Hodogaya, JP 74 21 285, 1970.
[81] Daito Kagaku, JP 72 50 210, 1970.
[82] Hodogaya Chem., US 3 132 132, 1960.Sumitomo Chem., DE-OS 2 136 974, 1970 (M. Ohkawa et al.); JP 73 06 931, 1970 (M. Ohkawa et al.); JP 73 01 997, 1970.
[83] Hoechst, DE-OS 2 822 913, 1978 (M. Hähnke, R. Mohr, K. Hohmann).
[84] Sumitomo Chem., GB 1 186 753, 1966.
[85] Geigy, DE 1 044 023, 1954 (W. Bossard, J. Voltz, F. Favre); DE 1 050 940, 1954 (W. Bossard, J. Voltz, F. Favre).
[86] Geigy, FR 1 145 751, 1955.
[87] Hodogaya Chem., JP 51 56 828, 1974.Sandoz, DE-OS 2 509 095, 1974 (B. Henzi).
[88] Hodogaya Chem., DE-OS 2 104 624, 1970 (M. Otsuzumi et al.).
[89] Hodogaya Chem., DE-OS 2 022 624, 1969 (M. Otsuzumi et al.).
[90] Bayer, DE-OS 2 811 258, 1978 (H. Fürstenwerth).
[91] Geigy, BE 666 693, 1964. Prod. Chim. Ugine Kuhlmann, EP 1 731 1977 (M. Champenois).
[92] BASF, DE-OS 2 234 348, 1972 (M. Eistert, K. Schmeidl).Yorkshire Chem., GB 1 591 532, 1977 (K. Bramham, M. Robert).
[93] Ciba-Geigy, DE-OS 3 544 574, 1984 (W. Stingelin, P. Moser).
[94] Sandoz, DE-OS 2 819 197, 1977 (B. Henzi).Bayer, DE-OS 2 837 908, 1978 (H.-P. Kühlthau).Sandoz, DE-OS 2 927 205, 1978 (B. Henzi, U. Blass).
[95] BASF, DE-AS 1 137 816, 1958 (H. Baumann, J. Dehnert).
[96] Bayer, DE-OS 2 837 953, 1978 (H.-P. Kühlthau); DE-OS 2 837 987, 1978 (H.-P. Kühlthau); DE-OS 2 908 135, 1979 (H.-P. Kühlthau).
[97] Bayer, DE 1 184 882, 1962 (A. Brack).

[98] Bayer, BE 671 139, 1964;BE 662 178, 1964.
[99] Ciba, DE 1 569 690, 1965.
[100] T. V. Stupnikowa et al., *Khim. Geterotsikl. Soedin.* **1980**, 959–964. T. V. Stupnikowa et al., *Khim. Geterotsikl. Soedin.* **1980**, 1365–1368.
[101] IG Farben, DE 614 325, 1933 (P. Wolff). DE 615 130, 1933 (P. Wolff).
[102] BASF, DE 1 163 775, 1955 (H. Pfitzner et al.).
[103] BASF, DE 1 228 586, 1956 (H. Pfitzner et al.); DE 1 205 637, 1960 (H. Pfitzner, H. Baumann).
[104] Hodogaya Chem., DE-OS 1 925 491, 1968 (M. Jizuka et al.); JP 73 00 851, 1970; JP 49 24 229, 1972.
[105] Hodogaya Chem., JP 72 50 208, 1970.
[106] Hodogaya Chem., US 3 822 247, 1969 (M. Ozutsumi, S. Maeda, Y. Kawada). Hodogaya Chem., JP 49 24 225, 1972.
[107] BASF, BE 660 201, 1964. Geigy, BE 683 733, 1965. Montedison, IT 1 110 641, 1979 (F. Merlo et al.).
[108] Ciba-Geigy, EP 55 224, 1980 (P. Loew).
[109] MLB Hoechst, DE 172 118, 1905; DE 175 034, 1906; DE 178 688, 1906; DE 189 942, 1906.

3.9 Acid Dyes (Anionic Azo Dyes)

3.9.1 Introduction

Anionic dyes include many compounds from the most varied classes of dyes, which exhibit characteristic differences in structure (e.g., azoic, anthraquinone, triphenylmethane, and nitro dyes) but possess as a common feature water-solubilizing, ionic substituents. The anionic azo dyes which are discussed here constitute the most widely used group of this class of dyes.

Most often sulfonic acid groups serve as hydrophilic substituents, because they are readily introduced and, as strong electrolytes, are completely dissociated in the acidity range used in the dyeing process. Almost invariably the products manufactured and employed are water-soluble sodium salts of the sulfonic acids.

In principle the anionic dyes also include *direct dyes*, but, because of their characteristic structures, these are used to dye cellulose-containing materials and are applied to the fiber from a neutral dye bath (see Section 3.3).

From the chemical standpoint the group of anionic azo dyes includes a large proportion of the *reactive dyes* which, in addition to the usual structural characteristics, also contain groups that can react with functional groups of the fiber during the dyeing process (see Section 3.1).

The anionic azo dyes also include formally the *metal-complex dyes*, which are derived from *o,o'*-disubstituted azo dyes and which can be metallized not only on the fiber but whose metal complexes are applied largely in the form of a complete dye (see Section 3.11).

For supplementary literature, see [1–4].

3.9.2 Chemical Constitution and Synthesis

Azo dyes with relatively low molecular masses and one to three sulfonic acid groups serve as acid azo dyes for dyeing and printing wool, polyamide, silk, and basic-modified acrylics and for dyeing leather, fur, paper, and food. The main area of application is the dyeing of wool and polyamide.

Disazo and polyazo dyes containing sulfonic acid groups are also frequently used in the above applications. Many of them are also sold as substantive dyes for cotton because of their pronounced affinity toward cellulosic fibers (see Section 3.3).

The term "acid dye" derives from the dyeing process, which is carried out in an acidic aqueous solution (pH 2–6). Protein fibers contain amino and carboxyl groups, which in the isoelectric range (ca. pH 5) are ionized mostly to NH_3^+ and COO^-. In the acid dyebath the carboxylate ions are converted to undissociated carboxyl groups owing to the addition of acid HX (sulfuric or formic acid), which causes the positively charged wool (H_3N^+–W–COOH) to take up an equivalent amount of acid anions X^- (hydrogensulfate, formate; Scheme 3.7) [5]:

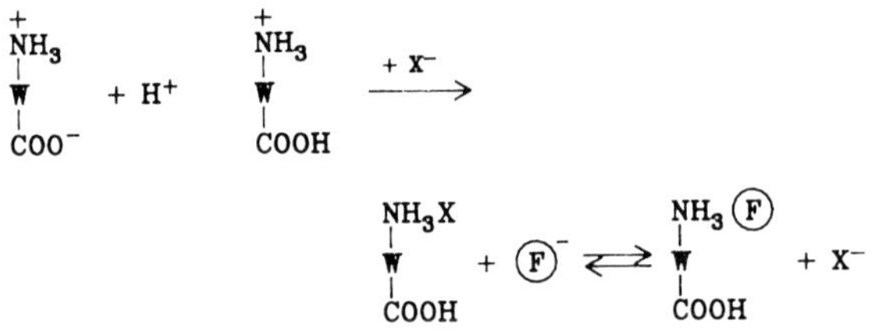

$$\begin{array}{c} \overset{+}{N}H_3 \\ | \\ W \\ | \\ COO^- \end{array} \quad + H^+ \quad \begin{array}{c} \overset{+}{N}H_3 \\ | \\ W \\ | \\ COOH \end{array} \quad + X^- \longrightarrow \quad \begin{array}{c} NH_3X \\ | \\ W \\ | \\ COOH \end{array} \quad + \text{\textcircled{F}}^- \rightleftarrows \quad \begin{array}{c} NH_3 \text{\textcircled{F}} \\ | \\ W \\ | \\ COOH \end{array} \quad + X^-$$

Scheme 3.7

The actual dyeing process consists of a replacement of the absorbed acid anions X^- by the added dye anions $\text{\textcircled{F}}^-$, since the latter exhibit a much greater affinity for the substrate than the much smaller acid anions. Thus the dye is bonded to the wool not only by electrostatic attraction (salt formation) but also by its affinity for the fiber.

Acid dyes are divided into three groups based on their differences in affinity, which is primarily a function of the molecular size:

1) Leveling dyes are relatively small molecules which form a saltlike bond with the protein fiber.
2) Milling dyes are large-volume dye molecules, for which salt formation with the fiber plays only a secondary role and the adsorption forces between the

hydrophobic regions of the dye molecule and those of the protein fiber predominate.
3) Dyes with intermediate molecular size not only form a saltlike bond with the wool fiber but are also bonded to the fiber by intermolecular forces and have properties lying in an intermediate position between those of the leveling and the milling dyes.

3.9.2.1 Wool Dyes

The standard dyes for wool are divided essentially into four groups according to their different dyeing behaviors.

Group A contains the low-priced, usually older leveling dyes with moderate lightfastness. These dyes are still used in larger quantities for cheap articles.

Group B also includes leveling dyes, which are applied in a strongly acid bath, but which exhibit very good lightfastness.

Group C contains the milling dyes, which are applied in a weakly acid to neutral bath.

Group D contains the acid dyes with good leveling properties, intermediate molecular sizes, and for the most part good lightfastnesses.

The product ranges of the various dye manufacturers do not completely fulfill the demands made on the dyes in respect of the individual fastnesses, and within a manufacturer's range the properties of the dyes can be individually specified.

In describing the separate dye types it is useful to distinguish between the monoazo and the disazo dyes, and to further break down the disazo dyes into primary and secondary types on the basis of chemicostructural principles.

In the industrial manufacture of acid azo dyes usually aniline derivatives are used as the diazo components. The coupling components for orange to blue shades are commonly aniline, naphthol, naphthylamine and aminonaphthol derivatives, whereas phenylpyrazolones are much used for preparing dyes in the yellow and orange shades.

With regard to the stability of the shade to changes in pH, care must be taken that hydroxyl or amino groups lie adjacent to the azo group in order to form a hydrogen bond with the latter. This hydrogen bond prevents a dissociation of the hydroxyl groups or a protonation of the amino groups and hence avoids an unwanted change in shade in response to moderate pH shifts. Groups that are not in the *ortho* position must be alkylated or acylated.

Acid Monoazo Dyes. Among the acid monoazo dyes are a number of much used wool dyes that possess no outstanding coloristic properties but that are distinguished by brilliance of shade, very good leveling power, and particularly low cost, while their wash- and lightfastness meet only low to medium requirements. Partly influenced by the introduction of the International Wool Label to label high-quality wool articles, the fastness requirements have risen considerably, and this has necessitated the development and manufacture of particularly fast dyes.

3.9.2 Chemical Constitution and Synthesis

For this reason many, and above all the older, acid dyes have lost all importance for the dyeing of wool and are used today only for the coloring of paper, soaps, food, and cosmetics.

In the following discussion the acid monoazo dyes are classified according to the type of coupling component.

In the case of *aromatic amines as coupling components*, one of the oldest azo dyes is *C.I. Acid Yellow 36*, 13065 (**1**) [587-98-4]. Because of the basic group suitable for salt formation the dye is not fast to acid, but it is still used today for dyeing wool and in special areas (leather and paper) primarily for price reasons.

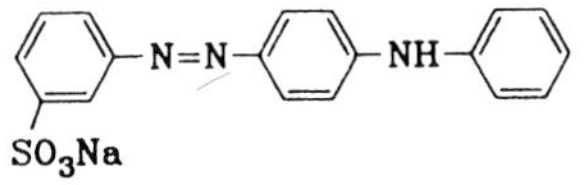

Coupling H acid to 1-(phenylamino)naphthalene-8-sulfonic acid (*N*-phenyl peri acid) yields *C.I. Acid Blue 92* (see Section 3.9.3). On wool it produces a very pure blue of good lightfastness and with moderate wetfastness and adequate leveling power.

Naphthols and Naphtholsulfonic Acids as Coupling Components. This series includes two important acid dyes with very similar structures: *C.I. Acid Red 88* (see Section 3.9.3), derived from diazotized naphthionic acid and 2-naphthol, and *C.I. Acid Red 13*, 16045 [2302-96-7] (**2**), from naphthionic acid and Schäffer's acid. Both are all-purpose dyes which, because of their attractive red shades, are still in use today in many areas of textile dyeing and also for leather and paper dyes. Wool dyeings produced with these dyes exhibit moderate fastness levels.

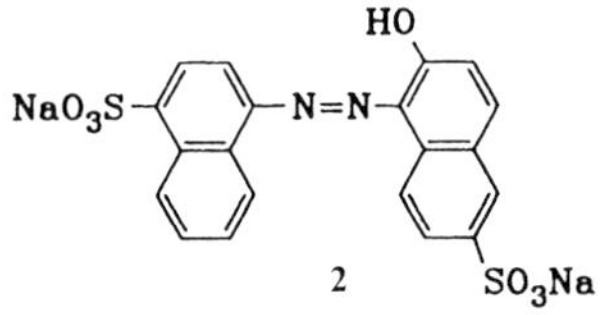

A long-known, inexpensive, but only moderately fast dye is *C.I. Acid Orange 7*, 15510 [633-96-5] (Orange II; **3**). As a wool dye it is now of secondary importance, but is used in special areas such as leather dyeing and paper coloration.

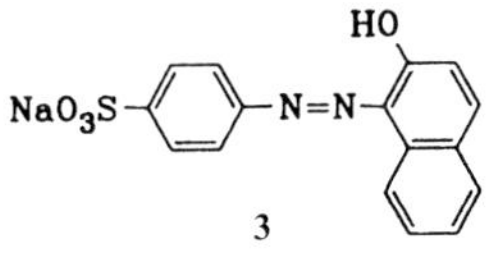

A previously much used red wool dye is *C.I. Acid Red 14* (see Section 3.9.3), which is now used to a limited extent in wool dyeing for inexpensive articles.

The usually very low wetfastness of these simple acid azo dyes can be improved at the expense of leveling power by the introduction of special, large-volume substituents. For example, *C.I. Acid Orange 19*, 14690 [3058-98-8] (**4**), is an important acid dye for wool and polyamide with good fastness properties and yet with good leveling power.

Aminonaphtholsulfonic Acid Coupling Components. Predominant here are γ acid (2,8-aminonaphthol-6-sulfonic acid) and H acid (1,8-aminonaphthol-3,6-disulfonic acid), from which many important wool dyes are derived, some of which exhibit very good light- and wetfastness. Especially in acid coupling γ acid yields dyes with very good lightfastness, which is presumably attributable to the formation of a hydrogen bond between the azo bridge and the hydroxyl group *peri* to the azo bridge. Two outstandingly lightfast dyes of good leveling power whose dyeings on wool exhibit moderate wetfastness are *C.I. Acid Red 42*, 17070 (**5**) [6245-60-9], and *C.I. Acid Red 37*, 17045 [6360-07-2] (**6**).

Through the choice of suitably substituted diazo components, bluish red to blue wool dyes with good light- and wetfastnesses can be produced, e.g., *C.I. Acid Violet 14*, 17080 (**7**) [4404-39-1].

Of interest is the manufacture of 4-amino-3-(4′-toluenesulfonyl)acetanilide, which is used as the diazo component: *p*-phenylenediamine is oxidized with iron(III) chloride to 1,4-benzoquinonediimine, to which *p*-toluenesulfinic acid is added. The final step is partial acetylation (Scheme 3.8).

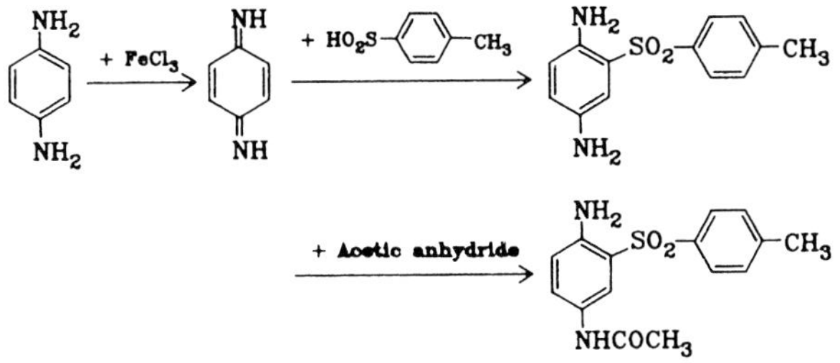

3.9.2 Chemical Constitution and Synthesis

[Scheme 3.8 reaction sequence: p-phenylenediamine + FeCl₃ → diimine + HO₂S-C₆H₄-CH₃ → 2-amino-4-amino-sulfonyltoluene intermediate + Acetic anhydride → acetylated product with NHCOCH₃, NH₂, SO₂-C₆H₄-CH₃ groups]

Scheme 3.8

Further examples are *C.I. Acid Red 32*, 17065 (**8**) [6360-10-7], and *C.I. Acid Blue 117*, 17055 [10169-12-7] (**9**). On wool both dyes exhibit quite good wetfastness and high leveling power.

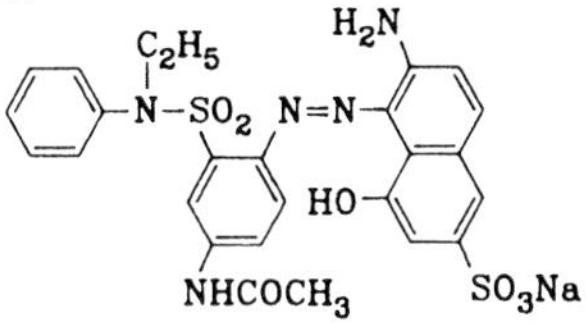

8 (structure: phenyl-N(C₂H₅)-SO₂- group with NHCOCH₃, linked via N=N to naphthalene bearing H₂N, HO, SO₃Na)

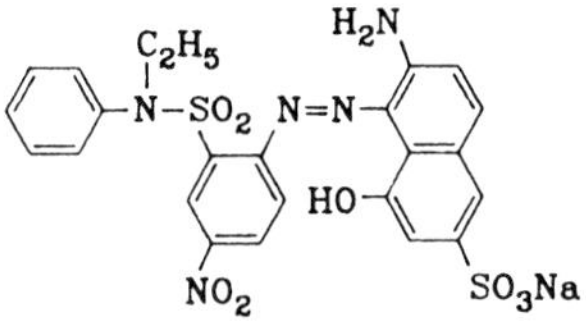

9 (analogous structure with NO₂ instead of NHCOCH₃)

To achieve valuable dyes with alkaline coupling to γ acid, the amino group must be acylated or arylated. Example: *C.I. Acid Red 68*, 17920 [6369-40-0] (**10**). On wool it exhibits high leveling power and yields light- and wetfast dyeings.

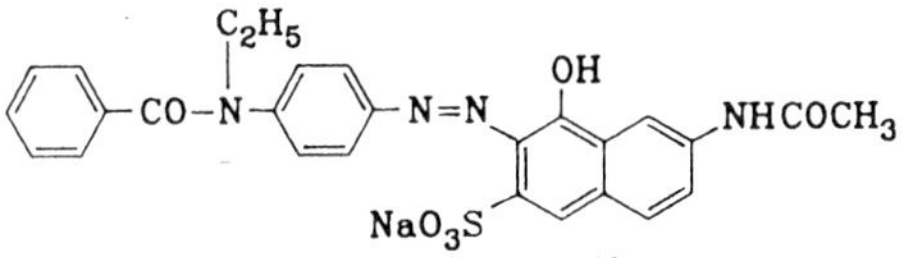

10 (structure: C₆H₅-CO-N(C₂H₅)-C₆H₄-N=N-naphthalene with OH, NaO₃S, NHCOCH₃)

N-Arylation of γ acid results in a strong deepening of the shade. The usually brown to black dyes thus obtained are fast to light and milling and go onto wool and silk from a neutral liquor. Example: *C.I. Acid Brown 20*, 17640 [6369-33-1] (**11**).

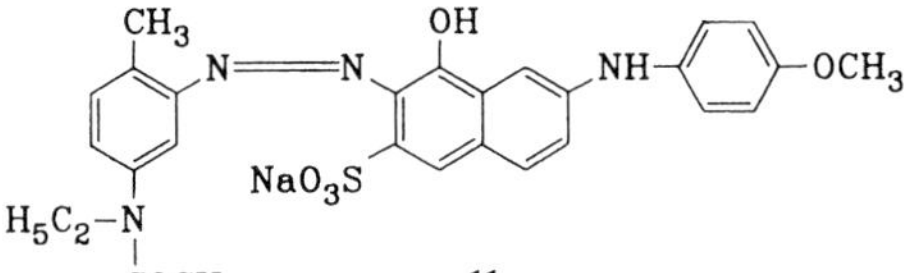

11 (structure: tolyl group with CH₃, H₅C₂-N(COCH₃)- substituents, N=N-linked to naphthalene with OH, NaO₃S, NH-C₆H₄-OCH₃)

An important coupling component is H acid. Realization that acylation of its amino group substantially improves the ligh fastness and in addition yields dyes with good leveling power led to many valuable light- and wetfast products that even today occupy a firm position on the market. The relatively simply structured wool dye *C.I. Acid Red 1*, 18050 [*3734-67-6*] (**12**) is obtained by the alkali coupling of diazotized aniline to *N*-acetyl H acid.

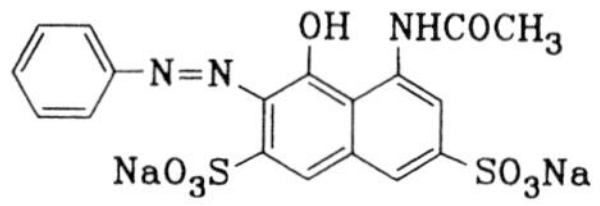

Ph—N=N—[naphthalene with OH, NHCOCH₃, NaO₃S, SO₃Na]
12

Valuable dyes can also be made with *N*-benzoyl H acid and *N*-toluenesulfonyl H acid.

An interesting development is achieved by incorporating long hydrophobic hydrocarbon radicals, which yields dyes with neutral affinity to wool and very good wetfastness properties [6]. Example: *C.I. Acid Red 138*, 18073 [*15792-43-5*] (**13**):

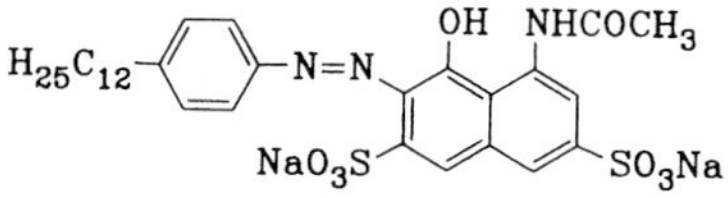

$H_{25}C_{12}$—[phenyl]—N=N—[naphthalene with OH, NHCOCH₃, NaO₃S, SO₃Na]
13

1-Phenyl-5-Pyrazolones as Coupling Components. Especially lightfast yellow shades are obtained by using *1-phenyl-5-pyrazolone coupling components.* The first representative of the class to appear was Tartrazine, *C.I. Acid Yellow 23*, 19140 [*1934-21-0*], which is prepared today from 1-(phenyl-4′-sulfonic acid)-3-carboxy-5-pyrazolone as the starting compound, obtained from oxaloacetic ester and phenylhydrazine-4-sulfonic acid and coupling with diazotized sulfanilic acid.

For price reasons the 1-aryl-3-methyl-5-pyrazolones and their derivatives are preferred to the corresponding 3-carboxypyrazolones. There is such a wide range of possible variations that dyes in this series extend from greenish yellow to reddish orange. An example is *C.I. Acid Yellow 17* (see Section 3.9.3), which on wool yields a clear, superbly lightfast yellow with good to very good general fastness properties, but relatively poor milling fastness.

C.I. Acid Yellow 76, 18850 [*6359-88-2*] (**14**), is obtained by coupling diazotized 4-aminophenol onto the pyrazolone component and then esterifying with *p*-toluenesulfonic acid chloride in an alkaline medium. The toluenesulfonic ester group substantially improves the fastness to milling and makes the shade obtained largely independent of pH; the lightfastness is not quite as good as that of *C.I. Acid Yellow 17*.

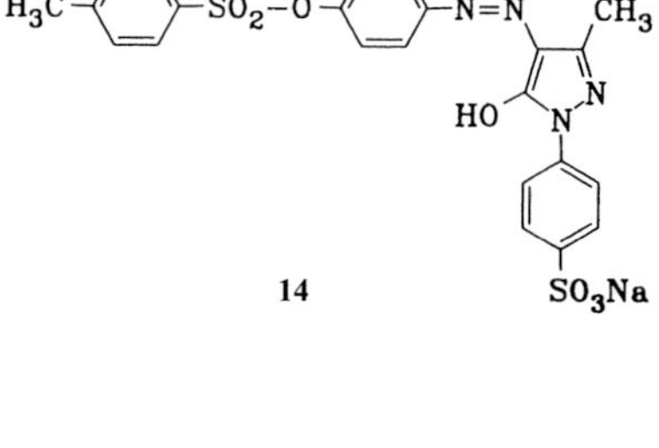

H_3C—[phenyl]—SO_2—O—[phenyl]—N=N—[pyrazolone with CH₃, HO, N-phenyl-SO₃Na]
14

3.9.2 Chemical Constitution and Synthesis

Acid Disazo Dyes. Dyes with two azo groups are divided on chemicostructural principles into primary and secondary disazo dyes. Primary disazo dyes are manufactured either according to the scheme $D_1 \rightarrow K \leftarrow D_2$ from a bifunctional coupling component K and two identical or different diazo components D_1 and D_2, or according to the scheme $K_1 \leftarrow D \rightarrow K_2$ by coupling a bis-diazotized diamine D to the coupling components K_1 and K_2, which may be the same or different. Secondary disazo dyes are manufactured according to the scheme $D \rightarrow M \rightarrow K$ (M = middle component) from a diazotized aminoazo dye ($D \rightarrow M$) and a coupling component K (series coupling).

Primary disazo dyes of the type $D_1 \rightarrow K \leftarrow D_2$. They are obtained by twofold coupling to, for example, H acid. Among the naphthalene derivatives H acid is one of the most important; it serves as a bifunctional coupling component K for the manufacture of many wool dyes now in use.

Dye synthesis following the above scheme generally permits little variation of the shade (black, dull brown, and blue), because completely conjugated chromophore systems are formed.

One of the most important acid dyes is *C.I. Acid Black 1* (see Section 3.9.3). It dyes wool in blue-black shades with very good lightfastness and exhibits high affinity and good leveling power but only moderate wetfastness. Nevertheless, because of its high affinity and good leveling power it has remained an important acid dye, which through shading with yellow, orange, or red forms the basis of most black acid dyes. To manufacture this type of dye, coupling is always first carried out in an acid medium with the more strongly electrophilic diazonium salt, here with diazotized 4-nitroaniline, then in the alkaline range with diazotized aniline:

$$\text{4-nitroaniline} \xrightarrow[\text{acid}]{1} \text{H acid} \xleftarrow[\text{alkaline}]{2} \text{aniline}$$

Usually the acid coupling of 4-nitroaniline is not continued to completion, because this requires too much time. Consequently, the commercial product normally contains a small proportion of red monoazo dye, which together with the greenish black disazo dye yields the desired neutral black shade.

Primary disazo dyes of the type $K_1 \leftarrow D \rightarrow K_2$ ($K_1 = K_2$ or $K_1 \neq K_2$). This series includes many milling dyes because of the molecular size achieved (group C). Depending on the type of coupling components K_1 and K_2, which may be phenols, pyrazolones, acetoacetic acid arylamides, or naphtholsulfonic acids, clear yellow to red shades are obtained. For a long time the preferred diazo components were diaminodiphenyl derivatives, such as benzidine, *o*-tolidine and *o*-anisidine. They are now of no further importance because of their carcinogenic potential.

An important discovery was the realization that diaminodiphenyls substituted in the 2,2'-positions are outstandingly suitable for the manufacture of very washfast and millingfast wool dyes, while the substantivity toward cellulosic fibers is reduced. A number of important acid wool dyes have been developed on this basis, for example, *C.I. Acid Yellow 44*, 23900 [2429-76-7] (**15**).

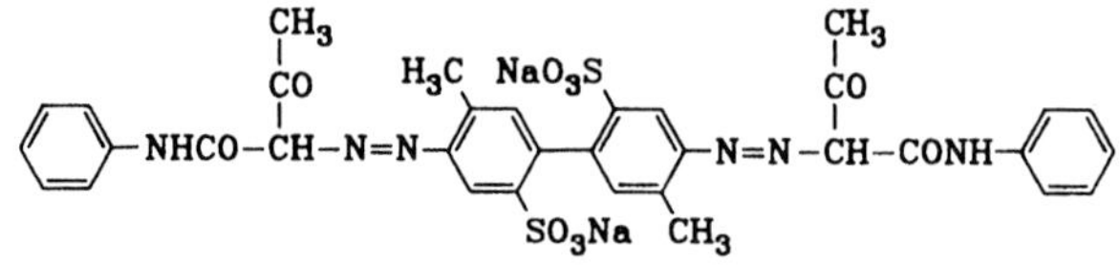

3.9 Acid Dyes (Anionic Azo Dyes)

15

In the manufacture of this dye bis-diazotized 5,5′-dimethylbenzidine-2,2′-disulfonic acid is coupled to two equivalents of N-acetoacetylaniline. The clear, intense, greenish yellow dyeing obtained on wool is millingfast and exhibits excellent wetfastness but only moderate lightfastness. *C.I. Acid Yellow 42* is a further member of this series (see Section 3.9.3).

Another class of substances of interest as bifunctional diazo components consists of special diamines that can be manufactured by the reaction of aromatic amines with benzaldehyde or cyclohexanone, e.g., **16** and **16a**.

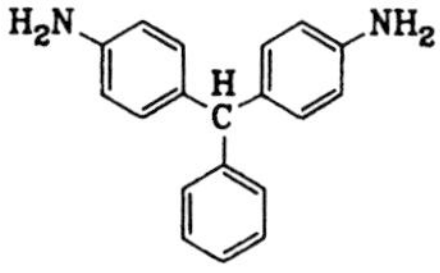

16

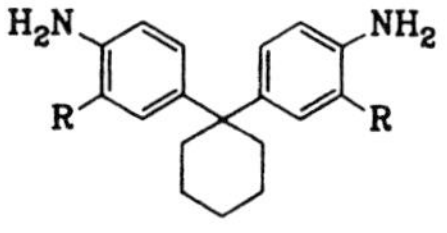

16a R = H, CH$_3$, OCH$_3$

Dyes based on these compounds possess, in addition to good lightfastness, excellent wetfastness and are usually neutral-dyeing on wool. This, although of no importance for dyeing pure wool, plays an important role in dyeing blended spun yarn and blended fabrics of wool and cotton or wool and viscose staple. The neutral-dyeing acid dye can be used in combination with direct dyes (union wool recipes). Examples are *C.I. Acid Yellow 56*, 24825 [6548-24-9] (**17**), and *C.I. Acid Red 154*, 24800 [6507-79-5] (**18**; R = CH$_3$).

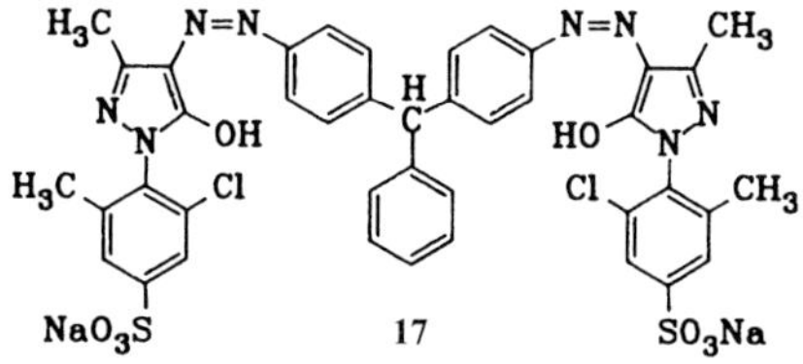

17

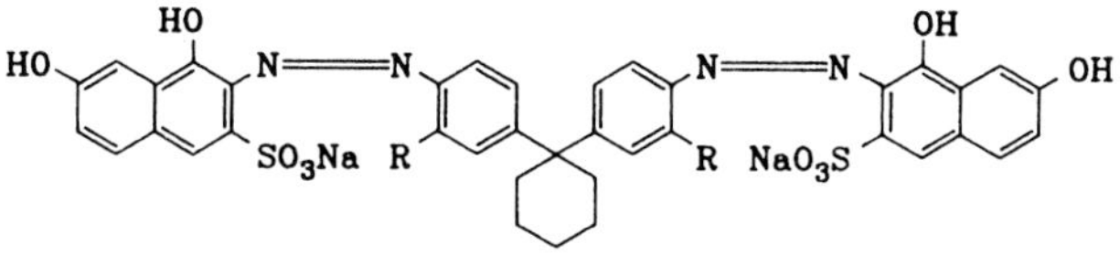

18

C.I. Acid Red 134, 24810 [6459-69-4] (**18**; R = OCH$_3$), is another interesting dye for wool and polyamide.

Particularly good milling fastness is achieved by the introduction of 4,4′-diaminodiphenylthio ether as the diazo component, for example, in the yellow wool dye *C.I. Acid Yellow 38*, 25135 [13390-47-1] (**19**).

3.9.2 Chemical Constitution and Synthesis 285

$$H_5C_2O-\text{[Ar]}-N=N-\text{[Ar]}-S-\text{[Ar]}-N=N-\text{[Ar]}-OC_2H_5$$
(NaO$_3$S, SO$_3$Na)

19

Secondary disazo dyes. These dyes can be manufactured by series coupling according to the scheme D→M→K (D = diazo component; M = middle component; K = coupling component), offer much greater possibilities for variation than the primary disazo dyes. This group includes red dyes and, above all, navy blues and blacks.

The most important dyes of the type D→M→K are black and navy blue wool dyes, which contain as the middle component M chiefly 1-naphthylamine or 1-naphthylamine-7-sulfonic acid and as the coupling component K *N*-phenyl peri acid (*N*-phenylaminonaphthalene-8-sulfonic acid), *N*-tolyl peri acid, and naphthol or 1-naphthylamine derivatives. Examples of these high-yield and very wash- and lightfast dyes are *C.I. Acid Black 24*, 26370 (see Section 3.9.3), which is manufactured by coupling diazotized 1-naphthylamine-5-sulfonic acid to 1-naphthylamine, further diazotization of the aminomonoazo dye, and coupling to *N*-phenylperi acid; and *C.I. Acid Blue 113*, 26360 [*3351-05-1*] (**20**). This blue dye is very important in the dyeing of wool and polyamide.

48

A very washfast black wool dye is *C.I. Acid Black 26*, 27070 [*6262-07-3*] (**21**).

21

It contains 4-aminodiphenylamine-2-sulfonic acid as the diazo component, 1-naphthylamine as the middle component, and 2-naphthol-6-sulfonic acid as the final coupling component.

Chrome Dyes [2, p. 464], [7]. Azo dyes with certain groupings can be converted on the fiber with chromium salts to chromium complexes that are more or less soluble (see Section 3.11). This "chroming" of largely acid monoazo dyes improves the lightfastness of the dyeings and above all their wetfastness, because complex formation blocks the hydrophilic groups in the molecule. It also results in a deepening and dulling of the shade, so that these dyes do not yield brilliant colors. Consequently, for combination dyeing the range of chrome dyes had been supplemented with brilliant acid dyes, which cannot form chrome lakes and survive the chroming process without change.

Chrome dyes are applied to the fiber chiefly by the afterchroming method and special products also by the one-bath chroming method (metachrome process). Today, for ecological reasons chrome dyes are only of very limited commercial importance.

In *afterchroming* the dye is allowed to go onto the fiber from an acid bath, and this is followed by treatment with dissolved alkali metal dichromate, which is reduced by the cystine group of the wool to trivalent chromium compounds. The chromium complex formed on the wool fiber contains two azo dye radicals for each metal atom (1:2 complex).

The afterchroming dyes have also lost their former importance because of the considerable amount of time required by the dyeing process and because of effluent problems, but they are still used in wool dyeing because of their low price.

In the yellow and orange types the salicylic acid grouping dominates. One of the most important chrome-developed yellow dyes is *C.I. Mordant Yellow 1*, 14025 [584-42-9] (**22**). After chroming it yields a somewhat dull yellow with good fastness properties. *C.I. Mordant Yellow 5*, 14130 [6054-98-4] (**23**), possesses two salicylic acid groups. After chroming the product is greenish yellow and exhibits very good general fastness properties.

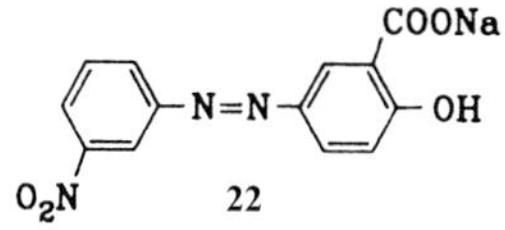

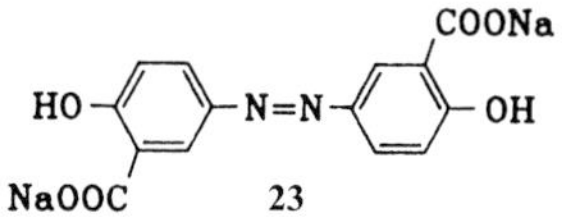

The complex formation on the salicylic acid group has little influence on shade and lightfastness, in contrast to *o,o'*-disubstituted arylazo compounds, in which complexing brings about a considerable deepening of shade and usually marked increase in lightfastness. Examples: *C.I. Mordant Red 7*, 18760 [3618-63-1] (**24**), dyes wool a dull orange, which through chroming becomes a clear red with very good fastness properties.

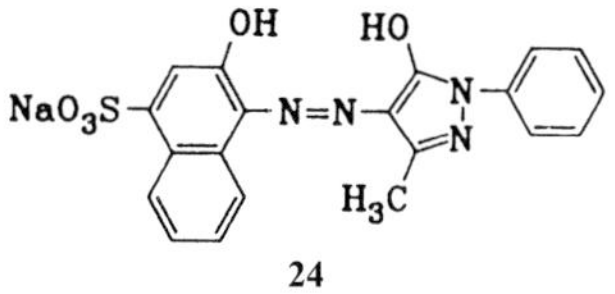

Red, blue, and primarily black afterchroming dyes contain naphthalene radicals in the diazo or coupling component (see Section 3.9.3).

Oxidation of the dye during chroming with dichromate creates a naphthoquinone structure in the molecule. Two other important black dyes are *C.I. Mordant Black 3*, 14640 [3564-14-5] (**25**, R = H), and *C.I. Mordant Black 11* (see Section 3.9.3).

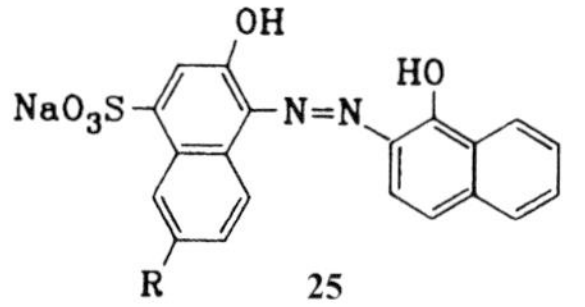

C.I. Mordant Black 11 is also used as an indicator in the complexometric titration of various bivalent metals with which it forms complexes [8].

The *o,o'*-dihydroxyaminoazo and *o*-hydroxy-*o'*-aminoazo compounds with phenols or aniline derivatives as coupling components yield largely brown shades on chroming; they include an important brown dye with very good fastness properties, namely, *C.I. Mordant Brown 33*, 13250 [*3618-62-0*] (**26**).

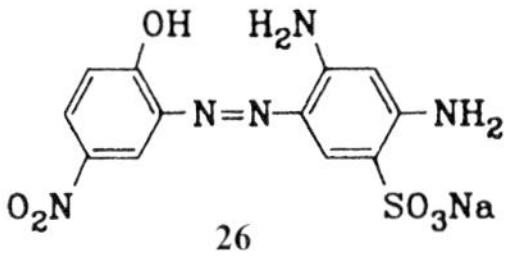

In the *metachrome process* (*one-bath chroming method*) the metal complex is formed on the fiber by the simultaneous action of chrome dye and dichromate. Compared with the afterchroming method metachrome dyeing has the advantage of simpler application, but the number of usable dyes for this process is more severely limited. Whether a dye is suitable for the metachrome process depends on its hydrophilic character and its rate of complex formation [9]. The method is used at present for wool dyeing primarily because of price considerations.

Examples of typical one-bath chroming dyes are: *C.I. Mordant Yellow 30*, 18710 [*10482-43-6*] (**27**), *C.I. Mordant Red 30*, 19360 [*6359-71-3*] (**28**), *C.I. Mordant Brown 48*, 11300 [*6232-53-7*] (**29**), and *C.I. Mordant Blue 7*, 17940 [*3819-12-3*] (**30**).

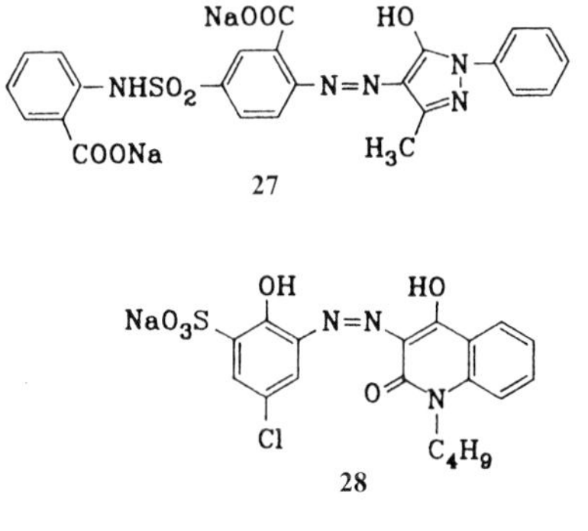

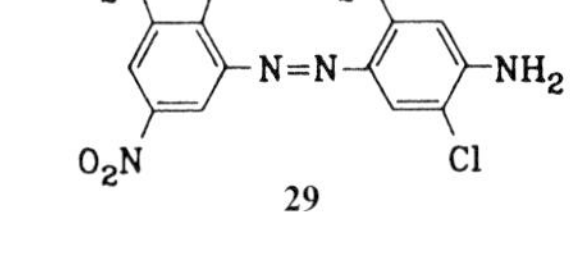

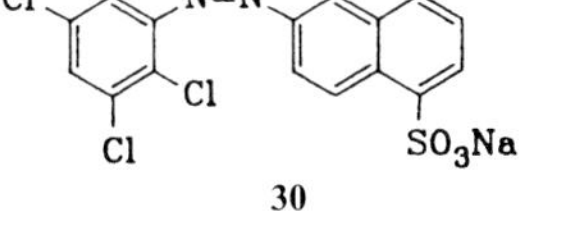

Metal-Complex Dyes. These dyes are manufactured in the complete dye form before the dyeing process and usually contain chromium or cobalt ions, rarely copper or iron ions, bound in the molecule (see Section 3.11). A distinction is made between 1:1 and 1:2 metal-complex dyes, depending on whether there is one metal ion for each dye molecule or for every two dye molecules. These products generally yield dyeings of high light- and wetfastness but of somewhat muted shades.

1:1 Metal-Complex Dyes. Among these dyes primarily the 1:1 chromium complexes containing sulfonic acid groups have achieved commercial importance. They must be applied from a strongly acid bath, which imposes certain limits on their range of applications. The 1:1 metal complexes are not suitable for polyamide, which is partially decomposed under the dyeing conditions for these products. Their main area of application is in the dyeing of wool, but they are also suitable for leather dyeing.

1:2 Metal-Complex Dyes. Because of their structure 1:2 metal-complex dyes exhibit anionic character. Those which have gained commercial importance are primarily the ones that are free of sulfonic acid groups and for which adequate water solubility is provided by nonionic, hydrophilic substituents, such as methylsulfone or sulfonamide groups [9, 10]. The introduction of 1:2 metal-complex dyes which are applied from a neutral to weakly acid bath, represented a significant technical advance over the strong-acid-dyeing 1:1 chrome complex dyes. It has led to better protection of the fiber material, simplification of the dyeing process, and improvement of the fastness properties.

Because these dyeings, like all metal-complex dyeings, usually exhibit subdued shades, the 1:2 metal-complex dyes are combined with small amounts of acid dyes or suitable reactive dyes to brighten the tint. Owing to their outstanding wetfastness, high lightfastness, and good fiber levelness, the 1:2 metal-complex dyes are of major importance in dyeing wool and polyamide.

The extent of variation is extremely large as a result of the great number of available diazo and coupling components, the choice of complex-forming metal (primarily chromium or cobalt), and the possibility of synthesizing 1:2 mixed-metal complexes.

For subdued shades the 1:2 metal-complex dyes with one or two sulfo groups continue to play an important role. Dyes with one such group contain the sulfonic acid group in one of the two parts bound to the azo group and complexed to chromium or cobalt. They are built up stepwise via the 1:1 metal complex. Compared with the 1:2 metal-complex dyes with nonionic hydrophilic groups they exhibit better solubility in water but poorer leveling power, and therefore they are always used together with specially selected leveling auxiliaries. The commercial ranges of these dyes contain primarily navy blue, black, dark brown, and olive shades.

The most recent development in this area is the 1:2 metal-complex dyes with two sulfonic acid groups. Contrary to the opinion frequently expressed in the literature that such dyes would exhibit inadequate levelness when applied in a weakly acid bath and inadequate stability to acid in a strongly acidic dyebath, they are very well suited for dyeing wool and polyamide if certain pH conditions are observed. These products are very soluble in water, have high tinctorial strength, and are very high yield dyes with a relatively simple structure.

3.9.2.2 Polyamide Dyes

Synthetic polyamides have a structure similar to those of wool and silk but differ in having a low acid-binding power and in their capacity to dissolve nonpolar compounds. Consequently polyamide materials can be dyed with disperse dyes and with selected acid dyes, including metal-complex dyes [11].

When choosing acid dyes for polyamides it must be borne in mind that owing to the lower acid-binding power dyes with two and more sulfonic acid groups in the molecule go onto the fiber much more slowly and to a much lower saturation value than dyes with one sulfonic acid group. Consequently, these dyes cannot be mixed or combined with one another, which has been taken into account in composing the product ranges.

Since acid dyes on polyamide behave as they do on wool in regard to leveling power, build-up, and fastness properties (see above), they fall into two classes: group A consists of acid dyes with good leveling power and low substantivity for polyamide; they yield dyeings with reasonably good wetfastness. Group B contains acid dyes with lower leveling power, higher substantivity, and high wetfastness on polyamide. Many of these acid dyes with a lower leveling power rather clearly reveal differences in fiber structure that may result from, for example, differences in the degree of drawing (so-called streakiness), so that it is usually necessary to add leveling and retarding auxiliaries (see Chapter 4).

The main area of application for the acid polyamide dyes of group A is in carpet dyeing, but they are also used in other areas of textile dyeing where the fastness requirements are not too stringent. A few structures are presented below to illustrate the types of azo dyes used: *C.I. Acid Yellow 25*, 18835 [6359-85-9] (**31**), *C.I. Acid Red 42*, 17070 [6245-60-9] (**32**), *C.I. Acid Red 32*, 17065 [6360-10-7] (**33**).

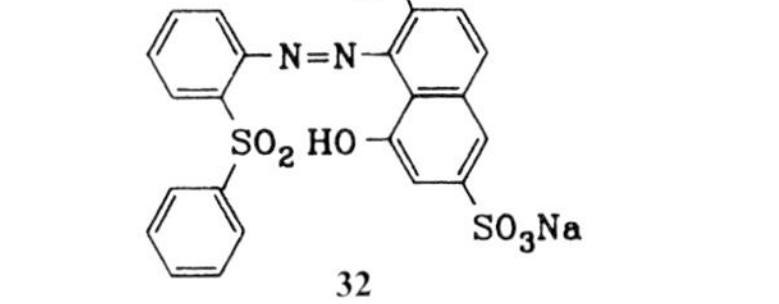

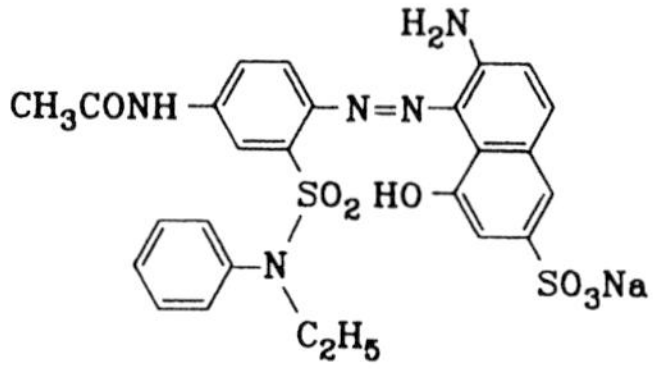

[Structure 33]

Group B dyes are used almost exclusively for clothing textiles, for which more stringent requirements are placed on wetfastness. Some of the typical azo dyes of this type are *C.I. Acid Yellow 65*, 14170 [6408-90-8] (**34**) and *C.I. Acid Blue 113*, 26360 [3351-05-1] (**35**).

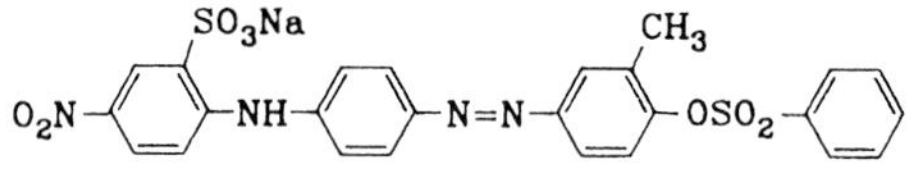

[Structure 34]

[Structure 35]

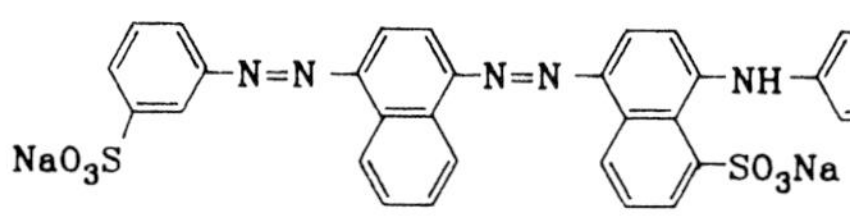

As regards the further development of acid disazo dyes for polyamide the more recent patent literature reveals a trend toward secondary disazo dyes (D→M→K) with one or two sulfonic acid groups bound to aromatic nuclei or with an external sulfate group which is situated mainly in the diazo component D. Aniline, 1-naphthylamine, and their derivatives preferably serve as the middle component M, whereas the coupling components of the phenol and arylamine series common in disperse dyes are used as the final hydrophobic component K [12–16], as illustrated by the general structure **36**. The patent literature indicates that polyamide dyes of this type are generally characterized by high tinctorial strength and lightfastness and a good capacity to compensate for nonuniformity in the material.

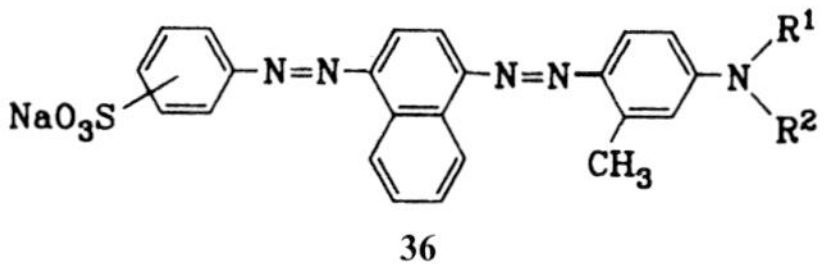

[Structure 36]

Because of increasing standards of wetfastness, lightfast *reactive dyes* are also being used in polyamide printing, especially to produce pastel shades. (see Section 3.1).

Because of their high fastness standards the *1:2 metal-complex dyes* have also acquired some importance in the dyeing of polyamide. They are used primarily in the clothing sector to achieve deep shades.

3.9.2.3 Silk Dyes

Because natural silk, as a protein fiber of animal origin, resembles wool in its chemical structure, it can be dyed with most of the classes of dyes used for wool. The choice of dyes depends essentially on the fastness properties required.

Of great importance for the dyeing of natural silk are selected members of the class of acid wool dyes (see above). The occasionally inadequate wetfastness of these dyeings can be substantially improved by the proper aftertreatment (e.g., with potassium sodium tartrate and tannin).

Selected acid 1:1 chrome complex dyes, because of their good wet- and lightfastness and their very good leveling power, and also 1:2 chrome complex and 1:2 cobalt complex dyes with hydrophilic groups, have successfully come into use as silk dyes.

Furthermore, direct dyes (see Section 3.3) are among the most important silk dyes. Their fastness properties can be improved by aftertreatment with metallic salts or formaldehyde.

Initially the development of synthetic fibers greatly reduced the importance of silk dyeing. Recently the processing of silk has undergone a marked increase owing to the growing quality consciousness of buyers, who appreciate the outstanding wear properties of silk.

3.9.3 Examples of Commercially Available Dyes

Monoazo Dyes
   C.I. Acid Blue 92, 13390 [3861-73-2] (**37**).

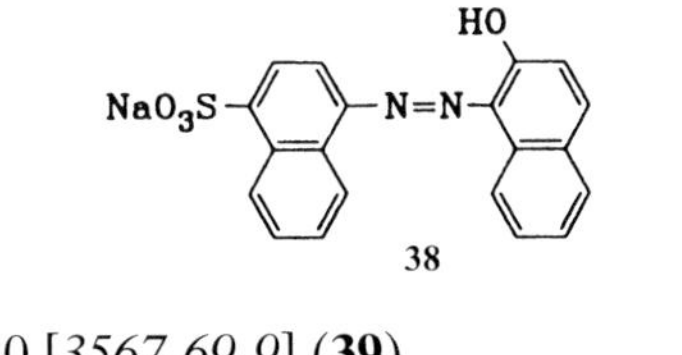

   C.I. Acid Red 88, 15620 [1658-56-6] (**38**).

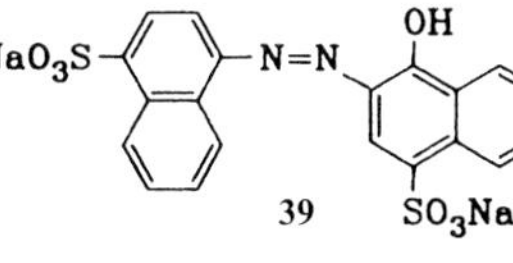

   C.I. Acid Red 14, 14720 [3567-69-9] (**39**).

3.9 Acid Dyes (Anionic Azo Dyes)

C.I. Acid Yellow 17, 18965 [6359-98-4] (**40**).

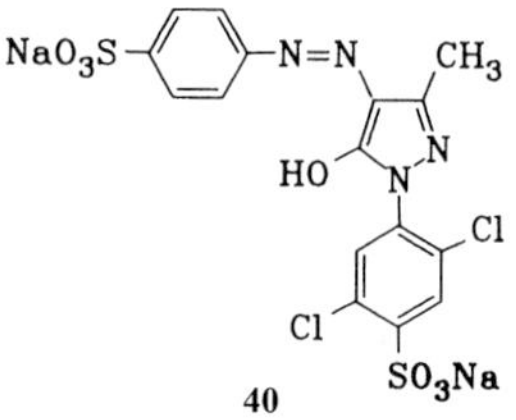

C.I. Acid Orange 67, 14172 [12220-06-3] (**41**).

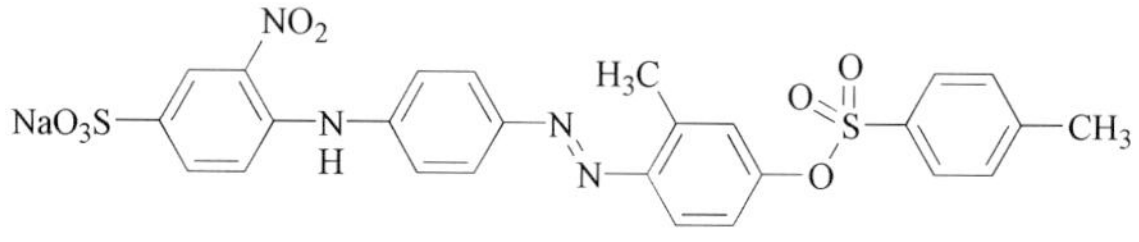

C.I. Acid Orange 10, 16230 [1936-15-8].
C.I. Acid Red 131 [70210-37-6].
C.I. Acid Red 249, 18134 [6416-66-6].

Mordant Dyes
C.I. Mordant Blue 13, 16680 [1058-92-0] (**42**).

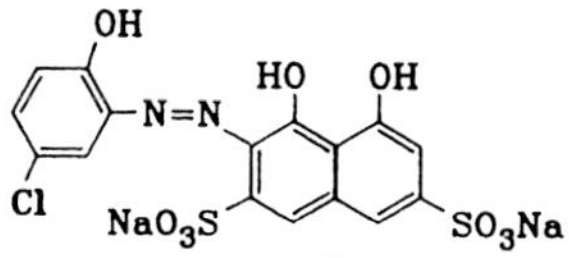

C.I. Mordant Black 9, 16500 [2052-25-7] (**43**).

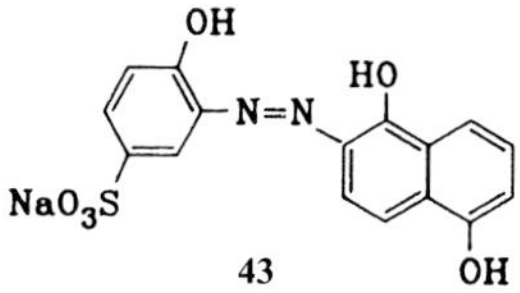

C.I. Mordant Black 11, 14645 [1787-61-7] (**44**).

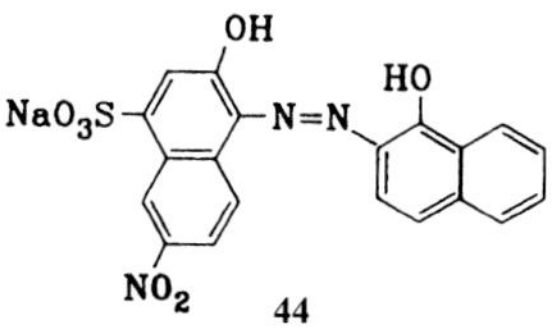

3.9.3 Examples of Commercially Available Dyes

C.I. Mordant Red 19, 18735 [*1934-24-3*] (**45**).

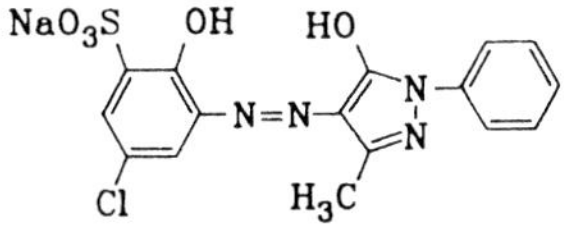

45

C.I. Acid Orange 65, 14170 [*6408-90-8*] (**46**).

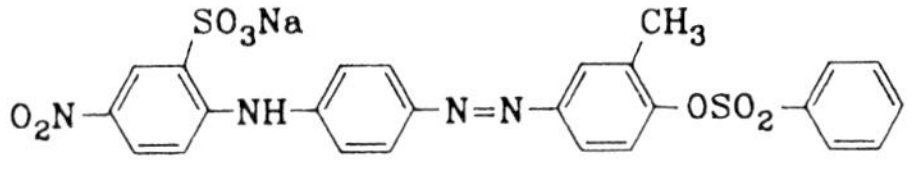

46

Disazo Dyes

C.I. Acid Black 1, 20470 [*1064-48-8*] (**47**).

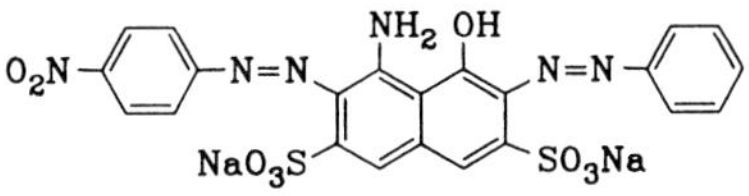

47

C.I. Acid Yellow 42, 22910 [*6375-55-9*].
C.I. Acid Black 24, 26370 [*3071-73-6*] (**48**).

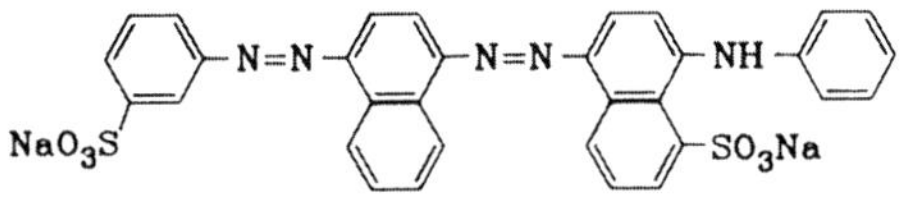

48

C.I. Acid Blue 113, 26360 [*3351-05-1*] (**49**).

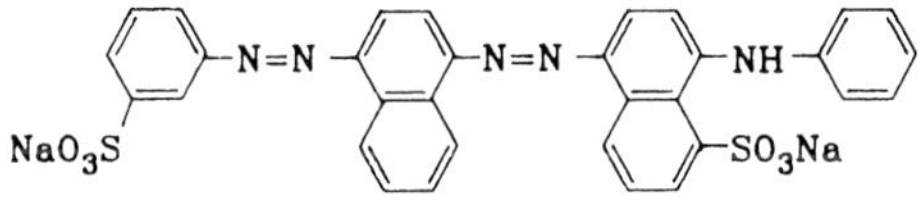

49

C.I. Acid Yellow 127, 18888 [*73384-78-8*].
C.I. Acid Red 119 [*70210-06-9*].
C.I. Acid Red 299 [*67674-28-6*].
C.I. Acid Brown 14, 20195 [*5850-16-8*].

Trisazo Dyes

C.I. Acid Black 210, 300825 [*99576-15-5*].
C.I. Acid Black 234, 30027 [*157577-99-6*].

Metal-Complex Dyes
  C.I. Acid Yellow 99, 13900 [10343-58-5].
  C.I. Acid Yellow 151, 13906 [12715-61-6].
  C.I. Acid Yellow 194 [85959-73-5].
  C.I. Acid Orange 74, 18745 [10127-27-2].
  C.I. Acid Blue 158, 14880 [6370-08-7].
  C.I. Acid Blue 193, 15707 [1239-64-2].

Various Structures
  C.I. Acid Orange 3, 10385 [6373-74-6].
  C.I. Acid Blue 25, 62055 [6408-78-2].
  C.I. Acid Green 16, [12768-78-4].
  C.I. Acid Green 25, 11359 [4403-90-1].
  C.I. Acid Green 27, 61580 [6408-57-7]
  C.I. Acid Green 28, [12217-29-7].
  C.I. Acid Brown 349 [72827-73-7].
  C.I. Acid Black 194 [61931-02-0].

3.9.4 References

[1] H. Zollinger, *Color Chemistry*, 2nd ed., VCH Weinheim, New York 1991. P. Rys, H. Zollinger: *Farbstoffchemie*, 3rd ed., Verlag Chemie, Weinheim, 1982.
[2] H. R. Schweizer: *Künstliche organische Farbstoffe und ihre Zwischenprodukte*, Springer Verlag, Berlin 1964.
[3] W. Seidenfaden: *Künstliche organische Farbstoffe und ihre Anwendungen*, Enke Verlag, Stuttgart 1957.
[4] *Colour Index*, Society of Dyers and Colourists and American Association of Textile Chemists and Colorists, 3rd ed., Bradford 1971, First Revision 1975, Second Revision 1982, 4th ed., Electronical release, 2002.
[5] T. Vickerstaff: *The Physical Chemistry of Dyeing*, Oliver and Boyd, London – Edinburgh 1954; K. Venkataraman: *The Chemistry of Synthetic Dyes*; Academic Press, New York, vol. I (1952), vol. II (1952), vol. III (1970), vol. IV (1971), vol. VI (1972), vol. VII (1974).
[6] K. Venkataraman: *The Chemistry of Synthetic Dyes*, Academic Press, New York, vol. III (1970), p. 268.
[7] H. Baumann, H. R. Hensel: "Neue Metallkomplexfarbstoffe. Struktur und färberische Eigenschaften", *Fortschr. Chem. Forsch.* **7** (1967) no. 4, 680.
[8] E. M. Diskant, *Anal. Chem.* **24** (1952) 1856.
[9] H. Baumann, H. R. Hensel: "Neue Metallkomplexfarbstoffe. Struktur und färberische Eigenschaften", *Fortschr. Chem. Forsch.* **7** (1967) no. 4, 689.
[10] G. Schetty, *J. Soc. Dyers Colour.* **71** (1955) 705.
[11] K. Venkataraman: *The Chemistry of Synthetic Dyes*, Academic Press, New York, vol. III (1970), p. 276.
[12] Sandoz, BE 717 458, 1967 (M. Studer).
[13] Bayer, DE-OS 1 923 680, 1969 (H. Nickel, F. Suckfüll).

[14] Toms River Chem. Corp., DE-OS 1 931 691, 1969 (H. A. Stingl).
[15] Crompton & Knowles, DE-OS 1 957 115, 1969 (J. F. Feeman).
[16] Bayer, DE-OS 1 960 816, 1969 (K. L. Moritz).

3.10 Solvent Dyes

3.10.1 Introduction

Solvent dyes [1] cannot be classified according to a specific chemical type of dyes. Solvent dyes can be found among the azo, disperse, anthraquinone, metal-complex, cationic, and phthalocyanine dyes. The only common characteristic is a chemical structure devoid of sulfonic and carboxylic groups, except for cationic dyes as salts with an organic base as anion. Solvent dyes are basically insoluble in water, but soluble in the different types of solvents. Organic dye salts represent an important type of solvent dyes. Solvent dyes also function as dyes for certain polymers, such as polyacrylonitrile, polystyrene, polymethacrylates, and polyester, in which they are soluble. Polyester dyes are principally disperse dyes (see Section 3.2).

For practical reasons the different solubility can be used as a basis for a classification of solvent dyes, although there is no strict differentiation. "Chemical constitution" is defined here as a structure which meets the corresponding solvent requirements.

3.10.2 Chemical Constitution and Application Properties

3.10.2.1 Alcohol- and Ester-Soluble Dyes

As far as chemical constitution is concerned, there is no precise delineation between alcohol- and ester-soluble dyes, nor between these two groups and the fat- and oil-soluble dyes described below.

With the exception of the blue copper phthalocyanine derivatives, these products are azo dyes that are soluble in polar solvents such as alcohols, glycols, esters, glycol ethers, and ketones. Dyes soluble in alcohols and esters are used in protective lacquers for the transparent coating of metal (aluminum) foils and other materials, such as wood (greening lacquers); in flexographic inks for the printing of metal foils, cellophane, and paper; as well as for the coloration of cellulose esters, celluloid, and poly(vinyl acetates), and, in the office supplies sector, for stamping inks and pastes for pressure recorders. In modern formulations for foil lacquers, dyes with better solubility in esters are given preference over the older products, which are mainly soluble in alcohols.

In accordance with chemical aspects, the most important alcohol- and ester-soluble azo dyes can be divided into three groups.

First, *1:2 metal complexes* of (mainly mono-) azo dyes, without sulfonic or carboxylic acid groups, and trivalent metals (see Section 3.11). The metals are preferably chromium and cobalt; nickel, manganese, iron, or aluminum are of lesser importance. Diazo components are mainly chloro- and nitroaminophenols or aminophenol sulfonamides; coupling components are β-naphthol, resorcinol, and 1-phenyl-3-methyl-5-pyrazolone. Formation of a complex from an azo dye and a metal salt generally takes place in the presence of organic solvents, such as alcohols, pyridine, or formamide. An example is *C.I. Solvent Red 8*, 12715 [*33270-70-1*] (**1**).

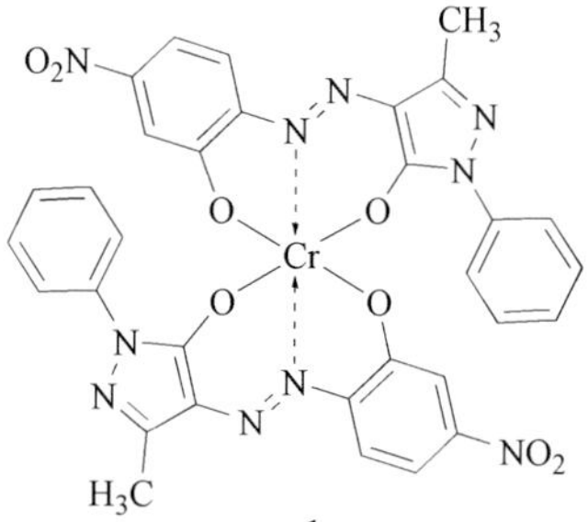

Second, *1:1 metal-complex azo dyes* that contain sulfonic acid or carboxylic acid groups and are present in the form of internal salts. Here, azamethine–metal complexes are also of importance. An examples is *C.I. Solvent Yellow 32*, 48045 [*61931-84-8*] (**2**).

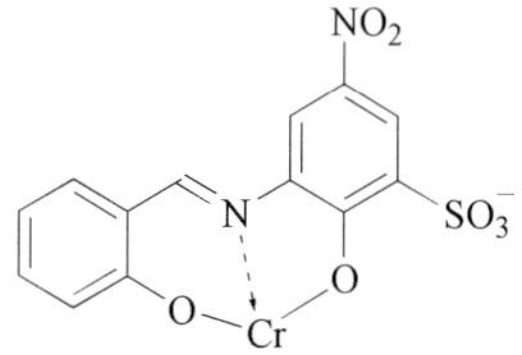

Other 1:1 metal complex dyes are *C.I. Solvent Orange 56*, [*12227-68-8*], similar to C.I. 518745:1, and *C.I. Solvent Yellow 82*, [*12227-67-7*], similar to C.I. 18690.

Third, *reaction products of acid azo dyes, acid 1:1 metal-complex azo dyes, or 1:2 metal complex azo dyes without acid groups, with organic bases or cationic dyes.* Cyclohexylamine, dodecylamine, and sulfonium or phosphonium compounds serve as bases, and derivatives of the xanthene range (rhodamines) are mainly used as cationic dyes. Example: *C.I. Solvent Red 109* [*53802-03-2*] is composed of *Solvent Yellow 19*, 13900:1 [*10343-55-2*] (**3**) and *Solvent Red 49*, 45170:1 [*81-88-9*] (**4**). These dyes are saltlike compounds of a metal-complex azo dye acid and a base.

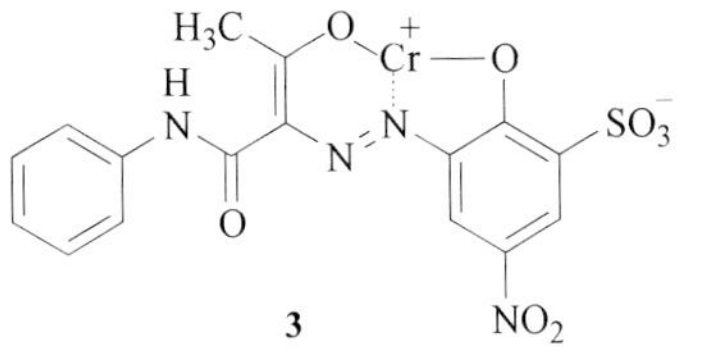

3.10.2 Chemical Constitution and Application Properties 297

3

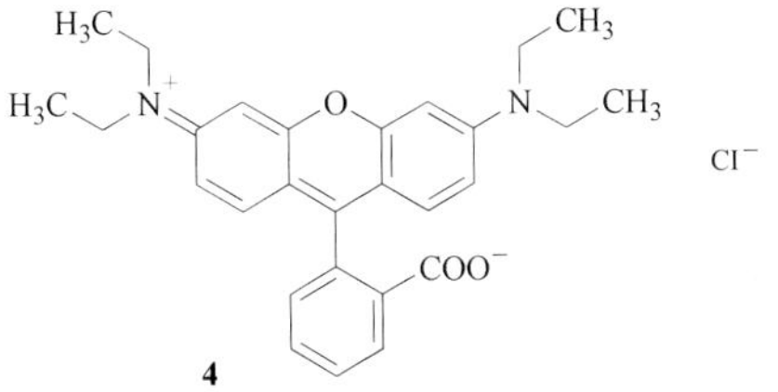

4

Anthraquinone dyes of relatively simple structure are used for coloring gasoline, oil, and plastics. For example, the highly soluble bis(alkylamino)anthraquinones are suitable additives for gasoline, whereas quinizarin and its derivatives are used for coloring heating oils. Note, that addition of alkali causes a color change. Examples are **5**, a blue dye for the coloring of hydrocarbons, including gasoline [2], and **6**, a mixture of blue dyes used to color gasoline and mineral oils [3].

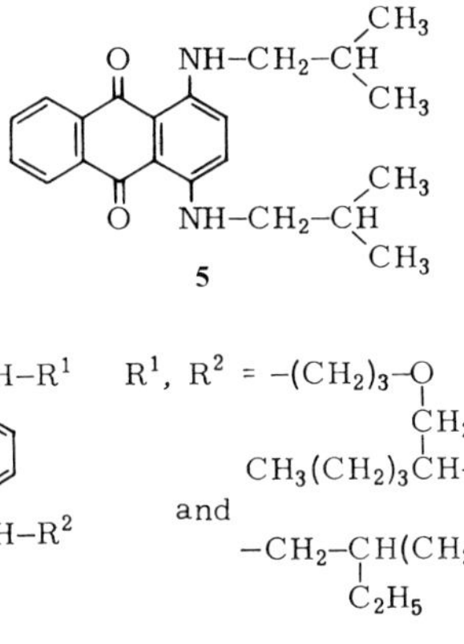

5

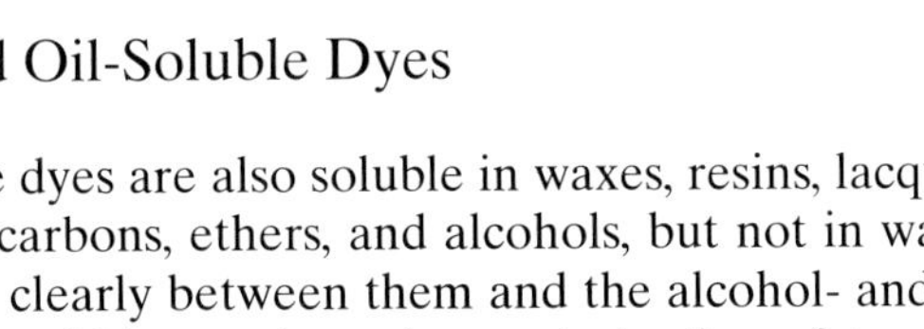

6 $R^1, R^2 = -(CH_2)_3-O-CH_2-CH(C_2H_5)(CH_2)_3CH_3$
and
$-CH_2-CH(C_2H_5)(CH_2)_3CH_3$

3.10.2.2 Fat- and Oil-Soluble Dyes

Fat- and oil-soluble dyes are also soluble in waxes, resins, lacquers, hydrocarbons, halogenated hydrocarbons, ethers, and alcohols, but not in water. It is not possible to differentiate clearly between them and the alcohol- and ester-soluble dyes. With the exception of blue anthraquinone derivatives, fat- and oil-soluble dyes are azo dyes, generally based on simple components. According to their degree of solubility they usually contain hydroxyl and/or amino groups, but not sulfonic acid and carboxylic acid groups. Examples of fat- and oil-soluble azo dyes are *C.I.*

Solvent Red 23, 26100 [*85-86-9*] (**7**; R = H) and *C.I. Solvent Red 24*, 26105 [*85-83-6*] (**7**; R = CH$_3$).

Further examples of commercial solvent dyes are given in Section 3.10.3.

The products are sold in powder form, as granules, or as flakes, and some dyes also as liquid brands. The liquid brands are highly concentrated solutions of fat-soluble dyes in aromatic hydrocarbons, in some cases also solvent-free 100 % liquid products.

Fat- and oil-soluble dyes are used on a large scale in a wide variety of industrial sectors. The main fields of application are the coloration of products in the mineral oil and plastics industries, as well as of wax products (e.g., candles, shoe polishes, floor polishes).

Mineral oil products (fuels, fuel oil, lubricating oils, and greases) are colored as a means of distinguishing between different grades (e.g., of gasoline) or for compulsory identification, e.g., of diesel and fuel oils, for duty purposes. Interestingly, the lightfastness of fat and oil-soluble dyes is highly dependent on the medium colored. Whereas, for example, 0.05 % colorations of the dye *Solvent Yellow 56*, 11021 [*2481-94-9*] (see 3.10.3) in candle materials possess only moderate lightfastness (step 3 on the eight-step Blue Scale), transparent colorations in polystyrene are distinguished by outstanding lightfastness (step 8).

Other fields of application are the pyrotechnics industry (pronounced sublimation tendency of the dyes), the office supplies industry (inks for felt-tip pens, oil-based stamping inks), the lacquers industry (especially coloration of transparent lacquers on aluminum foil), and the cosmetics industry (approval restricted to a few products).

3.10.2.2 Dyes Soluble in Polymers

Although these products belong to the group of fat- and oil-soluble dyes, they are differentiated here for the sake of clarity. The plastics industry values the fat- and oil-soluble dyes for the highly transparent colorings, usually with very good lightfastness, that are obtainable with them. Products in this dye class are most frequently used for the coloration of polystyrene, as well as for incorporation in polymethacrylate and unsaturated polyester casting resins.

Polymer-soluble dyes of very different chemical constitutions were developed for the coloration of *polystyrene* and *other transparent plastics*. These dyeings

3.10.2 Chemical Constitution and Application Properties

require dyes soluble in the plastics materials so that transparency can be retained.

Apart from *C.I. Solvent Yellow 56* other examples are *C.I. Solvent Orange 60*, 564100 *[61969-47-9]* and *C.I. Solvent Red 135*, 564120 *[71902-17-5]*.

Several anthraquinone dyes are used in dyeing thermoplastics such as polymethacrylate, (modified) polystyrene, and polycarbonate. The dyes also are used in combination with titanium dioxide or other materials to provide body colors for the thermoplastics. Initially compounds with relatively simple structures were used, many of them drawn from stocks of existing intermediates. Since then new products designed to satisfy the specific requirements have become available.

In contrast to disperse dyes solvent dyes for spin dyeing of *polyester fibers* are initially soluble in the polymer, thus avoiding the dipersing processes for the application of disperse dyes or pigments.

In addition to *C.I. Solvent Red 135* other examples of spin-dyeing dyes are C.I. Solvent Yellow 133, *48580*; *C.I. Solvent Red 212*, *48530 [61300-98-9]*; *C.I. Solvent Blue 122*, 60744 *[67905-17-3]*; *C.I. Solvent Brown 53*, 48525 *[64969-98-6]*; and *C.I. Solvent Green 28*, 625580 *[71839-01-5]*.

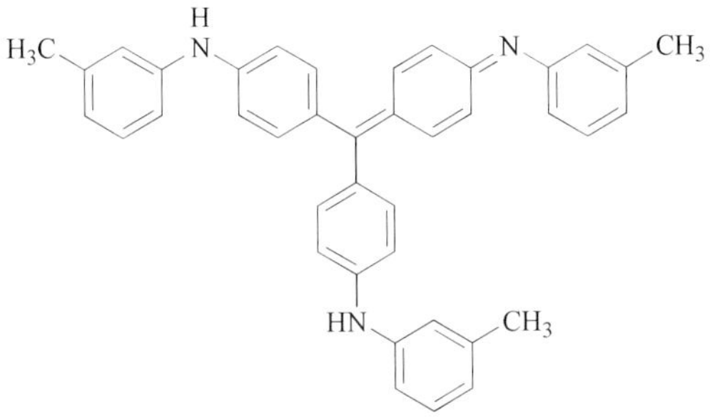

8

Dyes for the coloration of polyacrylonitrile fibers must be soluble in the solvents used for the PAN fiber production process, such as DMF, dimethylacetamide, and ethylene carbonate. Examples are: *C.I. Solvent Yellow 147*; *C.I. Solvent Yellow 163*, 58840 *[13676-91-0]*; *C.I. Solvent Red 202*; and *C.I. Solvent Blue 66*, 42799 *[58104-34-0]* (**8**); and *C.I. Solvent Blue 131*.

3.10.2.4 Solvent Dyes for Other Applications

Many relatively simple amino- and hydroxyanthraquinone derivatives show a high degree of order in liquid-crystalline systems and are therefore suitable as dichroic dyes for guest–host displays. Several new dyes for this application have been described. For example, **9** is a blue dichroic dye [4].

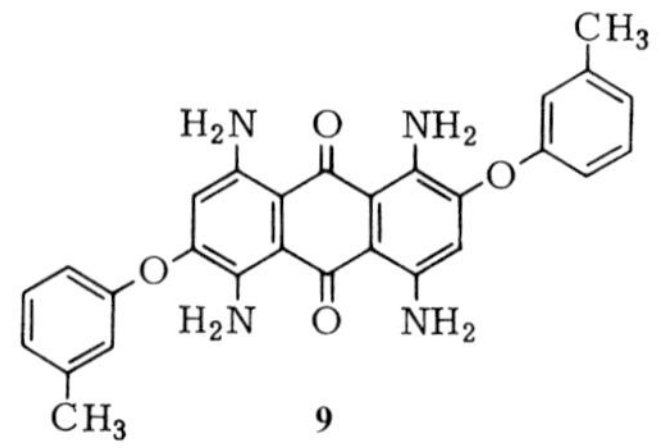

3.10.3 Examples of Commercially Available Dyes

Alcohol- and Ester-Soluble Dyes

C.I. Solvent Orange 56, [12227-68-8], simliar to C.I. Solvent Orange 5, 518745:1, [13463-42-8], which is a 1:1 chromium-complex dye, synthesized by diazotizing 2-hydroxy-3-amino-5-nitrobenzenesulfonic acid and coupling to 1-phenyl-3-methyl-5-pyrazolone.

C.I. Solvent Yellow 82, [12227-67-7], similar to C.I. Solvent Yellow 21, [5601-29-6], 18690, which is a 1:1 chromium complex of the azo dye obtained from anthranilic acid and 1-phenyl-3-methyl-5-pyrazolone.

Fat- and Oil-Soluble Dyes

C.I. Solvent Yellow 56, 11021 [2481-94-9] (**10**).

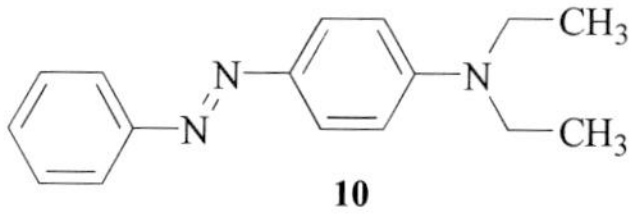

C.I. Solvent Yellow 14, 12055 (Orange) [842-07-9] (**11**).

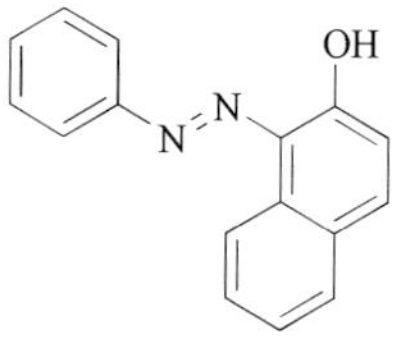

C.I. Solvent Yellow 16, 12700 [*4314-14-1*] (**12**).

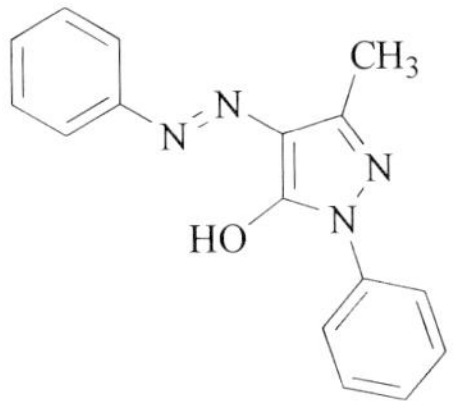

12

C.I. Solvent Black 3, 26150 [*4197-25-5*] (**13**).

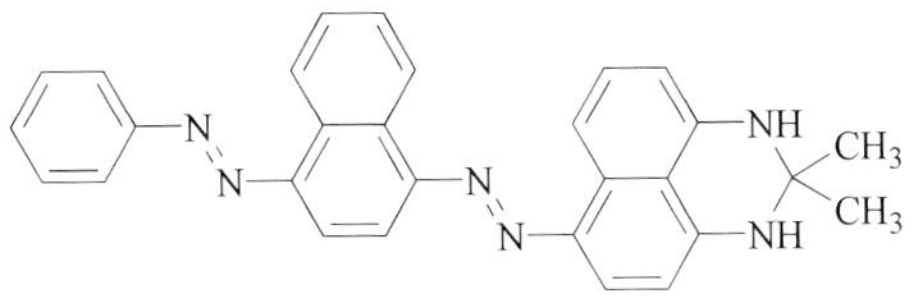

13

Dyes Soluble in Polymers
C.I. Solvent Red 111, 60505 [*82-38-2*] (**14**).
C.I. Solvent Red 52, 68210 [*81-39-0*] (**15**).
C.I. Solvent Violet 13, 60725 [*81-48-1*] (**16**).
C.I. Solvent Green 3, 61565 [*128-80-3*] (**17**).

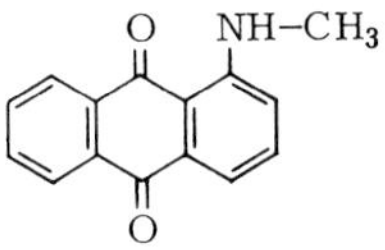

14 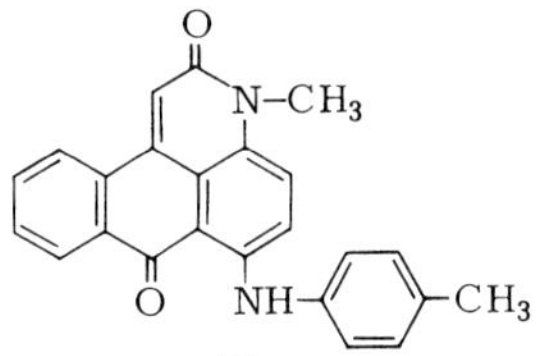   **15**

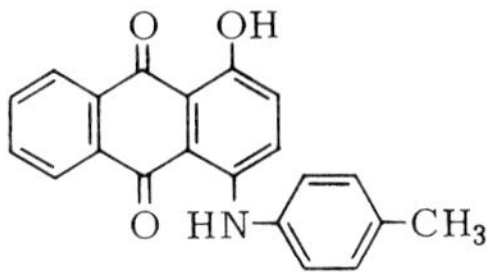

16 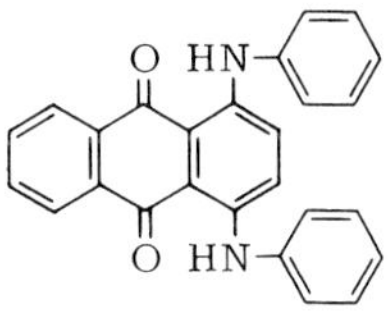   **17**

3.10.4 References

[1] *Colour Index International — Pigments and Solvent Dyes*, The Society of Dyers and Colourists, Bradford, UK, Fourth ed. online, 2002.
[2] Bayer, DE 849 158, 1944 (K. Bähr).
[3] BASF, DE-OS 1 644 482, 1967 (F. Graser, G. Riedel).
[4] Bayer, EP 93 367, 1982 (M. Blunck).

3.11 Metal-Complex Dyes

3.11.1 Introduction

Metal-complex dyes are very versatile in terms of applications. Virtually all substrates, apart from a few synthetic fibers, can be dyed and printed with this class of dyes. Countless shades from greenish yellow to deep black can be generated, depending upon the metal, the dye ligand, and the combination of dye ligands in mixed complex dyes. In commercial terms the most important chelated metals are chromium, cobalt, copper, iron, and nickel. Nickel complexes have gained commercial significance primarily as organic pigments. For textile fibers, only the chromium, cobalt, and copper complex dyes achieve desired technical effects. The resulting dyeings of chromium and cobalt complex dyes are in general dull but exhibit a high standard of fastness, particularly lightfastness. Because of dullness, these metal-complexes are chiefly used to produce deep colors for which a large amount of dye must be applied. Their use is restricted to nitrogen-containing substrates, such as wool, nylon, and leather, since they have only a little affinity to cellulosic fibers. However, with the advent of reactive dyes, chromium and cobalt complex dyes containing a fiber-reactive group are also finding application in cellulose dyeing. Copper complex dyes, in contrast, have little or no application as wool and nylon dyes due to their insufficient stability to acid treatment. Leather, however, which has milder dyebath conditions, can also be dyed with copper complexes in addition to chromium, cobalt, and iron complexes.

By far the most important and widely used metal-complex dyes are derived from azo compounds. Although they deliver a multitude of shades, only a few basic components are necessary to produce metal-complex azo dyes. The most useful starting materials are the amines **1–4**, which serve as diazo components for reaction with suitable coupling components to give tridentate azo ligands. In addition, the amines **1** and **2** can be condensed with salicylaldehyde or arylazo-substituted salicylaldehydes to give tridentate azomethine ligands.

1 $R^1, R^2 = H, Cl, CH_3, NO_2, SO_3H, COOH$

2

3: $R = H$
4: $R = NO_2$

Suitable coupling components are acetoacetanilides for preparing yellow, pyrazolones for red, naphthols for blue and black, naphthylamines for green, and phenols for brown chromium and cobalt complex dyes. A combination of acetoacetanilides or salicylaldehydes with pyrazolones in unsymmetrical chromium complexes or of chromium with cobalt complexes leads to orange shades, of pyrazolones with naphthols to brown shades, and of naphthols with acetoacetanilides or salicylaldehydes to green shades.

With divalent metals such as Cu^{II}, Ni^{II}, and Co^{II}, bidentate azo and azomethine dyes form 1:2 metal-complexes with the basic structure **5** [1].

$M = Cu^{II}, Ni^{II}, Co^{II}$
$X = O, NH$

5

In general, these complexes exhibit too little stability and washfastness to be of value as dyes. Except for Co^{III} [2], trivalent metal salts do not form complexes with bidentate o-hydroxy- and o-aminoazo dyes.

Another class of bidentate azo dyes are terminally metallizable dyes derived from bidentate coupling components such as salicylic acid, catechol, salicylaldoxime and 8-hydroxyquinoline. They are mainly used as mordant dyes and rarely as premetallized dyes [3, 4].

Forerunners of the preformed metal-complex dyes are the chrome or mordant dyes, which can be applied only on wool fibers.

o,o'-Dihydroxyazo and o-hydroxy-o'-aminoazo dyes were first prepared by E. Erdmann and O. Borgmann [5] in 1883. Because of alkali sensitivity and color instability they had no coloristic significance until ca. 1900, when they were adapted to the long-known mordant dyeing process. From now on it was possible to create dyeings on wool with high fastness to potting and light. In the ensuing years a multitude of metallizable mordant dyes were developed and marketed, and the afterchrome process became the most important mordant method. Typical dyes used in the afterchrome dyeing are *C. I. Mordant Black 11*, 14645 *[25747-08-4]* (**6**) and *C. I. Mordant Red 19*, 18735 *[25746-81-0]* (**7**).

6

7

The high wetfastness has been attributed firstly to the complex formation within the fiber [6] and secondly to chromium coordination to wool. The interaction of chromium with wool is not elucidated in all detail. Hartley suggested reduction of Cr^{VI} to Cr^{III} by Cr^{IV} and Cr^{II}. The last-named forms a complex with a carboxyl group, and is then rapidly oxidized by air to a Cr^{III}–wool complex [7], and this damages the wool. In recent years large efforts have been devoted to minimize the chromium concentrations in dyehouse effluents and to achieve accurate color matching and reproducity [8].

The mordant dyes are preferred for deep colors in black, navy blue, and brown shades. They represent because of their economy and high fastness to the present day the most important dye class for dyeing wool, with a market share of more as 35 % [9].

3.11.2 Chemical Constitution and Synthesis

To provide an overview chemical constitution and synthesis are combined in this section. The various application media for metal-complex dyes are a further ordering principle, which generates overlap of some sections with reactive dyes (Section 3.1), leather dyes (Section 5.1), and paper dyes (Section 5.3), demonstrating the typically complex interrelationship of constitution and application of dyes.

3.11.2.1 Chromium and Cobalt Complexes for Wool and Polyamides

Efforts to overcome the drawbacks of mordant dyeing resulted in the manufacture of the so-called premetallized dyes, which are 1:1 chromium, 1:2 chromium, and 1:2 cobalt complexes.

1:1 Chromium Complexes Containing Sulfonic Acid Groups. The production of premetallized complex dyes began in 1912, when R. Bohn of BASF synthesized 1:1 chromium complexes and BASF and CIBA introduced ranges of sulfonic acid containing 1:1 chromium complex azo dyes. No metal other than chromium is suitable for this class of dyes.

The synthesis of the 1:1 complex **8** is an example of [10] *demethylative chromation*. It is prepared by diazotization of 5-amino-2-chlorohydroquinone dimethyl ether and coupling onto 1-hydroxynaphthalene-5-sulfonic acid. The reaction product and Cr_2O_3 in formic acid are heated in an autoclave at 130 °C. The chromium complex **8** [*80004-31-5*] is obtained as a black powder that gives grayish blue dyeings on wool and leather.

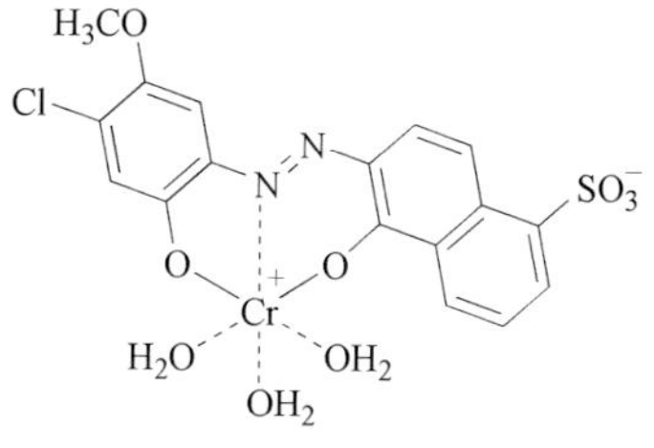

8

For stability reasons a pH of ca. 2 is required for the dyeing process. However, the resulting excellent leveling of dyeings is accompanied by hydrolytic wool damage and serious impairment of its soft handle. Use of higher pH values in the dyebath produces uneven dyeings due to unprotonated nucleophilic groups in the wool keratin, which displace ligand water from the 1:1 complex and thus form coordinative bonds to the chromium atom. Nevertheless, because of high migration and leveling capacity, the brighter hues compared with mordant dyes, and the single-bath dyeing process, the 1:1 chromium complexes are of commercial interest to the present day. Moreover, they are essential intermediates for manufacture of unsymmetrical 1:2 chromium complex dyes.

1:2 Metal Complexes without Water-Solubilizing Groups. In 1951 a dye range which contains no hydrophilic groups entered the market [11]. These dyes are insoluble in water and can only be used in dispersed form to dye nylon in aqueous dyebaths. However, they are soluble in organic solvents and can be applied in organic aqueous solutions to spray dyeing of leather. In practice, they are manufactured in the presence of an organic solvent such as formamide.

Compound **9** is prepared by 1:2 chromation in organic medium [12]: Diazotized 1-amino-2-hydroxy-4-nitrobenzene is coupled with 1-phenyl-3-methyl-5-pyrazolone. The resulting azo dye is heated at 110 °C in a mixture of formamide and a aqueous solution of chromium(III) formate. After completion of chromation, the dye **9** [*64560-69-6*] is precipitated with water.

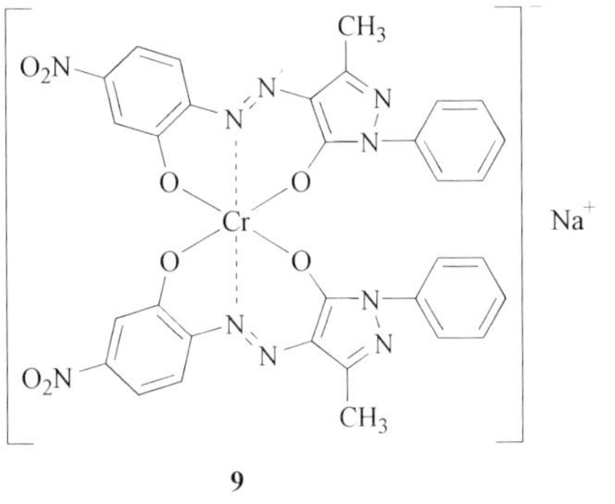

1:2 Metal Complexes with Hydrophilic Groups. In 1949 the water soluble 1:2-chromium complex **10** [*12218-94-9*] was introduced [13].

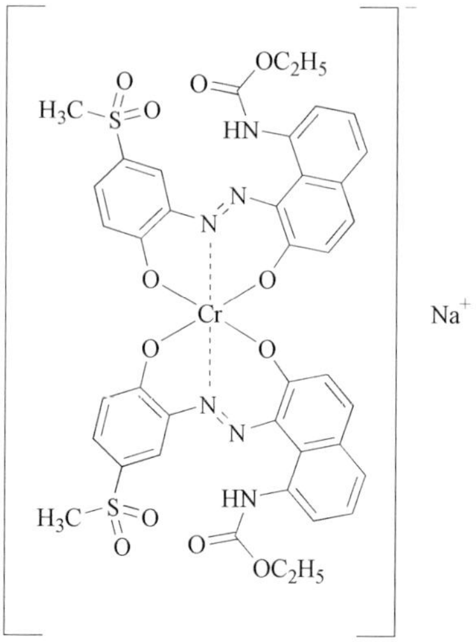

10

The commercial success of this dye and of the following range was the incentive for a worldwide research and development of corresponding product lines. Their water solubility is achieved by nonionic hydrophilic substituents, such as the methylsulfonyl group in **10** or the sulfonamide group, which attained the greatest industrial importance. Further reports concern mono- and dialkylated sulfonamide [14], ethylsulfonyl [15], ureido [16], alkylsulfoxide [17], and cyclic sulfone groups [18]. 1:2 Metal complexes with hydrophilic groups can be applied under neutral to weakly acidic conditions to give even dyeings with outstanding wet- and lightfastness. The preparation is usually carried out by the chromium(III) salicylic acid or the dichromate–glucose method.

As an example of 1:2 chromation by the *chromium(III) salicylic acid method* [19], the sodium salt of the azo dye obtained from diazotized 1-amino-2-hydroxy-5-nitrobenzene and acetoacetic acid 3-sulfonamidophenylamide in water and a solution of chromium(III) sodium potassium salicylate are refluxed for several hours. The 1:2 chromium complex **11** is precipitated at below pH 4. It is a yellow brown powder, which is soluble in hot water with a greenish yellow color.

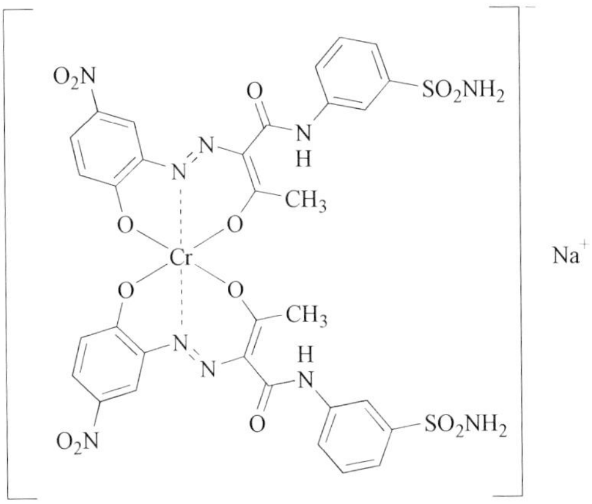

3.11.2 Chemical Constitution and Synthesis

[Structure **11**]

As an example of 1:2-chromation by the *dichromate–glucose method* [20], the azo dye obtained by coupling of diazotized 1-amino-2-hydroxy-5-nitrobenzene with 1-methylsulfonylamino-7-hydroxynaphthalene is dissolved in dilute NaOH and boiled after addition of $K_2Cr_2O_7 \cdot 2\,H_2O$ and glucose. Chromation is complete after a few minutes. The resulting 1:2 chromium complex **12** [*83748-22-5*] is precipitated with NaCl to give a black powder that imparts bluish olive dyeings on wool in a weakly acidic dyebath.

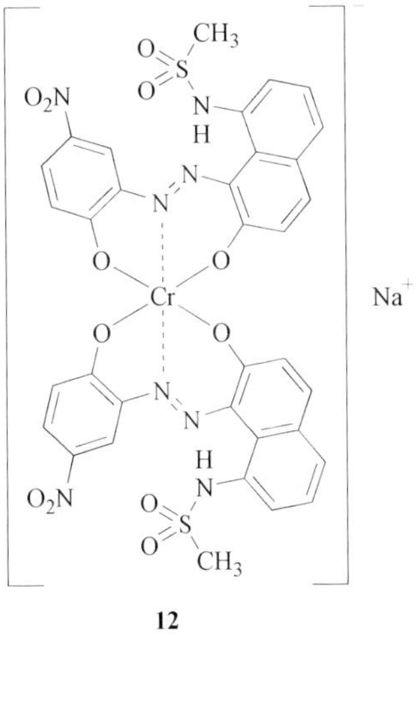

[Structure **12**]

1:2 Chromium Complexes with One Sulfonic Acid Group. By definition, 1:2-metal complexes with one sulfonic acid group must be unsymmetrical, because only one of the two azo dye ligands bears a sulfonic acid group. In 1962 the first metal-complex dyes of this type were brought onto the market [21]. The introduction of the sulfonic acid substituent extends the available diazo and coupling components and thus the possibilities for variations. This dye class possesses the same favorable properties as the unsulfonated 1:2 complex dyes and is moreover distinguished by higher brilliancy, better solubility, and simpler preparation. In general, the dyes are prepared in two steps with intermediacy of the sulfonated 1:1 chromium complex.

In the stepwise synthesis of the unsymmetrical complex dye **13** [*70236-60-1*] [10], the azo dye made from diazotized 1-amino-2-hydroxy-5-nitrobenzene and 1-phenyl-3-methyl-5-pyrazolone and the 1:1 chromium complex obtained from 6-nitro-1-diazo-2-hydroxynaphthalene-4-sulfonic acid and 2-naphthol are heated together at 80 °C for 5 h. The adduct is salted out with NaCl. A black powder is obtained that dyes wool and leather in dark brown shades. The resulting colors are fast, particularly on shrink-resistant wool.

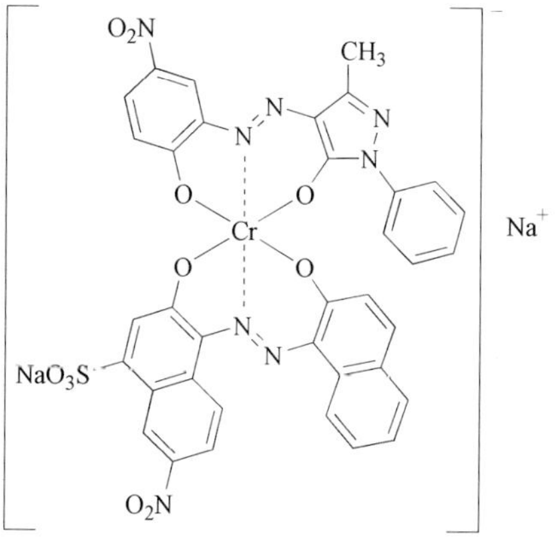

13

Unsymmetrical 1:2 chromium complexes based on an azo and an azomethine dye ligand can be prepared directly from a mixture of a 1:1 chromium complex and equimolar quantities of an 2-aminophenol, an anthranilic acid, or an aliphatic amino acid, and of salicylicaldehyde or a substituted salicylaldehyde [22].

In the synthesis of the unsymmetrical 1:2 chromium complex **14** containing an azo and an azomethine dye ligand [23], the azo dye formed from diazotized 4-chloro-2-amino-1-hydroxybenzene-6-sulfonic acid and 4′-hydroxynaphtho-(2′,1′:4,5)-oxathiazole-*S*-dioxide is heated in water at 140 °C under pressure together with $CrCl_3 \cdot 6 H_2O$. The suspension is filtered, distributed in water, and heated at 70 °C together with 4-nitro-2-amino-1-hydroxybenzene and salicylaldehyde while maintaining the pH within a range of 7–8 with sodium hydroxide solution. The mixed chromium complex **14** [*86158-66-9*] is precipitated by adding NaCl to give a dark green powder that imparts green dyeings on wool and polyamide with good fastness properties.

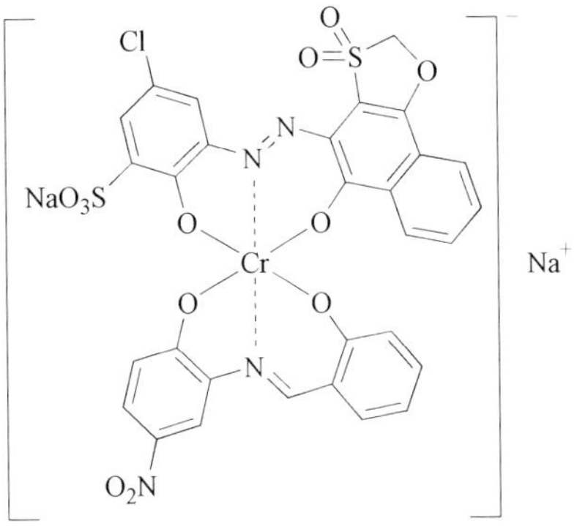

3.11.2 Chemical Constitution and Synthesis

[Structure 14]

14

1:2 Chromium Complexes with Two or More Sulfonic Acid Groups. The monosulfonated 1:2 chromium complex dyes were followed by those with two or more sulfonic acid groups. These more highly sulfonated complex dyes had been known since 1932 [24], but were introduced only in 1970 after detailed studies [25]. An example is *C.I. Acid Blue 193*, 15707 [75214-58-3] (**15**).

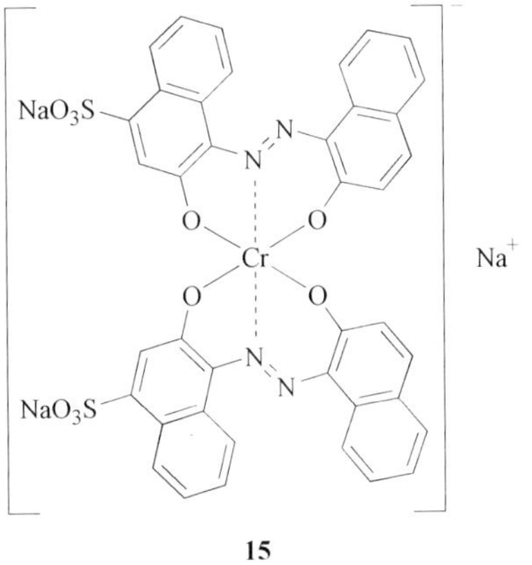

[Structure 15]

15

Because of its strongly hydrophilic character this dye type needs specific dyeing auxiliary agents to perform even and deep dyeings [26]. Wetfastness can be increased by using polyquarternary ammonium compounds to give machine-washable wool products, particularly in the case of chlorinated shrink-resistant wool [27]. Due to the sulfonic acid group the metal-free precursor dyes are slightly water soluble and can be chromed conveniently in weakly acidic medium without use of Cr^{VI} compounds or supporting agents like salicylic acid . Economy and high fastness resulted in a relatively large market share within a short time.

In the synthesis of the disulfonated 1:2 chromium complex dye **16** [*82269-28-1*] [28], a mixture of the azo compound made from diazotized 2-amino-1,4-dichlorobenzene-5-sulfonic acid and salicylaldehyde is heated to 50 °C with anthranilic acid, water, and NaOH/sodium acetate. A chromium(III) formate solution, prepared from Cr_2O_3, formic acid, and water is added, and the mixture is heated under reflux until conversion to the 1:2 chromium complex is complete. The pH is then adjusted to 7–8 with NaOH, and the dye is isolated by spray drying. The resulting brown powder gives light- and wetfast reddish yellow dyeings on wool and polyamides.

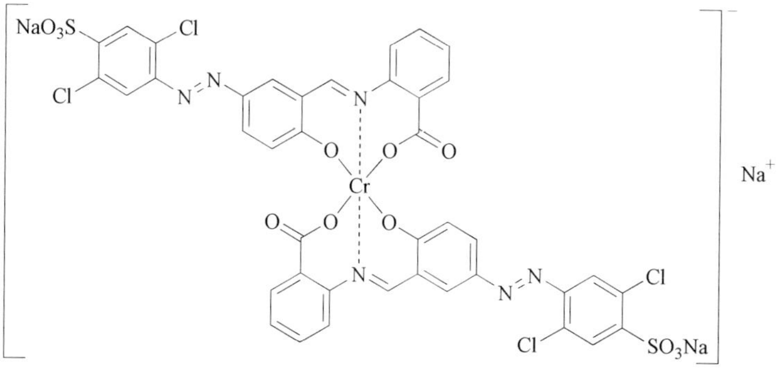

16

Sulfonated Cobalt Complexes. Unlike triaqua 1:1 chromium complexes, the corresponding cobalt complexes can not be prepared in acidic solution. The 1:1 cobalt complexes are only stable in the presence of a surplus of nitrogen donor groups. Sandoz has developed sufficiently stabilized 1:1 cobalt complexes by using nitrite ions as N ligands [29]. The coordinatively bound nitrite ions can easily be displaced by another tridentate azo dye [30].

In the synthesis of the 1:1 cobalt complex **17** by the *nitrite method*, a mixture of $CoSO_4 \cdot 7\,H_2O$ and $NaNO_2$ in water and the dye formed by acid coupling of diazotized 2-amino-4,6-dinitrophenol with 2-phenylaminonaphthalene-4′-sulfonic acid is subjected to dropwise addition of dimethylformamide. Metallization commences immediately and is complete within 2 h. The 1:1 cobalt complex **17** is precipitated by addition of NaCl and isolated. It is a suitable intermediate for producing unsymmetrical 1:2 cobalt complex dyes such as **18**.

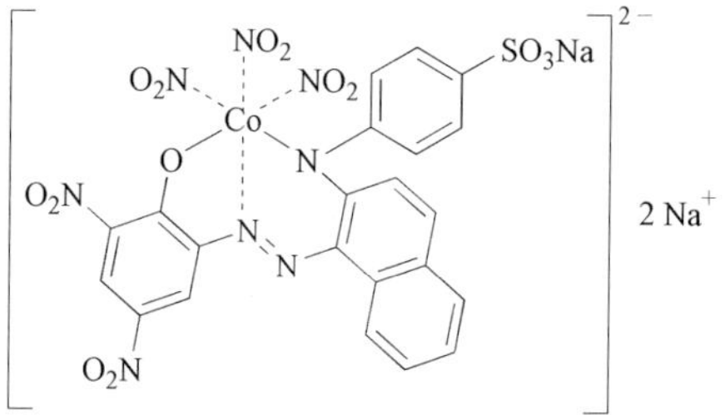

17

In the synthesis of the unsymmetrical 1:2 cobalt complex **18** [*68928-31-4*] [30], the 1:1 cobalt complex **17** and the azo dye made by coupling of 1-acetoacetyl-amino-2-ethylhexane with diazotized 2-aminophenol-4-sulfonamide are introduced into water and ethyl alcohol. The suspension is stirred and converted in solution by maintaining pH 12 with NaOH until the 1:1 cobalt complex has completely disappeared. The resulting mixed cobalt complex **18** is precipitated by adding KCl. It provides greenish olive, fast dyeings on polyamide.

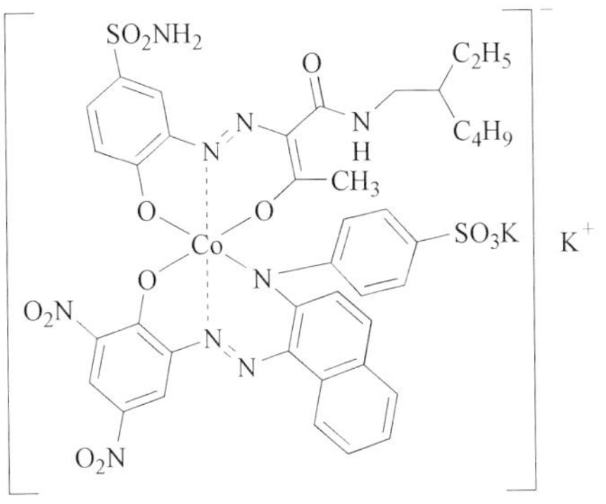

18

In the synthesis of the symmetrical 1:2 cobalt azomethine complex **19** [*41043-60-1*] [31], a mixture of 2-amino-6-nitrophenol-4-sulfonic acid and salicylaldehyde is heated to boiling in sodium acetate solution. The suspension is cooled and mixed first with $CoCl_2 \cdot 6\,H_2O$ and then with 30 % H_2O_2. The resulting solution is salted out with NaCl and KCl. The dye yields exceptionally lightfast yellow colors on wool and polyamides.

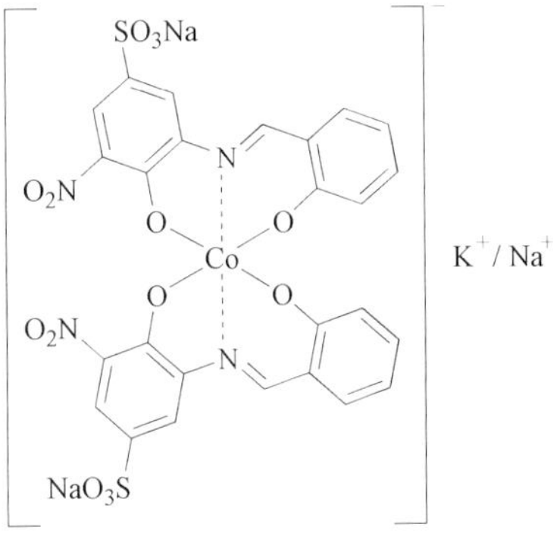

19

3.11.2.2 Metal Complexes for Cotton (see also Section 3.1)

As opposed to copper complex dyes, 1:2 chromium and cobalt complexes can only be bound to cotton if a fiber-reactive group is incorporated in the dye backbone [32]. For example, the reactive 1:2 chromium–cobalt complex **20** *C.I. Reactive Black 8*, 18207 M = Cr/Co = ca. 3:1; M = Cr [*79828-44-7*]; M = Co [*79817-89-3*] provides black prints on cotton, whereby the corresponding pure 1:2 chromium complex contributes the blue color component, and the 1:2 cobalt complex the reddish one [33].

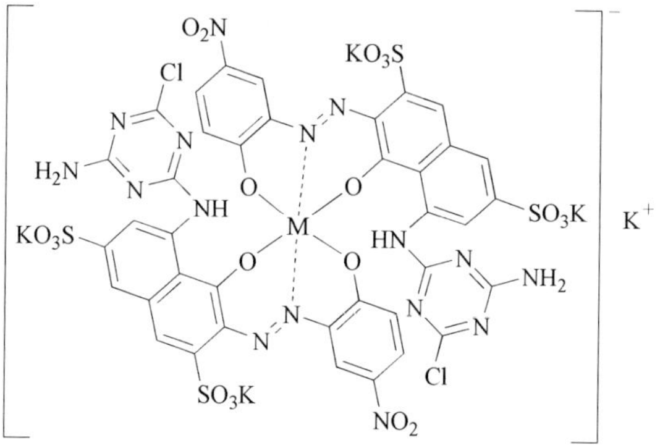

20

Copper-containing substantive dyes such as **21** C.I. Direct Blue 93, 22810 [*13217-74-8*] have long been used for coloring cotton [34].

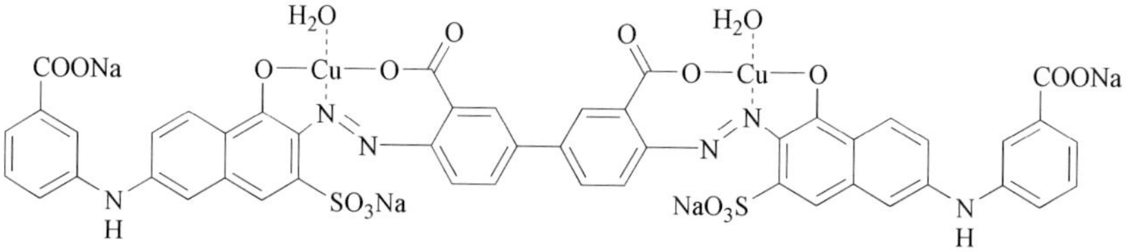

21

The Benzo Fast Copper (I.G. Farbenindustrie) and the Cuprophenyl (Geigy) dyes are converted to insoluble copper complexes on the fiber by treating the dyeings with a copper salt. Copper complex formation on the fiber has strongly lost technical significance due to the copper content in dyehouse effluents. The mostly copper-containing Indosol direct dyes recently launched by Sandoz are cross-linked on and with the fiber by aftertreatment with a special reactive agent [35].

In *demethylative coppering* to give the bis-copper complex **22** [*74592-99-7*] (3 Na, Li salt) [36], the sodium salt of the disazo compound made by coupling of bis-diazotized 3,3'-dimethoxy-4,4'-diaminodiphenyl with two equivalents of 8-amino-1-hydroxynaphthalene-3,6-disulfonic acid in alkaline medium is dissolved in water by adding diethanolamine. An ammonia alkaline copper(II) sulfate solution made from $CuSO_4 \cdot 5\,H_2O$ and of ammonia are added. The mixture is heated at 80–90 °C for 14 h. The solution is then cooled to 40 °C and the bis-coppered dye **22** salted out with sodium chloride. It dyes cotton in lightfast blue shades.

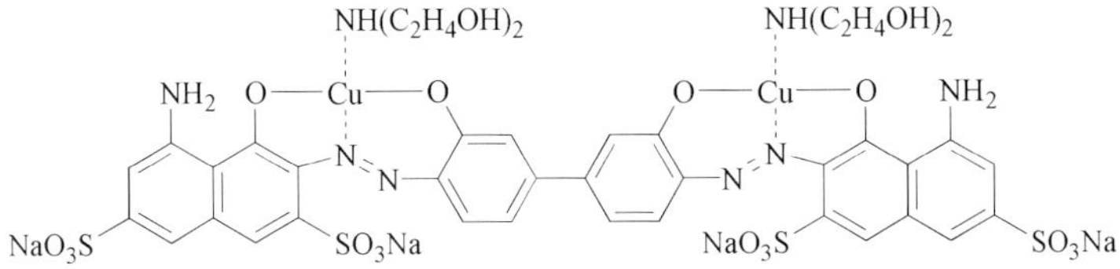

22

3.11.2.3 Metal Complexes for Leather (see also Section 5.1)

The 1:1 chromium and copper and 1:2 chromium, cobalt, and iron complex dyes can be used for dyeing of various leather substrates, such as chromium-tanned leather or leather that has been retanned with various vegetable or synthetic tanning agents [37]. These metal complexes are a indispensable group of dyes, particularly with respect to lightfastness and dye fixation. Color can be imparted by drum dyeing, usually with solid dye preparations, or by spray and dip dyeing with liquid dye formulations. The application of pumpable liquid dyes free of dusting problems is gaining increasing significance. Important leather dye ranges containing metal-complex dyes exist in powder form. The most important liquid dye ranges consist mainly of 1:2 chromium complex dyes.

The dyeing behavior is governed by the hydrophobic–hydrophilic balance. This property is basically affected by the type and number of hydrophilic groups and the molecular mass or the number of azo/azomethine groups the dye possesses. Increasingly hydrophilic character leads to higher solubility, penetration, and fastness to penetration into PVC and dry cleanability, whereas enhanced hydrophobicity better meets the requirements of various wetfastnesses and fixation rate [38]. Tanning agents used in the widely varying retannages strongly influence the depth of shade of the dyeings. On purely chrome tanned leather deeper shades are achieved than on retanned leathers, but with several hydrophilic 1:2 metal-complex dyes devoid of sulfonic acid groups, for instance, **23**, deep and well-covering dyeings are also attained on retanned leathers.

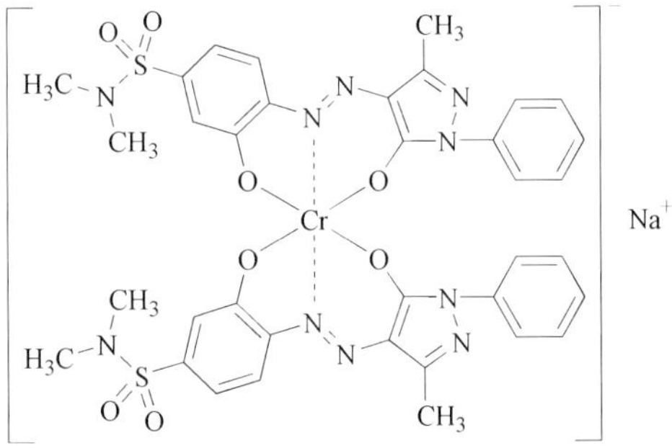

23

Due to the appropriate hydrophilic–hydrophobic balance, the bulky disazo 1:2 chromium complex **24** [*119654-54-5*] containing two sulfonic acid groups besides two hydrophilic sulfonamide groups affords leather dyeings with high resistance to migration into PVC. This so-called PVC fastness is a particularly important property for sport shoes.

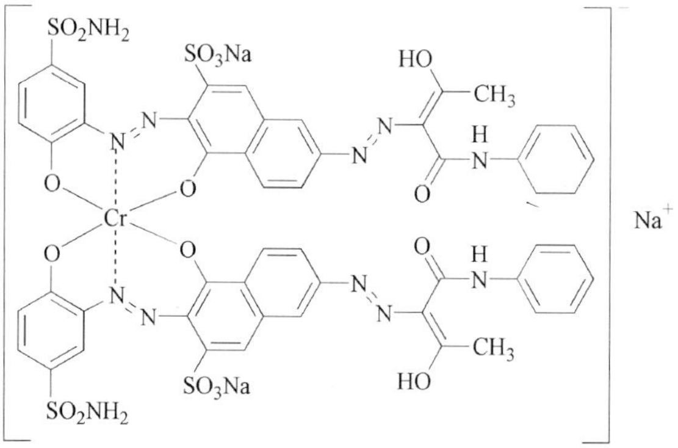

24

There is a increasing demand for liquid dye formulations, not only for spray and dip dyeing but also because of easy handling and use in automatic dosing equipment and through-feed dyeing machines. The chief requirements of liquid formulations are high storage stability, high dye concentration, and infinite miscibility with water. These properties can be met by adding water-soluble solvents such as polyglycols, glycol, or polyglycol monoalkyl ethers [39] or cyclic lactams [40]. Further approaches to storage-stable metal-complex dye solutions are liquid-phase separation [41], exchange of the countercation sodium for lithium or hydroxyalkyl-substituted ammonium species [42], and membrane-separation techniques [43].

In the synthesis of the liquid dye **25** for dip dyeing in a through-feed dyeing machine [44], the 1:1 chromium complex from diazotized 2-amino-1-hydroxy-4-nitrobenzene-6-sulfonic acid and 2-hydroxynaphthalene is suspended in water, and 1-ethoxy-2-propanol, 2-amino-1-hydroxybenzene-4-sulfonamide, and the azo dye prepared by coupling of diazotized 1-amino-2-choro-4-nitrobenzene with salicylaldehyde are added. The reaction mixture is adjusted to pH 6.5 with LiOH and heated to 75 °C until the starting materials are no longer detectable. The resulting brownish black solution of **25** is cooled and filtered. It dyes leather in brown shades with good resistance to water spotting.

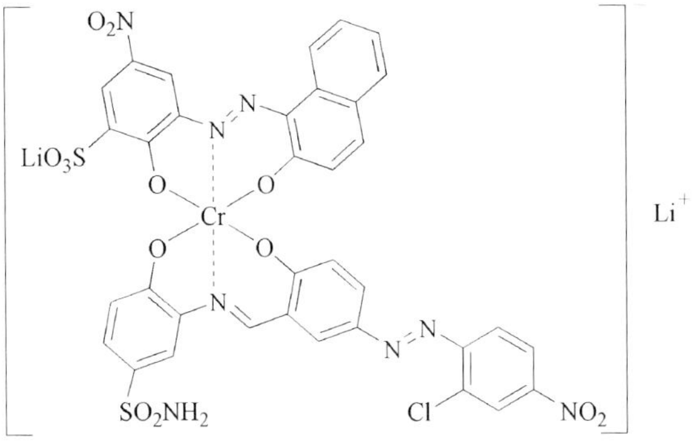

3.11.2 Chemical Constitution and Synthesis 315

25

3.11.2.4 Copper Complexes for Paper (see also Section 5.3)

Among metal-complex dyes only the planar 1:1 copper complexes exhibit sufficient substantivity to dye paper, whereas the octahedral 1:2 chromium and cobalt complexes are unsuitable for this purpose. Cationic substantive dyes, such as the bis-copper complex **26** [45] provide the most colorless effluent waters and superior fast paper dyeings that do not bleed out upon contact with milk, fruit juice, or alcohol [46].

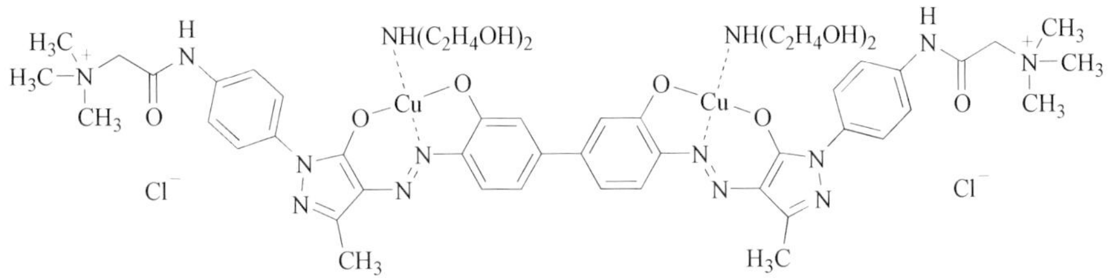

26

In the synthesis of the bis(1:1) copper complex **27** [*119103-25-2*] (copper-free) by oxidative coppering [47], the disazo compound made by coupling of bis-diazotized 4,4′-diaminostilben-2,2′-disulfonic acid with two equivalents of 6-acetylamino-1-hydroxynaphthalene-4,8-disulfonic acid is dissolved in water at 60 °C. After adjusting the pH to 6 with acetic acid a solution of $CuSO_4 \cdot 5 H_2O$ in water is added. Then 3 % H_2O_2 is dropped in, and the pH is maintained at 5–6 by addition of aqueous NaOH. When coppering is complete, the mixture is heated to 80 °C and precipitated with NaCl. The dye is isolated by filtration and dried. It dyes paper and cotton in lightfast blue shades.

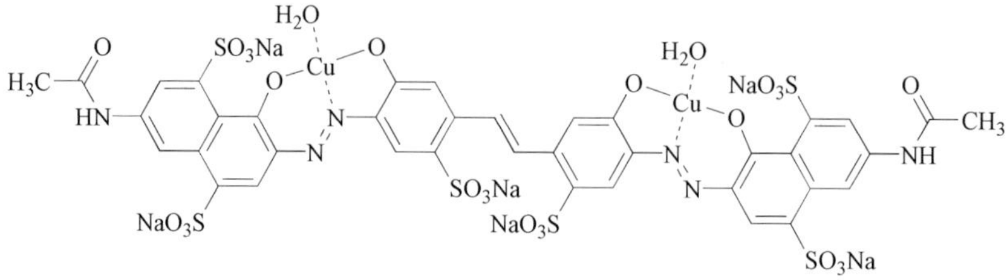

27

3.11.2.5 Metal-Complex Dyes for Polypropylene

Tridentate azo dyes having a ring nitrogen atom in the position β to the azo group and an hydroxyl group in *ortho* to the azo group are claimed to dye synthetic fibers including polyester, polyamide, and cellulose acetate and to form metal complexes on these fibers by aftertreatment with a suitable metal salt [48]. In a similar manner color may be imparted also to polypropylene. This material contains no functional groups providing anchor sites for anionic dye molecules, but to improve ligthfastness, propylene is polymerized with nickel salts. The resulting nickel-modified polypropylene produces colored nickel complexes when dyed with the above-mentioned heterocyclic dyes.

In the synthesis of the heterocyclic dye **28** [*83156-84-7*] [49], 2-amino-4-chlorophenol is diazotized. A solution of 1-methylimidazo[1,5a]pyridine in 10 % HCl is then added dropwise to the diazotation mixture while maintaining the pH at 4–5 with sodium acetate solution. The resulting azo dye is isolated by vacuum filtration and consists of a brown powder that dyes nickel-modified polypropylene fibers in greenish blue shades.

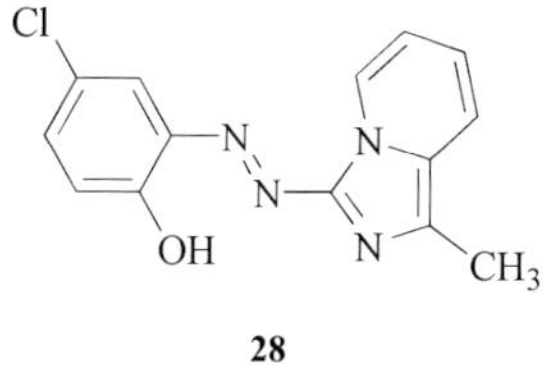

28

3.11.2.6 Formazan Dyes

Many metallized formazans are well suited as dyes for textile materials on account of their coloring strength and good fastness. The blue to blue-green copper complexes of quadridentate formazans are particular important because of their color clarity, which is similar to that of anthraquinone dyes. They can also be combined easily with yellow, orange, and red azo dyes and therefore present a low-priced alternative to blue anthrachinone dyes for trichromatic dyeing [50]. Copper complexes from quadridentate formazans are preferred on account of their high stability. Some dicyclic copper complexes of tridentate formazans also show good fastness and good application properties [51].

Wool and Polyamide Dyes. Water-soluble formazan complexes having sulfonamide, alkylsulfonyl, or sulfonic acid groups possess a high affinity to nitrogen-containing fibers. As with acid dyes, they can be used to dye wool and polyamide in neutral to weakly acid baths. Commercial dyes of this type are *C.I. Acid Blue 267* and *297*, and the 1:2 cobalt complex *C.I. Acid Black 180*, 13710 [*11103-91-6*] (**29**).

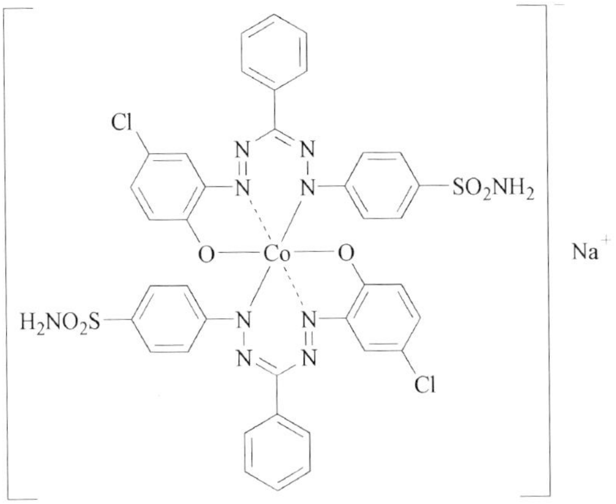

29

As environmentally friendly alternatives to the Cr^{III}- and Co^{III}-complexed acid dyes a range of 1:2 iron complex dyes based on tridentate 1,3,5-triarylformazans **30** was synthesized [52], which produce violet, blue, black, and brown shades, furnishing dyeings on wool and polyamide with good wet- and lightfastness. Iron complexes derived from tridentate 1,5-diaryl-3-cyanoformazans were also investigated [53].

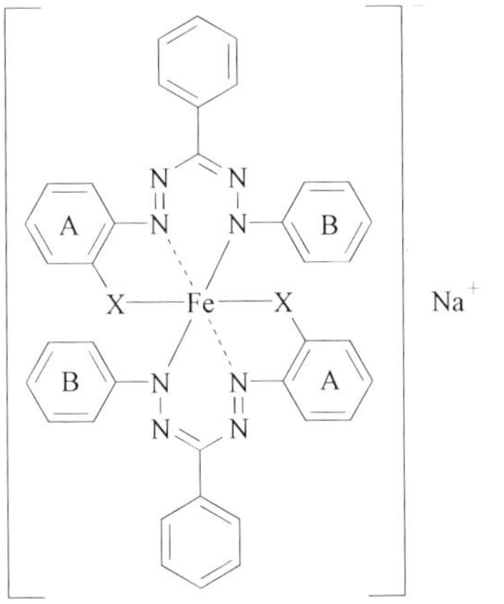

X = O, COO
Rings A and B may contain other substituents

30

3.11 Metal-Complex Dyes

Synthesis of the copper complex **31** [*167780-70-3*] (H, Na salt) involves coupling of a diazonium compound with a hydrazone [54]: 2-(3-amino-4-hydroxyphenylsulfonamido)chlorobenzene is dissolved in hot water, to which dilute NaOH solution is then added. After addition of sodium nitrite, the solution is cooled to –5 °C and then added to a well-stirred mixture of HCl, water, and ice. The diazonium salt suspension is neutralized with NaHCO$_3$ and combined with the weakly acidic solution of the hydrazone formed with 2-hydrazino-5-sulfobenzoic acid and benzaldehyde in water. The reaction mixture is kept at 0–10 °C, and at pH 10–12 by dropwise addition of diluted NaOH. When the coupling is complete, the pH of the formazan solution is reduced to 8 with HCl, a dilute aqueous solution of copper tetraaminosulfate is added and the solution is stirred for several hours. The copper complex **31** is precipitated at 85–90 °C with NaCl, then filtered and dried. The resulting powder colors wool in clear, fast blue tones.

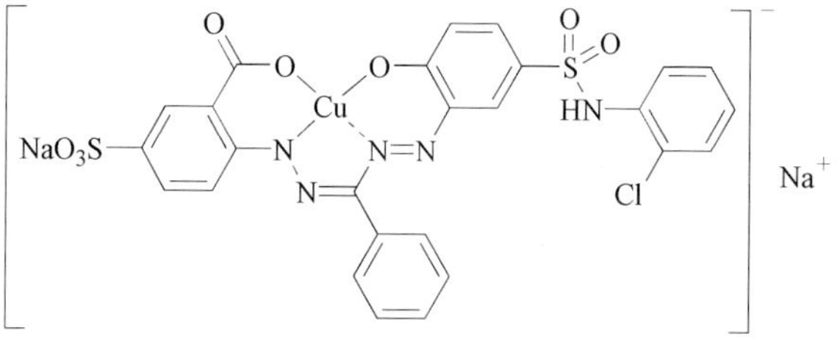

31

In the multistep synthesis of the copper complex **32** [55], 0.1 mol of ethyl phenylacetate and 14 mL of 10 mol/L NaOH are stirred together at 0 °C in 300 mL of water in the presence of an emulsifier. 14 g of Na$_2$CO$_3$ and the diazonium salt from 0.1 mol of 2-amino-6-chloro-4-nitrophenol are added and coupling is carried out below 25 °C. When the reaction is complete, the solution is neutralized with 10 mol/L HCl, diluted with a equal volume of water, and brought to the boil. Then 80 mL of 10 mol/L NaOH are added. The mixture is saponified for 10 min at 95–100 °C and neutralized again with 10 mol/L HCl. After cooling to 20 °C, the second coupling is carried out with the diazonium salt of 0.1 mol of 1-amino-6-nitro-4-sulfo-2-naphthol in presence of 0.1 mol of copper sulfate and 26 g Na$_2$CO$_3$. The coupling reaction is completed by heating to 60 °C. The resulting coppered formazan **32** is precipitated at 80 °C with NaCl, filtered, and dried. The dark brown powder imparts on wool, natural silk, and polyamide fibers an intense green tone with a blue cast.

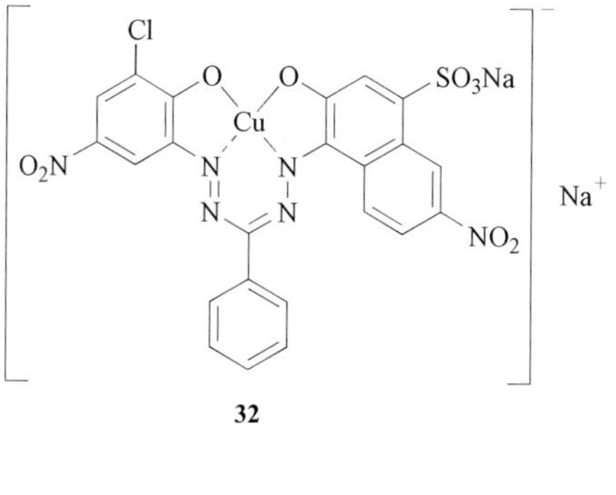

32

Cotton Dyes. The most important and best etablished formazan dyes are those containing reactive groups for dyeing and printing cellulose fibers. They are usually blue, occasionally green, [56], copper complexes derived from 1-(2-hydroxyphenyl)-3-phenyl-5-(2-carboxyphenyl)formazans containing sulfonic acid and reactive groups. The following commercial products belong to this group: *C. I. Reactive Green 15 and Reactive Blues 70, 83, 84, 104, 157, 160, 182, 202, 209, 212, 216, 218, 220, 221, 226, 228, 235, 260, 263* (see Section 3.1). The number of patents issued involving formazan dyes and trichromatic dyeing with formazan dyes was nearly 100 between 1985 and 1989 and nearly 65 between 1990 and 1994. A typical copper formazan reactive dye is exemplified by *C.I. Reactive Blue 216* [89797-01-3] (**33**), shown in the tricyclic, less strained 6/5/6-membered ring arrangement.

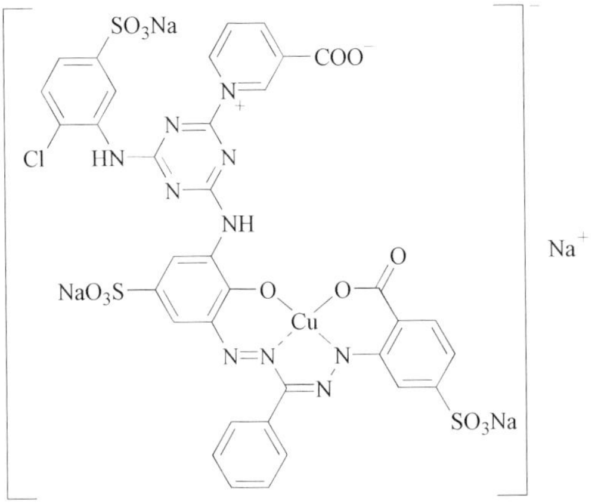

3.11.3 Miscellaneous Uses

3.11.3.1 Azo Metal-Complex Dyes

Solvent Dyes (see also Section 3.10). The 1:2 chromium and cobalt complex dyes devoid of any hydrophilic substituent have a considerable solubility in organic solvents, especially alcohols, ketones, and esters. Enhanced solubility can be achieved by converting the metal-complex sodium salts into salts of organic cations [57]. Such cations may be cationic dyes, long-chain aliphatic ammonium ions, or protonated guanidines. For example, the bluish red solvent dye **34** reaches a solubility in organic solvents of up to 1000 g/L [58].

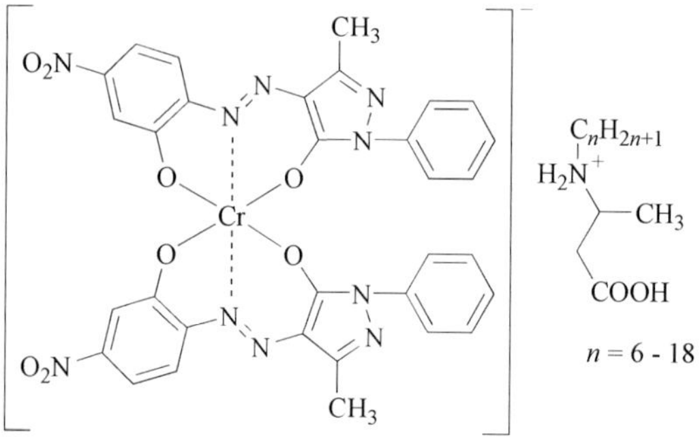

3.11 Metal-Complex Dyes

34

These laked 1:2 metal-complex dyes belonging to the generic class of solvent dyes have a broad application range, including transparent lacquers, wood stains, numerous office products, foil printing, and recently ink-jet printing [59].

Ink-Jet Dyes (see Section 5.6). Printing inks based on solvents such as methyl ethyl ketone or alcohol contain laked metallized solvent dyes, for example the statistically mixed 1:2 chromium complex **35** [60]:

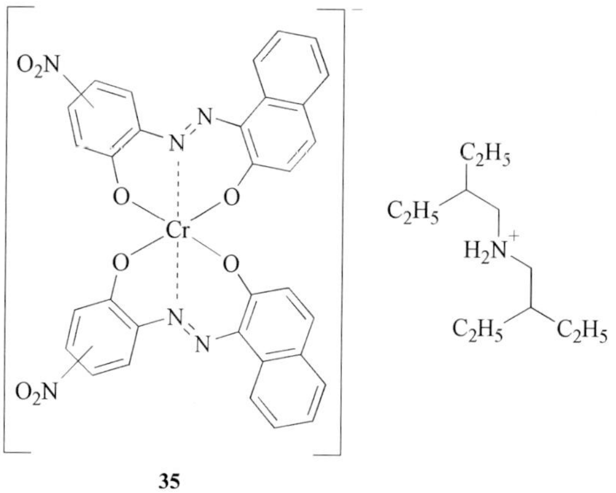

35

The chief prerequisite for printing inks is the absence of inorganic salts, because otherwise the nozzles of the printer heads might be plugged or degraded by corrosion. This requirement can easily be met in the case of laked metal-complex dyes, since they can completely freed from salts by washing with water due to their water insolubility. A further requirement is the stability of ink-jet solutions based on water, glycols, or cyclic aliphatic amides [61]. This property is fulfilled by the soluble metal complexes **36** M = Cr : Co = 70 : 30, M = Cr [*104815-63-6*], M = Co [*104815-64-7*] [59] and **37** [*113989-79-0*] [62].

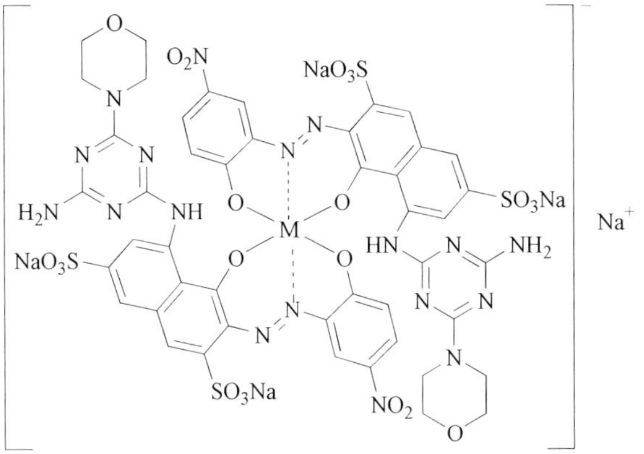

36

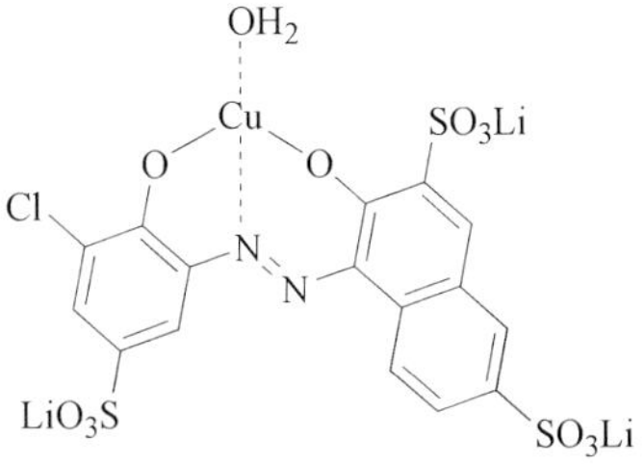

37

Anodized aluminum is a good acceptor for water soluble 1:2 chromium and cobalt complex dyes. For example, the cobalt complex **38** [*112144-04-4*] imparts a black color to this substrate [63].

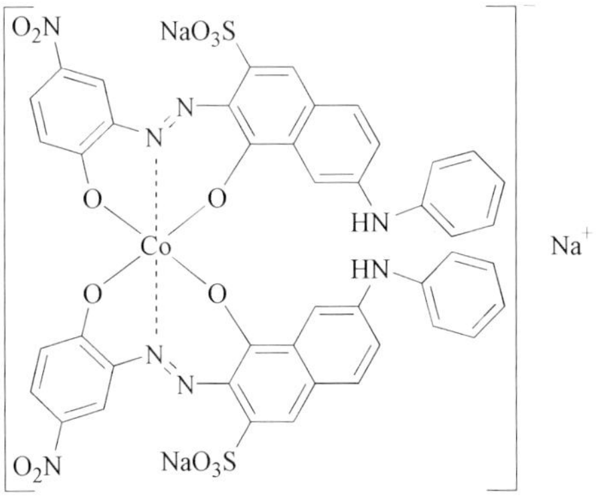

38

Redox Catalysts. Metallized redox catalysts are closely related to organic metal-complex pigments and distinguished, in general, by a polycyclic structure. The well-known cobalt complex salcomin (**39** [*14167-18-1*]) reversibly binds oxygen and serves, like the corresponding iron complex, as a model for biological redox catalysis. Further redox catalysts are outlined in [64].

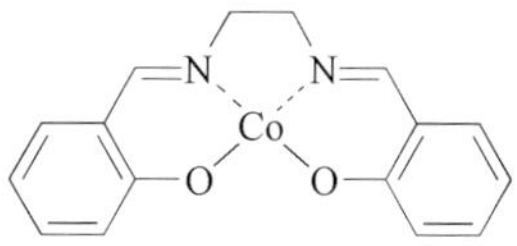

39

Color Photography (see Section 5.7). Chromium 1:1 complexes such as **40** [*79230-32-3*] [65], and 1:2 chromium complexes, both containing a hydroquinone residue as developer group, are used as photo dyes [66]. Nickel complexes of 2-amino-6-arylazo-3-pyridinol dyes can also be used for photographic purposes [67].

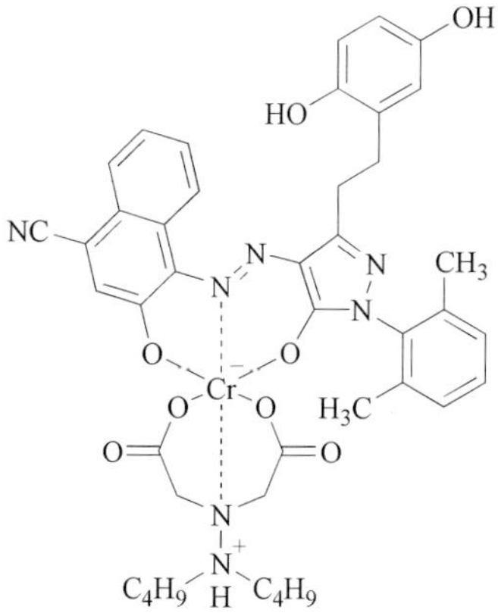

40

Electrophotography (see Chapter 6) comprises the ubiquitous technologies of photocopying and laser printing, both involving the photoconductor and toner as key components. Toner particles are composed of a resin with a low softening point, a pigment, and a charge-control agent. In black toners the pigment is usually carbon black, and colored toners contain yellow, magenta, and cyan pigments. The basic processes and the requirements for the dyes are treated in [68]. The charge control agents are basically 1:2 metal complexes of azo dyes, for example, the negative-charge-controlling 1:2 iron complex **41** [*163669-55-6*] [69], the positive-charge-controlling 1:2 chromium complex **42** [*100012-90-6*] [70], and the nearly colorless 1:2 chromium complex **43** [*119604-58-9*] [71]. The intensely colored complex dyes **41** and **42** can only be used in black toners, whereas the colorless complex **43** is required for colored toners so as not to affect the shade of the color copy.

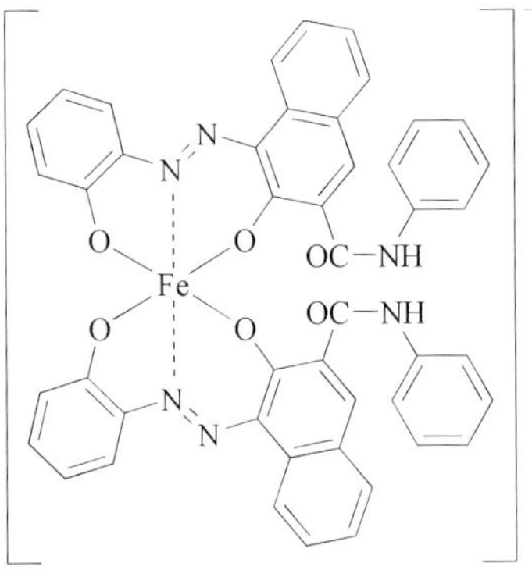

41

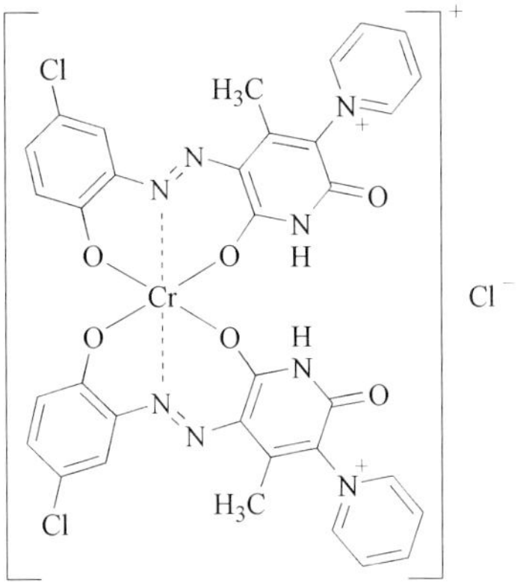

42

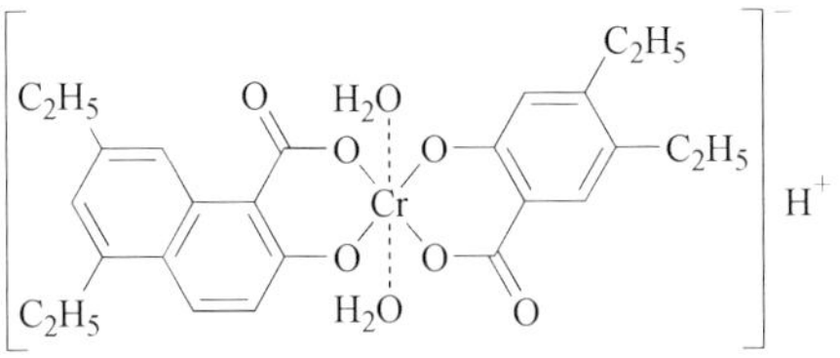

43

IR-Absorbing Dyes (see Chapter 6). Binary nickel, palladium, and platinum complexes of dithiol derivatives are distinguished by high extinction coefficients between 750 and 950 nm. They can be used as IR-absorbing dyes in optical recording media, laser components, and sunglasses, or as antioxidants, corrosion inhibitors, or protective agents for film and textile colors [72]. They are produced by treating an α-hydroxyketone with phosphorous pentasulfide and subsequent metallization to give, for example, the anionic palladium complex **44** [*105892-91-9*].

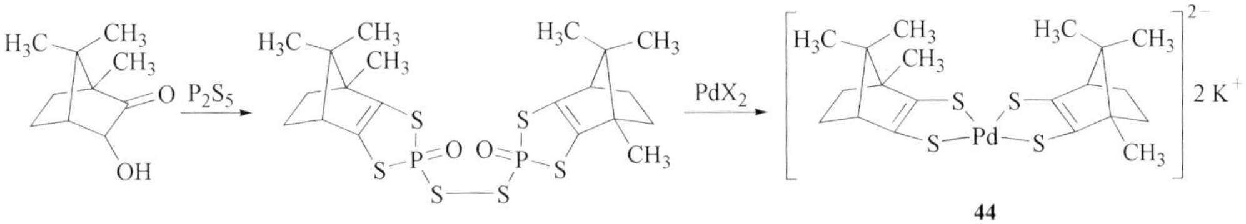

44

Liquid Crystals (see Chapter 6). Certain bicyclic copper complexes, such as **45** [*78452-82-1*], show discotic mesophases and are therefore attractive as liquid crystals for electro- and thermo-optical display elements [73].

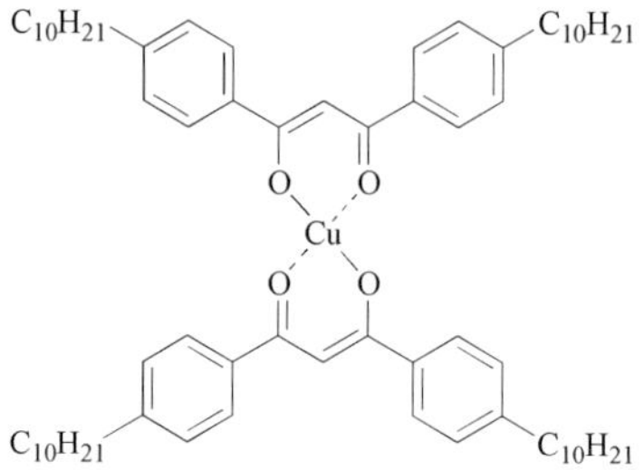

In the synthesis of **45** [73], a solution of CuCl$_2$ in ethanol is added to a solution of bis(4-decylbenzoyl)methane in THF at 40 °C. The mixture is stirred at this temperature for 30 min, then slowly mixed with 20 % ammonia solution and stirred for a further 2 h. The resulting precipitate is isolated by filtration and recrystallized from 2-propanol to yield greenish bronze crystals. They transform into liquid crystals at 87 °C and enter the isotropic fluid state at 130 °C.

Analytical Reagents. Various chelators give specific color changes on combining with different metal salts, such as vanadium, iron, cobalt, nickel, copper, and palladium salts and thus identify the corresponding metal ions [74]. Other chelators, such as *C.I. Mordant Black 11* (Eriochrome Black T) (**2**; see Section 3.11.1), are employed as indicators for complexometric titrations.

3.11.3.2 Formazan Dyes

Analytical Reagents. As a result of their fast and intense color reaction, formazans can be used for the spectroscopic determination of a variety of metal ions. In a sensitive blue color reaction, Zn and Cu can be determined with the commercial reagent Zinkon (**46** [*135-52-4*]). The detection limit for Zn (λ_{max} = 625 nm) is 0.02 ppm.

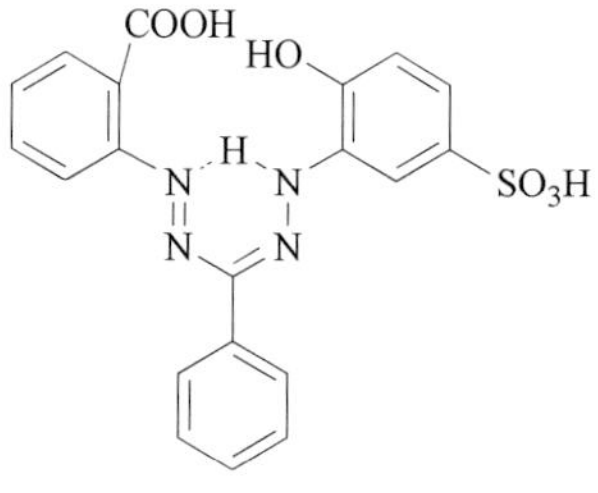

A concise survey regarding very specific reagents based on heterocyclic-substituted formazans is compiled in [75]. The cyclic formazans **47** are lithium-specific indicator dyes, useful for the quantitative colorimetric determination of lithium ions in biological fluids such as blood [76].

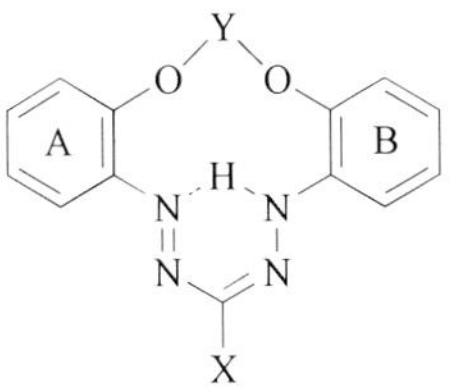

X = CN, NO$_2$
Y = H$_2$C−CH$_2$, H$_2$C−CH$_2$−CH$_2$, H$_2$C−CH$_2$−O−CH$_2$−CH$_2$

47

Bioindicators. Water-soluble, colorless tetrazolium salts can be reduced to water-insoluble, deeply colored formazans. The reduction of tetrazolium salts in plant tissue at pH 7.2 was first demonstrated in 1941 [77]. Tetrazolium salts have since been used in biochemistry, cytochemistry, and histochemistry because of the great sensitivity of this reaction. They can be used to detect biological redox systems in blood serum, in living cells, tissues, tumors, and bacteria. Tetrazolium Blue [*167429-81-7*] (**48**) is a particularly sensitive reagent.

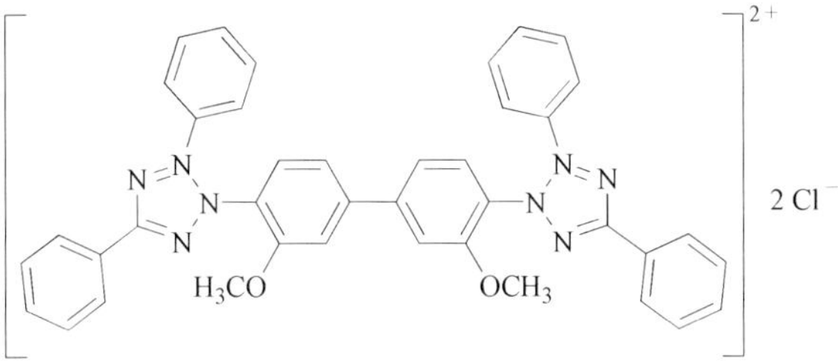

48

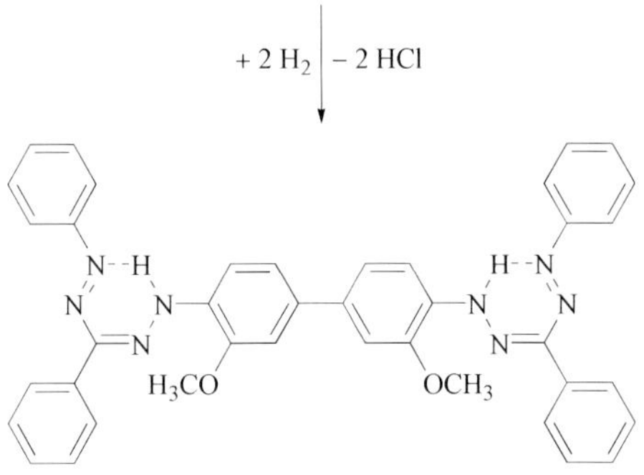

+ 2 H$_2$ | − 2 HCl

Indicator reactions are reviewed in [78], [79]. Another area of application is the fast determination of bacterial capacity in industrial wastewaters [80] and the determination of dehydrogenase activity in sludges [81].

Photoreagents (see Section 5.7). The conversion of tetrazolium salts into metallizable formazans dyes can also be employed in photographic development processes [82], [83]. In another development process [84], the silver produced on the film is amplified with formazan dyes. Formazan dyes are also effective in photographic emulsions as antifogging agents [85].

Ink-Jet Dyes (see Section 5.6). Formazans **49**, shown here in the open bicyclic form, are used as dyes in inks for ink-jet printing [86].

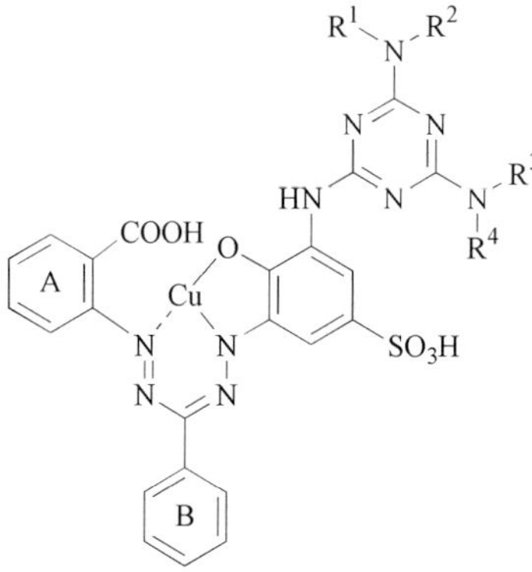

49

Optical Recording Dyes (see Chapter 6). Mixtures of the nickel-complexed tridentate formazans **50** are claimed to provide good light stability to optical recording layers and elements [87].

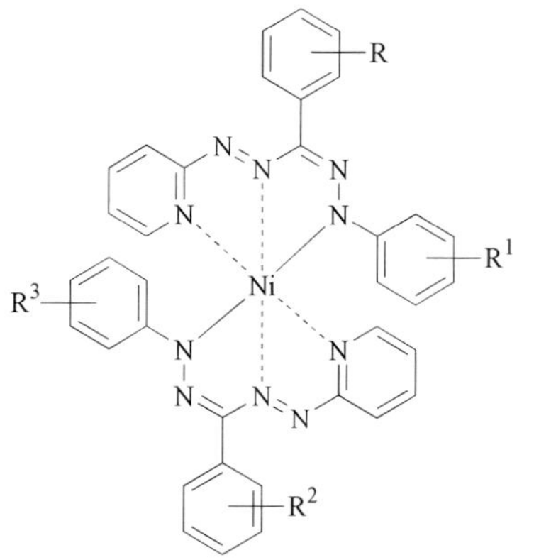

50

3.11.4 References

[1] H. D. K. Drew, J. K. Landquist, J. Chem. Soc. (1938) 292; J. A. J. Jarvis, *Acta Cryst.* 14 (1961) 961.
[2] R. Price, *Chimia* **28** (1974) 221.
[3] R. Price in: K. Venkataraman: *The Chemistry of Synthetic Dyes*, Academic Press, New York, vol. III (1970) 318.
[4] H. Baumann, H. R. Hensel: "Neue Metallkomplexfarbstoffe, Struktur und färberische Eigenschaften", *Fortschr. Chem. Forsch.* 7 (1967) 739–749.
[5] DE 78 409, 1893 (Erdmann, Borgmann).
[6] E. Race, F. M. Rowe, J. B. Speakman, *J. Soc. Dyers Colour.* **62** (1946) 372; F. R. Hartley, *J. Soc. Dyers Colour.* **85** (1969) 66.
[7] F. R. Hartley, *J. Soc. Dyers Colour.* **86** (1970) 209.
[8] G. Meier, *Textil Praxis* **31** (1976) 898; G. Meier, A. Overhoff, *Textil Praxis* **33** (1978) 293; G. Meier, *J. Soc. Dyers Colour.* **95** (1979) 252; W. Langmann, G. Meier, *Melliand Textilber.* **57** (1976) 580; R. Schaffner, W. Mosimann, *Textilveredlung* **14** (1979) 12.
[9] A. C. Welham, *J. Soc. Dyers Colour.* **102** (1986) 126.
[10] BASF, DE-OS 3 010 872, 1980 (H. Baumann, G. Grychtol).
[11] BASF, DE 734 990, 1940 (J. Müller, H. Pfitzner); DE 748 970, 1942 (G. von Rosenberg, K. Anaker, H. Pfitzner).
[12] I. G. Farbenindustrie, DE 715 082, 1938 (K. Holzach, H. Pfitzner).
[13] G. Schetty, *J. Soc. Dyers Colour.* **71** (1955) 705.
[14] Ciba, FR 836 297, 1938; Allied Chemicals, US 2 499 133, 1945 (K. F. Conrad); Ciba, DE 850 038, 1950 (W. Widmer, A. Bühler); DE 932 979, 1952 (A. Bühler, C. Zickendraht); Geigy, US 2 623 871, 1951 (G. Schetty, F. Beffa); Sandoz, US 2 753 336, 1953 (P. Maderni); Bayer, FR 1 111 970, 1954; General Anilin, US 2 809 962, 1953 (H. B. Freyermuth, A. F. Strobel); ICI, GB 748 421, 1953 (C. H. Reece); BASF, DE-AS 1 032 865, 1955 (H. Pfitzner, O. Kaufmann); Bayer, GB 796 723, 1955.
[15] Geigy, BP 730 036, 1951; BP 736 034, 1952.
[16] Geigy, BP 796 350, 1954.
[17] Geigy, BP 820 502, 1955.
[18] BASF, DE 846 142, 1950; Bayer, GB 796 759, 1955; GB 823 742, 1956; DE 940 483, 1952 (G. Dittmar, R. Pütter, F. Suckfüll).
[19] Ciba, DE 842 981, 1949 (W. Widmer, C. Zickendraht).
[20] Bayer, DE 929 567, 1952 (G. Dittmar, R. Pütter).
[21] Ciba, BP 765 355, 1953.
[22] Geigy, DE 1 079 247, 1957 (G. Schetty, W. Biedermann, F. Beffa); DE 1 544 580, 1964 (F. Beffa, E. Steiner); Bayer, DE 2 210 260, 1972; EP 66 230, 1982; EP 72 501, 1982 (W. Mennicke).; BASF, DE 2 461 481, 1974, (H. Baumann); Ciba-Geigy, DE 2 620 981, 1975 (F. Beffa, G. Back, E. Steiner); BASF, DE 2 735 287, 1977 (H. Baumann, K. Grychtol).
[23] Bayer, EP 73 951, 1982 (W. Mennicke).
[24] Ciba, DE 600 545, 1932.
[25] BASF, DE 1 008 253, 1956 (H. Baumann, W. Federkiel); Ciba, DE 1 133 846, 1959 (A. Bühler, A. Fasciati, C. Zickendraht).
[26] K. R. Schneider, *Textilveredlungspraxis* **2** (1972) 151.
[27] K. Reincke, *Melliand Textilber.* **67** (1986) 191; Hoechst, DE-AS 2 845 536, 1978 (K. H. Röstermundt).
[28] BASF, EP 47 427, 1981 (H. Baumann, K. Grychtol).

[29] Sandoz, DE-AS 2 153 548, 1971 (J. Dore).
[30] Sandoz, DE-AS 2 153 549, 1971 (J. Dore).
[31] Bayer, DE-OS 2 123 453, 1971 (W. Mennicke).
[32] D. M. Lewis, *J. Soc. Dyers Colour.* **98** (1982) 165; "Reactive Dyes" in: K. Venkataraman: *The Chemistry of Synthetic Dyes*, Academic Press, New York, vol VI (1972).
[33] ICI, GB 985 481, 1960 (A. H. Berrie, A. Johnson, C. E. Vellins).
[34] I. G. Farbenindustrie, DE 553 045, 1929 (J. Hilger, K. Wiedemann).
[35] H. Zollinger: *Color Chemistry*, VCH Verlagsgesellschaft, Weinheim 1987, p. 135.
[36] Ciba US 2 536 957, 1947 (H. Riat, H. Mayer).
[37] J. F. Feeman in: K. Venkataraman: *The Chemistry of Synthetic Dyes*, Academic Press, New York, vol VIII (1978) 44–49.
[38] C. S. Tysoe, *J. Soc. Leather Technol. Chem.* **79** (1995) 67.
[39] Sandoz, DE-OS 2 816 983, 1978 (J. v. Diest).
[40] Bayer, DE-OS 3 208 141, 1982 (K. Linhart, H. Gleinig); DE-OS 4 124 451, 1991 (R. Gerlach, W. Mennicke, W. Müllers).
[41] Bayer, EP-A 24 609, 1980 (W. Scholl, K. H. Schündehütte, W. Mennicke).
[42] Bayer, DE-AS 2 443 483, 1974 (W. Mennicke, P. Suchanek, P. Scotch, P. Vogt); Ciba-Geigy, EP-A 44 805, 1981 (R. Lacroix).
[43] Ciba-Geigy, EP-A 88 727, 1983 (P. Erzinger).
[44] Bayer EP 066 230, 1982 (W. Mennicke).
[45] Sandoz, DE-OS 3 117 127, 1981 (P. Doswald, E. Miriconi, H. Moser, H. Schmid).
[46] Sandoz, DE-OS 3 313 695, 1983 (J. Dore, R. Pedrazzi); EP-A 92 520, 1983 (M. Greve, H. Moser, R. Pedrazzi, R. Wald); DE-OS 3 440 777, 1984 (J. Dore, R. Pedrazzi); DE-OS 3 609 590, 1986 (H. Moser, R. Wald); DE-OS 3 715 066, 1987 (H. Moser, R. Wald).
[47] Bayer, DE-OS 3 710 077, 1987 (K. Kunde).
[48] Eastman Kodak, US 2 835 661, 1956 (J. M. Straley, J. G. Fisher); US 2 822 359, 1954 (J. M. Straley, J. Sagal); US 2 871 231, 1955 (J. M. Straley, D. J. Wallace, J. G. Fisher).
[49] Agfa-Gevaert, EP-A 54 230, 1981 (V. Runzheimer, G. Wolfrum, H. Heidenreich, P. Bergthaller, G. Schenk).
[50] Bayer, DE-OS 3 329 193, 1983 (R. Weitz, R. Neef, H. Hugl).
[51] Hoechst, DE-OS 3 323 638, 1983 (G. Schwaiger); DE-OS 3 440 265, 1984 (G. Schwaiger).
[52] J. Sokolowska-Gajda; H. S. Freeman, A. Reife, *Dyes Pigm.* **30** (1996) 1.
[53] M. Szymczyk, W. Czajkowski, R. Stolarski, *Dyes Pigm.* **42** (1999) 227.
[54] Geigy, DE-OS 1 419 841, 1960 (F. Beffa, G. Schetty).
[55] Soc. Carbochimique, DE-AS 1 062 285, 1952 (H. Ziegler).
[56] Geigy, DE-OS 1 644 254, 1966 (P. Dussy, H. Meindl, H. Ackermann).
[57] BASF, DE 743 848, 1939 (H. Pfitzner, B. Eistert); DE 746 839, 1939 (H. Pfitzner); DE-AS 1 260 652, 1960 (H. Baumann et al.).
[58] Williams (Hounslow) Ltd., US 4 204 879, 1980 (N. K. Paskins, D. C. Redman, I. D. Turner).
[59] ICI, EP-A 177 138, 1985 (A. G. Baxter, S. B. Bostock, D. Greenwood).
[60] BASF, EP-A 163 113, 1985 (H. Löffler, R. Dyllick-Brenzinger).
[61] ICI, EP-A 246 763, 1987 (A. Quayle, C. V. Stead).
[62] Canon, EP-A 233 769, 1987 (T. Eida, Y. Suga, K. Shirota).
[63] VEB Chemiekombinat Bitterfeld, DD 246 549, 1986 (G. Knöchel et al.).
[64] *Ullmann's Encyclopedia of Industrial Chemistry*, Fifth Edition, Wiley-VCH, Weinheim 1996, vol. A16, pp. 318–319.
[65] Polaroid Corp., US 4 267 252, 1980 (E. M. Idelson).

[66] Polaroid Corp., US 3 970 616, 1973 (H. Bader, M. H. Feingold); US 4 206 115, 1977 (E. M. Idelson); US 4 231 950, 1979 (E. M. Idelson); US 4 656 117, 1982 (E. Chinoporos, E. M. Idelson, P. F. King); EP-A 149 158, 1984 (P. F. King, S. G. Stroud).
[67] Agfa-Gevaert, DE-OS 3 322 058, 1983 (R. Stolzenburg, P. Bergthaller, G. Wolfrum, J. Strauss); DE-OS 3 329 774, 1983 (P. Bergthaller, G. Wolfrum, H. Heidenreich); DE-OS 3 337 118, 1983 (G. Wolfrum et al.); Eastman Kodak, EP-A 95 127, 1983 (J. A. Reczek, J. K. Elwood).
[68] K. L. Birkett, P. Gregory, Dyes Pigm. **7** (1986) 341.
[69] Ricoh, JP 95 114 218, 1995; JP 96 123 096, 1996; Kao Corp., JP 96 262 800, 1996.
[70] Hodogaya, JP 60 106 859, 1983.
[71] Konishiroku Photo, JP 63 217 362, 1987 (T. Okuyama, S. Ikeuchi, A. Matsubava, M. Avai); Derwent Report 88-296 629/42.
[72] Fuji Foto, JP 61 011 704, 1984; Derwent Report 86-059 321/09; JP 61 032 003, 1984; Derwent Report 86-085 065/13; JP 61 057 674, 1984; Derwent Report 86-116 734/18; Secretary of Navy, US 4 593 113, 1985 (M. H. Kauffman).
[73] Cent. Nat. Rech. Scientifiq, FR 2 486 946, 1980 (A.-M. Giroud-Godquin).
[74] Agency of Ind. Sci. Tech., JP 57 136 570, 1981; Derwent Report 82 659.
[75] *Ullmann's Encyclopedia of Industrial Chemistry*, Sixth Edition, 2002 Electronic Release, "Metal-Complex Dyes", Section 6.3.
[76] Eastman Kodak, EP-A 325 389, 1988 (M. H. Delton, S. F. Eiff).
[77] R. Kuhn, D. Jerchel, *Ber. Dtsch. Chem. Ges.* **74** (1941) 949.
[78] W. D. Hooper, *Rev. Pure Appl. Chem.* **19** (1969) 231.
[79] F. Wohlrab, E. Seidler, K. H. Kunze: *Histo- und Zytochemie dehydrierender Enzyme*, Verlag J. A. Barth, Leipzig 1979.
[80] V. Zotov, *Vodosnabzh. Sanit. Tekh.* **1972**, 16.
[81] J. Gajdusek, J. Kupec, J. Svancer, *Acta Hydrochim. Hydrobiol.* **2** (1974) 521.
[82] Eastman Kodak, US 3 642 478, 1972 (A. T. Brault, V. L. Bissonette).
[83] M. Lelental, H. Gysling, *J. Photogr. Sci.* **26** (1978) 135.
[84] *Res. Discl.* **217** (1982) 139.
[85] *Res. Discl.* **183** (1979) 393.
[86] ICI, EP-A 219 232, 1985 (A. Baxter, S. Bostock, D. Greenwood).
[87] Eastman Kodak, EP-A 832 938, 1998.

3.12 Naphthoquinone and Benzoquinone Dyes

3.12.1 Introduction

A large number of naphthoquinone and benzoquinone dyes occur naturally as plant constituents, and in the past many of these have found use as colorants. The chemistry of synthetic analogues began with the discovery of naphthazarin (**1**) [*475-38-7*] by Roussin in 1861. Naphthazarin and its derivatives first found use as black chrome mordant dyes for wool and silk. In contrast to the naphthoquinones, synthetic benzoquinone dyes are few in number and are generally of the

2,5-diarylamino-1,4-benzoquinone type, suitable for application as vat dyes, e.g., *C.I. Vat Yellow 5*, 56005 [*4370-55-2*] (**2**).

From about 1930 onwards, developments in the field of naphthoquinone dyes concentrated on the use of naphthazarin and intermediates for the preparation of violet, blue, and green acid and disperse dyes [1]. More recently there has been interest in the synthesis and color and constitution properties of simple colored naphthoquinones, stimulated by the fact that such dyes have similar tinctorial properties to the anthraquinones but a smaller molecular size. The naphthoquinones provide a useful alternative to the anthraquinones for certain specialized applications, e.g., as pleochroic dyes with improved solubility for liquid-crystal displays. As a result, research interest in these chromogens remains unabated, even though they have failed to make any major impact as textile dyes [2–8].

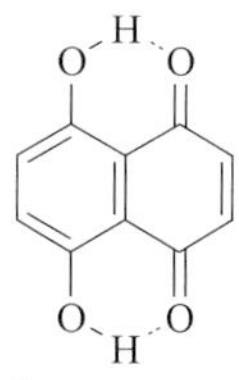

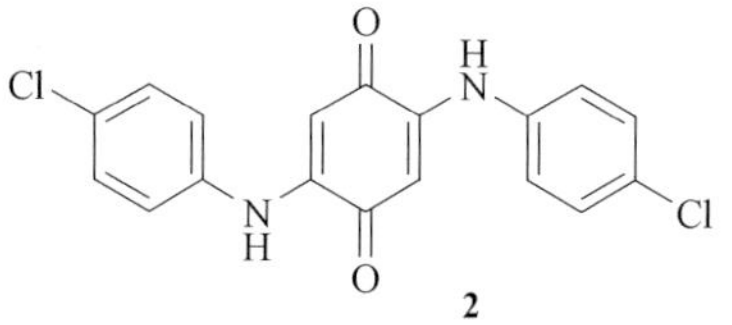

3.12.2 Benzoquinone Dyes

From the commercial point of view, only the 2,5-diarylamino-1,4-benzoquinones have been of any significance. They were used as vat dyes for wool and cotton, but are no longer in commercial production. Other examples of early dyes of this type are listed in the Colour Index under numbers 56000–56050. The inadequate color properties of the amino-substituted benzoquinone chromogen are exemplified by 2,5-bis(dimethylamino)-1,4-benzoquinone [*1521-02-4*], which absorbs at 530 nm in ethanol with an absorption coefficient of only ca. 500 L mol^{-1} cm^{-1}. This low intensity is characteristic of the 2,5-disubstitution pattern. Although other substitution patterns have higher intensities (e.g., the 2-amino derivatives have an ε_{max} of ca. 5000 L mol^{-1} cm^{-1}), dyes based on these have not been exploited.

3.12.3 1,4-Naphthoquinone Dyes

3.12.3.1 Simple 1,4-Naphthoquinones

Color can be produced in the 1,4-naphthoquinone chromogen by introduction of amino and hydroxyl groups into the quinonoid ring (i.e., positions 2 and 3), into the benzenoid ring (more particularly positions 5 and 8), or into both rings. The spectroscopic effects of such substitutions have been investigated in detail [9–12].

2-Amino-1,4-naphthoquinones are generally yellow to red, with ε_{max} values of ca. 2000–5000 $L\,mol^{-1}\,cm^{-1}$, depending on the nature of the substituents attached to the amino group and to the 3-position of the ring. In general, substituents in the 3-position cause a decrease in intensity due to steric crowding, whereas N-aryl groups induce a higher intensity. Typically, 2-phenylamino-1,4-naphthoquinone [6628-97-3] has λ_{max} 472 nm, ε_{max} 4800 $L\,mol^{-1}\,cm^{-1}$. A now obsolete vat dye of this type is **3** [3144-89-6]. Fluoro- and trifluoromethylphenylamino analogues have been described for coloring wool and polyester [13].

The more intensely colored 1,4-naphthoquinones with amino and/or hydroxyl groups in the 5- and 8-positions are analogous to the commercial anthraquinone dyes, but are generally more bathochromic [10]. For example, 5-methylamino-1,4-naphthoquinone has λ_{max} 529 nm in cyclohexane, whereas 1-methylaminoanthraquinone has λ_{max} 495 nm. This is also true for the 5,8-disubstituted naphthoquinones, but this potential advantage has not proved of commercial significance, and few such compounds have been considered as textile dyes. An exception is 5-amino-2,3-dichloro-8-hydroxy-1,4-naphthoquinone (**4**) [68217-33-4], which has been claimed as a violet disperse dye for acetate fibers [14]. Naphthazarin (**1**) is an example of a 5,8-disubstituted naphthoquinone of greater value as an intermediate than as a dyes.

When auxochromes such as NHR or OH are present in both rings of the naphthoquinone system, the absorption spectrum consists of two overlapping bands, and this can result in dull colors, typically blue-grays. A 3-cyano group appears to counteract the dulling effect, and **5** [79469-30-0] has been described as a blue dye for polyester [15].

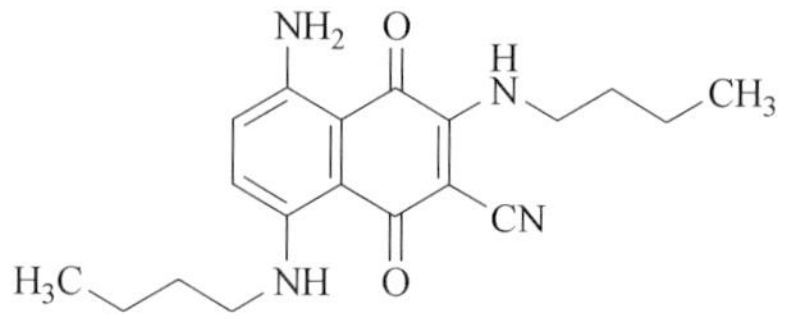

In recent years there has been considerable interest in 1,4-naphthoquinones that absorb in the near-infrared, particularly for application in optical recording materials and related high-technology areas. The first simple near-infrared 1,4-naphthoquinone to be described was the 5-amino-2,3-dicyano-8-phenylamino derivative **6** (X = H) [*68217-41-4*], which despite its relatively small molecular size has λ_{max} at 770 nm in dichloromethane [10]. The derivative **6** (X = $C_{12}H_{25}$) has been produced as Langmuir–Blodgett films which can be written on with 830-nm laser light with high sensitivity [16]. Other derivatives related to **6** have been patented extensively for optical recording discs [17] and have found commercial use.

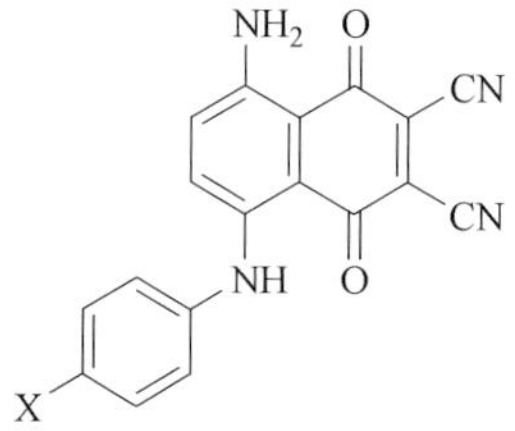

3.12.3.2 Heteroannelated 1,4-Naphthoquinones

A large number of quinones of the general structure **7** are known, where X corresponds to a heterocyclic residue containing one or more rings, and the chemistry of these has been reviewed extensively [6–8]. If X consists of a benzenoid or azabenzenoid ring system, then such compounds are best regarded as anthraquinone analogues. In the case of other heterocyclic residues, they may be considered as 1,4-naphthoquinone derivatives.

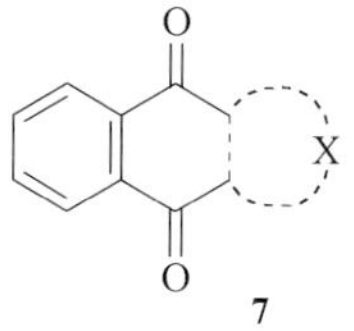

The simplest members of this group are those in which X is a monocyclic residue, such as furan, thiophene, and pyrrole ring systems. Color may then be produced in the usual way by attaching auxochromes to the 5- and/or 8-positions. Few dyes of this type have been suggested for textile applications, presumably because their chemistry is generally more complex than that of comparable anthraquinones. It appears that the color properties of such dyes differ only slightly from those of related anthraquinone structures, as exemplified by studies on imidazole dyes of general structure **8** [18]. However, triazole dyes of formula **9**, where R^1–R^3 are auxochromes, have been patented as disperse dyes for polyester [19].

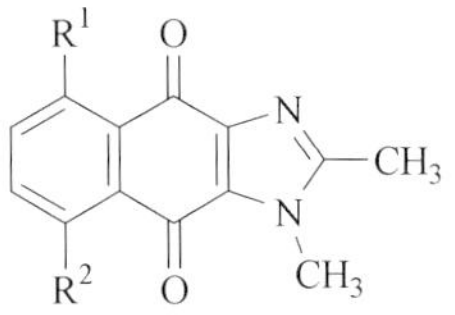

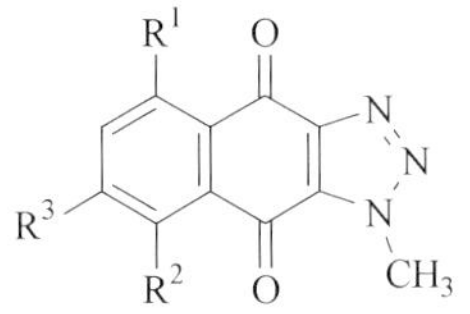

Of greater potential are dyes of general structure **7** in which X corresponds to a polycyclic heterocyclic system, and a large number of these have been examined as possible dyes and pigments. Perhaps the most closely investigated are the brazanquinones **10**, which have been known for more than eighty years. They can be prepared readily by condensing 2,3-dichloro-1,4-naphthoquinone [*117-80-6*] with substituted phenols and naphthols. For example, arylamides of 2-hydroxy-3-naphthoic acid, which are commercially available as coupling components for azoic dye combinations, condense to give yellow and orange vat dyes **11** with good substantivity for cotton [20].

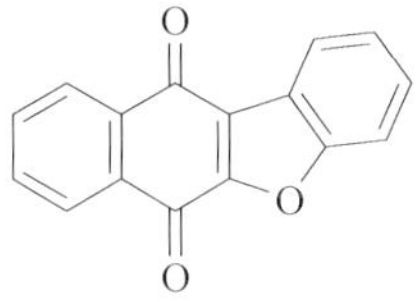

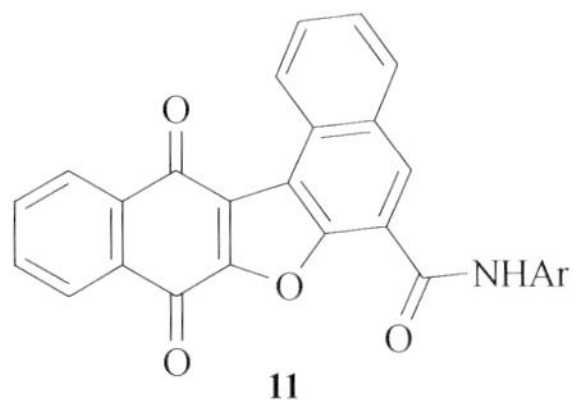

The brazanquinone system has also been used to provide disperse dyes by attaching auxochromes to the benzene ring of the naphthoquinone moiety. Thiophene analogues of the brazanquinones have received much less attention. Disperse dyes of the type **12** proved to be rather more bathochromic than corresponding anthraquinones, but of lower intensity [21]. Pyrrole analogues have also been neglected, but it is interesting to note that the old dye C.I. 56070 [*6451-05-4*] (**13**), a vat dye for cotton, is the parent compound of this type.

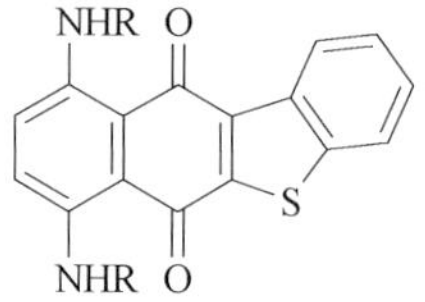

12

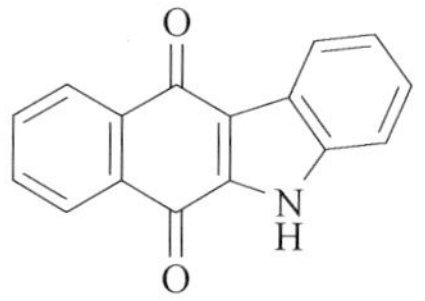

13

A different class of polycyclic naphthoquinone dyes is based on the naphth[2,3*b*]indolizine-6,11-dione system **14**, which is readily accessible by condensation of 2,3-dichloro-1,4-naphthoquinone with active methylene compounds in the presence of pyridine [22], or by reaction of 2-methoxy-3-pyridino-1,4-naphthoquinone with an active methylene compound [23]. In **14**, the bridgehead nitrogen atom acts as an effective auxochrome, and hence orange to red colors are observed without further substitution. Derivatives of **14** (R = amide group) are of particular value as vat dyes and pigments. Related isomeric heterocyclic structures have also attracted interest, e.g., **15**, a yellow disperse dye for polyester [24].

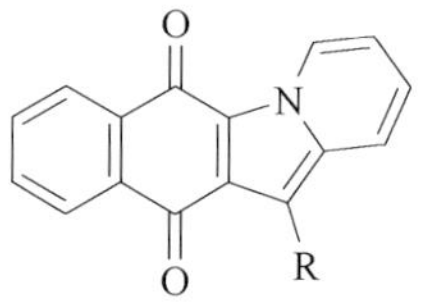

14

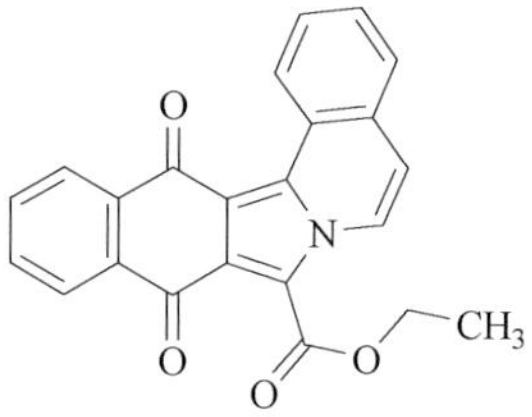

15

As with the simple 1,4-naphthoquinones, most recent research activity in the heteroannelated compounds has been directed towards the development of dyes that absorb in the near-infrared. Noteworthy is the dithiadinaphtho-1,4-naphthoquinone system **16** (X = S), which can be prepared by condensing 2,3-dichloro-5,8-dihydroxy-1,4-naphthoquinone with 2-aminothiophenol [25]. The parent compound **16** (X = S) [96692-25-0] shows multiple absorption bands between 400 and 800 nm, with the most intense peak at 725 nm in chloroform (ε_{max} 15 200 L mol^{-1} cm^{-1}) [25]. Oxidation of the sulfur atoms to SO groups produces a large bathochromic shift, and **16** (X = SO) shows a peak at 827 nm (ε_{max} 17 600 L mol^{-1} cm^{-1}) in chloroform [25].

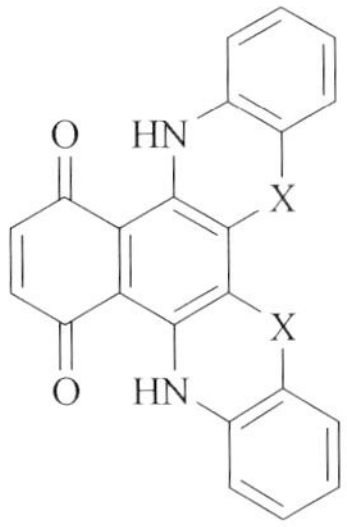

16

3.12.4 1,5-Naphthoquinones

Dyes of this class have the general formula **17**, in which R^1 and R^2 can be hydrogen, alkyl, or aryl groups. Historically and structurally, naphthoxidine (**17**, R^1 = R^2 = H) can be regarded as the parent compound; it is a violet-blue compound possessing strong intramolecular hydrogen bonding and exhibiting solubility in organic solvents and in water. This substance and its N-substituted derivatives have been represented variously as 1,5-naphthoquinones **17** and as the tautomeric 1,4-naphthoquinone imines **18**. Naphthoxidine itself, for example, is listed under two CAS registry numbers, i.e., [27823-83-2] and [6259-68-3] for the 1,5-quinone and the 1,4-quinone imine forms, respectively. Although definitive structural evidence has yet to be provided, the generally preferred formulation is the former, and the exceptionally low IR carbonyl frequencies of these compounds (i.e., below 1600 cm^{-1}) support this suggestion. Such a low frequency would be characteristic of a 1,5-naphthoquinone [26], particularly if intramolecular hydrogen bonding were present. Hydrogen bonding in a 1,4-quinone structure such as **18** would not account for such low frequencies, because, for example, 5-amino-1,4-naphthoquinones show absorptions for hydrogen-bonded carbonyl groups near 1630 cm^{-1}. In the following account, the 1,5-quinone structure **17** is assumed.

3.12 Naphthoquinone and Benzoquinone Dyes

17

18

Naphthoxidine is a very useful intermediate for dyes of this group, and it can be prepared conveniently by treating 1,5-dinitronaphthalene with sulfur in oleum [27], [28]. A solution of sulfur in 20 % oleum is added slowly over 35 min to a slurry of 1,6-dinitronaphthalene in concentrated sulfuric acid, maintaining the temperature at 50–55 °C. The dark brown mixture is stirred for a further 10 min at this temperature and then for 30 min at 20 °C. The mixture is then poured onto ice and the suspension filtered. The blue filtrate is made alkaline with sodium hydroxide solution, employing efficient external cooling in order to keep the temperature below 40 °C. The resultant blue precipitate of the sodium salt of naphthoxidine is filtered off and washed with a little water. The solid is extracted with boiling acetic acid, and concentration of the extracts gives lustrous red needles of naphthoxidine.

The most successful naphthoxidine-derived dye was **19** [26846-51-5], which is prepared by bromination of naphthoxidine in acetic acid in the presence of a catalytic amount of iodine or iron. The commercial dye is in fact a mixture of various brominated products, of which **19** is the predominant component. The product is a clear blue disperse dye suitable for acetate and polyester fibers.

Condensation of naphthoxidine with arylamines results in replacement of one or both primary amino groups by the amine, and the resultant *N*-aryl derivatives are greener than **17** ($R^1 = R^2 = H$) and can be used as disperse dyes. An example is **17** (R^1 = 4-ethoxyphenyl, R^2 = H) [1]. Arylamines react directly with **17** ($R^1 = R^2 = H$) in acetic acid, but the reaction is unsuccessful with alkylamines. The latter can be introduced by first reducing **17** ($R^1 = R^2 = H$) to a stable leuco compound with hydrosulfite and then condensing this with the alkylamine. The product is then oxidized to give the quinone dye. It appears that monocondensed dyes have greater affinity for cellulose acetate fibers than the bis-condensation products. Interest in the latter has been stimulated by the discovery that such compounds can provide useful dyes for liquid-crystal displays and exhibit high-order parameters (i.e., contrast ratios). For example, the compound **19** with $R^1 = R^2$ = 4-methoxyphenyl is a blue-green dye providing an excellent contrast ratio in liquid-crystal media [29].

3.12.4 1,5-Naphthoquinones

Brominated condensation products can be obtained by using appropriately halogenated naphthoxidine intermediates such as **19**. For example, **19** can be condensed with aniline in acetic acid at 90 °C to give **20** [1]. Naphthoxidine and its brominated derivatives can also be condensed with aminoazo compounds to give mixed azo–quinone chromogens, e.g., **21**. These are of interest as homogeneous green dyes [1].

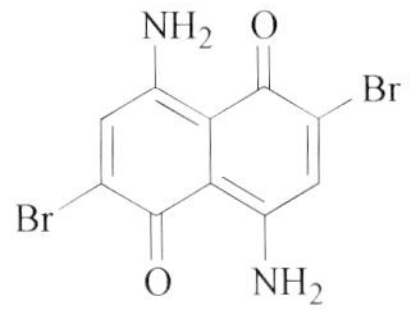

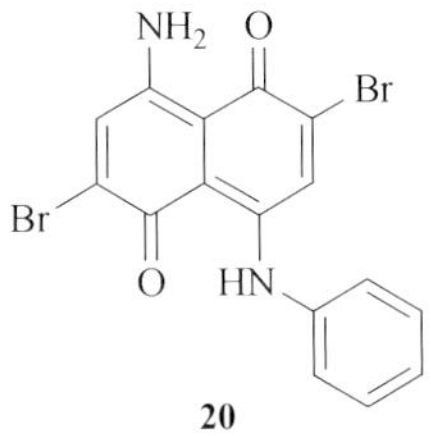

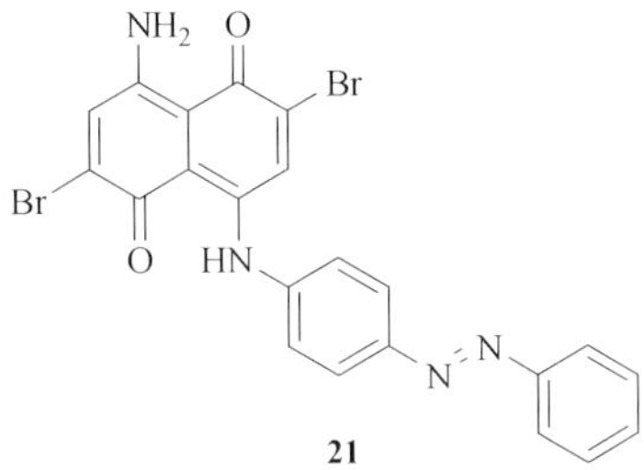

It is possible by modification of reaction conditions to effect the introduction of amino groups into the 2-position of naphthoxidine, but this generally produces dyes with reduced purity of shade and of lower lightfastness [1].

As in the case of the 1,4-naphthoquinones, there has been considerable interest in developing near-infrared dyes of this class. The dye **22** [*100012-51-9*] is essentially the 1,5-quinone analogue of **16** (X = S) and shows a long-wavelength peak at 750 nm in trichloromethane (ε_{max} 32 000 L mol^{-1} cm^{-1}) [30]. Thus, although **22** has the same molecular mass as **16** (X = S), it is both more bathochromic and more intensely absorbing.

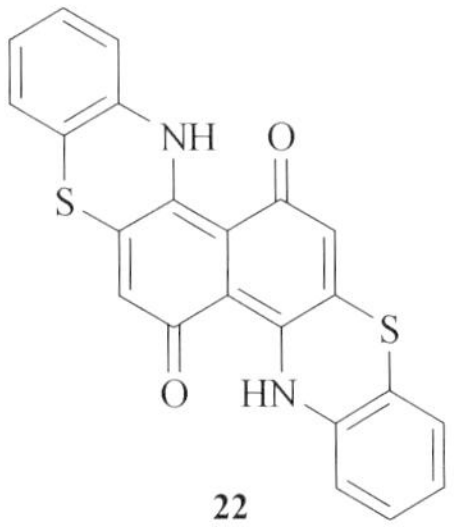

3.12.5 References

[1] E. Merian, *Chimia* 13 (1959) no. 6, 181; *Am. Dyest. Rep.* **48** (1959) 31.
[2] *Colour Index*, 1st ed. (1924), 2nd ed., Vols. 1–6 (1956), 3rd ed., Vols. 1–6 (1971), The Society of Dyers and Colourists, Bradford, England.
[3] R. H. Thomson: *Naturally Occurring Quinones*, Butterworths, London 1957.
[4] *Houben-Weyl*, **7** (**3a**).
[5] S. Patai (ed.): *The Chemistry of the Quinoid Compounds*, Parts 1 and 2, J. Wiley & Sons, London–New York, 1974.
[6] M. F. Sartori, *Chem. Rev.* **63** (1963) 279.
[7] I. Baxter, B. A. Davis, *Q. Rev. Chem. Soc. (London)* **25** (1971) 239.
[8] B. D. Tilak in K. Venkataraman (ed.): *The Chemistry of Synthetic Dyes*, Vol. V, Academic Press, New York–London, 1971.
[9] K. Y. Chu, J. Griffiths, *J. Chem. Soc. Perkin Trans. 1* **1978**, 1083.
[10] K. Y. Chu, J. Griffiths, *J. Chem. Res. Synop.* **1978**, 180; *J. Chem. Res. Miniprint* **1978**, 2319.
[11] K. Y. Chu, J. Griffiths, *J. Chem. Soc. Perkin Trans. 1* **1979**, 696.
[12] C. Blackburn, J. Griffiths, *J. Chem. Res. Synop.* **1983**, 168; *J. Chem. Res. Miniprint* **1983**, 1556.
[13] L. K. Vinograd, S. M. Shein, A. P. Cherepivskaya, G. V. Shalimova, *Zh. Prikl. Khim. (Leningrad)* **38** (1965) 208; *Chem. Abstr.* **62** (1965) 11941e.
[14] Sandoz, US 2 764 600, 1956 (E. Merian).
[15] Mitsubishi Chemical Industries, JP 57 145 157, 1982.
[16] D. Heard, G. G. Roberts, J. Griffiths, *Thin Solid Films* **180** (1989) 305.
[17] NEC, JP 60 149 490A, 1983. NEC, JP 60 150 241A, 1983. NEC, JP 60 150 242A, 1983, NEC, JP 60 161 192A, 1984. NEC, JP 60 190 388A, 1984.
[18] G. Green-Buckley, J. Griffiths, *J. Chem. Soc. Perkin Trans. 1 1979*, 702.
[19] American Cyanamid, US 2 967 863, 1961 (M. Scalera, A. S. Tomcufcik, W. B. Hardy).
[20] B. D. Tilak, V. K. Dikshit, B. Suryanarayana, *Proc. Indian Acad. Sci. Sect. A* **37** (1953) 92.
[21] A. T. Peters, D. Walker, *J. Chem. Soc.* 1956, 1429.
[22] E. F. Pratt, R. G. Rice, R. W. Luckenbaugh, *J. Am. Chem. Soc.* **79** (1957) 1212.
[23] J. A. VanAllan, R. G. Reynolds, R. E. Adel, *J. Org. Chem.* **28** (1963) 3502.
[24] Sandoz, DE 2425662, 1974 (B. L. Kaul).
[25] K. Takagi, M. Kawabe, M. Matsuoka, T. Kitao, *Dyes Pigments* **6** (1985) 177.
[26] L. K. Horst, H. Kratzin, P. Boldt, *Liebigs Ann. Chem.* 1976, 1560.
[27] Du Pont, US 2238959, 1941 (M. S. Whelen).
[28] K. Y. Chu: Ph.D. Thesis, University of Leeds, 1977.
[29] Merck, DE 3126108, 1983 (G. Haas, G. Weber).
[30] K. Takagi, S. Kanamoto, K. Itoh, M. Matsuoka, S. H. Kim, T. Kitao, *Dyes Pigments* **8** (1987) 71.

4 Textile Dyeing

4.1 Introduction

Production of colored textiles is one of the basic technologies in human civilization. Textile consumption is steadily increasing worldwide, following the growth of world population and stimulated by a growing GDP in many countries, primarily in Asia. Synthetic fibers increased their share, particularly polyester fibers see (Table 4.1).

Table 4.1 World fiber consumption (in 10^6 t) in 1998 and average growth rate for 1991–1998 (in %)

Fiber	Consumption	Growth
Wool (Wo)	1.5	–1.8
Cellulosics (CEL)	21.2	0.1
Polyamide (PA)	3.4	1.8
Polyacrylic (PAC)	2.5	0.6
Polyester (PES)	16.0	8.7
Others, primarily olefin	3.0	5.2
Total	47.7	2.5

Cellulosics and polyester together account for 78 % of world textile consumption. Following this trend, disperse dyes and dyes for cellulosic fibers reclaimed the principal market share (Table 4.2).

Table 4.2 World textile dye market in 1998

Dye class	Market share, 10^9 DM	Market share, %
Disperse dyes	2.3	28
Reactive dyes	2.3	27
Vat dyes	0.8	10
Indigo	0.3	4
Sulfur dyes		6
Direct dyes		10
Naphthols		4
Others (anionic, cationic, etc.)		12
Total	8.5	100

In preindustrial times, textiles were dyed primarily with plant dyes. Dyeing with animal dyes (kermes, Tyrian purple) was restricted to special cases. Today, synthetic dyes are used almost exclusively. However, use of natural dyes is again increasing [1, 2, 3, 4].

4.1.1 Dyeing Technology

For details, see [5, p. 355–390], [6]. The goal of every dyeing is a colored textile in the desired shade, homogeneous in hue and depth of shade, produced by an economic process and which exhibits satisfactory fastness properties in the finished state.

Colored textiles are produced today on a large industrial scale. Although modern automation techniques have been introduced for color measurement, metering of dyes and auxiliaries, and automatic control of the dyeing process, much human intervention is still required. Fibers can only be standardized to a limited extent, due to biological and environmental factors, e.g., in growing cotton or raising sheep. And new developments in fashion and application of textiles require constant modifications of finishing procedures. To remain flexible with regard to fashion and fastness properties, dyeing is carried out at the end of the production process whenever possible.

The textile material generally needs a pretreatment before dyeing. Wool must be washed to remove wax and dirt and is sometimes bleached, cotton must be boiled and bleached to remove pectins and cotton seeds and is mercerized. Sizes and spinning oils must be eliminated [5, p. 253–266].

4.1.1.1 Principles of Dyeing

Basically there are three methods of dyeing textiles:

1) Mass dyeing: dyeing of a synthetic polymer before fiber formation.
2) Pigment dyeing: affixing an insoluble colorant on the fiber surface with a binder.
3) Exhaustion dyeing from an aqueous bath with dyes that have an affinity for the fiber.

Mass and pigment dyeing are not considered here.

In exhaustion dyeing, the dye, which is at least partially soluble in the dyebath, is transported to the fiber surface by motion of the dye liquor or the textile. It is then adsorbed on the fiber surface and diffuses into the fiber. Finally, depending on the dye–fiber interaction, it is fixed chemically or physically. The dye can be applied to the textile discontinuously from a dilute solution (exhaustion dyeing from a long liquor) or continuously by immersing the textile in a concentrated bath and squeezing-off excess liquor (padding), followed by separate steps for diffusion and fixation in the fiber.

Water is almost exclusively the liquid phase which transports the dye in textile dyeing. For most dyeing processes demineralized water should be used; if not available, complexing agents must be added, where necessary, to avoid negative interference of metal ions.

For exhaustion dyeing, recipes are made up in percent relative to the weight of the textile goods. For continuous processes the amounts of dyes and auxiliaries used are given as g/L of the dye liquor and have to be based on a predetemined liquor add-on per unit weight of the textile material. The speed of exhaustion (dyeing rate) of individual dyes can vary widely, depending on their chemical and physical properties and the type of textile material. The dyeing rate depends on factors such as temperature, liquor ratio, dye concentration, and the chemicals and auxiliary products in the dye bath.

High dyeing rates bear the danger of unlevel dyeings. Therefore, dyes have to be carefully selected when used together in one recipe. The dye producers readily communicate their knowledge on dyeing characteristics of their dyes and on the particular requirements for all textile fibers.

The end of the dyeing process is characterized by the equilibrium phase, when dye concentration in the fiber and in the liquor do not change any more. Under standard conditions, the distribution coefficient of the dye between liquor and fiber is constant; in other words, the rates of adsorption and desorption are equal. The equilibrium phase is frequently necessary and may be extended to level out by migration inhomogeneities incurred during the exhaustion phase. When the dyeing is carried out continuously, it is of prime importance to achieve perfectly homogeneous dye application and to avoid migration during subsequent steps, e.g., when drying. Leveling out a dyeing after fixation of the dye is tedious and time-consuming. Stripping a faulty dyeing is difficult in most cases and should be avoided.

4.1.1.2 Bath Dyeing Technology [7, 8, 9]

High dyeing rates and good leveling properties of the dyes are normally reached at high temperatures. To achieve short dyeing cycles modern equipment is enclosed to be as near as possible to the boiling temperature of the dye liquor. Most discontinuous dyeing equipment even permits dyeing under pressure at temperatures up to 135 °C. Two types of machines exist: *circulating machines*, in which the goods are stationary and the liquor circulates, and *circulating-goods machines*, in which the textile material and normally also the liquor are in motion.

Circulating Machines. The goods — flock, card sliver, tow, yarn, or fabric — are packed loose, on cones, or perforated beams, and the liquor is pumped through the goods. The pump characteristics and the density of the material determine the circulating speed of the liquor and hence the necessary dyeing time. The geometry of the machine and homogeneous packing of the material govern the levelness of dye absorption and migration.

Circulating-Goods Machines. Traditional dyeing equipment belongs to this group, such as winch becks, where fabric is drawn through a dye beck by way of a winch; or the jigger, where fabric is mounted in whole width on rolls and guided back and forth through a dyebath. There exists a large variety of modern modifications of these principles. As a rule, the machine should be closed, and even be pressurized to permit dyeing above the boiling temperature. Fabric is moved as a rope or in open width by mechanical means or by a liquor jet produced by a circulating pump, or just by pumping the liquor in one direction determined by the geometry of the dyeing vessel. Also air jets are used to transport the fabric, permitting a very short liquor ratio and thus saving energy and wastewater. For particular articles, such as hosiery, carpets, finished garments, specially adapted dyeing machines have been developed [10].

Process Control in Bath Dyeing. Technical developments in the construction of dyeing machinery and process control are geared to maintaining the pristine properties of the goods, to reduce dyeing time, and improve productivity. Since bath dyeing runs discontinuously, automatic process control must work in cycles. A completely automated dyeing process is almost impossible to achieve, because so many variables determine the result of a dyeing, and a wide range of operating factors interact with each other during dyeing. Most standard dyeing equipment today provides a time–temperature program. Also other functions are widespread, such as pH and concentration control, filling and empting the dyeing vessel, flow rate control, and automatic metering of dyes and chemicals [11]. Process development based on local data of equipment, dyes, and chemicals is beginning.

A precondition for automatically controlling the dyeing process is detailed knowledge of the characteristics of the fiber to be dyed, the dyes and auxiliaries to be used, and the equipment available. To assure level dyeing from the beginning the exhaustion curves for the dyes combined in one recipe must be controlled. This requires constant color measurement of dye concentration [12, 13, 14].

4.1.1.3 Continuous and Semicontinuous Dyeing

Continuous dyeing means treating fabric in a processing unit in which application of the dye to the fabric and fixation are carried out continuously. As a rule, continuously working units are assembled into lines of consecutive processing steps, sometimes including pretreatment of the fabric. The following principles apply to all continuous dyeings:

– Fabric is treated in open width.
– Any unevenness in the equipment across the width of the goods leads to unlevel dyeings.
– The width of the goods and longitudinal tension influence each other (streching, shrinking).
– The running speed determines the dwell time in the treatment unit.
– Interruptions cause spoiled fabric.

Application of the Dye. Two types of dye application techniques exist:

1) Direct application of dye liquor by spraying, foam application, or printing.
2) Continuous immersion of the fabric in a dyebath and removal of the excess of liquor by squeezing or suction (padding).

The most important technique is the *pad process*, in which the dry fabric is passed through a pad trough. Good wetting of the fabric is imporant. The dwell time in the padder is adjusted according to the speed of swelling of the fiber (e.g., cotton). The amount of liquor remaining on the fabric after squeezing depends on the type of fiber and the construction of the fabric. A low pickup is often desirable to minimize migration and to save energy during drying. Preferential adsorption of dyebath components can lead to unevenness over the length of the batch to be dyed. It may be caused by substantivity of dyes, which leads to depletion of the dye content in the padding liquor; other factors may include reaction of reactive dyes with alkali, precipitation and mutual interactions of dyes, and transfer of impurities by the fabric from a pretreatment step into the padder.

Intermediate Drying. To assure uniform fixation of the dyes, the fabric is often dried before the fixing step. During drying, the dye liquor migrates to the surface of the fabric, which can increase the visual color yield. This process ends when the liquor content of the fabric is below 30 %. The drying equipment normally works with infrared heat or with a hot air stream or a combination of both. Drying should start contact-free to avoid smearing of the fabric and soiling of the equipment. Dryers should be optimized for energy consumption, in particular to prevent losses of hot air.

Dye Fixation. On the dried fabric, the dye is only deposited on the fiber surface. It must penetrate into the fiber during a fixation step and be incorporated in the fiber by chemical reaction (reactive dyes), aggregation (vat, sulfur dyes), ion-pair formation (acid, cationic dyes), or in the form of a solid solution (disperse dyes).

Fixation is performed by steaming under various conditions, depending on the fibers and dyes involved. Generally, saturated steam at ca. 100 °C is applied. Pressure steamers permit a reduction of the steaming time. Disperse dyes are fixed in polyester fibers by the Thermosol process; the fabric is heated in a infrared unit by radiation, by hot air, or with contact heat to 210–220 °C for 30–60 s. Superheated steam facilitates rapid fixation at lower temperatures.

Afterwards, fabrics are usually washed to remove unfixed dye and chemicals. Vat and sulfur dyes must be reoxidized.

Continuous Dyeing Plants. A continuous production line must be designed for a particular type of textile material. A continuous line must provide for

- synchronous throughput,
- adequate longitudinal tension,
- maintenance of open width of the fabric,
- no longitudinal creases,
- continuous metering of the products.

An exemple is a thermosol pad steam plant for dyeing PES–CEL blends with vat and disperse dyes:

- padder for dyes and chemicals,
- infrared predrying shaft,
- hot-flue dryer,
- thermosol unit,
- chemical padder for vatting,
- steamer,
- cold rinsing bath, followed by oxidizing and neutral washing in a washing equipment,
- squeezing and drying in hot air.

Instead of a continuous procedure the dye can be fixed on the fiber after application by batching on a roll at a controlled temperature. For reactive dyes a cold batch process is widely used.

There exists a highly developed technology for the automation and process control of different stages of continuous dyeing. It differs from the control of bath dyeing in that constant conditions must be maintained and no time program is required.

4.1.1.4 Printing

A wide variety of techniques exists for applying dyes by printing. For details see [5 p. 498–512] and [15].

Direct printing is the application of a printing paste containing dyes, thickeners, and auxiliaries directly to the fabric by *roller printing* (on engraved cylinders) or *screen printing* (with flat or rotating screens). The dominant technique is screen printing (ca. 80 %).

Recently methods were developed for computer-controlled *ink-jet printing* [16, 17]. Production speed is still low (less than 20 m/min). On the product side, no problems arise with soluble dyes. However, disperse dyes and pigments must be very finely distributed to avoid blocking of the jets. Ink-jet processes are already well established in carpet printing (space dyeing).

In *transfer printing* the design is first printed on paper with dyes sublimable in the range of 180–230 °C. The paper can be stored, and the pattern may later be transferred by contact heat to textiles made of synthetic fibers. (see Section 4.12.2.7) No afterwash is necessary.

Two-phase printing is the separate application of dyes and fixing agent in reactive and vat printing.

In *discharge printing* the fabric is dyed with a dischargeable dye and then printed with a discharge paste in the desired pattern. The discharge paste may contain a discharge-resistant dye (color discharge printing).

Special printing techniques were developed e.g. for floor coverings (spray printing, space dyeing), and for yarn or wool slubbing (Vigoureux printing).

The most important fibers for printing are cellulosics (70 %) and polyester (20 %). All others account for less than 10 %.

4.1.1.5 Dispensing Dyes and Chemicals

The dyer is obliged to follow precisely the dyeing formulation worked out previously for a particular color. In critical shades like gray or beige, variations of 2–3 % in color strength and hue can be seen by the human eye. The amount of dye required for certain shades can vary from ca 0.001 % to 10 %, relative to the amount of textile material; and for most shades more than onr dye is required. In addition, the dyer must consider a great number of dyes and auxiliaries necessary for different textile materials, and a large scale of colors to be dyed. Automatic dispensing is a considerable help. For this purpose either liquid preparations of dyes are used, supplied by the dye manufacturer, or the dyer prepares masterbatches which can be dispensed with metering pumps or flow meters [18].

Similar strict requirements for accuracy apply to chemicals and dyeing auxiliaries. Most auxiliaries show a distinct affinity for fibers or dyes. Their amount must therefore be strictly calculated and dispensed to the requirements of each individual dyeing. Today metering devices exist in the dyehouse which permit the dispensing of widely differing volumes and weights.

4.1.2 Standardization of Textile Dyes

The dyers task is to give a visual impression to the human eye. The eye is a very sensitive organ that can detect very slight color differences. Since dyes are industrial products, they may contain impurities and colored or colorless byproducts, depending on their production process. Therefore, after production dyes must be standardized with regard to their later use. Dyes are standardized not to a certain chemical composition, but to a defined coloristic value. Consequently, they may be composed of mixtures of dyes and may contain shading components. For example, all disperse and cationic blacks are mixtures of dyes. In addition, dyes contain various amounts of dispersing agents, as in the case of vat or disperse dyes, neutral salts in direct or acid dyes, solvents for liquid preparations, etc. Hence, dyes which, according to their listing in the Colour Index, should be chemically identical can, as commercial products, vary significantly in their overall composition, including dye strength, solubility, dispersion stability, etc. [5, p. 459–461].

A committee formed by specialists of dye producers has set up testing procedures for textile dyes [19]:

I. Relative color strength in solution.
II. Relative color strength and residual color difference by reflectance measurements.
III. Solubility and solution stability.
IV. Electrolyte stability of reactive dyes.
V. Viscosity of liquid dyes.
VI. Dispersion behavior and dispersion properties.

Color Strength, Hue, Chroma [20, p. 99–105]. The coloring properties of a dye are assessed by preparing a dyed test sample whose color is evaluated. This must, in principle, always be done by the human eye because color perception, being a subjective sense impression, is not accessible to direct measurement. However, with the aid of colorimetry this visual perception can be represented more or less closely by measurable quantities. Since colorimetry is an objective method and is therefore more accurate and reproducible than subjective visual assessment, it is very widely used today. Color is a three-dimensional quantity and must therefore be expressed by a set of three numbers (color coordinates). In practice, these are typically the values of color strength, hue, and chroma.

Color can be described with any accuracy only in a comparative way (i.e., in terms of deviations from a standard). Although the three-dimensional color coordinates can be determined by colorimetry (as absolute values), this is in practice not sufficient to provide an exact specification. In the case of a dye, the reference is a standard dye sample provided by the producer against which all further supplies are controlled and standardized. To be used as a basis for assessment the standard sample must be extremely stable.

To test a dye, dyed test specimen from both the sample dye and the standard are compared under conditions that are as identical as possible and assessed visually or colorimetrically and then calculated in relative terms to the standard. A widely used method to calculate color strength is a determination based on the weighted sum of the K/S values (K is the coefficient of absorption, S the coefficient of scatter), which is used in standard specifications [21].

Dyers can in practice compensate for differences in color strength by adjusting the concentration. Therefore, the hue and chroma of a dyeing are always judged against a standard dyeing of equivalent color strength. This is carried out by visual assessment in which the colorist estimates the remaining differences in hue and chroma for a strength-corrected dyeing. In a colorimetric evaluation, the residual color difference is determined by a mathematical correction to give equivalence in color strength. The result is expressed in CIELAB color difference units (DIN 6174) and is usually separated into components for hue (ΔH^*) and chroma (ΔC^*) [22, 23]. In a recent development, neural networks have been used as a reliable tool to predict the result of a visual assessment by reflection measurement [24].

Frequently, color strength is determined by the simpler and more easily reproducible measurement of the extinction of the dye solution. However, in many cases this does not agree with the values determined by dyeing because dyes in solution may not follow the Lambert–Beer law and are not always absorbed quantitatively by the fiber [25].

Solution and Dispersion Behavior. For the dyeing process in aqueous liquor, the dye must have adequate solubility or dispersibility. In general, good solubility is necessary for good application properties. If the solubility is poor (i.e., if any of the dye is present in the dye liquor in the form of undissolved particles), local coloration (specks), spots, uneven effects, and poor fastness can be produced, leading to serious defects and costly complaints.

In addition to solubility properties, the stability of the liquor during preparation and use (i.e., the *solution stability*) is important. For reactive dyes the solution stability in the presence of electrolytes is important (see Test Methods ISO 105-Z07-Z09). For disperse dyes the dispersion properties are of similar importance, but they are not easy to describe because of the complex processes and unstable dispersion states. Therefore, a large number of test methods exist, the results of which are usually limited in their application. A critical review can be found in [26].

Dusting. The dust produced from dyes can be a health hazard, and severe soiling of the textile material can be caused by very small amounts of dust particles. The content of fine dust (1–10 µm) must be reduced by adhesive bonding or prevented by the use of special formulations (granulation) of the dye. The amount of dust depends not only on the dusting behavior of the dye, but also on the nature and intensity of handling operations that generate dust. Therefore, the dusting behavior of a product is significant only for comparison with other products and cannot be used to predict dust exposure. In a widely used test method, the dust produced when a defined amount of dye is allowed to fall is determined quantitatively or qualitatively (ISO 105-A03).

4.1.3 Colorfastness of Textiles

Dyed textiles are exposed to a variety of treatments during subsequent manufacturing steps and later in daily use. Standard test methods have been developed to evaluate the colorfastness, i.e., resistance of dyed textiles to the conditions they may endure. These test methods are maintained by the International Organisation for Standardization (ISO), Technical Committee 38, Subcommittee 1 (TC/SC1). The Standard Test Methods are published as ISO Standard 105 [27].

Change in color and staining of undyed adjacent fabric are assessed visually or colorimetrically by comparison with a gray scale of dyed fabric (ISO 105-A02-A04). The results are recorded on a numbered scale:

Lightfastness and colorfastness to weathering: 1 (poor) to 8 (optimum).
All other fastness tests: 1 (poor) to 5 (optimum).
The color fastness tests are grouped according to different treatments of dyed textiles:

ISO 105-A	General
ISO 105-B	Colorfastness to Light, Weathering, and Photochromism
ISO 105-C	Colorfastness to Washing and Laundering
ISO 105-D	Colorfastness to Dry Cleaning
ISO 105-E	Colorfastness to Aqueous Agencies
ISO 105-G	Colorfastness to Atmospheric Contaminants
ISO 105-J	Measurement of Color Differences and Whiteness
ISO 105-N	Colorfastness to Bleaching Agencies
ISO 105-P	Colorfastness to Heat Treatments
ISO 105-S	Colorfastness to Vulcanization
ISO 105-X	Colorfastness to Miscellaneous Agencies
ISO 105-Z	Colorant Characteristics

Standard fastness properties of dyes on various textiles are given in the Colour Index [28].

4.1.4 Laboratory Dyeing Techniques

A modern dyehouse is not operable without support from laboratory work. The laboratory's tasks are among others:

- Working out formulations for new shades, supported by colorimetry and computer color matching.
- Production tests on the textile material to be dyed; fastness testing.
- Maintaining consistency of production, adjudicating claims.
- Establishing tolerance ranges.
- Dye selection in view of compatibility, fastness properties, dyeing rate, time requirements, and cost.
- Development of appropriate dyeing techniques, suitable for new types of textiles and novel styling requirements.

Laboratory dyeing technique should simulate actual production conditions as closely as possible. It also should allow for a multiplicity of tests in a short period of time. Laboratory equipment permits dyeing of small (5 g) to larger (1 kg) textile samples. Most lab dyeing machines work batchwise, but installations for continuous operation are also available [29].

4.2 Reactive Dyes on Cellulose and Other Fibers

Reactive dyes are the newest class of dyes for cellulose fibers (see Section 3.1). ICI introduced the first group of reactive dyes for cellulose fibers (with 2,4-dichloro-1,3,5-triazine anchor) in 1956. In the dye molecule, a chromophore is combined with one or more functional groups, the so-called anchors, that can react with cellulose. Under suitable dyeing conditions, covalent bonds are formed between dye and fiber. One-third of the dyes used for cellulose fibers today are reactive dyes. The range of available reactive dyes is wide and enables a large number of dyeing techniques to be used. Shades ranging from brilliant to muted can be obtained. They have better wetfastness properties than the less expensive direct dyes. Chlorine fastness is slightly poorer than that of vat dyes, as is lightfastness under severe conditions.

In addition to cellulose, many other fibers can be dyed with ractive dyes, provided they have chemical groups capable of forming a chemical bond with the reactive dye, e.g., wool or polyamide fibers (see Section 4.2.4).

4.2.1 Fundamentals [30, 31]

In general, dyeing with reactive dyes is very similar to the well-known process of direct dyeing (see Section 4.3). The major difference is that the reactive dye forms a chemical bond to the fiber. Important factors determining the dyeing properties of reactive dyes include affinity of the dye for the fiber (substantivity), its diffusibility, reactivity, and the stability of the dye–fiber bond.

The anchor components of reactive dyes determine both the fixing properties and the wetfastness of the dyed material. More than 300 electrophilic anchors for reactive dyes are described in the patent literature. Of this large number, fewer than ten are currently of practical importance for dyeing cellulose fibers (Figure 4.1).

Hot dyers	Warm dyers	Cold dyers
80 °C	60 °C	40 °C

		Dichlorotriazine
		Difluorochloropyrimidine
	Monofluorotriazine	
	Dichloroquinoxaline	
	Vinylsulfone	
Monochlorotriazine		
Trichloropyrimidine		

───── Reactivity ─────▶

Figure 4.1 Reactivities of established anchor systems and optimum dyeing temperatures in exhaustion dyeing.

Hydrolysis and Reactivity. Two possible reaction mechanisms exist: addition and substitution.

Addition anchors include, for example, vinylsulfones [Eq. (1)].

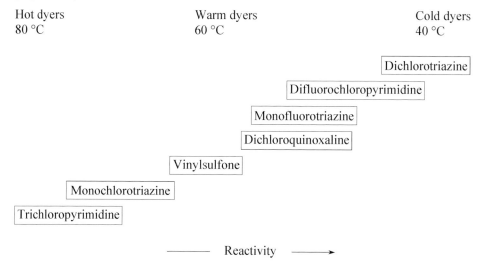

Substitution anchors include, for example, monochlorotriazines [MCTs; Eq. (2)].

$$\text{Dye}-\underset{\underset{Cl}{N}}{\overset{\overset{NHR}{N}}{\underset{\|}{N}}}\! + \text{Cellulose}-\text{O}^- \longrightarrow \text{Dye}-\underset{\underset{O-\text{Cellulose}}{N}}{\overset{\overset{NHR}{N}}{\underset{\|}{N}}}\! + \text{Cl}^- \quad (2)$$

As *substitution anchors*, heterocycles with a reduced π-electron density in the heterocyclic ring are used. The rate of hydrolysis or alcoholysis (i.e., the reaction with the cellulose anion) is influenced by the nature of the heterocyclic nucleus, the substituents, and the leaving groups. In the reaction with the cellulose anion, the substitution anchors listed in Figure 4.1 are bonded by ester linkages. In contrast, cellulose adds to the vinylsulfone anchor in a nucleophilic reaction, and an ether link is formed.

Hydrolysis and alcoholysis are competitive processes, both being first-order reactions [32]. The ratio of the rate constants of the reactions with cellulose and with hydroxide ions (ratio cellulose-O$^-$/OH$^-$) is termed selectivity and can be determined easily in a homogeneous medium by using model substances. However, the differences in selectivities are slight. The preference of the anchor for the fiber is explained by the greater nucleophilicity of cellulose compared to hydroxide. At high pH, the cellulose-O$^-$/OH$^-$ ratio decreases [33]. Furthermore, hydrolysis has a higher activation energy than alcoholysis (i.e., it is favored at higher temperature). As a result, the selectivity with respect to alcoholysis decreases with increasing pH and temperature. The rate of hydrolysis increases as a rule by a factor of 10 per pH unit. The rates of both reactions increase by a factor of approximately 2–3 per 10 °C. Therefore, pH and temperature can be varied in practice only within narrow limits. Depending on the optimum dyeing temperature, established anchor systems are classified as cold, warm, and hot dyers (see Figure 4.1).

In addition to the fastness and application properties the *fixation yield* is most important, i.e., the percentage of dye chemically bound to the cellulose at the end of the dyeing process relative to the total amount of dye employed. The fixation yield depends on the chemical nature of the anchor and many other factors [34]. It determines the economics of the dyeing process.

Substantivity is given by the ratio of the amount of dye in the fiber to the amount in the dye liquor after dyeing equilibrium has been reached (without chemical reaction). In analogy to direct and vat dyes, the adsorption properties of reactive dyes can be represented by the Freundlich relation. Substantivity is a necessary precondition for reaction with the fiber. Dye that has not become attached to the fiber is hydrolyzed when the alkali is added and lost to the dyeing process. For monoanchor dyes, high substantivity correlates closely with attainable fixation yields. Substantivity is influenced by the chemical constitution of the dye and the process parameters. In general, high substantivity is obtained with low dyeing temperature, a low liquor ratio, the addition of electrolytes (e.g., NaCl or Na$_2$SO$_4$) to the dyebath, and low dye concentration. An extended planar structure of the dye molecule is believed to favor high substantivity.

The dyer must also consider the effect of the substrate. Substantivities for mercerized cotton and regenerated celluloses are higher than those for unmercerized cotton.

Limits to substantivity exist in practice. If the substantivity is too high, the final shade may be uneven; it may be difficult to wash off unfixed dye; and the wetfastness of the dyeing may suffer. In continuous processes, tailing can occur (see Section 4.2.2).

Diffusion Rate is the rate at which dye is transported to the interior of the fiber material. A high diffusion rate leads to rapid establishment of dyeing equilibrium and rapid leveling out of irregularities in dye absorption. It also facilitates washing off the hydrolyzed dye. Increased dye substantivity is frequently combined with lower diffusibility. The diffusion rate doubles with a temperature increase of 10–20 °C.

The substrate also has an important influence on diffusion of the dye. The diffusion rate increases with caustic soda treatment or mercerization of cotton. However, with regenerated cellulose fibers, which have a marked skin–core structure, the outer parts of the fiber can act as a diffusion barrier.

Stability of the Dye–Fiber Bond. Because of the large variety of reactive dyes, generalizations about colorfastness are difficult. While wetfastness is determined mainly by the anchor system used, most other fastness properties depend on the dye as a whole or the chromophore present. Most reactive dyes are azo or anthraquinone derivatives whose standard of fastness varies greatly. Phthalocyanine, formazan, and triphenodioxazine derivatives are also very important. In addition, application conditions and finishing processes of the dyed goods can affect fastness properties. Thus, with some resin-finished textiles (dimethylolpropyleneurea finish) a decrease in lightfastness is observed.

Significant differences in the stability of the bond toward hydrolysis can be observed in the anchor systems listed in Figure 4.1. The ether bond to cellulose formed by addition anchors (vinylsulfone dyes) is extremely stable to acid hydrolysis. Its instability in alkaline solutions leads to poor fastness to hot sodium carbonate solution. In general, the stability of substitution anchors is significantly higher in alkaline solution. However, fastness can be reduced considerably by the presence of electron-withdrawing substituents in the anchor. An example is provided by dichlorotriazines (DCTs), which are more reactive than MCT dyes. The hydroxyl group formed by the second chlorine atom after reaction of the first with cellulose leads to a significant increase in the rate of cleavage of the dye–fiber bond [Eq. (3)].

$$\text{Dye--}\underset{\underset{Cl}{N}}{\overset{\overset{Cl}{N}}{\underset{\|}{C}}}\! \xrightarrow[-Cl]{Cellulose-O^-} \text{Dye--}\!\!\!\!\!\!\!\!\!\!\!\!\!\!\! \xrightarrow[-Cl]{+OH^-} \text{Dye--}\!\!\!\!\!\!\!\!\!\!\!\!\!\!\! \tag{3}$$

One aim of the introduction of heterobifunctional anchor systems has been to combine the good fastness profiles of different anchor systems (see Section 3.1).

4.2.2 Dyeing Techniques for Cellulose

Currently, a wide range of reactive dyes of varying constitution is available which are suitable for many different applications [33], [6, p. 510–516]. In most cases, dye is fixed to the substrate under alkaline conditions. However, dyes with phosphonate groups are applied in an acid dyeing bath, and quaternary nicotinic acid derivatives require neutral conditions. These dyes are used mainly for dyeing PES–CEL blends.

The suitability of a reactive dye for a particular application depends not only on its reactivity, but also on properties that are determined by its overall structure, such as solubility, substantivity, and diffusion. Special recommendations are therefore made by manufacturers for the application of each type of dye.

About 20 % of reactive dyes are used in textile printing, 30 % in pad dyeing, and 50 % in exhaustion dyeing processes.

Exhaustion Dyeing. Equipment for exhaustion dyeing consists, e.g., of package dyeing or hank dyeing machines, jiggers, winch becks, or dye-spraying machines (Section 4.1.1). The general trend is toward the use of concentrated (short) liquors. With woven textiles, liquor-to-goods ratios of 3:1 can be achieved today; with knitted goods, 5:1.

In the exhaustion dyeing process, highly or moderately substantive dyes are usually employed. The optimum dyeing conditions depend on the reactivity of the dyes. *Cold dyers* are dyed at 30–50 °C and a pH of 10–11; *hot dyers* at 70–90 °C and a pH of 11–12. For pH control normally mixtures of sodium carbonate and sodium hydroxide are used. To enhance substantivity of the dye sodium sulfate or sodium chloride is added. For less than 0.5 % dye based on textile weight, salt concentrations of 10–30 g/L are recommended. For deep shades (more than 4 %), ca. 50 g/L is used; with vinylsulfone dyes having low substantivity, up to 80 g/L.

In the *all-in process*, the dye, the salt, and the alkali are all added to the dyeing bath at the start of dyeing. The *stepwise process* consists of an exhaustion and a fixing phase, i.e., alkali is added only after the dye has been absorbed on the fiber. The stepwise process has advantages with goods that are difficult for the liquor to penetrate (e.g., dyeing wound packages in a circulating-liquor machine). A preceding exhaustion step also aids in preventing an uneven final appearance. The exhaustion step is usually started at 30 °C. After the dye absorption phase, and sometimes also an additional leveling step (at temperatures sometimes higher than the fixing phase), alkali is added. With hot dyers, a uniform dye–fiber reaction can be obtained by controlled heating. Heating rates are between 1 and 2 °C/min. In the case of cold dyers, fixing can be controlled smoothly by means of pH. Alkali can be added either in portions or by metering.

After dyeing, the liquor is usually drained off; the material is rinsed and then washed off with addition of a surfactant. The washing intensity depends on the substantivity of the dye hydrolysate. With alkali-sensitive vinylsulfone dyes, soaping at the boil must be carried out in a neutral or weakly acidic liquor to prevent cleavage of the dye–fiber bond.

Pad Dyeing Processes. The dye, additives, and alkali are applied to the textile by dipping followed by squeezing out in a padder. Fixing of the dye on the textile is carried out in a subsequent step.

In these processes, dye and alkali can be applied together or separately. In the *one-bath process*, the so-called pad liquor stability, the stability of the dye to hydrolysis as a function of time, is important. With increasing reactivity of the dye, the risk of a loss in color strength with too long a dwell time in the pad box or replenishing tank increases. For this reason, metering equipment is commonly used. Dye and alkali are metered separately into the padder, which means that the dwell time in the pad box can be kept to a minimum. In addition, pad boxes are constructed with very small liquor volume, so that the liquor is replaced within 5 min.

When selecting dyes, the dyer must consider the possibility of so-called tailings. This refers to differences in shade and depth of shade between the beginning and end of the batch of fabric. With highly substantive dyes, the liquor running back into the padder from the squeezing rollers has a lower dye concentration than the initial liquor. This causes a gradual depletion of dye in the liquor.

In general, pad dyeing can be a semicontinuous or continuous process.

Special *semicontinuous processes* were developed for reactive dyes. The *cold pad-batch* process is by far the most important; it is of special interest from an economic point of view. The textile is padded with dye and alkali. To obtain a level fabric appearance, the fabric must be uniformly absorbent. If necessary, the solubility of the dyes in the padding liquor can be increased by adding urea. The material is then either plaited down in a box wagon or rolled up into batches. Fixing takes place during storage. The amount of alkali and the dwell times required for dye fixation depend on the reactivity and quantity of the dye. With cold dyers, adequate fixation can be achieved with sodium carbonate or bicarbonate; with DCT and MCT dyes, sodium carbonate or even trisodium phosphate can be used. With less reactive dyes, mixtures of sodium carbonate and sodium hydroxide are employed. The batches are wrapped, because atmospheric carbon dioxide causes a pH decrease and hence a lighter final shade at the edges. A high process reliability can be achieved by utilizing the buffering action of waterglass (sodium silicate). Usually, highly or moderately reactive dyes are used for the cold pad-batch process. For a complete dyeing reaction, 2–8 h are required. After dyeing, the material is rinsed and washed off in open-width, e.g., on perforated beams, or on rope-washing machines. The material is then dried. Recent developments see [34a].

In *continuous dyeing processes*, padding, fixing, washing-off, and drying are carried out in consecutive steps. Two different types of processes exist: dry-heat and steaming processes. The *dry-heat process* is either a single-stage pad dry process or a thermofixation process at elevated temperatures.

Hot dyers are preferred for use in *thermofixation*. Advantages of the thermofixation process include good color yields and covering on immature cotton [35]. The dye and alkali are padded at the same time. The preferred alkali is sodium carbonate. For less reactive dyes, sodium bicarbonate is recommended. In addition, urea (for ecological reasons often cyanoguanidine) is added, which increases the solubility of the dye in the liquor, acts as a solvent for the dye during fixation (m.p. 133 °C), and prevents browning of the cellulose fiber. After pad dyeing, the material is dried, leaving a residual moisture content of 10–15 % because further drying can lead to loss of color strength. Subsequent fixation takes place in hot air (1–2 min at 150–200 °C, depending on the reactivity of the dye).

In the *pad dry process*, fixing and drying occur simultaneously. Here, dyes of medium to high reactivity are preferred. The fixing time is 30 s to 5 min.

In the first stage of the *pad steam process*, dye is applied by padding. Intermediate drying is then carried out at 110–140 °C. Alkali is applied by subsequent padding. High salt concentrations (sodium chloride) are used to prevent detachment of the dye and to give a high degree of fixation. The solubility of the dye can be increased by adding urea. Also here, the amount of alkali employed must be adapted to the dyeing system used. Dye fixation is carried out with saturated steam at 103 °C for ca. 30–90 s. A new process (Eco-flash) combines drying and dye fixation by a treatment with superheated steam (20–30 s at 180 °C). For dyes that are sensitive to reduction, a weak oxidizing agent can be added to the pad liquor.

In all heat-fixing processes, optimum fixing times must be adhered to. If the times are too short, loss of color strength occurs due to incomplete dye–fiber reactions. With too long a fixing time, cleavage of the dye–fiber bond is a risk, especially for vinylsulfone dyes. After fixation, the material is washed off and dried.

Special Processes and Development Trends [36, 37]. Many attempts have been made to remedy the weak points of the reactive dyeing system. Lightfastness could be significantly improved over a wide range of shades by development of effective chromophores (e.g., triphenodioxazine and copper formazan for blues). However, in the red range, reactive dyes are still inferior in lightfastness to vat and naphthol dyes.

The salt load and discoloration of wastewater from dyeworks using reactive dyes often attract criticism on environmental grounds. With monofunctional dyes, the fixation yield in exhaustion dyeing processes is in the region of 60 %.

An important step forward was achieved with the introduction of double- or multiple-anchor dyes. These can be homo- or heterofunctional (mixed-anchor systems). When dyes are provided with anchors of different reactivity, a wide range of dyeing properties can be realized. At the same time, process reliability and dyefastness are improved. Thus far, combination of the triazine and vinylsulfone anchors has had the best effect.

In terms of dyeing methods, the economical padding process is favored. In the exhaustion dyeing process, the trend is toward the short-liquor technique.

Drying with superheated steam at 180 °C guarantees dye fixation in a very short time (Babcotherm).

Trials were made to dye cellulose with reactive dyes from supercritical CO_2.

4.2.3 Reactive Dyes on Wool, Silk and Polyamide Fibers

Reactive Dyes on Wool and Silk. Reactive dyes produce brilliant shades on wool with good fastness. They differ from reactive dyes for cellulose fibers because the reactivity of the chemoactive groups in wool is considerably higher than that of the hydroxyl groups in cellulose. To achieve level dyeing on wool, dyes with reduced reactivity must be used, and an auxiliary agent added.

Among the reactive groups in wool the most important are amino, imino, and hydroxyl groups. Reactions occur in a weakly acidic medium (pH 3–5) and include nucleophilic substitution of leaving groups (usually Cl, F, and, rarely, sulfonate or ammonium groups) or addition reactions to polar aliphatic double bonds.

Anchor groups such as *N*-methyltaurine ethylsulfone and β-sulfatoethylsulfone have the advantage that their functional groups are masked at the start of the process; consequently, no premature reaction occurs below boiling temperature [38]. At the same time, level dyeing is made possible by the increased solubility. Other dyes are bifunctional and can have a cross-linking effect on wool [39].

In the acidic range, ionic bonds are formed between dye and fiber, and the dye is still capable of migration. At pH 5, covalent binding to the fiber predominates.

In the *dyeing process for wool*, a nonionic or weakly cationic leveling agent is added to the liquor, and the pH adjusted to 3–4 with formic or acetic acid. The process is started at 40 °C, and after some time the pH is adjusted to 5–6 with sodium dihydrogenphosphate. Dyeing is conducted at boiling temperature for 1 h. To eliminate hydrolyzed dye, an aftertreatment is performed at 80 °C with ammonia (pH 8.5–9.0). The last rinsing bath is weakly acidified.

On *silk*, reactive dyes are only applied when brilliant shades are needed and when the colorfastness produced with acid dyes does not meet the required level [40, 41, 42]. Dyeing is performed on silk with the addition of sodium sulfate. The system is heated up to 50–70 °C, and soda is added. Dyeing is completed at 70 °C. An afterwash at 80 °C increases wetfastness.

Reactive Dyes on Polyamide. In principle, the reactive dyes used for wool are also suitable for PA. Covalent binding of dyes to the terminal amino groups produces dyeings that have excellent wet- and rubfastness. The lightfastness corresponds to that of analogous acid dyes. The depths of color attainable are limited by the number of end groups and are higher for PA 6 than for PA 66.

In the dyeing process, the dyeing bath is made weakly acidic (pH 4.5–5). The process is started at 20–45 °C, followed by heating at a rate of 1 °C/min and dyeing at near boiling temperature. Aftertreatment (soaping) is performed with a nonionic surfactant and sodium bicarbonate or ammonia at 95 °C.

Reactive dyes are well suited to dye *blends of cellulose and PA fibers*. Clear shades with very good fastness are obtained. Like with vat dyes, the depth of shade of reactive dyes depends relatively strongly on the type of PA and structural differences. Dyeing is carried out in a three-step process with appropriately selected products. First, the reactive dyes in a weakly acidic liquor are allowed to absorb on the PA component. Salt is then added to improve the yield on the cellulose component. Finally, the liquor is made alkaline for reaction with the cellulose fiber. Dyes (e.g., with MTC anchor) that dye PA from a neutral liquor in the presence of salt are applied in a two-step process, as in the case of cellulose. In the reversal of this dyeing process, the cellulose component is dyed first at alkaline pH, followed by neutralization with acid, and the PA component is then covered at elevated temperature.

For reactive dyes on *cellulose–PAC fiber blends* see Section 4.14.8.

4.2.4 Reactive Dyes for Printing on Cellulose

The reasons for the increasing importance of reactive dyes [5, p. 523-527] in printing are the brilliance and wide range of available colors, the ease with which the dyes are applied, high fixation yield and good build-up properties, easy removal of hydrolyzed dye, and high stability in the printing paste. Mainly the moderately reactive MCT dyes are used for printing. Sulfatoethylsulfone dyes are used for two-phase applications, with alkali treatment in a second step. In general, dyes with low substantivity for the cellulose fiber are used to avoid bleeding of the unfixed dye on the white ground during final washing. To obtain clear, brilliant colors, the fabric must be carefully desized, bleached, and given good wetting properties; cotton should be mercerized. The thickener should not react with the dye; therefore sodium alginates are used. Oil-in-water type "half-emulsions" with alginates produce brilliant colors with good dye leveling, and are easily removed by washing, but environmental problems can arise with the naphtha used. Synthetic thickeners (Na polyacrylates) are sensive to electrolytes, which influence the viscosity, and are difficult to wash out. Frequently, urea is added to the print paste; because of its negative heat of solution, it binds steam to the fabric, helps to dissove the dye in the print paste, and functions as a dyeing medium. Sodium carbonate is an appropriate alkali for the fixation of MCT dyes. The dyes are usually fixed by steaming in saturated steam at 102 °C or by hot air treatment at 160 °C. Dyes with high reactivity are applied in a two-phase process: The dyes are printed alkali-free; after drying alkali is applied in a second step, and the print is then fixed without drying by steaming. Finally, thickener is removed by cold rinsing; remaining unfixed dye is washed out. Prints with reactive dyes should not be stored in an acidic atmosphere to avoid splitting of the dye–fiber bond, which may cause bleeding of the dye onto the white ground.

Most reactive dyes are easily discharged, which facilitates the elegant technique of printing vat dyes on reactive grounds: The reducing agent and alkali required for the development of the vat dye discharge the reactive ground.

Recent developments show how vinyl sulfone reactive dyes can be applied on cotton by transfer printing [43].

Reactive dyes can easily be printed by the ink-jet technique.

4.3 Direct Dyes on Cellulosic Fibers

In contrast to some naturally occurring dyes like indigo or kermes, which must be vatted or mordanted to be applied in textile dyeing, direct or substantive dyes can be used on cellulosic fibers "directly". Their use is widespread because of their easy handling. Today, direct dyes still account for ca. 10 % of the world textile dye consumption. For use on paper, see Section 5.3. Recent research on direct dyes concentrated on the replacement of possibly carcinogenic benzidine dyes [44].

Only in individual cases do direct dyes achieve the brilliance of reactive dyes. The lightfastnesses of substantive dyes covers the entire scale from 1 to 8, meeting the highest requirements. With increasing depth of color the wetfastness decreases to such an extent that direct dyeings must generally be aftertreated (see Section 4.3.4). Most direct dyes are characterized by excellent affinity to the fiber, leaving dye liquors almost colorless at the end of the dyeing process.

4.3.1 Dyeing Principle

The affinity (substantivity) of direct dyes for cellulose fibers depends on the type of chromophore and can be influenced by the choice of dyeing parameters (see Section 4.3.3). Chromophores are azo compounds (Sections 2.2 and 3.2), stilbenes, oxazines, and phthalocyanines (see Section 2.8). They always contain solubilizing sulfonic acid groups that are ionized in aqueous solution. The dye molecule is present in the dyebath as the anion. As a result, an electrolyte must be added to the dye liquor because cellulose fibers have a negative surface charge in water, which would repel the dye anion. The cations of the electrolyte neutralize the negative charge and favor the aggregation of dye ions on the fiber (salting-out effect). Exothermic adsorption can be described with the help of Freundlich or Langmuir adsorption isotherms [45]. After adsorption, dye molecules diffuse from the surface into the amorphous areas of the cellulose fiber. The rate of diffusion can be controlled by the dyeing parameters. In general, Fick's law can be applied to diffusion [46]. Dyes are bound to the fiber by hydrogen bonds or van der Waals forces [47]. Without additional measures (see Section 4.3.4), the strength of these bonds is low, so that the dyes can be washed out of the fiber again. At dyeing equilibrium, the rate of absorption is equal to the rate of desorption.

4.3.2 Dyeing Parameters [48]

Cellulose fibers require careful pretreatment before dyeing. Direct dyes are especially sensitive to differences in fiber affinity. Pretreatment involves desizing, bleaching, possibly subsequent mercerization and increasing the wettability.

The *liquor ratio* influences the dye solubility and the strength of the electrolyte effect. Lower liquor ratios are employed whenever possible, for ecological, economic, and technical reasons.

The *electrolyte* (i.e., sodium chloride or sodium sulfate), its concentration, and the speed of addition control the adsorptive behavior of the dyes and the degree of exhaustion. A high electrolyte content of the dyebath shifts the dyeing equilibrium toward the fiber.

The *pH value* influences solubility, substantivity of the dyes, and their stability in the dyebath. Lower pH values improve exhaustion; however, leveling proceeds preferably at higher pH values.

Temperature generally determines the position of the dyeing equilibrium, which at room temperature is shifted strongly toward adsorption on the fiber. At higher temperature (usually 80–95 °C) the equilibrium is attained more rapidly. The temperature dependence of the maximum degree of exhaustion is dye-specific. The same applies to hydrolytic stability, especially in high-temperature processes.

Water quality (metal-ion content) may be responsible for shifts in shade. Sequestrants are used as a safety measure.

The *leveling capacity* of dyes is an important characteristic that can be determined empirically. Unevenness results if dye adsorption occurs too quickly, and can frequently be traced to mistakes made in the addition of dye or salt. The leveling of uneven dye distribution proceeds via desorption of the dye and readsorption on the fiber (migration).

In combinations, dyes with similar properties are preferred. To help in the selection of combinable dyes, the Society of Dyers and Colourists has classified direct dyes according to their leveling capacity [49].

4.3.3 Dyeing Techniques

Exhaustion Process. In batch dyeing, the preferred mode depends on the type of dyeing equipment, the type of material to be dyed, and the solubility and affinity of the dyes. The dye is dissolved in hot water and added to the dyebath. The electrolyte is added either during heating of the bath or after attaining the optimum dyeing temperature. The material is dyed at 80–90 °C. Exhaustion is completed, if necessary, at somewhat lower temperature for a short time. After the dyebath has been drained, the fabric is quickly washed clear with cold water and generally subjected to aftertreatment (see Section 4.3.4).

High-Temperature Dyeing Process. With suitable dyes, dyeing can also be performed in a closed apparatus at up to ca. 130 °C (especially for polyester–cotton mixtures). As a result of the high rate of diffusion, high levelness is achieved in short dyeing times, even in the case of poorly penetrable materials. After the high-temperature phase, the dyebath is cooled to 80–90 °C and the dye continues to be adsorbed, resulting in the same depth of color as in normal dyeing processes.

Pad Processes. In a continuous or semicontinuous processes, the fabric is first padded with the dye solution. The substantive adsorption of dyes during padding is avoided by employing minimum amounts of pad liquor at ambient temperatures. For even penetration of the dye, the material is then steamed without intermediate drying for 1–3 min in saturated steam (*pad steam process*). Alternatively, it is heated to 80–85 °C and rolled up in a closed chamber where it is slowly rotated for 2–8 h (*pad roll process*).

In the *cold pad-batch process*, the padded material is rolled without heating and stored at room temperature for 8–24 h.

The *pad jig process* represents the easiest type of fixation of pad dyeings. The material padded with dye is passed through a salt liquor in a jigger at 80–90 °C, thus preventing desorption of the dye. In this process, additional dye can be added during jigger treatment for shading.

After dyeing the goods are rinsed with cold water and dried.

Printing with Direct Dyes is common only for inexpensive goods, because of the extended steaming time required and the limited wetfastness of the prints, even after treatment with cationic quaternary products [5, p. 514].

4.3.4 Aftertreatment

As a rule, the wetfastness of direct dyes does not meet the demands of daily use. It can be improved by aftertreatment [44], which results in dyeings that are fairly fast to water and perspiration and, in ideal cases, washable up to 60 °C. The improvement in wetfastness is achieved by reducing the solubility of the anionic dyes. This occurs by enlargement of the molecules by the formation of saltlike compounds with cationic aftertreatment agents [5, p. 288]. These are, e.g., quaternary (poly)ammonium compounds, polyamines, or poly(ethylene imine) derivatives. The fixation can be enhanced by cross-linking or by formation of covalent bonds to the fiber. This is performed with, e.g., formaldehyde condensation products with amines, polynuclear aromatic phenols, cyanamide, or dicyandiamide. For certain azo dyes, substantial improvement in lightfastness can be achieved by metal-complex formation with copper sulfate or potassium dichromate. A combination of these aftertreatments produces especially good fastness properties. With formaldehyde derivatives, as well as with metal salts, ecological aspects have to be considered.

4.3.5 Direct Dyes for Fiber Blends

Apart from cellulose, direct dyes have a strong affinity to wool and polyamide fibers. Blends of cellulose with wool ("half wool") used to occupy a considerable segment of the market, but are today without any significance. However polyamide (PA) fibers are included in articles made of cellulose fibers to improve dimensional stability, ease of care, and durability, e.g., in sportswear and knitwear, corduroy fabrics, or plush articles, in which a PA pile is often anchored to a cellulose fabric base. Different dyeing methods are described in [50 pp. 433–437], [6, p. 570].

Dyeing is carried out in the presence of sodium dihydrogenphosphate without salt at boiling temperature. Solid shades from light to moderate can be obtained with high-leveling direct dyes (e.g., *C.I. Direct Yellow 44, Direct Red 81,* and *Direct Black 51*). The choice of suitable dyes is greater if dyeing is performed in a single bath with direct and disperse dyes at pH 8 in the presence of a leveling agent and a resist for the PA fiber to prevent the direct dyes exhausting mainly on the PA component.

If, instead of disperse dyes, wetfast acid or metal-complex dyes are used for the PA component, shades having good lightfastness and moderate wetfastness can be obtained. Dyeing is carried out at pH 5–8, depending on the depth of shade and the type of acid dye used. The dyeing temperature is raised to boiling to guarantee good leveling and firm fixation of the wetfast acid dyes in the PA fiber.

For direct dyes in *blends of cellulose with acrylic fibers* see Section 4.14.8.

4.4 Anthraquinone Vat Dyes on Cellulosic Fibers

Approximately 15 % of cellulose fibers and cellulose-containing fiber blends are dyed with vat dyes. The total consumption of vat dyes in 1993 was about 22 000 t/a of commercial products, excluding indigo.

4.4.1 Principles of Vat Dyeing

Vat dyes are water-insoluble, organic pigments that are used to dye cotton and other cellulose fibers. The principle of vat dyeing is based on chemical reduction of these dyes to the leuco compounds, which are soluble in aqueous alkali and exhibit fiber affinity, followed by reoxidation within the fiber to the water-insoluble starting dye.

Two main groups of vat dyes can be distinguished chemically:

1) *Indigoid vat dyes* are derivatives of indigo and thioindigo (see Section 3.5). Leucoindigo compounds have a comparatively low affinity for fibers. Thus, these dyes are used mainly in textile printing rather than dyeing (see Section 4.4.6).
2) The leuco compounds of *anthraquinoid dyes* have a high affinity for the cellulose fiber and give dyeings that meet the highest fastness requirements in use and processing.

Commercial anthraquinoid vat dyes are dye preparations that consist of a vattable colored pigment and a dispersing agent. Such *vattable pigments* are polycyclic quinoid compounds that contain two or more carbonyl groups in a closed system of conjugated double bonds (Section 3.4).

The pigment must be as finely dispersed as possible to achieve a satisfactory vatting rate and uniform dye application in pigmenting processes. For this reason, after the actual synthesis the dyes are wet ground in the presence of selected dispersing agents and then standardized for color strength and shade. The *dispersing agent* stabilizes the fine dispersion of pigment in the aqueous liquor.

The dyeing behavior of vat dyes, i.e., their absorptive behavior and leveling capacity, is determined basically by the substantivity and diffusion of the leuco compounds. The chemical stability and insolubility in water of the dye pigment that results from reoxidation in the fiber account for the generally unsurpassed light-, wet-, and weatherfastness of vat dyeings.

The dyeing process involves the following steps:

1) Vatting (reduction) of the dye
2) Absorption by the fiber
3) Oxidation in the fiber
4) Aftertreatment of the dyeing

4.4.2 The Vat Dyeing Process

4.4.2.1 Vatting

Using anthraquinone as an example shows that the dye molecule is converted to the sodium leuco form in alkaline solution by the gain of two electrons [Eq. (4)].

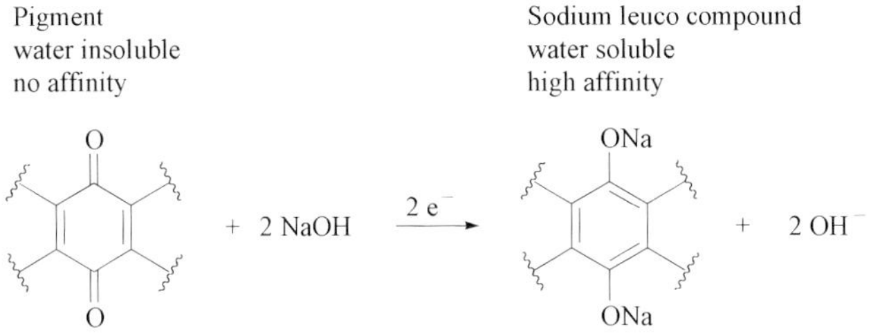

$$\text{Pigment (water insoluble, no affinity)} + 2\,\text{NaOH} \xrightarrow{2\,e^-} \text{Sodium leuco compound (water soluble, high affinity)} + 2\,\text{OH}^- \tag{4}$$

$$S_2O_4^{2-} + 4\,OH^- \longrightarrow 2\,SO_3^{2-} + 2\,H_2O + 2\,e^-$$

For a long time, *sodium dithionite* (hydrosulfite) has been the most important of the reducing agents [5, p. 288–289] for vat dyeing. It is inexpensive and, when produced by modern methods, virtually free of heavy metals (Zn, Hg). Its reduction potential (–970 mV) is high enough to vat all vat dyes quickly and completely.

The rate of vatting depends not only on the concentration of dye and reducing agent but also on the crystal form, surface, and dispersion of the pigment (i.e., on its finish quality [51]). Leuco compounds are soluble in alkali. In the case of anthraquinoid vat dyes, the pH of the vat is about 13. At lower values the risk of vat acid sediments exists. Reduction is usually performed at 50–60 °C. At higher temperature, over-reduction of certain dyes can occur (i.e., reductive destruction of the dye molecule).

Because hydrosulfite is consumed by atmospheric oxygen, the state of the vat should be monitored i.e., determination of the concentration of reducing agent, either qualitatively by using vat test paper, or quantitatively by measuring the redox potential, or by titrimetric methods. For this reason, in high-temperature dyeing less reactive, more stable reducing agents are used, e.g., derivatives of *sulfoxylic acid*.

The oxidation products of sulfur-containing reducing agents are sulfite and sulfate. In most countries, the concentration of these substances in wastewater must not exceed maximum permissible values stipulated by law.

Thiourea dioxide has a stronger reducing effect (–1100 mV) than hydrosulfite. Therefore, with sensitive dyes, a risk of over-reduction exists. In addition, the oxidation products of thiourea dioxide contribute to the nitrogen and sulfur contamination of wastewater.

Organic reducing agents such as *hydroxyacetone* do not contribute to the sulfur contamination of wastewater. However, the reducing effect of hydroxyacetone is much weaker (–810 mV), so it cannot replace hydrosulfite in all applications. The use of hydroxyacetone is restricted to closed systems because it forms strong-smelling condensation products in alkaline solution.

Systems have been described for *electrochemical vatting* by means of electric current in the presence of a mediator, an electron carrier such as an organic Fe^{2+}-complex, or anthrahydroquinones. This method is in the development stage [52].

The vat dyeing system is also supported by auxiliary agents. *Dispersing agents* prevent the agglomeration of pigment particles or the flocculation of slightly soluble components such as vat acids or the impurities in cotton. If the presence of alkaline earth ions is expected because of the use of hard water, the addition of *complexing agents* is required.

4.4.2.2 Dye Absorption in the Exhaustion Process [53, 54]

In the dissolved form, the vatted dye is present as the sodium leuco compound, either as a single molecule or as a complex of a few dye molecules. In the exhaustion process, the higher the affinity, the more complete is the absorption by the fiber. The degree of absorption of dyes depends also on dye concentration, liquor ratio, temperature, and the electrolyte present in the dyebath. It is generally between 80 and 90 %.

In the initial phase of dyeing, vatted dye is absorbed very quickly by the fiber. The final degree of exhaustion of the bath is achieved after 10 to 15 min. The dye is then present in a high concentration in the outer regions of the fiber, but its distribution may still be entirely uneven (unlevel). In the subsequent leveling phase, the dye diffuses into the fiber and the fiber becomes dyed uniformly. Dyes that are less substantive are absorbed more slowly and evenly in the initial phase and exhibit more favorable diffusion properties. Diffusion can be improved by increasing the temperature. *Leveling agents* can favorably influence absorptive behavior in the initial phase because they themselves possess a high affinity for the dye and release it in a controlled manner during the course of dyeing [5, p. 275–276], [55]. However, a certain amount of dye is retained by the leveling agent, which results in a slightly weaker shade.

4.4.2.3 Oxidation

After absorption by the fiber, the leuco dye is converted to the original pigment by oxidation and, in this way, fixed to the fiber. Oxidation can be achieved with atmospheric oxygen or by the addition of oxidants, such as hydrogen peroxide, perborate, or 3-nitrobenzenesulfonic acid, to the liquor.

After the dyeing process, the material is usually rinsed with water to remove unfixed dye and decrease the alkalinity of the liquor. Depending on the oxidizing agent, oxidation is carried out at pH 9–12 and 50–60 °C.

The leuco dye present in wastewater is also converted to water-insoluble pigment, which can easily be separated mechanically or adsorbed on clarification sludge. Thus, colorization of the wastewater by vat dyes is negligible.

4.4.2.4 Aftertreatment ("Soaping")

After oxidation, all vat dyeings are subjected to heat treatment in the presence of water, generally in a weakly alkaline detergent liquor at boiling temperature. By boiling, the final shade is obtained, and the fastness ratings (e.g., light- and chlorine-fastness) are improved substantially. Thus, this soap treatment is not restricted to the removal of pigment particles adhering to the surface but is an important step in vat dyeing. The processes occurring during soaping are unknown. Crystallization of amorphous dye particles or coarsening of primary particles by aggregation is assumed to occur. If soaping takes too long, migration of the aggregates to the surface of the textile can cause deterioration of the rub-fastness.

4.4.3 Dyeing Techniques [56]

Batch Processes. In batch processes, dyeing is conducted in various circulating-goods machines (e.g., jigger, winch beck, jet, and overflow machines) and circulating-liquor machines (yarn, piece, and package dyeing machines) (see Sections 4.1.1).

Vat dyes are divided into groups according to their affinity for the fiber and the amount of alkali required for dyeing: IK (I = indanthrene, K = cold), IW (W = warm), IN (N = normal).

The IK dyes have a low affinity for the fiber; they are dyed at 20–30 °C and require little alkali. Salt is added to increase the absorption of these dyes. The IW dyes have a higher affinity than IK dyes. Dyeing is performed at 40–45 °C with more alkali and little or no salt. The IN dyes are highly substantive and applied at 60 °C. They require much alkali for vatting but no addition of salt.

Dyeing with vat dyes is performed according to the following methods:

In the *leuco process*, auxiliaries, dye, sodium hydroxide solution, and reducing agent are added to the dyebath. The textile is introduced into the bath after vatting. Depending on the depth of shade, dyeing is performed for 30–60 min and completed in the usual manner (rinsing, oxidation, soaping).

In the *preimpregnation process*, the dye liquor initially contains only the dye as a water-insoluble dispersion. After the temperature is increased and, if necessary, sodium hydroxide solution and salt are added, the pigment partially deposits on the fiber. After the reducing agent is added, vatting occurs rapidly and the dye is absorbed by the fiber. As a result of partial deposition of the dye in the nonsubstantive form, the levelness of dyeing is improved. Prepigmentation of piece goods is conducted in a jigger and can also be used in dye packages (cheese, warp, yarn, and piece dyeing beams).

The *semipigmentation process* is used preferentially for dyeing lighter shades. Dye, sodium hydroxide solution, and reducing agent are added to the dyebath, and dyeing begins immediately at room temperature. Depending on the vatting rate of the dyes, only a small amount of substantive leuco dye is present intially. For this reason, absorption proceeds slowly and evenly. With increasing dyeing temperature, more and more dye is converted to the leuco form and can be absorbed by the fiber.

In the *high-temperature process*, the leveling-out properties of vat dyes are improved by increasing the dyeing temperature to 90–115 °C; exhaustion may be completed at lower temperature.

Continuous Processes. Vat dyes are especially suitable for continuous application. Dyeing is performed according to the *pad steam process* or the *wet-steam process*. The continuous processes are used almost exclusively for dyeing woven fabrics and only to a small extent for knitwear.

In the *pad steam process*, which is most frequently applied, the textile is impregnated in a padder with the aqueous dye dispersion in the presence of an antimigration agent and a wetting agent, if required. The material is squeezed to a constant residual moisture content and then dried. Subsequently, the fabric is passed through a chemical padder, which contains the required amounts of alkali and reducing agent, and is fed immediately to a steamer. In the steamer, the dye is fixed for 45–60 s under saturated steam conditions at a wet-bulb temperature of 99 °C. After steaming, the material is rinsed, oxidized, and soaped in an open-width washing machine.

To achieve maximum fixation yield and depth of shade, dyes must be very finely dispersed. However, fine dispersion also increases the migration of dye particles during intermediate drying. For this reason, antimigration agents are used (polyacrylates, alginates, etc.), which have an agglomerating effect on the pigment particles and thus restrict their mobility during the drying process [57].

The *wet-steam process* is used mainly for voluminous open fabrics with a high liquor-carrying capacity (e.g., toweling, cord, velvet). Unlike the pad steam process, this process does not require drying after dye application in the padder.

Printing with Vat Dyes [5, p. 516–521]. For printing, vat dyes must be used in a carefully finished form with a very small particle size to guarantee rapid vatting; liquid (paste) forms are preferred. Vat dyes are printed by an all-in or a two-phase printing process.

In the *all-in*, also known as Rongalit C potash process, the print paste contains the dye and the chemicals necessary for fixation. The reducing agent hydroxymethanesulfinate (Rongalit C), reaches its maximum reactivity under steaming conditions, but decomposes slowly already at room temperature [5, p. 296–298]. Potassium carbonate (potash) is the preferred alkali. The dyes are fixed with saturated steam at 100–102 °C for 5–15 min. The prints are then rinsed, reoxidized, neutralized, and soaped at the boil.

In the *two-phase process* dyes and chemicals are applied in separate steps: first dyes and thickener ("solid phase") are printed, then chemicals and auxiliaries are added as an aqueous solution by padding. Sodium dithionite (Hydrosulfite) is used as reducing agent, and steaming time is kept short (20–40 s). For stabilized reducing agents, see [5, p. 296–298].

4.4.4 Vat Dyes for Fiber Blends

Polyester fibers are dyed to a considerable extent by selected low-molecular vat dyes with satisfactory fastness properties. They can be applied on *CEL–PES fiber blends* by the Thermosol process (see Section 4.12.4).

Vat dyes are used for *PA–CEL blends* when the highest possible wetfastness is required [5, p. 457]. The coloring strength on the PA component depends on the type of PA fiber and on dyeing conditions (the depth of shade on PA 6 is deeper than on PA 66). The lightfastness of some vat dyes can be lower on PA than on cellulose fibers. The advantage of using vat dyes is that in a single bath with selected products, dyeings of the same shade are obtained on both components. For instance, *C.I. Vat Orange 26* [12237-49-9], *C.I. Vat Blue 4*, 69800 [81-77-6], *C.I. Vat Blue 14*, 69810 [1324-27-2], *C.I. Vat Brown 55*, 70905 [4465-47-8], *C.I. Vat Blue 9*, *C.I. Vat Blue 25*, 70500 [6247-39-8] are widely applicable (in the exhaust and pad steam process). The dye yield on the PA component increases with increasing temperature. With a dyeing time of 60 min, the most favorable temperature for the dyes mentioned above is 80–90 °C. The additives to the dyebath and the dyeing processes are the same as those used for dyeing cellulose fibers alone (see Section 4.4.6).

For vat dyes used in *CEL–PAC fiber blends* see Section 4.14.8.

4.5 Leuco Esters of Vat Dyes on Cellulosic Fibers

The leuco esters of vat dyes are anthraquinoid or indigoid vat dyes (Section 3.4) that have been made water soluble by reduction and esterification of the hydroxyl groups with sulfuric acid. These dyes are used mainly for dyeing high-quality articles of cellulose fibers in light colors and polyester–cellulose blends to moderate depths of shade [58], [59, p. 568].

The advantages of dyeing with the leuco esters of vat dyes are not only good levelness and penetration but also the excellent fastness of vat dyes. The low affinity of the leuco ester for the cotton fiber and the relatively high dye costs are obstacles to wide application. After application to the fiber, the ester is hydrolyzed, usually with sulfuric acid at room temperature or slightly elevated temperature (up to 70 °C), and the vat dye is returned to the insoluble form by oxidation, usually with sodium nitrite.

Batch Dyeing. Because of the low substantivity of leuco esters of vat dyes for cellulose, batch processes (exhaustion processes) are rather unimportant.

Continuous processes yield more favorable dye yields [60]. The most important dyeing methods are:

- Processes without intermediate drying: padding (dye, soda, nitrite), air passage, development (dipping in acid at 70 °C), air passage, and aftertreatment.
- Processes with intermediate drying: padding (dye, soda, nitrite), drying, development (dipping in acid at 70 °C), air passage, and aftertreatment; this process gives a higher fixing yield than the first method.
- Pad steam development process without intermediate drying: padding (dye, soda, nitrite), steaming (1 min, saturated steam, 103 °C), development (dipping in acid at 70 °C), air passage, and aftertreatment.

As a result of their high price, low substantivity, and toxicological problems during production, the importance of the leuco esters of vat dyes is decreasing. They can be replaced easily by pigment coloring or by reactive or vat dyes.

For leuco esters of vat dyes on *CEL–PES fiber blends*, see Section 4.12.4.

4.6 Dyeing with Indigo (see 2.3, 3.5)

Indigo is probably the most important single textile dye worldwide. The demand for indigo depends greatly on fashion. In comparison with other dyes, indigo occupies a special position because it is used almost exclusively for dyeing warp yarn in the production of blue denim. Being identical to the natural material, indigo leads to environmentally friendly clothing.

Beside indigo itself, the indigoid colorant thioindigo and its derivatives are also important vat dyes, but because of the low affinity of vatted thioindigo dyes for the fiber, these dyes are used mainly in textile printing.

4.6.1 Dyeing Technique on Cotton

Indigo is available as a powder, granulate, or liquid preparation. The dyeing technique has not changed over the years and, like vat dyeing, involves the following steps (see Section 4.4): (1) dissolving of the dye by reduction in alkaline medium, (2) dyeing in the vat, and (3) oxidation on exposure to air.

The first step can be omitted by using solutions of pre-reduced indigo (leucoindigo). In this way, about half of the amount of reducing agent (depending on the type of machine) required for dyeing can be saved, which means lower discharge of sulfate in the wastewater. In addition, using 20 % or better 40 % solutions of leucoindigo facilitates replenishing exhausted baths in continuous slasher or rope dyeing machines.

Recent developments have shown how indigo and vat dyes can be reduced electrochemically to the soluble form. An organic iron complex (Fe^{2+}–amine, Fe^{2+}–sugar acid, or anthraquinone–Fe^{2+}–amine complex) is used as a mediator for dye reduction on the cathode. When using leucoindigo solution, the dye is electrochemically maintained in the reduced form. Considerable savings in cost for reducing agents and wastewater are accomplished [61]

As a result of the relatively low affinity of indigo for the fiber, only about 10–20 % of dye offered in the dyebath is taken up by the fiber; therefore, deep blue shades are attained only if dyeing and oxidation are repeated several times. Therefore, in the past, indigo dyeing took several days. In modern indigo dyeing plants for dyeing warp yarn, several dyebaths and oxidation units (usually up to five units) are arranged in a row for continuous operation [62, 63]. For blue denim articles, pronounced ring dyeing of the yarn is desirable.

Indigo is dyed continuously on rope dyeing machines and slashers. In *rope dyeing machines*, 300–400 warp threads are combined into a rope and, in this form, passed through the dyeing system. *Slashers* were developed in the 1970s from sizing machines. In these machines, the warp threads are dyed side by side, parallel to each other [64]. Both types of machines have advantages and disadvantages and should be assessed according to the specific requirements (e.g., small or large batches, required fastness of dyeing). A detailed working procedure has to be developed under local conditions.

4.6.2 Indigo on Wool

In the past, vat dyes played an important role in dyeing wool. Indigo was considered to be "king of dyes" because of its unsurpassed fastness on wool. Until the mid-1950s, navy cloth was dyed with indigo. The application was problematic with regard to the negative effect of reducing agents and alkali on wool. Today, indigo, thioindigo, and their derivatives have been replaced in the dyeing of wool by classes of dyes that are easier to handle.

4.7 Sulfur Dyes on Cellulosic Fibers [65, 66]

The dyeing of cellulose and its blends with synthetic fibers, is the main field of application of sulfur dyes (see Sections 2.9 and 3.6). They are also used to a limited extent to dye polyamide fibers, silk, leather, paper, and wood. These dyes are used for deeper, muted shades, such as black, dark blue, olive, brown, and green, where their favorable price has its full effect. With respect to fastness, sulfur dyes are close to vat dyes but not equal to them. They have varying lightfastness; the washfastness at 60 °C is good, and their fastness to boiling water is moderate. Wetfastness can be improved by aftertreatment of the dyeing with quaternary (poly)ammonium compounds or formaldehyde condensation products. Most sulfur dyeings are not fast to chlorine. Therefore, faulty dyeings can be stripped by treatment with sodium hypochlorite.

The main applications are *piece dyeing*, which is important for the dyeing of the CEL component of PES–CEL blended fabrics, *yarn dyeing*, and *dyeing* of *flock, card, sliver*, etc., for blended yarns with wool and man-made fibers.

4.7.1 Types and Mode of Reaction [67, 68, 69]

Sulfur dyes contain in their molecule S—S bridges. For technical and applicatory reasons, sulfur dyes are available in various modifications, which are classified in the Colour Index under the general generic names:

Sulfur Dyes (C.I. Sulphur). Two types of sulfur dyes exist: *amorphous powders*, which are insoluble or partially soluble in water. They must be brought into solution by reduction with sodium sulfide, according to Equation (5).

$$\underset{\substack{\text{Water-insoluble} \\ \text{sulfur dye}}}{DS-SD} + 2\,S^{2-} \rightleftharpoons \underset{\substack{\text{Reduced} \\ \text{(leuco) dye}}}{2\,DS^-} + S_2^{2-} \qquad (5)$$

With polysulfide melt dyes (i.e., quinoneimine sulfur dyes), the reducing agent can attack not only the disulfide groups but also the quinoneimine group (as in the case of vat dyes). Dispersible pigments are used particularly for pad dyeing.

The dispersible, *partly reduced pigments* bearing the name "Sol" contain a dispersant and a certain amount of reducing agent.

Leuco Sulfur Dyes (C.I. Leuco Sulphur). These liquid dyes already contain the reducing agent required for dyeing. They must simply be diluted with water before application. Ecologically more favorable, low-sulfide types have come onto the market recently.

Water-Soluble Sulfur Dyes (C.I. Solubilised Sulphur). Water-soluble sulfur dyes are available in the form of Bunte salts (*S*-arylthiosulfuric acid salts) of sulfur dyes and can be dissolved in hot water. The addition of alkali and reducing agent gives them an affinity for the fiber. If NaSH or another reducing agent is used, alkali must be added to trap H^+ ions resulting from reduction.

The leuco form is absorbed by the substrate and then oxidized to the insoluble dye. Since several mercapto groups are present per dye molecule, oxidation causes not only dimerization but also cross-linking of dye "monomers". For polysulfide melt dyes, the quinonimine group is also involved in the reaction. This is apparent in the change in shade that occurs during reduction.

With respect to both application and fastness, a distinction is made between sulfur dyes and sulfur vat dyes. The latter include dyes of the type Hydron Blue (*C.I. Vat Blue 43*, 53630 [*1327-79-3*]) and Indocarbon (*C.I. Sulphur Black 11*, 53290 [*1327-14-6*]).

4.7.2 Additives to the Dye Bath

Reducing Agents. The reducing agents generally used to bring the dyes into the soluble form with affinity for the fiber are sodium sulfide or sodium hydrogensulfide:

- 100 parts of crystalline sodium sulfide ($Na_2S \cdot 9\,H_2O$; ca. 30–32 wt % Na_2S).
- 50 parts of concentrated sodium sulfide (ca. 60–62 wt % Na_2S).
- 65 parts of liquid sulfhydrate F 150 (31 wt % NaSH).
- 25 parts of concentrated sodium sulfhydrate powder (95 wt % NaSH).

In the dye liquor, the amount of concentrated Na_2S should be greater than or equal to 2.5 g/L. In aqueous solution, the sulfide ion is almost completely hydrolyzed. If NaSH and soda are used istead of NaOH, the lower alkalinity offers:

- Better color yield; the dyeing can be rinsed clear substantially faster.
- During dyeing and rinsing, the dyeing is less sensitive to oxidation (bronzing, rubbing-fastness).
- In the case of regenerated cellulose, embrittlement of the material is avoided.

Sodium dithionite (hydrosulfite) is a proven reducing agent for *sulfur vat dyes*, and is used together with alkali. The color yield of some sulfur dyes is remarkably increased by the addition of hydrosulfite. However, as a result of possible "overreduction" or destruction of the dye chromophore, hydrosulfite and similar reducing agents can be used only with selected dyes.

For the reduction of *water-soluble sulfur dyes*, glucose and hydroxyacetone are also used [70]. Glucose-based combinations with hydrosulfite, formamidinesulfinic acid, and hydroxyacetone (binary reducing-agent systems) have gained acceptance for ecological, technical, and applicatory reasons [71]. Mixtures of sodium hydroxide solution and soda are used as alkali. Glucose prevents or buffers possible overreduction or destruction of the dye chromophore.

Alkali metal polysulfides are added to stabilize alkaline sulfur dye liquors against atmospheric oxygen. This is common for Hydron Blue, especially on the winch beck. An excess of sulfide and reducing agent in the wastewater has to be avoided; therefore, low-sulfide dyeing methods were introduced [71].

Recently, the electrochemical reduction of sulfur dyes was studied. For continuous processes a mediator is not necessary. In discontinuous application it is possible to reuse dyebaths [72].

Wetting Agents. As wetting agent or padding auxiliary usually sulfonated oleic acid amides are used; they exert a leveling effect and confer softness on the material [5, p. 268–270]. Products based on ethoxylates are unsuitable because they have a retarding, and sometimes precipitating, effect on the dye.

Water Softeners. Sequestrants (EDTA or polyphosphates) are used, especially in circulating-liquor dyeing with hardness-sensitive dyes [5, p. 273–274].

4.7.3 The Dyeing Process

Absorption of Dyes. In the exhaustion process, the dye is generally absorbed at 60–110 °C in 30–60 min, and in the pad steam process at ca. 102–105 °C in 30–60 s. Dyeing with reducing agents is always performed by the exhaustion process. The fixing yield depends greatly on the liquor ratio. Optimal fixation and dye penetration are achieved with the pad roll process, in which the dye is applied from a concentrated liquor.

Oxidation of Dyes [73]. Sulfur dyes, like vat dyes, are fixed on the substrate by oxidation. First, the remaining reducing agent, salts, and alkali, as well as the unfixed dye, must be removed by rinsing or washing to attain optimal rubbing fastness. The following methods of oxidation are common.

In *air oxidation and acid treatment*, the dyeings are rinsed and acidified with acetic acid (exception: sulfur black). Since after-oxidation can occur on storage, the shade may change.

In *hydrogen peroxide oxidation*, bright shades are obtained in the alkaline to neutral pH range. However, wetfastness is reduced somewhat compared with other oxidation processes because oxidation is not specific. Thus, especially with a high concentration of H_2O_2, high temperature and pH (>50 °C, pH>7), and excessively long oxidation times, over-oxidation occurs; i.e., the mercapto groups of the sulfur dyes are oxidized not only to disulfide groups, but also in part to sulfonic acid groups. Wetfastness can be greatly improved by aftertreatment of the dyeings with cationic auxiliaries, which are used in combination with oxidizing agents [74]. A deterioration of wetfastness can largely be avoided by oxidation in acetic acid (pH 4–4.5) at 70 °C.

Other oxidizing agents such as bichromate, iodate, bromate, chlorite, bromite, chloramine T, and peroxosulfate are not equally suitable for all dyes and are seldom used, for price reasons or because of contamination of the wastewater with heavy metals. Sodium hypochlorite requires careful metering and can be used only with sulfur dyes that are relatively fast to chlorine.

Alkylation. With special alkylating agents, sulfur dyes can be converted to thioethers. In this way, the dye is stabilized against oxidative and reductive effects [74]. Bi- or polyfunctional alkylating agents are especially effective because they lead to cross-linking and enlargement of the dye molecule. In some cases, the wetfastness of vat dyeings can be achieved.

In alkylation, care must be taken that only the mercapto groups and not the quinonimine groups of dyes are alkylated; otherwise these compounds are fixed in their leuco form, which can lead to losses in coloring strength. Thus, in practice, the leuco vat form is oxidized by air before alkylation, or, the alkylating agent is used in combination with oxidizing agents such as chlorite.

Rinsing or Washing. After oxidation and alkylation, the dyeing is rinsed or washed at 60–95 °C to remove excess oxidation or alkylating agent and, possibly, detached dye.

4.7.4 Dyeing Techniques [75]

Batch (Exhaustion) Processes. *In jigger dyeing*, the dyebath is charged with all the chemicals, and the dye and salt are added in steps at the desired temperature. After about six passages (or 60 min), dyeing is completed by rinsing and oxidation.

In *jigger dyeing with preimpregnation*, with water-soluble sulfur dyes, the material can be impregnated at boiling temperature with dye, salt, and wetting agent in the shortest possible liquor ratio. The dyebath is then filled with water, and the reducing agent added. In comparison with the normal jigger process, better penetration is achieved.

In the *pad jig process*, after the material is padded with dye and wetting agent, it is treated in a jigger at the dyeing temperature. The liquor contains the reducing agent, salt, and some padding liquor.

In *winch beck dyeing*, the material is pretreated in a bath containing the chemicals and wetting agent at 40–50 °C. The dissolved, reduced dye is then added, the bath is brought to dyeing temperature, and salt is added. After 20–30 min, the material is rinsed clear with cold water

In *circulating-liquor dyeing*, the dyebath contains wetting agent, chemicals, and the dissolved, reduced dye. After a prerun at 30–40 °C, the dyeing temperature is adjusted, salt is added, and the process is terminated after about 30 min by rinsing clear and oxidation.

In *jet dyeing* [76], foam control is important in the case of semiflooded jets. After wetting the material, reducing agent and chemicals are added at ca. 40 °C, followed by the dissolved dye. Since the air in the jet consumes reducing agent, hydrosulfite and NaOH must be added accordingly. After heating to dyeing temperature, more hydrosulfite is added, if necessary, and the material is dyed for 30 min. Dyeing is completed by rinsing, oxidation, and re-rinsing.

Continuous Pad Dyeing Processes. Ready-for-use liquid or water-soluble sulfur dyes and dispersible pigments are employed preferentially in continuous pad dyeing. As a result of the varying substantivity of the dyes in pad processes with reducing agents, tailing can occur. This can be avoided by taking into account both the feeding liquor factors of the individual dyes and the substrate factors. The recipe of the starting bath and of the feeding liquor can be calculated [77].

In the *one-bath pad steam process*, the material is impregnated with dye, reducing agent, and wetting agent and subjected to air-free steaming at 102–105 °C for 30–60 s. Rinsing is followed by oxidation and re-rinsing in a washing machine. The *two-bath process* is preferred especially for mercerized material to obtain a more even appearance. The material is padded with soluble or dispersed dyes, wetting agent, and possibly antimigration agent. After intermediate drying, if necessary, it is impregnated with reducing agent, followed by steaming and aftertreatment, as in the one-bath process.

In the *pad air-passage process*, for light to moderate shades, the material can be impregnated with dye, reducing agent, and wetting agent at elevated temperature (70–80 °C). During air passage for ca. 1 min, sufficient fixation is achieved. Dyeing is completed in washing machines as usual.

The *hot pad dry process* is a variant of the pad air-passage process. After impregnation, the material is subjected to hot drying on a drying cylinder. For final fixation of the dye, the material is impregnated with a cationic alkylating agent and redried.

In the *semicontinuous pad roll process*, after impregnation with the pad liquor, the material is batched up and stored in an air-free steam chamber (dry and wet temperature ca. 98 °C) where it is rotated slowly for about 1 h. By this procedure, the dye is almost completely fixed. Excellent penetration is obtained.

The *pad thermofixing process* permits dyeing with water-soluble sulfur dyes without reducing agents [78]. After padding and drying, the dye is fixed by heat treatment in the presence of acid-donating catalysts, urea, and thiourea. In this way, the water-solubilizing Bunte salt groups are split off. An excellent appearance is obtained in this process.

In a specially designed process, warp yarn dyeings with indigo can be imitated [79].

4.7.5 Combination with Other Dyes

For stretch cord and similar articles made of PA–CEL blends, metal-complex dyes that are resistant to reducing agents are added to dye the PA fibers, because only a few sulfur dyes stain PA sufficiently. Water-soluble sulfur dyes can be combined with selected reactive dyes. In this manner, more brilliant shades are obtained.

4.8 Azo (Naphtol AS) Dyes on Cellulosic Fibers
[5, p. 417–418]

An azo dye is formed from a Naphtol AS coupling component and a diazotized base (developing component) under suitable conditions (see Section 3.7). This coupling is carried out on the fiber itself. If components without solubilizing groups are used, the dye formed is an insoluble pigment, and dyeings with excellent fastness properties are achieved. Today, azoics are used primarily for deep clear shades, like red and purple.

4.8.1 Application of Azo Dyes

The dyeing process is carried out in steps: First, a cotton fabric is impregnated with a solution of sodium naphtholate and, after drying, is passed through a solution of a diazotized aniline derivative to give an attractive deep dyeing. Besides bases that need to be diazotized before use in dyeing, diazotized diazonium compounds in stabilized form are also being marketed as fast color salts (see Section 3.7). Cellulosic fibers in all processing states can be dyed with Naphtol AS combinations. Handling is safe and relatively simple.

On cellulose acetate, developing dyes are still used, but for black only. Although dyeings are produced that are especially fast to wet treatments, the process is more tedious than dyeing with direct disperse dyes.

4.8.2 Dyeing Processes on Cellulosic Fibers

In the *impregnating process* (hot solution process) the cotton fiber is treated first with Naphtol AS, in the presence of a protective colloid [5, p. 272]. Most Naphtol AS types can also be solubilized with alcohol and caustic soda by using the *cold solution process*. The concentrated sodium naphtholate solution is added to the impregnating bath. For certain naphthols formaldehyde is added to stabilize them on the fiber (air stability). By addition of sodium chloride a rather substantive attachment of the naphthol to the fiber is achieved. On dyeing from a long liquor, a lower temperature is used for impregnating the fabric, followed by rinsing and development. When using a padder, a higher temperature and, in general, intermediate drying is required.

Hank yarn is impregnated in a long liquor in a trough or on hank yarn dyeing machines. Wound yarn can be impregnated in the usual package dyeing machines. Piece dyeing is carried out on the winch beck and the jigger or, preferably, continuously by padding.

For *developing*, the base must first be diazotized in the cold, if necessary with the addition of ice, by using a solution of sodium nitrite in hydrochloric acid. To increase the coupling capacity, excess mineral acid must be neutralized or buffered by the addition of sodium acetate, disodium phosphate, or similar compounds. The textile impregnated with Naphtol is passed through the cold developing bath containing the diazotized base so that the dye is formed rapidly. In dyeing machines the developing solution circulates through the stationary textile. Alkali carried over from the padding bath must first be neutralized (e.g., with acetic acid, aluminum sulfate) to prevent decomposition of the diazonium salt.

The use of naphthols and bases in solution was introduced in the 1970s. These formulations are safer and simpler to apply. Diazotization of the liquid bases generally does not require the addition of ice.

After development, the textile is rinsed and subjected to careful *afterwash* in which excess dye lake is removed from the fiber surface. Otherwise, it severely impairs the rubbing-fastness of the dyeing. In addition, through the change in pigment crystal structure occurring during aftertreatment, a stable final shade is achieved with enhanced fastness.

Naphtol AS combinations that can be used at low temperature are suitable for dyeing cellulose fiber materials that are partially coated with wax (*batik*). After dyeing the wax is washed out by boiling.

The use of Naphtol AS dyes for the production of *denim articles* is of some interest. The cotton warps are simultaneously sized and padded with a coupling component. After weaving the fabric is developed with a "fast salt", i.e. a diazotized base and subjected to the usual aftertreatment.

4.8.3 Printing with Azo (Naphtol AS) Dyes on Cellulosic Fibers

Fabric is treated with the coupling component 2-hydroxy-3-naphthoic acid anilide (Naphtol AS), dissolved with alkali and a dispersant in a padder, and dried. It is then printed with a print paste containing a diazonium salt, acetic acid, sodium acetate, and a suitable thickener. The process is simplified by using diazonium salts stabilized with zinc chloride, 1,5-naphthalenedisulfonic acid, or tetrafluoroboric acid. Coupling takes place immediately after printing, so after drying the print can be washed to remove unconsumed Naphtol. Since Naphtols allow only for a limited range of shades, unavailable colors can be supplemented with selected reactives for yellow or Phthalogen dyes for brilliant blues [5, p. 521].

4.9 Dyeing Cellulosic Fibers with Other Dye Classes

4.9.1 Mordant Dyes on Cellulosic Fibers [59, p. 578], [80, p. 649]

The fiber is first treated with metal salts (mordanted). Highly adhesive, basic metal compounds are formed on the fiber. These compounds are capable of producing insoluble colored complexes (lakes) with certain azo and anthraquinone derivatives. Alizarin is the best-known anthraquinone derivative for this process (see Section 2.3). It used to be isolated from the root of the madder plant but has now been replaced by the synthetic product. Suitable azo dyes contain, e.g., hydroxyl or carboxyl groups in the position *ortho* to the azo group on one or both of the aromatic nuclei. The shade of the dyeing depends on the type of metallic mordant used. Alizarin with aluminum or calcium salts produces the well-known Turkey red.

The mode of operation in alizarin dyeing is relatively tedious. The old Turkey red dyeing process required at least ten working steps: boiling, treatment with rancid olive oil, washing, treatment with a tanning agent, mordanting with alumina, dyeing in hard water, softening, acidification, and soaping. In the new process, treatment with Turkey red oil (sulforicinoleic acid) has replaced the oiling step. The light- and wetfastnesses obtained are excellent.

In the past, this method was used to dye, e.g., cambric and bunting. Today, easier methods using developing dyes and vat dyes have replaced this process.

4.9.2 Acid Dyes on Plant Fibers [59, p. 580]

Acid dyes, i.e., azo or anthraquinone dyes containing sulfonic acid groups, were occasionally used for dyeing cotton; however, they are no longer important. They do not have sufficient substantivity for cellulose. Only those dyes are usable that can form a metal complex on the fiber when applied together with a metal salt. Excess heavy metal contaminates the wastewater.

Acid dyes can attach directly to vegetable hard fibers (jute, sisal) by salt formation because the fiber companion substances contain basic groups.

4.9.3 Basic Dyes on Cellulose [59, p. 580, 80, p. 548]

Cationic (basic) dyes show no substantivity for cellulose. Therefore, pretreatment of the fiber with tannic acid (which contains phenolic OH groups) is required. The tannin mordant is insolubilized with antimony salts (tartar emetic). Synthetic tanning products are also suitable. Thus, the fiber is dyeable in a weakly acidic medium, a saltlike bond being formed with the acidic phenolic hydroxyl groups. Basic dyes attach to bast fibers without a mordant. Suitable dyes are to be found in the azo, diphenylmethane, and triphenylmethane series, and among thiazine, azine, oxazine, thiazole, and quinoline derivatives. Bright colors are obtained at low cost with rhodamine, auramine, fuchsin, or methylene blue, among others. Since the fastness, especially lightfastness, is poor, this method is of little significance.

4.9.4 Oxidation Dyes on Cellulosic Fibers [80, p. 437]

In the application of oxidation dyes, aromatic amines form insoluble polyazine derivatives in the fiber (azine dyes [81]). Synthesis proceeds in several steps in a hydrochloric acid medium by oxidation (e.g., with dichromate). For the chief representative of this group, aniline black, the chromophore consists of dibenzopyran rings. Oxidation dyes are rapidly decreasing in importance because aniline and other aromatic amines as well as the bichromate used for oxidation are toxic. The colors produced by aniline black are characterized by a full bluish black shade and excellent fastness. Since they are easily reserved, they are still used occasionally for printing grounds. For individual processes, see [82].

4.9.5 Phthalogen Dyes on Cellulosic Fibers

As with oxidation dyes, the actual dye in the Phthalogen process—the insoluble phthalocyanine pigment—is formed in the fiber itself. The dyeing is performed with precursors. There are two main processes.

Aminoiminoindolenine and its derivatives are *low-molecular Phthalogen developers*, applied together with heavy-metal donors (preferentially Cu or Ni) by padding. The pigment is formed on heating (150 °C) in the presence of weak reducing agents (e.g., glycols). This is followed by aftertreatment with hydrochloric acid, with the addition of sodium nitrite to remove secondary products. The most important aminoiminoindolenine is *C.I. Ingrain Blue 2:1*, 74160.

Polyisoindolenines, which are *high-molecular Phthalogen developers* and are complexes of heavy metals and indolenine, are developed with reducing agents in a wet treatment. The polyisoindolenine complexes are fiber-affinitive and can be applied by using the exhaustion process. Their substantivity is low but can be increased by pretreating the cellulose with anionic products. Reductive development is performed in an alkaline medium with hydrosulfite. For details, see [82, 83].

Dyeings made with Phthalogen developers are characterized by high brilliance and excellent fastness. Thus, they are suitable for washfast and weatherproof articles. No danger of photochromism exists if a finishing operation is performed with synthetic resin products.

For *printing*, the Phthalogen dye is mixed in a printing paste with solvents, and an ammoniacal solution of a copper or nickel complex. After printing, the fabric is dried immediately above 80 °C, and fixation is accomplished by steaming for 8–5 min at 100–102 °C or by hot-air treatment at 135–145 °C for 3–5 min. Excess heavy metal is removed by an acid treatment, and thickener and loosely adhering dye pigment are washed off.

4.9.6 Coupling and Diazotization Dyes on Cellulosic Fibers [59, p. 547, 84, p. 285–286]

Direct dyes (see Section 3.3) that contain aromatic amino groups can be diazotized on the fiber after dyeing and then coupled with a "developer" (a phenol, naphthol, or aromatic amine). Wetfastness, in particular, is improved by such an enlargement of the molecule, and the shade also changes (see Section 2.2). Conversely, water-soluble, substantive azo dyes which bear amino or hydroxyl groups capable of coupling can also be used, followed by aftertreatment with a diazonium compound. The resulting polyazo dye shows excellent wetfastness (see also Section 4.8).

4.9.7 Pigments and Mineral Dyes on Cellulose

Organic and inorganic pigments are not considered as true textile dyes, since they have no affinity to any textile fiber. However, organic pigments as well as inorganic colored oxides are widely used for coloring textiles, primarily consisting of cellulosic fibers.

Pigment dyeing and printing [82], p.625–630, [5], p. 418–419. Being insoluble products, pigments can only be fixed to the surface of the fiber. For pigment coloring, very fine (particle size ≤0.5 µm) inorganic or organic pigments are used in a nonionic preparation. They are fixed with the help of a binder. The binder is an aqueous dispersion of cross-linkable mixed polymers (basis polyacrylate, polystyrene, polyurethane). A film, in which the pigment particles are embedded, is formed on the fiber. Cross-linking occurs during heating. The binder film must adhere firmly to the substrate without the material becoming too stiff or sticky. The type and amount of binder used depend on the amount of pigment and determine not only the feel, but also the rubfastness and wetfastness of the dyeing or print [180]. Special polyurethane-based binders from elastic films that can bind large amounts of pigment, making deeper coloration possible.

Since pigments show no substantivity for textile fibers, piece goods are impregnated continously with a liquor that contains the pigment, the binder, an antimigration agent, a cross-linking agent if necessary, an acid donor, and a softener. A typical print paste contains, in addition to the pigment preparation, a nonionic detergent, a resin-forming binder, a high molecular thickener, an antifoam and other components for particular requirements. The pieces are dried at 90–120 °C and fixed at 160–180 °C. No afterwash is required. A resin finish can be carried out at the same time, e.g., a crease-resistant and antishrink finish, hydrophobing or a soil release finish.

Pigment dyeing and printing offer the advantages of universal applicability to all fibers (including glass fibers). It is a single-stage process, also for fiber blends, i.e., an energy-saving process.

Aftertreatment is not required. Raw cotton is covered well. The resistance to light is very good; the disadvantages are limited depth of color, stiffer feel of the product, and limits to the rub- and wetfastness.

Pigment dyeing is used commonly for heavy textiles (e.g., canvas), light printing grounds, dress materials, shirting, bed linen and furnishing articles.

Dyeing with mineral dyes [59] p. 584, [80] p. 520, [84] p. 395. In a pad process, inorganic metallic salts can be applied to cotton and then converted in an alkaline medium to the corresponding oxide by treatment with steam. Brown and khaki shades (mineral khaki) are obtained with chromium and iron salts. This dyeing is inexpensive, lightfast, and weather- and rotproof. The disadvantages are wastewater contamination and definite hardening of the fabric. This method is still important in some countries for dyeing tarpaulin and uniform materials.

4.10 Acid and Metal-Complex Dyes on Wool and Silk

4.10.1 Principles of Dyeing of Wool and Silk [5, p. 421–428].

The dyeing properties of wool vary depending on the origin, breed, age, food, season, and habitat of the sheep. Even fleece from a single sheep reveals pronounced differences in quality. In fact, differences are also found within a single fiber due to biological and environmental factors. By sorting of wool provenances and fleece components, batches are put together that largely correspond with each other with regard to fineness and to processing and application properties.

The wool fiber is composed of protein filaments and consists mainly of keratin with a very complex structure. The amino groups of keratin are of decisive importance for the dyeing process. The amount of basic groups titratable with acid is 850 µmol per gram of wool fiber. In the acidic and neutral range, carboxyl groups are present largely in the undissociated state.

The number of ionic groups in the fiber depends on the pH of the medium. Keratin is most stable when the number of negative and positive ions is identical. The isoelectric point of wool is pH 4.9 at room temperature.

The *morphological structure* of the fiber determines the pathway that dyes take during dyeing and is critical for the rate and extent of dye uptake. In some way, the dye has to penetrate the more or less hydrophobic layer on the fiber surface, formed by the epicuticle and the exocuticle. The strong swelling capacity of the intercellular cement is important for the penetration of dyes into the fiber. Only then are the sulfur-rich keratins also penetrated by the dye molecules. In general terms, Fick's law can be applied to the diffusion phenomena [46].

Bonding Forces Between Dye and Fiber. Dye anions can participate in ionic interactions with fibers that possess cationic groups. However, the formation of ionic bonds is not sufficient to explain dye binding, because compounds that can dissociate are cleaved in the presence of water. Secondary bonds (dispersion, polar bonds, and hydrogen bonds) are additionally formed between dye and fiber [47]. Close proximity between the two is a prerequisite for bond formation. However, this is counteracted by the hydration spheres of the dye and of wool keratin. On approach, these spheres are disturbed, especially at higher temperature, and common hydration spheres are formed. The entropy of the water molecules involved is increased in this process (hydrophobic bonding). In addition, coordinate and covalent bonds can be superimposed on secondary and ionic bonds.

Technological Aspects of Wool Dyeing. Wool is dyed in all states of production, i.e., flock, slubbing, yarn, and piece. High-quality articles woven from fine wool require gentle treatment to avoid shrinkage. Jig dyeing is only possible with shrinkproof material. Wool suffers degradation when dyed at the boil; therefore, dyeing is normally conducted at 80–95 °C. With polyester–wool blends dyeing temperatures of up to 120 °C are necessary for the polyester part. In this case formaldehyde is added to the liquor, which stabilizes the wool keratin against hydrolytic degradation [85, p. 318].

Dye selection depends on the type of textile material and the shades to be dyed. For light an brilliant shades disulfonic acid dyes are preferred; for deep dyeings, such as navy and black, chrome or metal-complex dyes are used. However 1:2 metal complex sulfonic acid dyes should not be used for piece dyeing, because of their poor leveling properties. Weekly acidic 1:2 metal complex dyes are the easiest to use. Leveling is good, and shading after dyeing is possible. The weekly acidic dyeing medium is also most favorable to maintain feel, elasticity, and strength of the wool. Excellent fastness to milling and potting is achieved with chrome and metal complex dyes.

Slubbing is frequently dyed by a continuous process. Here, a thickener, a special auxiliary combination, and an acid or acid donor are added to the pad liquor. The padding auxiliary assures homogeneous distribution of the dyes on the fiber, enhances diffusion in the fiber, and helps to remove the thickener in the afterwash. The material is steamed for 15–60 min after padding, depending on the dyes used.

Chlorinated (shrinkproof) wool showes a different coloristic behavior. Dye uptake and release is facilitated; wetfastness is reduced. In contrast, an antifelt treatment with a synthetic resin hardly leads to any change in coloristic behavior.

Dyeing Silk. Being chemically similar to wool, silk can in principal be dyed with the same dye classes as wool. However, the number of amino groups is considerably lower (230 μmol per gram); the mechanical stability is also much lower, because silk has no cystine cross-links. Therefore acid is added slowly during the dyeing process, and dyeing is conducted at 70–80 °C. To protect the fiber, silk is frequently dyed in the soap bath (pH 8–8.5), which otherwise precedes dyeing and is used for degumming. Acid dyes are the most important dyes for silk; but direct, metal-complex, and reactive dyes are also used.

4.10.2 Acid Dyes on Wool

Acid dyes (see Sections 3.4.2 and 3.9), are made water-soluble by the introduction of sulfonic acid groups. Dissociation gives rise to dye anions that interact with the ammonium groups of the fiber

From a coloristic standpoint, acid dyes are subdivided according to their affinity. Affinity increases from *leveling dyes* and weakly acidic absorbing (moderately leveling) *milling dyes* to neutral absorbing (poorly leveling) *super-milling dyes*. This sequence approximates the increase in size of the dye molecule. The presence of aliphatic groups in the dye molecule contributes to a substantial increase in binding to wool, converting leveling dyes to types that are fast to fulling. The sulfonic acid groups determine not only the number of possible ionic bonds to the fiber, but also hydration, which counteracts binding.

Dyes with similar absorptive behavior should be selected for combined application. The combination values for acid dyes established originally for polyamide were later transferred to wool [86, 87].

In dyeing, addition of salt has a retarding and leveling effect. Higher concentrations of sulfate ions are assumed to compete with dye anions for ammonium groups of the fibers, and inorganic salts exert an aggregating effect on dye molecules.

Dyeing Processes. Dyeing instructions for different types of acid wool dyes vary, especially with regard to the pH range used (see Table 4.3). The greater the affinity for the fiber, the more strongly must the ionic binding component be repressed, i.e., the pH value at the beginning of dyeing must be higher.

The starting temperature is 60 °C for leveling dyes, 50 °C for milling dyes, and 30 °C for super-milling dyes. Usually, the acid is added first to the dyebath, then salt followed by the dissolved dye. The system is heated slowly and dyed at 95 °C. Addition of acid toward the end of the dyeing process completes bath exhaustion. Warm and cold rinsing follows. With selected dyes dyeing can be performed at 80 °C. Milling dyes require the addition of a leveling agent [5, p. 276–277].

Table 4.3. Dyeing with acid dyes on wool

Addition	Leveling dyes	Milling dyes	Super-milling dyes
Na_2SO_4, cryst.	2–4 g/L	2–4 g/L	2–4 g/L
Acid (donor)	1–2 g/L H_2SO_4 (96 %)*	1–3 g/L acetic acid (30 %)	1–3 g/L $(NH_4)_2SO_4$
pH value	2.5–3	4–5	6–7
Leveling agent		0.5 g/L	0.5 g/L
Final addition		1 g/L H_2SO_4*	1–2 g/L acetic acid (30 %)

* Or formic acid.

Trichromic dyeing (matching shades with a standard combination of three dyes) can be conducted with leveling dyes. In the case of milling dyes, the desired shade is better achieved by selection of a dye with a hue similar to the required and shading.

Printing with Acid Dyes on Wool and Silk. For printing on wool, suitable acid dyes are selected to combine sufficient light- and wetfastness with high brilliance. Both 1:1 and 1:2 metal complex dyes give a higher degree of fastness, but they show rather subdued, dull colors. Dissolving acid dyes for printing requires the use of urea, a solubilizing agent such as thiodiethylene glycol or a dispersing agent, and an acid donor (to promote fixation) in hot water. Guar products together with British gum are used as thickeners. Fixation on wool or silk requires prolonged steaming (30–60 min) with saturated steam at 100–102 °C. Overheated steam leads to wool damage and must be avoided. Afterwashing must be carried out under minimum mechanical stress and at low temperature (<40 °C).

Fastness Improvement. The wetfastnesses of dyeings on wool that has been chlorinated to prevent felting and has received a finish with synthetic resin (polyamide–epichlorohydrin or polyurethane; superwash wool) can be increased by means of methylol amide compounds. Both fastness and antifelting finish can be improved through the application of a polyquaternary compound [88]. Anionic condensation products such as aromatic sulfonic acids with formaldehyde can form a barrier at the surface of the fiber and thus diminish the bleeding of anionic dyes.

4.10.3 Chrome Dyes on Wool

Chrome dyes (see Section 3.11.2) are selected acid dyes that form complexes with chromium ions. With complex formation a strong bathochromic shift of shade occurs. In addition, as a result of the superposition of several excited states, a marked dulling of the hue is observed.

Complex formation occurs during the dyeing process in a strongly acidic medium with participation of electron donors (ligands) from the chromophore and the fiber. Chromium, the central atom of the complex, acts as a link between dye and fiber. This results in a very strong bond, which is reflected in the excellent fastnesses obtained. The binding of chromium occurs by substitution of the H in neighboring COOH or OH groups and through electron lone pairs from C=O, NH_2 or —N=N— groups (dative bonds). Thus, the dyes must contain suitable functional groups, such as salicylic acid or alizarin type or o,o′-dihydroxyazo groups.

Six-coordinate chromium(III) acts as the central atom. It is formed from dichromate, which is reduced by the fiber. The reducing effect of wool can also be enhanced by organic acids (tartaric, lactic, or formic acid). Thiosulfate also acts as a reducing agent, increasing the rate and degree of conversion in chroming and decreasing fiber damage [89]. Lowering the temperature from boiling to 90 °C contributes to fiber protection [90]. Because pollution of wastewater by chromium must be avoided, dyeing methods have been developed which require no excess potassium dichromate [91, 92]. Individual chrome factors are given in pattern cards.

Auxiliary agents, such as ethoxylated fatty alcohols, alkylphenols, and fatty amines, can improve level dyeing.

In the *afterchroming method* with chrome-developing dyes, the dyeing liquor is prepared with formic acid (pH 3.5–3.8), sodium sulfate, and an amphoteric leveling agent as a wool protectant. The process is started at 40 °C, and dissolved dye is added. The system is heated, and dyeing is performed at 90 °C for 30–

45 min. If the exhaustion of the bath is too low, more formic acid is added toward the end. For subsequent chroming the bath is cooled to 70 °C, followed by the addition of potassium dichromate and heating. The temperature is maintained at 90–100 °C for 30–45 min, then sodium sulfate is added to detach the saltlike bound chromate from the wool and make it accessible to complex formation. Other variants use reducing and complexing agents to support the chemical reactions [93]. Finally, the bath is neutralized with ammonia (pH 8).

In the *one-bath chroming method* (*metachrome process*), complex formation must be preceded by dye diffusion. For this reason, the release of Cr^{VI} must be delayed. The dye liquor is prepared with metachrome mordant (a mixture of sodium chromate and ammonium sulfate) and sodium sulfate. After a prerun at 40–50 °C, the system is heated slowly and dyed at boiling temperature for 45–90 min. Shortly before dyeing is completed, the bath is exhausted by the addition of acetic acid.

The boiling time is shortest in the afterchroming method, and the most level and fast colors are produced. Poor shading possibilities represent a disadvantage. The one-bath chroming method is most important.

4.10.4 Metal-Complex Dyes on Wool

Metal-complex dyes (Sections 2.10 and 3.11) are chemically similar to chrome dyes. The risk of fiber damage during dyeing is reduced because the complex is formed during dye production.

4.10.4.1 1:1 Metal-Complex Dyes

Dyeing is conducted in a strong sulfuric acid medium at pH 1.9–2.2. Amino groups in the fiber are then present in the ammonium form, and ionic bonds to the dye anion are formed. Under these conditions, amino groups are not available as ligands. Only with increasing pH during rinsing can they be included in the complex in exchange for aquo ligands. The addition of auxiliaries (alkanolethoxylates) allows a decrease in the amount of acid added (pH 2.5–3) because they form addition compounds with the dye. The dyeing process is slowed down and exhaustion is made uniform. With synergistic amphoteric mixtures of auxiliary agents, dyeing can be performed at even higher pH (3.5–4) [94].

In the *dyeing process*, the pH is adjusted with sulfuric acid to 1.9–2.2 (pH 2.5 in the presence of auxiliary agents). After addition of sodium sulfate, the dyebath is heated to 40–50 °C; then the dissolved dye is added. Dyeing is carried out at boiling temperature for 90 min. The material is then cooled and rinsed; ammonia or sodium acetate can be added to the last rinsing bath. A lowering of the dyeing temperature to 80 °C is possible in the presence of an ethoxylated fatty amine (pH 1.9–2.2).

4.10.4.2 1:2 Metal-Complex Dyes

Metal complexes in a molar ratio of 1:2, i.e., two chromophores coordinated to one central Cr or Co atom, owe their high affinity for the fiber to the large size of the dye molecules, their compact, often almost spherical shape, and their negative charge. To achieve sufficient levelness, the incorporation of sulfonic acid groups into the dye molecules was initially avoided so additional ionic interactions between fiber and dye could be excluded. Instead, water solubility was achieved by incorporation of methylsulfone ($-SO_2CH_3$) or sulfonamide ($-SO_2NH_2$) groups.

With the development of suitable dyeing auxiliaries, dyes containing sulfonic acid groups were introduced. Their production is inexpensive; they give a high dye yield and are soluble in the cold. In general, the wetfastness is somewhat higher than that of dyes containing methylsulfone and sulfonamide groups. Unfortunately, dyes containing two sulfonic acid groups, in particular, are susceptible to nonlevel dyeing. To avoid tippy dyeing, auxiliary agents, especially ethoxylated fatty amines, are used [5, p. 276–277], [95], which form adducts with the dye. These adducts break down at higher temperature. The leveling effect is increased by addition of Glauber's salt.

Since the 1:2 dye complexes are formed in a weakly acidic medium, they can be applied to wool at a higher pH. This enables a gentle mode of dyeing.

In the *dyeing process*, the pH is set at 4.5–5.5 (acetic acid, ammonium sulfate, or ammonium acetate), and the dyeing auxiliaries are added to the liquor. The dyes are added at 30–50 °C; the system is heated and dyed at boiling temperature. After rinsing, formic acid is used for acidification to improve the feel and wetfastness.

4.11 Acid and Metal-Complex Dyes on Polyamide

4.11.1 Chemical and Physical Structure of the Fiber

For further information, see [96, pp. 569–579].

Chemical Composition. In *polyamide 6* (PA 6, polymerization product of ε-caprolactam) and *polyamide 66* (PA 66, adipic acid polymerized with hexamethylenediamine) one chain end consists of an amino group, which can be present in the free state or in the acylated form. Amino groups are of special importance for dyeing because they form ammonium groups in an acidic dyebath by addition of protons. The lower dye uptake in comparison to wool is caused by the comparatively low number of amino groups. The depth of color achieved on PA 6 is somewhat less than that on PA 66.

During dyeing a pH below 2.4 should be avoided, because fiber damage occurs at low pH.

Modified polyamides contain a varying number of amino groups [96, p. 576–577]. The dyeing characteristics of different PA fibers as a function of the content of amino groups can vary in a wide range from light to ultradeep dyeing types. In addition, polyamide fibers dyeable with cationic dyes can be produced by the incorporation of sulfonic acid groups.

Supramolecular Structure. The state of order of the polymer in the fiber, i.e., the degree of crystallinity and degree of orientation are critical for its dyeing behavior. However, the amorphous remainder is of greater importance in dyeing because the chain segments in this part have a mobility that permits penetration of swell water and dye molecules. The system can be regarded as a uniaxial stretched dynamic network in which the ordered regions act as knot points [97].

The amorphous regions are present in the glassy state at lower temperature, and the mobility of chain segments begins only above the glass transition temperature T_g. In the presence of water, T_g lies below the dyeing temperature for PA and hence has no influence on the course of dyeing.

The fiber exhibits certain structural differences in the radial direction (skin–core structure); the more compact surface layer in PA 66 obstructs dye uptake slightly. Technological operations change the fibrous structure. Stretching leads to increased orientation and impeded dye uptake. Thermal treatments also result in changes in structure. Dyeing behavior is not changed essentially by *hydrosetting*, but the structure is loosened. Thus, the size of the dye molecules has a smaller effect on dye uptake. Free shrinkage of the material results in a faster rate of dyeing. On the other hand, on heat setting under tension the critical range of absorption is shifted upward by 15 K. The required penetration time is almost doubled. In *texturing*, the internal forces acting during torsion of the thread change the inner structure so greatly that dye molecules can diffuse more easily into the fiber. Any unevenness in operations that changes the supermolecular structure of the polymer leads to disturbances of dye uptake (stripiness or Barré effect). In general, PA 6 is dyed more evenly than PA 66 because it has a more open structure.

4.11.2 Interactions Between Dye and Fiber

The binding of dyes to polyamide is based on various interactions [98, p. 270 ff.].

The chromophore system of the dye, with its aromatic rings (delocalized π-electron systems), easily polarizable azo groups (electron lone pairs), and other donor and acceptor groups, forms many polar bonds, dispersion interactions, and hydrogen bridges with the polymer. These *secondary bonds* are superimposed on all other bonds. The larger the dye molecules and the fewer the hydrophilic groups carried by the dye, the greater is the superposition.

Acid dyes are present in the fiber in ionic form. There is a stoichiometric relationship between the content of amino end groups in the fiber and dye uptake. This can be ascribed to electrical neutrality by formation of *ionic bonds*. Dye anions are able to displace counterions of the ammonium groups in the fiber because they are supported by numerous secondary bonds (ion-exchange mechanism). Additional dye uptake beyond the ionic binding capacity is possible only in the form of purely secondary adsorption. Dissociation and hydration counteract this effect. The dyeings obtained by overdyeing have poor wetfastness. The simultaneous action of ionic bonds and secondary bonds determines the degree of dye uptake in equilibrium. This is frequently reflected in a superposition of the Langmuir and Nernst isotherms [99].

The amount of dye bound at equilibrium decreases with increasing number of sulfonic acid groups in the dye molecule due to increasing hydration. For this reason, mono- and polysulfonated acid dyes should not be used together in combination dyeings.

Amino groups of the fiber can participate as ligands in the *coordinate bonding* of 1:1 metal-complex dyes. This is, however, of little practical importance. In addition, they can form *covalent bonds* to reactive dyes.

4.11.3 Dyeing Processes with Different Classes of Dyes

Because of their good leveling properties, acid dyes are used for medium shades up to ca. 1/1 standard depth of shade. For deep and dark shades metal-complex dyes are preferred, because of their high fastness level, though they tend to nonlevel absorption and structure-related stripiness.

The fastness of colors produced by various classes of dyes varies widely. Whereas acid dyes produce dyeings with medium to good lightfastness, metal-complex dyes attain an excellent lightfastness level. Wash- and perspiration-fastness with acid dyes is medium; good to very good values are obtained with metal-complex (and also with reactive and some disperse) dyes. The values for PA 66 are slightly better than those for PA 6. Improvements can be obtained by aftertreatment with synthetic tanning products. The lightfastness of dyeings obtained with acid dyes on PA is slightly lower than on wool. Resistance to the chlorinated water of swimming pools plays a special role for bathing suits. PA 66 is usually superior to PA 6 in this respect.

4.11.3.1 Acid Dyes

Basically, the same sulfonated acid dyes which are used to dye wool are also suitable for polyamide (see Sections 4.10.2–4.10.4). In a weekly acidic dye liquor the sulfonate anion of the dye binds to the ammonium cation, formed from the amino end groups of the fiber. Affinity to the fiber can be increased by introducing additional substituents into the dye molecule. The affinity increases along the series: olefinic<aliphatic<aromatic<cycloaliphatic.

Dye selection depends on technical requirements. *Monosulfonated acid dyes* comprise leveling dyes and neutral absorbing dyes; they can easily be combined with each other. *Disulfonated acid dyes* produce dyeings with good wetfastness. They are used as self-shades but are not recommended for three-color combinations [100].

The dyeing behavior of acid dyes can be described with the help of the combination values K, which are derived from combination dyeings at 70, 80, and 90 °C. The K values of dyes from different dye manufacturers cannot be compared [101].

Dyeing Conditions. The *liquor pH* is important for level dyeing, for exhaustion of the bath, for wastewater contamination, and for the effort involved in the afterwash after dyeing. With increasing dye affinity, the ionic interaction in the initial phase must be repressed to achieve uniform absorption, i.e., the liquor must be neutral at the start of the dyeing process, and only in the course of dyeing is the liquor slowly acidified by one of the following methods [102, 103]:

1) Acid is added at the beginning of dyeing or continuously, preferably by using pH-controlling instruments.
2) Acid donors release acid during the dyeing process. Substances used are ammonium sulfate, which liberates NH_3 during dyeing; sodium pyrophosphate ($Na_4P_2O_7$), which is converted to the acidic disodium hydrogenphosphate; esters of organic acids; and heterocyclic compounds, which slowly release organic acids when exposed to heat.

In the case of low-affinity acid dyes, addition of acid toward the end of dyeing is recommended for full exhaustion of the dyebath.

Instead of pH control for regulation of the dyeing process temperature control can be more convenient [104].

Electrolytes retard liquor exhaustion. The competition of sulfate anions with the dye for dye-affinitive positions is responsible for this effect.

To improve leveling a nonionic or amphoteric *auxiliary agent* is used. In the case of fiber-affinitive agents, the sum of the dye anions and auxiliary anions should not exceed the binding capacity of the fiber to avoid blocking effects [105]. The use of computerized dosing pumps for dyes and acid permits reliable control of dye sorption [105a].

Aftertreatment. The wetfastness of dyeings produced with acid dyes is often unsatisfactory. Formerly, tannic acid and tartar emetic were used, which form a 1:1 adduct on the material by hydrogen bonding. This process is expensive and said to be carcinogenic. Synthetic tanning agents (Synthanes) are preferred. They are condensation products of aromatic sulfonic acids with formaldehyde [106] or of phenols with formaldehyde, which are made water-soluble by reaction with bisulfite. These products deposit on the fiber surface and form an electrostatic barrier that counteracts the release of dye anions from the fiber. However, Synthanes are poorly resistant to dry heat and steam; treatment with Synthanes results in slight changes in shade, reduced lightfastness, and a hardening of the feel. In the presence of cationic softeners, the fastness to rubbing is decreased.

Dyeing Processes. In *dyeing with temperature control*, the pH of the dyebath is adjusted to 4.5–7, and leveling agent is added, followed by the dissolved dye; the temperature is increased according to a predetermined program, and dyeing is carried out near boiling temperature.

In *dyeing with pH control*, dye is dissolved in the liquor, the pH is adjusted to 8.5–10 with ammonia, and the auxiliary agent is added as described above. The material is placed in the boiling liquor, and acid is added slowly. The final pH of 4.5 is attained after 45–60 min. Alternatively, a pH regulator is used.

In *aftertreatment*, the exhausted bath or fresh liquor is adjusted to pH 4.5 with formic acid, and Synthane is added. Aftertreatment is started at 40 °C, and the material is treated at 70–80 °C; it is then rinsed.

4.11.3.2 1:2 Metal-Complex Dyes

The metal-complex dyes used for polyamide dyeing are acid dyes that are bonded to transition metals, especially chromium and cobalt, in the molar ratio of 1:2. These dyes are negatively charged and can form a ionic bond with the fiber. The amino groups of the fiber are not included in the complex. Water solubility of commonly recommended 1:2 metal complex dyes for polyamide is based on the presence of sulfonamide and methylsulfone groups. However, also dyes containing sulfonic groups, which are critical for their limited leveling capacity, are commercially available (see below). The high affinity of metal complex dyes for the fiber is due to secondary bonds. Their high fastness level makes them suitable for articles that are exposed to intense use, such as floor coverings and car seat covers. The tendency of these dyes to mark structural differences in the material is a disadvantage. For this reason, they are recommended only for dark shades. The disulfonated acid dyes give good light- and wetfastness on PA and are frequently used in PA dyeing despite the poorer build-up properties. The high affinity of 1:2 metal-complex dyes makes the addition of amphoteric or nonionogenic leveling agents necessary.

The absorption of dyes by PA increases with decreasing pH. This tendency is especially pronounced in the case of disulfonic acid dyes. They are dyed in the presence of acetic acid. Dyeing with dyes free of sulfonic acid groups depends less on the charge on the fiber and more on the extent of formation of free complex acid, which is more readily absorbed than the anion.

In the *dyeing process*, dye and auxiliary are added to the liquor. It is then made weakly acidic with ammonium sulfate and acetic acid, if necessary. High-affinity dyes can be applied in a neutral or weakly alkaline medium. The process is started at 30–40 °C, followed by heating, and dyeing at boiling temperature. A prerun with auxiliary agent at boiling temperature reduces the stripiness of the dyeing.

4.11.4 Technology of Dyeing Polyamide

Dyeability of Polyamide Material. As outlined in Section 4.11.1, the dyeability of polyamide materials can vary over a wide range. The dye-binding capacity of a given PA material can be characterized by a saturation value by comparing dyeings at pH 4.7 with reference material, the amount of dye taken up being given as the fiber sum number S_F. The values lie between 1 and 3.

Pretreatment. Before dyeing, fabrics generally must be prefixed to compensate for material-related differences in affinity and to reduce the sensitivity to creasing during the dyeing process. Prefixing can be performed in hot air as *thermofixing* (15–20 s at 190 °C for PA 6 or 200–230 °C for PA 66) or as *hot-water fixing* (45 min at 130 °C). Soiled material should also be washed before dyeing.

Dyeing of Special Materials. Exceptional features in dyeing exist for certain articles.

Stockings, *tights*, and *socks* are dyed either in drum-dyeing machines or on automatic apparatus (for stockings). In these machines the stockings or tights are drawn over aluminum forms, sprayed with dye liquor, and fixed.

Belts for technical purposes (*safety belts*) can be dyed most efficiently by the pad steam process in continuous systems.

In the piece finishing of *carpets*, stringent demands are made on the evenness of the shade. Dyeing on special, open-width winch becks (4–5 m wide) is frequently performed at 90 °C by the pH shift process, starting at pH 8. Within 30–45 min, the pH is lowered to 4.5–6.5, depending on the depth of color, by using pH regulators or acid dosing [107]. Dyeing in a cold batch process lowers consumption of water and energy [108]. In continuous dyeing of carpets thickening agents are added to the pad liquors to prevent dye migration. The dye liquor is applied to the voluminous material with help of a minimal application process by using foamed liquors. The liquor uptake is between 120 and 500 % [109, 110]. Frothing auxiliaries are added to prevent dye from being washed out of the fiber tips by the formed condensate (frosting effect). Dye fixation is effected by steaming. Differential dyeing materials require a specific selection of dyes, auxiliaries, and pH values.

For *printing* with acid and metal-complex dyes on polyamide material see Section 4.10.2.

Aftertreatment. Frequently, textiles manufactured from polyamide fibers (hosiery, underwear) come into close contact with the skin. Therefore, the material has to be carefully rinsed or washed after dyeing. For the treatment with synthetic tanning agents to improve the wet fastness, see above (Acid Dyes).

4.12 Disperse Dyes on Polyester and Other Man-Made Fibers

For information on polyester (PES) fibers, see [96, p. 464–473, 579–600]. For properties, see also [111]. The PES fiber is quantitatively the most important synthetic fiber (see Table 4.4). At the end of 2000, world capacity was 21.2×10^6 t/a (cf., PAC 2.7×10^6 t/a, PA 3.8×10^6 t/a).

Table 4.4 World polyester fiber production (10^6 t/a)

Year	Total	Staple fiber	Filament
1970	1.6	0.9	0.7
1980	5.1	3.0	2.1
1990	8.7	4.7	3.9
1998	16.0	7	9

Their inexpensive production from petrochemical raw materials and excellent textile properties alone and in combination with natural fibers guarantee PES fibers universal applicability. Beside technical applications, uses include clothing, especially mixed with wool for suiting and trouser materials and mixed with cotton for shirts, raincoats, trousers, and casual wear. Pure PES fibers, especially in textured form, are employed in the knitwear sector. Polyester has found a wide field of application in the microfiber sector.

Since PES fibers are hydrophobic, water-soluble dyes do not attach. In contrast, PES fibers can be dyed easily with water-insoluble, small molecular dyes originally developed for dyeing cellulose acetate. Since the preferred dyeing medium is an aqueous liquor, the poorly water-soluble dyes must be dispersed before application (Section 3.2).

More than 50 % of disperse dyes are simple azo compounds, about 25 % are anthraquinones, and the rest are methine, nitro, and naphthoquinone dyes (see Sections 2.2, 2.3, 2.6, 3.12).

4.12.1 General Aspects

4.12.1.1 Dyeing in Aqueous Liquor

Since the fundamental work of Meyer and Kartaschoff in 1925 [112] and the studies of Vickerstaff [54] in the early 1950s, a large number of studies have been published on the kinetics and thermodynamics of dyeing cellulose acetate and syn-

4.12.1 General Aspects

thetic fibers with disperse dyes in aqueous dyebaths. For a survey, see [80, p. 67, 147]. Disperse dyes must be molecularly dispersed in the dyebath, i.e., the dye must be dissolved in the aqueous bath before it can adsorb to the fiber surface and then diffuse into the fiber.

Thus, the form in which the disperse dye is present in the dye liquor is decisive for the dyeing process. The water solubility and rate of dissolution are influenced by many factors, including temperature, particle size, crystal modification and crystal growth, melting behavior, type of dispersion and dispersing agent used (fineness, stability, agglomeration, aggregation of dispersion), presence of salts, and behavior in dye mixtures [113, 114, 115].

The water solubility of pure disperse dyes is a few milligrams per liter and increases strongly with temperature. It is also increased manyfold by dispersing agents [116].

The state of the dye in the fiber is often compared to a solid solution. Thus, the thermodynamic dyeing equilibrium of a disperse dye between water and fiber follows the Nernst distribution law, i.e., $C_F/C_L = K$ at infinite time. At constant temperature, the distribution coefficient K [i.e., the ratio of the concentration of dye dissolved in the fiber (C_F) and in the liquor (C_L) at equilibrium] is constant. At normal dyeing temperatures the rate of dissolution of dispersed dyes in the dye liquor is assumed to be high, so that C_L is constant as long as undissolved dye is present. Although not always valid in practice, the motion of the liquor is also assumed to be high enough that molecularly dissolved dye is always transported to the fiber surface in sufficient amounts and the adsorption of dye proceeds rapidly. Then, diffusion in the fiber is the step that determines the rate of dyeing. The amount of dye g that diffuses in a certain time follows Fick's first law [Eq. (6)].

$$g = \frac{D}{h} \cdot (C_L - C_F) \qquad (6)$$

The diffusion coefficient D is 10^{-10} to 10^{-12} cm^2/s and can be increased only by an increase in temperature. The diffusion path h is predetermined by fiber geometry; the concentration gradient C_L-C_F, by C_L.

Many experiments were carried out to develop an equation for the kinetics of dyeing. However, only approximations result because of the complexity of the fiber state and of the processes occurring in the dye liquor. The following phenomena determine the course of dyeing:

First, the polyester fiber contains varying proportions of crystalline and amorphous regions, depending on the degree of drawing and the fixing state. Only regions that are amorphous are accessible to dye diffusion above the glass transition temperature. During dyeing, the amorphous portion can change as a function of stress and shrinkage. In addition, the dyeing medium (water) and the dye already diffused into the fiber change the fiber structure and thus the dyeability.

Second, in the dyebath, the dye goes through a large number of stages, from solid particle to a single molecule, which interact with each other. These include (1) Dissolving in the liquor and possibly melting, (2) Adsorption of dispersing agent and association to dispersing agent, (3) Incorporation into dispersing agent micelles, (4) Diffusion through the laminar liquor layer on the fiber surface and adsorption to the fiber, (5) Agglomeration, aggregation, and crystal growth in the dyebath, (6) Mutual influencing of dyes, isomorphism of different dyes (formation of mixed crystals with different solution behavior than pure components), (7) chemical changes, e.g., reductive destruction.

Third, the levelness of dyeing, that is, the basic requirement of coloring, is endangered by differences in temperature and concentration. In practice, such differences are unavoidable because of irregularities in the velocity and direction of the liquor flow at various positions in the dyeing aggregate and gradual depletion of dye in the liquor during passage through the dyebath. Hence, a constant supply of dye at the fiber surface is not guaranteed, and diffusion no longer determines the rate of dyeing.

Despite many careful studies of model systems and individual processes, which permit a semiquantitative interpretation of individual steps under standardized conditions, dyeing results still cannot be mathematically predicted unambiguously. Thus, an industrial dyeing process must be supported by dyeing experiments. Recipes are drawn up on the basis of trial dyeings with the help of color measurements.

When elaborating a dyeing process in practice, it is important to realize that, with regard to the rate of absorption, the system consisting of hydrophobic fiber and disperse dye occupies an intermediate position compared to other classes of dyes. The rate of diffusion of disperse dyes into hydrophobic fibers is moderately temperature dependent; therefore, absorption can be easily regulated by controlling the temperature. Furthermore, this intermediate position between fast- and slowly diffusing dyes is characterized by adequate migration capacity of the dyes, especially at high temperature. Thus, unlevel dyeings can be leveled out within manageable periods of time. To achieve satisfactory dyeings, both methods—temperature control during exhaustion and high-temperature leveling—are basically available for dyeing with disperse dyes.

In the former method, as a result of rapid absorption of the dyes, the dyeing at the end of the exhaust phase is uneven and becomes level through subsequent adequate migration at high temperature. In the latter, The dyeing at the end of the exhaust phase is largely level due to controlled absorption with a bath exhaustion rate that is below a previously determined limit. The subsequent high-temperature phase serves only to ensure good penetration.

Irrespective of the method selected, a series of basic rules with regard to the time requirement, safety, and quality can be deduced from knowledge of the optimization of the absorption phase. First, in one formulation, dyes should be combined that, in the amounts applied, have exhaustion characteristics that are as similar as possible. Second, while the bath is being heated, the main exhaust range of the dyes, which covers about 30 °C, should be passed through sufficiently slowly. Third, the heating rate in the critical temperature range should be adjusted to the type and design of the textile and to liquor circulation or the circulation rate of the material [117, 118].

Information about the combinability of dyes is obtained from dye suppliers. Moreover, dye producers offer premixed disperse dyes whose components are selected in such a manner that the mixture is absorbed evenly at relatively low temperature and, consequently, makes fewer demands on the dispersion stability. At the same time, the depression of the melting point of mixtures should not be overlooked [119]. When using mixed dyes, differences in the fastness of the components must be accepted.

Empirical values are used to elaborate apparatus-specific control programs that guarantee optimization of the dyeing process with regard to the time course and dyeing levelness. Balancing the critical dyeing rate of the apparatus system (V_{CRIT}) with the significant dyeing rate of the formulation (V_{SIG}) is helpful. Programs of this type are promoted as fast-dyeing processes.

The dyeing process can be also controlled with dyeing accelerants (carriers). Carriers are aromatic compounds that exert a swelling effect on PES fibers and a dissolving effect on disperse dyes [5, p. 283–284]. They increase the rate of absorption and diffusion into the fiber and improve level dyeing.

4.12.1.2 Thermosol Process

In dyeing PES fibers, the dyeing time can be shortened significantly by increasing dyeing temperature. The rate of dyeing increases exponentially above the glass transition temperature; a temperature increase of 10 °C doubles the rate. For this reason, at ca. 200 °C fast colored fibers can be obtained with disperse dyes within seconds. Especially for PES–cellulose mixtures, this property is used to dye PES fibers continuously in a pad-fix process, which was introduced by Du Pont as the thermosol process [80, pp. 191–202, 50, pp. 141–142]. Application of the dye from an aqueous dispersion is followed by drying and fixing.

During drying, the padded dye liquor on the fiber surface can migrate to positions with a higher rate of drying; this is counteracted with an antimigration aid [5, p. 286]. To obtain a dye distribution on the fiber surface that is as homogeneous as possible, additives are used that wet the fiber and, at the same time, have a dissolving effect on dyes (e.g., ethoxylated fatty alcohols or fatty acids).

After drying, the dye is present on the fiber surface embedded in a film of dyeing aids and dispersing agents. In the case of PES–CEL fiber blends, the dye is also present on the surface of the cellulose fibers. At fixing temperatures of 210 °C, most disperse dyes have melted or are dissolved in the liquid film of auxiliary agents. Disperse dyes diffuse into the fiber only in the molecular form. Sorption on the fiber proceeds rapidly. Buildup of a high dye concentration on the fiber surface layer and diffusion are competing steps that depend on the prevailing temperature and determine the rate of dyeing.

In heat fixation of disperse dyes, hot air and contact heat are used most commonly. Steam has a swelling effect on the fiber. If superheated steam is used for fixation, the temperature can be lowered in comparison with hot air, the fixing times being the same. This is an advantage especially with articles made of textured fibers.

In the case of PES–CEL fiber blends, the larger part of the liquor is absorbed by the CEL fiber during drying, because of its higher absorption capacity. Nevertheless, in the subsequent thermofixing step, disperse dyes are almost completely fixed to the PES fiber. Transfer from the CEL to the PES fiber occurs both by migration via direct fiber contact and through the gas phase. The type of transfer depends on the sublimation behavior of the dyes [120].

4.12.2 Dyeing Processes for Polyester Fibers with Disperse Dyes

PES fibers are almost exclusively colored with disperse dyes. For some time, diazotization dyes were used for dark shades. They are coupled in the fiber with suitable components (e.g., 2-hydroxynaphthoic acid). Although fast colors are obtained, the dyeing process is time-consuming and susceptible to failure, and dyeing reproducibility is poor. Vat dyes are also occasionally used for PES dyeing. Small molecular thioindigo derivatives, in particular, diffuse into PES fibers if they are applied in the thermosol process.

4.12.2.1 Suitability of Disperse Dyes for Different Applications

A large variety of disperse dyes are available for coloring PES fibers. The Colour Index provides information on dyes made by various producers that correspond chemically with each other. As in the case of vat dyes, however, chemically identical disperse dyes can exhibit marked differences not only in coloring strength and shade, but also in their preparation and finish. Especially in modern dyeing processes, exacting standards are set for fine dispersion and dispersion stability. Disperse dyes are supplied as powder and liquid products. Powdered dyes contain a high percentage of dispersing agent, which is required to maintain a very fine dispersion when drying a dye that has been ground and dispersed in an aqueous slurry. Liquid products contain much less dispersing agent. Depending on concentration, the yields obtained with liquid products in various dyeing processes can deviate from those given by powders.

Criteria for dye selection are the required fastness properties of finished goods and the suitability for certain dyeing techniques. A strict standard is applied to the lightfastness of articles for car seat covers, furnishings, etc., and to the wetfastness of textiles that are washed in commercial laundries. Fastness to thermofixing is important for all articles that are heat-set, pleated, or resin-finished after dyeing. Thermal stability plays also an important role in the selection of suitable products for thermosol dyeing; easily subliming dyes contaminate the fixing aggregate. To facilitate selection of a proper dye for a certain dyeing process, dyes are classified into groups according to their rate of dyeing and the main temperature range in which they exhaust [121]. To determine combinability, the

American Association of Textile Chemists and Colorists (AATCC) recommends an easy-to-perform immersion test [122].

Another classification is based on covering up material differences in textured PES fibers. Dyes that have a low fastness to thermofixing often hardly mark differences in texture; they dye at relatively low temperature and are suited to carrier dyeing processes. Dyes that are very fast to thermofixing diffuse slowly into the fiber and must be dyed at high temperature; they are not suited for carrier dyeing under atmospheric pressure; they tend to mark texture and fixing differences in the fiber.

According to the recommendation of the Society of Dyers and Colourists, four tests are used to characterize the dyeing properties of disperse dyes [123]:

1) Migration test for leveling capacity.
2) Building up capacity to deep shades.
3) Exhaustion test to determine the temperature range in which the main amount of dye attaches.
4) Diffusion test (BASF Combi Test [124]) for rate of diffusion in the fiber.

Many dyeworks illustrate the coloring behavior of dyes by exhaustion curves (e.g., [85, pp. 384–501]).

For dyeing aids and carriers especially suited to dyeing PES fibers, see [5, pp. 278–285].

4.12.2.2 Dyeing from Aqueous Dye Baths

Articles made of pure PES fibers are dyed almost exclusively in exhaustion processes. Disperse dyes are most stable at a pH of 4–5 (acetic acid). A dispersing agent is also added to the dyebath. Dyes sensitive to heavy metals (e.g., anthraquinoid red products) are dyed in the presence of a complexing agent (EDTA type).

Dyeing under High-Temperature Conditions. Because of the slow diffusion of disperse dyes at boiling temperature, PES fibers are dyed whenever possible at 125–135 °C under pressure (HT dyeing). To develop a course of dyeing that is as quick and efficient as possible, studies have been performed on optimizing the temperature–time program, using dyes together that have similar exhaustion properties and adapting the circulation of the liquor or goods to the type of material to be dyed (see Section 4.12.1). In favorable cases, the entire dyeing cycle can thus be shortened to less than 60 min. The same goal can be approached by using complex dye mixtures [125, 126, 127]. HT dyeing can be made still more economical by using specially designed dyeing equipment to reduce the liquor ratio. The evenness of dye absorption is improved by dyeing in the presence of a leveling agent. Carriers can also be used as HT leveling agents. They are, however, technically outdated and ecologically harmful. For this reason, the carrier dyeing technique is recommended only when dyeing temperatures of 130 °C are impossible (e.g., with PES–wool blends).

Dyeing below 100 °C. The highest possible dyeing temperature is of advantage for coloring PES fibers. However, if no pressure-stable dyeing aggregates are available, dyeing can be carried out at atmospheric pressure with the addition of dyeing accelerants. The boiling temperature must be approached as closely as possible by using covered dyeing equipment. This also prevents inhalation problems due to the carriers, which are generally steam-volatile and can be ecologically harmful. Carriers must be accurately dosed. The amount of carrier to be applied is determined by the type of carrier, the dyes, the type of fiber, the dyeing temperature, the depth of shade, the liquor ratio, and the dyeing aggregate. Since they can also affect fastness, especially lightfastness, carriers must be removed after dyeing by hot rinsing and washing or by drying at high temperature or heat setting.

4.12.2.3 Special Dyeing Processes

Very light shades are sometimes dyed without the use of pressure and without a carrier. Only those disperse dyes that diffuse sufficiently quickly into PES fibers even at 100 °C are suited to this purpose (e.g., *C.I. Disperse Yellow 7*; *Orange 1, 33*; *Red 4, 11, 60*; *Blue 56, 81*). The dyeing equipment must be tightly closed, and top heating in vats and jiggers is advantageous. Dyeing time is extended to several hours. Since, in a process of this type, a portion of the dye adheres to the fiber surface without being fixed, dyeing must be followed by a reductive clearing step.

Dyeing from Foam. For saving energy, attempts were made to entirely dispense with an aqueous dye liquor and to apply disperse dyes to PES articles from a foam medium. Problems can arise with regard to the levelness of dyeing. When considering the energy balance of the foam dyeing technique, it is important that although water is saved in the dyeing process, considerable amounts of water are required e.g., for pretreatment and aftercleaning.

Air as Transport Medium. A clear economic advantage is observed when air is used as the transport medium for the textile material (e.g. Then Air Jet). Here, efforts are made to prevent foam formation. The small amount of aqueous dye liquor present carries dyes and auxiliaries.

Dyeing in Alkaline Medium. Oligomeric fiber components migrate out of the fiber during dyeing and can cause problems due to formation of agglomerates with dyes and deposition on the textile material or in the dyeing equipment. The cyclic trimer of ethylene terephthalate is especially harmful because of its low water solubility (<0.5 mg/L at 100 °C). The oligomer problem is counteracted by dyeing in alkaline medium. In this process, oligomers leaving the fiber are hydrolyzed to water-soluble components before their deposition. To achieve this the pH is adjusted to ca. 10.5 with NaOH at the start of dyeing; it then decreases gradually to about 7.5. Alkali-stable dyes must be used. Many blue azo dyes are unsuitable for this process. The reductive alkaline aftertreatment can be omitted.

Dyeing from Organic Solvents. Chlorinated hydrocarbons, in particular, have been recommended as a medium for continuous or batch dyeing of acetate and PES fibers. For continuous dyeing, dye application from either chlorinated hydrocarbons or an aqueous liquor with fixation in solvent vapor is possible. This method is disfavored for ecological and toxicological reasons [82, p. 676].

Dyeing from Supercritical Carbon Dioxide. According to studies performed in the early 1990s, dyeing from supercritical CO_2 is possible because of its adequate solubilizing power for disperse dyes [128]. Temperatures up to 150 °C and pressures near 30 MPa are used. Short dyeing times, good levelness, and low disposal costs are expected. The CO_2 is recovered. The equipment is naturally very complex. Dyeing to a predetermined shade is difficult.

Dyeing via the Gas Phase. Dyeing with dye vapor in the vacuum or in a gas stream as the medium has also been investigated. Although advantageous from an energy standpoint, the problems encountered on a practical scale have not been solved [129].

4.12.2.4 Continuous and Semicontinuous Dyeing Processes

Thermosol Process. The most important continuous dyeing process for PES fibers is the thermosol process. It is applied primarily to PES–CEL blends (see Section 4.12.4). The thermosol process consists of four individual steps: (1) Padding of the dye liquor on the fabric, (2) Drying, (3) Fixing of the dyes in the fiber, (4) Aftertreatment. The four individual steps usually follow one another in one pass. Systems also exist that include the subsequent overdyeing of the cellulose component in blended fabrics [e.g., by using the pad steam technique (see Section 4.1.1)].

The quality of thermosol dyeing is assessed according to its levelness over large yardages. For good quality, uniform pretreatment of the material and a very uniformly operating thermosol system are required.

Pretreatment ensures, above all, that the material is free of burls and wrinkles, contains no dirt or preparation stains, and is uniformly wettable. It includes careful singeing, washing, and bleaching.

In *padding*, a long dipping section is used so that the liquor penetrates the material completely and is applied evenly over the width of the material. Wetting agents are avoided if possible because they foam, and strong foam formation interferes with the process. Silicone-based foam depressants can negatively affect the stability of the dispersion. The pH of the pad liquor should not exceed 6–7 because disperse dyes can be sensitive to alkali. A padding auxiliary, e.g., a polyacrylate or an alginate thickener [5, p. 286], is used to prevent migration of dye liquor on the fiber surface during drying. Furthermore, dye migration is counteracted by low liquor application (concentrated pad liquor, high squeezing pressure in the padder), low drying temperature, and low air circulation.

Drying is carried out contact free up to a residual moisture of 20–30 % and then on any equipment available.

The *fixing time and temperature* depend on the aggregate, the dyes used, and the depth of shade (see Table 4.5 for guide values).

Table 4.5 Fixing times and temperatures in different dyeing aggregates

Dyeing aggregate	Temperature,* °C		Time, s
	Dyes with average rate of diffusion	Dyes with low rate of diffusion	
Hot flue, frame	200–215	215–225	60–30
Suction drum	200–215	215–225	45–20
Cylinder fixing	215	220–225	30–15

* Lower dyeing temperature corresponds to longer dyeing time.

In one formulation, dyes with similar rates of diffusion should be used. Very quickly diffusing dyes sublime easily, contaminate the fixing aggregate, and produce dyeings that have a low fastness to thermofixation. The color yield of very slowly diffusing dyes is sensitive to temperature variations.

After the dyes are fixed, dyes and auxiliaries that adhere superficially must be washed out, if necessary, by an alkaline reductive treatment. In the case of PES–CEL fiber blends, this washing can be combined conveniently with afterdyeing of the cellulose component, e.g., simply by overdyeing with vat dyes (see Section 4.4.3). For details of the thermosol process, see [85, pp. 122–131].

Pad Roll Process (see Section 4.1.1). Disperse dyes are fixed only incompletely to PES fibers under the conditions of the pad roll process (padding and subsequent rotation of the batch in saturated steam atmosphere). Only light colors can be produced economically with adequate fastness. Carrier additions are not useful because very large amounts would be required. In the pad roll process, rapidly diffusing dyes must be used.

Continuous Dye Fixation with Microwaves. The moist fabric padded with dye liquor is irradiated with microwaves in the presence of vapor. The addition of urea or carrier is recommended for rapid diffusion of the dyes [130, 131].

4.12.2.5 Dyeing of PES Microfibers

Fibers with a fineness of less than 1 dtex are called microfibers. Articles made of microfibers are characterized by a low weight, soft feel, weather resistance, and good breathing activity. The large fiber surface leads to rapid dye absorption and thus to the risk of unevenness. For physical reasons, the colors also appear to be lighter than on normal fibers. Therefore, greater amounts of dye are required to achieve a certain shade. The large fiber surface and the large amounts of dye used also lead to lower light- and wetfastness. Preliminary tests are useful [132, 133, 134].

PES fibers that have been subjected to a surface peeling process with alkali behave similarly to microfibers [135].

4.12.2.6 Dyeing of Modified PES Fibers

Since disperse dyes diffuse very slowly into PES fibers, efforts have been made to increase the rate of dye strike by chemical or physical alteration of the fiber. The fiber is also modified to reduce the pilling tendency, to increase shrinkage and elasticity, and to reduce flammability. Such modified fibers exhibit improved dye receptivity. Fibers with improved dyeability can be dyed with disperse dyes at boiling temperature without a carrier or with basic dyes when they are modified with acidic components (5-sulfoisophthalic acid). Fibers of this type are used if dyeing cannot be carried out easily above 100 °C (e.g., in the case of floor coverings, articles made of PES–wool blends, stretch materials, and cord). Strongly crimped PES bicomponent fibers are produced for special purposes. These fibers are normally also dyeable at the boil and without a carrier [136, 137, 138].

Modified PES fibers are usually more sensitive to hydrolysis than normal fibers. The lightfastness of the dyeings is often lower than on normal fibers. Thus, dyes for coloring carpeting, upholstery, and drapery must be carefully selected. On modified PES fibers dyes start exhausting at low temperatures (ca. 60 °C) and the dyebath is exhausted after a short time, so problems with levelness may arise.

The joint use of anionically modified and normal PES types can be exploited for differential dye effects (joint application of different classes of dyes that differ in their affinity to the fiber components).

4.12.2.7 Printing wth Disperse Dyes on Man-Made Fibers
[5, p. 528–543]

Practically all synthetic fibers can be printed with disperse dyes. Cationic dyes are used preferentially for acrylic fibers, and acid dyes and metal-complex dyes can be used for prints on polyamide fibers. The importance of printing with disperse dyes and the relative amount of different man-made fibers used for prints varies according to fashion and local requirements. Polyester fabrics alone or in combination with cotton are the most important. After precleaning, fabrics made from synthetic fibers must be heat-set to achieve dimensional stability and crease resistance. The usual setting conditions are 20–30 s at 190–210 °C, and for texturized articles about 30 °C lower.

In addition to the normal direct printing process disperse dyes can be applied by the so-called *transfer printing* technique, in which disperse dyes are first printed on paper and then transferred to the textile by sublimation. With suitable dyes discharge and resist printing are also possible. See also [139].

Disperse dyes are applied by direct printing on flat or rotary printing screens or by the roller printing method. Dye selection is limited by the high-pressure steaming conditions, under which azo dyes sensitive to reduction may undergo color change, and dyes which are easily sublimable may stain white grounds. Liquid (paste) products are preferred for their fine distribution and low content of dispersants. To obtain prints with optimal pattern sharpness high-solids thickeners such as starch ethers and dextrin are used, but they lead to a brittle film. Low-viscosity alginates or seed-flour derivatives produce an elastic film, that can easily be removed by washing.

Standard fixation conditions for disperse dyes on polyester fibers:

- 30 min at 102 °C in saturated steam; only 30–50 % color yield.
- 30 min at 0.25–0.4 MPa in saturated steam; up to 100 % dye fixation.
- 7 min at 175 °C in superheated steam; high dye yield; continuous treatment in high-capacity steamers is possible.
- 40–60 s at 200–220 °C in dry heat; dye yield 50–70 %.

Discharge printing on dyed grounds is not recommended, because of yellowing of the white ground. However *discharge resist printing* leeds to satisfactory results. In this case the fabric is padded with dischargeable dyes and dried only. It is then printed with a reductive discharge paste containing discharge resist dyes. The amount of reducing agent required is a function of color intensity and the fixing process. For lower consumption of reducing agent, a two-stage fixing process is sometimes applied, e.g., steaming first at 100 °C in saturated steam, followed by pressure steaming. Instead of discharge resist also alkali resist printing is possible, which uses dyes in the print paste which are sufficiently stable under alkaline conditions.

Blends of polyester and cellulosic fibers are printed with disperse and reactive dyes only in cases, when pigment printing is unsatisfactory. Dye selection follows basically the rules given for continuous dyeing polyester and cellulose fibers.

For printing disperse dyes on *cellulose acetate*, dyes are selected according to their colorfastness on 2.5-acetate and triacetate. Fixation is carried out for 2.5-acetate in saturated steam at 102 °C for 20–40 min; for triacetate 20–30 min at 0.25 MPa (127 °C) or in superheated steam 8–6 min at 165–185 °C.

Disperse dyes are commonly printed on polyamide swim wear, umbrella fabrics, and carpets. Presetting of the fabric is always required. For good color yield the printing paste should contain thiodiethylene glycol and urea. Best fixing results are obtained with saturated steam under pressure (20–30 min at 0.15 MPa). Discharge printing is possible. Color-discharge printing with vat dyes leeds to interesting effects.

4.12.3 Aftertreatment

In the case of light and medium PES dyeings, as in the coloring of other fibers with disperse dyes, the material need only be rinsed thoroughly or soaped after dyeing. In the case of dark shades, nonfixed dye components are held firmly on the fiber surface and impair fastness properties. Oligoesters (oligomers) leaving the fiber can also interfere with spinning or winding because they deposit on yarn guiding devices. Therefore, loose material and yarn, in particular, must frequently be subjected to reductive alkaline clearing after dyeing. For this purpose, hydrosulfite (sodium dithionite) is used in the presence of an emulsifying washing agent. In the case of loose stock, reductive aftertreatment also improves spinnability because the fiber becomes smoother. Good spinnability is especially pronounced when organic reducing agents are used for aftertreatment. Thickeners used for printing pose no additional problem during aftertreatment.

To improve the lightfastness of dyes used to color car fittings, benzophenone or triazole compounds are often employed in the dyebath or as aftertreatment [140].

Stripping of faulty dyeings is difficult and only possible under high temperature conditions at 130 °C. Dyeings can be brightend (partial dye removal) by using high concentrations of nonionic leveling agents and carriers, but some fiber damage can occur. Some dyes can be destroyed by chlorite bleaching with addition of carrier [5, pp. 290–291].

4.12.4 Dyeing Blends Containing Polyester Fibers

The favorable technological properties of synthetic fibers are often combined with the pleasant feel and wearing comfort of natural fibers. Hence, the introduction of man-made fibers has resulted in rapid expansion of the mixed processing of various types of fibers.

Since efforts are made to place the dyeing step as close as possible to the end of the finishing process, the problem of coloring fiber blends arises frequently. In the joint dyeing of two components, a compromise often must be made with regard to the dyeing process, tone-on-tone dyeing, or fastness of the finished product. For this reason, not every fiber blend is advantageous from the point of view of finishing.

4.12.4.1 Polyester–Cellulose Blends

The most important fiber blends consist of polyester and cellulose fibers or wool. Of the worldwide consumption of PES fibers, ca. 45 % is used in blends with cellulose fibers. In polyester–cellulose blends the cellulose component is usually cotton or viscose staple fibers The preferred mixing ratio is PES:CEL 67:33 and, for textiles worn close to the skin, 50:50 to 20:80.

General Dyeing Information [5, pp. 452–455], [85, pp. 209–289], [50, pp. 453–457], [6, pp. 592–600], [141]. As in the case of cellulose articles, the dyeability of PES–CEL mixtures is influenced greatly by the methods used for pretreatment, the boiling and bleaching of the cellulose component, and the heat setting of the PES component. Pretreatment methods depend especially on the type of cellulose fiber employed. In addition to the shades to be dyed and the scheduled dyeing method, tests must also be conducted to determine the required degrees of whiteness and hydrophilicity [5, pp. 254–264].

In dyeing PES–CEL mixtures, disperse dyes are used for the PES component. The CEL portion can be dyed with practically all classes of dyes suited to cellulose (see Sections 4.2–4.8). The selection of suitable dyes depends not only on the equipment available, the desired shade, and the batch size, but also on economic factors, with the required fastness taken into account. Thus, *reactive dyes* (see Section 4.2) are used preferentially for all types of clothing textiles, but *vat dyes* (see Section 4.4) are employed where extreme demands are made on washfastness or weatherproofness.

Dyeing with premixed disperse and vat dyes is especially easy. The mixed dyes can be applied in a single bath by the exhaustion or continuous process. The dyes are formulated in such a way that the same shade is obtained on both fibers, which facilitates shade matching [85, pp. 62–63, 221–246]. They are normally intended for 80:20 to 50:50 PES–CEL blends.

In dyeing the PES and CEL components with two dye classes, the staining of one fiber by the dye used for the other fiber should be taken into account. *Disperse dyes* stain cellulose fibers only slightly. The soiling can be removed easily by washing or, in stubborn cases, by alkaline reductive treatment, which is required anyway in vat dyeing. Most of the dyes used for CEL stain PES only slightly or not at all. Selected *vat dyes* that can be fixed to PES by a special application process and *sulfur dyes*, which often dye PES deeply, exhibit good fastness in use and produce a shade that corresponds with the tint on cellulose.

In the case of light shades, a slight staining of the partner fiber can give an attractive appearance to the entire article. Therefore, for pastel shades, depending on the article and its design, only one component may be dyed, provided that the staining on the other is fast enough. *Leuco esters of vat dyes* are also suitable for the simultaneous fast dyeing of PES and cellulose in pastel shades. The mode of dyeing is simple (see Section 4.5). To obtain good fastness on the PES component, drying should be carried out at 170–180 °C. *Pigment dyeing* is also commonly used for light shades.

For matching off the two fibers, the components often must be inspected separately. To this effect, the cellulose in PES–CEL blends can be dissolved out by treatment with sulfuric acid (72 %, 45 min at room temperature or 5 min at 70 °C) or orthophosphoric acid (85 %, 3–5 min at 80–90 °C), which is facilitated by ultrasonic treatment.

Dyeing Processes. Since safe and efficient continuous dyeing methods are well known, and large yardages are often dyed in the same color, PES–CEL mixtures are frequently dyed in continuous processes [142]. Nevertheless, batch dyeing is of major importance for dyeing yarn as well as piece goods, especially in the knitwear sector.

Disperse and Vat Dyes. In *batchwise dyeing*, a single bath is used in two stages. A suitable choice of dyes produces dyeings with unsurpassed fastness to washing (e.g., for work clothes) and thermofixing (e.g., for colored weaving yarn). The PES component is dyed at ca. 130 °C without a carrier since carriers can hold back vat dyes in the dyebath (for mode of dyeing, see Section 4.12.2). During the high-temperature step, the CEL portion is uniformly covered with vat dye. After the PES component is dyed, the bath is cooled to 80 °C and made alkaline, the vat dye is vatted with hydrosulfite, and dyeing is completed according to the procedure for vat dyeing (see Section 4.4.3). The disperse dye attached to the CEL portion is removed by means of the vat so that a reductive clearing step to improve fastness becomes unnecessary.

For the *continuous process*, the materials must be carefully pretreated to guarantee rapid wetting and uniform absorption of liquor [5, pp. 254–264]. The padding liquor should contain an antimigration agent [5, p. 286]. After dye application, the fabric is uniformly and gently dried to minimize migration of the insoluble dye, and the disperse dye is fixed in the PES component by the thermosol process (see Section 4.12.2). Subsequently, the vat dye on the CEL portion is developed either continuously according to the pad steam process or batchwise by the pad jig method. The most elegant method is *pad steam development*, which frequently follows the thermosol process in one pass. In pad steam development, the fabric is padded with reducing agent and alkali, and the vat dye is fixed by steaming in saturated steam (103–105 °C dry thermometer). Finely dispersed vat dyes must be used in this process because of the short vatting time.

Disperse and Reactive Dyes (see also Section 4.2.2). Reactive dyes are often used to dye the cellulose portion of PES–CEL blends because the coloring possibilities of reactive dyes are especially diverse, and a large number of relatively easy application processes produce dyeings that are sufficiently fast for the clothing sector and, frequently, for household textiles.

In the *two-bath batch method* the PES component is predyed with disperse dyes, and the CEL portion is topped with reactive dyes after reductive intermediate clearing, if necessary.

In a *one-bath two-stage batch process*, the bath contains all the dyes from the start. After the PES component is dyed at 130 °C, the bath is cooled to ca. 80 °C, and alkali is added to fix the reactive dyes to the CEL component. Careful selection of dyes is required because the reactive dyes must be stable at pH 4–5 and 130 °C.

In a *reverse process*, the CEL component is predyed at alkaline pH, the pH is then lowered, and the PES component is dyed at 130 °C. In this case, alkali-stable disperse dyes must be chosen if the disperse dye is added at the start. In addition, large amounts of salt in the bath can interfere with dispersion stability. Hence, especially fast exhausting disperse dyes are used.

Alkali dosing processes with controlled addition of alkali are of great importance for dyeing the CEL component, particularly when vinylsulfone-anchor dyes are used. The control program can also allow a progressive increase in pH. Thus, the rate of absorption, which is responsible for leveling, and the rate of fixation are optimized. Occasionally, the pH is controlled by cleavable alkali-releasing chemicals.

The most important *continuous dyeing* methods are the thermosol pad steam process and the thermosol pad batch process. In all continuous processes, the substantivity of the reactive dyes should be taken into account, i.e., the reactive dye can exhaust from the pad liquor or accumulate to a greater or lesser extent even at room temperature depending on the pH value and salt content.

In the *thermosol pad batch process*, a semicontinuous process, alkali-sensitive disperse dyes can be used if the alkali is applied after the thermosol process. Conversely, if the thermosol process is carried out after the pad batch step, the pH must be lowered before the thermosol process so that the disperse dye is not destroyed.

The *thermosol pad steam process* is especially important. Disperse and reactive dyes are padded together, and alkali and salt are padded after the thermosol passage. If a steamer is not available, the reactive dye can be fixed continuously by means of "alkali shock" in a roller vat with a relatively high alkali concentration.

Although the *thermosol thermofixing process* is especially easy and economical, it is less reproducible and therefore of little importance. In this process, disperse and reactive dyes are padded together with alkali and fixed by dry heat at ca. 210 °C. This is followed by washing in an open-width washing machine and soaping. Disperse dyes used for this process must be alkali-stable, and the dispersing agent contained in the dye must not react with the reactive dye. Suitable reactive dyes are hot dyers that are stable and poorly substantive in the pad liquor.

Disperse and Sulfur Dyes (see also Section 4.7.3). The *thermosol pad steam process* is the primary method for application of disperse and sulfur dyes. The sulfur dyes that are used preferentially in the ready-for-dyeing soluble form, and are especially inexpensive for muted shades, must be applied after the thermosol step because of their incompatibility with disperse dyes.

Disperse Dyes and Leuco Esters of Vat Dyes (see also Section 4.5). If a thermosol passage is provided under conditions that cleave the ester, so that the vat acid can diffuse into the PES fiber, leuco esters of vat dyes can produce up to medium shades with a high fastness level without the additional use of disperse dyes.

Only one class of dyes is also needed when small molecular vat dyes are used. Fast dyeings are obtained in lighter shades. Padding and the thermosol process are, if necessary, followed by vatting and normal aftertreatment to improve fastness on the cellulose component.

Disperse and Direct Dyes (see also Section 4.3). If no great fastness demands are made on the dyed article, PES–CEL blends can be dyed in a one-bath, one-step or two-step process with disperse and direct dyes. The direct dyes selected here must be stable under HT dyeing conditions. Dyes whose wetfastness can be improved by aftertreatment are of advantage. The subsequent resin finishing must also be taken into account in dye selection.

In a two-bath, two-step dyeing process, other dyes suited to dye CEL fibers may also be used for PES–CEL blends. For instance, in spite of its tediousness, Naphtol AS dyeing is still employed (see Section 4.8), especially for wetfast red shades, and the Phthalogen process is used for turquoise (see Section 4.9.5). Pigment dyeing is also applied to PES–CEL blends.

4.12.4.2 Polyester–Wool Blends

Mixtures of PES fibers and wool are used widely, especially as woven goods and knitwear in the outerwear sector. They offer the advantages of dimensional stability, easy care, and fashionable design possibilities. PES is generally used as staple fiber and frequently as a low-pilling type [143]. The ratio PES:wool of 55:45 is widely employed. In methods used for pretreating and dyeing PES–wool blends, the relatively high sensitivity of wool must be taken into account. Difficulties are circumvented by dyeing the components separately as loose stock or slubbing (see Sections 4.10.2–4.10.4 and 4.12.2). Especially fast dyeings are obtained in this manner.

Dyes. In dyeing PES–wool mixtures, disperse dyes are used for the PES component, and acid or metal-complex dyes for the wool. Disperse dyes can soil wool to a great extent. Since they produce poorly fast dyeings on wool, the dyes selected must stain wool as slightly as possible or must be easily removable by a washing step, which may be reductive if necessary. Frequently used dyes are C.I. *Disperse Yellow 23, 54, 64; C.I. Disperse Orange 30, 33; C.I. Disperse Red 50, 60, 73, 91, 167, 179*; and *C.I. Disperse Blue 56, 73, 87*. Premixed dyes consisting of disperse and wool dyes are occasionally available.

Wool is stained less at high temperatures and longer dyeing times because the disperse dye that was initially on the wool component migrates to the PES fiber. Carrier-free dyeable PES fibers should preferably be employed; these fibers are easily dyeable even at boiling temperature (see Section 4.12.2). The nature of the wool also plays a role, fine wool and reclaimed wool being more strongly soiled.

Dyeing Processes. For practical information, see [85, pp. 291–328], [144]. Wool cannot be dyed at the high temperatures normal in the HT dyeing of PES fibers or PES–CEL blends. The dyeing time should also be as short as possible so that the wool is not damaged. When normal polyester fibers are involved, the following dyeing processes are usually employed: 1) At boiling temperature with a carrier, 2) At 103–106 °C with little carrier, 3) At 110–115 °C with the addition of formaldehyde as a wool protectant and without carrier (HT dyeing). The significance of HT dyeing has increased in more recent times. The dyeing is carried out in one or two baths.

In the *one-bath process* preferred in practice, PES and wool are dyed simultaneously at pH 4–5, which is favorable for wool. Reductive aftertreatment is possible only if the wool dyes used can withstand it.

The *two-bath process* is applied for deep shades and for stringent fastness requirements. In the first step the PES part is dyed with disperse dyes, followed by reductive intermediate clearing and wool dyeing. However, while dyeing the wool component, disperse dye must always be expected to detach from PES and retint the wool.

Prebleached wool becomes somewhat yellow on dyeing. To obtain clear pastel shades bleach and dye are used preferably in the same bath. Mildly bleaching reducing agents (especially stabilized bisulfite or hydroxylamine salts) can be employed because they do not affect the dyes used.

Continuous or semicontinuous dyeing processes are uncommon because dry heat treatment (thermosol) hardens wool. Besides, application of the wool dye requires dyeing in aqueous liquor or dye fixation by steaming.

Auxiliaries. At dyeing temperatures above 100 °C, formaldehyde (5 % on weight of fabric) is added to protect the wool. Complexing agents should be used with care in the dyebath if metal-complex dyes are present. Ethoxylated fatty amines or ethoxylated fatty alkyl sulfates are suitable leveling dyeing aids. Condensation products of naphthalenesulfonic acid and formaldehyde, which hardly foam, have proved especially useful dispersing agents. A leveling agent is required primarily if tippy-dyeing grades of wool are present. In the selection of a carrier, technical as well as ecological factors must be considered [5, p. 283–284].

If possible, yarn is dyed at 106 °C or even higher temperatures because in this way the production fastness (e.g., fastness to setting) required for further processing is easier to guarantee, and thermally stable disperse dyes that exhaust only at higher temperatures can be used.

Aftertreatment. After dyeing, an afterwash is always indicated to remove the disperse dye attached to the wool. The materials are treated with an ethoxylated fatty amine in a weakly acidic liquor at 60 °C. Reductive aftertreatment is possible only in the case of appropriately stable wool dyes.

4.13 Disperse Dyes on Other Fibers

4.13.1 Disperse Dyes on Cellulose Acetate

For properties of cellulose acetate fibers, see [96, pp. 464–472], [145]. Cellulose 2.5-acetate (CA) and cellulose triacetate (CT) are preferentially processed as filaments. Because of their silklike feel, dull gloss, and pleasant wearing quality, they are especially popular for dress and blouse materials, scarves, and linings. Cellulose acetates are being increasingly replaced by synthetic PA and PES fibers. In contrast to the other regenerated cellulose fibers, CA and CT are hydrophobic. Therefore, they have little in common with viscose or cuprammonium rayon from the viewpoint of dyeing. As hydrophobic fibers, they can be dyed with disperse dyes. The dyeability of CA and CT with disperse dyes is very similar to that of PES and other synthetic fibers. The rules described for dyeing PES fibers with disperse dyes (see Section 4.12) basically apply to CA and CT as well. Tests have been drawn up by the Society of Dyers and Colourists for the selection of suitable dyes [146]. Dye producers offer special ranges of dyes selected for the particular requirements of cellulose acetate. In addition, the ranges of disperse dyes intended for PES fibers contain numerous products that are very well suited to CT in particular.

The choice of dye depends, on the one hand, on the desired use and production fastness and, on the other hand, on the suitability for certain dyeing techniques. Dyes that produce especially washfast dyeings require higher dyeing temperatures for level absorption. An important criterion for the selection of dyes for CA can be fastness to exhaust gases, i.e., nitrogen oxides and other acidic, oxidizing exhaust and industrial waste gases which are absorbed by the fiber. Products which are sensitive to gas fading are found especially among the red and blue anthraquinone dyes. Damage can be avoided by aftertreatment with special cationic products. Articles made of CT are subjected to thermal treatment in the course of finishing. The fastness of disperse dyes to thermofixing or pleating recommended for CT is usually at least as good as on PES fibers.

4.13.1.1 Dyeing Processes for Cellulose 2.5 Acetate
[147, 50, pp. 76–128]

For the theoretical aspects of dyeing CA and CT with disperse dyes see Section 4.12.1. CA is dyed by the exhaustion method in the presence of a nonionic or anionic dispersing and leveling agent in a weakly acidic bath (pH 5–6). A series of less wetfast disperse dyes are taken up by acetate at temperatures as low as 50–60 °C. For this reason, the process must be started at a correspondingly low temperature to obtain level dyeings. Dyeing is normally done at 80–85 °C. Dyes that are fast to wet treatments require temperatures up to 90 °C. If CA is dyed at boiling temperature, the fibers may become delustered.

Piece goods made of CA are usually dyed in a jigger, less often in a star frame or winch beck, with gentle material guidance. The piece beam is suitable for warp materials. Yarn was previously dyed in hanks. Since suitable winding machines are available, CA can also be dyed on cylindrical cheeses. Woven CA fabrics are occasionally dyed semicontinuously in a pad roll system at 80–90 °C. Dyeing auxiliaries are subsequently washed out.

4.13.1.2 Dyeing Processes for Cellulose Triacetate [50, 147]

In terms of dyeing and finishing, CT is more similar to purely synthetic fibers than CA. It can be permanently pleated. For stress relaxation, articles made of CT, like those made of PES, are heat set (thermofixed) after dyeing [80, pp. 92–100]. CT, like PES, can be dyed by the thermosol process (see Section 4.12.1).

The absorption of disperse dyes from a long liquor by CT fibers is slower than CA and somewhat faster than PES fibers. With a few disperse dyes, acceptable dyeings are obtained on CT even at boiling temperature. As a rule, however, dyeing is conducted at 120 °C, especially to obtain dyeings with adequate fastness properties. If this is not possible, a dyeing accelerant (based on butyl benzoate or butyl salicylate) is used. As in the case of CA, CT is dyed in a weakly acidic liquor (pH 5–6) in the presence of a leveling dispersing agent. The lower the dyeing temperature, the more must leveling be supported with a suitable auxiliary agent.

Dyeing equipment suitable for CA is also suited to dyeing CT. In accordance with the higher dyeing temperature used for CT, pressure dyeing aggregates are preferred (e.g., piece beam dyeing machines, jet dyeing machines, or HT winch becks).

4.13.2 Disperse Dyes on Polyamide Fibers [50, pp. 273–277]

Though originally developed for dyeing acetate fibers, disperse dyes also proved very useful for polyamide (PA) fibers. The shades of colors obtained with disperse dyes depend on the substrate. Thus, in comparison with acetate or PES, many shades on PA are bathochromically shifted (orange to red, red to violet). This is attributed to interaction of the amide groups with the chromophores. The colors are generally vivid, except for the red shades. Disperse dyes excel in their good leveling properties, even when structurally nonuniform material is to be dyed. Therefore disperse dyes are applied especially in dyeing lighter shades. Wetfastness deteriorates with increasing depth of color.

Dyeing Process. A dispersing agent is added to the liquor, and the pH is adjusted to 5 with acetic acid. The process is started with the material at 40 °C and followed by addition of dye. Then, the system is heated up, and dyeing is conducted in 60 min near boiling temperature.

For the technological aspects of polyamide dyeing, see also Section 4.11.4.

4.13.3 Disperse Dyes on Other Fibers [5, pp. 449–451]

Disperse dyes can be used to produce light to medium deep shades on *acrylic and modacrylic fibers* [96, p. 639]. The dyeing mechanism and process correspond to those used on PES and CA fibers (see Section 4.12). However, dyeing can be performed below 100 °C. Addition of carriers is not required. The good migration properties of disperse dyes result in problem-free level dyeing.

Poly(vinyl chloride) fibers (PVC) [96, pp. 642–645], are characterized by their flame retardance. They are dyed preferably with disperse dyes [50, p. 404], [6, pp. 611]. As with modacrylic fibers, high temperatures must not be used because of shrinkage of the PVC fiber. Hence, some fibers are dyed at 60–65 °C with dyeing accelerants. Other PVC fibers can be dyed at 100 °C without a carrier and a few even at 110 °C. Dyes must be selected with regard to the lightfastness desired.

Elastomeric polyurethane fibers [96, pp. 609–615], are contained in stretch articles and in knitted fashion materials. Light shades can be dyed tone-on-tone on polyamide–polyurethane mixtures with disperse dyes at 95–98 °C and pH 6–7. However, the wetfastness of these dyeings on polyurethanes is lower than on polyamide. Because of the temperature sensitivity of polyurethane fibers, mixtures of elastomeric and polyester fibers must be dyed with small-molecular, rapidly diffusing disperse dyes in 30 min at 120 °C according to the HT process [148]. Modified PES fibers that are dyeable at 100 °C without a carrier are often used in mixtures with elastomeric fibers. In all dyeing processes for elastomeric fibers, dyeing equipment that permits low-strain guidance of the material and the lowest possible thermal stress are important.

Polypropylene fibers are derived from an inexpensive raw material that produces fibers with a series of desirable properties such as dimensional stability, low weight, low water absorption, resistance to chemicals, and rotproofness [96, pp. 615–623], [149]. Consisting of pure aliphatic hydrocarbons, these fibers have no groups with a special dye affinity that can interact with dyes normally used for textile dyeing. For this reason, numerous attempts have been made to modify polypropylene to make it dyeable from aqueous liquor [150, 151, 152]. In one method, the polypropylene structure is loosened by a polyether or a polyamide, making the fiber accessible to disperse dyes. Another recommends etching the fiber with a low-temperature plasma in the presence of O_2 [153]. The depth of shade and fastness properties of disperse dyes are sufficient for many textile purposes including floor coverings. In all dyeing processes, the low melting point of the polypropylene fiber must be taken into account. The fiber becomes plastic at 130 °C, and treatment temperatures must be kept below 120 °C because of shrinkage. Before dyeing an unknown type of polypropylene, preliminary tests are required, to determine the type of modification.

4.14 Cationic Dyes on Acrylic Fibers

4.14.1 General Aspects

Acrylic fibers (PAC) are, together with PES and PA, the most important synthetic fibers [154]. For methods of production and properties of PAC fibers see [96, pp. 629–642]. To obtain fibers with satisfactory dyeing properties, anionic comonomers are used. In this way, the glass transition temperature T_g is lowered and anionic groups are available that can act as dye sites for cationic dyes. Thus, acrylic fibers are reliably and economically dyeable wth cationic dyes.

The main advantages of acrylic fibers are their similarity to wool in feel, heat retention, and processing, as well as good dyeability. In Western Europe and the United States, ca. 60 % are used for clothes and 30 % for household textiles.

Dyeing is performed mainly by exhaustion processes (batchwise). Another dyeing method with increased importance is the *gel dyeing process*. Dyeing in the gel state takes place during production of the fiber, i.e., after the fiber has been extruded and the solvent has been washed out, but before the fiber is stretched and dried (see Section 4.14.6). Continuous dyeing of stock, cable, and tops is also possible according to the pad steam process.

4.14.2 Cationic Dyes

Cationic dyes (see Sections 2.5, 2.6, 3.7, and 3.8) are used predominantly for acrylic fibers [155] because practically any depth of color can be obtained and they yield brilliant shades with good fastness properties. In choosing the dyeing conditions, the dyeing mechanism and relevant properties of the dyes and fibers should be taken into account. This is relatively easy because despite the great variety of fibers and dyes, uniform and simple laws are valid for dyeing with cationic dyes. A detailed description is presented in [50, pp. 359 ff.], and a discussion with a literature survey of the developments in certain areas (e.g., pore and surface structure, fiber modifications, thermal yellowing, effect of carriers and solvents during dyeing, and transfer print) in [156]. The physical chemistry of dyeing in general is presented in [157, 158].

Acrylic fibers can be dyed wih cationic dyes only above the glass transition temperature T_g, which has a characteristic value for each type of fiber and for most fibers lies between 70 and 80 °C [96, p. 469]. Cationic dyes form a heteropolar bond with anionic groups of the fiber. The number of anionic sites in the fiber determine the saturation value S_F, i.e. the maximum amount of dye wich can be taken up by the fiber. T_g and S_F are indicated by the fiber producer.

The dyeing process is governed by the affinity of the dyes to the fiber and their diffusion in the fiber. Since cationic dyes have a very high affinity to acrylic fibers, the equilibrium distribution between dye liquor and fiber lies on the side of the fiber and yields a normal bath exhaustion of 97–100 %. The dye diffusion depends on the difference in dye concentration between fiber and dye liquor according to Fick's law (see Section 4.12.1.) [46]. For standard conditions, relative diffusion coefficients can be calculated. Since the dye concentration at the fiber surface can be considered constant, at least during the early stages of dyeing, diffusion is the rate-determining step, unless very light dyeings are considered. The speed of exhaustion based on dye diffusion is a fiber characteristique and is strongly temperature dependent. Above T_g it doubles with a temperature increase of 2.5–3 °C. Therefore, it is very important for dyeing PAC with cationic dyes to carefully control the temperature during dyeing. Consequently, the optimum time–temperature program for each dyeing should be established before a production run.

When the dyeing equilibrium is reached, the relative saturation value S_{rel} indicates the degree of levelness which can be expected during dye absorpion; the higher S_{rel} the better the levelness. For similar S_{rel} for different dyeings a similar course of the dyeing process can be assumed.

High affinity for the fiber and rapid diffusion mean short dye cycles, independence from liquor ratio, no dyeing accelerants, and good reprocibility. However it also means poor migration properties of the dyes after they have formed a heteropolar bond with the anionic sites on the fiber, so that inequalities in the dyed material are difficult to level out. Therefore, it is very important to aim for homogeneous bath exhaustion.

Cationic dyes influence each other during absorption. If several dyes exhaust tone-in-tone from one bath (i.e., the depth of color, but not the shade, changes in the course of absorption), the dyes are referred to as compatible. In the case of cationic dyes on PAC, the property of being compatible is dye specific and also applies to dyeing under different conditions and in different shades. It depends on the product DA (D = diffusion coefficient, A = dye affinity) and can be described by the compatibility value K, which ranges from 1 to 5.

In acrylic dyeing, combinations of dyes with the same K value exhibit considerable advantages in level dyeing. In combinations with different K, dyes with a lower K absorb before those with a higher K. Cationic retarders can also be assigned K values [159]. The K values change with the use of anionic retarders.

4.14.3 Retarders and Auxiliaries

Some low molecular cationic dyes migrate to a certain extent when dyed in the presence of a *migration auxiliary*, e.g. Glauber's salt or organic salts with large cations [5, pp. 279–281].

With light shades, the amount of dye applied is often not enough for an adequate S_{rel}. Sometimes even saturating the fiber surface is not sufficient. To meet the danger of nonlevelness in such cases, *cationic retarders* are used; these are colorless quaternary ammonium salts which compete with dyes for dye sites on the fiber and thus reduce the speed of bath exhaustion for dyes. Retarders are usually quaternary amines with an aromatic substituent and an aliphatic chain. The longer this chain is, the greater is the affinity for the fiber. Retarders also improve migration and homogeneous penetration of the fiber. Retarders are used mainly to increase S_{rel}, thereby improving levelness. Retarders can be assigned a saturation factor S_F and treated formally like a dye. The amount of retarder necessary for a particular recipe is calculated from S_F of the fiber to be dyed and the type and amount of dyes used.

Depending on their affinity and penetrating ability, some *cationic softeners* when used in the dyebath act like cationic retarders and contribute to S_{rel}. Thus, less cationic retarder is required.

Polycationic retarders, i.e., compounds (usually polymeric) with numerous cationic groups, have been recommended as leveling agents. They occupy the fiber surface without penetrating into the fiber and counteract dye adsorption, thereby reducing the rate of dyeing [160]. Depending on their adsorption characteristics, they also decrease the migration capacity and the bath exhaustion in the dyeing equilibrium. However, a favorable effect on leveling has not been found.

In some cases, anionic auxiliaries are also used as *anionic retarders*. These contain two or more anionic groups (sulfonic acid groups) per molecule and form a 1:1 addition compound with the cationic dye that is soluble in the dye liquor. To ensure this solubility, at least two anionic groups must be available from the start for each dye cation. The minimum amount of anionic retarder required can be calculated. In the course of dyeing, a growing excess of auxiliary anions is formed, and the concentration of free dye cations in the liquor falls below that required for saturation of the fiber surface. Dyeing rate and bath exhaustion at dyeing equilibrium are decreased. Anionic retarders promote the migration of cationic dyes. They reduce the danger of precipitations when anionic dyes are used in addition to cationic dyes (for fiber blends). The higher the affinity of the cationic dyes for the fiber, the greater is their effect. The main disadvantages are the lower bath exhaustion and poorer reproducibility.

4.14.4 Exhaustion Process

Normal PAC textile material is dyed as cable or stock in packages, as yarn in hanks or packages (direction of liquor: inside to outside), as pieces on beam, overflow, paddle machines (knitwear), or drums (socks). High-bulk (HB) yarns are dyed as muffs [161]. Strongly soiled textile material is prewashed, preferably in a weakly alkaline bath.

Because of the strong temperature dependence of exhaustion and the poor migration properties, dye absorption must be as level as possible from the start. Correspondingly, attention should be paid to uniform packaging, package density, liquor flow or movement of textile material, and controlled machine conditions.

As far as possible, dyes with the same K value are selected; the largest choice of dyes and most suitable retarders are available at $K = 3$. In cases where sufficient levelness during absorption is critical, dyes with a relatively low affinity (with high K) are recommended [162]. According to the dyes present in the actual recipe the amount of retarder is adjusted to give a favorable S_{rel}. The dyes are dissolved in acetic acid, the dyebath is made up with all auxiliaries (see Table 4.6) at 60–70 °C, and the textile material is added. Dyeing is conducted by controlling the temperature, as shown in Tables 4.7 and 4.8. This is followed by cooling (ca. 0.3 °C/min to 60 °C, i.e., well below the glass transition temperature), rinsing, and aftertreatment.

Table 4.6 Additives to the dye bath, general formulation

Additive	Light hues	Medium hues	Deep hues
Nonionic dispersant, g/L	0.2–0.5	0.5	0.5
Acetic acid, final pH	3.6	3.6	4.5
Sodium acetate anhydrous, g/L			0.5
Sodium sulfate anhydrous, g/L	5	5	
Cationic retarder,* %	1.5–3	0.3–1.5	

* Retarder with a K value of about 3, in percent based on mass of material (see Section 4.14.3).

Table 4.7 Temperature range of controlled heating (general method; T_{start}–T_{end}) and residence time (t_{max}) at maximum temperature T_{max}

S_{rel}	Low	Medium	High
T_{start}, °C	75	80	85
T_{end}, °C	95	98–100	98–100
t_{max} at T_{max} = 98 °C, min	20	40	60
t_{max} at T_{max} = 102 °C, min	15	30	40

Table 4.8 Heating rate (general method)

Temperature range*	Stock, top	Yarn (pack.)	Hank, piece
up to T_{start} (until absorption of dye starts)	quick	quick	quick
T_{start}–T_{end} (usually linear)	0.5 °C/min	0.3–0.4 °C/min	0.25 °C/min
T_{end}–T_{max} (until absorption of dye is complete)	quick	quick	quick

* Numerical data for the temperature can be obtained from systematic optimization processes.

The details given in Tables 4.6–4.8 refer in a general way to the principles explained in Section 4.14.1, especially the need to achieve sufficiently level dyeings from the start and to have the shortest possible dyeing time. The liquor movement or circulating speed of the material can be taken into account [163, 164]. As a rule, satisfactory results are obtained in this manner.

Dyeing can be performed more economically and reliably by using systematic methods for the optimization of formulations and dyeing conditions [165]. Here, dyes with the same K are selected and the following factors are determined for each dyeing: (1)The amount of retarder that leads to the most favorable relative saturation value S_{rel}, (2) The temperatures at which exhaustion starts or ends (the critical temperature range outside which rapid heating will not cause unlevelness), (3) The most favorable rate of heating in this temperature range, depending on the dyeing machine and the textile material, and (4) The dwelling time at the maximum (final) dyeing temperature required for penetration of the dyes used. This optimization can be performed with the help of tables [166] or with a computer [167]. It leeds to level dyeings in short dyeing times, good reproducibility and savings in laboratory work.

4.14.5 Dyeing of Special Fiber Types with Cationic Dyes by the Exhaustion Process

High-Bulk Material. The bulking of high-bulk (HB) material proceeds either as a pretreatment with steam continuously for yarn, batchwise for yarn and knitwear, or in the dyebath before dye absorption at ca. 80 °C for 5–10 min without liquor movement [168]. To avoid nonlevelness in this process, retarders are employed to ensure that very little dye is absorbed during the bulking time. With HB bicomponent fibers, a high S_{rel} can interfere with the reversible shrinkage. For this reason, anionic retarders and dyes with a K value of 4 or more are preferred [166].

Pore Fibers. For the surface and pore structure of PAC fibers, see [169]. As a result of their greater light scattering, pore fibers (light fibers with high water absorption capacity) require up to three times more dye than normal fibers [170]. Navy and black shades are hard to dye; for dyeing other deep hues, an ammoniacal reductive aftercleaning step is required. The diffusion of dyes proceeds mainly in the pores of the fiber, changing the K values of the dyes. With the exception of a higher softener requirement, the rules of normal PAC dyeing apply.

Microfibers. PAC microfibers (single titer less than 1 dtex) behave differently in some respects from normal acrylic fibers [171]. First, more dye is required for the desired shade in each case because of the greater light scattering of the fiber. This requirement increases more than twofold with increasing depth of color (similar to pore fibers). For this reason, the fiber saturation is usually adjusted to a higher value by the fiber producer. The characteristic dyeing rate of the fiber is also higher. Corresponding to these increased values, more cationic retarder is required and the starting temperatures are lower. Above an S_{rel} of about 0.75, the rub- and wetfastness can be lower; reductive aftercleaning is sometimes necessary. Some dyes exhibit increased sensitivity to boiling down (lower dyebath stability) when the pH deviates from 4.5. For this reason, dye producers recommend a special selection of dyes for microfibers. In the dyeing of stock, top, or yarn, the liquor flow can be lower and its uniformness is more problematic than in the case of normal acrylic fibers. For this reason, packaging and winding must be especially even and the rate of heating should be slow. In piece dyeing, longitudinal tension should be kept as low as possible and creasing must be avoided. Cooling to below 60 °C should not proceed more than twice as fast as heating.

Modacrylic Fibers [96, p. 639]. Modacrylic fibers containing more than 20 % of vinyl chloride or vinylidene chloride as comonomer are used because of their reduced flammability. Depending on the comonomers, various types of modacrylic fibers exist that differ especially in the range of dyeing temperatures. In all cases, the critical temperature range is greater (25–50 °C) than with normal acrylic fibers (15 °C), and absorption starts in most cases at lower temperature. Modacrylic fibers generally exhibit increased plasticity and a strong tendency to shrink at higher dyeing temperature. The fiber types and their dyeing properties change frequently. The lightfastness on these fibers is considerably poorer than on normal PAC fiber, making an appropriate dye selection necessary. Dyeing is performed at pH 4.5 with acetic acid, sodium acetate, and a nonionic dispersing agent. Depending on the type of fiber, carriers or cationic retarders and different temperatures are recommended [50, pp. 384–387]. Recommendations of the fiber producers should be consulted.

Modacrylic fibers can also be dyed with *disperse dyes*, similarly to normal PAC fibers, but the fastness to sublimation is poor (see Section 4.13.3).

4.14.6 Continuous Processes with Cationic Dyes

Stock, cable, and top can be dyed on special machines according to the *pad steam process* [172], preferably with pressurized steam at temperatures above 100 °C to obtain short fixing times. To avoid drying and condensed water spots, the steam should be neither superheated nor wet. A typical pad liquor contains acetic acid (pH 4.5) and dye solvent in addition to steam-resistant, readily water soluble cationic dyes (usually liquid brands) [173].

Space dyeing can be regarded as a variant of this mode of dyeing. A new suggestion involves simultaneous bulking and dyeing in a continuous process [174].

The pad steam process with *fixing by saturated steam* at 100–102 °C is used for piece goods, especially upholstery material (velour). This requires longer fixing times (10–20 min). Rapidly diffusing cationic dyes and dye solvents, which also exhibit a carrier effect [5, p. 279–281], are required. If direct dyes are used simultaneously (for cotton backs), the dye solvent should also have a good dispersing action.

Dry heat fixing (*thermosol process*, see Section 4.12.2) is unimportant for PAC dyeing because the fiber tends to yellow and harden in this process and cationic dyes are fixed slowly and incompletely.

Gel dyeing, i.e. dyeing wet-spun PAC fibers during the production process in the *gel state* is especially economical. It is estimated that ca. 25 % of PAC fiber production is dyed in the gel state by the fiber producer. The gel state is defined as the state of production of the fiber after extrusion, when the solvent has been washed out, either before or after stretching, and before drying (sintering). Due to the high porosity of the fiber in the gel state, cationic dyes are adsorbed very rapidly (within seconds) on the fiber. Therefore, adsorption, which is insensitive to temperature variations, is the rate-determining step of dyeing. Fixation of the dye follows later during drying at 105–150 °C. There is almost no difference from normal PAC dyeing with regard to saturation, shade, and fastness. Liquid brands, which are miscible without precipitation in the dye applicator, are preferably used. Mixtures, even diluted, must be stable on storage [175, 176].

Although acrylic fibers can also be printed with disperse dyes, *printing with cationic dyes* is preferred because of their high brilliance, spreading power and good lightfastness [5, pp. 534–537]. Preliminary tests should be carried out, since acrylic fibers differ in their dye absorpion rates for different groupes of dyes. Best fixation results are obtained with high-pressure steam at 0.12–0.14 MPa (104–110 °C) for 30-min.

Transfer printing is not suitable without either elaborate pretreatment or special swelling agents or solvents [160]. Disperse dyes produce only very light shades.

4.14.7 Cationic Dyes on Aramid Fibers

Aramid fibers, i.e. polyamide textile fibers made from aromatic amines and dicarboxylic acid [177] are similar to polyamide and polyester fibers and are highly heat resistant and flame retardant. Aramid fibers must be heat set by steaming before wet finishing and washed before dyeing for good leveling.

For dyeing, cationic dyes are used exclusively. Suitable products are, e.g., *C.I. Basic Yellow 15, 21, 23, 25, 29, 49, 53, 79*; *Basic Orange 22, 35*; *Basic Red 29, 46*; *Basic Blue 41, 54*; and *Basic Green 6*. A lightfastness of 3–4 is achieved, with C.I. Basic Green 6 giving 5–6. To attain a good color yield, dyeing is carried out at pH 3–4 and 120 °C for 2 h. In addition, a carrier is used, preferably benzyl alcohol in amounts of 10–40 g/L, depending on the depth of color. A carrier-free technique recommends dyeing at 180 °C under pressure [178]. After dyeing, thorough afterwashing is required with the addition of sodium hydrogensulfite, if necessary, to remove superficially adhering dye [80, pp. 675–676].

In the case of faulty dyeing, the dye can be largely stripped by treatment at 120 °C with benzyl alcohol in an aqueous bath or with a carrier and a retarder which has a leveling effect in the presence of hydrosulfite in an alkaline liquor.

4.14.8 Cationic Dyes in Fiber Blends

Most textile dyes are anionic in nature or they are prepared with anionic finishing products. Therefore, whenever cationic dyes are used together with other dye classes, possible interactions between different dye classes should be considered.

PAC–CEL Blends are used for household textiles and imitation fur. In these plush and fur materials, the pile consists of PAC fibers while the back is made of cellulose. Furthermore, they are used for leisure wear, sport stockings, drapery, and table linen. The percentage of PAC in the mixtures varies widely. Cationic dyes reserve CEL well, because of their high affinity for acrylics. For a survey of dyeing methods, see [50, pp. 474–476] and [6, pp. 608–610].

Continuous dyeing of PAC–cotton plush with *cationic* and *direct dyes* by the pad steam process plays an important role. The choice of dyes must take into account liquor stability, reservation of PAC or CEL fiber, and solubility. Precipitation of cationic and anionic dyes present in the pad liquor at relatively high concentrations cannot be avoided solely by dye selection. Suitable auxiliary systems have been developed. Differently charged dyes are kept in solution separated from each other in two phases by the combination of anionic and nonionogenic surfactants. With the help of fixing accelerators, good penetration of PAC fibers can be achieved in 10–15 min with saturated steam at 98–100 °C.

With *cationic and vat dyes*, dyeing can be conducted batchwise according to the pigment process in one bath and two steps. Here, the PAC component is dyed with cationic dyes and the CEL portion is simultaneously pigmented. After cooling to 70 °C (max.) and vatting, the vat dye is fixed (see also Section 4.4.3). Another possibility is to complete the dyeing of the PAC component first with cationic dyes and then add the vat dye to the same bath after cooling.

PAC–CEL mixtures can also be dyed in a one-bath, two-step process with *reactive and cationic dyes*. Here, too, the PAC component is predyed at pH 5 with cationic dyes, and after the pH is raised, the cellulose component is covered with reactive dyes. For the dyeing of yarn, see [179].

PAC–Wool Blends. No other synthetic fiber achieves the wool-like character of PAC fibers. For this reason, PAC fibers are especially suited to the production of blends with wool. The high water-absorbing capacity of wool and the high strength of PAC fibers complement each other and confer good wearability. The felting tendency of the blend is lower than that of wool alone. The blending ratio of PAC to wool varies widely. The blends are preferred for outer clothing, especially knitwear, and for household textiles.

In principle, three possibilities exist for dyeing PAC–wool mixtures [50, pp. 473–474], [6, p. 610]:

1) Two-bath method, i.e., dyeing of wool and PAC in succession in two baths.
2) One-bath, two-step method, i.e., dyeing of wool and PAC in succession in the same bath.
3) One-bath, one-step method, i.e., dyeing of wool and PAC simultaneously.

The *one-bath, one-step method* is preferred for economic reasons. However, problems arise due to the simultaneous presence of cationic and anionic dyes in the dyebath. Precipitation can be prevented by dispersing auxiliaries and precise dye selection, but very deep shades are normally dyed in a one-bath, two-step process.

Although only cationic dyes can be used for dyeing the PAC component, metal-complex dyes, chrome dyes, or reactive dyes can be used for wool. In the case of 1:1 metal-complex dyes, almost no danger of precipitation exists because of intramolecular compensation of the negative charge by the positively charged metal atom. The pH is adjusted to 2.5–3 (see also Section 4.10). Most cationic dyes can be applied at this low pH value. Only the auxiliaries and the acid are added while the dyebath is being heated. After the initial dyeing temperature (ca. 70 °C) is reached, the cationic dyes are added first. After balancing concentration differences, the anionic dyes are then added. A pH of 4.5–5.5 is maintained for acid, chrome, reactive, and 1:2 metal-complex dyes.

In the choice of dyes, attention must be paid not only to the prevention of precipitation but also to the reserve of PAC fibers and wool. Cationic dyes attach first to wool, producing colors of very poor fastness, and then boil over onto the PAC fiber. Even if well-reserving dyes are selected, dyeing must be conducted for a sufficiently long time (60–90 min) until the boilover is complete.

Mixtures of Acrylic with other Synthetic Fibers. A frequently processed mixture (e.g., for outerwear or furnishing fabrics), consists of *PES and PAC fibers*. No basic difficulties are involved in dyeing, both union shades and bicolor dyeing being possible. Selected disperse and cationic dyes are used. Although disperse dyes color PAC more weakly than PES, the shade produced is the same. In the case of bicolor dyeing, the PAC component must always be kept darker. A dyeing temperature of 106 °C should not be exceeded because of the PAC component. Thus, carriers must be used as a rule. In one-bath dyeing, attention must be paid to the compatibility of the various dyebath additives with each other. Cationic retarders are not used in one-bath dyeing [85, pp. 346–355], [6, pp. 601–602].

Problems with the compatibility of the dyes can also arise in dyeing mixtures of *PA and PAC fibers*. Such blends are produced, e.g., for skiwear. The cationic dyes used for the PAC component do not dye normal PA fibers. Acid and metal-complex dyes leave PAC fibers practically white. Sulfo group containing 1:1 metal-complex dyes possess a zwitterionic character and therefore can be applied in low concentration in the same bath with cationic dyes, without the occurrence of precipitation. In this manner, one-bath dyeing can be carried out. The stability of the bath can be improved by addition of a nonionic auxiliary. Cationic retarders should be excluded in all cases. With 1:2 metal-complex dyes dyeing is carried out in two steps: first with cationic dyes, because they exhaust almost quantitatively, and then with metal-complex dyes for the PA component [50].

4.15 References

General References

R. H. Peters, *Textile Chemistry*, Vol. 3, "The Physical Chemistry of Dyeing", Elsevier, Amsterdam 1975.
C. L. Bird, W.S. Boston (ed.), *The Theory of Coloration of Textiles*, The Dyers Comp. Publ. Trust, Bradford, 1975.
H. K. Rouette, C.-H. Fischer-Bobsien, Lexikon für Textilveredlung, Dülmen, 48249 Dülmen bei Recklinghausen, 1995.
Ullmann's Encyclopedia of Industrial Chemistry, 5th ed., vol. **A 26**, 1995, pp. 351–487.

Specific References

[1] Ullmanns Encyklopädie der technischen Chemie, 4th ed. 1976, **11**, 99–134.
[2] H. Schweppe, *Handbuch der Naturfarbstoffe, Vorkommen, Verwendung, Nachweis*, Ecomed Verlag, Landsberg 1993.
[3] Kirk-Othmer, 3rd ed., **8**, 351–373.
[4] D. J. Hill, *Rev. Progr. Col. Relat. Top.* 27 (1997), 18–25.
[5] Ullmann's Encyclopedia of Industrial Chemistry, 5th ed., VCH, Weinheim 1995, **A 26**.
[6] M. Peter, H. K. Rouette, *Grundlagen der Textilveredlung, Handbuch der Technologie, Verfahren, Maschinen*, 13th ed., Deutscher Fachverlag, Frankfurt 1989.
[7] G. Grüninger, *Melliand Textilber.* **80** (1999) 820–832; P. Owen, ITMA 99, *Text. Chem. Color Am. Dyest. Rep.* **1** (1/1999) 22–25; A. Barthold, *Textilveredlung* **34** *(1999) no. 7/ 8, 21–34.*
[8] ITMA Show-view, *Intern. Text. Bull. Veredlung* 4/1999, 77–80.
[9] J. Hilden, Intern. *Text. Bull. Veredlung* 4/1999, 77–80.
[10] M. White, *Rev. Progr. Col. Relat. Top.* **28** (1998), 80–94.
[11] Alison Gilchrist, *Rev. Progr. Col. Relat. Top.* **25** (1995) 35–43.
[12] B. Drechsel, F. Hoffmann, S. Ottner, *Melliand Textilber.* **79** (1998), 328–232.
[13] B. C. Burdett, A. J. King, *Rev. Progr. Col. Relat. Top.* **29** (1999) 29–36.
[14] H. Beckstein, *Intern. Text. Bull. Veredlung* 4/1999, 95–98; M.R. Chaney, J. H. Nobbs, Text. Chem. Color. **31** (2/1999), 35–39.
[15] K. Spitzner, *Textildruck*, 3rd revised ed., VEB Fachbuchverlag, Leipzig 1980.
[16] T. L. Dawson, *Jet printing, Rev. Progr. Col. Relat. Top.* **22** (1992), 22–31.
[17] B. Siegel, K. Siemensmeyer, M. Dorer, *Melliand Textilber.* **79** (1998) 864–865; W. C. Tincher, *Textile World* 11/1999, 27–32.

[18] J. R. Knapton, *Rev. Progr. Col. Relat. Top.* 21 (1991), 7–10; J. Hilden, *Int. Text. Bull., Veredlung* 4/1999, 92–94.
[19] *Text. Chem. Color.* I **19** (2/1987) 25–28; II **19** (3/1987) 32–35 and **19** (4/1987) 21–22; III **19** (5/1987) 33–34 and **19** (6/1987) 14; IV **19** (8/1987) 21–22; V **20** (9/1988) 63–67; VI **21** (2/1989) 30–32.
[20] Ullmannn's Encyclopedia of Industrial Chemistry 5th ed., VCH 1995, **A9**.
[21] ISO/CD 787-26 (Draft).
[22] W. Baumann et. al., *Melliand Textilber.* **67** (1986), 562–566; *J. Soc. Dyers Colour.* **103** (1987), 100–105; R. H. Warmann, *Rev. Progr. Col. Relat. Top.* **24** (1994), 55–75.
[23] R. G. Kuehni, *Text. Chem. Color.* **31** (6/1999) 11–16; B. Schick, *Textilveredlung* **35** (2000) no. 1/2, 12–16.
[24] G. Faes, *Melliand Textilber.* **79** (1998), 462–465.
[25] R. Broßmann et al., *Melliand Textilber.* **67** (1986) 499–502; *J. Soc. Dyers Clour.* **103** (1987) 38–42.
[26] T. v. Chambers et al. *Melliand Textilber.* **69** (1988), 755–758; *J. Soc. Dyers Colour.* **105** (1989), 214–218.
[27] P. J. Smith, *Rev. Progr. Col. Relat. Top.*, **24** (1994) 31–40.
[28] Colour Index vol. 6, First suppelemnt to vol. 1–4 of the 3rd ed. 1975, *The Soc. of Dyers and Colourists and Amer. Assoc. of Textile Chemists and Colourists* vol 7 (1982), vol. 8 (1987).
[29] ITMA 91, *Melliand Textilber.* **73** (1992) 180–183; C. Schukei, *Int. Text. Bull. Veredlung* 4/1999, 100–103.
[30] K. Venkatarman: *The Chemistry of Synthetic Dyes*, vol. 6: Reactive Dyes, Academic Press, New York 1972.
[31] M. R. Fox, H. H. Sumner in C. Preston (ed.): *The Dyeing of Cellulose Fibers*, Dyer's Company Publication Trust, Bradford 1986, pp. 142–195.
[32] J. P. Luttringer, P. Dussy, *Melliand Textilber.* 62 (1981, 84–94.
[33] M. R. Fox, H. H. Sumner in C. Preston (ed.): *The Dyeing of Cellulose Fibers*, Dyer's Comp. Publ. Trust, Bradford 1986, pp. 142–195.
[34] S. Abata, K. Akahor, U. Meyer, H. Zollinger, *J. Soc. Dyers Colour.*, **107** (1991), 12–19.
[34a] R. Tüxen, *Melliand Textilber.* **81** (2000) 842–846.
[35] R. F. Hyde, Rev. Progr. *Color. Relat. Top.*, **28** (1998) 26–31.
[36] A. H. M. Renfrew, J. A. Taylor, *Rev. Progr. Col. Relat. Top.*, **20** (1990), 1–9; J.A.T. ib. **30** (2000), 93–107.
[37] D. M. Lewis, *J. Soc. Dyers Colour.* **109** (1993) 357–364.
[38] D. M. Lewi, *Melliand Textilber.* **67** (1986), 717–723.
[39] D. M. Lewis, *J. Soc. Dyers Colour.* **98** (1982) 165–175.
[40] K. Y. Chu, J. R. Provost, *Rev. Progr. Colour. Relat. Top.* **17** (1987), 22–28.
[41] M. L. Gulrajani, *Rev. Progr. Colour. Relat. Top.* **23** (1993) 51–56.
[42] W. Zuwang, *Rev. Progr. Colour. Relat. Top.* **28** (1998), 32–38.
[43] A. W. Peddersen, *Melliand Textilber.* **75** (1994) 54–56.
[44] J. Shore, *Rev. Progr. Colour. Relat. Top.* **21** (1991), 23–42.
[45] J. J. Porter, *Text. Chem. Color.* **25** (no. 2/1993) 27–37.
[46] P. R. Brady, *Rev. Progr. Colour. Relat. Top.* **22** (1992), 58–78.
[47] D. M. Lewis, *Rev. Progr. Colour. Rel. Top.* **28** (1998) 12–17.
[48] J. R. Aspland, *Text. Chem. Color.* **23** (11/1991), 41–45.
[49] Commitee on Direct Dyes, *J. Soc. Dyers Colour.* **62** (1946) 280–185; **64** (1948) 145–146.
[50] D. M. Nunn (ed.), *The Dyeing of Synthetic-polymer and Acetate Fibres*, Dyers Comp. Publ. Trust, Bradford 1979.
[51] U. Baumgarte, *Textilveredlung* **2** (1967) 896–907.

[52] T. Bechtold et al., *Melliand Textilber.* **55** (1974), 953–962.
[53] U. Baumgarte, *Melliand Textilber.* **55** (1974) 953–962.
[54] T. Vickerstaff, *The physical Chemistry of Dyeing*, Oliver & Bayd, London 1954.
[55] U. Baumgarte, *Melliand Textilber.* **59** (1978) 311–219.
[56] BASF-Manual, *Cellulosic Fibers, Sizing, Pretreatment, Dyeing*, B 375e, Ludwigshafen 1977, p. 193–201.
[57] U. Baumgarte, H. Schlüter,, *Melliand Textilber.* **62** (1981) 555–561.
[58] H. U. von der Eltz, H. Walbrecht, *Dtsch. Färber-Kal.* **77** (1973), 692–697.
[59] H. Rath, *Lehrbuch der Textilchemie*, 3rd ed., Springer Verlag, Berlin, 1972.
[60] U. Baumgarte, *Rev. Progr. Colour. Relat. Top.* **5** (1974), 17–32.
[61] Th. Bechtold, E. Burtscher, G. Kühnel, F. Boble, *J. Soc. Dyers Colour.* **113** (1997) 135–144; Th. Bechtold et al.##, *Text. Res. J.* **67** (1997) 635–642.
[62] P. Richter, *Textilveredlung* **10** (8/1975), 312–316. BASF Technical Information, Ludwigshafen 1985.
[63] J. A. Greer, G. R. Turner, *Text. Chem. Col.* **15** (1983) no. 6, 101–107.
[64] L. Haas, *Int. Text. Bull. Veredlung* **2** (1990), 45–50.
[65] C. Heid, K. Holoubek, R. Klein, *Melliand Textilber.* **54** (1973) 1314–1327.
[66] O. K. Venkataraman (ed.), *The Chemistry of Synthetic Dyes* vol. VII, Academic Press New York 1974, p. 26–33 (D. G. Orton, Sulfur Dyes); p. 56–68 (C. D. Weston, Bunte Salts).
[67] C. Heid, *Melliand Textilber.* **45** (1964) 648–652, 833.
[68] W. Prenzel, *Chemiefasern Textind.* **24/76** (1974), 293–302.
[69] C. Heid, *Z. Gesamte Textilind.* **70** (1968) 626–630.
[70] W. Titzka, *Text. Prax. Int.* **33** (1978), 1387–1390.
[71] M. Hähnke, *Dtsch. Färber-Kal.* **96** (1992) 154–156.
[72] T. Bechtold, E. Burtscher, A. Turcanu, *Text. Chem. Color.* **30** (8/1998), 72–77; T. Bechtold et al., *Textilveredlung* **32** *(1997), 204–209.*
[73] L. Tigler, *Am. Dyest. Rep.* **61** (1972) no. 9, 46, 107.
[74] C. Heid, *Melliand Textilber.* **59** (1978), 247–254.
[75] C. Heid, *Text. Prax. Int.* **33** (1978) 285–287; W. Titzka, *Melliand Textilber.* **60** (1979) 254–256; G. Kreuzpaul, *Text. Prax. Int.* **42** (1987) 140–142.
[76] H. M. Tobin, *Am Dyest. Rep.* **68** (no. 9/1979), 26–28.
[77] H. Bernhardt, *Int. Text. Bull.* **33** (1987) 5–19.
[78] C. Heid, *Z. Gesamte Textilind.* **70** (1968) 626–630.
[79] R. Klein, *Textilveredlung* **10** (1975) 112–117.
[80] R. H. Peters, Textil Chemistry vol. 3: *The Physical Chemistry of Dyeing*, Elsevier, Amsterdam 1975.
[81] Ullmann's Encyclopedia of Industrial Chemistry, 5th ed., **A 3**, p. 213.
[82] Ullmanns Encyklopädie der technischen Chemie, 4th ed. 1982, **22**, 656.
[83] H. Vollmann in K. Venkataraman (ed.): *The Chemistry of Synthetic Dyes*, vol. 5, Academic Press, New York, 1971, p. 283–-311.
[84] A. Schäffer, *Handbuch der Färberei* vol. 1, Konradin-Verlag R. Kohlhammer, Stuttgart 1949.
[85] BASF Manual: *Dyeing and Finishing of Polyester Fibers* B 363e, 1975.
[86] F. Hoffmann, *Rev. Progr. Col. Relat. Top.* **18** (1988) 56–64.
[87] H. Flensberg, W. Mosimann, H. Salathé, *Melliand Textilber.* **65** (1984), 472–477.
[88] K. Reincke, *Melliand Textilber.* **74** (1993) 408–417.
[89] D. M. Lewis, G. Yan, *J. Soc. Dyers Colour.* **109** (1993), 193–197.
[90] P. A. Duffield, R. D. D. Holt, *Textilveredlung* **24** (1989) 40–45.
[91] L. Benisek, *J. Soc. Dyers Colour.* **94** (1978), 101–105.
[92] G. Meier, *Text. Prax. Int.* **31** (1976) 898–901.

[93] A. C. Welham, *J. Soc. Dyers Colour.* **102** (1986), 126–131.
[94] D. de Meulemeester, I. Hammers, W. Mosimannm, *Melliand Textilber.* **71** (1990) 69.
[95] J. Cegarra, A. Riva, *J. Soc. Dyers Colour.* **104** (1988), 227–233.
[96] Ullmann's Encyclopedia of Industrial Chemistry, 5th ed. **A 10.**
[97] G. Kühnel, H. J. Flath, R. Gärtner, *Melliand Textilber.* **72** (1991), 288–290, 360–362.
[98] H. Zollinger, *Color Chemistry*, 2nd ed., VCH, Weiheim, 1991.
[99] E. Sada, H. Kumazawa, T. Ando, *J. Soc. Dyers Colour.* **98** (1982), 121–125.
[100] D. Schwer, *Textilveredlung* **23** (1988) 296–301.
[101] F. Hoffmann, *Rev. Progr. Col. Relat. Top.* **18** (1988), 56–64.
[102] B. C. Burdell, *Rev. Progr. Col. Relat. Top.* **13** (1983) 41–49.
[103] F. Hoffmann et al., *Melliand Textilber.* **81** (2000), 284–290.
[104] O. Annen, J, Carbonell, *Textilveredlung* **15** (1980) 296–302.
[105] J. R. Aspland, *Text. Chem. Color.* **25** (1993), 19–23.
[105a] W. J. Jasper, R. P. Joshi, *Text. Res. J.* **71** (2001), 57–64.
[106] C. C. Cook, *Rev. Progr. Col. Relat. Top.* **12** (1982) 73–89.
[107] H. Schiedegger et al., *Textilveredlung* **13** (1978), 302–304.
[108] H. Thielemann, *Melliand Textilber.* **75** (1994) 50–54.
[109] W. Hartmann, *Am. Dyest. Rep.* **69** (6/1980), 21–22.
[110] M. Mitter, *Textilfasern Text. Anwendungstech.* **31** (1981) 55–56, 558.
[111] Faserstoff-Tabellen according to P. A. Koch, *Chemiefasern/Textilind.* **43** (6/1993), 508–522.
[112] K. H. Meyer, *Melliand Textilber.* **6** (1925) 737–739; V. Kartaschoff, Helv. Chim. Acta **8** (1925) 928–942; **9** (1926) 152–173.
[113] H. Braun, *Rev. Progr. Col. Relat. Top.* **13** (1983), 62–72.
[114] H. Leube, *Text. Chem Color.* **10** (1978) no. 2, 32–46.
[115] F. Jones, *J. Soc. Dyers Colour.* **100** (2/1984), 66–72.
[116] G. L. Baughman, T. A. Perenich, *Text. Chem. Color.* **21** (1/1989) 33–37.
[117] H. Leube, W. Rüttiger, *Melliand Textilber.* **59** (1978), 836–842.
[118] T. M. Baldwinson, Rev. Progr. Col. Rel. Top. **15** (1985) 6–14.
[119] A. Keil, H. Noack, H. J. Flath, *Textiltechnik* **40** (1990), 264–267, 317–318, 382–384, 432–437.
[120] H. Gerber, *Textilveredlung* **8** *(1973) 449–456.*
[121] A. Maier, AATCC Intern. Dying Symp. Washington 1977, p. 66–69.
[122] R. A. Walsh, *Text. Chem. Color.* **7** (10/1975) 184/35–187/38.
[123] Disperse Dye Comm., *J. Soc. Dyers Colour.* **93** (1977), 228–237.
[124] M. Hammoudeh, E. Schönpflug, *Melliand Textilber.* **52** (1971) 1063–1068.
[125] J. Park, *Rev. Progr. Col. Relat. Top.* **15** (1985), 26–27.
[126] A. D. Cunningham, H. Burchardi, *Melliand Textilber.* **77** (1996) 468–375.
[127] B. Drechsel, F. Hofmann, S. Ottner, *Melliand Textilber.* **79** (1998), 328–332.
[128] D. Knittel et al., *Melliand Textilber.* **75** (1994) 388–391; D. M. Lewis, *Rev. Progr. Col. Relat. Top.* **29** (1999) 23–28.
[129] F. Jones, J. Kraska, *J. Soc. Dyers Colour.* **82** (1960), 333–338.
[130] J. -H. Chiao-Cheng, B. M. Reagen, *Text. Chem. Color.* **15** (1983) 12–19.
[131] C. A. Li, Chem. Abstr. 100 (24/1984), 193 415.
[132] W. Griesser, H. Tiefenbacher, *Textilveredlung* **28** (4/1993) 88–96.
[133] M. Laufer, *Textilveredlung* **28** (4/1993), 96–101.
[134] D. Fiebig et al., *Textilveredlung* **33** (7/8/1998) 20–23; Z. Uddin et al., *Text Res. J.* **72** (2002), 77–82.
[135] P. Richter, *Chemiefasern/Textilind.* **41/93** (9/1991) 1118–1125, E 129–133.
[136] J. Cegarra et al., *Melliand Textilber.* **65** (1984) 405–409.

[137] C. Grardel, *Ind. Text.* (Paris) **1115** (5/1985), 519–520.
[138] H. Beiertz, K. H. Röstermundt, *Text. Prax. Intern.* **48** (2/1993) 135–137.
[139] C. E. Vellins in K. Venkataraman, *The Chemistry of Synthetic Dyes*, vol. VIII, Academic Press 1978, p. 191–220.
[140] G. Reinert, V. Misun, *Melliand Textber. Intern.* **74** (1993) 1007–1014.
[141] F. Fellner, P. richter, *Melliand Textilber.* **80** (7/8/1999), 608–610.
[142] A. D. Moorhouse, *Rev. Progr. Col. Relat. Top.* **26** (1996) 20–28.
[143] K. H. Rostermundt, Text. Prax. Intern. **47** (7/1992), 649–454.
[144] S. M. Doughty, *Rev. Progr. Col. Relat. Top.* **16** (1986) 25–38.
[145] Ullmann's Encyclopedia of Industrial Chemistry, 5th ed. **A 5**, 438–444.
[146] Report of Committee: Dyeing Properties of Disperse Dyes, I Cellulose Acetate, *J. Soc. Dyers Colour.* **80** (1964) 237–242, II Cellulose Triacetate, **81** (1965) 209–210; J.-H. Choi, A.D. Towns, *Color. Technol.* **117** (2001), 127–133.
[147] BASF Manual: *Dyeing and Finishing of Acetate and Triacetate and their Blends with Other Fibers*, S 388e (1970.
[148] A. Berger-Schunn et al., *J. Soc. Dyers Colour.* **103** (1987) 128–139, 140–141, 272–274.
[149] Polypropylen-Fasertabelle (nach Koch), *Chem. Fibers Intern.* **50** (5/2000), 233–255.
[150] H. Leube, *Melliand Textilber.* **46** (1965) 743–749.
[151] J. Shore, *Rev. Progr. Col. Relat. Top.* **6** (1975), 7–12.
[152] B. D. Gupta, A. K. Mukherjee, *Rev. Progr. Col. Relat. Top.* **19** (1989) 7–19.
[153] H. Thomas et. al., *Melliand Textilber.* **79** (1998), 350–352.
[154] Stanford Res. Inst.: Chemical Economics Handbook, Fibers 5433000 D, Menlo Park 1993.
[155] R. Raue, *Rev. Progr. Col. Relat. Top.* **14** (1984), 187–203.
[156] I. Holme, *Rev. Progr. Colour. Relat. Top.* **13** (1983) 10–23.
[157] F. Jones, *Rev. Progr. Col. Relat. Top.* **4** (1973), 64–72.
[158] H. Sumner in A. Johnson (ed.): *The Theory of Colouration of Textiles*, chap. 2, Soc. Dyers Colour. Bradford, 1989.
[159] AATCC Metropolitan Section, *Text. Chem. Color.* **8** (1976), 165.
[160] S. Shukla, M. Mathur, *J. Soc. Dyers Colour.* **109** (1993) 330.
[161] J. Park, J. Shore, *J. Soc. Dyers Colour.* **97** (1981), 223.
[162] W. Biedermann, *Rev. Progr. Col. Relat. Top.* **10** (1979) 1.
[163] F. Hoffmann, *Text. Chem. Colour.* **22** (1990), 11.
[164] A. Kretschmer, *Text. Prax. Intern.* **44** (1989) 1098, 1316, **45** (1990) 225, **46** (1991) 1220.
[165] F. Hoffmann, *Melliand Textilber.* **59** (1978), 239.
[166] W. Beckmann, K. Jakobs, *Bayer Farben Rev.* **16** (1968) 1.
[167] F. Hoffmann, H. Schubert, J. Fiegel, *Text. Prax. Int.* **47** (1992), 233.
[168] J. Park, J. Shore, *J. Soc. Dyers Colour.* **97** (1981) 223.
[169] I. Holme, *Rev. Progr. Col. Rel. Top.* **13** (1983), 10.
[170] R. Rohner, H. Zollinger, *Text. Res. J.* **56** (1986)1.
[171] U. Reinehr, G. Häfner, A. Nogaj, *Chemiefasern/Textilind.* **35** (no. 87/1985), 588.
[172] J. Shore, *Rev. Progr. Col. Relat. Top.* **10** (1979) 33.
[173] L. Kostova, R. Iltscheva, R. Detcheva, *Textilveredlung* **27** (1993), 398.
[174] I. Hardelov, J Mikhailova, *J. Soc. Dyers Colour.* **109** (1993) 369.
[175] I. Holme, *Rev. Progr. Col. Relat. Top.* **7** (1976), 1.
[176] H. Fleischer, VII. Intern. Chemiefaser Symp., Siofok, Hungary, 1987.
[177] J. E. McIntyre, *Rev. Progr. Col. Relat. Top.* **25** (1995), 44–56.
[178] DUP, US 5232 461 (1993).
[179] W. Haertl, *Textilveredlung* **24** (no. 6/1989), 214–218.
[180] W. Schmidt, G. Faulhaber, A. J. Moore, *Rev. Progr. Col. Relat. Top.* **2** (1971), 33.

5 Nontextile Dyeing

5.1 Leather Dyes

5.1.1 Introduction

Leather is a very complex and heterogeneous substrate. Once the animal has been slaughtered, the hide or skin has to be removed, cleaned, depilated, and tanned in the so-called "beam house" process. The term "hide" is used for the outer covering of animals with a body surface area of more than 1 m^2 like bovines, and the term "skin" for that of smaller animals like sheep, goats, and pigs. Large proportions of sheepskins are used for fur production.

Tanning is the process of stabilizing the collagen matrix to retain the fiber structure of the hide and to raise the hydrothermal stability to around 100 °C. Today, the most important process is chrome tanning with trivalent chromium compounds. Chrome-tanned leather is greenish blue and traded in the "wet blue" state or dried and slightly retanned as "crust" leather. "Wet white" is chrome-free semifinished leather that is tanned with aluminum or glutaraldehyde. Vegetable-tanned leather is chiefly used for soles and its color is yellowish to reddish brown. Oil tanning is performed mainly with fish oils to give "chamois leather". The leather obtained is yellow due to reaction with acrolein from the unsaturated fats.

After the wet blue or wet white has been sorted and shaved, it is subjected to wet finishing. The purpose of the so-called "wet end" process is to make the leather more uniform regarding fullness, softness, and color. Fullness is achieved by retanning with synthetic tanning chemicals (syntans), vegetable tanning agents, or special filling compounds, mostly resins and polyacrylate derivatives. The application of fat liquors provides the required softness. Dyeing with water-soluble synthetic organic dyes results in "aniline leather", known as best quality leather providing the grain quality is appropriate. On drying, leather can shrink and harden. Therefore the aim is to avoid shrinking by controlling the drying process and mechanical stretching (called "staking") to break up the fiber–fiber adhesion. "Dry milling", rotating in a drum, is also used but results in a coarse and high grain pattern, and the area yield is reduced.

In most cases the leather surface has to be protected against humidity and any kind of dust. Casein and other proteins as well as wax have been used for centuries. In the 1930s polymer finishing was introduced in combination with pigment dyes. "Semi-aniline" leather receives only a small amount of the transparent polymer finish to maintain the breathability of the material. Stronger covers containing pigments are applied to leather from raw hides with surface defects, so making the "pigmented leather" more uniform and suitable for the manufacture of end products.

Today, leather is mostly used for shoes, upholstery, clothing, gloves, bookbindings, handbags, soles and work protection, or orthopaedic purposes. At the consumer level, the outstanding stability of leather and its colors ensure a long service life for leather goods. Nevertheless the method of care depends on the finish and surface condition of the leather in question. For perfect leather care the surface must be free from residual dirt particles. Furthermore leather should not be wet through or moist in parts. Stains are wiped off with a moist cloth soaked in neutral soap solution. Heavy stains need to be treated with great care. Aniline and semi-aniline leather should be protected from direct sunlight [1].

5.1.2 Color Selection

Dye selection and classification used to be largely empirical and was based primarily on specific tannages and on the requirements to be met by the leather end product. Drawing on this fund of knowledge, supplemented by scientific findings and conclusions, the dye-manufacturing industry now creates essentially tailor-made ranges. By and large, the principle of grouping together dyes with identical dyeing characteristics and marketing them as specialized ranges has paid handsome dividends. This includes implementation of trichromatic dyeing philosophy which teaches that all shades can be matched with the three primary colors: yellow, blue, and red.

The dyes used for leather are mostly selected from ranges designed for textiles, paper, or foods, and pigments for printing, paints, plastics, and other applications.

5.1.2.1 Aniline Leather

Color is one of the key attributes by which the consumer judges the quality of leather goods. High-grade leathers nowadays are through dyed so that they have the same shade on the cut edge and on the back (flesh side) as on the more compact grain side. The dyes are applied not to the skin (collagen) but to the tanned leather, which differs according to the tanning process.

Furthermore, because leather is of heterogeneous animal origin it does not have a definite composition. It also has inherent grain characteristics, which change within animal species according to breed, age, and nutritional status of the animals. All this makes it difficult to attain uniformity of shade throughout the piece and within a batch. Therefore dyeing agents play an important role in improving levelness, penetration and fixation of dyeing.

Besides shade, gloss is also an essential attribute, and special products like cationic fat liquors are used. Silicones give a very appropriate silky sheen often seen as a two-way (often called writing) effect on buffed leather. Most coat finishing processes also provide a glossy appearance.

Dyes for leather are selected according to the application process. In the wet end process, dyeing is normally conducted in wooden drums. Flawless dyeing in these vessels depends on good mixing of float and leather. Bars and pegs prevent the leather from rolling up at the bottom of the drum. The requirements for the drum dyeing process vary for the various end products. *Surface (top) dyeing* for leveling and correcting defects on the grain side uses high molecular weight anionic direct dyes or 1:2 metal-complex dyes with moderate solubility. *Semi-penetrated dyeing* to reduce the obtrusiveness of damaged patches use medium molecular weight anionic acid or direct dyes or 1:2 metal-complex dyes with high solubility. For *through-dyed leather*, to avoid edges on the goods, low molecular weight acid dyes, 1:1 metal complex dyes, or soluble sulfur dyes are used.

Alternative procedures are paddle dyeing for sensitive skins in an open vessel, and immersion dyeing, in which the leather passes through a dye solution and is subsequently squeezed by a roller. Spraying and roller machines have mostly replaced other techniques such as traditional brush dyeing, which colors the surface only.

5.1.2.2 Pigmented Leather

The final color of *pigmented leather* is achieved by mixing and shading inorganic or organic pigments, which are incorporated in a binder film system, also called "pigment coats". Dissolved or homogenized dyes and finely dispersed pigment pastes are used. Pigment coat binders are polymer dispersions which form films on drying. They can be solvent-, solvent–water-, or water-based polyacrylates, polybutadienes, or polyurethanes. Spraying of the coats with compressed air is the most common application method, followed by curtain and roll coaters. Screen printing using a doctor blade and fine-meshed negative stencil, like that used in textile printing, is also sometimes practised.

The most frequently used pigments are organic pigments (water- and solvent-insoluble metal-complex and other pigments for good fastness qualities), inorganic pigments (titanium dioxide, carbon blacks, iron oxides, ultramarine, heavy-metal compounds), and special pigments.

Anionic dyes (brilliant acid dyes) and solvent dyes (unsulfonated metal-complex and other solvent-soluble dyes for brilliant shades) are used for shade correction.

5.1.2.3 Colour Index

The Colour Index (C.I.) is almost exclusively a listing of textile dyes (and pigments), and dyes for leather appear only in a supplementary chapter and are classified mostly as acid, direct, and mordant dyes [2]. About 1500 leather dyes are listed in the Colour Index (Table 5.1), with more then 1000 acid and more than

250 direct dyes. However, an even larger number, including many mixtures, are suitable for leather. Conversely, not all the dyes listed are manufactured and commercially available nowadays. The chemical structure of many dyes recorded in the Colour Index has not been reported; ca. two-thirds of leather browns and blacks have an undisclosed structure.

Table 5.1: Leather dyes found in the Colour Index*

Class	Number	Aniline leather	Pigmented leather
C.I. Acid Dyes	1031	regularly	occasionally
C.I. Direct Dyes	283	regularly	occasionally
C.I. Mordant Dyes	76	seldom	seldom
C.I. Reactive Dyes	53	occasionally	seldom
C.I. Sulfur Dyes	25	occasionally	seldom
C.I. Solvent Dyes	4	occasionally	regularly
C.I. Basic Dyes	2	occasionally	occasionally
C.I. Natural Dyes	*	seldom	seldom
C.I. Pigments	*	seldom	regularly
Total	1474		

* Natural dyes and pigments are not listed in the Colour Index for leather dyeing purposes. Natural dyes are mostly of historical interest. Pigments are the most common finishing colorants.

The most important leather dyes identified by typical chemical nomenclature are azo, stilbene, triarylmethane, quinoline, azine, oxazine, thiazine, sulfur, and phthalocyanine.

The C.I. Name refers to the colorant only and not to the commercial preparation, purity of the main component, balance of synthesis related byproducts, shading elements and their diluents, and presence of dispersants and other chemicals. Consequently the dyes can differ in technical properties, concentration, and ecotoxicological parameters.

The C.I. Constitution number is given in this paper as a guide to references for synthesis in the Colour Index. Any duplication of these syntheses should be attempted only under the supervision of an experienced chemist who is aware of the latest ecological, toxicological, and safety requirements. These requirements must be constantly observed even if attention is not explicitly drawn to them. A further point to note is that dye chemists often write the free sulfonic acid form, even if they have obtained a salt. The reason is that the dyes are often not completely neutral or may not be a defined salt. This commonly arises when potassium or sodium salts are involved, and therefore the free sulfonic acid formula is frequently used as a general term. A sulfonate salt does not influence the color as such, but water solubility is affected. Consequently, for pigments the cation is always given in the chemical formula.

5.1.3 Natural and Mordant Dyes

The oldest method of dyeing vegetable-tanned leather involved treating it with solutions of metal salts that formed complexes with tanning agents [3]. Iron salts yielded a gray to black shade, copper salts a dark brown shade, titanium salts orange shades, and dichromate deep brown shades. The so-called "beer black" was based on this dyeing principle. Old nails and iron turnings were added to fermenting beer to form the corresponding iron salts. The resulting organic iron complexes were somewhat less corrosive than an iron salt and could be used to yield slightly deeper shades.

5.1.3.1 Tanning Agents

Vegetable tanning agents are extremely complex mixtures of low to high molecular aromatic compounds, with several hydroxyl compounds for solubility and binding on skin collagen. Tanning agents used to be classified according to the main cleavage products pyrogallol and pyrocatechin, which form on heating at 180–200 °C.

Of the numerous other chemical principles for classifying tanning agents, subdivision into hydrolyzable and condensed types should be mentioned. Hydrolyzable tanning agents are high-molecular esters that mostly form gallic acid when hydrolyzed. The condensed tanning agents are derivatives of flavan, generally types of catechin which are held together via carbon linkages. In simplified terms, the pyrogallol tanning agents represent the hydrolyzable variety and the pyrocatechin tanning agents the condensed variety.

C.I. Natural Brown 6 (constitution not available), basic structure gallic acid (gallotannin; **1**), is based on a large variety of vegetable products having a high content of hydrolyzable tannin, which mostly form gallic acid when hydrolyzed.

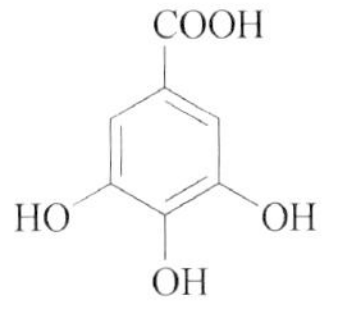

C.I. Natural Brown 3, 75250 (catechu) has the basic structure of catechin (**2**) and is a condensed tannin, e.g., gambier from leaves and twigs of the *Uncaria gambier* bush. Many of these tanning agents are based on a catechin flavanoid.

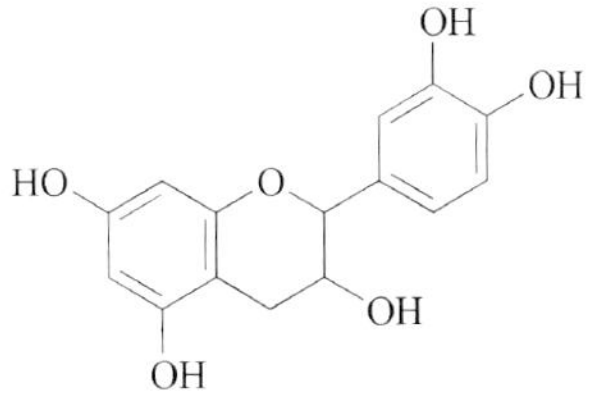

5.1.3.2 Dyewood

Before the synthetic dyes started to dominate the coloration of leather, extracts of dyewoods, chiefly logwood, redwood, and fustic, were used. These dyewoods exhibit a tanning action, dye the flesh and grain sides of leather equally well, and have no tendency to accentuate grain defects. These favorable properties arise from the fact that the dyewoods contain low-molecular, less strongly aggregating constituents. Also dyewood extracts can be considered as mordant dyes of the flavonoid* and neoflavanoid** classes. Typical shades obtained by laking with potassium alum are listed in Table 5.2.

Table 5.2 Lake of dyewood

Dyewood	Color	Dye	C.I. name
Fustic	Yellow	Maclurin*, Morin*	C.I. Natural Yellow 8, 11, resp.
Brazilwood	Red	Brazilein**	C.I. Natural Red 24
Logwood	Violet	Haematein**	C.I. Natural Black 1, 2, (3, 4)

* Flavanoids. ** Neoflavanoids.

C.I. Natural Yellow 8, 11, 75660, 75240 (fustic)

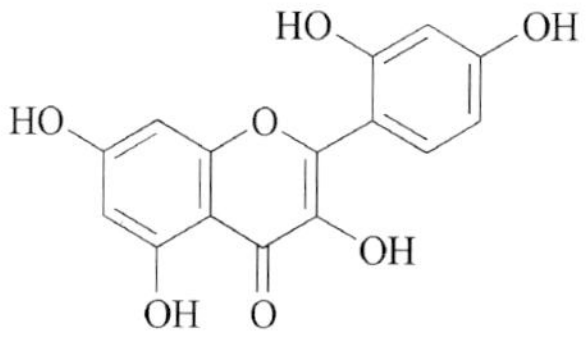

Morin

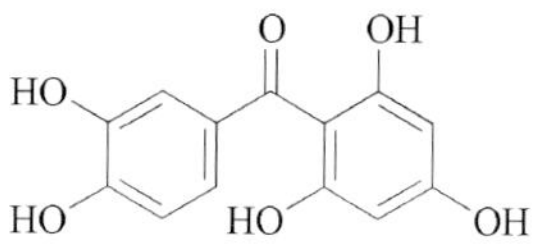

Maclurin

C.I. Natural Red 24, 75280 (brazilwood)

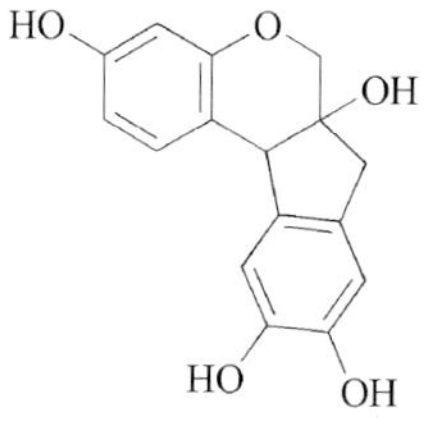

Brazilin

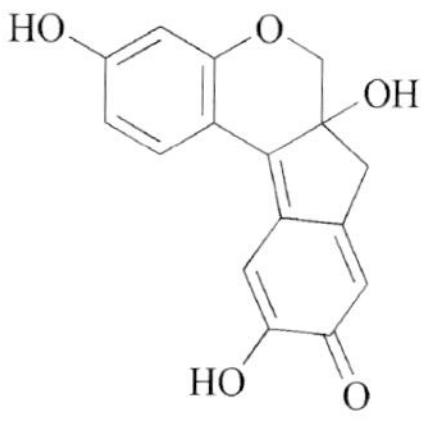

Brazilein

C.I. Natural Black 1, 2, (logwood), *3, 4* (Al, Cr, Fe, or Sn lakes), 75290, 75291

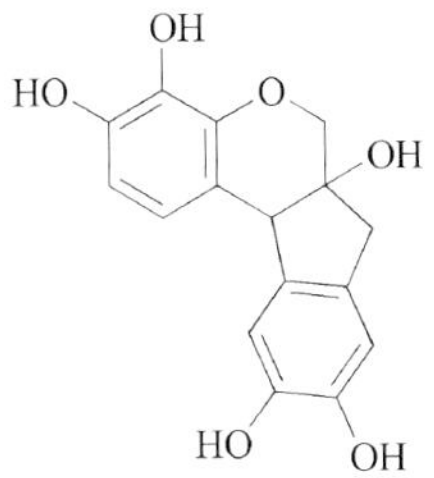

Haematoxylin

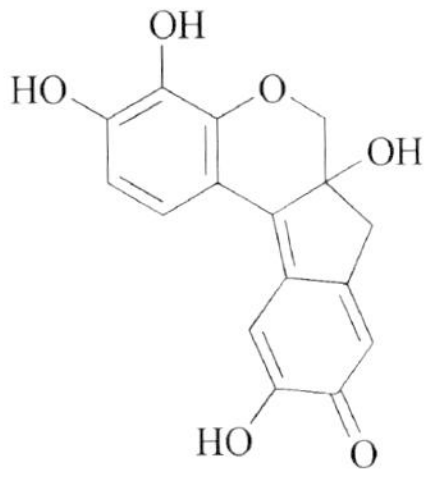

Haematein

5.1.3.3 Synthetic Mordant Dyes

The use of both natural and synthetic mordant dyes has long passed its zenith. Synthetic mordant dyes are still listed as leather dyes in the Colour Index. These are chiefly monoazo dyes, which are used especially for prechromed wool. For leather, the operations involved in the metallization step are too complex and time-consuming. Nowadays, modern tanneries prefer premetallized dyes.

5.1.4 Basic Dyes

The first synthetic leather dyes were cationic. Mauveine, discovered by W. H. Perkin, marked the start of the synthetic dye industry. The cationic dyes were characterized by a previously unknown brightness. Nevertheless, the introduction of the new synthetic dyes was not a simple matter. The lightfastness properties were poor. However, the presence of Perkin at the first International Leather Congress in 1897 in London, undoubtedly ensured the breakthrough in the use of synthetic dyes for leather.

5.1.4.1 Azine Dyes

A large number of azines were produced in the early decades of industrial dye manufacture. The synthesis was an all-in operation involving a sequence of oxidation reactions. Mauveine (**3**), for example, was prepared by oxidizing aniline containing *o*- and *p*-toluidines with potassium dichromate in cold dilute sulfuric acid solution.

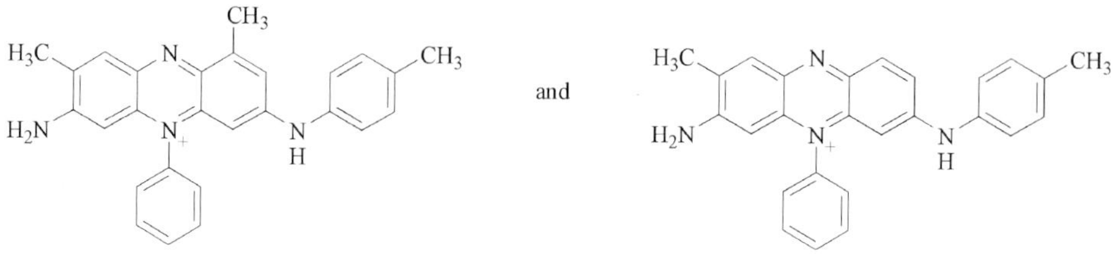

3

Basic azines are predominantly of historical interest. Yellow, red, brown, blue, and black shades can be obtained with these dyes. Today, Nigrosine Spirit Soluble is still used in shoe polish and creme. The sulfonated variety is applied as a leather dye. It is a polymeric dye containing phenazine ring systems, similar to Aniline Black. *C.I. Solvent Black 5*, 50415 [*11099-03-9*] (Nigrosine, spirit-soluble); *C.I. Solvent Black 7*, 50415:1 [*8005-02-5*] (**4**) is the free base.

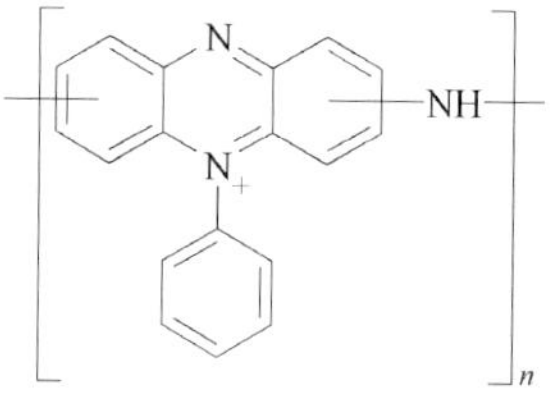

4

5.1.4.2 Other Cationic (Basic) Dyes

The triarylmethane dyes were also very important leather dyes in the past. The central carbon is sp^2-hybridized (triarylmethine) and can add a hydroxyl group to form the respective carbinol base with sp^3-hybridized carbon. The dyes react like Lewis acids and some species act as pH indicators (e.g., phenolphthaleine). The term cationic dyes is therefore recommended for this dye class instead of basic dyes. Nowadays only the sulfonated triphenylmethine dyes are used for leather.

Basic dyes are used almost exclusively for overdyeing, e.g., to achieve deep black shades by laking. Recent research indicates that new cationized sulfur dyes for leather will shortly become commercial products. In the 1970s dicationic azo dyes for paper were introduced by azo coupling of aromatic diazonium ions with heterocylic components. The resulting cationic azo dyes are also applied on leather e.g. *C.I. Basic Black 11*. The constitution is not disclosed.

5.1.5 Acid Dyes

These are, with the exception of some metal complexes, low molecular weight wool dyes that usually have monoazo, disazo, or simple anthraquinone systems. They exhaust from a strongly acid to more neutral bath. The relatively small molecules create exceptionally bright shades, with the exception of some premetallized dyes. Given their small size, they diffuse rapidly into the interior of the leather without fast binding. They are therefore classified as penetrating dyes.

5.1.5.1 Amphoteric Dyes

In 1862 Nicholson sulfonated basic dyes and thus created the first amphoteric and acid leather dyes. Triarylmethane (triarylmethine) dyes (Table 5.3) and their different uses are typical examples of this development.

Table 5.3 Triarylmethane dyes

Technical name	Form	C.I. name	Use
Aniline blue/blue T base	methine	*Solvent Blue 23*	solvent
Alkaline blue	carbinol base, one sulfo group	*Acid Blue 119*	silk
Sea blue/ reflex blue B	two sulfo groups	*Acid Blue 48*	wool, paper
Ink blue	three sulfo groups	*Acid Blue 93*	ink, leather

An example is *C.I. Acid Blue 93*, 42780 [*28983-56-4*] (**5**; simplified structure).

5

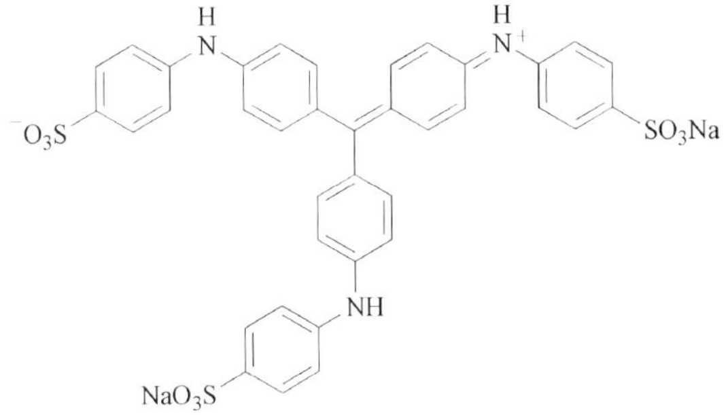

Soluble sulfonated triarylmethane compounds are still used today as single dyes. They are brilliant blue, but with medium lightfastness. There are therefore often used in leather dye mixtures as shading components or as brightening dyes.

Almost all basic dyes have been sulfonated, including Nigrosine Spirit Soluble (see Section 5.4.1) The sulfonated variant *C.I. Acid Black 2*, 50420 [*8005-03-6*] is still applied to leather, mainly as a shading component for deep black.

5.1.5.2 Anthraquinone Dyes

Alizarin, a naturally occurring anthraquinone, exists as coloring matter in madder. In the middle ages the red color of the famous cordovan leather (leather from Cordoba, Spain) was created by mordanting with alum. Nowadays, anthraquinone dyes are the classical source for vat dyes. However, vat dyes are not applied to leather. The alkali used for the reduction of the vat dyes would destroy the leather substrate.

In 1869 Caro succeeded in performing the sulfonation step, and the anthraquinone dyes could be used to color leather. These bright acid dyes have excellent lightfastness and a bright, mostly blue to reddish shade. The arylaminoanthraquinone types are chiefly utilized for coloring leather. The sulfo groups can be present in the anthraquinone or/and in the substituent. An example is *C.I. Acid Blue 25*, 62055 [6408-78-2] (**6**).

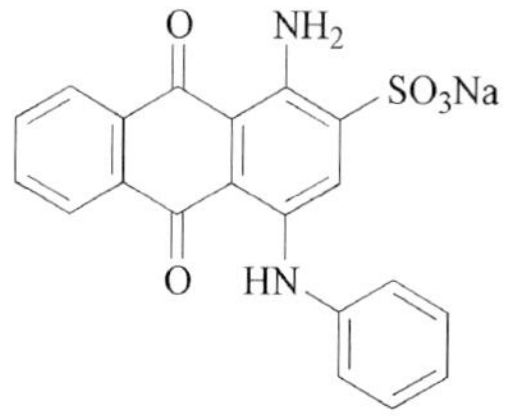

6

5.1.5.3 Low-Molecular Azo Dyes

Yellow, red, and navy blue for leather can be obtained with monoazo dyes. Generally speaking, azo dyes with naphthalene moieties give deeper shades than those with phenyl residues. These classical acid dyes were developed for wool, silk, and polyamide fibers, and suitable ranges are applied to leather. The molecular weight for penetration dyes is below 500. Typical leather penetration dyes are *C.I. Acid Yellow 11* 18820 [6359-82-6] (**7**) and *C.I. Acid Red 7*, 14895 [5858-61-7] (**8**; Orange II), which are still used today.

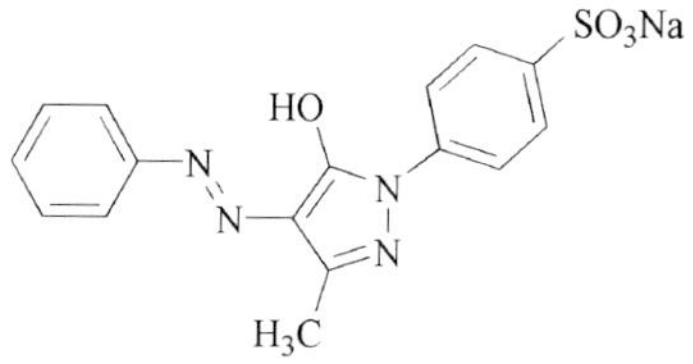

7

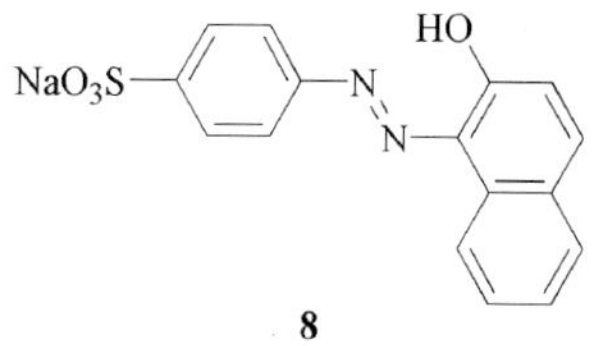

8

5.1.5.4 Resorcinol Azo Dyes

Homogeneous brown dyes are, unfortunately, difficult to synthesize. For the first time in the 1930s, research started to target synthetic leather dyes. The readily synthesized polyazo brown dyes have given good results, particularly on leather though they could scarcely meet textile requirements in terms of stability and fastness properties. This was particularly the case with azo dyes having resorcinol or diaminobenzene as coupling component. Multiple azo coupling is feasible on resorcinol at three points, depending on reaction conditions.

Browns are the most important shades for leather after blacks. Despite their higher molecular weight and sole use on leather, these dyes are normally also classified as acid dyes. *C.I. Acid Brown 123*, 35030 [6473-04-7] (**9**), for example, although a tetrakisazo compound, is classified as an acid dye. The dyeing is more on the surface.

(structure 9)

The mono and disazo resorcinol dyes can act as indicators and change color very markedly, depending on pH. Monoazo and disazo varieties are often found as byproducts with higher molecular weight resorcinol dyes. This property makes them less obtrusive on leather. Because of their strong intermolecular forces of attraction, resorcinol dyes are very well fixed on leather, despite their low to medium molecular weight. Their indicator properties are scarcely noticeable, in contrast to the case with textiles. This explains why these dyes are not necessarily suitable for textiles, particularly polyamides.

Although the major part of research on resorcinol dyes was actually completed in the 1930s, an increasing number of patent applications were filed into the 1980s. The more recent work was conducted to enhance the scope of available structures, to optimize synthesizing procedures, and to meet new fastness requirements by minimizing byproducts.

5.1.5.5 Azo Metal-Complex Dyes [4] (see Section 3.11)

Currently, tanneries use the premetallized 1:1 and 1:2 metal complexes. The 1:1 complexes with their good penetrating properties contain one metal ion and one dye ligand. The 1:1 complexes with a trivalent, six-coordinate chromium ion contain further ligands (e.g., water), depending on the method of synthesis. The most important metal complex dyes are *o,o'*-dihydroxyazo and *o,o'*-carboxyhydroxyazo, in which a nitrogen atom of the azo group participates in the metallization, e.g., *C.I. Acid Red 183*, 18800 (1:1 Chrome) [6408-31-7] (**10**).

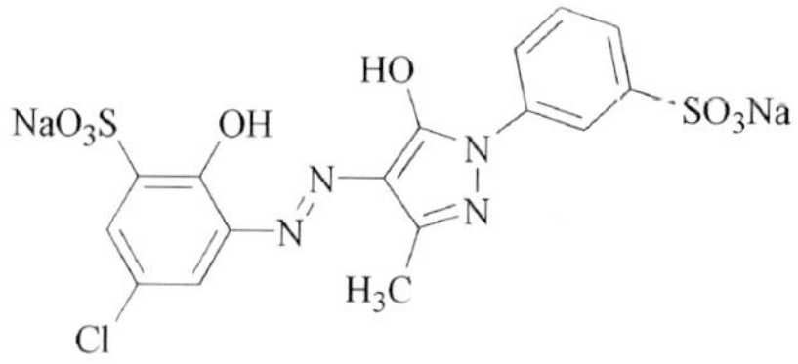

10

In contrast, 1:2 complexes contain one metal ion and two dye ligands and tend more towards dyeing the surface. This system is characterized by five- or six-membered rings, with the metal in the center of the octahedral structure. Metals are the trivalent, six-coordinated chromium, cobalt, and iron ions. Today, the principles underlying the structure and syntheses of metal-complex dyes are well known. Examples are *C.I. Acid Blue 193*, 15707 (1:2 Chrome) [*12392-64-2*] (**11**) and *C.I. Acid Yellow 151*, 13906 (1:2 Chrome) [*12715-61-6*] (**12**).

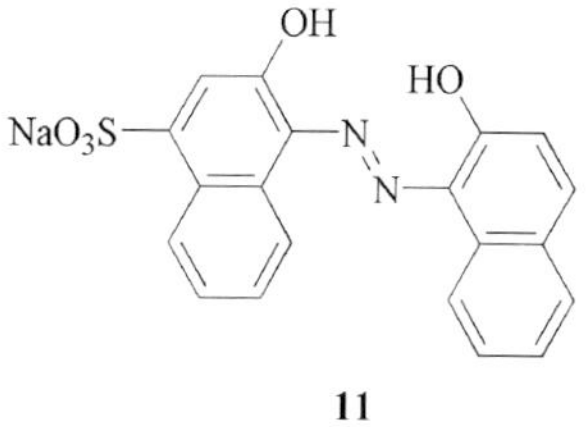

11 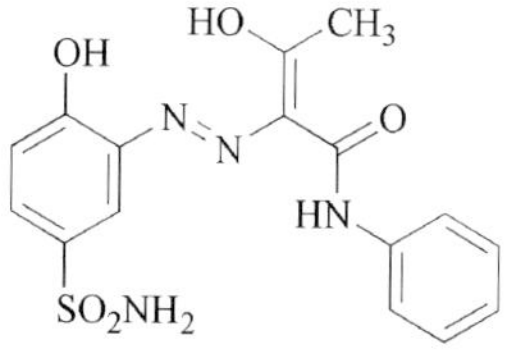   **12**

Since the 1:1 chromium complexes can easily be converted to 1:2 complexes with complexable azo dyes, it is also possible to produce unsymmetrical chromium complexes. This opens up an extraordinary number of combination options for producing brown and olive shades, along with navy blue, gray, and a variety of black shades, enormously extending the shade spectrum of chromium complexes. Yellow, red, blue, and black are also obtainable with symmetrical complexes.

The 1:1 cobalt complexes are almost impossible to isolate because they convert too rapidly to the 1:2 complexes. Since the bathochromic shade shift accompanying cobalt complex formation is not as marked as with chromium, the cobalt complexes are chiefly encountered in the yellow, light brown, orange, and bluish gray shades. The lightfastness of the cobalt complexes is often somewhat better than that of the corresponding chromium complexes.

Soluble iron complexes are extremely unstable and therefore hardly suitable for textile dyeing. On leather, however, they have given very good results with mostly brown shades. Homogeneous iron complexes are almost impossible to isolate.

Since copper is divalent and four-coordinate, it does not give 1:2 complexes. Also, 1:1 azo copper dyes are unstable in solution but are used for dyeing cotton (direct dyes) and occasionally for leather. The important phthalocyanine complexes are regarded as direct dyes (see Section 5.1.6).

5.1.6 Direct Dyes

The direct or substantive dyes, which were designed for cotton, generally have a molecular weight in excess of 500, and elongated, planar structures. Their water solubility is less than that of the acid dyes. Hard water or acid can speed up their natural tendency to undergo aggregation; this tendency has a number of disadvantages. The aggregated dye exhausts very rapidly onto the damaged areas of the grain side of leather and onto the larger surface of the flesh side. These parts of the leather are therefore frequently dyed a duller, deeper shade than the others. Under unfavorable dyeing conditions, direct dyes can also flocculate out of solution and smear the leather surface, a phenomenon also referred to as bronzing. Nevertheless, direct dyes have a decisive advantage over acid dyes. Although they fix at a faster rate and hence tend to dye the surface more, tanners can produce wetfast, penetrated, and even through-dyeings with them. Desirable results are achieved with the aid of special techniques, e.g., by using a short float, low temperature, fat liquors, and/or dyeing auxiliaries.

5.1.6.1 Condensation Dyes

Walther obtained the first direct dyes by self-condensation of 4-nitrotoluene-2-sulfonic acid. So-called Sun Yellow is a mixture of different components, depending on the concentration of sodium hydroxide, the temperature, and the duration of the reaction. Oxidation of the intermediate dye and subsequent reduction with iron and hydrochloric acid gives 4,4' diaminostilbene-2,2'-disulfonic acid, which is used for fluorescent whitening agents and azo dyes. The shades are mostly yellow to red. The structure of *Direct Yellow 11*, 40000 [*65150-80-3*] (**13** is one of the main components) probably contains a mixture of stilbene, azo, and/or azoxy groups.

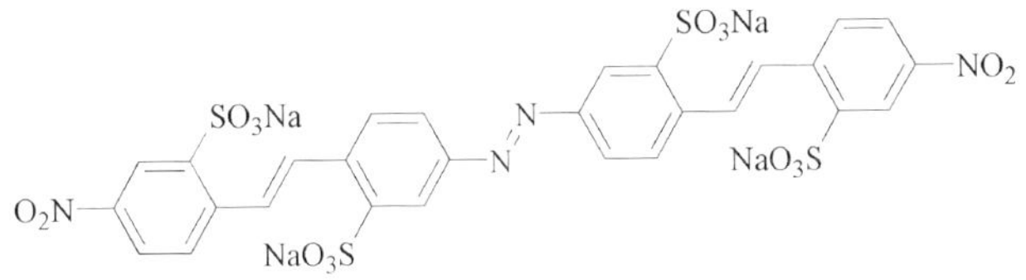

13

Some bright blue direct dyes with good lightfastness are sulfonated dioxazine compounds, mostly synthesized by condensation of chloranil with amines followed by ring formation and sulfonation, e.g., *C.I. Direct Blue 106*, 51300 [*6527-70-4*] (**14**).

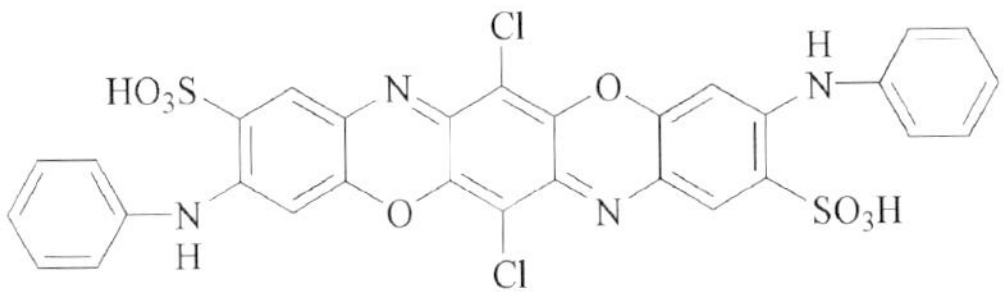

14

For leather, the quinoline chromophore, prepared by condensation of quinoline derivatives with phthalic anhydride or similar substances, is of some interest. The mainly yellow or red compounds are sulfonated and yield valuable dyes for leather, e.g., *C.I. Acid Yellow 2*, 47010 (**15**).

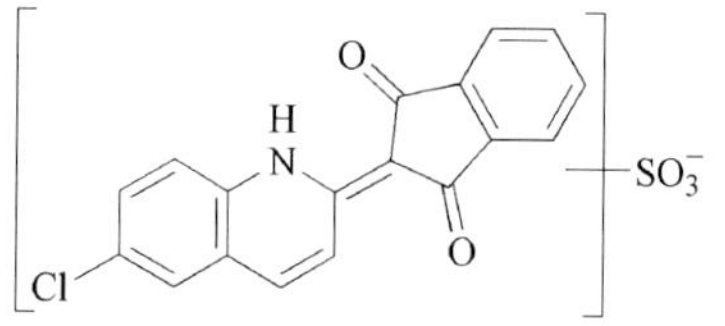

15

5.1.6.2 Polyazo Dyes

The direct dyes that are most suitable for leather have the following features: predominantly long-chain dye molecules, sulfonate groups at the ends of the molecule, additional non-dooble-bonded groups, a balanced ratio of double bonds to solubilizing groups, and predominantly *meta*-substituted azo coupling components. These empirical observations have been used to enhance the technical properties of anionic polyazo leather dyes.

While the development of direct dyes was essentially completed in the 1930s, it was reactivated in the 1960s and 1970s, when it was found that benzidine, an important direct dye intermediate at the time, is carcinogenic. Benzidine dyes are used for some yellows and reds and mostly for black. In view of the ensuing concern, the leading dye producers withdrew benzidine-based dyes preemptively to protect their workers, dyers, and consumers. Studies to find alternative dyes were undertaken, and this work led eventually to many of today's most important black dyes for leather, e.g., *C.I. Direct Black 168*, 30410 [85631-88-5] (**16**).

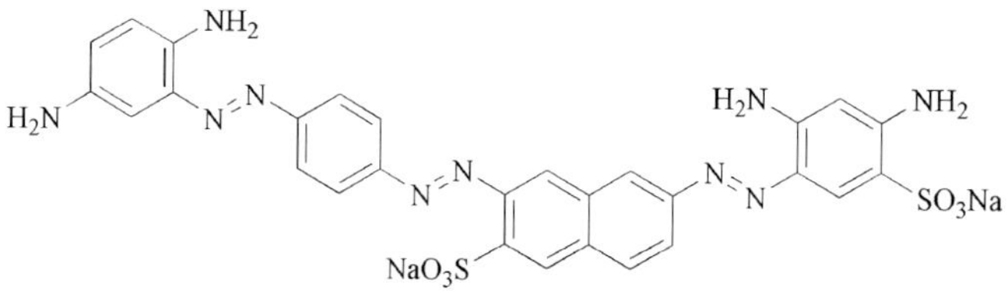

16

5.1.6.3 Phthalocyanine Dyes

The best lightfastness and brilliant blue and green (turquoise) shades are obtained with copper phthalocyanine dyes. Phthalocyanines are synthesized by heating phthalodinitrile with copper chloride and then sulfonated, e.g., *C.I. Direct Blue 86*, 74180 [1330-38-7] (**17**).

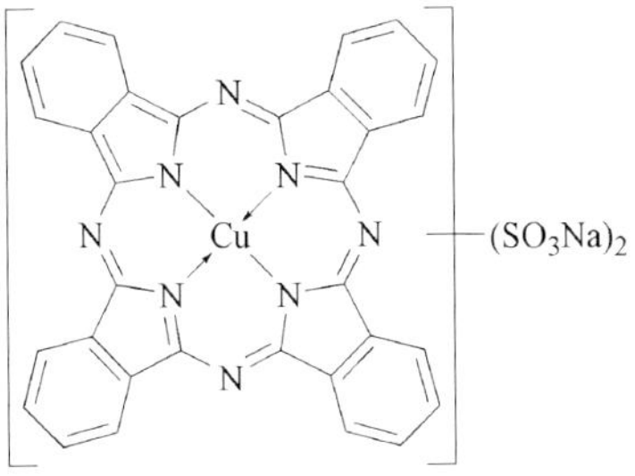

17

5.1.7 Sulfur Dyes

Initially, sulfur dyes were water-insoluble, macromolecular, colored compounds formed by treating aromatic amines and aminophenols with sulfur and/or sodium polysulfide. R. Vidal developed these dyes in 1893 but they only became attractive for leather with the introduction of water-solubilizing groups. Today, the sulfur dyes can be divided into three classes: conventional water-insoluble, leuco, and solubilized sulfur dyes. Most sulfur dyes are synthesized by condensation of aromatic amines with sulfur or sodium polysulfide in the so-called bake process, or else in water or under pressure as a solvent-reflux reaction.

Many sulfur dyes contain benzothiazole, thiazone or thianthrene groups, and almost the whole range of shades can be obtained. The number of commercially available sulfur dyes is small but their production volume is large. Sulfur Black T with all conceivable variations may be the biggest synthetic dye which is used as a penetrating dye on leather: *Sulphur Black 1*, 53185 [1326-82-5] (**18**; most probable structure).

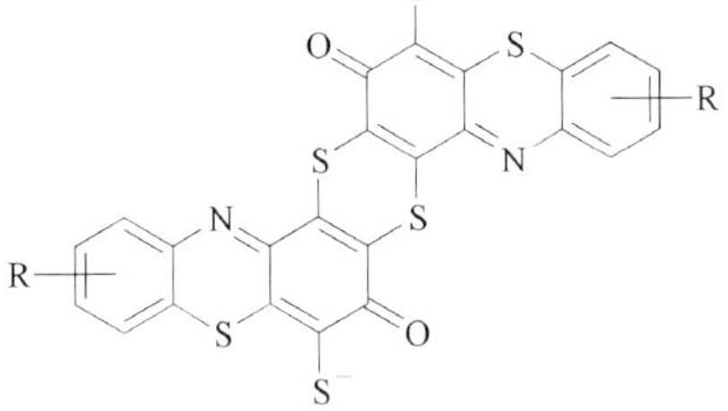

18

Sulfur Black T is produced by boiling 2,4-dinitrophenol with sodium polysulfide and is presumably a phenothiazonethianthrene macromolecule. The water-insoluble forms of these dyes can be treated with various reducing agents to yield the water-soluble leuco sulfur dyes, which are fairly unstable. More stable water-solubilized sulfur dyes can be obtained by the action of sodium sulfite or bisulfite on the parent sulfur dyes to produce thiosulfonic acid groups. The thiosulfate groups can also be found in azo dyes, e.g., *C.I. Condensed Sulphur Orange 2*, 18790 (**19**).

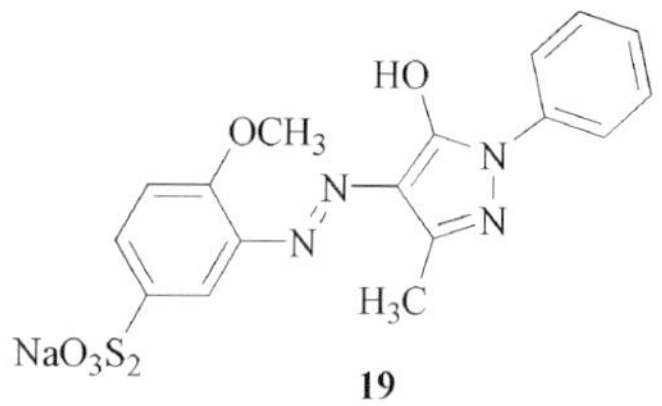

This dye is not commercialized for leather; however, recent research and development have produced a true azo–sulfur hybrid and also cationized sulfur dyes for leather. These dyes will shortly become commercial products.

Sulfur dyes are eminently suitable for through-dyeing leather, and because of the recent upsurge in demand for such leather this class of dyes has experienced a renaissance. In spite of numerous advantages, ecological problems regarding sulfur byproducts have to be taken into account by the manufacturer of these dyes and by tanneries applying them on leather.

5.1.8 Reactive Dyes

Reactive dyes are colored compounds that contain groups capable of forming covalent bonds between dye and substrate. Approximately 80–90 % of reactive dyes are azo dyes. The other chromogenic classes are anthraquinones, dioxazines, phthalocyanines, and some 1:1 copper azo complexes. Reactive 1:2 complex leather dyes were also commercially manufactured for a short time. Constitution and producer have not been disclosed.

The characteristic structural features of a reactive leather dye are water solubilizing group(s), chromogen, bridging link (often amino) between chromogen and electrophilic group, and an electrophilic substituent with a nucleophilic leaving group.

Although reactive dyes have the potential to provide the best fastness to wet treatments, this potential has only been realized in some small segments. Since 1951, when the first reactive dyes were designed for cotton, many reactive dyes including complete ranges have been offered for leather. However, no major breakthroughs have been achieved to date.

The trichlorotriazine molecule was the first reactive compound that was found to be able to form a reactive bridge between dye and substrate. One chlorine atom reacts with the amino group of the dye, and the other two chlorine atoms can then react bifunctionally with the substrate (or water) to form covalent bonds. An example is *C.I. Reactive Yellow 4*, 13190 [1222-45-8] (**20**).

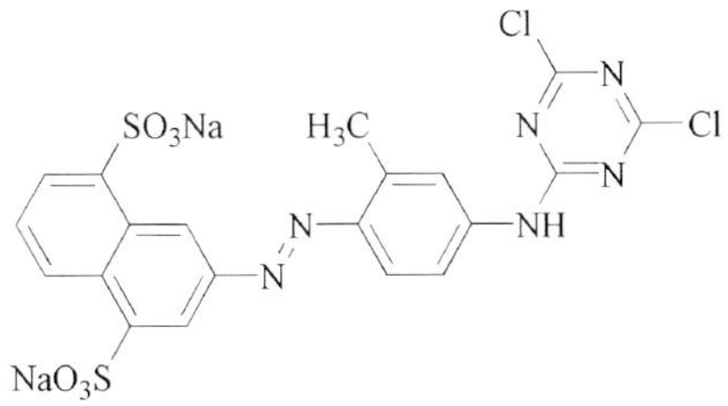

20

Another reactive component for leather (major use dyeing wool) is the α-sulfatoethylsulfonyl group, as in e.g., *C.I. Reactive Violet 4*, 17965 [*12769-08-3*] (1:1 Copper; **21**), as well as acryloamido, e.g., α-bromoacryloamido groups.

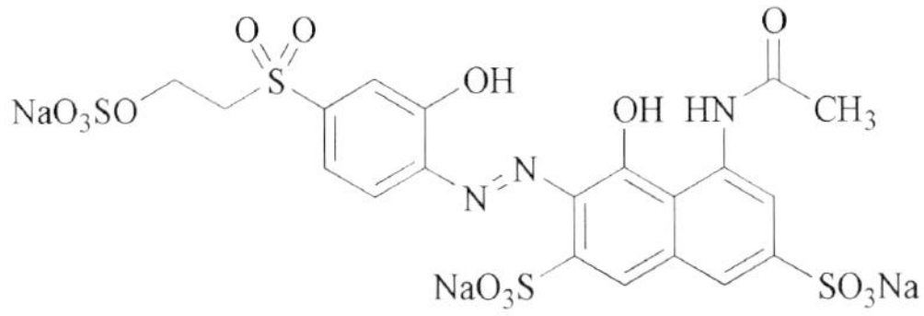

21

On leather, reactive dyes attach to the amino group of lysine and hydroxylysine moieties of collagen. A tanning effect may occur if one reactive group reacts with leather. However, the reaction of the electrophilic group of reactive dyes with water (hydrolysis) competes with the fixation reaction of forming a covalent bond between the dye and the substrate. The hydrolyzed dye cannot react with the fiber. Leather absorbs the noncovalently bound dye like a conventional anionic dye. Unlike on textiles, these hydrolyzed dyes cannot be easily washed off. That is the reason why sometimes no decisive wetfastness improvement can be achieved.

5.1.9 Solvent Dyes

A distinction is made in pigmented leather between the transparent finishing preparation, which contains solvent dyes or small-particle organic pigments, and the covering finishing preparations containing inorganic pigments.

Commercial solvent dyes for leather are employed in a great variety of solvents. It is common practice to use mixed solvents to obtain the requisite physical properties. They are also applied to correct off-shade dyeings and improve the brilliance of shades. Alcohols or glycols are the most common solvents but esters and ketones are also convenient. The colors are normally applied to dyed or undyed tanned leather by spraying from solvent solution or aqueous/organic emulsions. Nowadays, water-based finish recipes, which are combinations of solvent with water, are commonly preferred as a result of ecological and toxicological requirements.

The features of chemical constitution associated with the special requirement of solvent solubility include a number of chemical groups on the chromophores. Sulfo groups are often absent, and only hydroxy or amine groups are present. There are mostly cationic and neutral and sometimes also anionic azo, 1:2 azo metal-complex, and a few anthraquinone dyes. An example is *C.I. Solvent Yellow 21*, 18690 [*5601-29-6*] (**22**, 1:2 Chrome; also *C.I. Acid Yellow 121*).

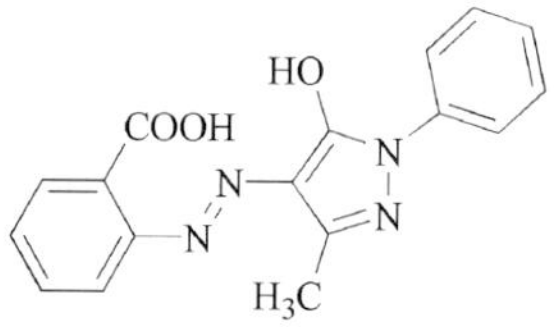

22

The dyes should not contain any inorganic byproducts or diluents. The preferred anionic dyes include salts formed by using ammonia as cation, or special tetraalkylamines, e.g., *C.I. Solvent Orange 49*, (Orange II; see Section 5.3) with an amine condensate as cation.

5.1.10 Pigments

Pigments must be insoluble in water and solvents and chemically unaffected by the vehicle or substrate in which they are incorporated. The physical form and crystallographic structure of pigment particles are of primary importance for their application and properties. Physical finishing methods, such as grinding and milling or solvent treatment are necessary for the raw pigment after synthesis in order to achieve applicable products. Depending on the treatment, different coloristic and application properties are obtained even from compounds with the same chemical structure. Finally, all pigments must be dispersed and embedded in auxiliaries, such as dispersing agents, to avoid instability in application.

Certain inorganic pigments, such as carbon black (*C.I. Pigment Black 7, 8, 9, 10*, 77265-77268), brown iron oxides (*C.I. Pigment Brown 6, 7*, 77492) and white titanium oxide (*C.I. Pigment White 6*, 77891) are most commonly used in leather finishing. Additionally, ultramarine (*C.I. Pigment Blue 29*, 77007) chromium oxide (*C.I. Pigment Green 17*, 77288), iron red (*C.I. Pigment Red 101*, 77491) and pearlescent pigments are frequently used. For these and organic pigments, see [5, 6].

5.1.11 References

Specific References

[1] H.S. Freeman, A.T. Peters, *Colorants for Non-Textile Application*, Elsevier, Amsterdam, 2000.
[2] *Colour Index International*, Third Edition, 1999, CD-Rom, Clarinet Systems Ltd., SDC and AATCC, 1999.
[3] H. Schweppe, *Handbuch der Naturfarbstoffe*, ecomed, Landsberg, 1992.
[4] H. Baumann, H. Hensel, "Neue Metallkomplexfarbstoffe", *Fortschr. Chem. Forsch.* Bd. 7/4,
1967, 643.
[5] W. Herbst, K. Hunger, *Industrial Organic Pigments*, VCH, Weinheim, 1997.
[6] G. Buxbaum, *Industrial Inorganic Pigments*, VCH, Weinheim, 1998.

General References

H. Zollinger, *Colour Chemistry*, 2nd ed., VCH, Weinheim, 1991.
K. Eitel, "Das Färben von Leder", *Bibliothek des Leders*, Band 5, Umschau Verlag, Frankfurt am Main, 1987.
R. Schubert, "Lederzurichtung", *Bibliothek des Leders*, Band 6, Umschau Verlag, Frankfurt am Main, 1982.
G. Otto, *Das Färben des Leders*, Roether Verlag, Darmstadt, 1962.
J. F. Feeman, *Leather Dyes*, Vol. VIII, Academic Press, New York, 1978.
H. E. Fierz-David, *Farbenchemie*, 8. Aufl., Springer Verlag, Wien, 1952.
H. R. Schweizer, *Künstliche organische Farbstoffe und ihre Zwischenprodukte*, Springer Verlag, Berlin, 1964.
K.H. Schündehütte, *Houben Weyl, Methoden der organischen Chemie*, 4. Aufl. Band X/3, Thieme, Stuttgart, 1965, 213.
W.R. Dyson, A. Landmann, "Dyeing of Leather", *Rev. Prog. Coloration*, **4** (1973) 51 ff.
A. Hudson, P. A. Britten, "Leather Coloration", *Rev. Prog. Coloration*, **30** (2000) 68 ff.
E. Heidemann, *Fundamentals of Leather Manufacturing*, Eduard Rother Verlag, 1993.

Websites

General

www.emag.ch
www.leathernet.com
www.leatherbiz.com
www.leatherxchange.com

Schools, Institutes and Associations

www.lgr-reutlingen.de
www.blcleathertech.com
www.northampton.ac.uk
www.leatherinstitute.com
www.vci.de
www.veslic.ch
www.leatherchemists.org
www.sltc.org
www.leatherusa.com
www.leder.ch
www.souzkogevnikov.ru
www.aimpes.com
www.ichslta.org
www.tannerscouncilict.org
www.valles.com/aqeic
www.cotance.com

5.2 Fur Dyes

5.2.1 Introduction

5.2.1.1 Origin of Fur

Fur is one of the oldest forms of clothing we know. From the very start humans were forced to protect themselves from the weather, and they used skins of animals for this purpose. Throughout history humans have valued fur for warmth, protection, adornment, and prestige. The earliest skins may have hung loosely over the body, probably without preservation. After prolonged wear the natural fat present in skin, possibly in combination with perspired salt, would have had a certain preserving and tanning effect, resulting in supple leather. One of the oldest recovered fur garments, found in the Austrian/Italian Alps and worn by the man now known as "Ötzi", might have been made durable in this way. This type of preservation, called "Leipzig dressing", is still practised today. It consists merely of pickling with common salt and sulfuric acid or, preferably, an organic acid, followed by oiling. However this cannot be regarded as true tanning. Real tanning requires the presence of reactive double bonds in the fatty substances. Tanning with oxidizable fats such as fish oil is called "chamois tannage".

The term "fur" refers to any animal skin or part that has hair, fleece, or fur fibers attached, in either the raw or processed state. Skins of fur-bearing animals are also called peltries or pelts. For the production of furs the skins of more than one hundred different species of hunted, trapped, fur-farmed, or domestic mammals are used.

The wild animal fur trade was responsible for the exploration and development of North America and Siberia. It played a prominent role in the economy of colonial America as well as in the expansion of Russia. As the demand for furs increased, enterprising trappers began raising the more valuable fur bearers in captivity and fur farming soon became an important agricultural industry. The processing of wild animal pelts into fur still remains a craftsman's trade, the fur being converted into luxury garments.

Fur farming, or raising animals in captivity under controlled conditions, started in Canada in 1887 on Prince Edward Island. Fur farmers customarily crossbreed and inbreed animals to produce furs with desirable characteristics. The silver fox, developed from the red fox, was the first fur thus produced. Today, so-called mutation minks, ranging from white to near black and from bluish to lavender and rosy-tan colors, each with exotic trade names, are raised on thousands of fur farms, as are chinchilla, nutria, and fox. Fur-farmed animals provide a steady supply of fine-quality, well cared for pelts. The main suppliers are North America, the former Soviet Union, Scandinavia, and Poland.

Domestic woolly sheep are bred for wool and meat in large quantities in Australia, New Zealand, and other agricultural countries. The skins are byproducts, and their conversion into high-quality fur or leather can be regarded as beneficial waste management. Lambskins and sheepskins have grown markedly more popular in recent years for clothing, automotive seat covers, medical, and other applications. Woolly sheepskins account for about half of total fur production and are processed on an industrial basis.

In recent times fur hunting, trapping, farming, and retailing have been the target by protests from animal rights activists who at the same time promote woolly sheepskins and "bioleather" from happy sheep and cows grazing in green meadows. This has led to the increasing popularity of woolly sheepskins.

5.2.1.2 Animal Rights

Various groups, concerned that certain animal species are threatened with extinction or that using furs to make apparel constitutes cruelty to animals, have sought to protect them. Efforts by such organizations as the World Wildlife Fund, Friends of Animals, and the Fur Conservation Institute of America resulted in the enactment of the Endangered Species Conservation Act of 1973. Under the terms of an added convention in 1977 nearly 80 nations established procedures to control and monitor the import and export of endangered species covered by treaty. The act and convention define as endangered any species that is in danger of extinction, and as threatened, any species that is likely to become endangered within the foreseeable future. Covered by the act and convention are some seals, many cats, otters, badgers, Columbus monkeys, some rabbits, non-fur-farmed chinchilla, flying squirrels, and wolves.

5.2.1.3 Fur Hair and Classification

The hair of fur skins varies widely in form and structure. Flayed fur pelts are dried or preserved by salting. Prime furs, those caught during the coldest season, are labelled as firsts. Unprime furs, caught earlier or later, are labelled as seconds, thirds, or fourths. The shorter "fluffier" fur keeps the animal warm, and the longer, "coarser", harsher guard hairs shed water and protect the fur fiber. These are the beauty marks of most fur, and they receive special treatment. Lambskins and sheepskins possess only underhair. From the auction house, pelts are sent for tanning or dressing, as the process is called within the industry. A distinction has to be made between the term "fur dressing", which does not significantly change the hair but the skin, and "fur finishing", which modifies the hair's appearance, e.g., by dyeing. In colloquial English, "fur" is a luxury item often associated with the wild or farmed animal species, while "woolly sheepskins" are considered more a commercial-grade commodity.

5.2.1.4 Fur Dressing

Dressing entails carefully scraping the skins to remove fat, washing them, and treating them with a series of chemicals that soften and preserve them. This process is one of the most critical steps in the making of a fur garment, adding to its durability. Dressing further includes pickling to remove soluble protein and to prepare the skin portion for tanning. Tanning converts the skin into leather by raising the shrinking temperature. One of the oldest methods, still often used, is treatment with aluminum salts. In contrast, dressing with oils has lost its importance, as stability is quite unsatisfactory. The switch from formaldehyde to glutaraldehyde has improved interest in aldehyde tanning. Tanning with chromium(III) salts produces the most stable furs, chiefly used for woolly lamb- and sheepskins. Fur skins are also fat-liquored at all states of processing, the purpose being to render the leather portion light, soft, and supple.

5.2.1.5 Fur Finishing

Naturally white, but slightly yellowed fur skins from sheep and lambs are reduction bleached, most commonly with dithionite or/and in combination with hydrogen peroxide in a weakly alkaline medium. A true bleach in the sense of destroying the natural pigments in the hair can only be achieved by means of oxidation with, e.g., hydrogen peroxide, catalyzed by iron(II) salts. This exothermic process is very difficult to control and must be monitored carefully. In addition, the skin must not be chrome tanned, as this could cause serious damage.

There are many variations of fur dyeing depending on the kind of fur. Bath treatments include "blending" to adjust the natural shade, "reinforcing" to intensify natural colors, and "blueing", or overdyeing yellowish hair. Other techniques are tipping, brushing, or spraying to dye the tips of guard hairs or striping for an application to dye the central or dorsal region. In general, a dyeing temperature higher than 50–60 °C cannot be used for tanned fur because of the limited hydrothermal resistance of the leather portion. Therefore "carriers" or/and "killing" chemicals are used to open the hair. The chemicals used for killing include ammonia, soda, or caustic soda solution together with wetting detergents. Carriers are leveling auxiliaries that improve penetration of the dyes and deepen the shade.

Normally, on wild or farmed skins, the hairs are dyed only if unavoidable. The leather portion, known as "suede", is seldom dyed. In contrast, sheep- and lambskins are colored according to fashion trends. They are called "double-face" and are dyed "tone-in-tone", with the same shade for suede and wool. "Two-tone" dyeing is also common. It is frequently named "bicolor" because suede and hair have different colors. For example, black-dyed hair looks much more exquisite if the suede is dyed blue. Dyed suede with white wool can be found, and conversely undyed suede with dyed hair, too.

Fur dyes are applied in a bath of color, by rolling the dye onto the fur, stroking it on with a feather, or simply by touching up the tips of the guard hair. For batch production, the vessels most frequently used for wet treatment are paddles in which the fur skins float in long (dilute) liquors.

Today, color effects for double face such as "snow top", in which the dyed hair is brushed with special reductive chemicals to bleach it, are in fashion. Finally, it is also usual to print some lamb- or woolskins to imitate more valuable furs. Imitations are produced not only with sheep or lambskin but with rabbitskin, too. These various starting materials may be stencilled to resemble leopard or other spotted furs.

5.2.1.6 The Final Stage

Repeated tumbling in sawdust removes unwanted residue. A final glazing, ironing, or spraying with a chemical and air blowing puts a sheen on the finished fur. The wool of sheep and other skins with dull and crimped hair can be stretched and made more lustrous by ironing on a special machine. It is brushed with water and some acetic acid and alcohol then ironed for a short time at ca. 200 °C.

When the dressing process is complete, the pelts emerge as shiny, silky fur and soft, supple leather. Some furs go through additional beautification steps. Coarse guard hair, for example, from beaver and Alaska fur seal, is removed by plucking. To brighten furs, fluorescent dyes are used.

The remaining fur fiber is then sheared with revolving blades to a velvety texture. Nutria, some rabbit, and muskrat may also be sheared to imitate seal. "Pointing", a process of gluing either badger or monkey guard hair into furs, adds thickness and beauty to the fur by adding contrasting colors. Furs that, after glazing, have a rich color are sold in their natural state. The hair side is normally worn outside to show the beautiful hair structure and the natural looking coloration. Chrome-tanned, buffed, and suede dyed wool sheepskins, known as "suede shearling", is worn with the hairs inside, the wool being sheared to a length of 5–8 mm.

5.2.1.7 Garments and Fashion

The processed skins are then made into fur garments. To keep fur looking nice it should always be hung and kept dry. Since the mid-1970s furs have been made in more varied, sporty, and exotic ways as designers have created new, dramatic styles. Good-quality garments are made from the choice parts of the skin, which excludes the belly (flanks), paw, and head sections. Less costly garments are made from this waste fur.

Large skins, such as mouton lamb or sheep, may have the garment pattern cut from them. The garment parts are then joined together. Most animals, however, have smaller skins that must be joined in various ways to create a garment. The skin-on-skin method, commonly used with muskrat, squirrel, rabbit, small lamb, and some chinchilla, joins the trimmed skin lengthwise to other skins. This method leaves a straight, zigzag, or rounded joining mark, visible in all but curly-haired skins. Garment parts are then stitched together, linings are inserted, and the garment is tailored for fit and drape. Identical shade and appearances are essential requirements. Waste parts of the fur, cut from the skins of these quality garments, are assembled into sections, called plates, that are later cut, as is fabric, to make less costly garments.

5.2.1.8 Labels

After processing, dyeing, shearing, and construction, furs are often difficult to identify. To protect sellers from others who may falsify their products and to protect consumers against misrepresentation, labels have been created. In the USA, for example, the Fur Products Labeling Act was passed, effective August 9, 1952, with minor amendments added in 1961, 1967, 1969, and 1980. Under this law, furs must be invoiced, advertised, labeled, and sold under their accepted English names. Waste fur and used fur articles must be so labeled, too. In addition, if any dye, color alteration, or change has been made that affects the fur's appearance, it must be so noted. If furs have been pointed, that fact must be disclosed on the label.

5.2.2 Fur Dyeing

5.2.2.1 History and Outlook

Documents tell us that furs were dyed in ancient Egypt, and Pliny reported in 60 A.D. that the Romans had imported purple-dyed furs from China, claiming they were as valuable as gold. Nearly 400 years ago, when North America and Siberia were just beginning to be settled, fur was in great demand in Europe. The tremendous demand lured the early frontiersmen to explore the interior of the continent in search of pelts. Trappers and explorers sold their pelts to trading posts and purchased provisions from them. To speed up drying they used open fire. The smoke darkens the hair, resulting in more costly painted furs. Commercial dyeing started on a regular basis in the early 1800s. Even today, the more valuable furs are processed without dyeing. However, the amount of woolly sheepskins which are dyed has increased markedly in recent years. Synthetic dyes, especially acid dyes, are in high demand.

5.2.2.2 Color Selection and Colour Index

A few dyes can be found in the Colour Index under the designation "Fur". Five of them are acid dyes and some are oxidation dyes. These dyes are selected to dye the hair. However, an even larger number, including many mixtures, are suitable for dyeing fur hair. For dyeing suede almost the whole range of leather dyes are suitable (see Section 5.1). Nevertheless, they must be carefully selected in line with the desired requirements.

For some typical dyes the C.I. constitution number is given in this book. Any duplication of dyes or application process should be attempted only under the supervision of an experienced chemist who is aware of the latest ecological, toxicological, and safety requirement. This guidance must be constantly observed even if attention is not explicitly drawn to it.

5.2.2.3 Vegetable Dyes

The industrial fur dyeing business started around 200 years ago with vegetable dyes such as logwood, redwood, fustic, or sumach extracts (see Section 5.1). Mordanting is carried out with iron, chromium, or copper salts. The dyeing process is repeated several times, and between the treatments the fur skins are hung up in a humid room to ensure satisfactory oxidation and laking. Vegetable dyes are seldom used today. Occasionally, metal salts such as those of copper or iron are used for brightening natural hair.

5.2.2.4 Oxidation Bases

These dyes were very popular initially until the mid-1900s and are still commonly applied. Aniline black, discovered by J. Lightfoot (1863), is the oldest known oxidation dye. The polymeric dye is obtained by impregnating the hair with aniline hydrochloride and oxidizing with sodium chlorate in the presence of copper and/or vanadium salt as catalyst at pH 1–2. Aniline black, *C.I. Oxidation Base 1*, 50440 [*13007-86-8*] (**1**) has a polymeric structure.

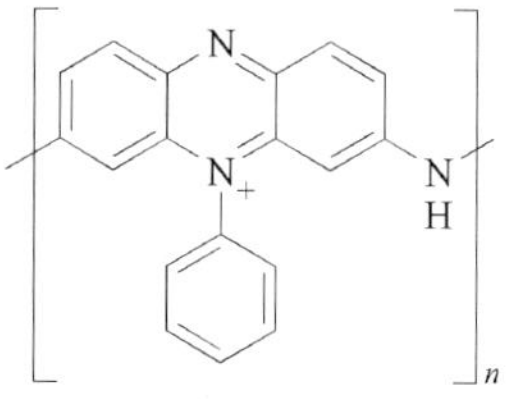

1

During the period 1888–1997 H. Erdmann and E. Erdmann patented a number of *para*-diamines for dyeing human and animal hair. *p*-Phenylenediamine has become the most important oxidation fur dye. The reason for their popularity is that they can be applied much more easily and with less fiber damage than the aniline analogues. The precursor group, including *p*-aminophenols and their *o*-isomers, as well as analogous naphthalene derivatives, generate black, gray and brown colors when oxidized. These amines are called oxidation bases, developers, or primary intermediates.

Synthesizing blacks such as *C.I. Oxidation Base 10*, 76060 [*13007-86-8*] (*C.I. Developer 13*) is very complex and depends on the oxidation conditions. The first step with *p*-phenylenediamine is oxidation to *p*-benzoquinonediimines, then to a trinuclear species called Bandorowski's base (**2**). The final polymeric dye structure is unknown.

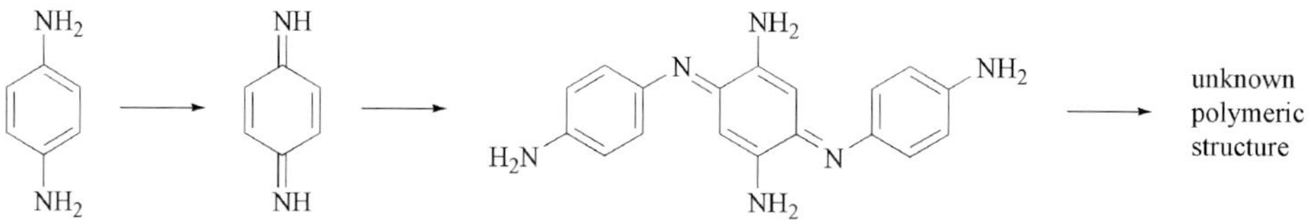

2

The use of primary intermediates as sole dye precursors severely limits the range of shades; some couplers are therefore sometimes added. They do not themselves develop a significant color effect as such but if present in primary intermediates they modify the resulting colors. The usual couplers are 1,3-diaminobenzene (blue), 3-aminophenols (red), and resorcinol (yellow-green). This technique is mostly used for human hair (see Section 5.4.2), the oxidation steps being performed at around pH 9.5 with hydrogen peroxide.

Usually, the fur colors are obtained with only one or two primary intermediates, but in combination with a metal salt in order to convert the dyes into a color lake by complexing or by insoluble salt bonding. This improves their wetfastness and the depth of the dyed shade.

Before dyeing with oxidation dyes, the furs are treated with the appropriate killing agents and then mordanted with metal salts. Iron, chromium, and copper salts, alone or in combination, are used for mordanting, and the uptake process requires several hours. Adjustment of the pH is effected with formic, acetic, or tartaric acid. The final dyeing process is carried out in paddles with the precursors and hydrogen peroxide until the actual dye lake is developed and adsorbed within the hair fiber. It takes quite a few hours at room temperature until the dyeing process is finished.

After dyeing with oxidation dyes, the fur skin must be washed and rinsed thoroughly to remove all loose and incompletely oxidized dye. In working with oxidation dyes, certain precautions have to be taken, as some people are allergic to specific amines.

5.2.2.5 Disperse Dyes

Disperse dyes are almost but not completely insoluble in water. There are utilized as fine aqueous disperse suspensions obtained by grinding the dye crystals in water down to 1–3 μm. A stable dispersion is achieved by adding an anionic dispersing agent. The dyes can be applied to the fur hair with an appropriate carrier such as aryl phosphate esters, which promote uniform dyeing. Disperse dyes yield dyeings with average lightfastness. Dye fixation on the hair protein is not very strong due to weak physical bonding and is mostly unsatisfactory for ironing. They are particularly suitable for shading and blending.

Disperse and other synthetic dyes can only be taken up by the hair at elevated temperatures, and for this reason the skins must be chrome-tanned. The shrinking temperature of suede has to be at least 20 °C higher then the dyeing temperature. No mordant is required, unlike with oxidation dyes, and killing need be far less intensive, too. In actual fact the operation is more hair cleaning than a killing process. The suede portion is not dyed and can be cleaned very easily with standard washing auxiliarics.

An example is *C.I. Disperse Red 17*, 11210 [*3179-89-3*] (**3**).

3

Primarily azo disperse dyes are used to obtain the so-called snow-top effects. After dyeing at a pH of approximately 3.5 and a temperature of 50–60 °C for 1–2 h, the dyed hair top is given a reduction bleach with hydrosulfite. The bleaching solution is applied by brush or spray.

5.2.2.6 Acid and Direct Dyes

The general wetfastness properties of acid dyes are normally enhanced in comparison to those of the disperse dyes due to the fact that the hair protein forms additional salt linkages with the anionic dyes. Moreover, the lightfastness is in most cases better than with disperse dyes. The application technique and killing pretreatments for acid dyes are similar to those for disperse dyes.

In dyeing the hair and the suede portion a number of factors must be allowed for. The keratin of the hair contains basically the same amino acids as the collagen of the skin but in a different ratio. The keratin of the hair includes cysteine, which cross-links the polypeptide chain and imparts stability. The collagen of the skin does not have these substances, and the cross-links are made by the tanning agent. On the other hand, only L-hydroxyproline can be found in the collagen. As a result the thermal stability is different, and in addition the isoelectric points of the two polypeptides diverge.

These facts play an important factor in the dyeing behavior. The isoelectric point of hair is at a pH of ca. 5, whereas untreated collagen has its isoeletric point at a pH of ca. 6, which varies with the kind of tannage. For pure chrome suede it is close to pH 7 or about pH 6 after slightly anionic retanning. Consequently, for good dye fixation the acidification at the end of a dyeing is around pH 3 for fur hair and ca. pH 4 for chrome-tanned suede.

One method of classifying acid dyes is to divide them into groups according to their application behavior. The traditional wool/nylon/cotton classification is customarily employed by the Colour Index and is therefore used here, too.

Acid leveling dyes provide an uniform appearance and their lightfastness is generally good. They have a low molecular weight and usually require a highly acidic dyebath for good exhaustion on the wool. They are particularly suitable for strong, deep shades, including blacks on fur hair. They dye the suede only weakly and can easily be stripped. Blue is obtained with anthraquinone, yellow with nitro species, and for the other colors including yellow, orange, navy, and red monoazo compounds are commonly used.

C.I. Acid Blue 25, 62055 [6408-78-2] (**4**).

4

C.I. Acid Orange 3, 10385 [6373-74-6] (**5**).

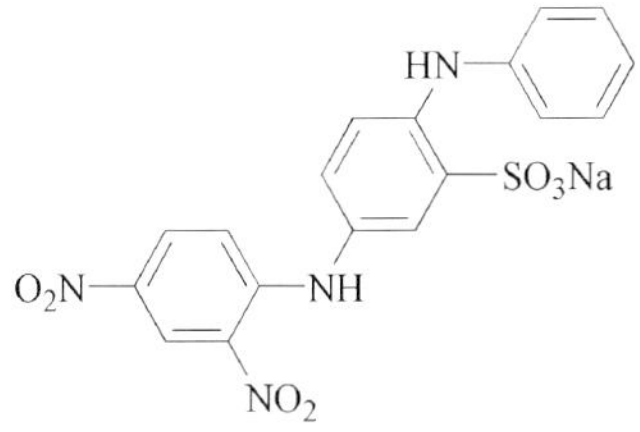

5

C.I. Acid Red 88, 15620 [1658-56-65] (**6**).

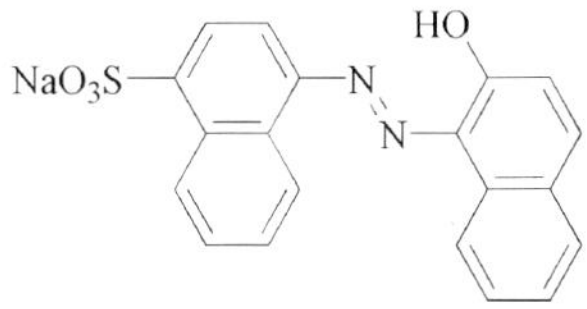

6

In practice, a less acidic pH and a lower temperature are used for fur than for wool. A pH of 4.5 and a temperature of 50–60 °C are sufficient, if used in combination with leveling agents. These are normally ethoxylated fatty alcohols or amines. The auxiliaries promote leveling and regulate the uptake of the dyes. Combination with an alkyl phosphate ester as carrier is recommended.

Milling dyes were originally used on wool fabrics which were to be subjected to severe wet treatments in milling to improve the fabric density. Some of them are also used to dye nylon. Their affinity for wool is higher than that of leveling dyes, and the application pH is higher (5–6). These dyes have even better washing fastness then the leveling acid dyes because of their higher affinity. The shades often lack brightness and the good leveling power of the leveling acid dyes.

Wool milling as such has nothing to do with fur hair, although the dyes are also suitable for dyeing fur hair. They are used where improved wash and lightfastness are requested. In general the dyes are of higher molecular weight than the leveling dyes. The dyeing process can be carried out at a lower temperature of 40–50 °C. For uniform dyeing a combination of a leveling agent and a carrier must be used, as adsorption is faster but less level.

In practice, the fur dyer does not differentiate between these two dye classes. Dye manufacturers offer a selected color range with compatible anionic dyes including the suitable auxiliaries. Consequently the recommended optimized dyeing conditions should be monitored closely. As a rule, the trichromatic dyeing technique is applied, in which the standard colors blue, red, and yellow are mixed to obtain the desired shade.

Direct Dyes. Direct dyes are designed to dye cotton. A few of them dye wool in a similar way to certain acid dyes with higher molecular weight. When these acid and direct dyes are used for wool, they are often called "super milling" or neutral dyeing anionic dyes. Wet and lightfastness properties are considered excellent.

Some of these dyes have more then one sulfo group, and their application requires considerable care, e.g., leveling, because of their comparatively high molecular weight and blocking effect, especially on nylon. Fur hair is difficult to dye. However, dyeing suede is mostly simple. Therefore, certain dyes are used for dyeing suede shearling with hair reservation.

C.I. Direct Blue 78, 34200 [*34200-73-3*] (**7**).

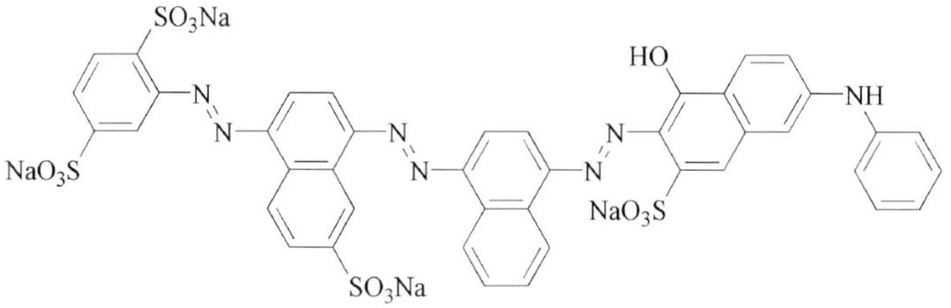

7

5.2.2.7 Metal-Complex Dyes

Iron, chromium, or cobalt 1:2 metal-complex dyes are usually regarded as acid dyes despite their high molecular weight. They contain one equivalent of metal combined with two equivalents of dye. They possess outstanding light and wetfastness but are not as bright as true small azo acid dyes. In application the solubilizing groups are decisive for the dyeing behavior.

Specific 1:2 metal complexes dye hair and suede tone-in-tone in one bath. The 1:2 metal-complex dyes without ionized substituents (e.g. sulfonamides) in the dye moieties are used to dye suede and hair. A number of selected unsymmetrical 1:2 metal complex dyes in which only one ligand carries one ionic group (e.g. sulfo) show similar behavior.

C.I. Acid Yellow 151, 13906 [*127-15-61-6*], 1:2 Co complex (**8**).

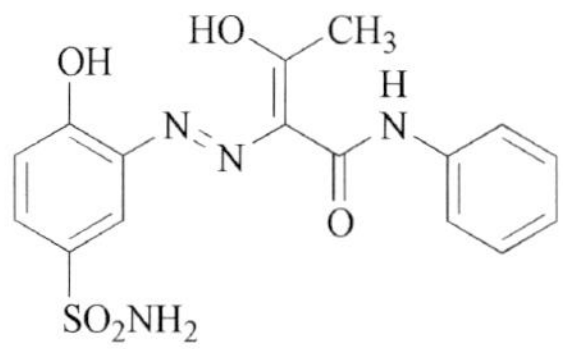

8

If the metal complex dyes carry two sulfo groups, they are suitable for dyeing suede with hair reservation. For full penetration on suede, dyeing starts at a weakly alkaline pH of 8 (mostly adjusted with ammonia), and anionic dye-penetration agents such as naphthalenesulfonic acid–formaldehyde condensate. Finally, the dyes are fixed with an organic acid, preferably formic acid, at a pH of ca. 3.5.

C.I. Acid Blue 193, 15707 [*12392-64-2*], 1:2 Cr complex (**9**).

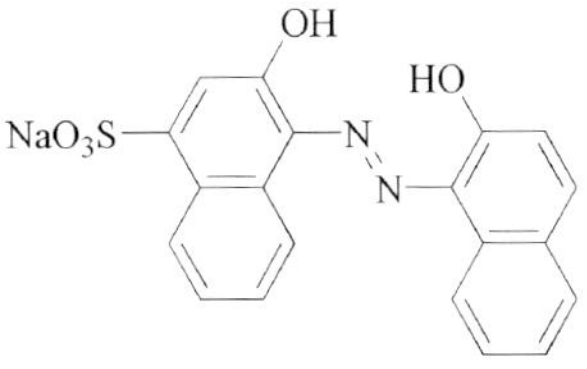

9

For light to medium shades on suede 1:1 metal complex dyes can be applied. Chrome dyes with 1:1 equivalent ligand have good leveling and penetrating power as well as excellent lightfastness.

C.I. Acid Red 186, 18810 [*52677-44-8*], 1:1 Cr complex (**10**).

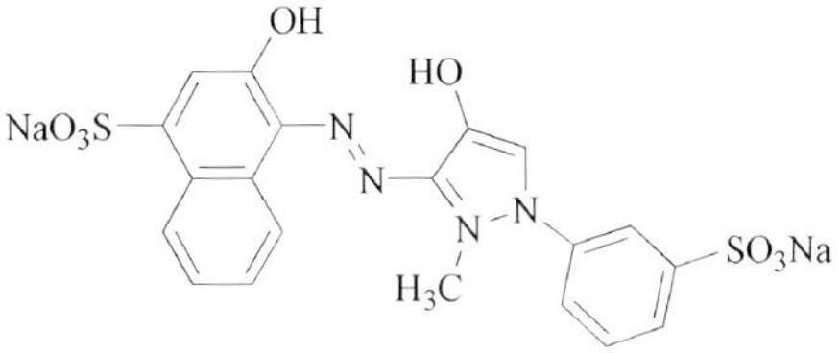

10

Bicolor shades are achieved by dyeing the hair with acid dyes and fixation at pH 3–3.5. Neutralization to pH 5.5–6 is followed by suede scouring with ammonia and auxiliaries. Dyeing with metal-complex or direct dyes is effected at neutral to weak acid pH. Finally, the dyes are fixed with formic acid at pH 4.

5.2.2.8 Other Synthetic Dyes

Cationic dyes are occasionally used fur blueing or brightening the hair, frequently in combination with a fluorescent dyes. P. Krais discovered the brightening effect in 1929 by impregnation of rayon with a horsechestnut extract known as esculin, a fluorescent glycoside of 6,7-dihydroxycoumarin. Coumarin derivates are still used for wool and fur hair.

C.I. Fluorescent Brightener 130, 501101 [*87-01-4*] (**11**).

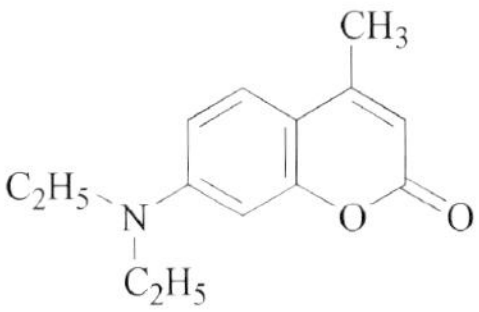

11

Synthetic mordants are rarely applied. Reactive dyes for hair or suede and soluble sulfur dyes for suede are chosen only for special purposes. Finally organic solvent dyes of the 1:2 metal-complex dye type are used for printing together with thickening agents, replacing the oxidation dyes.

C.I. Solvent Yellow 21, 18690 [*5601-29-6*], 1:2 Cr complex (**12**).

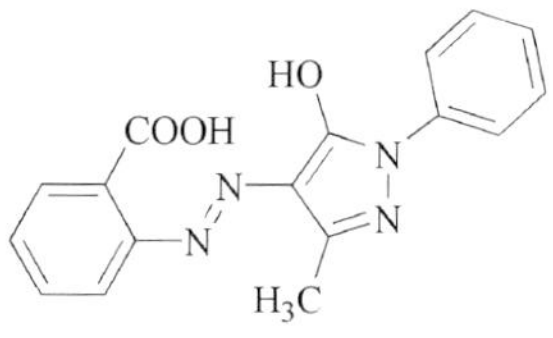

12

5.2.3 References

General References

Colour Index International, 3rd ed., 1999, CD-ROM, Clarinet Systems Ltd. (SDC and AATCC), 1999.

H. Zollinger, *Colour Chemistry*, 2nd ed., VCH, Weinheim, 1991.

H. Schweppe, *Handbuch der Naturfarbstoffe*, ecomed, Landsberg, 1992.

H. S. Freeman, A.T. Peters, *Colorants for Non-Textile Application*, Elsevier, Amsterdam, 2000.

W. Graßmann, *Handbuch der Gerbereichemie*, 3. Band/2.Teil, Springer-Verlag, Wien, 1955.

Websites

http://www.fur.org
http://www.furcommission.com
http://www.iftf.com
http://www.fur.ca
http://www.ffs.fi
http://www.woolskin.au
http://www.deutsches-pelzinstitut.de
http://www.pelzfachverband.ch
http://www.wwf.org

5.3 Paper Dyes

5.3.1 Introduction

Paper has always been dyed with dyes which were developed for other industries, particularly since the discovery of the new synthetic organic dyes in the second half of the 19th century. Usually these were textile dyes [1]. It was not until the 1960 and the beginning of the 1970s that a few dye manufacturers began to develop specific paper dyes, because many of the dyes used no longer satisfied the growing requirements. The main reasons were the increasing speed of the paper machines and the increasingly stringent regulations regarding cleanliness of the wastewater. Specific research on paper dyes produced new dyes which are superior in many respects to the textile dyes that were often used before and which enable simpler, more economical dyeing processes.

Today a whole series of different types of dyes are used to dye paper, which can be divided into a wide variety of dye classes. This is because paper is not just paper [2]. The main raw material for producing paper, wood, consists mainly of cellulose and other components such as lignin, hemicellulose, and resins [3]. During chemical pulping of the wood to obtain the cellulose, the lignin components and the hemicellulose are largely dissolved out, and during bleaching other interfering substances are removed. However, apart from bleached pulp, unbleached pulp or mechanical wood pulp is also used depending on the final use and scope of application.

The cellulose molecule

consists on average of 6000–8000 glucose radicals joined in the 1,4-positions to form a chain. The most important properties for dyeing are mainly due to the long drawn out structure of the cellulose molecule, the capillary structure of its smallest regions, and the free hydroxyl groups of the glucose structural elements. Hemicellulose has a similar structure but is far less polymeric and contains other components such as mannose or xylose apart from glucose. Although hemicelluloses are relatively unimportant for the dyeing of paper, they do contribute significantly to a wide variety of properties. Lignin, on the other hand, which is an amorphous substance which contains phenolic and aliphatic hydroxyl, carbonyl, and methoxyl groups, has a complicated composition and varies from one type of wood to another. Lignin has a considerable influence on the dyeability.

The dyeing procedure for paper can be described basically by two processes: the penetration of the dye molecule into the capillary spaces of the cellulose and then its adsorption on the surface of the fiber. The bonding forces are due to the effects charge (ionic bonds), precipitation, and intermolecular forces.

Since unbleached cellulose and mechanical wood pulp have a distinctly negative charge due to the acid lignin groups, products of opposite polarity, i.e., basic or cationic, can interact with the fiber directly by ionic bonding.

Bleached pulp, which has only a few residual negative charges, exhibits insufficient dyeing behavior, especially with low molecular basic dyes. Here the so-called substantive dyes are used, i.e., long molecules which have a conjugated π-electron system and only a few water-solubilizing groups. Intermolecularly such dyes can attach themselves on a plane to the cellulose fiber and form a charge-transfer complex between their π-electrons and the electrons of the OH groups of the cellulose.

For the precipitation effects, acid dyes with little affinity to the fiber are rendered sparingly soluble (e.g., precipitated with alum) or are bound to the cellulose fiber with a polymer, generally a cationic fixative.

When dyeing with pigments the situation is completely different from that described with water-soluble dyes. Here physical effects are important. The smaller the pigment particles, the greater their specific surface and the more strongly the adsorption forces come into action. Suspended in water, most pigment particles have a weakly negative charge and are repelled by the negative charge of the cellulose fiber. The fibers are therefore recharged with the aid of aluminum ions, so that the fibers and pigment particles attract each other.

5.3.2 Classification of Paper Dyes

Due to the multitude of paper compositions and the different dyeing methods, dyes are generally classified according to the principles of the Colour Index. For dyeing paper the following main classes of dyes are used: direct dyes, acid dyes, basic dyes, pigments, and sulfur dyes. This classification describes not only the chemical structure, but also the dyeing behavior and type of charge. For this reason azo dyes, for example, can be found in all the classes mentioned except sulfur dyes.

5.3.3 Direct Dyes

This is an important class of dyes for the dyeing of paper. Direct dyes are also called substantive dyes [4, 5] because they tend to have a high affinity to cellulose fibers due to their linear molecular structure and a system of conjugated double bonds and usually also exhibit good wetfastness properties with the addition of a fixative.

5.3.3.1 Anionic Direct Dyes

which are specially suitable for dyeing cotton, is also used for dyeing paper. The oldest dye which was recognized as substantive is congo red (**1**), which was discovered in 1884 by Böttiger.

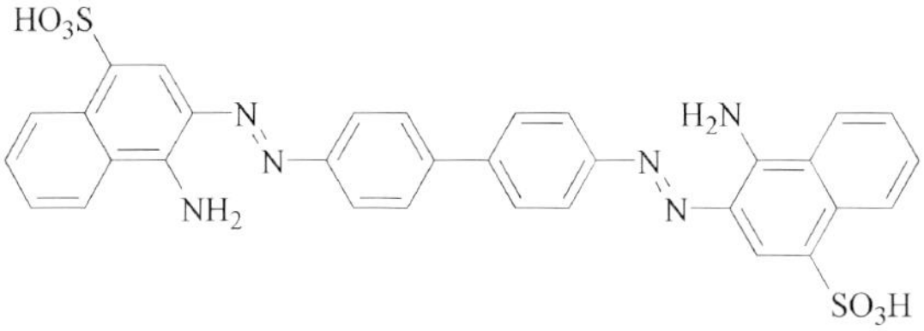

Benzidine and its homologues give the molecule a planar structure and exhibit good conjugation. Due to the proven carcinogenicity of benzidine, dyes of this type have not been produced for over 30 years. The homologous compounds 3,3′-dimethylbenzidine and 3,3′-dichlorobenzidine have not been used in the production of paper dyes since about 1985. However, 3,3′-dimethoxybenzidine is still in use because it enables demethylative coppering which leads to lightfast substantive blue dyes.

4,4′-diaminostilbene-2,2′-disulfonic acid exhibits similar behavior to benzidine. For example, when it is coupled with phenol and the phenolic hydroxyl groups are etherified, chrysophenine, *C.I. Direct Yellow 12*, 24895 [*2870-32-8*] (**2**) is obtained, which is a highly substantive yellow paper dye.

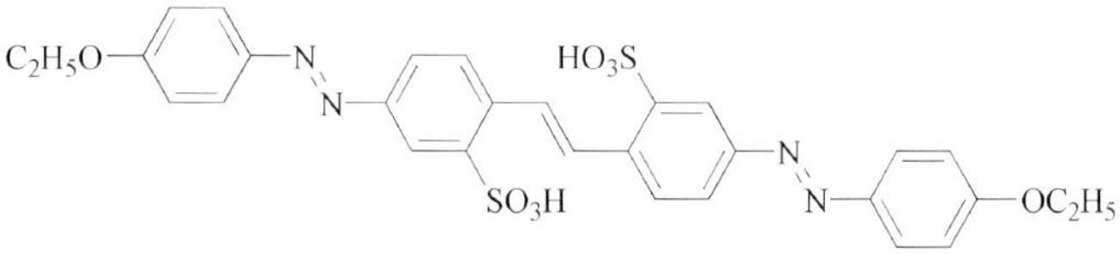

However chrysophenine is fairly difficult to dissolve and is no longer widely used for dyeing paper.

Another important group in the yellow to orange region, but which is not based on diazotization and coupling, are the so-called stilbene dyes. They are obtained by alkaline self-curing of 4-nitrotoluene-2-sulfonic acid and aromatic chains of different length with azoxy and azo groups and stilbene groups. The most important dye of this class is *C.I. Direct Yellow 11* which has very good affinity to paper fibers. However, its constitution is not yet totally clarified [6]. In any case, the dye is a mixture of stilbene azo and azoxy compounds with different chain lengths, the shortest of which is probably **3**.

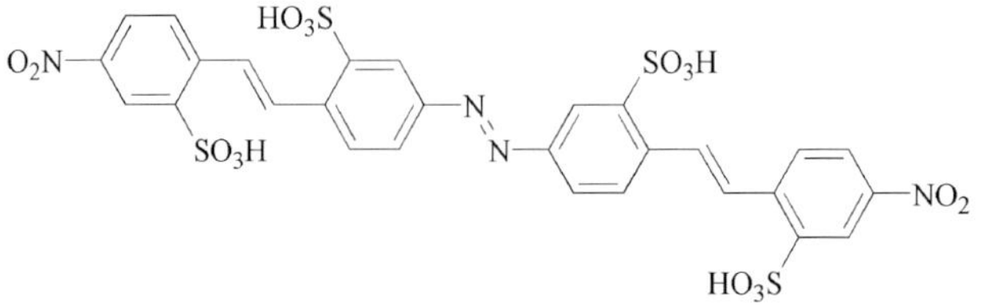

5.3 Paper Dyes

3

With the increasing popularity of the continuous method of dyeing in the paper industry, new direct dyes were developed with particularly good cold-water solubility without any impairment to the substantivity. Improvements of this kind with regard to substantivity and solubility can be seen in the further development of the widely used, but only moderately soluble, cotton dye *C.I. Direct Red 81*, 28160 [*2610-11-9*] (**4**).

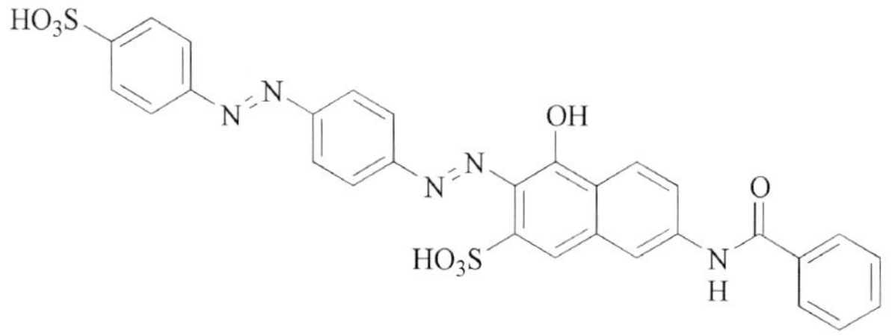

4

If the benzoyl group is exchanged for 1,3,5 trichlorotriazine and the residual chlorine atoms are treated with, e.g., monoethanolamine, *C.I. Direct Red 253*, [*142985-51-1*] (**5**) is obtained, which has both higher substantivity and better solubility.

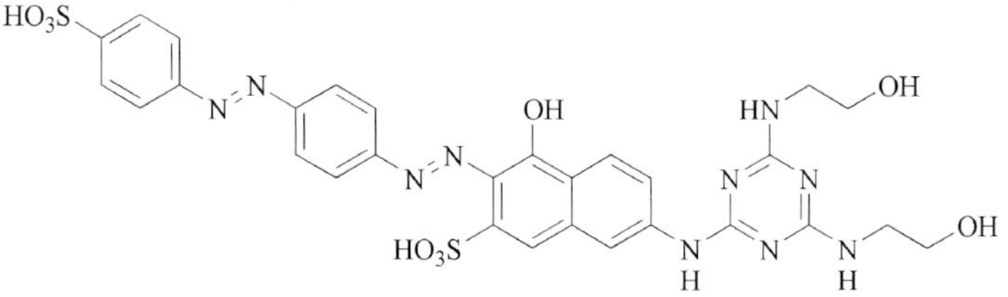

5

Most of the anionic direct dyes which were developed for dyeing paper are dis-, tris- and polyazo dyes, which have proved to have particularly good substantivity due to the strong conjugation power of the double bonds present. Only a few anionic monoazo dyes have sufficient affinity for paper. However, dehydrothio-*p*-toluidine (**6**), obtained in a baking process from *p*-toluidine and elemental sulfur, can be sulfonated to give **7**, which coupleswith pyrimidine derivatives to give the paper dyes *C.I. Direct Yellow 147*, [*35294-62-3*] (**8**) and *C.I. Direct Yellow 137*, [*71838-47-6*] (**9**), which have relatively good substantivity.

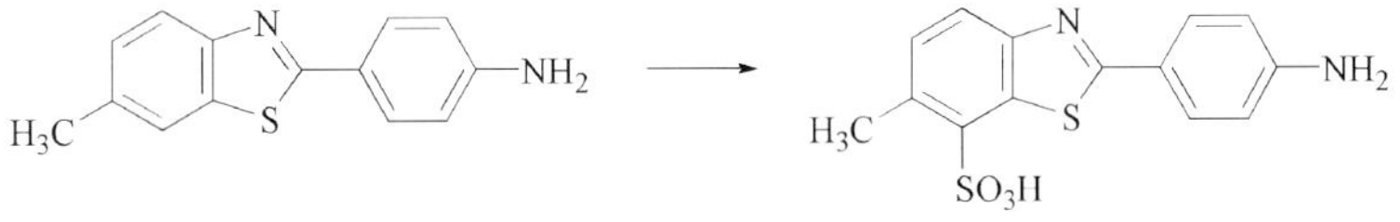

6

7

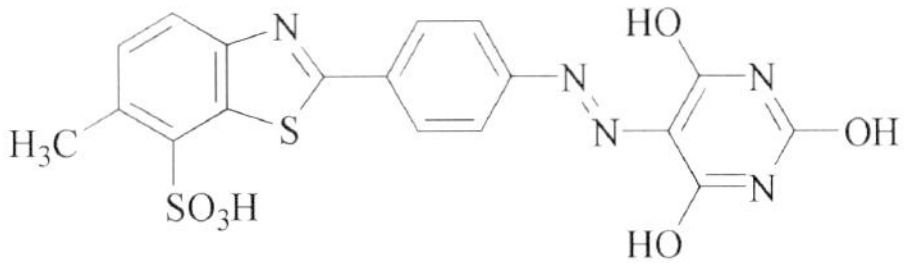

8

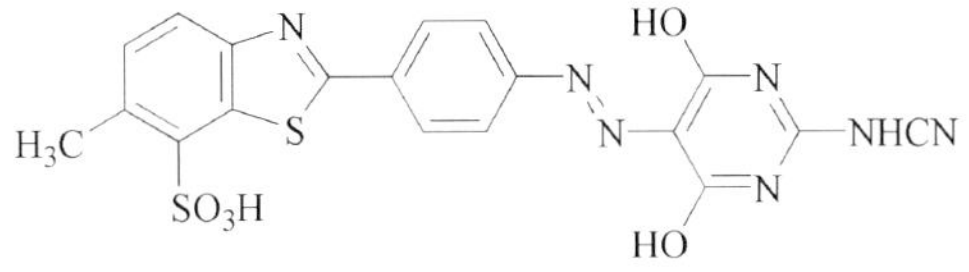

9

Oxidation of dehydrothio-*p*-toluidinesulfonic acid with sodium hypochlorite [7] gives the yellow azo paper dye *C.I. Direct Yellow 28*, 19555 [8005-72-9] (**10**) without diazotization and coupling.

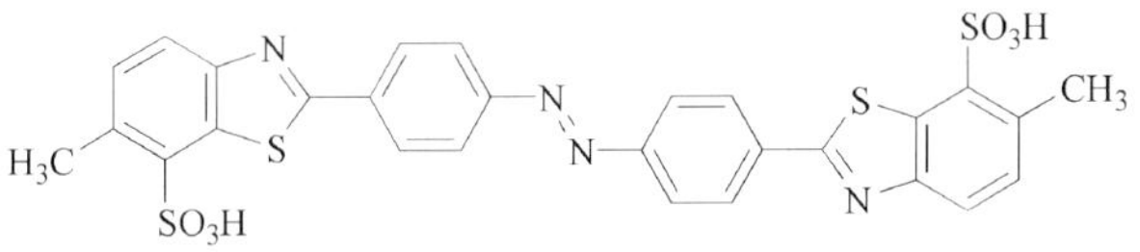

10

The bridging of mono- or disazo compounds with a link capable of conjugation, such as phosgene, also leads to the substantivity necessary for paper dyes, e.g., *C.I. Direct Yellow 51*, 29030 [6420-29-7] (**11**).

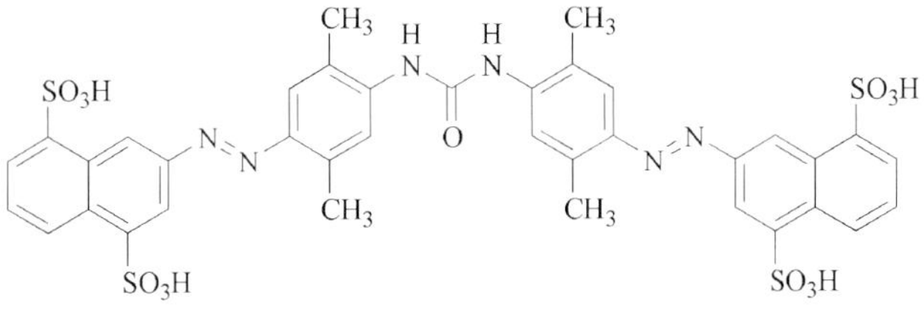

11

An important coupling component for paper dyes is 2-amino-5-naphthol-7-sulfonic acid (**12**) and its derivatives (R= H, acetyl, benzoyl, aryl, imidazoyl, thiazolyl, triazolyl, etc.).

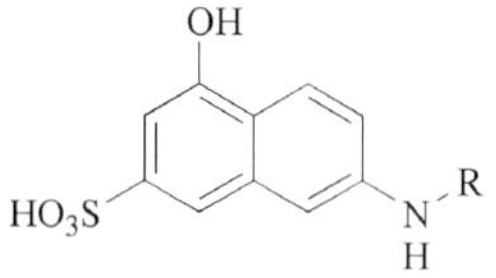

5.3 Paper Dyes

12

For example, if 2-amino-5-naphthol-7-sulfonic acid is bridged with phosgene, highly substantive orange to red paper dyes such as *C.I. Direct Orange 102*, 29156 [6598-63-6] (**13**) and *C.I. Direct Red 239*, [28706-25-4] (**14**) are formed with suitable diazo components.

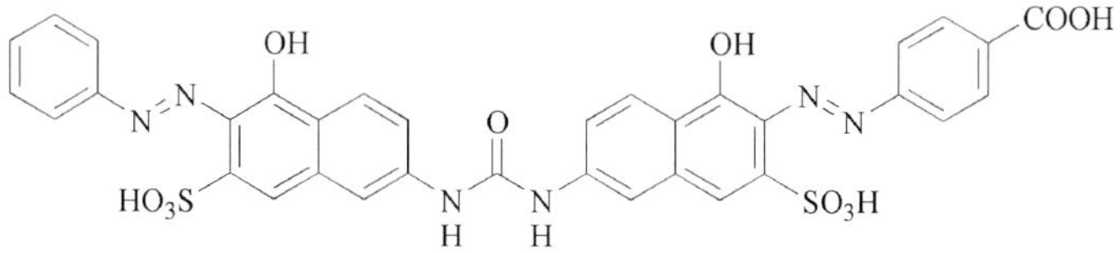

13

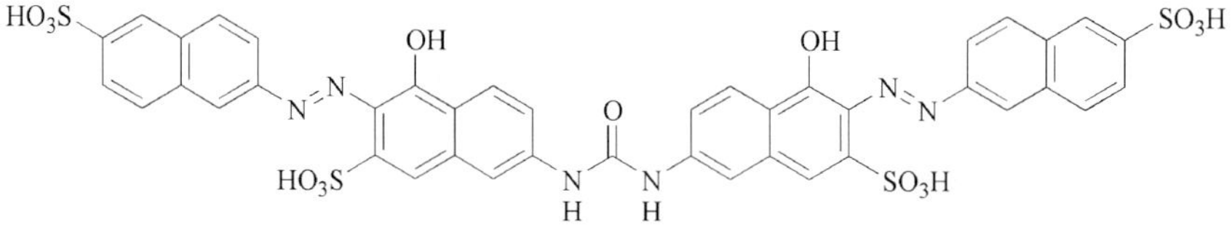

14

If suitable amino azo compounds are coupled to *N*-phenylamino-5-naphthol-7-sulfonic acid, violet, moderately lightfast paper dyes with good brilliance are obtained, e.g., *C.I. Direct Violet 51*, 27855 [6227-10-7] (**15**).

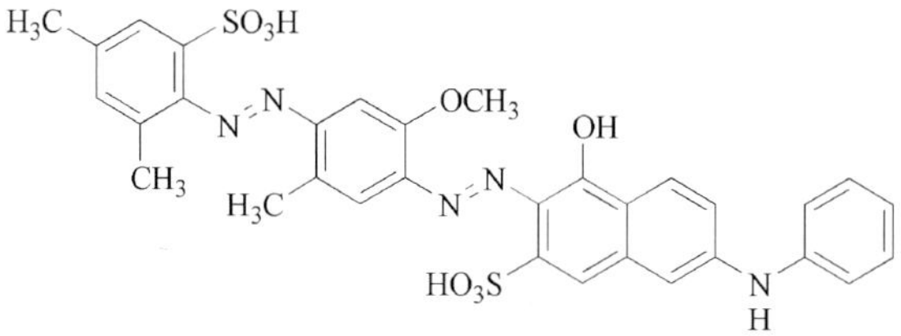

15

If amino azo compounds which contain 1-naphthylamine or 1-naphthylamine-6- or -7-sulfonic acid as their middle component are coupled with 2-amino-5-naphthol-7-sulfonic acid and their derivatives blue paper dyes such as *C.I. Direct Blue 71*, 34140 [4399-55-5] (**16**) are obtained, which have good lightfastness but generally a slightly dull shade.

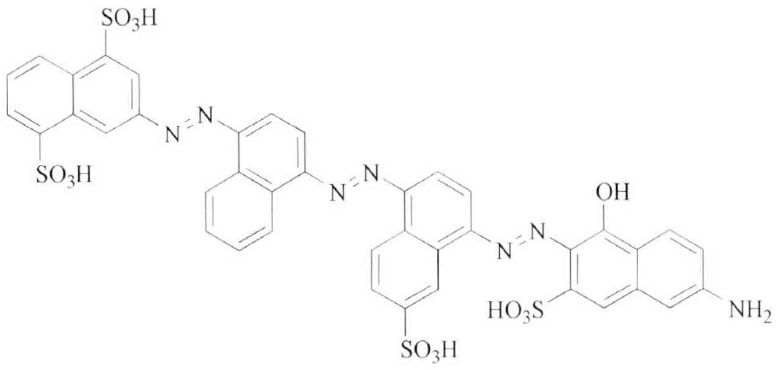

5.3.3 Direct Dyes

16

2,4,6-Trichloro-1,3,5-triazine (cyanuric chloride) has proved to be an interesting binding link because, in contrast to phosgene, it can selectively bind different aminoazo compounds. For example, homogeneous green dyes such as *C.I. Direct Green 26*, 34045 [*6388-26-6*] (**17**), which have better affinity to the fiber than a dye mixture, can be obtained by combining a yellow and a blue aminoazo dye.

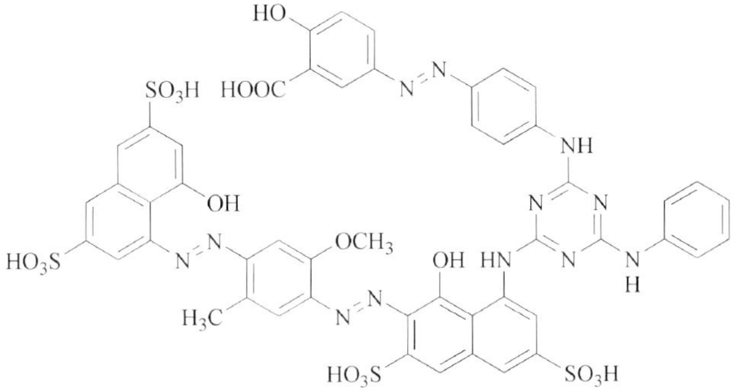

17

As a rule many of these dyes only have moderate lightfastness on paper. To improve this metal complexes are used, particularly with the blue azo dyes. Because planarity of the dye molecule is advantageous for a good dyeing, virtually only Cu^{2+} complexes such as *C.I. Direct Blue 218*, 24401 [*28407-37-6*] (**18**) are of interest for paper. 3,3′-Dimethoxy-4,4′-biphenyldiamine (*o*-dianisidine) has proved particularly suitable, because after coupling with a 1-naphthol derivative demethylative copperization can take place.

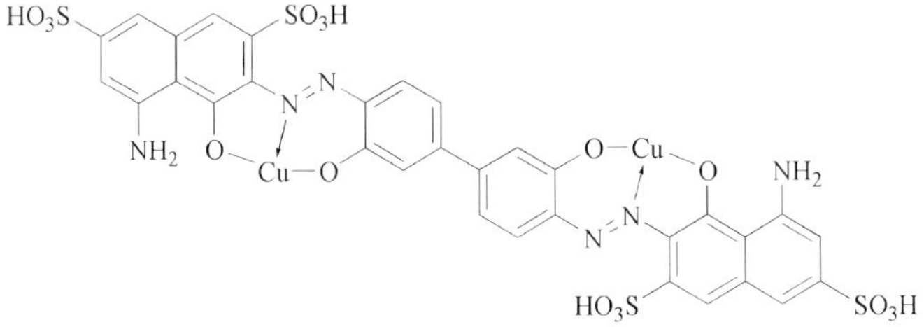

18

A whole series of such dyes based on dianisidine have been developed for the paper industry. In recent years, however, efforts have increased to replace *o*-dianisidine for metallized dianisidine dyes for toxicological reasons [8]. An example is *C.I. Direct Blue 273*, [76359-37-0] (**19**).

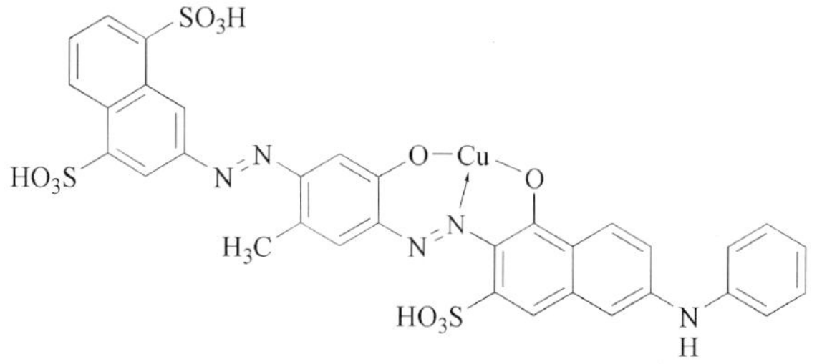

19

By linking heterocyclic ring systems such as triphenbisoxazine with 1,3,5-trichlorotriazine brilliant blue and lightfast dyes **20** were developed which are completely metal-free, contain no dianisidine, and have high substantivity on paper. [9].

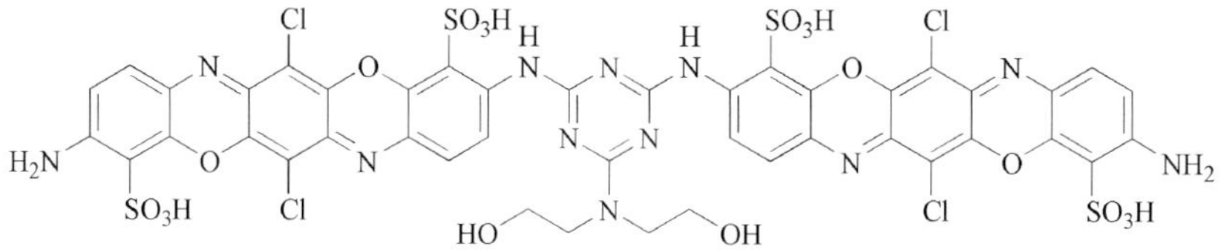

20

5.3.3.2 Cationic Direct Dyes

have extremely high affinity to bleached and wood-containing paper fibers due to their cationic charge and substantive properties and therefore do not require any additional fixative. The cationic direct dyes are therefore particularly suitable for dyeing unsized papers such as hygienic papers, serviettes, and kitchen papers in deep shades. In addition, the wastewater is clear after dyeing, and the bleedfastness properties are also significantly better than with the anionic substantive dyes. Interestingly, cationic direct dyes only retain their substantive character when the cationic charge is independent of the chromophore and not, as is the case with diazamethine dyes such as *C.I. Basic Blue 64*, [12217-44-6] (**21**), present in the chromophore in a delocalized form.

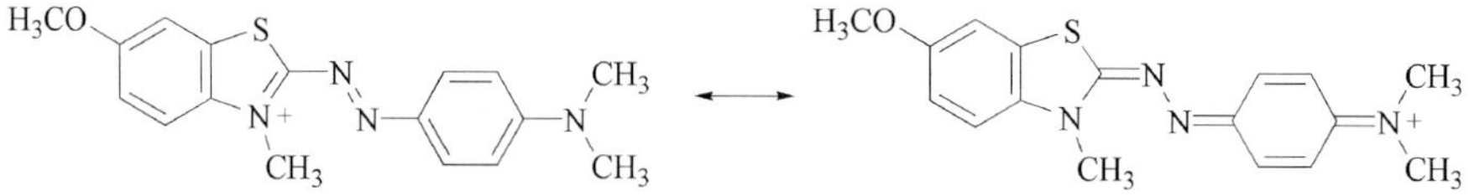

21

5.3.3 Direct Dyes

If aromatic diamines are coupled to substituted 4-methyl-6-hydroxy-2-pyridinones [10] which contain a quaternary ammonium group (**22**) or primary, secondary, or tertiary amines (**23**), brilliant yellow to orange dyes are obtained.

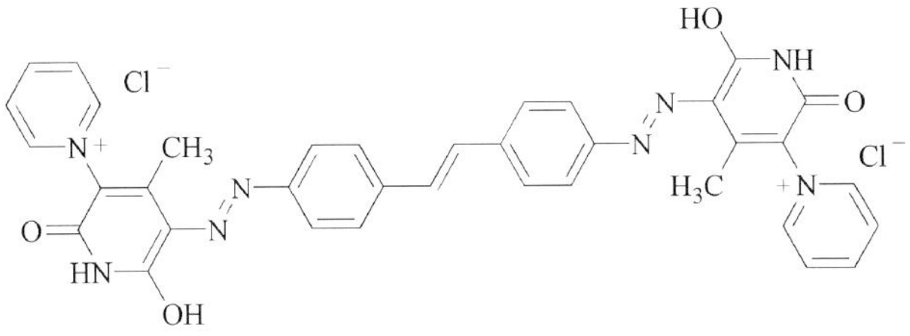

22

Usually it is necessary to exchange the chloride ions for organic acid anions to obtain adequate solubility. Dyes containing aliphatic amino groups (e.g., **23**) are more soluble because they can be protonated with organic acids such as acetic, lactic, or formic acid.

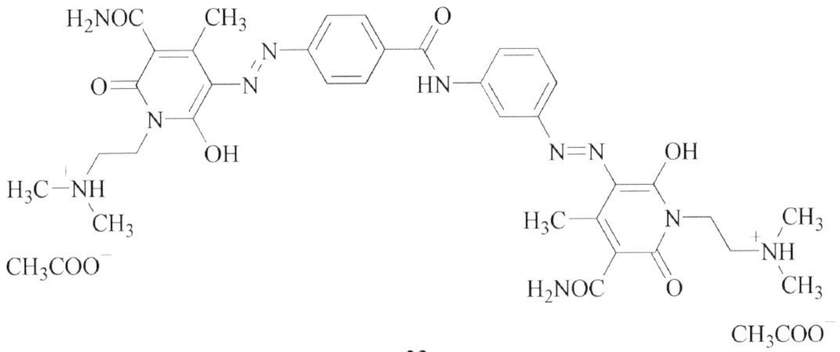

23

In polycationic dyes such as **24**, the presence of a conjugated molecular structure is only of secondary importance. The positive charge is sufficient to enable complete exhaustion onto paper fibers.

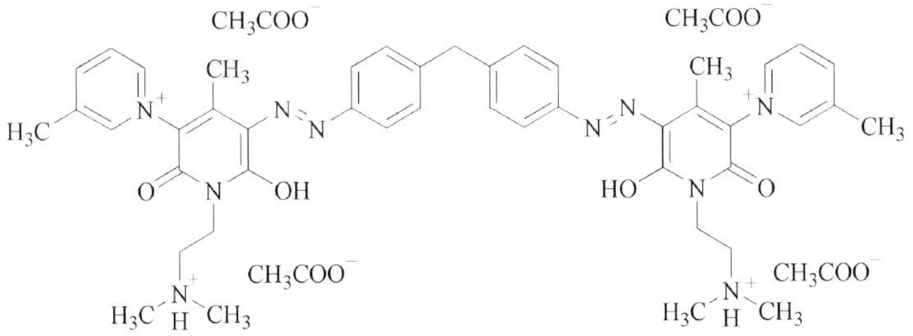

24

Substantive anionic direct dyes can be "repolarized" by incorporating an excess of cationic groups [11] without impairing the substantive properties. Several dyes of this kind have been developed for the paper industry, e.g., *C.I. Basic Red 111*, [118658-98-3] (**25**).

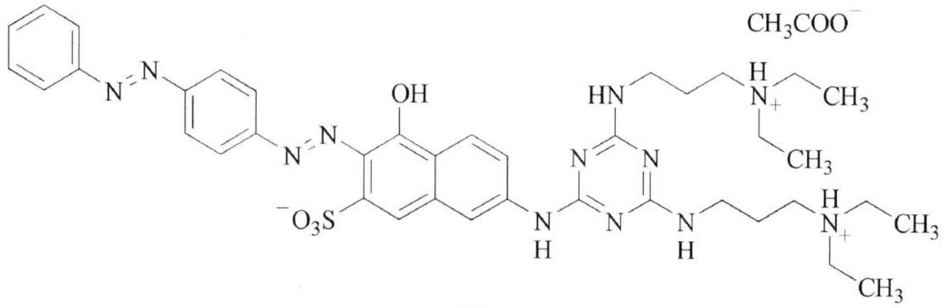

25

Attractive lightfast turquoise shades can be obtained with Cu phthalocyanines. Although these molecules are planar, they are insufficiently substantive in sulfited form so that the dyeings must be post treated with a fixative. By introducing cationic groups, e.g., by treatment with chlorosulfonic acid and subsequent amidation (**26**), very good affinity to paper is obtained, e.g., *C.I. Basic Blue 140*, [*61724-62-7*].

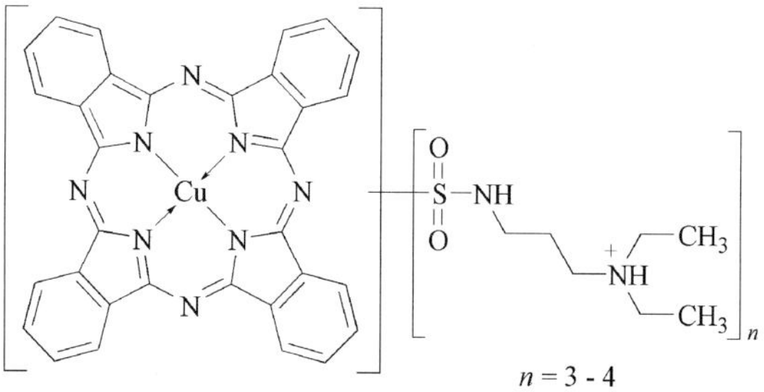

26

If amidation is not carried out to completion [12], Cu phthalocyanines **27** with a more anionic character depending on the degree of sulfonation are obtained, but still with adequate affinity to paper fibers.

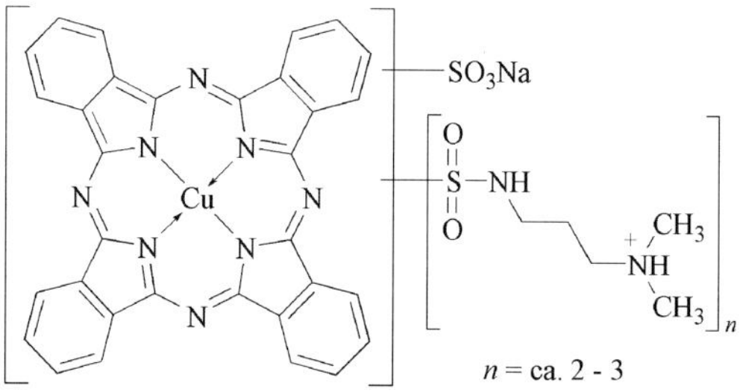

27

The cationic groups can also be introduced by other methods such as chloromethylation and quaternization [13] or be converted by curing suitable hydroxymethyl compounds [14] with a pigment. In this way, for example, the lightfast cationic blue paper dye **28** can be obtained with *C.I. Pigment Violet 23* 51319 [*6358-30-1*].

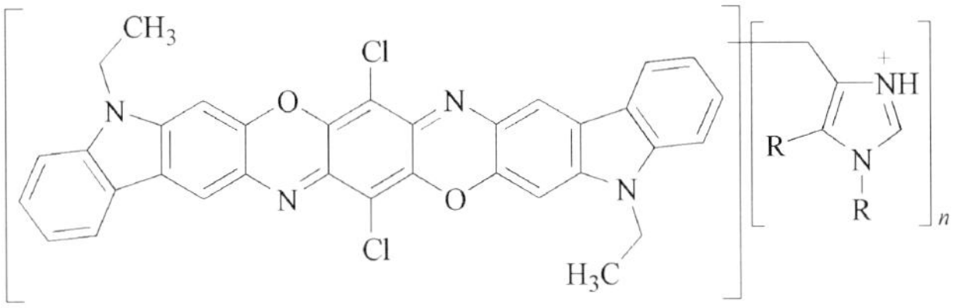

[Structure 28 shown with R = H, CH₃]

28

In general it can be said that the cationic direct dyes are the ideal complement to the anionic direct dyes. By combining selected elements from both dye groups it is even possible to achieve dyeing results which cannot be obtained with dyes from a single dye group.

5.3.4 Acid Dyes

Acid dyes differ from the anionic direct dyes in that they have smaller molecules and are less substantive due to the lack of a conjugated double-bond system. As a rule they are used to dye polyamide or wool, where they produce good dyeings thanks to their molecule size and good solubility. The acid dyes have no affinity to vegetable fibers. Although they penetrate well into the capillaries of the fibers, no fixed bond is formed and there is virtually no formation of a charge-transfer complex (**29**), (**30**). Only a few of these dyes are still used for dyeing paper. Examples are *C.I. Acid Orange 7*, 15510 [633-96-5] (**29**) and *C.I. Acid Violet 17*, 42650 [4129-84-4] (**30**).

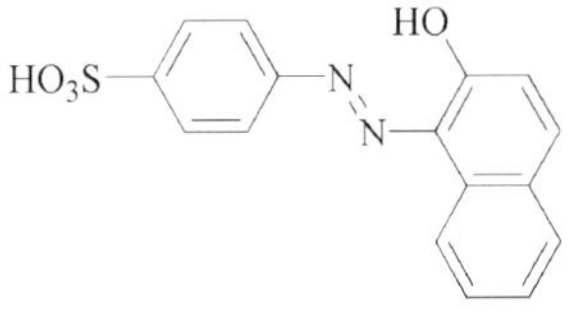

29

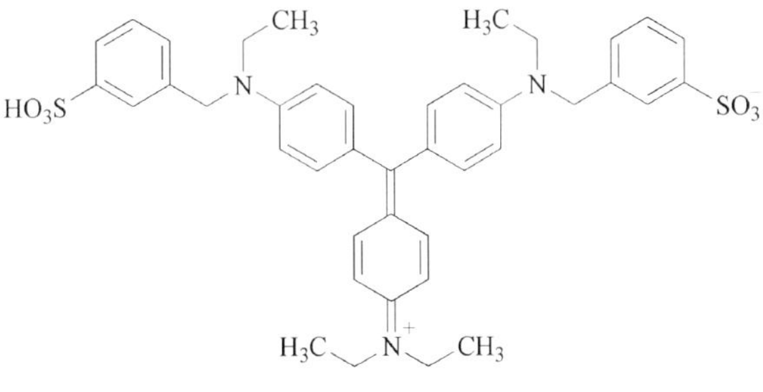

30

In order to achieve a satisfactory color yield, suitable cationic precipitants which exhaust straight onto the fiber must be used for deep dyeings. These products precipitate acid dyes and simultaneously have a fixing action as the binding link between the fibers and dye molecules.

5.3.5 Cationic (Basic) Dyes

Cationic dyes form positively charged dye ions by dissociation, and the positive electric charge is delocalized over the entire molecule. They are usually di- and triarylmethine dyes such as *C.I. Basic Green 4*, 42000 [*569-64-2*] (**31**), which are also known as triarylmethane dyes or triaryl carbenium ions [15], oxazine dyes such as *C.I. Basic Blue 3*, 51005 [*2787-91-9*] (**32**), or thiazine dyes.

31

32

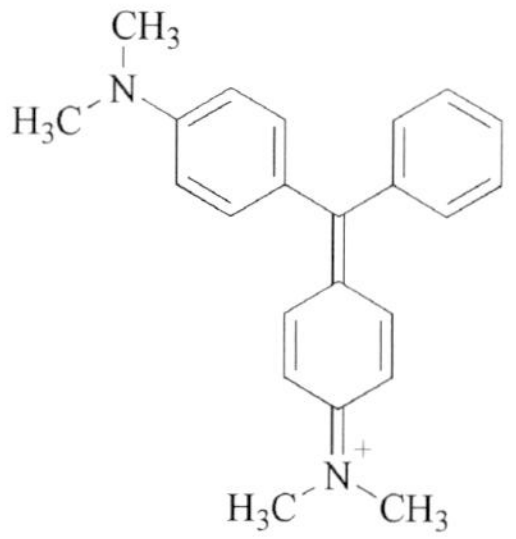

Cyclized di- or triarylmethine dyes are also known as acridine, xanthene, or thioxanthene dyes. Rhodamine GG (**33**), *C.I. Basic Red 1*, 45160 [*989-38-8*], is of the most interest for the paper industry. Rhodamine dyes dye paper in strongly fluorescent red shades.

33

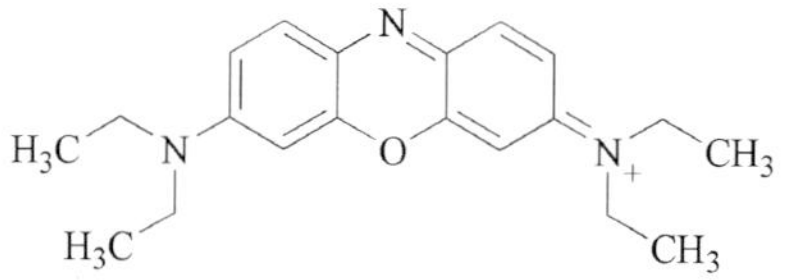

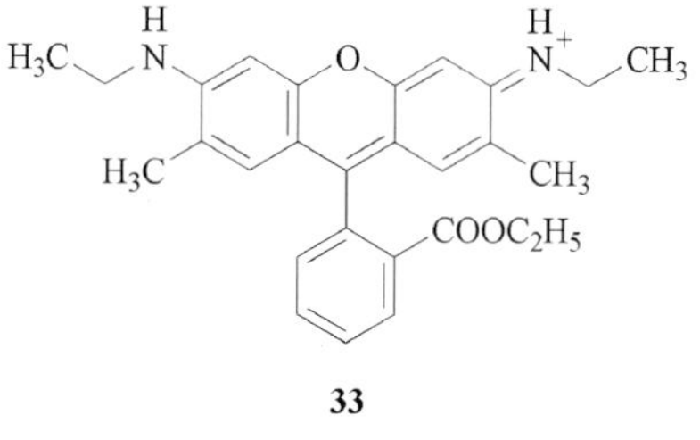

Low-molecular azo dyes, such as *C.I. Basic Brown 1* (**34**), which bear basic groups are also classified in this category.

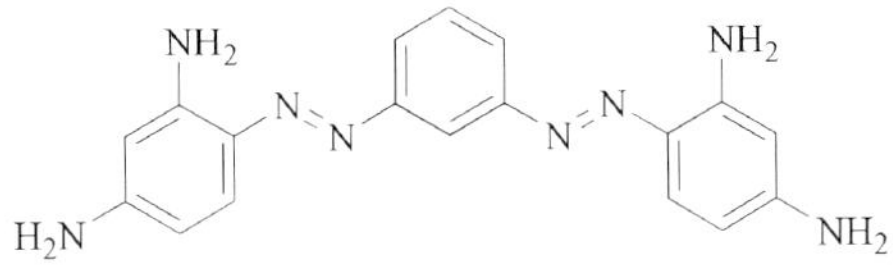

34

This styryl dye **35** has very high color strength and dyes paper in brilliant yellow shades.

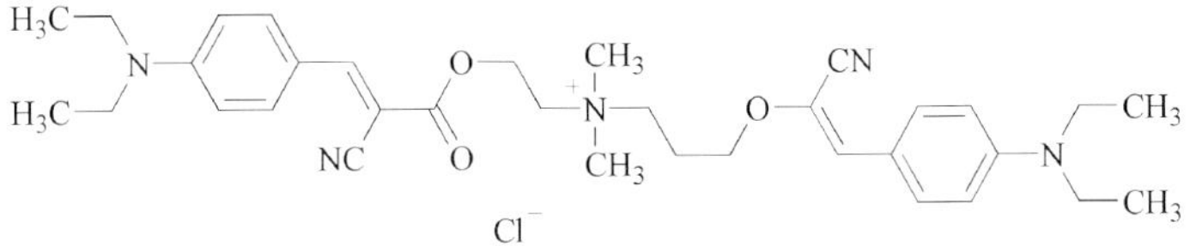

35

All these dyes are distinguished by the fact that they have very good exhaustion on lignin-containing, unbleached pulp qualities and wastepaper. However, as a rule they have very poor lightfastness, but as they are intense in color, brilliant, and cheap they are preferably used for dyeing wood-containing papers such as packaging, kraft, sleeve, and cheap envelope papers.

5.3.6 Sulfur Dyes

Sulfur dyes are water-insoluble, macromolecular, colored compounds which are produced by bridging aromatic amines, phenols, and aminophenols with sulfur and/or sodium polysulfide [16]. These dyes are of little interest for dyeing paper. Only C.I. Sulphur Black 1 (**36**), the most important dye of all in terms of volume, is used for special paper dyeings.

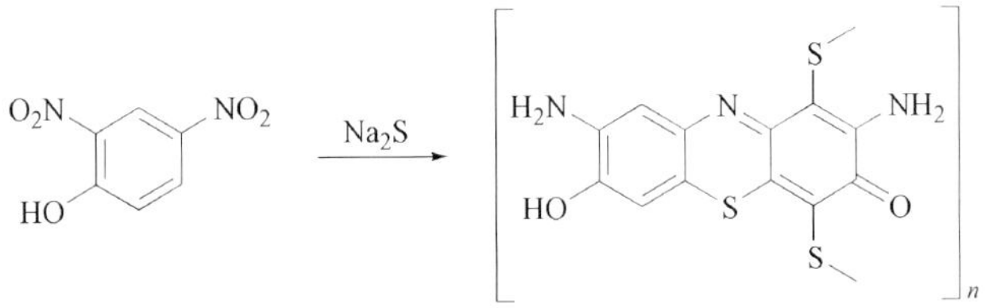

36

5.3.7 Organic Pigments

Pigments are water-insoluble colorants of organic and inorganic origin. Inorganic pigments are mainly iron, chromium, and titanium oxides. Organic pigments differ from textile and paper dyes in that they are completely insoluble in water. They do not react chemically with the fiber but are only physically attached. Their use in the paper industry is therefore limited and they are mainly applied because of their high lightfastness in laminating papers, as well as high-grade writing and printing paper (Examples are *C.I. Pigment Yellow 126* and *C.I. Pigment Red 112*). Blue and violet pigments are used for shading white (e.g., *C.I. Pigment Violet 23*) [17].

5.3.8 Special Requirements for Paper Dyes

The requirements of a dye for paper are greatly dependent on the end use of the paper. Packaging grades of paper and board, for which fastness properties may not be important, can be colored economically with basic dyes, whereas high-quality writing paper or tissue require dyes with higher fastness properties. To meet the requirements of the environment, dye producers are offering newly formulated product forms. Today a dust-free and easily dosed dye enables trouble-free handling. Thanks to new technologies [18] such as fluidized-bed spray granulation or compacting of powders, the dust problem can be reduced dramatically, although these drying techniques are more expensive than the conventional methods for producing powder forms. Further developments are imperative as modern trends, e.g., toward continuous dyeing, have led to use of dyes in concentrated liquid form. New techniques like ultrafiltration (or reverse osmosis) have been introduced to avoid the addition of auxiliaries by lowering the salt content of liquid dyes. The ultimate goal for a paper dye is a liquid product which should consist of the active substance which is adsorbed by the fiber material completely and water only.

5.3.9 References

[1] G. Martin, *Wochenblatt für Papierfabrikation*, **110** (1982) no. 13, 457.
[2] J. P. Casey, *Pulp and Paper — Chemistry and Chemical Technology*, 3rd edn, Vol. III, Wiley Interscience, New York, 1981.
[3] D. Fengel, G. Wegener, *Wood — Chemistry Ultrastructure and Reactions*, De Gruyter, Berlin, 1983.
[4] J. Wegmann, *Textil Rundschau*, **11** (1959) 631–642.

[5] H. Zollinger, *Chemie der Azofarbstoffe*, Birkhäuser Verlag, 1958, pp. 195 ff.
[6] H. E. Woodward, *The Chemistry of Synthetic Dyes and Pigments*, New York, 1955, p. 114.
[7] H. E. Fierz-David, L. Blangey, *Grundlegende Operationen der Farbenchemie*, 8. Aufl., Wien, 1952, p. 328.
[8] R. Pedrazzi, M. Golder, J. Geiwiz, R. Grimm, "Ecotoxicological Issues and their Influence on the Development of New Dyestuffs", *Chimia* **54** (2000) 525–528.
[9] Sandoz AG, DE 4 005 551, 1990, R. Pedrazzi.
[10] Sandoz Ltd, DE 2 627 680, 1977, M. Greve, H. Moser.
[11] Sandoz Ltd, DE 2 915 323, 1979, R. Pedrazzi; Sandoz Ltd, DE 3 030 197, 1980, R. Pedrazzi; Sandoz Ltd, DE 3 625 576, 1986, j. Dore; R. Pedrazzi.
[12] Nippon Kayaku Co Ltd, JP 01-297468, 1989 Masahiro Hiraki; Shimizu Oshiaki; Kojima Masayoshi.
[13] P. Crounse, P. J. Jefferies, E. K. Moore, B. G. Webster, "Dyeing Paper with a New Class of Cationic Colorants", *Tappi* **1** (1975) 120–123.
[14] BASF AG, EP 34 725, 1981, M. Patsch; M. Ruske.
[15] P. Rys, H. Zollinger, *Fundamentals of the Chemistry and Application of Dyes*, Wiley-Interscience, London, 1972.
[16] D. G. Orton in K. Venkataramen (Ed.): *The Chemistry of Synthetic Dyes*, Vol. 7, Academic Press, New York, 1974.
[17] W. Herbst, K. Hunger, *Industrial Organic Pigments*, 2nd edition, Wiley-VCH, Weinheim, 1997.
[18] H. Mollet, A. Grubenmann, *Formulierungstechik*, Wiley-VCH, Weinheim, 2000

5.4 Hair Dyes

This section deals with cosmetic colorants and agents used to modify hair color. However, changes due to intervention in biochemical processes (e.g., side effects of medication) are also known [1]. For example, the antimalarial drug chloroquine [*54-05-7*] and the spasmolytic mephenesin [*59-47-2*] brighten hair color. After discontinuation of drug treatment, the hair grows back in its original color. Chemicals such as hydroquinone [*123-31-9*] and its ether derivatives, also inhibit melanin formation [2]. These ways of altering color should not, however, be regarded as cosmetic treatment because of their mechanism of action and potential side effects (allergy, irritation, hair loss).

5.4.1 Bleaching

Bleaching of natural hair color can be done as a single cosmetic application but is also part of oxidative hair coloring. Bleaching, blonding, and lightening of human hair involve the irreversible destruction of melanin pigments by oxidation; partial or complete degradation is possible. The degree of lightening is controlled by the type and concentration of oxidant, the temperature, and the duration of treatment. Because melanin degradation also entails modification of the keratin in hair, the maximum possible extent of lightening is limited by hair quality.

Chemistry of Bleaching. The structure and formation of melanin are discussed in [3, Hair Preparations, Section 2.2.3]. Research has shown that the melanin in hair is relatively resistant to reduction and, in acid solution, to oxidation as well. In alkaline hydrogen peroxide, however, the melanin granules are attacked quickly. Evidently the polymer structure is partly destroyed by oxidation, with the formation of carboxyl groups that facilitate dissolution of the granule under alkaline conditions [4, pp. 67–79], [5]. If bleaching is complete, a void is left at the site of the melanin granule [6]. Peroxosulfates are generally added to hydrogen peroxide to intensify the bleaching action. Hydrogen peroxide can also be replaced by its addition compound with urea, carbamine peroxohydrate [*124-43-6*]. Other oxidants, such as sodium carbonate peroxohydrate ($2\,Na_2CO_3 \cdot 3\,H_2O_2$) or sodium peroxoborate tetrahydrate ($NaBO_3 \cdot 4\,H_2O$) have not met with acceptance; they are only used in speciality products. For a detailed description of peroxide bleaching, see [3, Bleaching, Section 2.2.4].

The undesirable reaction between bleach and the keratin in hair or wool has been well studied [7]. The principal point of attack is the disulfide bond of cystine, which is oxidatively cleaved via a series of intermediates [8]. The main degradation product is cysteic acid, which is responsible for the higher alkali solubility of bleached hair. In addition, hydrogen bonds and ionic bonds can be destroyed by hydrolysis. The hydroxyl and amino side chains of various keratin amino acids offer points of attack for oxidative degradation [9, pp. 216–221]. When a strong bleach is used, alteration of the chemical properties of hair also results in modified physical properties: a higher extensibility and thus lower mechanical strength; a rough, strawlike feel when dry; and a spongy feeling when wet. Because of its higher porosity, hair swells more rapidly in water, a permanent wave can be produced with more dilute solution, and the dyeing behavior is modified.

Photochemical bleaching of hair by sunlight or UV light has also been investigated with regard to bleaching of natural hair color [10, 11] and the damaging properties of light on hair keratin [12–14].

Bleaches. Bleach consists of an oxidant and a vehicle to prevent the product from running out of the hair. For sample formulations, see [15].

The most intensive bleaching action is achieved with paste bleaches, which are prepared before use by mixing hydrogen peroxide solution (6–12 vol %) with a bleach powder. The powder consists of a peroxodisulfate, an alkalizing agent, stabilizers, thickeners, and other additives. Sodium, potassium, or ammonium peroxodisulfate is used. The ammonium salt is most effective; when combined with an alkalizing component such as sodium carbonate or silicate, ammonia is formed.

The base readily penetrates the hair and promotes bleaching. The addition of stabilizers such as sodium pyrophosphate or sodium oxalate [16, 17] retards the decomposition of hydrogen peroxide in the alkaline preparation and thus enhances the bleaching action. The same holds for complexing agents (sequestrants) such as ethylenediaminetetraacetic acid, which hinder decomposition due to traces of heavy metals. Thickening additives include carboxymethyl celluloses, xanthine derivatives, and synthetic polymers. Certain dyes can also be added.

Dusting of bleaching powders has to be prevented to avoid inhalation of peroxosulfates. This can be achieved by: (1) Addition of oils to bind fine particles [18], (2) Granulation to enlarge particle size [19], or (3) Formulation of water-free, oil-containing pastes [20].

In spite of the importance of bleaches in oxidation dyeing, very few technical improvements apart from formulation aspects have occurred recently in this field. The selective adsorption of metal ions [21], especially of iron(II) salts [22], on melanin has been proposed for gentler bleaching of human hair. This process has achieved no more acceptance than the use of peroxocarboxylic acids or their precursors, which are important as bleaching intensifiers in textile bleaching.

5.4.2 Dyeing with Oxidation Dyes

Hair dyeing includes the use of permanent, semipermanent, and temporary dyes. A permanent dye lasts through any number of washings as well as permanent waving. A semipermanent dye is removed after two to ten washings, and a temporary dye is largely eliminated after one washing.

The best method of achieving a permanent hair color is the use of oxidation hair dyes. Only with these dyes can all the requirements, such as adequate color range, masking of white hair, and permanence, be satisfied. The principle of oxidation dyes is over 100 years old [23] and is used in a similar form for dyeing furs (see Section 5.2). In hair dyeing, however, toxicological considerations are more important. The dyes are produced inside the hairs from colorless precursors by oxidation with hydrogen peroxide in alkaline solution [24]. Given a suitable choice of intermediates, the hair is unifomly colored. The dyes themselves and the kinetics of their formation have been studied intensively [25].

Oxidation. A primary intermediate, i.e., a *para*-substituted aromatic compound such as a derivative of 1,4-diaminobenzene or of 4-aminophenol, is oxidized to a quinonediimine or quinonemonoimine, respectively. The imine then reacts with a secondary intermediate (coupler), which is a *meta*-substituted compound such as as a derivative of 1,3-diaminobenzene, 3-aminophenol, or resorcinol. Another oxidation step yields an indo, phenazine, or oxazine dye which is three times the size of the precursors. The mechanism of this coupling reaction is shown in Scheme 5.1.

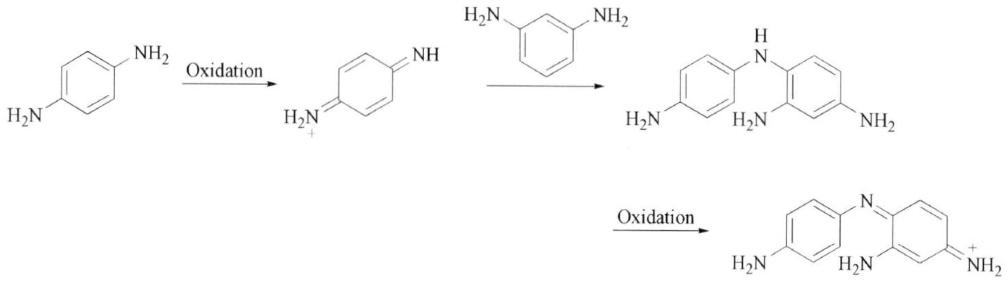

Scheme 5.1 The mechanism of oxidative coupling of *p*-phenylenediamine (primary intermediate) with *m*-phenylenediamine (coupler).

This size enlargement is largely responsible for fixing the dye in the hair. With the proper choice of substituents, any color from yellow to blue can be obtained [9, pp. 279–283], [26] and Scheme 5.2. If one or two primary intermediates are mixed with various (usually three to five) couplers, a range of colors is produced in the hair, giving the desired shade as an overall impression.

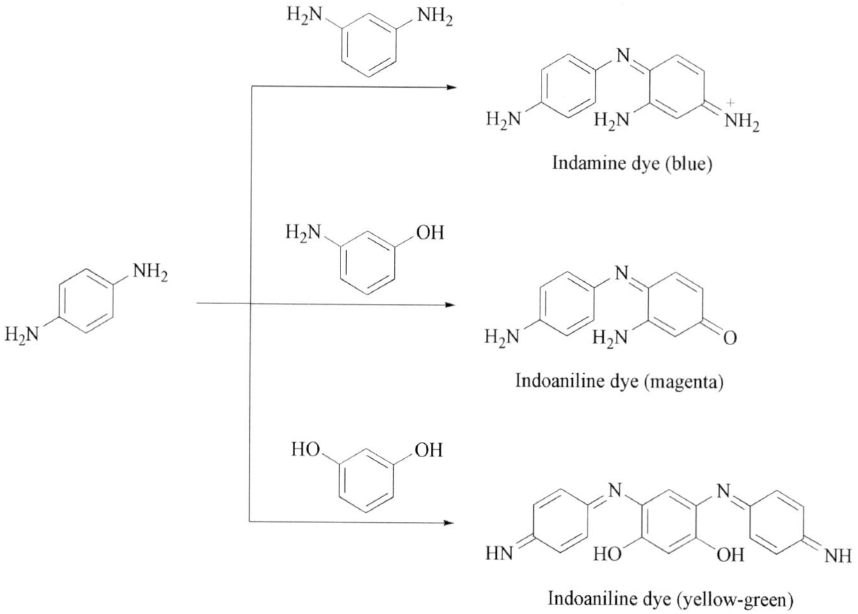

Scheme 5.2 Different types of pigments obtained by oxidative coupling.

Oxidants include hydrogen peroxide and its addition compounds; in special cases, atmospheric oxygen can also be used. Enzyme systems have also been proposed [27]. The alkali used to accelerate dyeing and swelling of the hair is ammonia or, less often, monoethanolamine. Under these conditions, lightening of the natural melanin pigment depends on the concentration of the oxidant and the primary intermediates. Lightening is desirable because it produces a more even background color; the synthetic dye would otherwise give rise to different shades when applied to gray hair, which is a mixture of naturally pigmented and white hair.

The simultaneous bleaching process also allows hair to be dyed a lighter color; only small amounts of dye intermediates are needed to offset the shade produced by bleaching. The degree of lightening is limited; the color corresponds to that obtained by cream bleaching without peroxodisulfate (see Section 5.4.1).

Advances in the field of hair dyeing by oxidation have related mainly to the areas of toxicology and application techniques.

Primary Intermediates. 1,4-Diaminobenzene [106-50] (*p*-phenylenediamine) has been the main primary intermediate used worldwide for more than a 100 years but has now been largely replaced by 2,5-diaminotoluene [*615-50-9*]. Other primary intermediates are tetraaminopyrimidine [*1004-74-6*] and *N,N*-dialkylated, ring-alkylated, or ring-alkoxylated 1,4-diaminobenzene derivatives (e.g., *N,N*-bis(2′-hydroxyethyl)-*p*-phenylenediamine [*54381-16-7*] and 2,5-diamino(hydroxyethylbenzene) [*93841-24-8*]). 4-Aminophenol and its derivatives, such as 4-amino-3-methylphenol [*2835-99-6*], yield red violet and orange indoanilines and indophenols with the usual couplers; they are therefore important for red shades, whereas diaminopyrazoles and their derivatives yield bright red, red-violet, and orange colors with a broad range of different couplers [28].

Couplers can be classified into three groups according to the color obtained with the aromatic 1,4-diamines: blue, red, and yellow-green couplers.

The usual blue couplers are 1,3-diaminobenzene derivatives with the ability to couple at the 2,4-positions relative to the amino groups. Examples in use include *N*- [29] and *O*-hydroxyethyl derivatives [30], dimeric couplers [31], and heterocyclic compounds [32].

The principal red couplers are 3-aminophenol [*591-27-5*], 5-amino-2-methylphenol [*2835-95-2*], and 1-naphthol [*90-15-3*]. Yellow-green couplers include resorcinol [*108-46-3*], 4-chlororesorcinol [*95-88-5*], benzodioxoles, and 2-methylresorcinol [*608-25-3*] and its derivatives. The importance of the yellow-green couplers lies in the broad-band absorption of the dyes produced, which makes natural-looking hair shades possible. Table 5.4 lists a range of colors obtained by reaction of primary intermediates with different couplers.

Table 5.4 Combinations of primary intermediates and couplers used in hair colorants

Primary intermediate	Coupler	Color on hair
2,5-Diaminotoluene	*m*-phenylenediamine	blue
2,5-Diaminotoluene	2,4-diaminophenoxyethanol	violet-blue
2,5-Diaminotoluene	*m*-aminophenol	magenta-brown
2,5-Diaminotoluene	5-amino-2-methylphenol	magenta
2,5-Diaminotoluene	1-naphthol	purple
2,5-Diaminotoluene	resorcinol	greenish brown
N,N-Bis-(2-hydroxyethyl)-*p*-phenylenediamine	*m*-phenylenediamine	green-blue
N,N-Bis-(2-hydroxyethyl)-*p*-phenylenediamine	2,4-diaminophenoxyethanol	green-blue
N,N-Bis-(2-hydroxyethyl)-*p*-phenylenediamine	*m*-aminophenol	green-blue
N,N-Bis-(2-hydroxyethyl)-*p*-phenylenediamine	5-amino-2-methylphenol	violet-blue
N,N-Bis-(2-hydroxyethyl)-p-phenylenediamine	1-naphthol	blue
N,N-Bis-(2-hydroxyethyl)-*p*-phenylenediamine	resorcinol	greenish brown
4-Amino-3-methylphenol	*m*-phenylenediamine	magenta
4-Amino-3-methylphenol	2,4-diaminophenoxyethanol	magenta
4-Amino-3-methylphenol	*m*-aminophenol	pale orange red
4-Amino-3-methylphenol	5-amino-2-methylphenol	orange red
4-Amino-3-methylphenol	1-naphthol	red
4-Amino-3-methylphenol	resorcinol	yellow-grey
4,5-Diamino-1-methylpyrazole	*m*-phenylenediamine	magenta-brown
4,5-Diamino-1-methylpyrazole	2,4-diaminophenoxyethanol	purple-brown
4,5-Diamino-1-methylpyrazole	*m*-aminophenol	bright orange red
4,5-Diamino-1-methylpyrazole	5-amino-2-methylphenol	intense red
4,5-Diamino-1-methylpyrazole	1-naphthol	purple
4,5-Diamino-1-methylpyrazole	resorcinol	red

Other Dye Intermediates. Besides the above-mentioned precursors other compounds such as 2-aminophenol [*95-55-6*] can be used. Trihydroxybenzenes, dihydroxyanilines, or diphenylamines are used chiefly in "autoxidative" systems, i.e., those that undergo oxidation with atmospheric oxygen without an added oxidant. The diphenylamines are oxidized to indo dyes (a general term including indamine, indoaniline, and indophenol dyes). The oxidation of 1,4-diaminobenzene alone leads to Bandrowski's base, a skin irritant [33], [3, Azine Dyes, Section 2.2.6]. Intensive work has been carried out since the first patent appeared [34] to produce indole precursors that mimic the biological melanogenesis process by using 5,6-dihydroxyindole [*3131-52-0*] or 5,6-dihydroxy-indoline [*29539-03-5*]. Until now, the relevance of these products has been low. "Progressive" coloration is obtained, i.e., the color intensity increases gradually with each treatment. For reviews of the patent literature on new intermediates and general progress in the field of hair dyeing generally, see [35–37].

5.4.3 Dye Classes

5.4.3.1 Direct Dyes

With direct dyes (see Section 3.3), as opposed to oxidation dyes, the actual dye (and not its precursor) is applied to the hair. It imparts a semipermanent or temporary color that lasts for a variable time. Lightening is not possible.

The first historically known hair dyes were semipermanent. The ancient Egyptians and Romans dyed hair and fingernails with henna, which contains lawsone [83-72-7] (2-hydroxy-1,4-naphthoquinone), a red-orange dye; walnut shells, containing juglone [481-39-0] (5-hydroxy-1,4-naphthoquinone), which gives a yellow-brown color; and indigo [482-89-3]. They also employed combinations and mordants with metal salts. To a slight extent, hair is still dyed with henna, juglone, indigo, and extracts such as chamomile (containing apigenin, 4′,5,7-trihydroxyflavone [520-36-5], which gives a yellow color). The use of these products is limited, however, because of poor selection of shades, uneven coloring, and laborious methods of application [9, pp. 236–241]. In addition natural dyes may be combined with oxidation or direct dyes to achieve a higher color intensity or stability [38].

Most of the chromophore systems common in dye chemistry (nitro, azo, anthraquinone, triphenylmethane, and azomethine) are currently used [37, 39, pp. 526–533]. Disperse, cationic (basic), and anionic (acidic) dyes are employed.

5.4.3.2 Nitro Dyes

Nitro dyes are the most important class of direct hair dyes; they are substituted derivatives of nitrobenzene or nitrodiphenylamine [3, Nitro and Nitroso Dyes]. By proper selection of donor groups and substitution site on the benzene ring, a spectrum of dyes from yellow to blue violet can be prepared [9, pp. 247–250], [40] (Scheme 3).

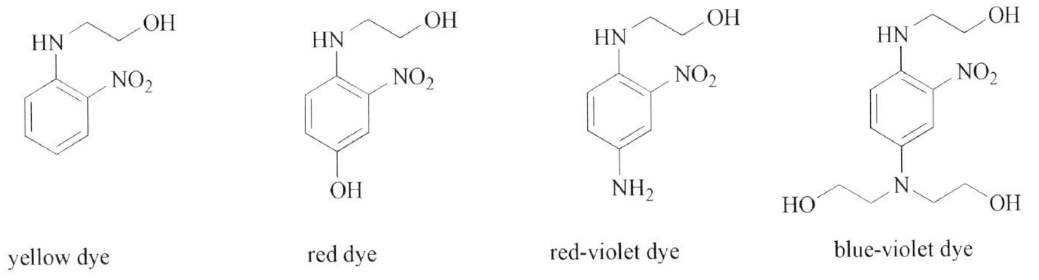

yellow dye   red dye   red-violet dye   blue-violet dye

Scheme 5.3 Influence of substituents on the color of nitro dyes

Nitro dyes are special because the small dye molecules penetrate the hair and impart color throughout the hair or in an annular pattern [41]. In this way, intense colors can be obtained even in hair with a large proportion of white. The dyes are not, however, fixed in dyed hair by size enlargement or ionic bonds (as in oxidation dyes), and so they slowly wash out.

The physical and chemical properties of dyes are very important for even coloring. First a hair dye should have comparable affinities for the roots, damaged areas, and tips. Second, the combined yellow-to-blue dyes used for shading must have similar properties so that color shifts will not take place when the hair is washed, for example. The importance of these problems is illustrated by the number of patent applications disclosing new, "custom-tailored" dyes [35] and suitable dye mixtures [42]. The relatively good colorfastness and stability of some nitro dyes allow them to be used in oxidation hair dyes as well, especially for the shading of brilliant (mainly red) fashion colors.

5.4.3.3 Cationic (Basic) Dyes (see Sections 3.7 and 3.8)

Dyes such as methyl violet [8004-87-3] and methylene blue [61-73-4] are used in rinses; cationic azine and azo dyes are also used in tints. Their molecular size and charge prevent them from penetrating the hair, so that the color pattern in cross section is annular [24]. The charge on the dyes, however, leads to a relatively stable ionic bonding of the dye to acid groups of the hair, which lasts through several washings.

5.4.3.4 Anionic (Acid) Dyes (see Section 3.9)

Anionic dyes are usually azo dyes and are employed only in special cases. Because the skin is more basic than the hair, these dyes are adsorbed preferentially on the skin, so that contact with the dyeing product leads to severe scalp staining. For this reason, the mainly anionic food colorings are not employed as hair dyes. Attempts to mitigate these disadvantages by using appropriate vehicles (see Seetion 5.4.4) have not been successful.

5.4.3.5 Disperse Dyes (see Section 3.2)

Disperse dyes of the azo and anthraquinone types, on the other hand, are used in hair tints. They also give annular color patterns, which are relatively durable because of the poor solubility of the dyes in aqueous systems. They can be combined with other classes of dyes.

5.4.3.6 Dyeing with Inorganic Compounds

Permanent hair dyes also include metal salt dyes, which produce a finely dispersed metal deposit on the hair. Metal oxides or sulfides may also be formed; this process can be promoted by the addition of sulfur compounds [9, p. 240]. A metallic gray, brown, or black coloration is obtained depending on the concentration and type of metal ion used.

Metal salts can be used in two ways. First, pretreatment with a trihydroxybenzene compound such as pyrogallol [*87-66-1*] is followed by treatment with an ammoniacal silver salt solution. This allows rapid dyeing due to the formation of metallic silver and oxidation products of the trihydroxybenzene derivative. If pretreatment is performed with thiosulfate instead of a benzene derivative, the process yields unstable silver thiosulfate and finally black silver sulfide.

In the second method, a metal salt solution (silver, lead, or bismuth; less often nickel, cobalt, or manganese) is applied; colloidal sulfur may also be added. Dyeing is based on (1) The reaction of the metal salts with the added sulfur and the sulfur in the hair keratin, which yields metal sulfides, and (2) The deposition of finely divided metals or metal oxides. With these products "progressive" coloration also is obtained.

Metal salt dyes are no longer important in professional hairdressing, for three reasons:

1) Their use involves toxicological problems.
2) In permanent-wave neutralization, bleaching, or oxidation dyeing, metal salts catalyze the decomposition of hydrogen peroxide used as oxidant. When the temperature rises, the liberated oxygen can make the hair brittle.
3) The selection of colors is limited, and the shades look metallic and unnatural.

In some countries, certain oxidation dyes may not be used on the lashes and eyebrows, and the above disadvantages of metal salt dyes are of no consequence. In this area, dyes based on silver nitrate are still used [43]. Dyes based on lead and bismuth salts are sold in small quantities as "color restorers".

5.4.3.7 Other Dyes

Metal-complex dyes are of minor importance. They can be used as such or can be formed on the hair, although the metal salt treatment entails problems. Brilliant fashion colors (pink, green, etc.) are mostly obtained with anionic dyes, which are often food colorants. Other methods for dye formation in situ [44] or with reactive dyes have not been accepted because of toxicological concerns.

5.4.4 Product Forms

The dyes discussed in Sections 5.4.3 can be used in a variety of coloring products [45] and forms. Many proposed formulations for the preparations discussed below can be found in [14, pp. 526–577, 649–676], and a more application-related review is given in [46].

Color rinses represent the simplest way of altering hair color. The hair is rinsed with a dilute aqueous or aqueous–alcoholic dye solution. The dyes are generally cationic and are adsorbed by the hair surface; they can mask the yellow color that results from bleaching or, given the proper selection of shade, can impart an attractive accent to the hair. Because of the short contact time with the hair, only slight color changes are possible.

The same holds for colored or tint setting lotions [3, Hair Preparations, Section 4.1] that contain added dyes (usually cationic, disperse, and/or nitro dyes). When the setting lotion is applied, the hair is tinted slightly. The colored film-forming polymers remaining on the hair also contribute to color impression. Changes are limited to refreshing the color, masking individual white hairs, and producing color effects as well as silver or gray shades in white or greying hair.

More pronounced color changes are possible with tints. Tints are formulated with direct dyes; intense colors can be obtained, especially with nitro dyes. Contents of 10–50 % white hair can be masked, depending on the type of dye used. Appropriate shading can produce fashionable colors or extreme color effects, especially with bleached streaks. Depending on the desired coloring effect, tints are left on the head for 3–20 min. Therefore, thickened products are generally employed. In foam tints, a surfactant solution is dispensed as a foam from an aerosol. The foam is distributed easily and does not run off the head. Tints can also be thickened with cellulose derivatives, natural mucilage, or synthetic polymers. Concentrated solutions with intense coloring action can be obtained by using cosolvents (e.g., alcohols and ethylene glycol ethers) and vehicles (e.g., urea derivatives or benzyl alcohol). In emulsion tints, the dye base consists of an emulsion.

Permanent hair colors can be achieved with tint shampoos. The shampoo base is adjusted to an alkaline pH and contains oxidation dye intermediates. Before application, it is mixed with hydrogen peroxide or a hydrogen peroxide addition compound. In comparison with oxidation hair dyes, tint shampoos employ lower concentrations of base and oxidant. This suppresses the simultaneous bleaching process that occurs during dyeing (see Section 5.4.2). As a result, damage to the keratin in hair is diminished, but the uniform coloring action is lost.

Early oxidation hair dyes were used in solution form; these have been replaced by cream- or gel-based formulas. The oil-in-water emulsions commonly used can be supplemented with auxiliary ingredients, such as polymers to improve combing ability, as well as other conditioning additives. Extensive patent literature is available on this point [35]. Gel formulations may be based on alcoholic solutions of nonionic surfactants or fatty acid alkanolamide solutions, which form a gel when mixed with the oxidant. The type (emulsion or gel) and the basic composition of the preparation strongly influence dyeing [47]. Different base formulations with the same dye content yield varying color depths and shading due to the distribution of the dye between the different phases of the product, interaction with surfactants, and diffusion from the product into the hair.

For use, the preparation is mixed with the oxidant and applied with a brush or squeeze applicator to the hair, first to the roots and then to the shafts. After a total contact time of ca. 30 min at room temperature, the hair is rinsed and washed, then generally treated with a conditioning product [3, Hair Preparations, Section 3.2, Conditioning Agents and Treatments]. The formation of acid groups during oxidation promotes the adsorption of cationic compounds, which improve the feel and combability of the hair. To prevent "delayed" oxidation, which may result from residual hydrogen peroxide in the hair, antioxidants such as ascorbic or glyoxylic acid may be added to the conditioning agent. If the conditioner is adjusted to an acidic pH, alkali residues are neutralized and the hair swelling is reversed.

In addition to cream and gel formulations, powder hair dyes are also marketed, but to a lesser extent. Besides the dye intermediates, they contain thickeners and oxidants in powder form. They are stirred with water before use.

5.4.5 Dye-Removal Preparations

The complete removal of oxidation hair colors was a problem for a long time. Older, more intense dyes, in particular, could not be removed entirely. Special reductive dye-removal products, mostly based on sulfites, reduce indo dyes to the leuco bases (diphenylamines), which are easier to remove by washing. This simple process has two deficiencies: it is not complete and the remaining diphenylamine is reoxidized. An oxidative treatment such as bleaching has therefore often been used for the removal of dyes, although certain colors are quite stable to oxidative cleavage. Recently mixtures of reducing agents (e.g., ascorbic acid, sulfite, and/or cysteic acid) together with organic acids have been described [48], which may be used for the complete removal of a broad range of oxidation dyes.

Metal salt dyes cannot be removed with these preparations because of the catalytic decomposition of hydrogen peroxide and their resistance to reducing agents.

Direct dye tints can be removed by washing with shampoo (the rate of removal depending on the type of dye) or, in some cases, by rubbing the hair with ethanol.

5.4.6 Testing Hair Dyes

Hair dyes must meet a number of conditions related to their end use. Color can be assessed by colorimetry [49]. The limits of precision are set by the substrate on which the measurement is performed. Studies on test subjects are difficult because of the uneven natural hair color and the background color of the scalp. Tresses are hard to prepare at a constant quality level. Measurements on wool cloth give reproducible results, but for oxidation dyes the shades are not identical to those produced on hair. Colorimetric methods are therefore useful only for comparative measurements on the same object, for example, in lightfastness tests. Because hair must be redyed after four to six weeks due to growth, the fastness required of hair dyes is generally less than that needed for textiles. However, stability is still a problem with many indo dyes (see Section 5.4.3). Some of them "fade" even in the absence of light; others exhibit poor light- [50], sweat-, or acid-fastness [51]. Results on the lightfastness of nitro dyes appear in [52]. The washfastness of dyes is also important [36, 53], as is wear resistance, especially for colored tints and setting lotions.

New dyes and formulations must be tested for toxicological acceptability [3, Hair Preparations, Chapter 10]. Cosmetic products are used extensively by most humans, at least in the economically developed countries. Many hair cosmetics are used daily or frequently over long periods of time. Consequently, consumer exposure to cosmetic ingredients may be considerable, so that current legislation, which defines far reaching safety requirements for cosmetic products, appears to be justified. Safety requirements from a regulatory point of view and consumer expectations are adequately summarised by the corresponding text within the EU Cosmetics Directive: "A cosmetic product put on the market within the community must not cause damage to human health when applied under normal or reasonably foreseeable conditions of use..." [54].

The history of cosmetology, past experience with the safety of cosmetic products, and in particular statistics published by poisoning advisory centers show that the cosmetic industry in general has met these criteria [55]. This is achieved by continuously evolving safety assessment strategies which take into account the actual state of the art in safety and toxicity testing.

5.4.7 References

[1] F. Herrmann, H. Ippen, H. Schaefer, G. Stüttgen: *Biochemie der Haut*, Thieme Verlag, Stuttgart 1973, pp. 100–104.
[2] S. S. Bleehen, M. A. Pathak, Y. Hori, T. B. Fitzpatrick, *J. Invest. Dermatol.* **50** (1968) 103.
[3] Ullmann's Encyclopedia of Industrial Chemistry, Sixth Edition, 2002, Electronic Release.
[4] C. R. Robbins: *Chemical and Physical Behaviour of Human Hair*, 3rd. Ed., Van Nostrand Reinhold Comp., New York, 1994.
[5] C. R. Robbins, *J. Soc. Cosmet. Chem.* **22** (1971) 339–348.
[6] I. J. Kaplin, A. Schwan, H. Zahn, Cosmet. Toiletries **97** (1982) 22–26.
[7] J. Cegarra, J. Gacen, "The Bleaching of Wool with Hydrogen Peroxide", *Wool Sci. Rev.* **59** (1982) 1–44. I. J. Kaplin, J. M. Marzinkowski, *J. Text. Inst.* **74** (1983) 155–160. H. Zahn, S. Hilterhaus, A. Strüßmann, *J. Soc. Cosmet. Chem.* **37** (1986) 159–175.
[8] H. Zahn, *J. Soc. Cosmet. Chem.* **17** (1966) 687–701.
[9] C. Zviak, *The Science of Hair Care*, Marcel Dekker, New York, Basel, 1986.
[10] Clairol, US 4 792 341, 1986 (S.D. Kozikowski, J. Menkart, L. J. Wolfram).
[11] Wella, EP 451 261, 1990 (R. E. Godfrey, T. Clausen, W. R. Balzer, G. Mahal, R. Rau); L'Oréal, EP 682 937, 1995, C. Caisey, C. Monnais, H. Samain).
[12] C. Dubief, *Cosmet. Toiletries* **107** (1992) 95–102.
[13] S. Kanetaka, K. Tomizawa, H. Iyo, Y. Nakamura in: *Preprints Platform Pres.*, Vol. 3, IFSCC Intern. Congress Yokohama 1992, Yokohama 1992, pp. 1059–1072.
[14] S. B. Ruetsch, Y. Kamath, H.D. Weigmann, *J. Cosmet.* Sci. **51** (2000) 103–125.

[15] K. Schrader: *Grundlagen und Rezepturen der Kosmetika*, Hüthig Verlag, Heidelberg 1979, pp. 575–577. J. S. Jellinek: *Kosmetologie, Zweck und Aufbau kosmetischer Präparate*, 3rd ed., Hüthig Verlag, Heidelberg, 1976, pp. 674–675.
[16] J. Cegarra, J. Ribé, J. Gacén, *J. Soc. Dyers Colour.* **80** (1964) 123–129.
[17] V. Böllert, L. Eckert, *J. Soc. Cosmet. Chem.* **19** (1968) 275–288.
[18] Goldwell, EP-A 560 088, 1993 (H. Lorenz, F. Kufner); Goldwell, EP 684 036, 1994 (H. Lorenz, W. Eberling); Clairol US 5 698 186, 1997 (G. Weeks).
[19] L'Oréal, DE 2 023 922, 1970 (F. A. Vorsatz, A. F. Risch); Wella, US 5 279 313, 1991 (T. Clausen, W. R. Balzer); Wella, EP-A 650 719, 1994 (T. Clausen, W. R. Balzer, V. Port, J. Kujawa); L'Oréal, EP 619 114, 1994 (J.-M. Millequand, C. Tricaud, A. Gaboriaud).
[20] Schwarzkopf, DE 3 844 956, 1988 (T. Oelschläger, W. Wolff); Wella, DE 195 45 853, 1997 (M. Schmitt, H. Goettmann, W. R. Balzer, H. Schiemann); Wella, DE 197 23 538, 1998 (M. Schmitt, U. Lenz, W. R. Balzer).
[21] A. Bereck, H. Zahn, S. Schwarz, *TPI Text. Prax. Int.* **37** (1982) 621–629.
[22] Shiseido KK, JP-Kokai 5 4129 – 134, 1978, (S. Kubo, 1. Yamada); Deutsches Wollforschungsinstitut an der RWTH Aachen, DE 3 149 978, 1981 (A. Bereck).
[23] H. Monnet, FR 158 558, 1883.
[24] H. Wilmsmann, *J. Soc. Cosmet. Chem.* **12** (1961) 490–500.
[25] J. F. Corbett, *J. Soc. Cosmet. Chem.* **30** (1979) 191–211 and references cited therein; H. Husemeyer, *J. Soc. Cosmet. Chem.* **25** (1974) 131–138.
[26] J. F. Corbett in K. Venkataraman (ed.): *The Chemistry of Synthetic Dyes*, vol. 5, Academic Press, New York, London 197 1, pp. 475–534.
[27] Kyowa Hakko Kogyo Kabushika Kaisha, EP 310 675, 1988 (Y. Tsujino, Y. Yokoo, K. Sakato, H. Hagino); Wella, DE 196 10 392, 1997 (M. Kunz, D. LeCruer), Novo Nordisk, WO 95/33 836, 1995 (R. M. Berka, S. H. Brown, F. Xu, P. Schneider, D. A. Aaslyng, K. M. Oxemboell).
[28] Wella, DE 38 43 892, 1990 (T. Clausen, U. Kern, H. Neunhoeffer); Wella, DE 42 34 887, 1994 (H. Neunhoeffer, S. Gerstung, T. Clausen, W. R. Balzer); L'Oréal, FR 9 505 422, 1995 (L. Vidal, A. Burande, G. Malle, M. Hocquaux).
[29] I.G. Farben, DE 559 725, 1930 (E. Lehmann).
[30] L'Oréal, FR 2 362 116, 1976 (A. Bugaut, J. J Vandebossche).
[31] Henkel, DE 2 852 156, 1978 (D. Rose, P. Busch, E. Lieske); DE-OS 3 235 615, 1982 (D. Rose, E. Lieske).
[32] Henkel, EP 106 987,1983 (N. Maak, P. Flemming, D. Schrader).
[33] J. F. Corbett, *J. Soc. Dyers Colour.* **85** (1969) 71–73.
[34] L'Oréal, US 2 934 396, 1958 (R. Charle, C. Pigerol); Henkel, DE 4 016 177, 1991 (G. Konrad, I. Matzik, E. Lieske).
[35] J. F. Corbett, **4** (1973) 3–7; **15** (1985) 52–65.
[36] J. C. Johnson: Hair Dyes, Noyes Data Corp., Park Ridge, N.J. 1973.
[37] J. F. Corbett, *Cosmet. Toiletries* **106** (1991) 53–57.
[38] Schwarzkopf, DE 196 07 220, 1997 (M. Akram, W. Wolff, S. Schlagenhoff, S. Schwartz, A. Kleen); Wella EP 898 954, 1998 (A. Sallwey, M. Schmitt, U. Lenz).
[39] J. B. Wilkinson, R. J. Moore, *Harry's Cosmeticology*, 7th ed., George Golswin, London, 1982.
[40] J. F. Corbett, *J. Soc. Dyers Colour.* **83** (1967) 273–276.
[41] S. K. Han, Y. K. Kamath, H.-D. Weigmann, *J. Soc. Cosmet. Chem.* 36 (1985) 1–16; G. Blankenburg, H. Philippen, *J. Soc. Cosmet. Chem.* **37** (1986) 59–71.
[42] L'Oreal, DE 3 131366, 1981 (R. de la Mettrie, P. Canivet).

[43] F. E. Wall in M. S. Balsam, E. Sagarin (eds.): *Cosmetics, Science and Technology*, 2nd ed., vol. 2, Wiley-Interscience, New York, 1972, p. 328.
[44] Ciba–Geigy, DE 2 807 780, 1978 (A. Bühler, A. Fasciati, W. Hungerbühler).
[45] P. Greß, D. Hoch, M. Schmock, D. Wanke: *Das Färben des Haares*, Wella AG, Darmstadt 1984.
[46] D. F. Williams, W. H. Schmitt (eds.): *Chemistry and Technology of the Cosmetics and Toiletries Industry*, 2nd ed., Blackie Academic & Professional (Chapman and Hall), London, 1996.
[47] R. L. Goldemberg, H. H. Tucker, *J. Soc. Cosmet. Chem.* **19** (1968) 423–445. K. Schrader, *Parfuem. Kosmet.* **63** (1982) 649–658.
[48] Wella, DE 196 47 493 1998 (M. Kunz, D. LeCruer); Wella, DE 196 47 494, 1998 (M. Kunz, D. LeCruer); Clairol, US 5 982 933, 1998 (G. Wis-Surel, A. Mayer, I.Tsivim).
[49] M. den Beste, A. Moyer, *J. Soc. Cosmet. Chem.* **19** (1968) 595–609. R. Feinland, W. Vaniotis, *Cosmet. Toiletries* **101** (1986) 63–66.
[50] H. H. Tucker, *J. Soc. Cosmet. Chem.* **18** (1967) 609–628; **22** (1971) 379–398.
[51] 1. Schwartz, J. Kravitz, A. D'Angelo, *Cosmet. Toiletries* **94** (1979) 47–50.
[52] J. F. Corbett, *J. Soc. Cosmet. Chem.* **35** (1984) 297–310.
[53] M. Y. M. Wong, *J. Soc. Cosmet. Chem.* **23** (1972) 165–170.
[54] Council Directive of 27 July 1976 on the approximation of the laws of the Member States relating to cosmetic products (76/768/EEC).
[55] A.Hahn, H. Michalak, K. Noack, G. Heinemeyer, Ärztliche Mitteilungen bei Vergiftungen nach § 16e Chemikaliengesetz (Zeitraum 1990–1995), Zweiter Bericht der "Dokumentations- und Bewertungsstelle für Vergiftungen" im Bundesinstitut für gesundheitlichen Verbraucherschutz und Veterinärmedizin.

5.5 Food Dyes

5.5.1 Introduction

An attractively colored food stimulates the appetite more than a discolored one. Apparently, there is a relationship between the eye and the gustatory nerves. Another purpose of food coloring is to provide a more variable range of products, which is especially important in the confectionery industry. Despite some controversy in this area, the coloring of food is unavoidable [1–3].

Coloring fresh food is not permitted as a matter of principle. As a rule, coloring is used only for processed food with no color of its own or in which only residual amounts of color remain. Food must not be colored to simulate a higher level of nutritionally important components or, worse, to mask poor quality or spoilage.

A synthetic dye is permitted for dyeing food only if thorough toxicological studies reveal no danger of toxic effects to the consumer. Food dyes are among the food additives that have been subjected to the most thorough toxicological examinations.

In terms of quantity and value food dyes play only a relatively minor role in the dye-making industry.

5.5.1.1 Specifications

Coloring is generally added to food at a very early stage in processing. Coloring agents must therefore be stable to heating, cooling, acid, or oxygen. In particular, they must remain stable during the storage of the food, during which they are often exposed to light. Because many natural colorants lack this stability, their use is limited, despite other advantages. Sulfur dioxide, which is still be used in a number of foods, can destroy many colors.

5.5.1.2 Uses and Individual Substances

Food colorants are distinguished on the basis of their solubility, i.e., water- or fat-soluble dyes and insoluble pigments. Water-soluble colorants are mainly used for food with a high water content, such as fruit products and beverages, fat-soluble colorants for fatty foods such as margarine and cheese, and pigments are used mainly for surface coloration of confectionery. Depending on their origin, coloring agents can also be classified as self-coloring food, natural coloring, or synthetic coloring. Examples of natural coloring are extracts of beetroot and turmeric, as well as carotene and other carotenoids, although as a rule the latter are produced synthetically. Among purely synthetic coloring agents, azo and triphenylmethane dyes are used most often because they are the most stable ones.

The main uses for food coloring are in cakes, fruit-based products, beverages, margarine, cheese, certain fish products, and confectionery.

Most synthetic dyes commonly used in food are summarized in Tables 5.5–5.7, which list dyes approved for food coloring in the EEC, the USA, and Japan. Some of the dyes are allowed for specific applications only.

For other areas of application, such as for drugs or cosmetics, different regulations apply. These regulations differ from one another with respect to the dyes permitted. This section deals with synthetic food dyes only.

5.5.2 Synthetic Dyes Approved for Coloring of Foodstuffs

European Union. In the EU colorants which may be used in foodstuff are regulated in 94/36/EC: On Colours for Use in Foodstuff. (Table 5.5). Food additives listed in EEC Directives are assigned E numbers. Approved dyes within the EU are also defined as food additives and thus carry E numbers. A substance with an E code and number is valid throughout the EU (E = EEC number of food additives).

Table 5.5 Synthetic dyes which can be used in foodstuff (excerpt from 94/36/EC, Annex 1)

E no.	Common name	CAS no.	C.I. Food	C.I. no.
E 102	Tartrazine	[1934-21-0]	Yellow 4	19140
E 101	Quinoline Yellow	[8004-92-0]	Yellow 13	47005
E 110	Sunset Yellow FCF	[2783-94-0]	Yellow 3	15985
E 122	Carmoisine (Azo Rubine)	[3567-69-9]	Red 3	14720
E 123	Amaranth [915-67-3]	[915-67-3]	Red 9	16185
E 124	Ponceau 4R (Cochineal Red A)	[2611-82-7]	Red 7	16255
E 127	Erythrosine	[16423-68-0]	Red 14	45430
E 128	Red 2G	[3734-67-6]	Red 10	18050
E 129	Allura Red AC	[25956-17-6]	Red 17	16035
E 131	Patent Blue V	[129-17-9]	Blue 5	42051
E 132	Indigo Carmine (indigotine)	[860-22-0]	Blue 1	73015
E 142	Green S (Acid Brilliant Green BS)	[3087-16-9]	Green 4	44090
E 151	Black PN (Brilliant Black BN)	[2519-30-4]	Black 1	28440

USA. The Federal Food, Drug, and Cosmetic (FD & C) Act provides that foods, drugs, cosmetics, and some medical devices are adulterated if they contain color additives that have not been proved safe to the satisfaction of the Food and Drug Administration for the particular use. A color additive is a dye, pigment, or other substance, whether synthetic or derived from a vegetable, animal, mineral, or other source, which imparts a color when added or applied to a food, drug, cosmetic, or the human body.

Regulations 21CFR, Parts 73, 74 (see Table 5.6), and 81 list the approved color additives and the conditions under which they may be safely used, including the amounts that may be used when limitations are necessary. Separate lists are provided for color additives for use in or on foods, drugs, medical devices, and cosmetics. Some colors may appear on more than one list.

Table 5.6 Color additives subject to certification (21CFR 74, Subpart A)

CAS no.	Chemical name	21CFR 74, Subpart A, Foods	C.I. Food	C.I. no./ CAS no.
2783-94-0	FD & C Yellow No. 6	74.706	Yellow 3	s. 5.5
3844-45-9	FD & C Blue No. 1	74.101	Blue 2	42090 [3844-45-9]
860-22-0	FD & C Blue No. 2	74.102	Blue 1	s. 5.5
2353-45-9	FD & C Green No. 3	74.203	Green 3	42053 [2353-45-9]
16423-68-0	FD & C Red No. 3	74.303	Red 14	s. 5.5
25956-17-6	FD & C Red No. 40	74.340	Red 17	s. 5.5
1934-21-0	FD & C Yellow No. 5	74.705	Yellow 4	s. 5.5
15139-76-1	FD & C Orange B	74.250	Acid Orange 137 [4548-53-2]	

Japan. In Japanese Food Law, synthetic and naturally occurring additives are treated differently. The latter, in particular naturally occurring flavors and vitamins, do not require any special permission for use. The Japanese Ministry of Health and Welfare (MHW) is responsible for the approval of color additives which are outlined in a positive list of approved colors. MHW allows petitions for the addition of a new colorant to the list. Synthetic dyes permitted for food coloring in Japan include food dyes approved in the EU or/and in the USA (Table 5.7).

Table 5.7 Examples of food dyes approved in Japan by MHW

Name in Japan	C.I. Name	C.I. no.	CAS no.
Food Blue 2	Food Blue 2	42090	s. 5.6
Blue No. 2	Food Blue 1	73015	s. 5.6
Green No. 3	Food Green 3	42053	s. 5.6
Red No. 2	Food Red 9	16185	s. 5.5
Red No. 3	Food Red 14	45430	s. 5.5 and 5.6
Red No. 102	Food Red 7	16255	s. 5.5
Red No. 504	Food Red 1	14700	[4548-53-2]
Yellow No. 4	Food Yellow 4	19140	s. 5.5
Yellow No. 5	Food Yellow 3	15985	s. 5.5 and 5.6

5.5.3 Examples of Chemical Structures

A brief survey is given here of the chemical structure of some food dyes that have been approved for food coloration. From the chemical viewpoint several of the most frequently used synthetic food dyes belong to the azo series.

C.I. Food Yellow 4, 19140 [1934-21-0], E 102, Tartrazine (**1**)

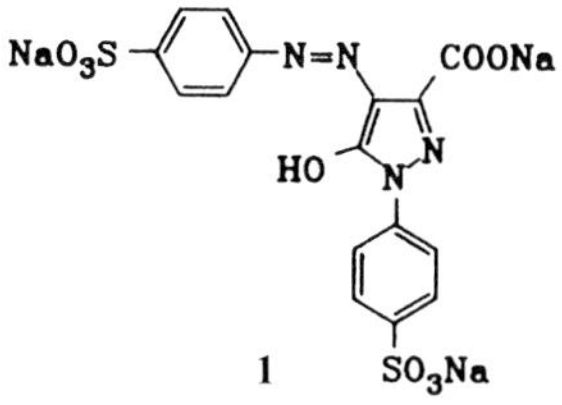

C.I. *Food Yellow 3*, 15985, [*2783-94-0*], E 110, Yellow Orange S (**2**)

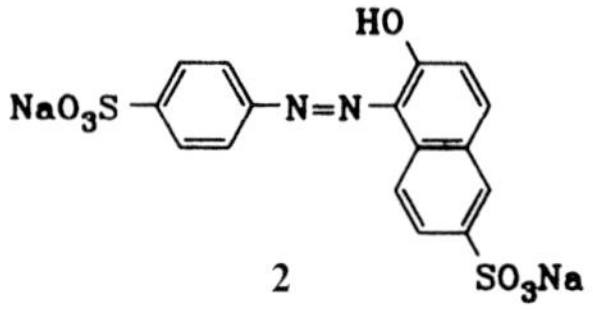

C.I. *Food Red 9*, 16185 [*915-67-3*], E 123, Amaranth S (**3**)

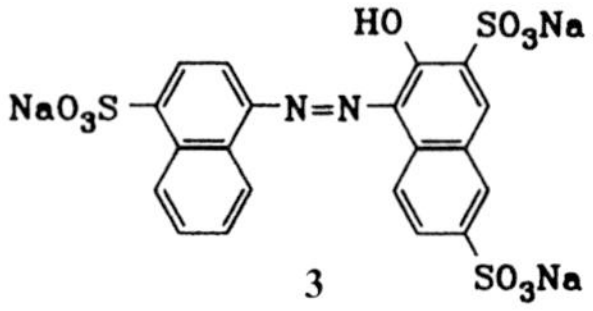

C.I. *Food Red 7*, 16255 [*2611-82-7*], E 124, Ponceau 4 R (**4**)

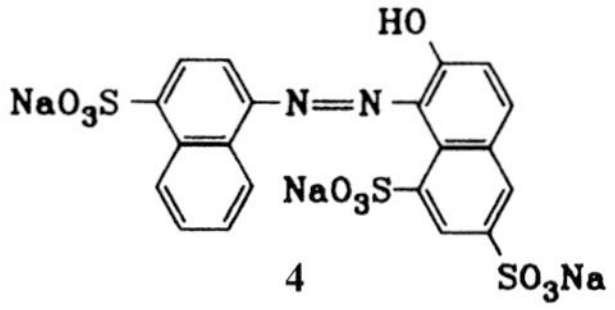

C.I. *Food Red 3*, 14720 [*3567-69-9*], E 122, Azo Rubine (**5**)

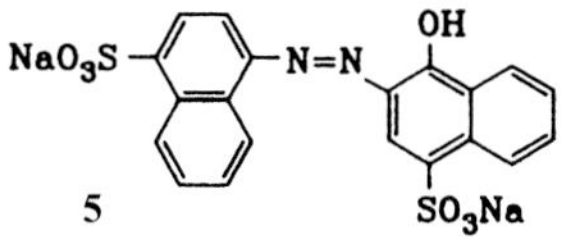

C.I. *Food Black 1*, 28440 [*2519-30-4*], E 151 Brilliant Black BN, (**6**)[1]

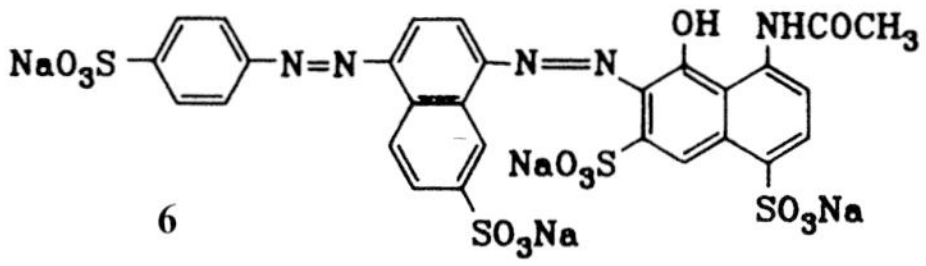

C.I. *Food Blue 5*, 42051 [*129-17-9*], E 131, Patent Blue V (**7**)

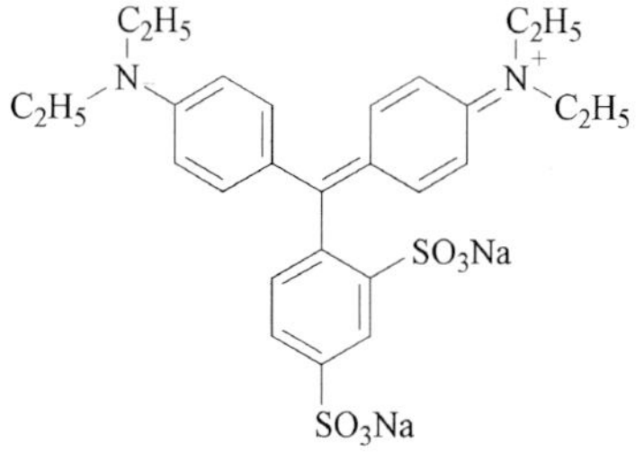

---

[1] In the Swiss List of Toxic Substances (Giftliste 1) the dye is classified as Toxic Cat. 4 (May 31, 1999).

C.I. Food Red 14, 45430 [*16423-68-0*], E 127, Erythrosine (8; as disodium salt).

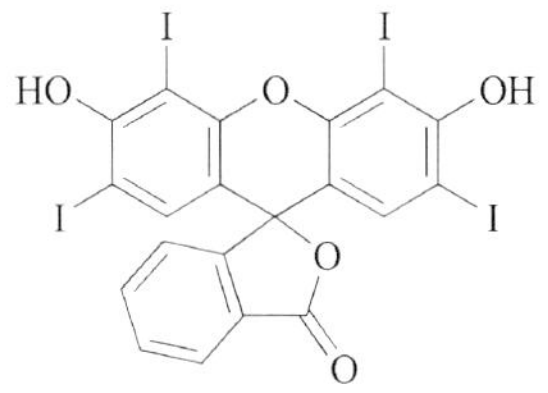

8

5.5.4 Purity Requirements

Purity criteria of colours for use in foodstuff are regulated in the EU under Directive 95/45/EC [4]. According to the legislation food dyes that are placed on the market, must meet special purity requirements
An example for *C.I. Food Yellow* 4 (tartrazine) is shown here:

Water-insoluble matter	not more than 0.2 %
Subsidiary coloring matter	not more than 1.0 %
Organic compounds other than coloring matter*	total not more than 0.5 %
Unsulfonated primary aromatic amines	not more than 0.01 % (calculated as aniline)
Ether extractable matter	not more than 0.2 % under neutral conditions
Arsenic	not more than 3 mg/kg
Lead	not more than 10 mg/kg
Mercury	not more than 1 mg/kg
Cadmium	not more than 1 mg/kg
Heavy metals (as Pb)	not more than 40 mg/kg

* 4-hydrazinobenzenesulfonic acid, 4-aminobenzene-1-sulfonic acid, 5-oxo-1-(4-sulfophenyl)-2-pyrazoline-3-carboxylic acid, 4,4′-diazoaminobis(benzenesulfonic acid), tetrahydroxysuccinic acid.

In the USA the regulations for the purity of food dyes are specified in the Code of Federal Regulations 21CFR Part 74 and Part 82. The Food and Drug Administration (FDA) is responsible for the registration and certification of food dyes.

5.5.5 Legal Aspects

Generally, food laws state which substances may be used for which foods. Sometimes, the maximum amount permitted in a food and other conditions are also fixed. Both authorization and maximum amount are based on technological necessities, with calculation of a sensible safety margin. In contrast to other food additives, major differences exist in the food-coloring laws of the EU countries, the United States and Japan.

5.5.5.1 Codex Alimentarius

The Codex Alimentarius Commission was established by the UN in 1961/1963 and is open to all member nations and associate members of FAO and WHO. In 1998, membership comprised 163 countries reepresenting 97 % of world population. One of the principle purposes of the commission is the preparation of food standards and their publication is the Codex Alimentarius [5].

In the Codex Alimentarius, *food additive* is defined any substance not normally consumed as a food by itself, and not normally used as a typical ingredient of the food, whether or not it has nutritive value, the intentional addition of which to food for a technological (including organoleptic) purpose in the manufacture, processing, preparation, treatment, packaging, transport or holding of such food results, or may be reasonably expected to result (directly or indirectly) in it or its byproducts becoming a component of or otherwise affecting the characteristics of such foods. The term does not include "contaminants" or substances added to food for maintaining or improving nutritional qualities [6].

5.5.5.2 EU and other European Countries

Colors for use in foodstuffs are regulated in the European Union by the following directives and amendments:

- European Parliament and Council Directive 94/36/EC of 30 June 1994 on colors for use in foodstuffs.
- Commission Directive 95/45/EC of 26 July 1995 laying down specific purity criteria concerning colours for use in foodstuffs.
- Commission Directive 1999/75/EC of 22 July 1999 amending Commission Directive 95/45/EC laying down specific purity criteria concerning colors for use in foodstuffs.
- Commission Directive 2001/50/EC of 3 July 2001 amending Directive 95/45/EC laying down specific purity criteria concerning colours for use in foodstuffs.

– Commission Decision of 21 December 1998 on the national provisions notified by the Kingdom of Sweden concerning the use of certain colors and sweeteners in foodstuffs (1999/5/EC).

The member states of the European Union are obliged to implement the EU Directives into their national laws.

In many countries, all additives permitted are summarized in a single list. Examples are relevant regulations in the Scandinavian countries, Germany, and Switzerland. In countries such as the United Kingdom, authorization of food additives is controlled by individual regulations, with reference to specified groups of additives. Some countries have authorizations within product regulations for certain foods. In addition, mixed systems exist, e.g., in the USA.

The international consensus is largely that, in principle, authorization for food additives should be given only if justified technologically. This means that without such additives, the required effect could only be achieved either uneconomically or not at all. In every country, food additives are subject to purity specifications which are fixed by law.

5.5.5.3 USA

In the USA, the term *additive* is more widely defined than anywhere else in the world, yet food coloring and GRAS (generally recognized as safe) substances are not classified as food additives. These GRAS substances include salt, sugar, vinegar, and baking powder, as well as many substances classified as food additives in other countries, which are considered to be especially safe, such as citric and sorbic acids. However, regulations covering the use of coloring and GRAS substances in the USA are very similar to those for food additives. In the USA, additives include practically all materials that come in contact with food in any way.

Distinction is made between three types of food additive:

– Food additives permitted for direct addition to food for human consumption are those substances which, in other countries, are considered food additives as such.
– Secondary direct food additives permitted in food for human consumption include polymers and polymer adjuvants for food treatment, enzyme preparations and microorganisms, solvents, lubricants, release agents and related substances, and other products that come in contact with food only temporarily.
– Indirect food additives are adhesives, components of coatings, food-packaging materials, components of paper and paperboard, polymers, processing aids, and sanitizers.

Requirements of Laws and Regulations Enforced by the U.S. Food and Drug Administration concerning *color additives* are:

– 21CFR Parts 73, 74, and 81
– 21CFR § 70.10 Color Additives in Standardized Foods and New Drugs
– 21CFR § 70.42 Criteria for the Safety of Color Additives

Testing and certification by the Food and Drug Administration of each batch of color is required before that batch can be used, unless the color additive is specifically exempted by regulation. Manufacturers who wish to use color additives in foods, drugs, devices, or cosmetics should check the regulations to ascertain which colors have been listed for various uses. Before using a color the person should read the label, which is required to contain sufficient information to assure safe use, such as "for food use only", directions for use where tolerances are imposed, or warnings against use, such as "do not use in products used in the area of the eye".

Manufacturers of certifiable colors may address requests for certification of batches of such colors to the FDA [7]. Certification is not limited to colors made by U.S. manufacturers. Requests will be accepted from foreign manufacturers if signed by both such manufacturers and their agents residing in the USA. Certification of a color by an official agency of a foreign country cannot, under the provisions of the Federal Food, Drug, and Cosmetic Act, be accepted as a substitute for certification by the Food and Drug Administration.

5.5.5.4 Japan

In Japan, the term "additive" means anything added to, mixed with, permeating, etc., food in the process of manufacturing, processing, or preserving it [8]. In Japanese food law, synthetic and naturally occurring additives are treated differently. The latter, especially naturally occurring flavors and vitamins, do not require any special permission for use. This explains, for example, why sweeteners isolated from plants must be specifically permitted as additives everywhere else in the world, while they can be used freely in Japan.

5.5.6 References

[1] J. N. Counsell (ed.): *Natural Colours for Food and Other Uses*, Applied Science, London 1981.
[2] *Food Technol. (Chicago)* **34** (1980) 77 – 84.
[3] *Farbstoffe für Lebensmittel*, Boldt, Boppard, 1978.
[4] Commission Directive 95/45/EC of 26 July 1995 laying down specific purity criteria concerning colors for use in foodstuffs.
[5] http://www.codexalimentarius.net/
[6] Joint FAO/WHO Food Standards Programme. Codex Alimentarius Commission: *Codex Alimentarius*, vol. 14, "Food Additives", Food and Agriculture Organization of the United Nations, Rome 1983.
[7] Food and Drug Administration, Office of Cosmetics and Colors (HFS-105), 200 C Street, S.W., Washington, D.C. 20204. For copies of regulations governing the listing, certification, and use of colors in foods, drugs, devices and cosmetics shipped in inter-

state commerce or offered for entry into the United States, or answers to questions concerning them, write to the Food and Drug Administration, at the above address.

[8] The Federation of Food Additives Associations: *Food Sanitation Law*, "Food Additives in Japan," Japan, 1981.

5.6 Ink Dyes

5.6.1 Introduction

The development of ink dyes is closely related to the progress in writing, drawing, and printing. Before the advent of synthetic dyes in the 19th century, dyes from natural resources were used for ink production. Typical examples are liquids from cuttlefish or molluscs. Recipes for iron–gallic inks were already used in the Middle Ages. Ink formulations containing iron salts of tannin and gallic acid are characterized by high fastness properties accepted for use on official documents [1]. Synthetic dyes are now widely used in inks for writing, drawing, and marking.

In recent years, ink jet printing has emerged as one of the main digital printing technologies and is challenging electrophotography and traditional printing technologies. Significant improvements have been achieved in all parts of ink-jet devices including software, print-head design, media, and ink. This progress resulted in the availability of high quality, low cost color printers for both personal and professional users. In designing ink jet printing devices, a great challenge is associated with the selection of dyes and formulation of inks. The choice of dye depends on the ink used for the specific technology (aqueous, solvent based, or hot melt). However, ink jet dyes have to meet specific requirements, irrespective of the solvent and printer system.

5.6.2 Application Principles

5.6.2.1 Ink-Jet Technology

In contrast to writing, drawing, marking, and conventional printing methods, ink-jet printing is a true primary, non-impact process. Liquid ink droplets are ejected from a nozzle under digital control and directed onto surfaces such as paper, plastics, metals, ceramics, and textiles to form a character or image [2, 3].

Generally, ink jet printers can be divided into two basic types, which can be subdivided further as shown in Figure 5.1 [4].

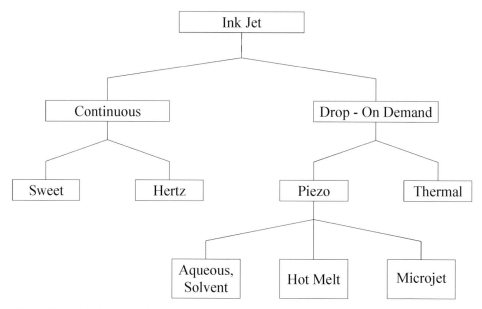

Figure 5.1 Ink jet printing technologies

Continuous Ink Jet Systems. In continuous ink jet systems a train of ink droplets continually emerges from a nozzle at high pressure and high frequency during the printing process. The direction of these ink jets is electrically controlled by charging the ink drops. There are three basic methods of deflecting the charged droplets passing through the high-voltage deflection zone.

In the *raster scan method* each droplet can be charged individually and deflected to the substrate. Uncharged ink drops are collected and returned to the ink reservoir.

In *binary continuous ink-jet systems* the uncharged ink droplets form the image and the charged drops are deflected to a gutter. Both raster scanning and binary continuous ink jets have their origins in the original *Sweet technology* and are mainly suited for monochrome printing and industrial applications.

The *Hertz method* [4] resembles the Sweet technology [4], since it also uses a continuous ink jet, the drops of which are charged by control electrodes. However, it uses charge intensity modulation instead of deflection and is especially suited for high resolution color printing.

The impulse or drop-on-demand ink jet is a system of fundamental simplicity. It differs from continuous ink jet systems in two aspects: (1) The image-forming ink droplets are not charged, so no deflection device is involved; (2) Electrical signals are only used to control the moment when an individual droplet is needed.

As seen in Figure 5.1, there are two main types of drop-on-demand technologies. The *thermal ink jet* or *bubble jet* has a heating element that causes a vapor bubble to eject an ink droplet from the nozzle. The *piezo ink jet* uses a piezoelectric transducer for ejecting ink droplets. The *microjet technology* is a further development of the piezo ink jet in which piezo ceramic walls of the ink jet channels vibrate to eject ink droplets. Advantages of this variation include high resolution and the potential to economically produce page-wide nozzle arrays. Furthermore, piezo-type systems are suited for both conventional aqueous and solvent based inks, as well as for hot melt or phase change inks.

5.6.2.2 Writing, Drawing and Marking Materials

There is a wide variety of inks for writing, drawing and marking devices [1, 5]. Modern *fountain pens* mainly contain aqueous inks. The writing paste for *ballpoint pens* is based on combinations of dyes and pigments. Water soluble dyes are used in ballpoint ink pens which combine the advantages of ballpoint pens and fiber-tip pens and ensure uniform continuous writing. *Fiber-tip pens* are available in many variations which can be used for fine writing, marking, and painting. As a result of manifold applications, various inks have been developed. Water based inks containing acid dyes are mainly used for writing and drawing on paper. For writing or marking on nonabsorbing materials like glass, metals, or plastics, inks containing volatile organic solvents and solvent or basic dyes have been developed. *Fluorescent marker pens* are widely used to highlight text passages. They are mainly based on water soluble dyes. A new type of marker pen is based on fluorescent basic or solvent dyes and uses resins in combination with surfactants and stabilizers [5].

5.6.3 Dye Classes

Generally, ink dyes for ink jet applications and writing, drawing, or marking materials are selected from food, acid, direct, sulfur, and reactive dyes. The choice of dye depends on the application and the ink used, whether it is aqueous, solvent based, or hot melt, and on the printer type: continuous ink jet or drop-on-demand, piezo or thermal ink jet.

5.6.3.1 Dyes for Ink-Jet Application

Irrespective of the printer and solvent system, ink jet dyes have to meet specific requirements [2], some of which are listed in Table 5.8.

5.6 Ink Dyes

Table 5.8 Properties required for ink-jet dyes

Colors	black, yellow, magenta, cyan
Fastness	high light, water, and smear fastness
Purity	low salt content, no insoluble materials
Solubility	high (5–10 %)
Optical density	high
Toxicology	nontoxic, nonmutagenic

In a first phase of developing ink-jet dyes, commercially available paper, leather, and textile dyes were selected and purified. However, considerable research has been invested in meeting the specific requirements of ink jets, and novel black dyes for monochrome printing and yellow, magenta, and cyan dyes for color printing have been introduced to the market [6].

Black Dyes. Water-soluble black ink jet dyes are selected from disazo, polyazo, metal complex, and sulfur dyes [7]. For ink-jet systems using the piezo technology, *C.I. Direct Black 19*, 35255 [*6428-31-5*] and *Direct Black 154*, 303865 [*54804-85-2*] are used. *Food Black 2*, 27755 [*2118-39-0*] was selected for thermal ink-jet printers due to its thermal stability, but this dye exhibits a poor waterfastness. Replacing the sulfo groups by carboxyl groups gave *C.I. Direct Black 195* (**1**), which has been introduced in commercial printers and shows much improved waterfastness due to pH-dependant differential solubility [3].

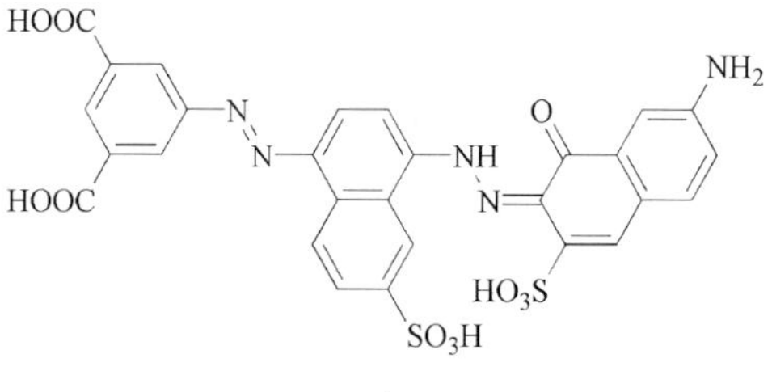

Recently, the structurally related disazo dye **2** was developed for commercial thermal ink jet printers [3].

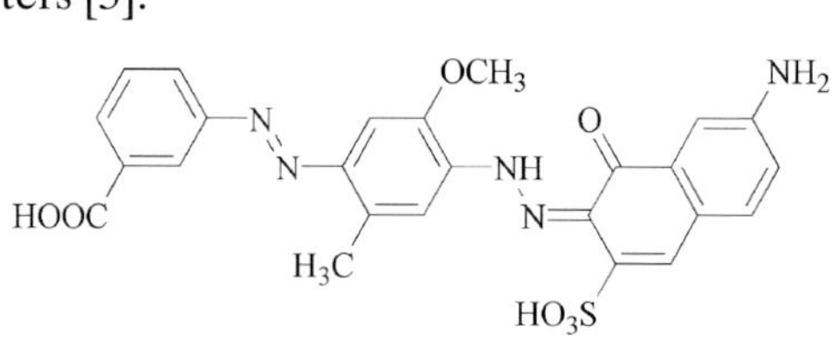

As a result of developing safe replacements for benzidine based dyes, *C.I. Direct Black 168*, Lithium Salt [*108936-08-9*] (**3**) was commercialized for thermal and piezo ink-jet printers; it has a neutral black shade and good fastness properties [7].

3

For newly emerging markets such as wide format printing and photographic imaging, there is a growing demand for lightfast dyes. It is well established that the fading resistance of dyes is considerably improved by copper-complex formation. This phenomenon has been explained by the formation of sheetlike aggregates on the substrate [8]. A representative example is *C.I. Reactive Black 31* (**4**), which was modified and adjusted to ink-jet requirements by addition of nucleophiles to the vinylsulfone group [9, 10].

4

Automatic identification is the machine readability of data by infrared lasers or LEDs. Consequently, *infrared absorbing dyes* are needed to provide readability. Black is the major color, and the predominant colorant is carbon black, which exhibits absorption throughout the visible and infrared regions of the spectrum. However, a disadvantage originates from its pigmentary and abrasive properties, especially in continuous ink jet printers. Therefore, infrared absorbing dyes are preferred for these applications. Representative examples of infrared absorbing dyes include *C.I. Solubilised Sulphur Black 1*, 53186 [*1326-83-6*] and the Nigrosine dye *C.I. Acid Black 2*, 50420 [*8005-03-6*], as well as combinations of these dyes with polyazo dyes, e.g., *C.I. Direct Black 19*, 35255 [*6428-31-5*] and *C.I. Direct Black 168*, 30410 [*85631-88-5*].

Black solvent dyes are used for *continuous* and *hot melt ink-jet printers*. Generally, there are two structural principles for these dyes: long-chain ammonium salts of lightfast acid, mostly metal complex dyes, or lightfast acid dyes with long-chain alkyl groups, e.g., sulfonic acid amides. Alkyl groups are normally branched and contain 8–10 carbon atoms. The most prevalent black solvent dyes are the 1:2 chromium(III) azo complex dyes *C. I Solvent Black 27*, 12197 [*12237-22-8*] and *35*, 12195 [*61931-53-1*]. Furthermore, ethanol soluble versions of *C.I. Direct Black 168* (see above) [3] are available, as well as infrared absorbing dyes from modified *C.I. Sulfur Black 1*, 53185 [*1326-82-5*] [7] and Nigrosine dyes *C.I. Solvent Black 5*, 50415 [*11099-03-9*] and *C.I. Solvent Black 7*, 50415:1 [*8005-02-5*].

Color Dyes. Color quality is determined by the dyes chosen for the primary colors yellow, magenta, and cyan. Important parameters for subtractive color reproduction are peak wavelength, breadth of absorption peaks and the absence of secondary absorptions [11].

Currently used water soluble yellow ink-jet dyes include monoazo dyes *C.I. Acid Yellow 17*, 18965 [6359-98-4] and *C.I. Acid Yellow 23*, 19140, [1934-21-0] which exhibit symmetrical narrow absorption bands, as well as the disazo dyes *C.I. Direct Yellow 86*, 29325 [50925-42-3] and *C.I. Direct Yellow 132* [61968-26-1]. Recently developed, improved yellow dyes include disazo dye **5** of the type *C.I. Direct Yellow 86*, 29325, but modified by introducing carboxyl groups. Similar to *C.I. Direct Black 195*, this dye shows a high waterfastness on uncoated paper due to differential, pH-dependant solubility [3, 12].

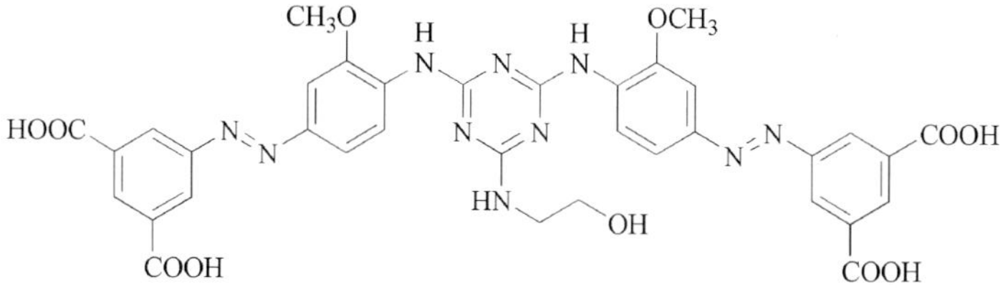

5

Magenta dyes exhibiting both brightness and high lightfastness are proving difficult to develop [13]. With xanthene dye *C.I. Acid Red 52*, 45100 [3520-42-1] ink-jet prints with excellent brightness are obtained in accordance with its narrow absorption band. However, the lightfastness is very low. Monoazo H Acid dyes of the type *C.I. Acid Red 149* and *C.I. Reactive Red 180*, 181055 [98114-32-0] are used commercially; they exhibit enhanced lightfastness and a combination of acceptable brightness and moderate waterfastness. Combinations of *C.I. Acid Red 52* and *C.I. Reactive Red 180* are used in commercial printers. Disazo dye *C.I. Direct Red 75*, 25380 [2829-43-8] has been selected for ink-jet inks and exhibits high lightfastness. Similar to yellow and black ink jet dyes, waterfastness of magenta dyes on plain, uncoated paper is improved by the introduction of carboxyl instead of sulfo groups [3].

For coated media, brilliant and lightfast monoazo dyes based on γ acid of the structural type **6** have been introduced to the photographic imaging market [14].

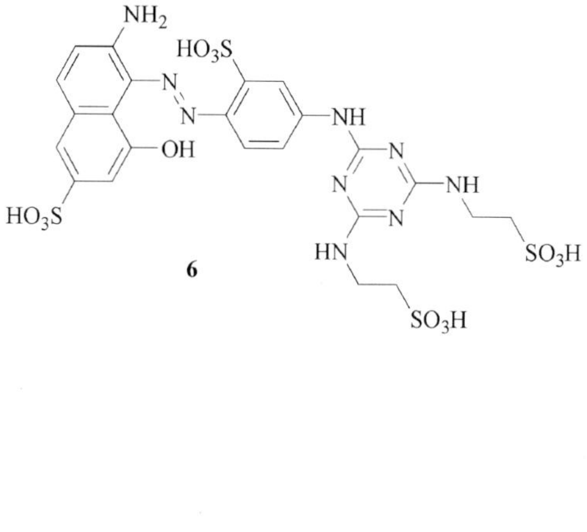

6

Lightfastness is further improved in copper complex dyes such as *C. I. Reactive Red 23*, 16202 [*12769-07-2*], albeit at the expense of brightness [13].

Standard *cyan ink-jet dyes* are *C.I. Acid Blue 9*, 42090 [*3844-45-9*] and *C.I. Direct Blue 199*, 74190 [*12222-04-7*]. *C.I. Acid Blue 9* is characterized by a superior brilliancy of shade and high color strength, but this triphenylmethane dye exhibits a low lightfastness. Copper phthalocyanine dye *C.I. Direct Blue 199* is suited for many applications due to its superior lightfastness and thermal stability. To improve its moderate waterfastness on plain paper, novel cyan dyes have been launched based on the differential solubility approach by introducing carboxyl groups into the phthalocyanine moiety [3, 12].

Solvent dyes for solvent based ink-jet inks and hot melt ink-jet systems are selected 1:2 chromium or cobalt complex azo (*C.I. Solvent Yellow 83:1* [*61116-27-6*], *C.I. Solvent Red 91* [*61901-92-6*]), anthraquinone (*C.I. Solvent Blue 45* [*37229-23-5*]), or phthalocyanine (*C.I. Solvent Blue 44* [*61725-69-7*]) dyes.

5.6.3.2 Dyes for Writing, Drawing, and Marking

Similar to ink-jet dyes, dyes used for writing, drawing and marking materials fall into two categories: water soluble dyes for aqueous inks and dyes for solvent inks [1, 5]. Dyes are listed according to application in Tables 5.9–5.13. Many of these dyes can also be used for ink jet applications. However, ink jet dyes have to satisfy more stringent criteria with regard to purity and salt content.

Table 5.9 Water-soluble dyes for fountain pen inks and fiber-tip pens

Color	C.I. name (no.)
Yellow	*C.I. Acid Yellow 3* (47005), 23 (19140), 36 (13065), 42 (22910), 73 (45350)
	C.I. Food Yellow 3 (15985)
	C.I. Direct Yellow 5 (47035), 11 (40000), 86 (29325)
Orange	*C.I. Acid Orange 7* (15510), 10 (16230), 56 (22895)
Red	*C.I. Acid Red 18* (16255), 51 (45430), 52 (45100), 73 (27290), 87 (45380), 92 (45410)
	C.I. Direct Red 75 (25380), 239, 254
Blue	*C.I. Acid Blue 1* (42045), 9 (42090), 15 (42645), 90 (42655), 91 , 91:3, 93 (42780), 104 (42735), 204
	C.I. Direct Blue 86 (74180), 199 (74190)
Violet	*C.I. Acid Violet 49* (42640)
Green	*C.I. Acid Green 1* (10020)
Black	*C.I. Acid Black 2* (50420), *C.I. Food Black 2* (27755) *C.I. Direct Black 19* (35255), *C.I. Direct Black 154* (303865)

Table 5.10 Solvent dyes for ballpoint pen inks

Color	C.I. name (no.)
Blue	*C.I. Solvent Blue 4* (44045:1)
	C.I. Solvent Blue 89 (74340)
Black	*C.I. Solvent Black 27* (Cr complex)
	C.I. Solvent Black 35 (12195 & 12197), *45* (Cr complex)

Table 5.11 Solvent dyes for fiber-tip pens

Color	C.I. name (no.)
Yellow	*C.I. Solvent Yellow 7, 82*
Orange	*C.I. Solvent Orange 3* (11270:1)
Red	*C.I. Solvent Red; 1* (12150), *122, 160*
Blue	*C.I. Solvent Blue 70*
Black	*C.I. Solvent Black 3* (26150), (Cr complex)

Table 5.12 Water-soluble dyes for fluorescent marker inks

Color	C.I. name (no.)
Yellow	*C.I. Acid Yellow 245*
Red	*C.I. Acid Red 52* (45100), *87 (45380)*
Blue	*C.I. Direct Blue 199 (59040)*

Table 5.13 Basic and solvent dyes for fluorescent marker inks

Color	C.I. name (no.)
Yellow	*C.I. Solvent Yellow 160:1*
	C.I. Basic Yellow 40
Red	*C.I. Basic Red 1* (45160)
Blue	*C.I. Basic Blue 3* (51004)
Green	*C.I. Solvent Green 7* (59040)

5.6.3.3 Fields of Application for Ink-Jet Printing

Ink jet printing has become a major printing technology in desktop, commercial and industrial applications. The growth of ink-jet technology into these markets has been well documented [15]. The reasons for this success can be attributed to the following factors:

1) Advances in computer and print head technology allow high printing speeds and digital color printing.
2) Tailoring of dyes to meet specific ink jet requirements.
3) Development of novel ink formulations to meet desktop and industrial printing requirements.
4) Development of novel ink-jet media matching both ink and print head parameters.

The diversity of ink-jet applications is summarized in Table 5.14.

Table 5.14 Applications of ink jet printing

Home/office	desktop printers, fax devices, color copiers, photographic imaging, large format and wide array printers
Commercial	graphic arts: computer aided design, indoor and outdoor signs, posters, canvas,
	pre-press printing: textile proofing, publication proofing,
	short run printing of textiles,
	carpet printing,
	digital color presses.
Industrial	bar code printing, marking, labeling

The colorant requirements for the different ink-jet applications differ widely. For example, high chroma is the key requirement for photorealistic printing, whereas high lightfastness is the key parameter for the wide format and outdoor markets. These widely differing properties provide opportunities for a variety of dye classes, as well as for pigments.

5.6.4 Inks
[16,17]

Generally, there are two basic types of inks. The first type contains finely dispersed carbon black or colored pigments in both solvent and water based systems. Pronounced improvements have been achieved in stabilizing aqueous pigment formulations for ink-jet applications by adding specific polymeric dispersing agents [18]. The second type of ink is based on water or solvent soluble dyes.

5.6.4.1 Ink-Jet Inks

In developing ink-jet printing devices a decisive task is associated with the formulation of inks. Basically, there are three types of ink-jet inks: aqueous, solvent, and hot melt. The choice of ink depends on the ink-jet technology used. Aqueous inks can be applied in both drop-on-demand and continuous printers, whereas solvent based inks are mainly used in continuous printers for industrial use. High boiling oily solvents exhibiting low vapor pressure are used in the microjet technology. Hot melt or phase change inks are based on low melting resins or waxes.

General features of ink jet inks influencing print quality and printer compatibility are summarized in Table 5.15.

Table 5.15 Performance requirements for ink jet inks

Physical properties	Printing properties	Functional properties
Viscosity	Optical density	Long shelf life
Surface tension	Lightfastness	Machine compatibility
Specific gravity	Waterfastness	Stable drop formation
Conductivity	Rubresistance	No nozzle crusting
Color	Smear resistance	Low corrosion
	Feathering	No biological growth
	Drying time	No particle formation
	Substrate sensitivity	Low toxicity
		No chemical hazards

Water based ink jet inks usually contain the following main components and additives: colorants; solvents, cosolvents, humectants; biocides; pH stabilizers; chelating agents; lightfast agents. Water miscible cosolvents play an important role: they act as humectants, prevent dye crystallization at the nozzles, improve solubility of dyes and storage stability of inks at low temperatures, and control surface tension and viscosity of inks. Commonly used cosolvents include glycols like diethylene glycol, triethylene glycol, thiodiglycol, 1,2-propylene glycol, butyl diglycol, butyl triglycol, glycerine, and N-methyl-2-pyrrolidone. Biocides are normally present to prevent bacterial growth. One of the most important production steps for inks are filtration procedures, including microfiltration to eliminate insoluble materials and membrane filtration to remove electrolytes such as chloride and sulfate ions. A typical formulation of an ink for drop-on-demand printers follows:

Acid or direct dye	4.0 %
Diethyleneglycol	10.0 %
N-Methyl-2-pyrrolidone	5.0 %
Triethanolamine	0.5 %
Biocide	0.1 %
Water	80.4 %

In conductive inks for continuous ink-jet printers salts such as lithium acetate and potassium cyanate are added to adjust the specific conductivity.

For commercial printing onto nonporous, hydrophobic materials such as plastics, metals, and glass, rapidly drying solvent inks which adhere to these substrates are needed. A typical solvent for these applications is methyl ethyl ketone (MEK). However, alternative alcoholic solvents like ethanol and N-propanol have been introduced for ecological and safety reasons. The predominant color for industrial marking is black, and solvent soluble 2:1 azo chromium(III) complex, polyazo, Nigrosine, and sulfur dyes are used. Typical formulation of an MEK based ink:

C.I. Solvent Black 27	5.0 %
Methyl ethyl ketone	80.0 %
Propylene glycol	10.0 %
Binder [poly(vinyl chloride)/poly(vinyl acetate) copolymer]	5.0 %

In *hot melt (phase change) inks* long-chain fatty acids, alcohols, or sulfonamides bearing 18–21 carbon atoms are used as vehicles exhibiting melting points in the range from 60 to 125 °C. These inks are melted in the printhead, and the ink droplets solidify after hitting the surface of the media. Good print and color quality is obtained. Since hot melt inks do not contain volatile solvents, the piezoelectric ink jet is the technology of choice, and the thermal ink jet cannot be applied. The colorants used are solvent dyes soluble in the particular wax, or pigment dispersions.

5.6.4.2 Writing, Drawing, and Marking Inks

Inks for writing, drawing, and marking materials are mainly based on water. They have a variety of compositions depending on the specific uses and materials.

Similar to ink-jet inks, aqueous drawing and writing inks contain water miscible cosolvents as humectants or moisturizers (e.g., glycols, glycerin, or sorbitol), which stabilize the ink in the pen and at the nib. Typically, biocides and tensides are added to the ink formulation.

Formulation of a blue fountain pen ink [1]:

C.I. Acid Blue 93 (42780)	0.5 %
Glycerine	1.7 %
Sulfuric acid (95 wt %)	0.2 %
Biocide	0.07 %
Nonionic surfactant	0.03 %
Water	97.5 %

Fountain pen inks are low in viscosity, exhibit a surface tension within small tolerances, and do not contain resins or pigments.

Composition of an aqueous fiber-tip pen ink [5]:

C.I. Acid Blue 9 (42090)	3.0 %
1,2-Propylene glycol	25.0 %
Polyvinylpyrrolidone	0.1 %
Nonionic surfactant	0.1 %
Deionized water	71.8 %

Highly concentrated dyes with very low salt content are preferred. For black inks, dye concentrations up to 8.0 % are necessary to achieve a high optical density. For fineliner pen inks the same ink formulations are used as in fiber-tip pen inks.

High concentrations of basic dyes or solvent dyes digested in oleic acid or resins in organic solvents are used for BALLPOINT WRITING PASTES. Typical ballpoint paste formulation [1]:

C.I. Solvent Blue 89 (74340)	30.0 %
Oleic acid	10.0 %
Benzyl alcohol	35.0 %
Polyacrylic acid (or polyvinylpyrrolidone)	25.0 %

Phthalocyanine based inks of this type are accepted for use in official documents in accordance with DIN 16554.

For felt-tip ink markers, combinations of volatile solvents like ethanol, 2-propanol, and high-boiling solvents such as diethylene glycol are used to prevent clogging of the writing point. Basic dyes and solvent dyes, as well as selected acid dyes, are suitable for these inks. Typical formulatioon of a solvent-containing marking ink [1, 5]:

Basic, solvent, or acid dye	15.0 %
Ethanol	25.0 %
2-Propanol	25.0 %
Diethylene glycol	15.0 %
Binder (natural or synthetic resin)	20.0 %

5.6.5 Properties of Ink-Jet Prints

Since the beginnings of ink-jet technology, permanence of ink-jet prints, in particular light stability and waterfastness, has been a major problem. Only recently were ink jet inks exhibiting both high light stability and good waterfastness successfully introduced to the market. The most important factors influencing image permanence are dye structure, ink formulation, and application environment.

The most important factor affecting image stability is the inherent functional property of the dye. Yellow disazo dyes of the type *C.I. Direct Yellow 86*, 29325 and *C.I. Direct Yellow 132* exhibit high light- and waterfastness. Copper phthalocyanine dyes and copper complex azo dyes show superior lightfastness and are suited for photographic and outdoor applications. Useful relationships between lightfastness of dyes and the physical and chemical structures are well established [8, 19]. Accordingly, besides the inherent photostability of the colorant the degree of aggregation on the substrate is among the most important factors [20]. Black dyes exhibiting excellent lightfastness are found in the range of azo copper(II)- and chromium(III) complex, Nigrosine, and sulfur dyes.

It is well documented that parameters such as print-head performance, print density, shade, print uniformity, drying time, and waterfastness are strongly influenced by the choice of solvents and additives in the specific ink formulation [21]. Solvents serve as solubilizers and humectants to prevent nozzle clogging. However, solvents can interfere with the permanence of the image in two ways: by influencing the aggregation behavior on the substrate and by acting as sensitizers that cause an increased fading rate. Special lightfastness additives have been developed for both inks and media to provide protection against light-induced fading [22]. For improving the waterfastness of ink-jet prints, reactive ink systems have been developed according to the following reaction principles [23]: (1) Coagulation of dyes with mordants on the substrate, (2) Reaction of ink components with materials incorporated in the paper, (3) Photopolymerization of UV curable inks, and (4) Formation of polymeric binders on the substrate.

A major improvement in permanence of ink-jet prints in recent years has been due to tailoring pigment dispersions for both solvent and water based ink jet inks [18]. These pigment containing inks are primarily used for outdoor applications.

The composition of the ink receiving medium represents an important parameter controlling both light fading and waterfastness of ink-jet prints [21]. Complex factors such as the nature of the organic polymer or mineral coating layer, nature of additives, layer structure, and layer surface influence the permanence of ink-jet prints. As shown by spectroscopy, dye aggregation within the receiving medium strongly affects the light fading performance [21].

5.6.6 References

[1] M. Colditz, E. Kunkel, *Ullmann's Encyclopedia of Industrial Chemistry*, 5th ed., Vol. A9, Wiley-VCH, Weinheim, 1987, pp. 37–47.
[2] P. Gregory, *High-Technology Applications of Organic Colorants*, Plenum Press, New York, 1991, pp. 175–205.
[3] R. W. Kenyon in P. Gregory (Ed.): *Chemistry and Technology of Printing and Imaging Systems*, Blackie Academic & Professional, London, 1996, pp. 113–138.
[4] J. Heinzl, C. H. Hertz, *Adv. Electronics Electron Physics* **65** (1985) 91–166.
[5] J. G. Pfingstag, *J. Soc. Dyers Colour.* **109** (1993) 188.
[6] W. Bauer, J. Ritter, *Am. Ink Maker* **75** (1995) 59–64.
[7] W. Bauer, J. Geisenberger, H. Menzel, R. Pedrazzi, *Ink World* **1999**, 40–43.
[8] H. Zollinger, *Colour Chemistry*, VCH, Weinheim, 1991, pp. 311–319.
[9] Hoechst, EP 312004, 1987.
[10] Hewlett Packard, DE 19727427, 1996.
[11] N. Ohta, *J. Appl. Photo. Engr.* **2** (1976) 75.
[12] R. W. Kenyon, IS&T's 9th International Congress on Advances in Non-Impact Printing Technologies, Yokohama, Oct. 4–8, 1993; Proceedings 279–281.
[13] W. Bauer, D. Baumgart, W. Zöller, *Am. Ink Maker* **76** (1998) 53–59.
[14] Ilford, WO 96/ 24636, 1995.
[15] R. N. Mills, IS&T's 10th International Congress on Advances in Non-Impact Printing Technologies, New Orleans, Oct. 30–Nov. 4, 1994; Proceedings, 410–414.
[16] *Kirk-Othmer, Encyclopedia of Chemical Technology*, 3rd Edition, Vol. 13, Wiley, New York, 1981, pp. 374–397.
[17] E. Kunkel, H. Roselieb, *Ullmann's Encyclopedia of Industrial Chemistry*, Vol. A23, Wiley-VCH, Weinheim, 1983, pp. 259–266.
[18] K. K. Keller, IS&T's NIP 15: International Conference on Digital Printing Technologies, Orlando, Oct. 17–22, 1999; Proceedings, 95–97.
[19] R. F. Gordon, P. Gregory, *Organic Chemistry in Colour*, Springer Verlag, Heidelberg, 1987, pp. 97–121.
[20] C. L. Bird, W. S. Boston, *Theory of Coloration of Textiles*, Dyers Company Publication Trust, White Rose Press, London, 1975, pp. 90–91.
[21] M. Fryberg, R. Hofmann, P. A. Brugger, IS&T's NIP 13: International Conference on Digital Printing Technologies, Seattle, Nov. 2–7, 1997; Proceedings, 595–599.
[22] C. Lee, J. Urlaub, A. S. Bagwell, M. G. MacDonald, R. S. Nohr, IS&T's NIP 13: International Conference on Digital Printing Technologies, Seattle, Nov. 2–7, 1997; Proceedings, 664–666.
[23] N. Noguchi, S. Fijii, IS&T's NIP 15: International Conference on Digital Printing Technologies, Orlando, Oct. 17–22, 1999; Proceedings, 129–132.

5.7 Photographic Dyes

Dyes are used in photography for two fundamental purposes. The first is to sensitize silver halide crystals from 400 to 1300 nm. The second purpose is the production of color images. Other uses for dyes in photography are for antihalation, color filtration, and opacification.

5.7.1 Cyanine Dyes

5.7.1.1 Introduction

Silver iodide bromide crystals have intrinsic sensitivity to blue light. In 1873 Vogel observed that certain silver halide plates had sensitivity to green light [1]. He determined that the antihalation dye corallin was absorbing green light and sensitizing the silver emulsion. This discovery led to a burst of activity to discover useful sensitizers for the new field of photography. "Cyanine" became a popular sensitizing dye for panchromatic emulsions before 1900. Work advanced significantly in the 1920s when the chemical structures were elucidated and correlated with photographic activity. The ability to correlate structure with activity resulted in an explosion of sensitizing dye syntheses. Many sensitizing dye classes were discovered. However, the cyanines and merocyanines have provided the majority of useful photographic sensitizers. Dyes have been developed which can sensitize silver halide emulsions throughout the visible spectrum from 400 to 700 nm and to ca. 1300 nm in the infrared [2–8].

5.7.1.2 Fundamental Aspects

The sensitizing dye extends the sensitivity of silver halide crystals by adsorbing to the surface of the silver halide crystal and forming dye aggregates which absorb light at the requisite wavelengths [9]. The excited state of the dye then injects an electron into the conduction band of the silver halide crystal. This photoelectron then reduces an interstitial silver atom at specially designed traps composed of silver sulfide and gold. Further reduction of silver atoms at this site produces a stable latent image. The most efficient photographic silver halide crystals can produce a stable latent image by absorbing as few as four photons [10, 11]. This latent image acts as a catalytic site which then allows the selective reduction of the silver grain by a variety of specially selected reducing agents in an alkaline medium [12]. By means of this selective reduction of silver atoms the latent image can be amplified by as much as 10^7. This is the basis of the high sensitivity of silver halide photographic systems.

5.7 Photographic Dyes

1 cyanine, $n = 0, 1, 2, 3$

2  merocyanine, $n = 0, 1, 2, 3$

3 oxonol, $n = 0, 1, 2, 3$

The fundamental structural element of cyanine dyes **1** is two nitrogen auxochromes linked by an odd number of sp^2-hybridized carbon atoms. This chromophoric moiety is characterized by high extinction coefficients. The nitrogen auxochromes are typically incorporated into a wide variety of heterocyclic bases to adjust the physical properties of the dyes. By varying the heterocyclic bases and the length of the connecting carbon chain, effective sensitizing dyes have been prepared with absorptions from 400 to beyond 1300 nm. Very simple modifications allow the facile preparation of dyes with solution absorptions ranging from 373 to 810 nm (Table 5.16). In the fundamental merocyanine chromophore **2** an oxygen atom replaces one of the nitrogen auxochromes of the cyanine. The oxonols **3** are characterized by two oxygen atoms as auxochromes.

Table 5.16 Variation of cyanine dye absorption as a function of chain length and heterocyclic base.

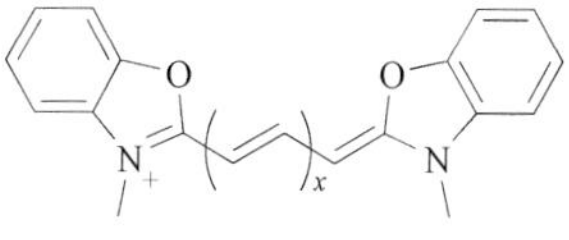

Benzoxazole 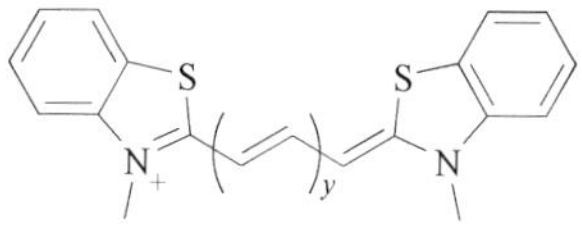 | Benzothiazole 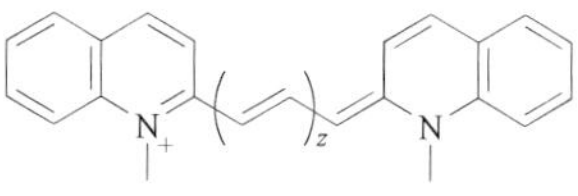 | Quinoline

x	λ_{max}, nm	y	λ_{max}, nm	z	λ_{max}, nm
0	373	0	426	0	523
1	495	1	560	1	606
2	580	2	650	2	709
3	697	3	765	3	810

Adsorption to the silver surface is one of the most important requirements for a sensitizing dye. Intimate contact with the silver surface facilitates electron transfer. To make stable films with predictable photographic properties it is also important that the dyes not be easily displaced from the silver surface. The formation of J-aggregates by the adsorbed dyes is often a critical factor for the performance of sensitizing dyes; J-aggregates are extensive bathochromically shifted aggregates [13]. In the case of a cyanine dye absorbing green light in solution (λ_{max} = 550 nm), the J-aggregate will have a peak absorbance approximately 100 nm to longer wavelength and sensitize silver halide to red light. H-aggregates are small aggregates of several dye molecules which are hypsochromically shifted relative to the solution spectrum of the monomer. The aggregation of sensitizing dyes is very sensitive to the size and position of substituents. Large substituents which inhibit either adsorption to the surface or J-aggregation can have an extremely deleterious effect on photographic performance [14].

The dyes must also have appropriate redox properties to function properly as sensitizers. If the dyes are oxidized too readily in the ground state, silver atoms can be prematurely and unselectively reduced, and this causes photographic fog. Fog is the unwanted, indiscriminate reduction of silver. Photographic fog causes a loss of signal-to-noise ratio and degrades image quality. The redox potentials of sensitizing dyes have also been extensively measured and correlated to performance [15].

5.7.1.3 Application of Sensitizing Dyes

Modern silver halide emulsions are generally prepared in the presence of photographic gelatin by a two-jet process in which silver nitrate and halide ion are added together with great precision. Rate of addition, pAg, concentrations of halide ions, sequence of addition, temperature, and various growth modifiers are combined to produce the wide variety of silver halide crystals of precise crystal habit, composition, and shapes that are used today. Once the crystal is fully formed to give the primitive emulsion, the silver halide crystals are chemically and spectrally sensitized. Chemical sensitization involves treating the silver halide with sulfur and gold to create low-energy traps for photoelectrons. Chemical sensitization can be performed either before or after spectral sensitization [16].

The spectral sensitizer is typically added as a solution in an alcoholic solvent such as methanol or trifluoroethanol. The optimum amount of dye is determined experimentally. A good carbocyanine sensitizer can be used at a concentration as low as 200 mg per mole of silver. Higher concentrations result in a loss of photographic sensitivity as a result of desensitization. It is common in modern photographic systems to use mixtures of two or more sensitizing dyes to optimize performance. A mixture of dyes allows the adjustment of the absorption envelope to allow better color reproduction [17, 18]. Sensitizing dyes can be neutral zwitterions or they can be positively or negatively charged.

5.7.1.4 Production of Sensitizing Dyes

The synthesis of sensitizing dyes has changed little since Hamer and Ficken's works cataloging the synthetic methods for making the various classes of sensitizing dyes [19,20]. The synthesis of symmetrical cyanine dyes is straightforward. Unsymmetrical dyes present a greater challenge. The total amount of sensitizing dyes used worldwide amounts to hundreds of kilograms. Ten million square meters of color negative requires about 20 kg each of blue, green, and red sensitizing dyes when coated at a coverage of one gram per square meter of silver sensitized with two milligrams of sensitizing dye per square meter for each color channel.

5.7.1.5 Cyanine Dyes as Sensitizers

Cyanines are the most widely used sensitizing dyes. Simple monomethine cyanine dyes **4–6** containing a variety of heterocyclic bases are effective blue sensitizers. These dyes adsorb on the silver halide surface and enhance the intrinsic blue sensitivity of the silver halide crystal by absorbing strongly from 400 to 500 nm. This is particularly important with the newly introduced high aspect ratio, tabular, silver grains used in many of today's color films [21]. These high aspect ratio grains are very thin, transparent, and have very low intrinsic sensitivity to visible light.

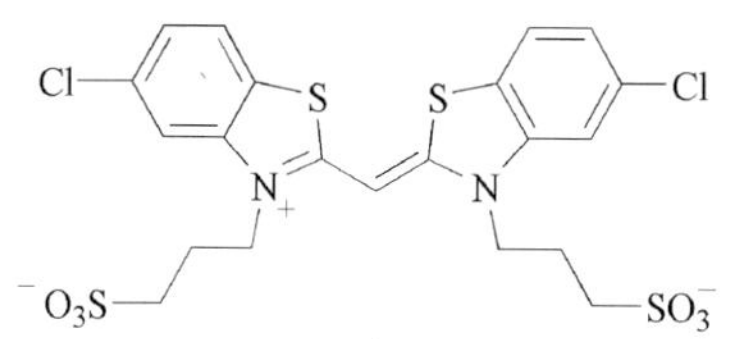

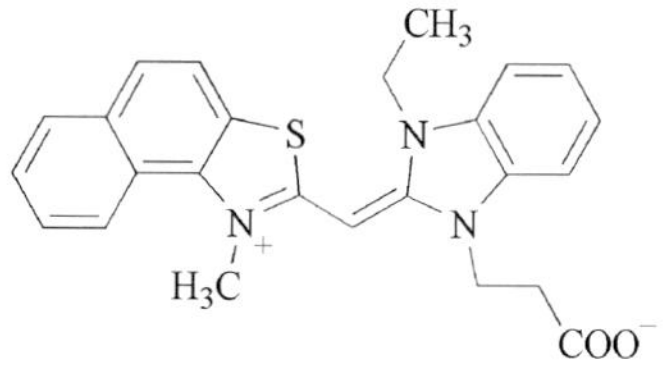

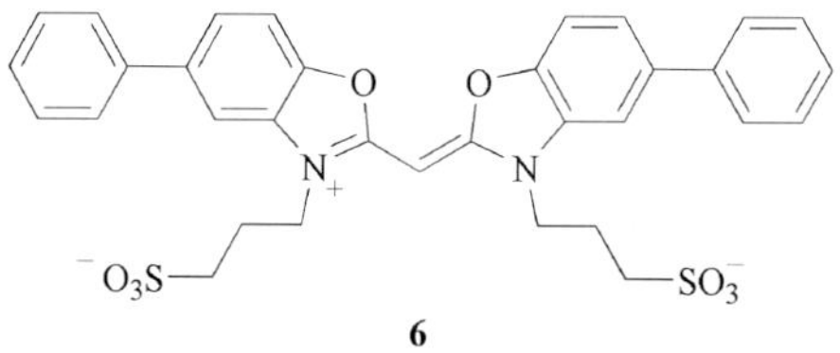

The most common cyanine dyes used as green sensitizers are symmetrical benzimidazole and benzoxazole dyes **7–9**. Dyes **7–9** also illustrate several common side-chain variations. The sulfobutyl and sulfopropyl groups are very common. The methylsulfopropyl side chain is used to improve solubility.

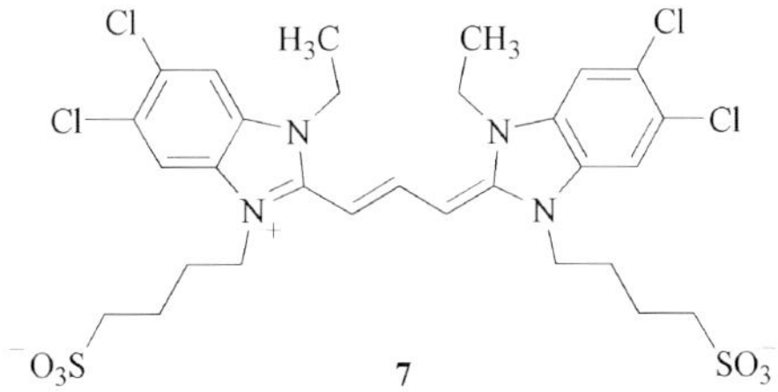

5.7.1 Cyanine Dyes

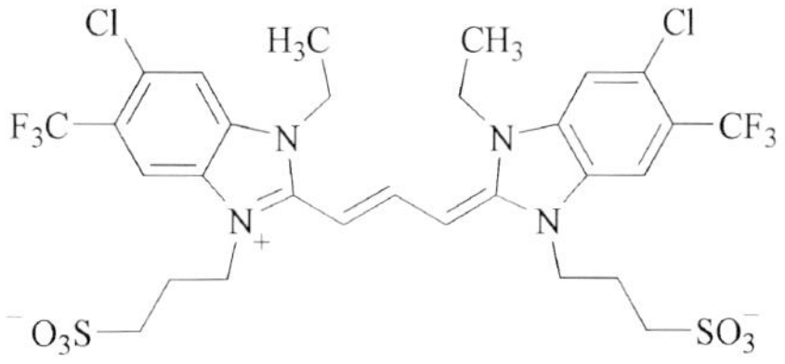

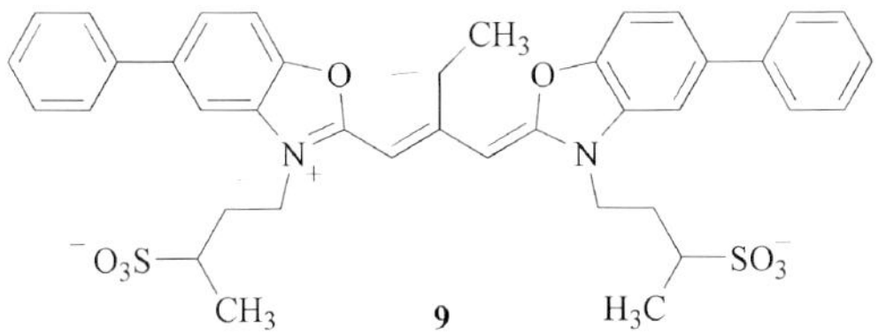

Unsymmetrical benzothiazole–quinoline dye **10**, benzothiazole–benzoxazole dye **11**, and benzimidazole–benzoxazole dye **12** are also effective green sensitizers. It is important that these dyes J-aggregate and absorb between 500 and 600 nm with a peak absorbance at ca. 550 nm.

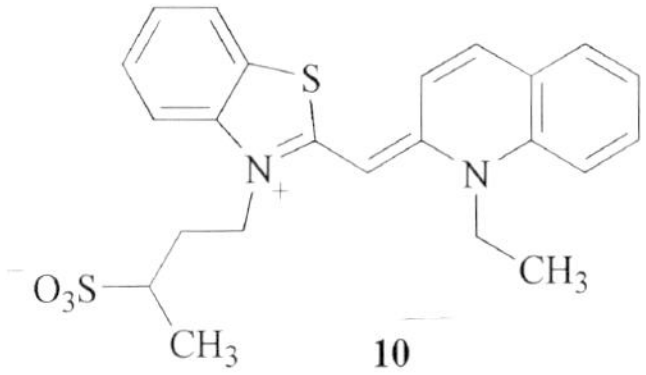

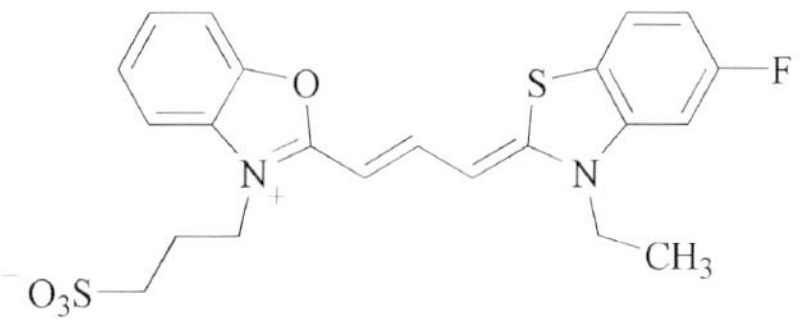

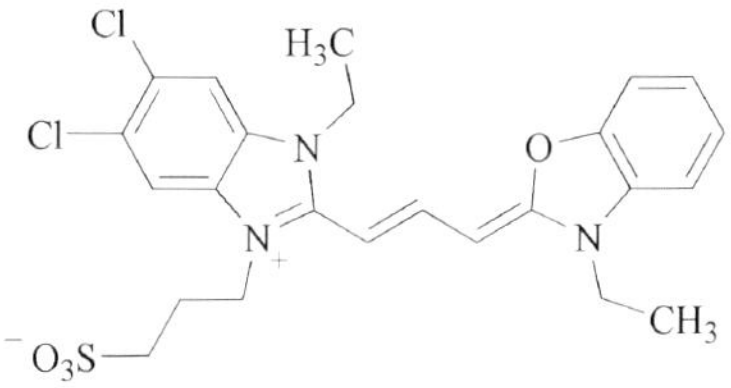

Red sensitizers typically use benzothiazole bases since these produce longer wavelength absorbances than the related benzoxazoles and benzimidazoles. Examples of effective red sensitizers are symmetrical dyes **13–15** and unsymmetrical dyes **16–18**. Red sensitizers are typically carbocyanines with a C_3 chain connecting the two heterocyclic moieties. An ethyl group on the central carbon of the chain of many carbocyanines is important for promoting the formation of strong J-aggregates.

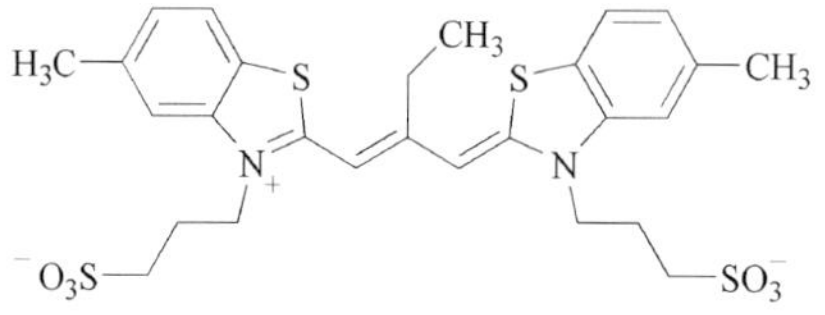

5.7 Photographic Dyes

13, 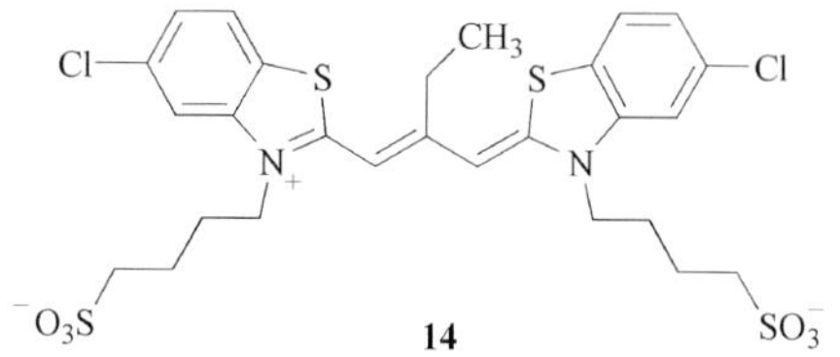 **14**, 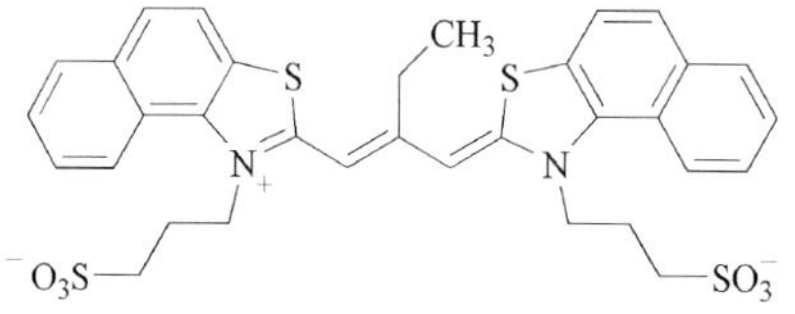 **15**, 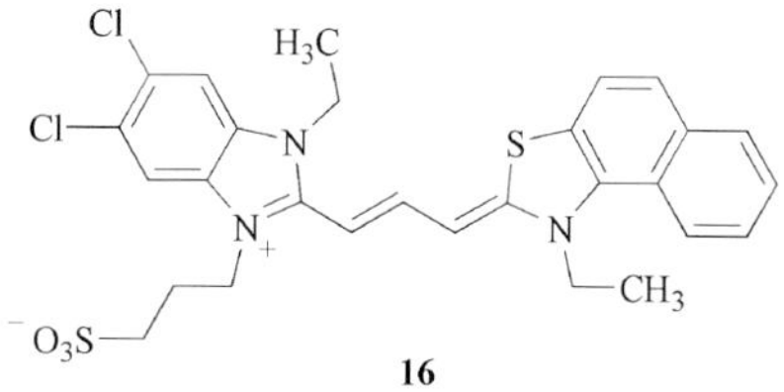 **16**, 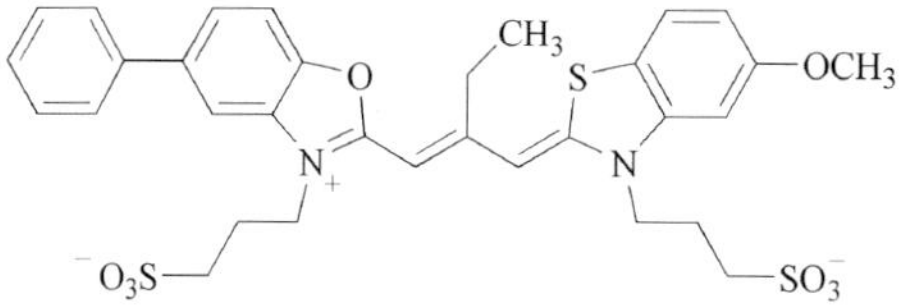 **17**, 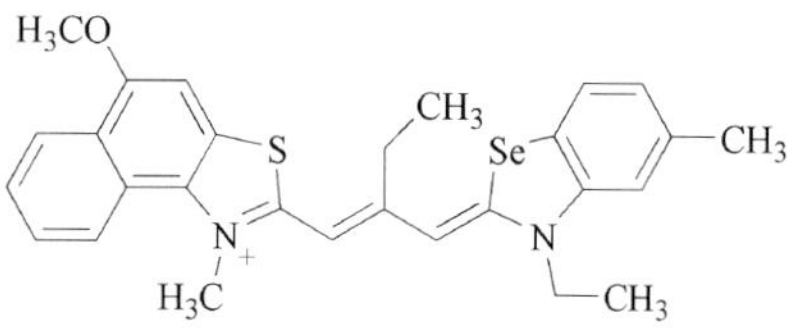 **18**

The use of cyanine dyes to sensitize silver halide grains to the near-infrared has been important for many years. This has particularly been true in recent years with the development of printers using light-emitting diodes (LEDs) or laser diodes (LDs) as the exposure devices. Early LEDs and LDs were most efficient in the red and near-infrared. Consequently, the photographic films for these devices had to be sensitized accordingly. For example, Fuji's Pictrography 3000 digital printer utilizes LDs emitting at 670, 750, and 810 nm to expose a false-sensitized color film. The Pictrography 3000 false-sensitized film uses the 670 nm red LD to control the amount of yellow dye in the final image, the 750 nm near-infrared LD to control the amount of magenta dye, and the 810 nm LD to control the amount of cyan dye [22]. In addition to these films manufactured for use with electronic printers, special photographic films have been manufactured for many years with sensitivity out to 1300 nm for recording images in infrared light [23]

Infrared sensitization presents special challenges. These dyes are subject to desensitization, and speeds are relatively low. They are relatively unstable and they do not tend to form J-aggregates. To improve the properties of dyes such as **19** and **20** the connecting chain is often included in one or more rings to improve stability and improve aggregation by making the chain more rigid and limiting the number of conformations [24–26].

5.7.2 Merocyanine Dyes

The merocyanines are also a valuable class of photographic sensitizers. Because of their nonionic structure they are more soluble in nonpolar solvents than the ionic cyanines. Merocyanines **21–23** can also sensitize from the blue to the near-infrared. They are most useful as blue and green sensitizing dyes.

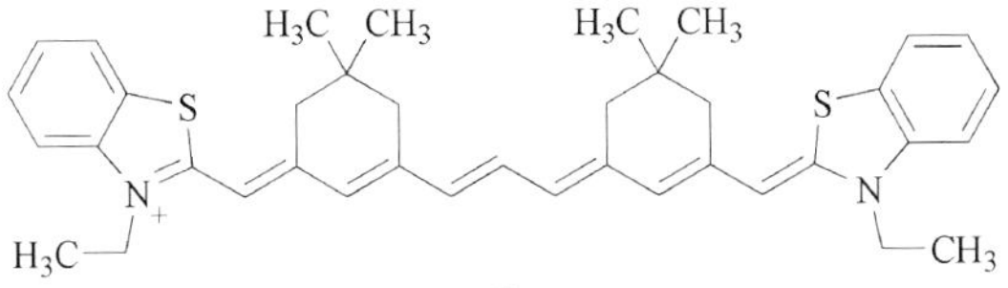

19

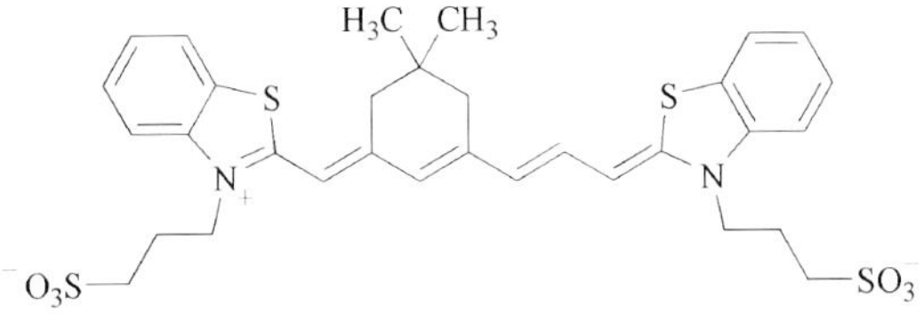

20

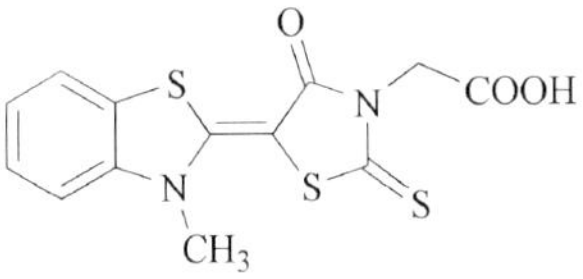

Blue sensitizer
21

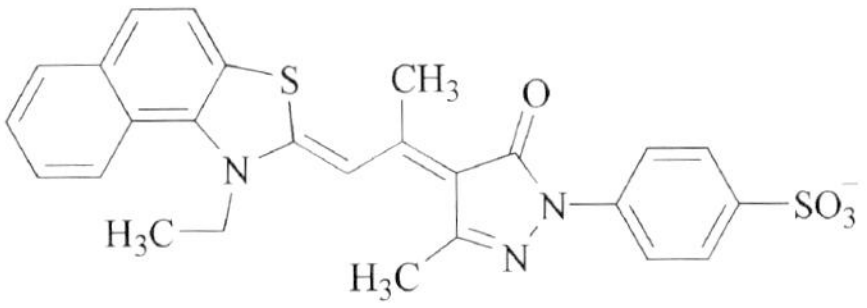

Green sensitizer
22

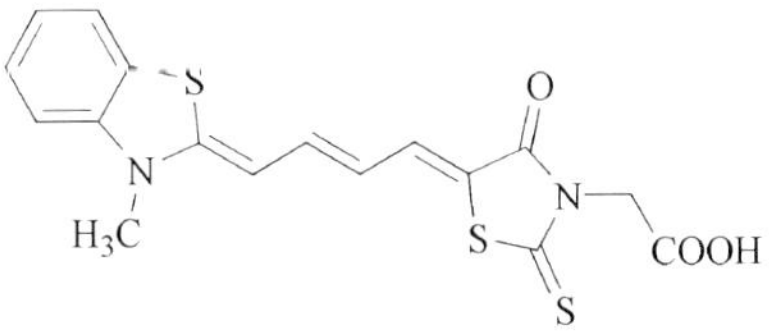

Red sensitizer
23

5.7.3 Oxonol Dyes

Oxonols are poor sensitizing dyes. However, they have found wide use as color filters and as antihalation dyes. Oxonol dyes are commonly used as microparticulate dispersions which are insoluble under neutral coating conditions. When processed with the alkaline photographic reagent, these filter layers dissolve and are removed from the final image. A combination of a monomethine oxonol dye **24** and the pentamethine oxonol dye **25** absorbs the entire range of the visible spectrum [27–29].

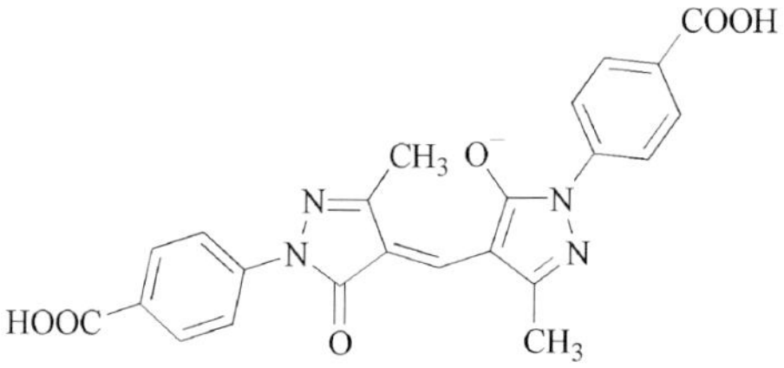

24

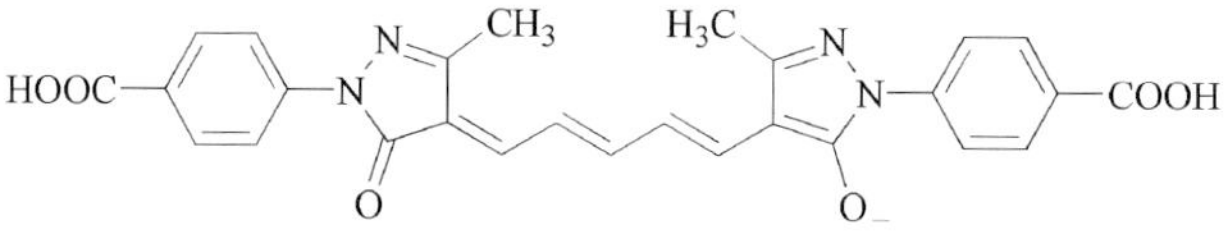

25

5.7.4 Azomethine and Indoaniline Image Dyes

5.7.4.1 Introduction

In 1912 Rudolph Fischer showed that the oxidation product of phenylenediamine, formed by reduction of exposed silver halide, reacts with nucleophilic compounds known as color couplers to give azomethine and indoaniline dyes [30]. This chromogenic reaction is the basis of conventional color photography. In color films and papers the blue-sensitive silver controls the formation of yellow dye, the green-sensitive silver controls the formation of magenta dye, and the red-sensitive silver controls the formation of cyan dye. The dye formed in each layer is proportional to the amount of light absorbed by each of the light-sensitive layers. Many thousands of kilograms of color developers and color couplers are used to produce the images in billions of rolls of color film and billions of color prints produced each year.

5.7.4.2 Color Developers

A limited set of phenylenediamine color developers are used. Kodak's CD-3, CD-4, and CD-6 (**26–28**) are the principle color developers used today [31]. They are directly incorporated into the alkaline processing fluid. The methyl group in the 2-position is important for preventing side reactions and enhancing the formation of the desired dye. The phenylenediamine undergoes a two-electron oxidation to quinonediimine, which then reacts with the color couplers to form the desired chromophores [32].

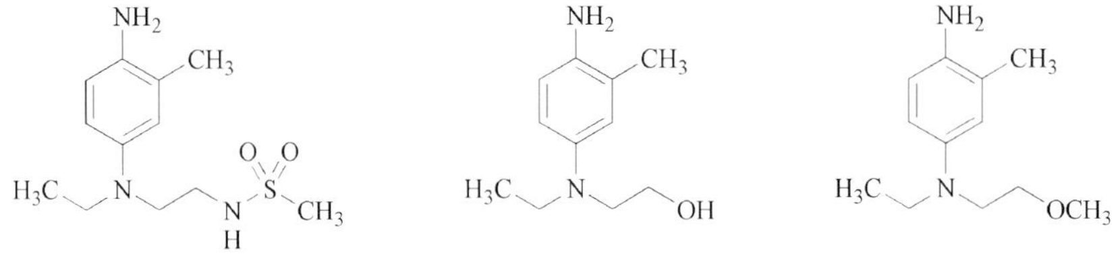

CD-3 CD-4 CD-5
26 **27** **28**

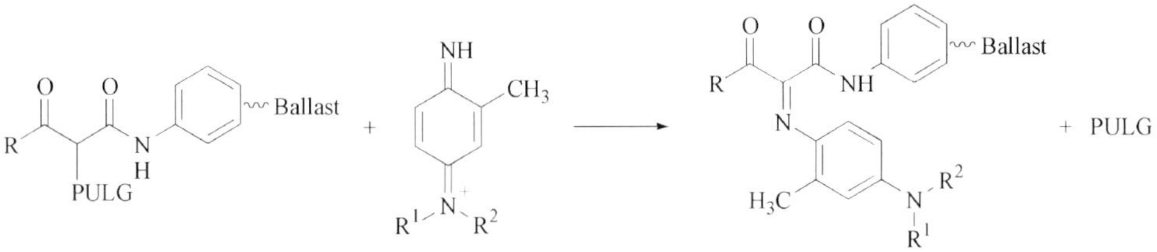

29 **31** **30**

R = *t*Bu, aryl
PULG = photographically useful leaving group

5.7.4.3 Yellow Azomethine Dyes

Simple 1,3-dicarbonyl compounds couple with quinonediimine to produce yellow chromophores. The most commonly used yellow color couplers are β-ketocarboxamides **29** [33–36]. Of the β-ketocarboxamides, the pivaloylcarboxanilides and the benzoylcarboxanilides are the most important. The pivaloylacetanilides are preferred for their improved light stability and low levels of undesired green absorption [37]. Color couplers usually contain a large hydrophobic group ("ballast") to prevent migration to adjacent layers and color contamination. This group can also inhibit aggregation and improve the chroma and light stability of the dye [38].

The coupler also includes a specially designed photographically useful leaving group (PULG) at the coupling position between the two carbonyl groups. This leaving group is important for several reasons. First, it results in the direct formation of the chromophore **30** from the coupling product with the quinonediimine **31**. Without the leaving group another two electron oxidation is required to form the chromophore from the colorless leuco dye. The leaving group is also used to perform other photographic functions. An example is the release of a heterocyclic development inhibitor such as phenyl mercaptotetrazole. This release of development inhibitors results in improved sharpness and improved grain. The released inhibitor can also migrate to adjacent layers and improve color reproduction. Although the color-forming reaction is quite simple and has been known since 1912, substantial molecular engineering of the ballast group and the PULG has occurred in the last 90 years to produce today's high-quality photographic products [39].

5.7.4.4 Magenta Azomethine Dyes

The most commonly used magenta couplers are 1-aryl-5-pyrazolones **32**, patented in 1934 [40]. The principles for their design and use are the same as that for yellow couplers. In recent years other classes of heterocyclic couplers have been developed. 1*H*-Pyrazolo[3,2-*c*]-1,2,4-triazole couplers **33** and the isomeric 1*H*-pyrazolo[1,5-*b*]-1,2,4-triazole couplers have become important because of their sharp absorptions in the green region and their ability to couple efficiently over a wide pH range [41–45].

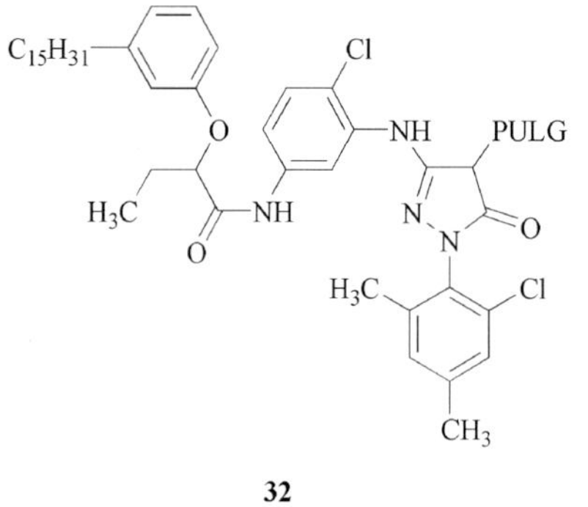

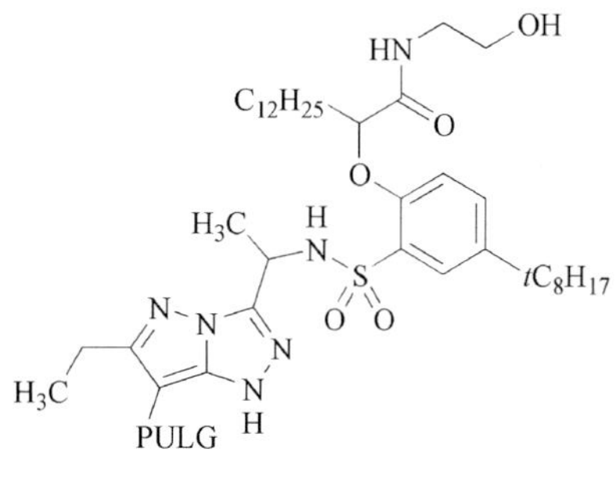

5.7.4.5 Cyan Indoaniline Dyes

The most common cyan couplers are phenols and naphthols which react with quinonediimine to give indoaniline dyes [46]. To obtain a good cyan chromophore, phenols must be highly substituted with electron-donating groups. Simple 1-naphthol-2-carboxamides **34** give excellent cyan dyes. However, the naphthol dyes tend to have less light and heat stability when compared to the best phenolic dyes [47]. Cyan dyes typically have maximum absorption at 685–700 nm for color negative film and from 630–670 nm for color print papers. For phenolic couplers

35 the design of the ballast side chain is particularly important for reducing the unwanted green and blue absorptions of the cyan chromophore. The selection of the acyl groups is important for adjusting the absorption maximum to the proper wavelength [48].

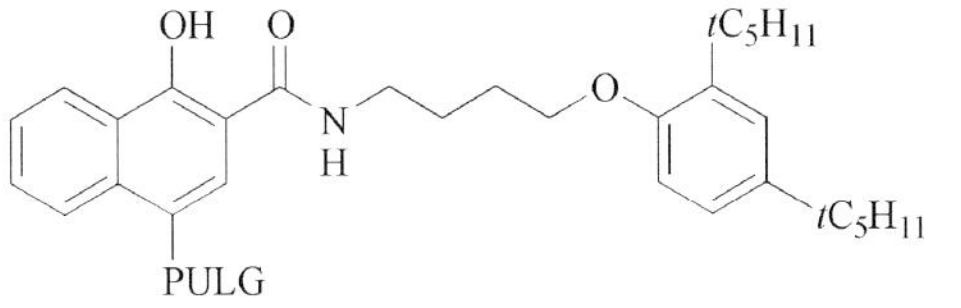

34

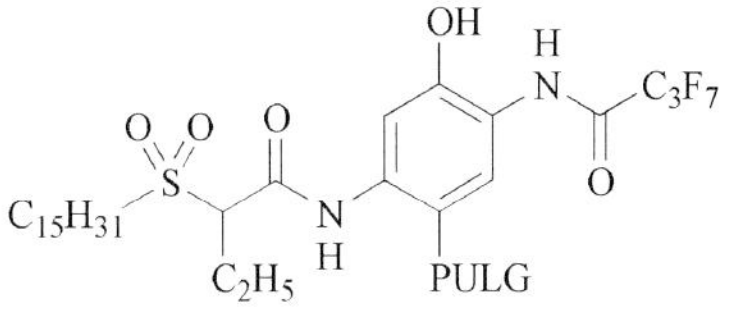

35

5.7.5 Azo Dyes

5.7.5.1 Diffusion-Transfer Imaging Systems

Azo dyes have found broad use as chromophores in diffusion-transfer imaging systems. Kodak's PR-10 and Fuji's instant films have exclusively used azo chromophores to generate images [49–52]. Polaroid's early instant films used yellow and magenta azo dyes, but these were replaced by more light stable metallized dyes [53]. Unlike conventional color photography where the chromophore is generated during processing with the phenylenediamine developer, the chromophores in instant negatives are fully formed and attached to a control moiety which mobilizes or immobilizes dye in proportion to the amount of silver development. The mobile dye then diffuses to an image-receiving layer to produce the instant color image. The chemistry of these control groups is of critical importance, but is beyond the scope of this article [54]. Pyrazolone yellow dyes **36** and naphthol-based magenta and cyan dyes **37** and **38** have been widely used in diffusion-transfer products [55].

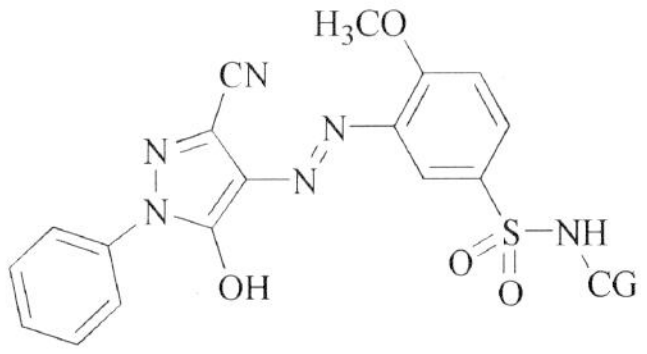

CG = control group

36

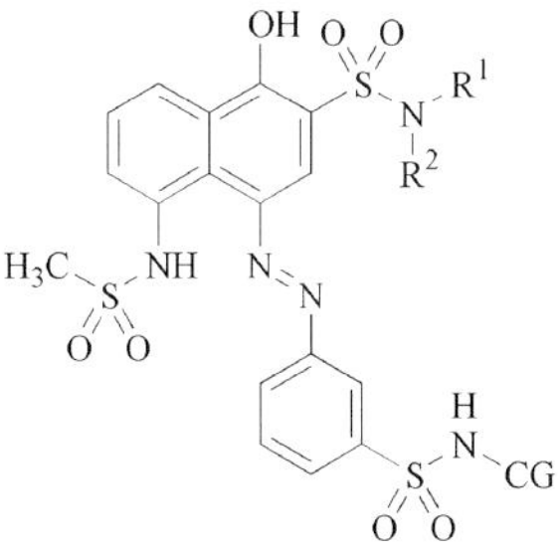

37

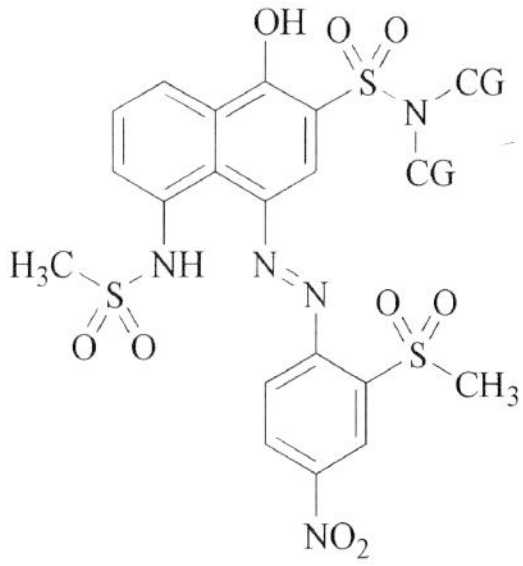

38

5.7.5.2 Silver Dye Bleach Processes

The archivally stable Cibachrome printing process is based on the use of developed silver metal in the negative image for the imagewise reduction of azo chromophores **39–41** to colorless anilines under highly acidic conditions. The anilines are then dissolved away to leave a positive image in dye. This process is also known as the silver dye bleach process [56, 57].

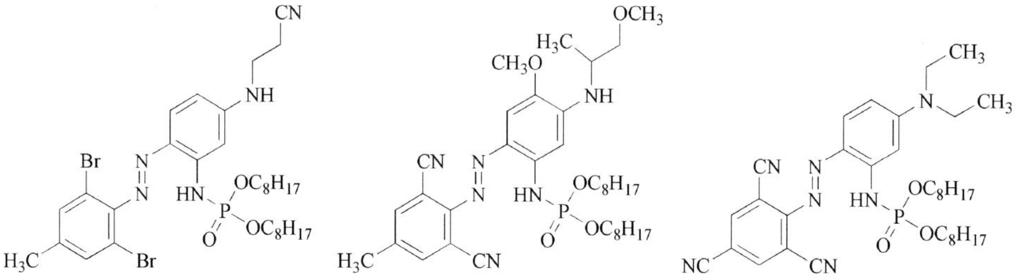

39   **40**   **41**

5.7.5.3 Color Masking

Another very sophisticated use of azo dyes in photography is for color masking [58]. Color masking with azo dyes gives color negative its distinctive orange color. Modern color negative is made with a small amount of magenta and cyan couplers which are themselves dyes. The coupler is carefully designed to be yellow in the case of the magenta coupler and red (a blue-green absorber) in the case of the cyan coupler. For example, during processing, the coupler **42** is destroyed in the formation of the magenta chromophore **43** in proportion to the amount of magenta dye produced [59]. This imagewise destruction of the coupler is accomplished by the loss of the azo group as the photographically useful leaving group. Ultimately, color masking compensates for the unwanted blue absorption of the magenta dye. Similarly, the unwanted blue and green absorptions of the cyan dye can be compensated for by a red coupler. The masking process leads to better color reproduction in the final image by affording brighter, cleaner colors.

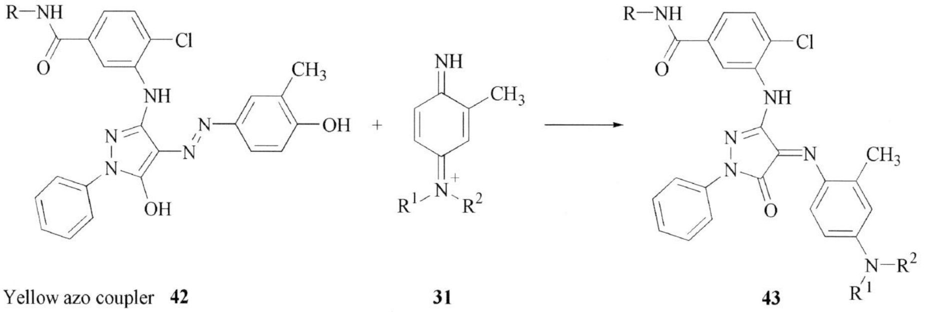

Yellow azo coupler **42**   **31**   **43**

5.7.6 Metallized Dyes

To improve the light stability of the original Polacolor film, Polaroid introduced a metallized set of image dyes in 1972. This set included a chromed azomethine yellow **44**, a chromed azo magenta **45**, and a highly solubilized derivative of copper phthalocyanine **46** [60].

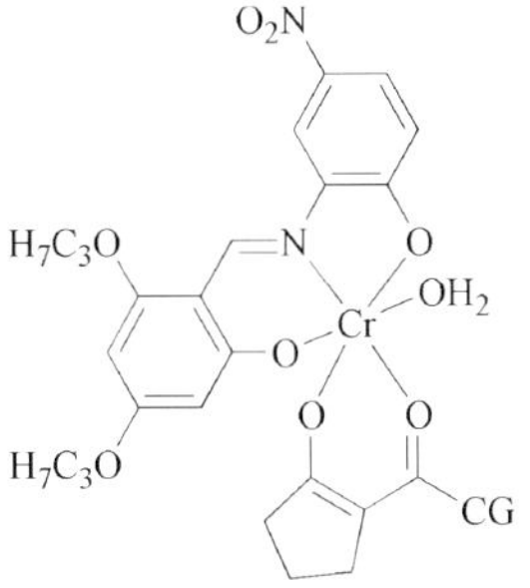

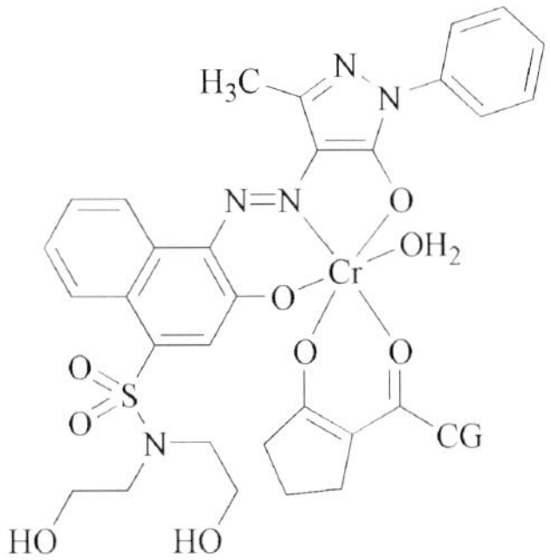

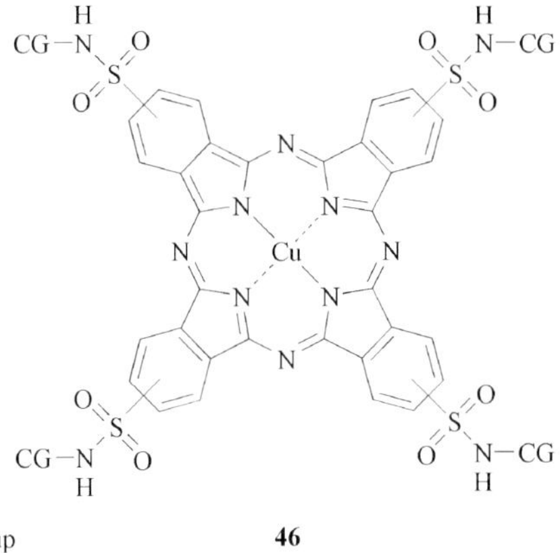

CG = control group

5.7.7 Xanthene Dyes

In the early 1980s Polaroid replaced the chromed azo magenta dye **45** with a xanthene magenta **47**, which has a far less blue absorbance and a much better chroma [61]. Polaroid has also pioneered the use of magenta xanthene **48** and cyan xanthene **49** as bleachable filter dyes which become colorless when treated with alkaline processing solutions. Changing the nitrogen auxochrome from an aniline to an indoline bathochromically shifts the xanthene from its usual magenta color to cyan [62–65].

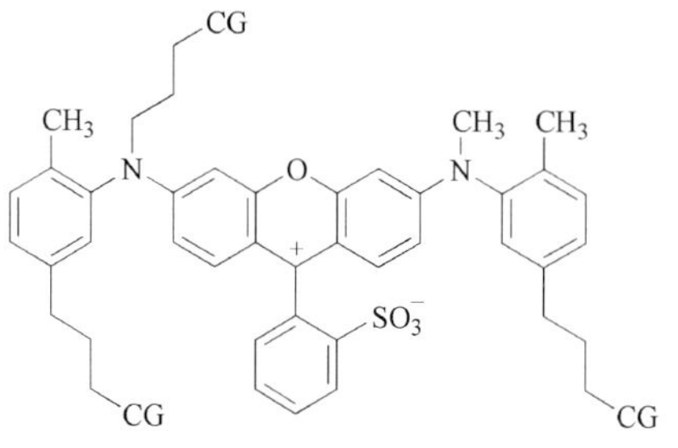

47

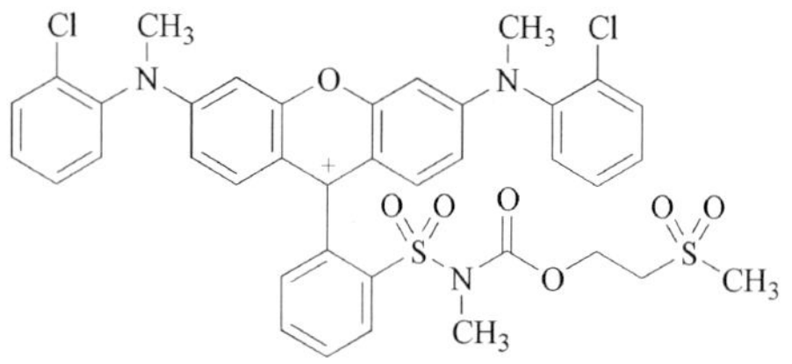

48

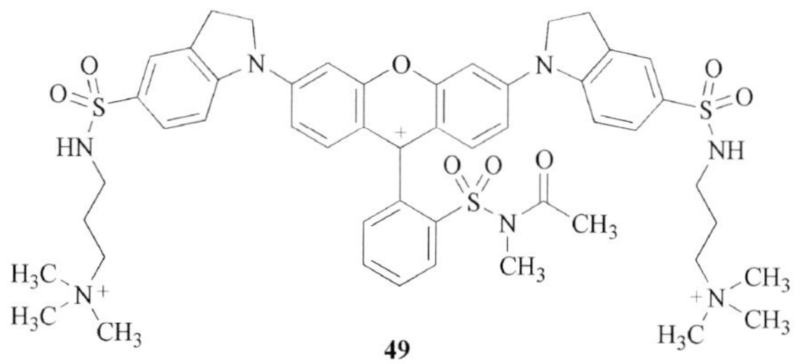

49

CG = control group

5.7.8 Triarylmethane Dyes

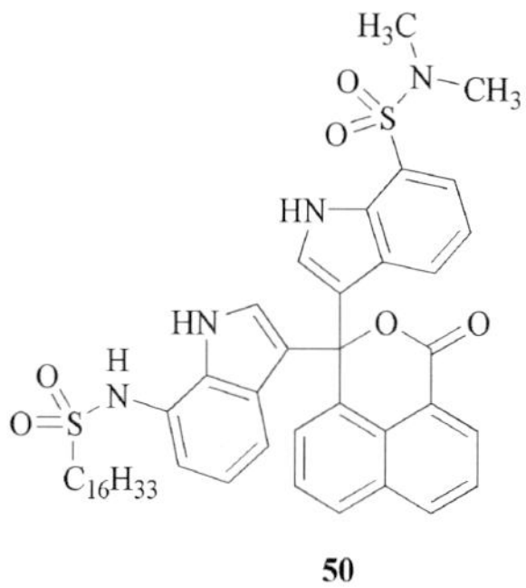

50

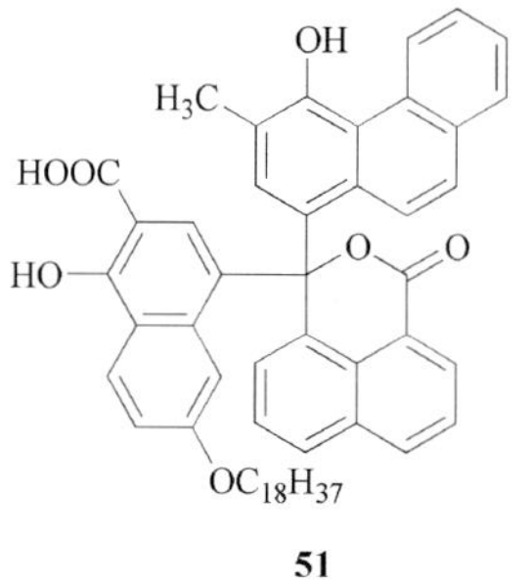

51

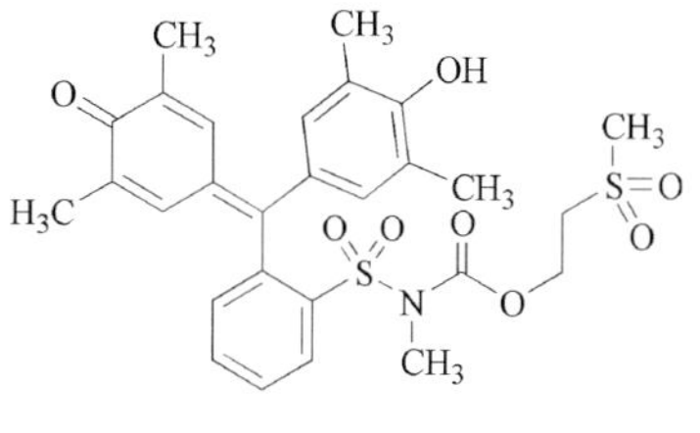

52

Polaroid's integral instant photography products would not exist but for the invention of highly alkaline stable, highly colored triarylmethane dyes. These dyes are incorporated into the alkaline reagent to protect the negative from ambient light during development after the film unit is ejected from the camera into the external environment. A short-wavelength opacifying dye **50** and a long-wavelength opacifying dye **51** are dissolved in a thick alkaline paste that includes the white, reflecting pigment titanium dioxide. The combination of the two dyes and the white pigment results in an optical density of approximately 6.0 which prevents ambient light from fogging the negative and destroying the photographic image. As the pH of the system is reduced by an acid neutralizing polymer the opacifying dyes decolorize, leaving a final dye image free of unwanted stain [66, 67]. Polaroid also commercialized the alkali-bleachable yellow triarylmethane filter dye **52** [68].

5.7.9 Anthraquinone Dyes

Anthraquinone dyes have not been widely used in photography. However, Polaroid's initial color film released in 1963 used the anthraquinone cyan **53**. This dye illustrates the dye-developer concept in which the control group for diffusion-transfer imaging is a pair of hydroquinone moieties. The branching of the side chain linking the control group to the chromophore is important for light stability [69].

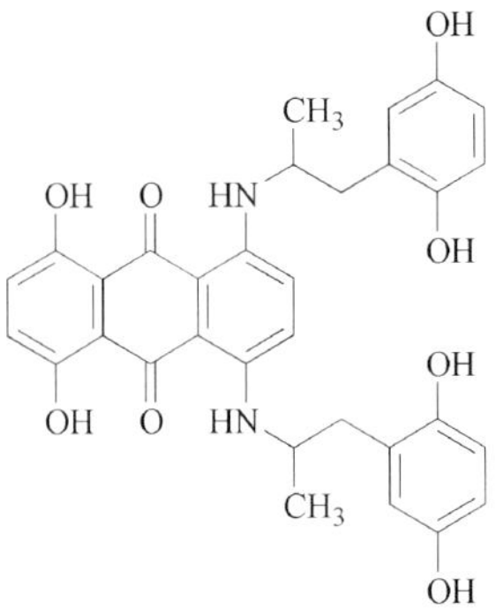

53

5.7.10 References

[1] H.W. Vogel, *Berichte* **6** (1873) 1302.
[2] P.B. Gilman., *Photochem. Photobiol.* **16** (1972) 221.
[3] C. B. Neblette (ed.), *Photography: It's Materials and Processes*, 6th ed., D. Van Nostrand Co., Princeton, NJ, 1962, pp. 196–204.
[4] W. West, *Photogr. Sci. Eng.* **18** (1974) no. 1, 35.
[5] F. M. Hamer, *The Chemistry of Heterocyclic Compounds*, Vol. 18, "The Cyanine Dyes and Related Compounds", A. Weissberger (ed.), Wiley Interscience, New York, 1964:, pp. 1–4.

[6] J. F. Thorpe, R. P. Linstead, *The Synthetic Dyestuffs*, Charles Griffin & Co. Ltd., London, 1933.
[7] C. E. K. Mees, T. H. James, *The Theory of the Photographic Process*, 3rd ed., The Macmillan Co., New York, 1966, pp. 198–232.
[8] J. M. Sturge (ed.), *Neblette's Handbook of Photography and Reprography*, 7th ed., Van Nostrand Reinhold Co., New York, 1977, pp. 73-112.
[9] T. H. James, *The Theory of the Photographic Process*, 4th ed., Macmillan Publishing Co., New York, 1977, pp. 194–234.
[10] R. K. Hailstone, J. F. Hamilton, *J. Imaging Sci.* **29** (1985) 125.
[11] R. K. Hailstone, *J. Phys. Chem.* **99** (1995) 4414.
[12] T. H. James, *The Theory of the Photographic Process*, 4th ed., Macmillan Publishing Co., New York, 1977, pp. 291–334.
[13] T. H. James, *The Theory of the Photographic Process*, 4th ed., Macmillan Publishing Co., New York, 1977, pp. 235–250.
[14] H. S. Freeman and A. T. Peters (ed.), *Colorants for Non-Textile Applications*, Elsevier, Amsterdam, 2000, pp. 109–124.
[15] T. H. James, *Advances in Photochemistry*, Interscience Publishers, New York, 1986, **13**, 388–395.
[16] T. H. James, *The Theory of the Photographic Process*, 4th ed., Macmillan Publishing Co., New York, 1977, pp. 88–104.
[17] US 3 672 898, 1972 (J. A. Schwan).
[18] US 6 140 035, 2000 (K. J. Klingman, B. E. Kahn, R. L. Parton, T. R. Dobles, D. A. Stegman, T. A. Smith, J. D. Lewis).
[19] F. M. Hamer, *The Chemistry of Heterocyclic Compounds*, Vol. 18, "The Cyanine Dyes and Related Compounds", A. Weissberger (ed.), Interscience, New York, 1964.
[20] G. E. Ficken, *The Chemistry of Synthetic Dyes*, Vol. 4, K. Venkataraman (ed.), Academic Press, New York, 1964.
[21] US 4 439 520, 1984 (H. S. Wilgus, C. G. Jones, J. T. Kofron, R. E. Booms, J. A. Haefner, F. J. Evans).
[22] T. Yokokawa, T. Inagaki and Y. Aotsuka, *Int. Cong. Adv. Non-Impact Print Technology/Jpn.*, Hardcopy 93, The Society for Imaging Science and Technology, Springfield, VA, 1993, pp. 493–496.
[23] US 3 690 891, 1972 (J. Spence, P. B. Gilman, C. G. Ulbing).
[24] T. H. James, *The Theory of the Photographic Process*, 4th ed., Macmillan Publishing Co., New York, 1977, pp. 196–197.
[25] US 5 807 666, 1998 (A. Adin, A. T. Wyand).
[26] US 6 245 499, 2001 (K. Suzuki, T. Kubo, Y. Inagaki, T. Arai).
[27] DE 2 700 651, 1977 (R. G. Lemahieu, H. Depoorter, J. Willy).
[28] US 5 326 687, 1994 (J. Texter).
[29] US 5 922 523, 1999 (M. J. Helber, W. J. Harrison, E. A. Gallo).
[30] DE 253 335, 1912 (R. Fischer).
[31] G. Haist, *Modern Photographic Processing*, Vol. 2, Wiley, New York, 1979, pp. 558–560.
[32] T. H. James, *The Theory of the Photographic Process*, 4th ed., Macmillan Publishing Co., New York, 1977, pp. 339–353.
[33] GB 848 558, 1960 (B. Tavernier, M. Desmit, A. DeCat).
[34] GB 808 276, 1959 (A. DeCat, R. VanPouke).
[35] US 2 875 057, 1959 (F. McCrossen, P. Vittum, A. Weissberger).
[36] US 3 265 506, 1966 (A. Weissberger, C. Kibler).

5.7.10 References

[37] J. Bailey, L. Williams, *The Chemistry of Synthetic Dyes*, Vol. IV, K. Venkataraman (ed.), Academic Press, New York, 1971, p. 362.
[38] S. Krishnamurthy, J. W. Harder, R. F. Romanet, 2000 International Symposium on Silver Halide Imaging-Silver Halide in a New Millenium, Final Program and Proceedings, The Society for Imaging Science and Technology, Springfield, VA, 2000, pp 260–262.
[39] H. S. Freeman, A. T. Peters (eds.), *Colorants for Non-Textile Applications*, Elsevier, Amsterdam, 2000, pp. 81–83.
[40] US 1 969 479, 1934 (M. Seymour).
[41] US 3 705 896, 1972 (J. Bailey).
[42] US 4 540 654, 1985 (T. Sato, T. Kawagishi, N. Furutachi).
[43] US 5 656 418, 1997 (T. Nakamine, M. Motoki, T. Kawagishi, N. Matsuda).
[44] US 5 609 996, 1997 (P. Tang, S. Cowan, D. Decker, T. Mungal).
[45] US 5 698 386, 1997 (P. Tang, D. Decker, S. Fischer, S. Cowan).
[46] T. H. James, *The Theory of the Photographic Process*, 4th ed., Macmillan Publishing Co., New York, 1977, pp.359–361.
[47] H. S. Freeman, A. T. Peters (eds.), *Colorants for Non-Textile Applications*, Elsevier, Amsterdam, 2000, pp. 78–79.
[48] US 5 681 690, 1997 (P. Tang, T. Jaozefiak).
[49] US 4 013 633, 1977 (J. Haase, C. Eldredge, R. Landholm).
[50] US 3 932 381, 1977 (J. Haase, R. Landholm, J. Krutak).
[51] US 3 942 987, 1976 (R. Landholm, J. Haase, J. Krutak).
[52] S. Fujita, *J. Synth. Org. Chem.* **40** (1982) 176–187.
[53] S. M. Bloom, M. Green, M. Idelson, M.S. Simon, *The Chemistry of Synthetic Dyes*, Vol. VIII, K. Venkataraman (ed.), Academic Press, New York, 1978, pp. 331–387.
[54] C. C. Van de Sande, *Angew. Chem. Int. Ed. Engl.* **22** (1983) 191.
[55] H. S. Freeman, A. T. Peters (eds.), *Colorants for Non-Textile Applications*, Elsevier, Amsterdam, 2000, pp. 96–104.
[56] T. H. James, *The Theory of the Photographic Process*, 4th ed., Macmillan Publishing Co., New York, 1977, pp.363-366.
[57] US 4 456 668, 1984 (J. Lenoir, G. Jan).
[58] W. Hanson, P. Vittum, *J. Photogr. Soc. Am.* **13** (1947) 94.
[59] US 4 163 670, 1979 (K. Shiba, T. Hirose, A. Arai, A. Okumura, Y. Yokota).
[60] S. M. Bloom, M. Green, M. Idelson, M.S. Simon, *The Chemistry of Synthetic Dyes*, Vol. VIII, K. Venkataraman (ed.), Academic Press, New York, 1978, pp. 371–380.
[61] US 4 264 701, 1981 (L. Locatell, H. Rogers, R. Bilofsky, R. Cieciuch, C. Zepp).
[62] US 4 304 833, 1981 (J. Foley).
[63] US 4 304 834, 1981 (R. Cournoyer, J. Foley).
[64] US 4 258 118, 1981 (J. Foley, L. Locatell, C. Zepp).
[65] US 5 187 282, 1993 (P. Carlier, M. Filosa, M. Lockshin).
[66] H. S. Freeman, A. T. Peters (eds.), *Colorants for Non-Textile Applications*, Elsevier, Amsterdam, 2000, pp. 105–107.
[67] J. M. Sturge (ed.), *Neblette's Handbook of Photography and Reprography*, 7th ed., Van Nostrand Reinhold Co., New York, 1977, pp. 268–269.
[68] US 4 283 538, 1981 (J. Foley).
[69] US 3 209 016, 1964 (E. Blout, M. Cohler, M. Green, M. Simon, R. B. Woodward).

5.8 Indicator Dyes

5.8.1 Introduction

Much work has been spent during the last 150 years to continously improve the stability of dyes. In nature, however, not only color-stable dyes (e.g., indigo) exist but also dyes which change their color. As early as 1660, Boyle observed that certain plant extracts changed their color on treatment with acid or base. This effect is well known to gardeners in the case of hydrangea (or hortensia): the color of its blossoms can be changed by watering it with slightly acidic or alkaline water.

The first indicator dye which came into use, was *litmus*, which was isolated from the orseille lichen, *Roccella tinctoria*. It changes its color from red (acid) to blue (alkaline).

The systematic investigation of natural colors led to the discovery of several dyes and dye classes which are able to change their color. Thus, it was found that the origin of the color of hydrangea blossoms is an *anthocyan* dye.

Interestingly, in nature these dyes are not always and necessarily used as indicators: the red color of roses and the blue color of the blue cornflower are based on the same molecule (also an anthocyan), which in vitro reacts as an indicator dye. In the plants however, this dye does not change color: the red rose does not change to blue due to the presence of an intrinsic buffer system.

With the detection and introduction of titration (more general: volumetry) into the arsenal of analytical methods, an increasing need arose for means which allowed the end-point of such an analysis to be detected. This need stimulated the search for indicator dyes, and consequently the methods of dye synthesis were also used for this purpose. As a result, most indicator dyes are now synthetic.

5.8.2 General Principles

In general terms, an indicator dye can be characterized as follows: (1) It shows different optical properties when the system of which it forms a part changes its status, and (2) This change is reversible.

The first property means that the indicator molecule reacts as a part of the system. When the number of $[H_3O]^+$ ions in this system is in a large excess compared to the number of indicator molecules, the influence of indicator molecules can be neglected. When, however, the numbers of $[H_3O]^+$ ions and indicator molecules become comparable, e.g., in very weakly buffered systems like raindrops, the influence of the indicator molecules is significant. This phenomenon is known as indicator error.

The second property in particular distinguishes indicator dyes from color-forming reagents. The latter are widely used in reactions in which colors are formed by chemical or biochemical (mostly enzymatic) reactions between the analyte and added reagents. These reactions are widely used in analytical and diagnostic test kits but are normally not reversible.

The different optical properties mentioned above are clearly linked to structural properties: a change in optical properties also indicates a change of molecular structure.

The most important structure changes for indicator dyes are (1) Reversible transition between acid and base form of a molecule (pH indicator), (2) Reversible transition between reduced and oxidized forms of a molecule (redox indicator), and (3) Reversible transition between the free molecule and its complex with a cation (metal indicator) [1].

5.8.3 Classes of Indicators

5.8.3.1 pH Indicators

In common laboratory use, the term *indicator* is often employed to refer to pH indicators.

A pH indicator molecule exhibits an equilibrium between the indicator acid In_A and the indicator base In_B [Eq. (1)].

$$In_A \rightleftharpoons In_B + H^+ \qquad (1)$$

To be suitable as an indicator molecule, the absorption spectra of In_A and In_B must be different.

The theoretical derivation of equations for equilibrium and pH is presented in [1]. Here, some more general observations shall be discussed. In the transition from the acidic to the alkaline form (and vice versa), a point occurs where In_A and In_B are present in comparable numbers. The consequence is a mixed color at this point. Important for the practical application is that the complete transition requires a certain pH range. Therefore a pH indicator does not change its color at a sharp pH value but within a range.

Table 5.17 lists ranges and colors of common pH indicators, and Figure 5.1 shows a graphical overview of the pH ranges and color changes.

Table 5.17 Commonly used pH indicators

Transition range, pH	Indicator	CAS no.	Color change	Composition of indicator solution, %
0.0–2.0	Malachite Green (C.I. 42 000)	[2437-29-8]	yellow–green-blue	0.1 (in water)
0.0–2.6	Brilliant Green (C.I. 42 040)	[633-03-4]	yellow–green	0.1 (in water)
0.1–2.3	Methyl Green (C.I. 42 590)	[82-94-0]	yellow–blue	0.1 (in water)
0.2–1.0	Picric acid (C.I. 10 305)	[88-89-1]	colorless–yellow	0.1 (in 70 % ethanol)
0.5–2.5	Cresol Red	[1733-12-6]	red–yellow	0.1 (in 20 % ethanol)
0.8–2.6	Crystal Violet (C.I. 42 555)	[548-62-9]	yellow–blue-violet	0.1 (in 70 % ethanol)
1.2–2.3	Metanil Yellow (C.I. 13 065)	[587-98-4]	red–yellow	0.1 (in 20 % ethanol)
1.2–2.8	m-Cresol Purple	[2303-01-7]	red–yellow	0.04 (in 20 % ethanol)
1.2–2.8	Thymol Blue	[76-61-9]	red–yellow	0.1 (in 96 % ethanol)
1.2–2.8	p-Xylenol Blue	[125-31-5]	red–yellow	0.1 (in 50 % ethanol)
1.2–2.8	Thymol Blue sodium salt	[62625-21-2]	red–yellow	0.05 (in water)
1.4–2.2	Quinaldine Red	[117-92-0]	colorless–pink	0.1 (in 60 % ethanol)
1.4–2.6	Tropaeolin OO	[3012-37-1]	red–yellow	0.1 (in water)
2.0–4.4	2,6-Dinitrophenol	[573-56-8]	colorless–yellow	0.1 (in 70 % ethanol)
2.2–3.2	Phloxine B (C.I. 45 410)	[18472-87-2]	colorless–purple	0.1 (in water)
2.8–4.0	2,4-Dinitrophenol	[51-28-5]	colorless–yellow	0.1 (in 70 % ethanol)
2.9–4.0	4-Dimethylamino-azobenzene (Methyl Yellow, C.I. 11 020)	[60-11-7]	red–orange-yellow	0.1 – 0.5 (in 90 % ethanol)
3.0–4.5	Bromochlorophenol Blue	[2553-71-1]	yellow–blue-violet	0.1 (in 20 % ethanol)
3.0–4.6	Bromophenol Blue	[115-39-9]	yellow–blue-violet	0.1 (in 20 % ethanol)
3.0–4.6	Bromophenol Blue sodium salt	[34725-61-6]	green-yellow–blue-violet	0.05 (in water)
3.0–5.2	Congo Red	[573-58-0]	blue–orange-red	0.2 (in water)
3.1–4.4	Methyl Orange (C.I. 13 025)	[547-58-0]	red–yellow-orange	0.04 (in 20 % ethanol)
3.5–5.8	2,5-Dinitrophenol	[329-71-5]	colorless–yellow	0.1 (in 70 % ethanol)
3.7–5.0	1-Naphthyl Red	[131-22-6]	red–yellow	0.1 (in 70 % ethanol)

5.8.3 Classes of Indicators

Transition range, pH	Indicator	CAS no.	Color change	Composition of indicator solution, %
3.8–5.4	Bromocresol Green	[76-60-8]	yellow–blue	0.1 (in 20 % ethanol)
3.8–5.4	Bromocresol Green sodium salt	[62625-32-5]	yellow–blue	0.05 (in water)
4.3–6.3	Alizarin (Red) S (C.I. 58 005)	[130-22-3]	yellow–pink	0.1 (in water)
4.4–6.2	Methyl Red (C.I. 13 020)	[493-52-7]	red–yellow–orange	0.1 (in 96 % ethanol)
4.5–6.2	Methyl Red sodium salt	[845-10-3]	red–yellow–orange	0.05 (in water)
4.7–6.3	Bromophenol Red	[2800-80-8]	yellow–purple	0.1 (in 20 % ethanol)
4.8–6.4	Chlorophenol Red	[4430-20-0]	yellow–purple	0.1 (in 20 % ethanol)
5.0–7.2	Hematoxylin (C.I. 75 290)	[517-28-2]	yellow–violet	0.05 (in 96 % ethanol)
5.0–8.0	Litmus (C.I. 1242)	[573-56-8]	red–blue	4 (in water)
5.2–6.8	Bromocresol Purple	[115-40-2]	yellow–purple	0.1 (in 20 % ethanol)
5.4–7.5	4-Nitrophenol	[100-02-7]	colorless–yellow	0.2 (in 96 % ethanol)
5.7–7.5	Bromoxylenol Blue	[40070-59-5]	yellow–blue	0.1 (in 96 % ethanol)
5.8–7.2	Alizarin (C.I. 58 000)	[72-48-0]	pale yellow–violet-red	0.5 (in 96 % ethanol)
5.8–7.6	Bromothymol Blue	[76-59-5]	yellow–blue	0.1 (in 20 % ethanol)
5.8–7.6	Bromothymol Blue sodium salt	[34722-90-2]	yellow–blue	0.05 (in water)
6.0–7.0	Nitrazine Yellow (C.I. 14 890)	[5423-07-4]	brownish yellow–blue-violet	0.05 (in 70 % ethanol)
6.4–8.2	Phenol Red	[143-74-8]	yellow–red-violet	0.1 (in 20 % ethanol)
6.5–8.0	Phenol Red sodium salt	[34487-61-1]	yellow–red-violet	0.05 (in water)
6.5–8.5	Cresol Red	[1733-12-6]	orange–purple	0.1 (in 20 % ethanol)
6.6–8.6	3-Nitrophenol	[554-84-7]	colorless–yellow	0.3 (in 96 % ethanol)
6.8–8.0	Neutral Red	[553-24-2]	blue-red–orange-yellow	0.1 (in 70 % ethanol)
7.1–8.3	1-Naphtholphthalein	[1301-55-9]	pink-brown–blue	0.1 in (96 % ethanol)
7.4–9.0	m-Cresol Purple	[2303-01-7]	yellow–purple	0.04 (in 20 % ethanol)
7.8–9.5	Thymol Blue sodium salt	[62625-21-1]	green–blue	0.05 (in water)
8.0–9.6	Thymol Blue	[76-61-9]	yellow–blue	0.1 (in 96 % ethanol)
8.0–9.6	p-Xylenol Blue	[125-31-5]	yellow–blue	0.1 (in 50 % ethanol)

5.8 Indicator Dyes

Transition range, pH	Indicator	CAS no.	Color change	Composition of indicator solution, %
8.2–9.5	o-Cresolphthalein	[596-27-0]	colorless–red-violet	0.02 (in 50 % ethanol)
8.2–9.8	Phenolphthalein	[81-90-3]	colorless–pink	0.1 (in 96 % ethanol)
9.3–10.5	Thymolphthalein	[125-20-2]	colorless–blue	0.1 (in 50 % ethanol)
9.4–12.0	Alizarin (Red) S (C.I. 58 005)	[130-22-3]	brown-orange–violet	0.1 (in water)
10.2–11.8	Alizarin (C.I. 58 000)	[72-48-0]	brownish red–violet	0.5 (in 96 % ethanol)
10.2–12.1	Alizarin Yellow GG (C.I. 14 025)	[584-42-9]	pale yellow–brownish yellow	0.1 (in water)
11.5–13.0	Alkali Blue (C.I. 42 765)	[568-02-5]	blue–pink-violet	0.1 (in 90 % ethanol)
11.6–13.0	Epsilon Blue	[73904-21-9]	orange–violet	0.1 (in water)
11.7–13.2	Malachite Green (C.I. 42 000)	[569-64-2]	blue–colorless	0.1 (in water)
11.7–14.0	Indigo Carmine (C.I. 73 015)	[860-22-0]	blue–yellow	0.25 (in 50 % ethanol)
12.0–14.0	Acid Fuchsin (C.I. 42 685)	[3244-88-0]	red–colorless	0.1 (in water)

5.8.3 Classes of Indicators 531

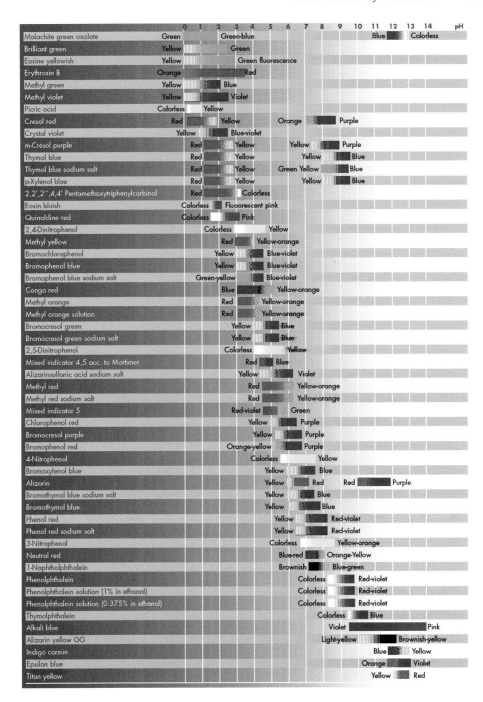

Figure 5.1 Color changes of commonly used pH indicators (courtesy of Merck). The pH ranges and color shades shown are approximations.

5.8 Indicator Dyes

Individual pH Indicators. The most commonly used pH indicators include azo dyes, nitrophenols, phthaleins, and sulfophthaleins. Varying the substituents on the chromophores yields pH indicators with different transition ranges, and the complete pH range between 0 and 14 can thus be covered, as shown in Table 5.17 and Figure 5.2. Introducing acid groups (sulfo or carboxyl) leads to water-soluble indicator dyes.

pH indicators are generally used as 0.1 % solutions in water, ethanol or water/ethanol mixtures. The special case of pH papers is described in Section 5.8.4; for pH indicators for nonaqueous systems, see below.

Structures of some of the most common pH indicators are shown in the following:

Methyl Orange (**1**)

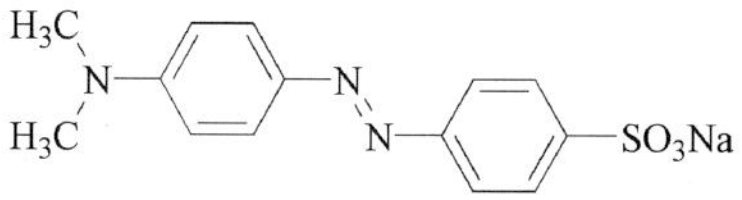

Methyl Red (**2**)

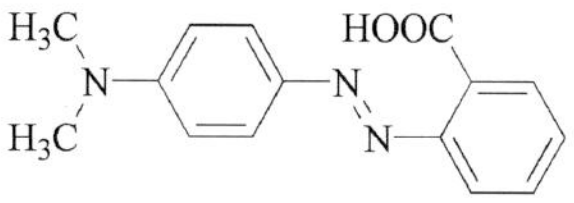

Phenolphthalein (**3**)

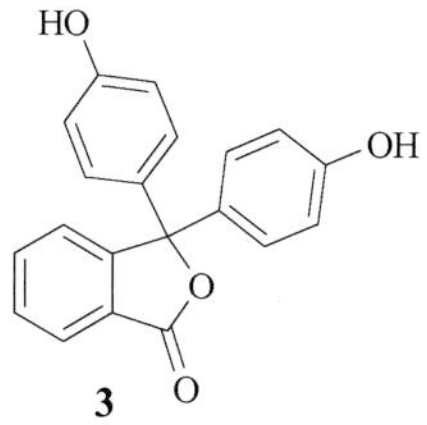

Phenol Red (**4**)

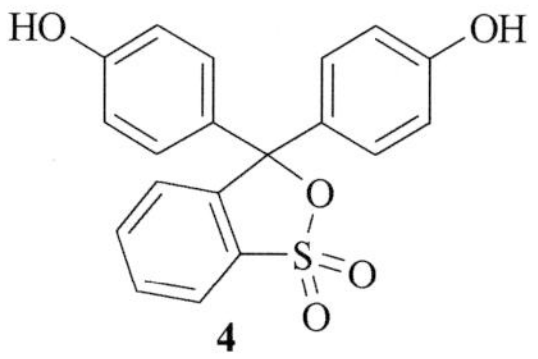

5.8.3 Classes of Indicators

Bromocresol Green (**5**)

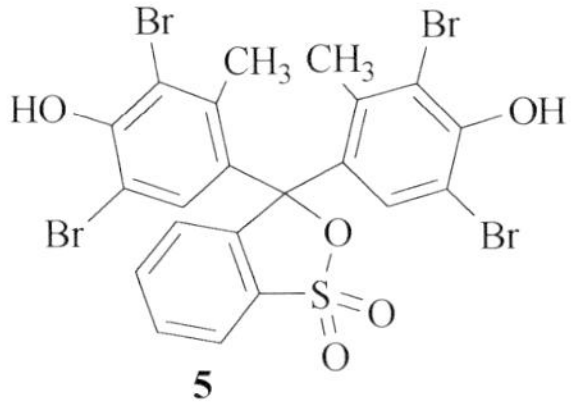

[Structure of compound 5]

Bromothymol Blue (**6**)

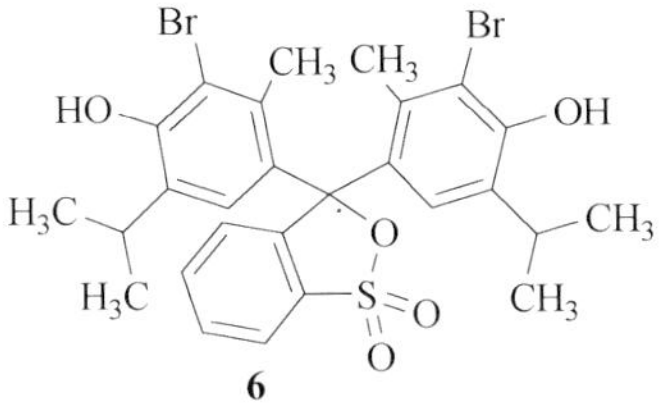

[Structure of compound 6]

Indicator Mixtures (Two-Component). An important criterion for the selection of an indicator is an transition interval which is clear, visually well detectable, and as sharp as possible. Ideally, the acidic and the alkaline forms should possess complementary colors, for example, red and green. Unfortunately, no individual indicator fulfills all these requirements. Furthermore, the mixture of colors in the transition range often deteriorates the quality of color change.

This unsatisfying situation can be improved in two ways. First, a pH-neutral dye can be added to the indicator molecule. This leads to a *screened indicator* (see Table 5.18).

Second, indicators with similar transition ranges can be combined to form a *mixed indicator* (Table 5.19).

Table 5.18 Screened pH indicators

Components of screened indicator mixture	Parts by volume	Transition point, pH	Color change Acid	Alkali
Dimethylaminoazobenzene (0.1 % in ethanol)	1	3.3	blue-violet	green
Methylene Blue (0.1 % in ethanol)	1			
Methyl Orange (0.1 % in water)	1	4.1	violet	green
Indigo Carmine (0.25 % in water)	1			

Components of screened indicator mixture	Parts	Transition	Color change	
Methyl Orange (0.2 % in water)	1	4.5–4.8	red	green fluorescence
Fluorescein (0.2 % in ethanol)	1			
Methyl Red (0.2 % in ethanol)	1	5.4	red-violet	green
Methylene Blue (0.1 % in ethanol)	1			
Neutral Red (0.1 % in ethanol)	1	7.0	violet-blue	green
Methylene Blue (0.1 % in ethanol)	1			
Phenol Red (0.1 % in ethanol)	1	7.3	green	violet
Methylene Blue (0.02 % in water)	1			
Phenolphthalein (0.1 % in ethanol)	1	8.9	green	violet
Methyl Green (0.1 % in ethanol)	2			
Phenolphthalein (0.1 % in ethanol)	1	10.0	blue	red
Nile Blue (0.2 % in ethanol)	2			
Nile Blue (0.2 % in water)	2	10.8	green	red-brown
Alizarin Yellow (0.1 % in ethanol)	1			

Table 5.19 Mixed pH indicators

Components of mixed indicator	Parts by volume	Transition point, pH	Color change	
			Acid	Alkali
Bromocresol Green (0.1 % in ethanol)	1	4.3	orange	green
Methyl Orange (0.02 % in water)	1			
Bromocresol Green (0.1 % in ethanol)	3	5.1	wine red	green
Methyl Red (0.2 % in ethanol)	1			
Bromocresol Green-Na (0.1 % in water)	1	6.1	yellow-green	blue-violet

5.8.3 Classes of Indicators

Components of mixed indicator	Parts	Transition	Color change	
Chlorphenol Red-Na (0.1 % in water)	1			
Bromocresol Purple-Na (0.1 % in water)	1	6.7	yellow	violet-blue
Bromothymol Blue-Na (0.1 % in water)	1			
Bromothymol Blue-Na (0.1 % in water)	1	7.5	yellow	violet
Phenol Red-Na (0.1 % in water)	1			
Cresol Red-Na (0.1 % in water)	1	8.3	yellow	violet
Thymol Blue-Na (0.1 % in water)	3			
Thymol Blue (0.1 % in 50 % ethanol)	1	9.0	yellow	violet
Phenolphthalein (0.1 % in 50 % ethanol)	3			
Phenolphthalein (0.1 % in 50 % ethanol)	2	9.6	slightly pink	violet
1-Naphtholphthalein (0.1 % in 50 % ethanol)	1			
Phenolphthalein (0.1 % in ethanol)	1	9.9	colorless	violet
Thymolphthalein (0.1 % in ethanol)	1			
Thymolphthalein (0.1 % in ethanol)	2	10.2	yellow	violet
Alizarin Yellow (0.1 % in ethanol)	1			

The difference in color change can clearly be seen by comparing the changes described in Tables 5.18 and 5.18 with those of Table 5.17: the changes of the mixtures are much more pronounced than those of the individual indicators.

Indicator Mixtures (Multicomponent). Mixing several indicators of different transition ranges leads to *universal indicators* which can be used over a wide pH range. Table 5.20 lists examples of such indicators.

Table 5.20 Some pH indicators with extended measuring ranges

Name	Measuring range, pH	Color change	Number of colors on evaluation scale	Accuracy
Indicator fluid, pH 0 – 5, Merck	0 – 5	orange red, ocher, lime green, light green, dark green, brown green, blue	11	± 0.25 pH units
Universal indicator fluid, pH 4 – 10, Merck	4 – 10	pink, yellow, yellow green, gray green, gray, bluish gray, gray violet, violet	13	± 0.25 pH units
Indicator fluid, pH 9 – 13, Merck	9 – 13	yellow green, lime green, gray green, gray violet, violet	5	± 0.5 pH units

pH Indicators for Nonaqueous Solutions. For the titration of weak acids and bases in a nonaqueous environment, special titrants are used, as well as special indicators. Some of them are listed in Table 5.21. As the pH ranges of these indicators differ from those which are found in an aqueous medium, they are not listed here. Indicator changes in nonaqueous media must be controlled and verified by potentiometric titration before routine use. Therefore, the use of indicator dyes in nonaqueous titration has been significantly declined during the last decade.

Table 5.21 Commonly used pH indicators for titrations in nonaqueous solvents

Substance to be titrated	Indicator	CAS no.	Color change
Weak bases			
In glacial acetic acid with perchloric acid	Crystal Violet	[548-62-9]	violet–blue-green–green–yellow
	Methyl Violet	[8004-87-3]	violet–blue–blue-green–yellow
	Malachite Green	[2437-29-8]	brown-green–green–yellow
	Nile Blue A	[3625-57-8]	blue–colorless
In acetic anhydride with perchloric acid	Crystal Violet	[548-62-9]	blue–green-yellow–yellow
	a-naphtholbenzein	[6948-88-5]	yellow–green
Weak acids			
In dimethylformamide			
With tetraalkylammoniumhydroxide	Thymol Blue	[76-61-9]	red–yellow–blue
With n-butylamine	Azo Violet	[74-39-5]	red–blue
With pyridine	2-nitroaniline	[88-74-4]	yellow–red

5.8.3.2 Redox Indicators

Redox indicators are organic molecules whose oxidized and reduced forms show different colors. The two levels of oxidation form a redox system which can be described by the standard potential or rH value. A short definition of the two terms is given in [1]. Here, some more practical properties are mentioned:

- The redox potential of an indicator depends strongly on pH because hydrogen ions are directly involved in the redox process, as the following example of Variamine Blue (**7**; C.I. 37255) shows.
- The color of redox indicators changes over a certain range. In addition, redox reactions often proceed slowly.
- The amount of redox indicator should be kept as small as possible because the indicator itself is a redox system which might influence the system under investigation. This parallels the indicator error of pH indicators mentioned above.

The redox potential of the indicator system must be chosen in such a way that the indicator does not react with the titrant: for an oxidizing titrant, the redox potential of the indicator must be greater than that of the titrant. For a reducing titrant, the reverse must be the case.

Examples of redox indicators are Indigo Carmine (C.I. 73015 [*860-22-0*]; **8**), which is frequently employed, and ferroin, an iron(II) complex with three 1,10-phenanthroline ligands (**14**), which is widely used in water and wastewater analysis in the determination of chemical oxygen demand (COD) in an internationally standardized procedure. For the structure of the organic ligand, see p. 539.

5.8.3.3 Metal Indicators

Introduction. Metal indicators are organic molecules which form specifically colored soluble complexes with metal ions in aqueous media. Here, the color of the complexes and of the free indicator must be different. These reactions can be used in two analytical procedures: volumetry (complexometry) and colorimetry/photometry. In both methods, the concentration of metal ions is determined, but with different techniques.

Colorimetry/photometry exploits the fact that the color intensity of the complex correlates to the metal ion concentration, at least within a certain concentration range. Therefore, sufficient indicator is added to the sample that the indicator forms a colored complex with all ions of the metal of interest in solution. The color intensity of this complex is then either compared visually against the color of reference samples (colorimetry) or the color intensity is measured at a defined wavelength with an instrument and then compared with a calibration curve (photometry).

Volumetry uses the formation of complexes between the metal ions and a complexing reagent (the titrant). This reagent is added in the same manner as in titration. Here, the metal indicator serves as a classical indicator: normally, a small amount of it is added before titration, and this leads to complex formation with a certain number of metal ions, but not with all, in contrast to colorimetry/photometry, and consequently to a colored solution. Addition of the titrant then leads to formation of complexes between this reagent and the metal ions which are colorless and more stable than the indicator–metal complexes. As a result, the color of the solution changes when all free metal ions have reacted with the titrant, because the next small addition of titrant forms a complex with the metal ions which were so far in the metal–indicator complex. This complex is destroyed, loses its color, and the color of the solution changes. The concentration of the metal ion is proportional to the added amount of titrant.

Titrants which are most frequently used are substances like N,N,N',N'-ethylenediaminetetraacetic acid (EDTA) and its salts. Due to their ability to form complexes the method is also called complexometry.

Due to the different requirements of the two methods, the indicators for metal determination are often to be different in colorimetry/photometry and volumetry/complexometry (see Table 5.22).

Table 5.22 Metal indicators: importance of parameters for different application areas

Parameter	Colorimetry/photometry	Volumetry/complexometry
Sensitivity	very important (detection of metals in the ppm and ppb range)	not very important
Selectivity	very important (ideally, an indicator indicates only one metal)	important
Complex stability as function of time	very important (ideally hours to days)	not too important (hours)
Complex stability against other reagents	important	must be lower compared to titrants
Color difference between free and complexed indicator	less important (photometry), quite important (colorimetry)	very important

The main issue with metal indicators is selectivity. Most of them form complexes with various metals. For a detailed review see [1, Table 5.10].

Generally, the selectivity can be improved by optimizing the reaction conditions, that is, by addition of buffers, masking reagents, etc. Nevertheless, many special reagents have been developed for very special applications in the last four decades, mainly by users of photometric methods, who had no or only limited access to modern sophisticated spectrometric methods.

Chemical Structure. The complexes formed by the metal indicator and the metal ions contain an electron donor and an electron acceptor part. If the donor molecule (*ligand*) contains two or more atoms with donor properties, the complex is known as a *chelate*. Most organic complexing agents can be regarded as *bidentate* ligands (they contain two donor atoms) and form five- or six-membered chelate rings. The electron-donating atoms in these chelates include oxygen, nitrogen, and sulfur. The other part of the molecule varies significantly, as the examples below show.

Tiron (**9**)

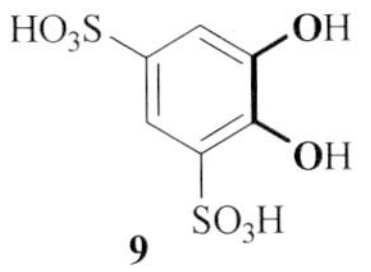

8-Hydroxyquinoline (**10**)

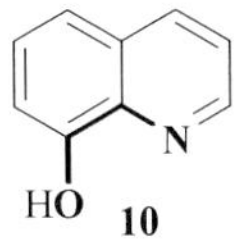

Dithizone (**11**)

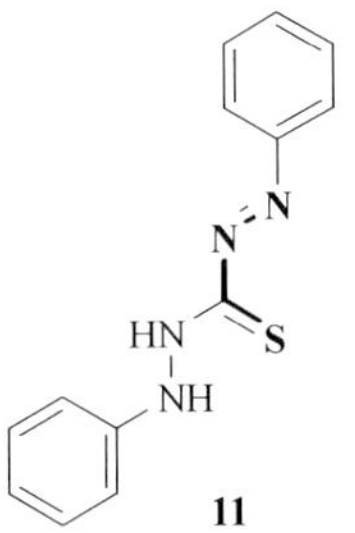

Chromotropic acid (**12**)

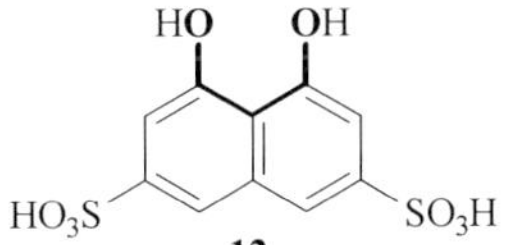

The two following examples demonstrate how the general structure of the molecule determines its properties:

Diacetyldioxime (dimethylglyoxime) (**13**)

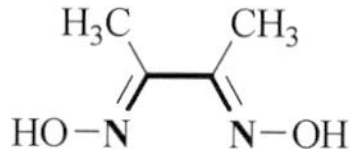

$$\begin{array}{cc} H_3C & CH_3 \\ \diagup\!\!\diagup & \diagdown\!\!\diagdown \\ HO-N & N-OH \end{array}$$
13

1,10-Phenanthroline (**14**)

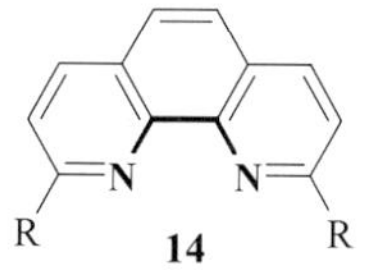

14

Both molecules contain the same donor atom configuration. Diacetyldioxime is highly specific for nickel and palladium. If the same donor atom configuration is incorporated into a heterocyclic system, as in the phenanthrolines, the ligands become specific for copper and iron. This specificity can be directed by choice of the substituents R in the molecule: if R = H, the ligand is specific for iron; with R = CH_3 and/or C_6H_5, the ligand is copper-specific.

Some of these metal indicators are used as reagents in photometric or visual test kits for the detection of metals. Here, these molecules act like color-forming reagents but are still indicators.

5.8.4 Indicator Papers

5.8.4.1 Bleeding Indicator Papers

These papers are obtained when uncoated unfilled paper is soaked with solutions of pH indicators and then carefully dried. Depending on the end use, one or more indicator dyes can be applied. Three types of indicator papers are commercially available.

Simple indicator papers only show if a solution is acidic, neutral, or alkaline. Examples of this type are litmus paper, phenolphthalein paper, and congo paper.

Universal indicator papers permit pH determinations over the total pH range from 0 to 14. The determination of the pH is made by comparing the color of the paper with a reference color scale. The accuracy is ca. 1 pH unit.

Special indicator papers cover ranges from 2 to 5 pH units. The determination method is the same as with universal indicator papers. The accuracy is better: 0.5 units is regularly attained; under favorable conditions, 0.2–0.3 units is possible.

The pH papers which are produced by impregnation are prone to "bleeding", that is, a certain portion of the indicator dyes dissolves in the test solution during immersion of the paper or the strip. This occurs particularly in alkaline medium, where the solubility of the dyes reaches a maximum, and in weakly buffered solutions, for which the papers must remain in the solution longer. Two negative effects result: the test solution becomes contaminated and colored, while the test paper loses dye and shows zones of different color intensity. Analysis becomes much more difficult and inaccurate. Despite all technical efforts, this intrinsic drawback has not been eliminated from pH papers produced with the impregnation technique.

5.8.4.2 Nonbleeding Indicator Papers

It can be prepared either by using appropriate direct dyes or reactive dyes. In the latter case, pure linters (the raw material for high-quality paper) are suspended in water, and the solution of the reactive dye is added. For example, the dye **15** reacts via the sulfonic acid group in the side chain. The reaction mixture is then made alkaline and the dye reacts with the linters. After completion of the reaction, the fiber pulp is centrifuged, washed electrolyte-free, and processed on a paper machine to form the pH paper. Paper produced in this way is mostly bonded onto a plastic material and used as pH test strip.

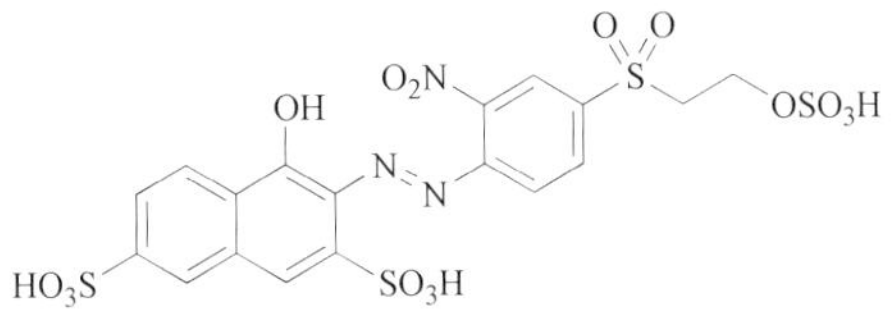

15

pH papers and strips with covalently bonded indicator dyes can remain almost indefinitely in solution without any migration or dissolution. Thus, pH determinations in hot water (up to ca. 70 °C), alkaline solutions, and weakly buffered solutions become possible. Also, pH measurements in turbid solutions or suspensions are possible, as the paper can be briefly rinsed with distilled water without changing the color.

Experiments have been performed with other carriers such as hydrophilic plastic foils and even textiles, but none of these products has come to market.

5.8.5 References

[1] E. Ross, J. Köthe, R. Naumann, W. Fischer, U. Jäschke, W.-D. Mayer, G. Wieland, E. J. Newman, C. M. Wilson, "Indicator Reagents", *Ullmann's Encyclopedia of Industrial Chemistry*, 5th ed., Vol. A14, VCH, Weinheim, 1989, pp. 127–148.
E. Ross, J. Köthe, R. Naumann, W. Fischer, W.-D. Mayer, G. Wieland, E. J. Newman, C. M. Wilson, "Indicator Reagents", *Ullmann's Encyclopedia of Industrial Chemistry*, 6th ed., Electronic Release, Wiley-VCH, Weinheim, 2002.

6 Functional Dyes

6.1 Introduction

The traditional applications for dyes are well known and form the main part of this book, since they represent high-volume uses. However, the last 20 years, and particularly the last decade, has witnessed a phenomenal rise in the growth of dyes for high-technology (hi-tech) applications. The advent and/or consolidation of new imaging technologies, such as electrophotography (photocopying and laser printing), thermal printing, and especially ink-jet printing; "invisible" imaging by using infrared absorbers in optical data storage, computer-to-plate and security printing; displays, such as liquid crystal displays and the newer emissive displays such as organic light emitting devices; electronic materials, such as organic semiconductors; and biomedical applications, such as fluorescent sensors and probes, and anti-cancer treatments such as photodynamic therapy, created the need for novel dyes to meet new and demanding criteria.Ced Dyes, and related ultraviolet (UV) and particularly infrared (IR) active molecules, which have been specifically designed for these hi-tech applications, are generally called functional dyes.

Functional dyes may be classified in several ways. Here, they are discussed according to their application. However, before doing this, it is pertinent to consider briefly the interactions of functional dyes with various agents. This enables a particular phenomena to be associated with each application.

6.2 Interactions of Functional Dyes

Functional dyes are designed to interact with electromagnetic radiation, pH, electricity, heat, pressure and even frictional forces. Each of these interactions is discussed, beginning with the most important interaction, that with electromagnetic radiation.

Electromagnetic Radiation. Functional dyes interact with electromagnetic radiation in the near-UV (300–400 nm), visible (400–700 nm), and near-IR (700–1500 nm) to produce a variety of effects required for hi-tech applications. The more important effects are described briefly, including their main applications.

Selective absorption of visible radiation produces *color* (hue). The important colors for hi-tech applications are black and the three subtractive primary colors yellow, magenta, and cyan. For displays, the three additive primary colors red, green, and blue are used. The main applications for colored dyes (and pigments) are ink-jet printing, photocopying, laser printing, thermal printing, liquid crystal displays, and organic light-emitting devices. Selective absorption of nonvisible radiation, such as UV and particularly IR, is also important in applications such as optical data storage, computer-to-plate, security, and printing.

Luminescence occurs when a molecule in an excited state, normally achieved by absorption of a photon, loses some or all of the excess energy as light rather than the normal relaxation mode of heat loss. Luminescence from the first excited singlet state to the ground singlet state is known as fluorescence. *Phosphorescence* is a much rarer spin-forbidden luminescence which occurs from the first excited triplet state of the molecule to the ground singlet state. Phosphorescence is normally only observed at low temperatures and/or when a dye is in a rigid matrix. Fluorescence and phosphorescence both occur at a longer wavelength than the absorption maximum, the difference being known as the Stokes shift. Fluorescence is important in biological applications, laser dyes, emissive displays, and in providing vivid, bright dyes, particularly for ink-jet printing.

A *color change* caused by electromagnetic radiation is called photochromism. The change in color may be from colorless to colored (or vice versa) or from one color to another. Photochromism is undesirable in traditional textile dyes, but is used in eyeglasses and optical data storage. Indicator dyes are probably the most familiar example of a color-change effect. In this case, a change in pH causes the color change.

Heat. As mentioned above most molecules lose energy from the excited state as heat. The most efficient molecules for converting electromagnetic radiation into heat are those that absorb in the near-IR region, i.e., infrared absorbers (IRAs). There has been much interest in IRAs because of their use in laser thermal transfer, optical data storage [the older write-once read-many (WORM) and the newer compact disc recordable (CD-R) and digital versatile disc recordable (DVD-R) systems], computer-to-plate printing, and as solar screens for car windscreens and windows.

Thermochromic dyes change color with temperature (heat). The effect may be due to a single dye or a composite system, and the color change may be reversible or irreversible. Thermochromic dyes find use in direct thermal printing and as temperature sensors, as well as in clothing and novelties.

Absorption of a photon to produce a triplet excited state (via intersystem crossing from the first excited singlet state) can lead to energy transfer to another molecule. The phenomenon is used in photography for *sensitization* of the red end of the visible spectrum. More recent uses are in photodynamic therapy, an anticancer treatment, and in photobleaching. In both these applications, a triplet-state dye molecule converts triplet oxygen to highly reactive singlet oxygen.

In the *photoelectrical effect* a photon removes an electron from a molecule to produce a radical cation or "hole". This effect is utilised in the key imaging step in photocopiers and laser printers, and in solar cells.

Electricity. Functional dyes have been designed to interact with electricity to produce a color change or to fluoresce. Electrochromic dyes change color, normally from colorless to colored, when an electrical voltage is applied. This phenomenon is used to provide electrochromic mirrors on luxury cars. The electrical injection of electrons and holes in organic semiconductors to produce fluorescence emission (red, green, and blue) is used in emissive displays such as organic light emitting devices (OLEDs). The process is essentially the reverse of that used in electrophotography, where light produces a hole and an electron.

Frictional Forces. Dyes have been designed to produce an electrostatic charge when frictional forces are applied. Such dyes, known as charge-control agents, are used in the toners for photocopiers and laser printers to both produce and regulate the triboelectric charge on the toner particles.

Pressure. Barochromic (pressure-sensitive) dyes change color with pressure. As with thermochromic dyes, the effect may be due to a single dye or a composite system. Barochromic dyes find use in imaging (carbonless paper) and for testing pressure points in, for example, aeroplanes and cars.

6.3 Functional Dyes by Application

6.3.1 Imaging

Over the past 30 or 40 years many technologies have attempted to take a share of the imaging market, particularly that for nonimpact printing [1]. Of these technologies, two now dominate, namely, laser printers and especially ink-jet printers, although thermal printers are also used, particularly in color printing.

6.3.1.1 Laser Printing and Photocopying

Laser printers and photocopiers use electricity and light to produce an image. The two main components are the photoconductor, now invariably an organic photoconductor (OPC), and a toner. The basic process, whether it is photocopying or the laser printing process described here, comprises six steps (Figure 6.1) [2].

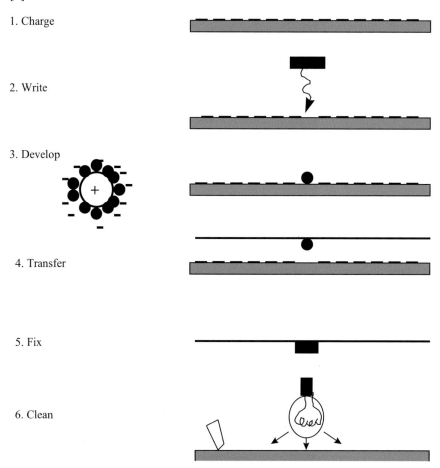

Figure 6.1 Laser printing process using an OPC

In the first step, the OPC drum or belt is given a uniform electrostatic charge of ca. 700 V. As the name implies, the photoconductor is a conductor of electricity in the presence of light but an insulator in the dark. In the second, imaging step, the laser, invariably a gallium-aluminium-arsenide semiconductor laser operating at 780 or 830 nm, writes on the photoconductor and dissipates the electrostatic charge. This produces a latent electrostatic image on the phtotconductor consisting of an uncharged image and a charged background. The third, development step consists of rendering this latent image visible by treating it with a toner. In laser printing, the toner particles have the same electrical charge (negative) as the background and are repelled onto the uncharged image areas. The fourth and fifth steps simply involve the transfer to and thermal fixation of the toner on the paper. This is the reason why prints/copies emerge warm from a laser printer or copier. The final step is simply a cleaning step to ensure the system is ready for the next print or copy.

Full color printing is essentially the same as for black and white but requires four steps, one each for the yellow, magenta, cyan, and black toners needed for full color reproduction [2, 3]. Consequently, color laser printers and copiers are significantly more expensive than monochrome laser printers and especially ink-jet printers. They are also slower than monochrome laser printers.

Organic photoconductors (OPCs) are dual-layer devices comprising a thin (ca. 0.1–1.0 μm) charge generation layer (CGL) on top of which is a thicker (ca. 20 μm) charge-transport layer (CTL) [2, 3] (Figure 6.2). Light passes through the transparent CTL and upon striking the CGL, which usually contains a pigment, forms an ion-pair complex. The electron passes to earth, leaving a positive hole, which is transported to the interface. Hence pigments of high crystallinity are required in the CGL to avoid crystal defects, which can trap the positive holes and hinder their transport to the interface. The CTL contains highly electron-rich compounds which readily donate an electron to the positive hole, forming a positive hole in the CTL; this is transported to the negatively charged surface by a hopping mechanism where it neutralizes the negative charge. Thus, the key chemicals in OPCs are the charge-generation materials (CGMs) in the CGL and the charge-transport materials (CTMs) in the CTL. The CGMs in the CGL are invariably pigments, and the CTMs in the CTL are electron-rich organic compounds, which are usually noncolored or only slightly colored, e.g., pale yellow.

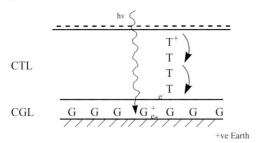

Figure 6.2 The photoconductive mechanism of an OPC (G = charge-generating molecule, T = charge-transport molecule)

A variety of organic pigments have been used for charge generation, including polyazo compounds, perylene tetracarboxydiimides, polycyclic quinones, phthalocyanines, and squariliums (Figure 6.3) [2, 3]. The pigments used as photoconductors must be extremely pure and possess the correct morphology, otherwise their performance is impaired. For example, traces of impurities can deteriorate the photoconductive characteristics of a compound. In some cases, the pigments are purified by sublimation, while in others less expensive crystallization is possible. The crystallinity of a pigment and its particle size are often important parameters in determining OPC performance.

Most modern laser printers now use titanyloxy phthalocyanine, type IV polymorph, as the CGM. This pigment has the best combination of properties and is unlikely to be surpassed.

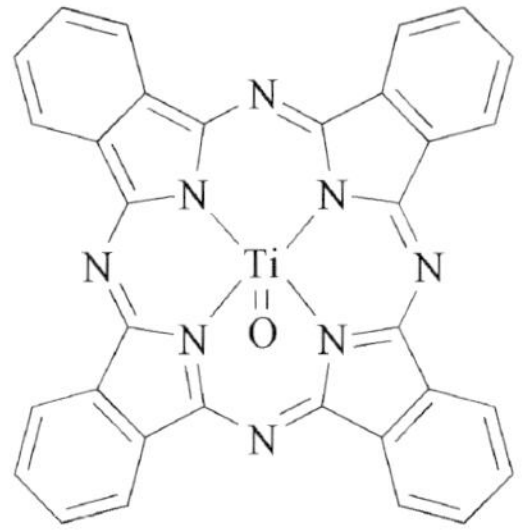

Titanyloxy phthalocyanine

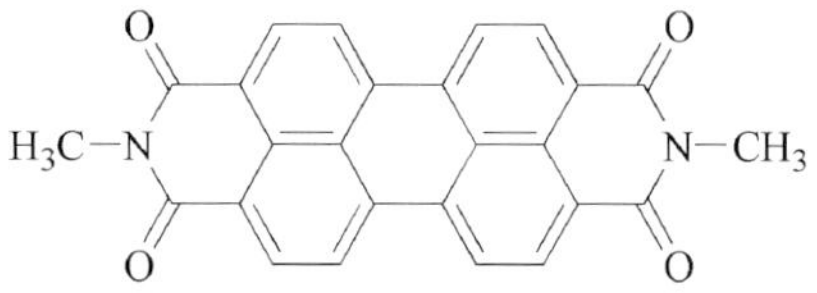

Perylene tetracarboxydiimide

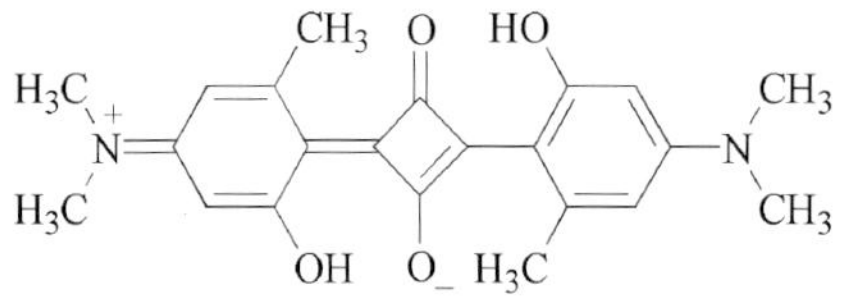

Squarylium

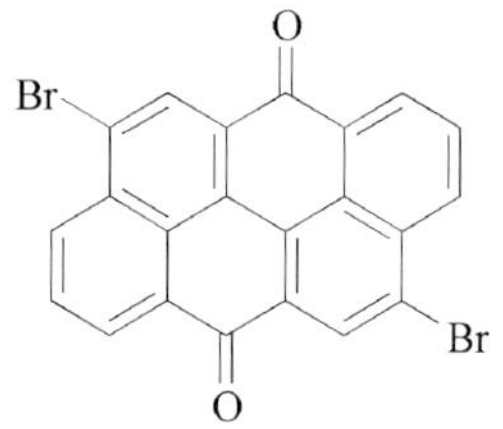

Dibromoanthanthrone

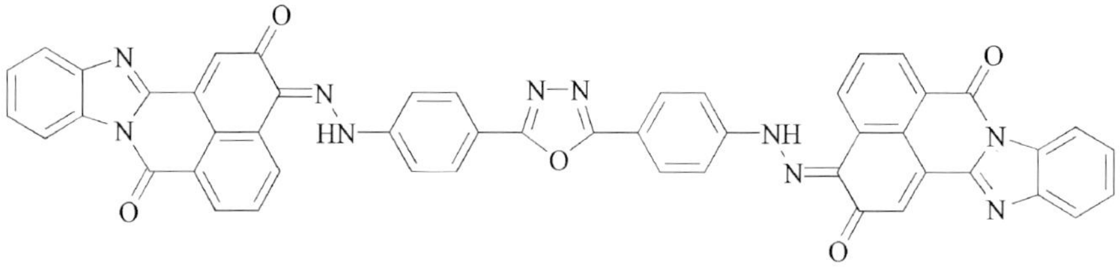

Azo

Figure 6.3 Typical CGMs for laser printers (left column) and copiers (right column)

The CTMs are p-type semiconductors (i.e., they transport a positive hole), the main types being aryl hydrazones, aminoaryl heterocycles such as oxadiazoles, and especially highly conjugated arylamines [2, 3]. Obtaining a good CTM necessitates a trade-off between good OPC performance on the one hand and good environmental stability, especially to light and atmospheric oxidation, on the other. The highly conjugated arylamines, such as the dimethyl diarylide shown in Figure 6.4, are currently the best and are used in most laser printers and copiers. However, a recently developed oligomeric arylamine is claimed [4] to be much faster than the arylamine shown in Figure 6.4.

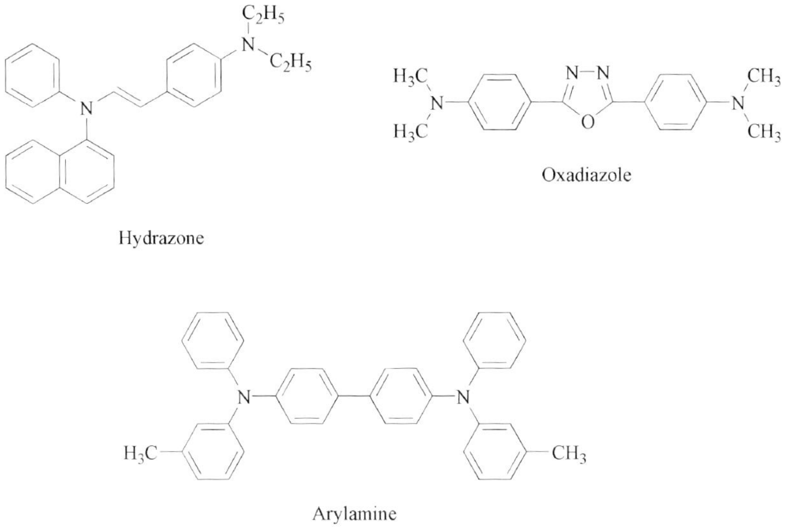

Figure 6.4 Typical CTMs for both laser printers and copiers

Toners. Two types of colored molecules are used in the toner, namely, the colorant and the charge-control agent (CCA). The colorant is used to impart color to the toner, while the CCA is used to help impart and especially to control the triboelectric charge on the toner particles. The color is normally black but can also be yellow, magenta, or cyan in colored toners. Typical colored pigments are the diarylide yellow *C.I. Pigment Yellow 12* (**1**), the quinacridone magenta *C.I. Pigment Red 122* (**2**), and the copper phthalocyanine cyan *C.I. Pigment Blue 15:4* (**3**) [2, 3]. Carbon black is the black pigment. Yellows such as *C.I. Pigment Yellow 155*

(**4**) and *C.I. Pigment Yellow 74* (**5**), which have higher transparency, have been proposed [5].

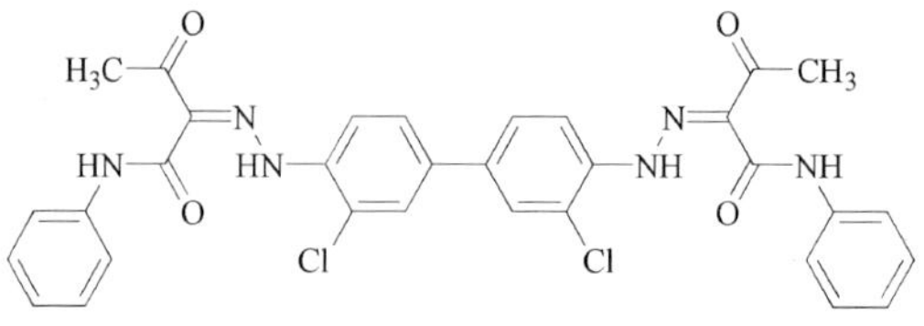

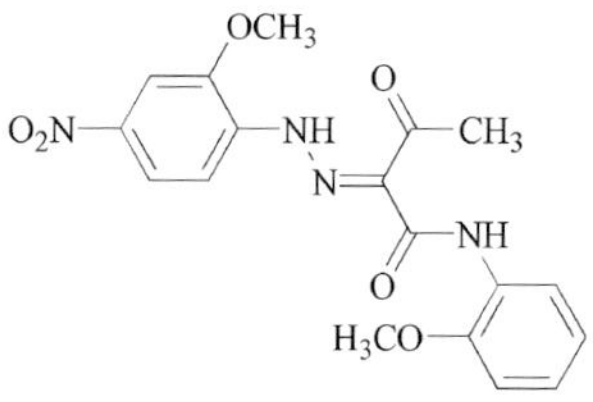

CCAs can be colored or colorless and are mainly dyes rather than pigments. Colored CCAs are more effective than colorless CCAs but, because of their color, are restricted to black toners. Colorless CCAs are needed for yellow, magenta, and cyan toners.

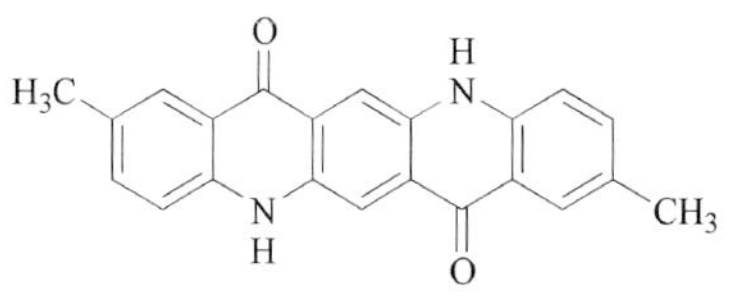

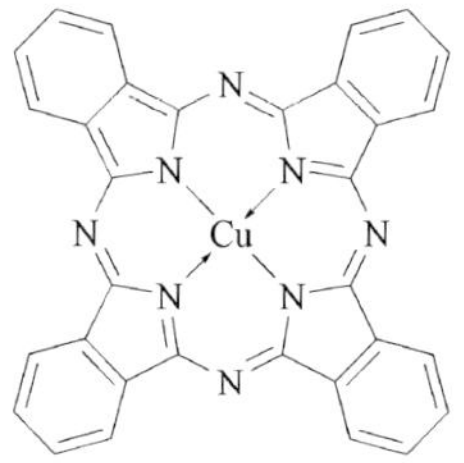

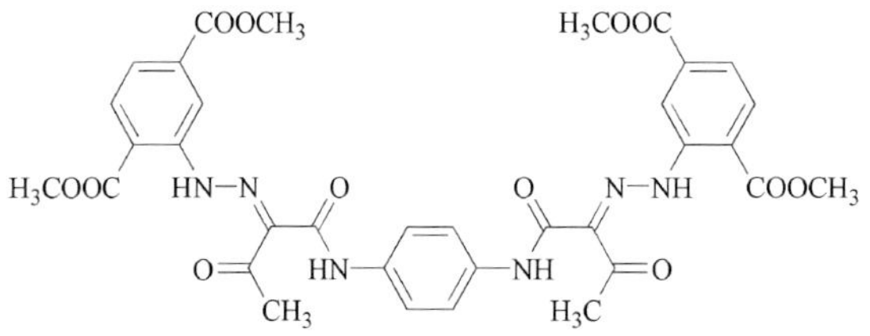

The CCAs which impart a negative charge to the toner are invariably metal complexes, particularly 2:1 Cr^{III} azo dye complexes such as (**6**), and colorless complexes such as (**7**), both of which have a delocalized negative charge [2, 3].

Positively charging CCAs are organic molecules with a positive charge. Colored positive CCAs tend to be the black Nigrosine dyes, which are mixtures of compounds in which highly arylated phenazines (**8**) are the major component. Colorless positive CCAs tend to contain a quaternary nitrogen atom: cetylpyridinium chloride (**9**), also found in antiseptic mouthwashes, is one of the most widely used [2, 3]. Triphenylmethane, benzimidazolium and beta di- and tri-carbonyl compounds are also used [6].

6.3.1.2 Thermal Printing

This process can be subdivided into direct thermal printing and thermal transfer printing. Direct thermal printing, based on color formers such as crystal violet lactone (**10**), which is converted to the cationic dye (**11**) on contact with an acidic compound [1, 3, 7], has declined in importance and is not considered further. Thermal transfer printing, particularly dye diffusion thermal transfer (D2T2) printing, is still important, especially for color images. Only D2T2 is discussed here, and not the older thermal wax-transfer printing based on pigments [3, 7].

                                    10   **11**

    The D2T2 process, the most recent thermal transfer technology, is shown in Figure 6.5. The dye-sheet is a thin plastic film, usually 6-μm polyester, coated on one side with a heat-resistant backcoat and on the other side with a color coat. This consists of an intimate mixture of a solvent-soluble dye and binder that attaches it to the substrate. The receiver sheet is normally white plastic which contains a thin, clear copolyester receiving layer. Heat supplied from the thermal heads, usually at high temperature (up to 400 °C) for short times (milliseconds), transfers the dye from the dye-sheet into the receiver layer [7, 8]. The dye actually dissolves in the receiver sheet (i.e., does not lie on the surface as in thermal wax transfer), and the amount of dye transferred is proportional to the quantity of heat supplied, so that gray scale is achieved. Three passes of the trichromatic color ribbon are required for a color print, one each for the yellow, magenta, and cyan dyes. Early reports that the process was one of dye sublimation, as in textile transfer printing, have been convincingly repudiated, and the D2T2 process is now accepted as being a melt-state diffusion process [9].

    Laser thermal transfer uses similar dyes, but a semiconductor infrared laser supplies the energy, which is converted to heat by an IR absorber in the transfer ribbon [7, 10].

    Since D2T2 is primarily used for producing full-color images, a precisely matched trichromat of yellow, magenta, and cyan dyes, and optionally a black for very high quality prints, is required. Inherently strong dyes are desirable to produce high print optical densities, which are essential for high-quality prints. Solubility in acceptable solvents is required for producing the trichromatic color ribbon by solvent coating and, as in thermal ink-jet printing, high-temperature stability is required to withstand the high temperatures reached by the thermal heads. Good fastness properties are also needed to give both good dye-sheet stability and good image stability (light- and heatfastness). Finally, the dyes must be nontoxic and nonmutagenic. The two key features which distinguish D2T2 from other nonimpact printing technologies are its ability to give continuous tones (gray scale) without loss in resolution, and its high print optical densities (up to 3.0). Both of these features are necessary for producing high-quality prints.

    The ultimate aim of D2T2 is application in digital photography [11]. This could provide a multibillion-euro outlet. However, this market is being captured by photorealistic ink-jet printers because of their lower cost, versatility, and higher speed.

6.3.1 Imaging

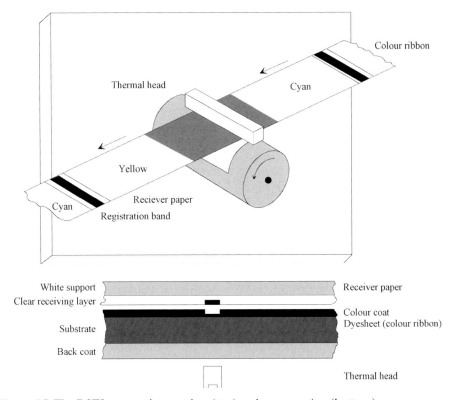

Figure 6.5 The D2T2 process in overview (top) and cross-section (bottom)

Yellow Dyes. A yellow D2T2 dye is probably the easiest to obtain both in terms of color and properties. The leading transfer-printing yellow dyes for textiles, such as quinophthalones and azophenols, are inadequate, as are typical photographic yellows such as azamethines. Two of the most widely used dye classes are methine (**12**) and especially azopyridones (**13**) [7, 11, 12].

Magenta Dyes. As for silver halide photography, a magenta dye is more difficult to produce than a yellow dye, both in terms of color and properties. Anthraquinones are the leading textile transfer-printing dyes, but these are red rather than magenta and are tinctorially weak (low print optical densities). However, they possess outstanding lightfastness, and dyes such as *C.I. Disperse Red 60* (**14**) are often used in mixtures with other dyes, such as heterocyclic azo dyes. Photographic magenta dyes such as azamethine pyrazolones are also poor D2T2 dyes. Again, novel dyes such as heterocyclic azos, especially those based on isothiazoles such as (**15**), had to be designed to produce D2T2 magenta dyes [7, 11, 12].

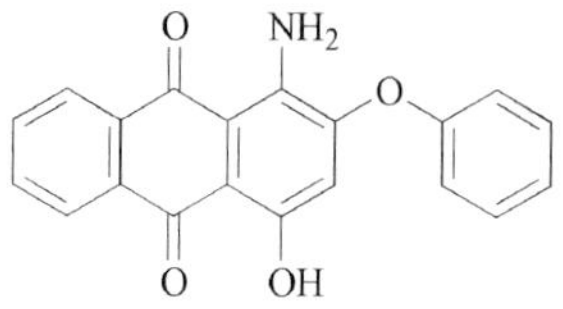

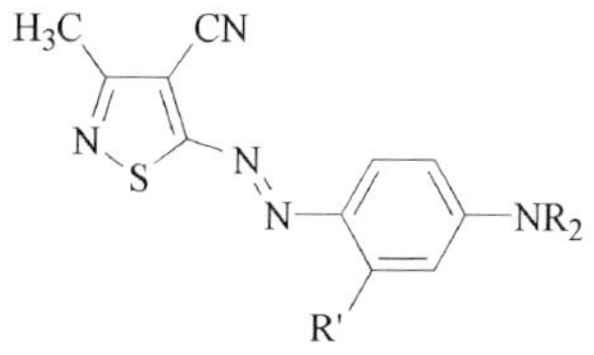

Cyan Dyes. True cyan dyes are scarce. The best cyan dyes are copper phthalocyanines, but these are unsuitable as D2T2 dyes because of their large molecular size. The leading textile transfer-printing cyan dyes are anthraquinones, but these have several disadvantages: they are blue rather than cyan, they are weak, and they have inadequate lightfastness. Photographic cyans (indoanilines) have the correct color, but they are also weak and generally exhibit poor lightfastness. Yet again, novel dyes had to be synthesized to produce satisfactory D2T2 cyan dyes.

Heterocyclic azo dyes such as (**16**), although promising in many respects, have poor thermal stability, probably associated with the nitro groups. Dyes of type (**17**), however, have overcome this instability problem. These dyes and novel indoaniline dyes (**18**), which have higher lightfastness than conventional indoaniline dyes, are currently the most prevalent cyan D2T2 dyes [12].

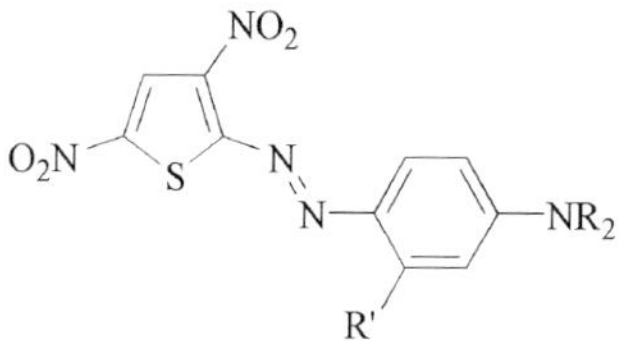

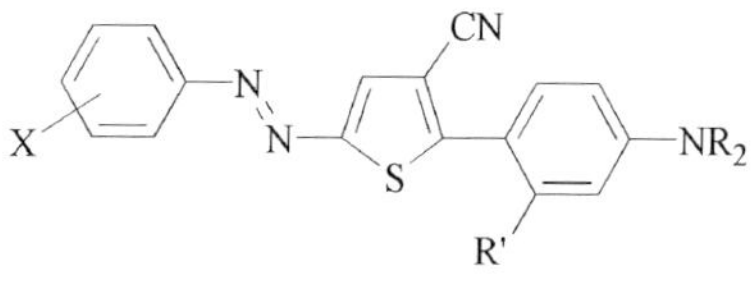

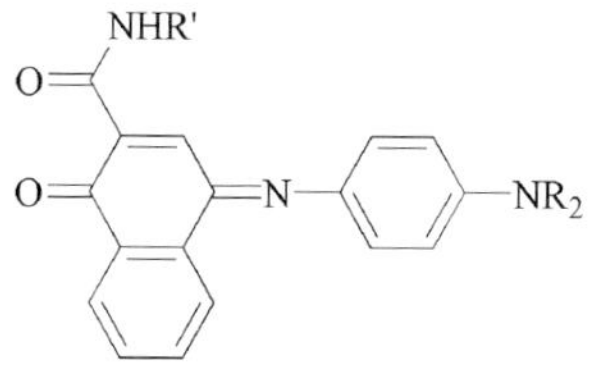

Black dyes are not normally used in D2T2. A black color is obtained by overprinting the yellow, magenta, and cyan dyes to produce a so-called composite black.

6.3.1.3 Dyes for Ink-Jet Printing

This topic is covered in detail in Section 5.6, but a brief mention of ink-jet printing is warranted in order to compare and contrast the colorants and technology to laser printing and thermal printing.

Ink-jet printing is now the dominant nonimpact printing technology. It has achieved this position from a platform of low cost, full color, very good (and ever-improving) quality, and reasonable speed. For example, good-quality color ink-jet printers currently retail for € 95, whereas the cheapest color laser printer costs ca € 1500 [13].

The main reason for the low cost of ink-jet printers is their simplicity. Unlike electrophotography and thermal transfer (D2T2), ink-jet printing is a truly primary process. There are no intermediate steps involving photoconductors or transfer ribbons. In ink-jet printing, a solution of a dye (an ink) is squirted through tiny nozzles, and the resulting ink droplets impinge on the substrate, usually paper or special media, to form an image (Figure 6.6).

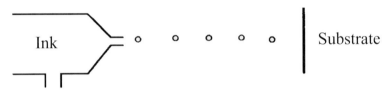

Figure 6.6 Principle of ink-jet printing

The simplicity associated with being a primary printing technology has enabled ink-jet printing to be used in many markets. From its humble beginnings in the small office and home office (SOHO) market of the mid-1980s, ink-jet printing has expanded into a diverse number of markets ranging from photorealistic, wide-format, textiles, and industrial to short-run printing and displays (e.g. printing of color filters for flat panel displays).

The predominant ink-jet inks consist of water-soluble dyes in an aqueous vehicle. The substrates are many and varied but are conveniently divided into two basic types: special media and (plain) paper. Special media include photographic-type media for photorealistic ink-jet printing, vinyl-type media for wide format, and overhead transparencies. All of these have in-built mechanisms to give high waterfastness. Hence, the main dye requirements are high chroma (vividness) and lightfastness. For plain paper, print quality (edge acuity), waterfastness, and optical density, especially for black, are the key requirements. For both special media and plain paper applications, excellent ink operability (reliability) is required.

Ink-Jet Dyes for Special Media. Anionic water-soluble dyes (**19**) are the dyes of choice for ink-jet printers. The properties required of these ink-jet dyes include high color strength, good stability in the appropriate solvent, good fastness properties (especially to light and water), and high purity. Obviously, the dyes must also be nontoxic. The high levels of purity are necessary to minimize nozzle clogging and printhead corrosion. High thermal stability is an additional requirement of dyes for thermal ink jets, since the localized ink temperatures can reach 350 °C.

$[\text{Dye}]\!-\!SO_3^-$

19

20

21

22

23

24

25

26

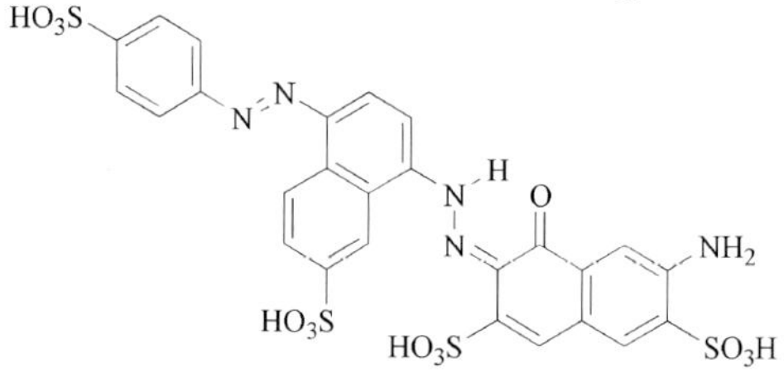

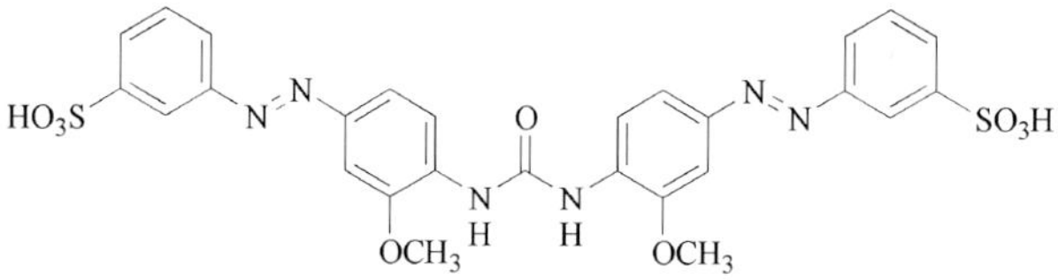

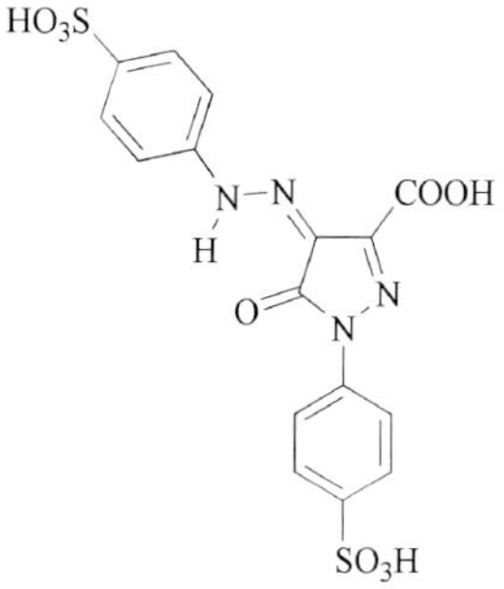

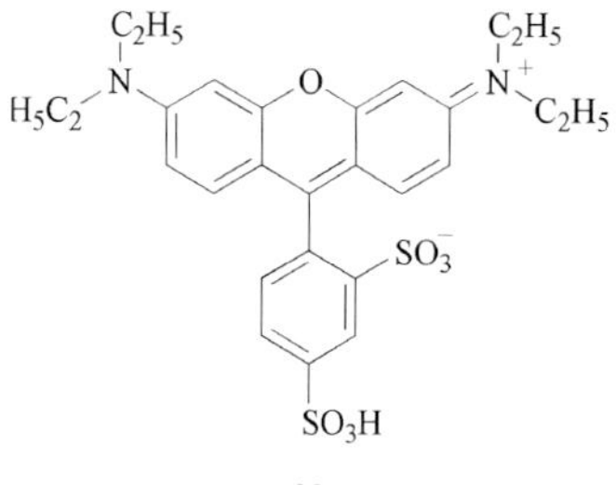

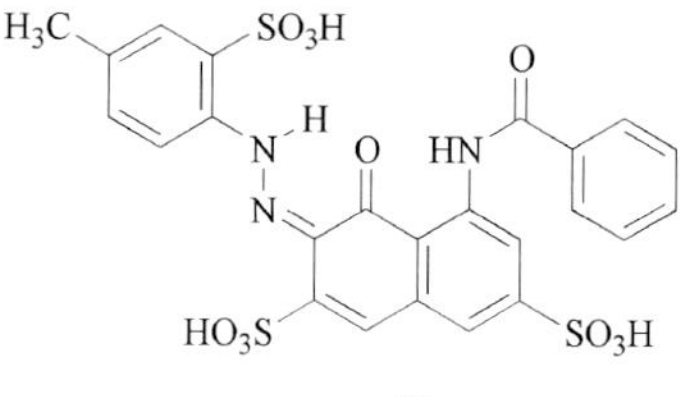

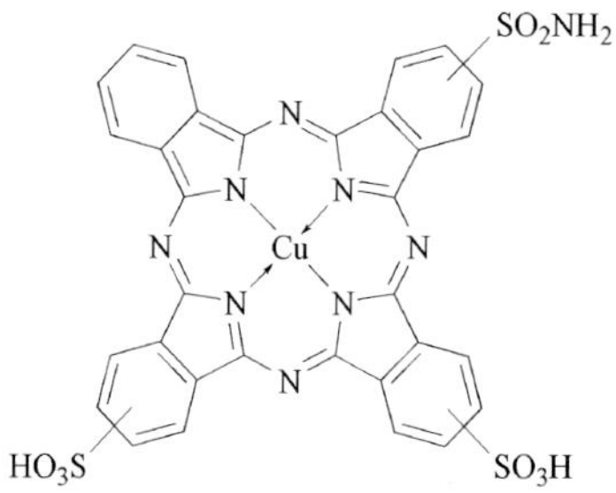

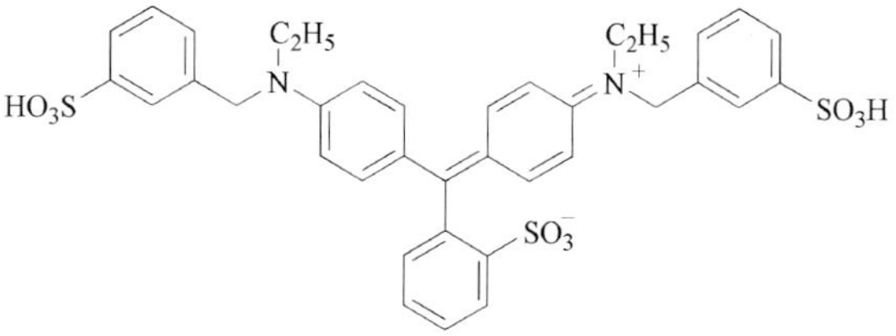

6.3.1 Imaging

Acid dyes, reactive dyes (passivated), and especially direct dyes provide the bulk of ink-jet dyes used currently. Azo dyes, including their hydrazone tautomers, are the most important and provide most of the yellows, magentas, and blacks. Copper phthalocyanines invariably provide the cyan dyes. Sometimes, very bright fluorescent dyes, such as xanthene magentas (e.g., *C.I. Acid Red 52*) and triphenylmethane cyans (e.g., *C.I. Acid Blue 9*) are also used, usually in admixture with other dyes to enhance the chroma or fine-tune the lightfastness. Typical dyes and structures are *C.I. Food Black 2* (**20**), the yellow dyes *C.I. Direct Yellow 132* (**21**) and *C.I. Acid Yellow 23* (Tartrazine; **22**), the magenta dyes *C.I. Acid Red 52* (**23**) and the hydrazone dye (**24**), and the cyan dyes *C.I. Direct Blue 199* (**25**) and *C.I. Acid Blue 9* (**26**). Dyes of this type are used extensively in special media applications such as photorealistic ink-jet printing [13].

Ink Jet Dyes for Paper are essentially the same water-soluble anionic dyes that are used for special media, with the addition of novel dyes tailor-made to give better waterfastness on paper. These are the black (**27**) plus trichromatic carboxy dyes introduced by Avecia in the early 1990s. They provide high, but not total or instantaneous, waterfastness, by using the principle of differential solubility and smelling salts, i.e., a volatile cation such as ammonium (Figure 6.7) [1,13-16]. Carbon black pigmented inks are also used [17].

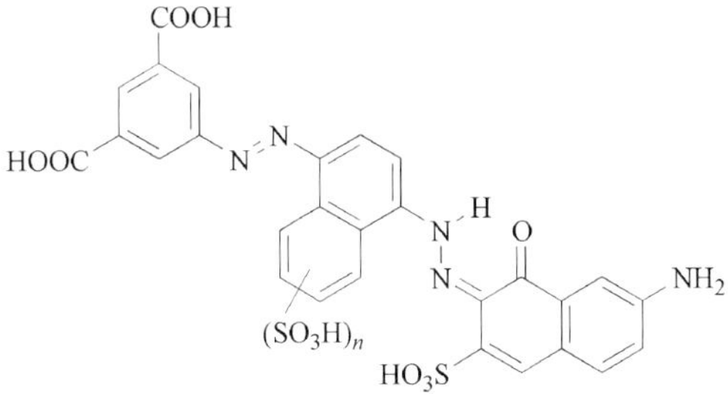

[Structure of compound **27**: a bis-azo dye with COOH, HOOC substituents on a phenyl ring connected via N=N to a naphthalene bearing $(SO_3H)_n$ where $n = 0$ or 1, further linked by N-N-H to a hydrazone with HO$_3$S and NH$_2$ groups and C=O]

$n = 0$ or 1

27

$$(NH_4)_2CO_3 \longrightarrow 2\,NH_3 + CO_2 + H_2O$$

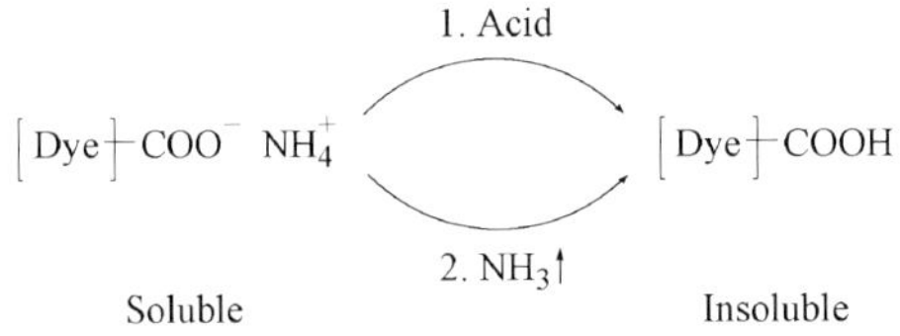

$$[\text{Dye}]\text{-COO}^-\,NH_4^+ \underset{2.\ NH_3\uparrow}{\overset{1.\ \text{Acid}}{\rightleftharpoons}} [\text{Dye}]\text{-COOH}$$

Soluble Insoluble

Figure 6.7 Principle of differential solubility and smelling salts

Ink Dyes for Textiles. The two main textile fibers are cellulose (cotton) and polyester. Ink-jet dyes for cotton are existing textile reactive dyes [18, 19] but purified

6.3 Functional Dyes by Application

to ink-jet standards. Both chlorotriazines and vinylsulfones are used (Figure 6.8). Smaller particle size disperse dyes are used for ink-jet printing polyester. Post-treatment and usually pre-treatment of the fibers are necessary to give good prints. Pigments are also used to ink-jet print textiles.

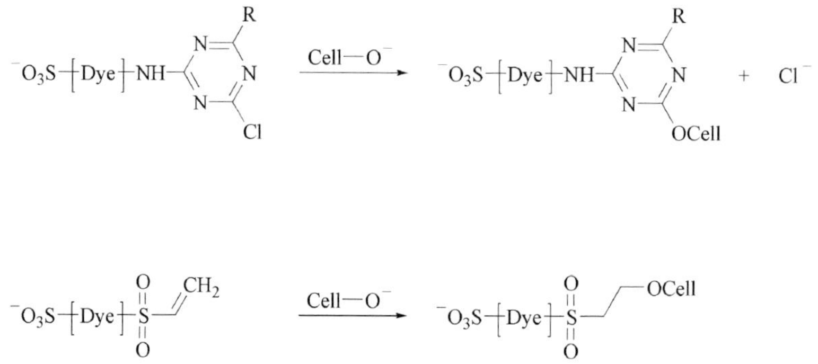

Figure 6.8 Chlorotriazine and vinylsulfone reactive dyes for ink-jet printing of cotton

Other Inks. In addition to dye-based inks, pigmented inks have also appeared in ink-jet printing. A carbon black pigmented ink, developed by DuPont [17], is used in some ink-jet printers and gives good quality black prints, although there are problems of rub- and smearfastness. Inks containing color pigments have not found widespread use for a number of reasons, including more complex and less stable inks, the inability to print on most special media, and duller colors. Therefore, they are mainly confined to applications in which high durability is required, such as outdoor posters and signs.

6.3.1.4 Other Imaging Technologies

These processes include *electrography* [3, 20] (similar to electrophotography but without light), *ionography* [3, 20] (ion deposition), and *magnetography* [3, 20]. These technologies, and the more recent *elcography* [21], use colorants similar to those in electrophotography.

Pressure-sensitive or carbonless copying utilizes encapsulated color formers similar to those used in direct thermal printing, such as crystal violet lactone (**10**), and an acidic compound [3]. Black is the most important color, and the black color formers (**28**) produce the black dyes (**29**) on contact with an acidic compound. The Meade Corporation developed a full-color imaging system (Cycolor) based on color-former technology [3, 22]. Although scientifically elegant, it failed commercially because of its complexity and expense compared to other technologies such as ink-jet printing.

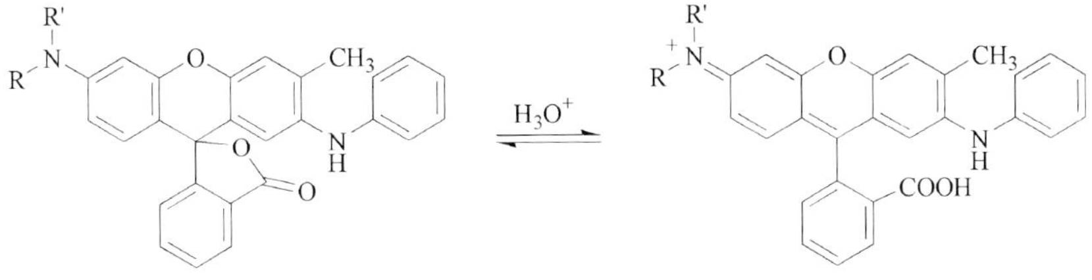

28 ⇌ (H₃O⁺) **29**

R, R' = CH₃, C₂H₅, —C₆H₁₁ (H), or —C₆H₄—CH₃

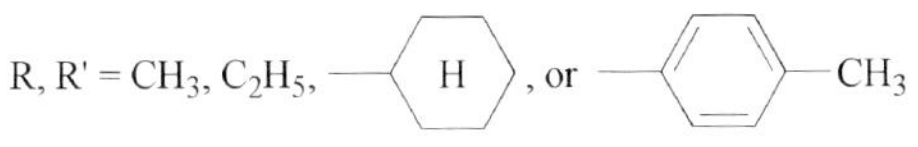

Because it is paramagnetic, triplet oxygen quenches fluorescence. This property has been used for testing vehicles such as airplanes and automobiles for high-stress points by using the fluorescent red dye platinum octaethylporphyrin. A solution of the dye is painted onto the test vehicle, which then fluoresces red. In a wind tunnel under UV illumination, the dye molecules luminesce in inverse proportion to the local pressure of oxygen over the surface during aerodynamic flow. The resulting image of the surface is brightest red in areas of low pressure (low oxygen levels) and dimmer red in areas of high pressure (high oxygen levels). Thus, the luminescence can be used to map pressure fields to help design aircraft and automobiles [23].

The older diazotype printing uses stabilized diazonium salts such as (**30**) and diazo couplers such as (**31**). When a diazonium salt contacts the diazonium coupler, which is present in diazo paper or film, an azo dye is produced [24].

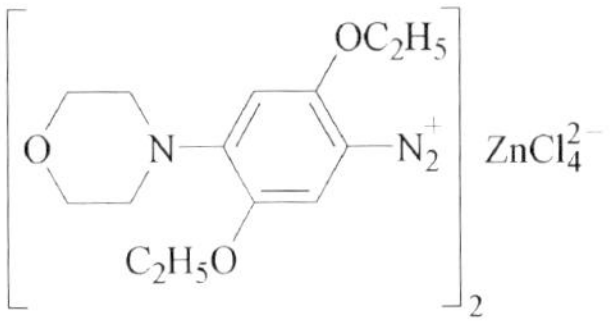

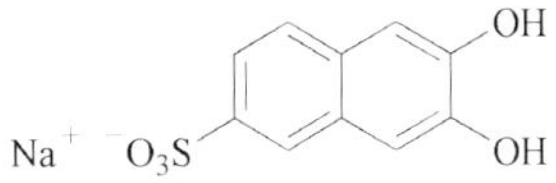

30   **31**

6.3.2 Invisible Imaging

The technologies described above produce visible images on a hard copy, usually paper or special media. However, there are important technologies which produce invisible images, usually for entertainment and publishing, using functional dyes.

6.3.2.1 Optical Data Storage

In its broadest definition, optical data storage involves a medium which functions by means of the application of near-infrared radiation or visible light for the recording and reading of information [25]. The use of a compact semiconductor laser, such as gallium-aluminium-arsenide, with its ability to produce a beam power of several milliwatts and which can be focused to a sub-micrometer spot size, makes it possible to record and replay data. One of the great benefits of the optical disc is that it provides large data storage capacity similar to a magnetic hard disk but is removable like a magnetic floppy disc. The other benefit is low cost for both the media and hardware [26]. The most common examples of optical data storage are compact discs (CDs) for audio playback, and the newer digital versatile discs (DVDs) for visual and sound playback (movies).

Rotating optical disk systems were first demonstrated in the 1960s and paved the way for the Philips Videodisc, introduced in 1973. Less than ten years later, following the successful development of the reliable, low-cost gallium-arsenide laser, the compact disc was launched, leading to the first penetration of laser technology into the consumer market.

Optical data storage media fall into three main classes [26] (Figure 6.9). Read only media comprise the CD in its audio and CD-ROM formats and DVD, also of the CD format but having about six to ten times the data capacity of a CD-ROM, depending upon whether one or two recording layers are used. These read only media are mainly dedicated to entertainment. As seen from Figure 6.9, the systems which use organic dyes are WORM (write-once-read-many), CD-R, and DVD-R.

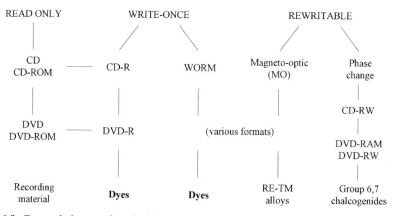

Figure 6.9 General classes of optical data storage media

The mechanism of WORM recording and indeed of all currently available optical disc media, whether erasable or write-once, depends upon the heat generated by a focused laser beam impinging on an absorbing medium. The energy, normally near-IR radiation, is transmitted to the absorbing material in a time much shorter than the time it takes for the heat generated to dissipate by thermal conduction.

6.3.2 Invisible Imaging

The basic structure of a CD-R disc is shown in Figure 6.10. The substrate contains a molded-in spiral pre-groove, which enables a tracking signal to be generated to guide the incident laser beam. Near-IR radiation at ca. 780 nm is absorbed by the IR absorber in the polymer layer in the groove of the CD-R and is converted rapidly into heat. The temperature rises quickly to ca. 250–300 °C, which is above the glass transition temperature of the polymer substrate and the decomposition temperature of most dyes [27]. The subsequent processes result in diffusion of the polycarbonate substrate, decomposition of the dye, and mechanical deformation of the reflective layer due to thermal contraction. At these points, the optical path length of the polarized readout beam is changed and this is interpreted as a mark.

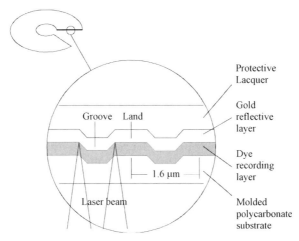

Figure 6.10 Recording process in CD-R

Most of the pioneering work on IR-absorbing dyes for optical recording media was carried out in the 1970s and 1980s, at which time the emphasis was on WORM media [26]. Many classes of dyes, based on patent applications filed by the late 1980s, are potentially suitable for WORM media [3, 25–30] (Table 6.1).

Table 6.1 Classes of IR-absorbing dyes for optical recording layers

Dye class	Producers
Cyanines	Fujitsu, Pioneer, Ricoh, TDK, Canon, Fuji Photo Film
Phthalocyanines	ICI (now Avecia), TDK, Mitsubishi, Xerox, Hitachi, BASF, Mitsui Toatsu
Naphthalocyanine	Mitsubishi, Yamamoto
Anthraquinone	BASF, Mitsui Toatsu
Nickel dithiol complexes	IBM, Kodak, TDK, Mitsui Toatsu, Ricoh

Dye class	Producers
(Di)imminiums	Philips, 3M, Ricoh
Squarylium	Philips, IBM, Fuji-Xerox
Pyrilium, thiapyrilium	Kodak, TDK
Triarylammonium	Sumitomo

The key requirements of infrared absorbers for optical data storage are:

- High absorption at the appropriate wavelength, i.e., ca 780 nm for WORM, ca. 715 nm for CD-R, and ca. 550 nm for DVD-R.
- High reflectivity in the range 775–830 nm for WORM and CD-R, and 630–650 nm for DVD-R.
- High solubility in organic solvents suitable for spin coating which do not dissolve the substrate, such as alcohols and aliphatic hydrocarbons.
- Low thermal conductivity and hence high sensitivity for heat-mode recording.
- Good oxidative and hydrolytic stability.
- Low toxicity.
- Reasonable cost with potential for cost reduction as the volume increases.

The phthalocyanine and cyanine classes of IR absorbers meet these criteria best and are therefore used widely in WORM and CD-R systems. Phthalocyanines have many desirable attributes for use in optical data storage. They meet most of the criteria outlined above but can have poor solubility. Indeed, the best blue and green pigments are phthalocyanines. The discovery [31] that the fully chlorinated green pigment (**32**) could be transformed smoothly into IR absorbers (**33**) with high solubility in organic solvents was a major breakthrough, and this resulted in dyes such as (**33**) (Ar = 2-naphthyl) for use in WORM systems. Naphthalocyanines (**34**) are also used.

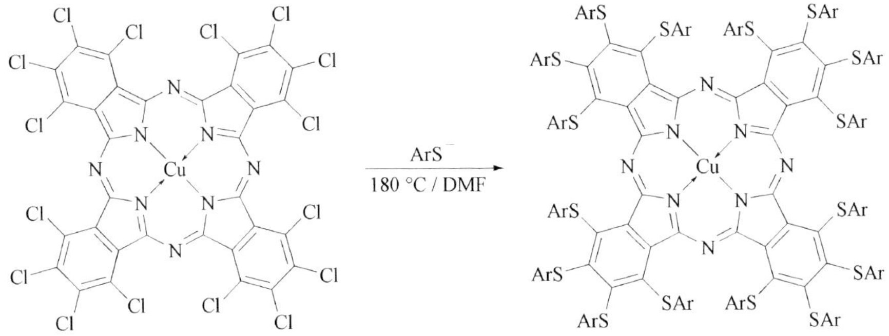

32 → (ArS⁻, 180 °C / DMF) → **33**

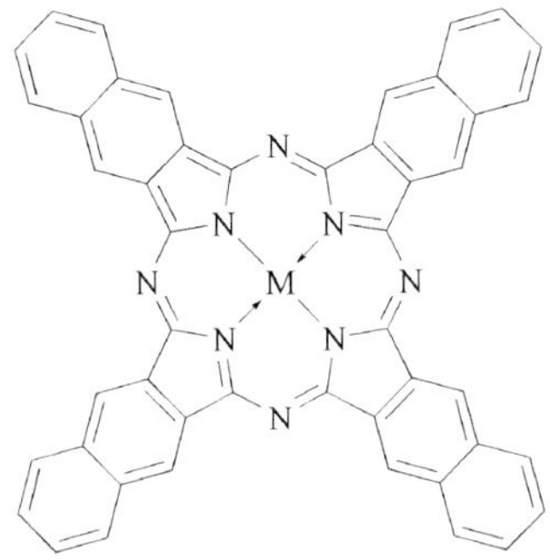

M = metal **34**

The palladium phthalocyanine (**35**), developed by Mitsui Toatsu and Ciba [26, 32], is one of the leading phthalocyanine infrared absorbers for CD-R. Bulky groups (**R**) reduce undesirable molecular aggregation, which lowers the extinction coefficient and hence absorptivity and reflectivity. Partial bromination allows fine-tuning of the film absorbance and improves reflectivity. The palladium atom influences the position of the absorption band, the photostability, and the efficiency of the radiationless transition from the excited state [26]. It is marketed by Ciba as Supergreen [33].

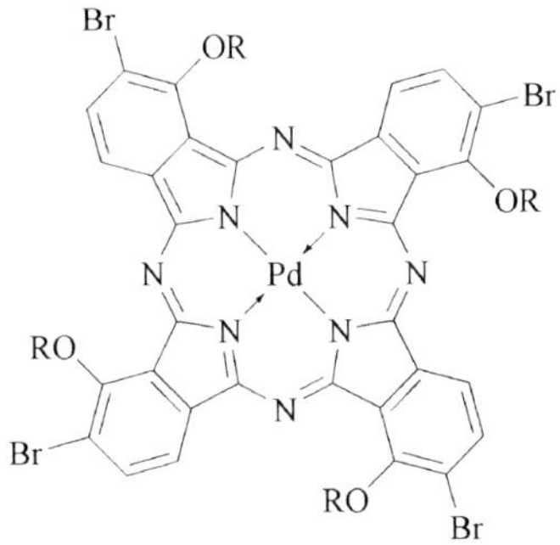

35

The two main cyanine infrared absorbers used in optical disc recording are the indoles (**36**) and benzindoles (**37**). These cyanine dyes tend to be more light-stable than other cyanines. However, they are still very much inferior to the phthalocyanines in this respect and must be used in combination with photostabilizers such as nickel dithiolates (**38**) and diiimminiums (**39**). The absorbance maximum is readily altered by the polymethine chain length. For example, in compounds (**36**) and (**37**), when $n = 3$ the absorption of the main band in the solid film is in the range 770–810 nm, which is ideal for WORM media. Reducing n by one shifts the peak to shorter wavelengths by ca 100 nm, making these dyes suitable for CD-R. A further reduction to $n = 1$ positions the band in the region suitable for DVD-R recording [26].

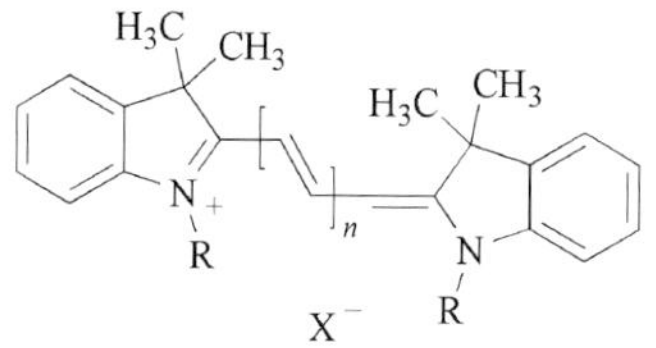

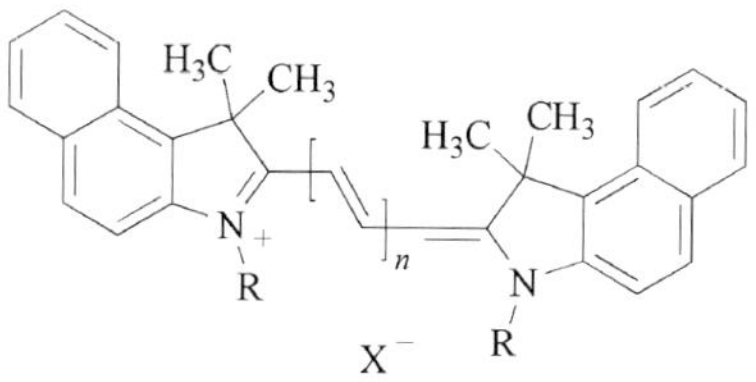

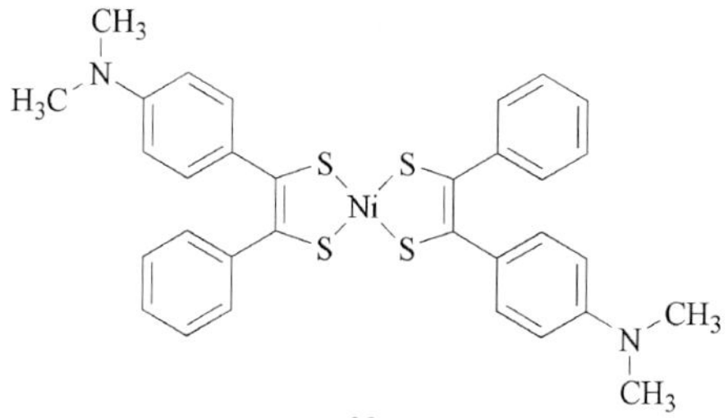

The application of cyanine dyes for optical storage media was developed primarily in Japan [30], and several dyes plus compatible stabilizers are commercially available in pure form from Japanese suppliers.

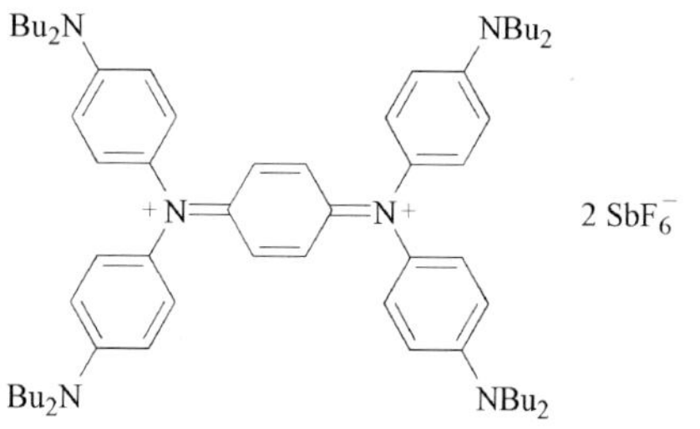

6.3.2.2 Other Technologies

Other dye-based technologies evaluated for optical data storage include photochromic dyes for rewritable systems and azo dyes for holographic data storage. Spirobenzothiopyran dyes such as (**40**) absorb in the red/near-IR in their colored form and are suitable for erasable optical data storage [34]. Dyes for holographic data storage, such as (**41**), are similar to those used in nonlinear optics [35] (see below).

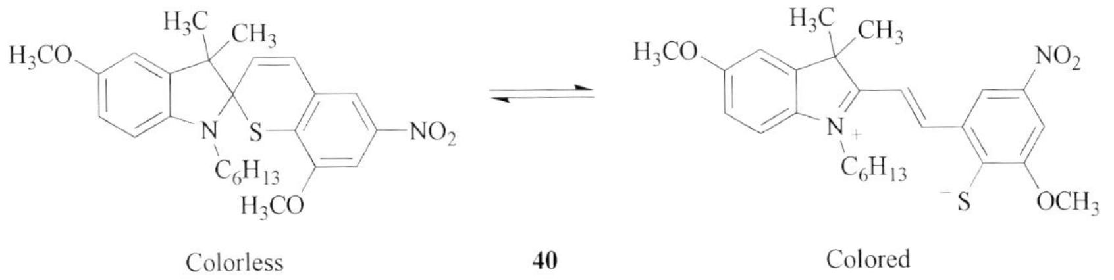

6.3.2 Invisible Imaging

40 Colorless ⇌ Colored

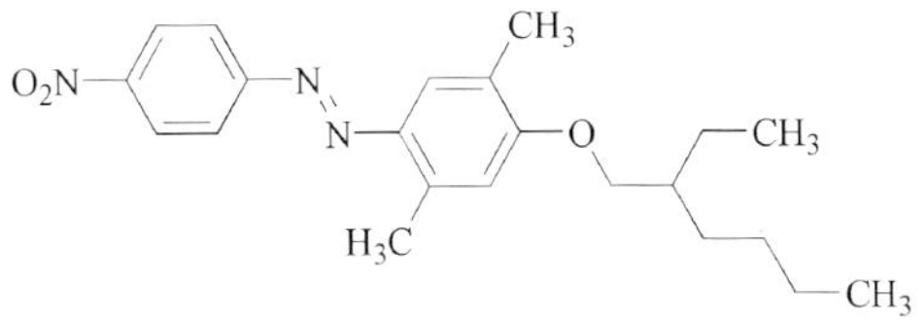

41

Modern imaging technologies are being used in traditional technologies. A good example of such a hybrid technology is *offset lithography* where electrophotography and especially thermal imaging using infrared absorbers are being increasingly used to produce lithographic plates. The most widely used infrared absorbers are cyanines, such as (**42**), which absorb around 830 nm [36].

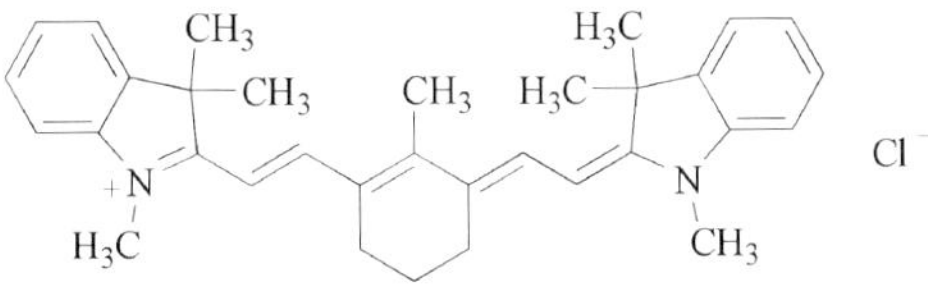

42

Infrared absorbers are used in *security printing*. Because of their durability and lower cost, phthalocyanines of type (**33**) tend be used [3, 37].

Infrared absorbers such as (**33**) and (**43**) [31] are being evaluated as *solar screeners* for car windscreens and windows to let in daylight but screen out the IR component which causes heating. Although phthalocyanines are renowned for their durability, it is proving difficult to meet the demanding requirements of ca. 10 years for cars and ca. 25 years for windows.

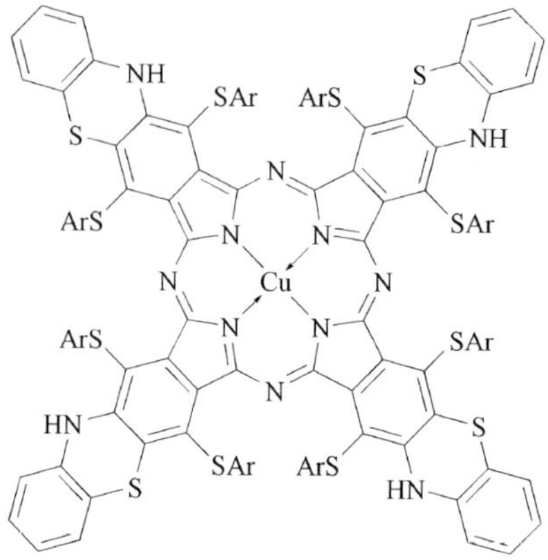

43

6.3.3 Displays

Most imaging is downloaded from a so-called soft copy on a display, such as a television (TV) or a visual display unit (VDU), on to hard copy by using laser or ink-jet printers. Therefore, the main types of displays are considered with focus on the functional dyes employed.

6.3.3.1 Cathode Ray Tube

The most familiar display is a television or visual display unit associated with a computer. Both are based on the mature cathode ray tube (CRT) technology, whereby an electron beam selectively activates red, green, and blue (RGB) inorganic phosphors. It is an emissive technology and therefore produces bright images.

6.3.3.2 Liquid Crystal Displays

Liquid crystal displays (LCDs) were developed from the discovery by Gray [38] that certain organic molecules, such as *para*-cyanobiphenyl, can be readily aligned in an electrical field. When the molecular axis of *para*-cyanobiphenyl is orthogonal to a beam of polarized light, the light is absorbed and a dark (gray/black) color results. When it is parallel to the direction of the light beam, no absorption occurs and a light-colored area results. This contrast allows monochrome imaging (Figure 6.11) [3].

6.3.3 Displays

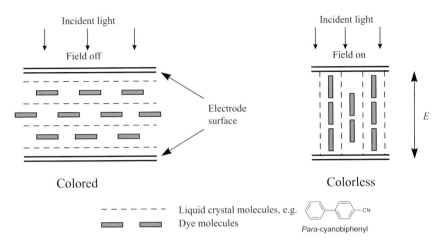

Figure 6.11 Operation of a liquid crystal display

Liquid crystal displays devoid of dyes are defective in several respects. For example, the color contrast is an unsatisfactory gray on dark gray, the angle of vision is limited, and troublesome polarizers are required. Incorporating suitable dyes into the liquid crystal overcomes the above disadvantages. Thus, a whole range of desirable color contrasts are possible, including black on white, from just three dyes, namely, yellow, red, and blue. Suitable dyes must be compatible with the liquid crystal, have high purity, produce the correct viscosity, have good light-fastness, and exhibit a high order parameter, i.e., the dye must align with the liquid crystal. An order parameter of greater than 0.7 is needed for a practical device.

Many solvent-soluble dyes have been examined, particularly azo and anthraquinone dyes. The latter tend to be used on account of their higher light stability. Typical dyes [3] are the yellow (**44**), the red (**45**), and the blue (**46**). Fluorinated disazo dyes such as (**47**) have good solubility and high order parameters (0.79) [39].

6.3 Functional Dyes by Application

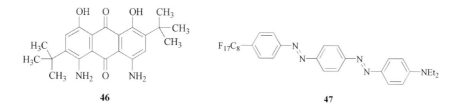

44 **45**

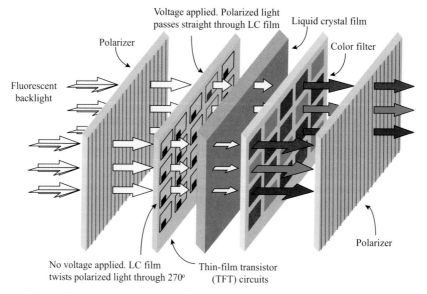

46 **47**

Figure 6.12 Configuration of a flat panel LCD

The main uses of LCDs are in digital watches and clocks, calculators, laptop computer and game displays, and instrument display panels for cars, airplanes, etc. Liquid crystals are also used as an on–off switch in color flat panel liquid crystal displays. They are interposed between the white backlight and the color filter and control the illumination of a matrix of RGB pixels in the color filter which produce the colored image (Figure 6.12) [40].

The colorants used in the color filter are generally pigments [41, 42]. Typical pigments are *C.I. Pigment Red 177* (**48**), *C.I. Pigment Green 36* (**49**), and *C.I. Pigment Blue 15:6* (ε copper phthalocyanine (**3**)]. Other pigments may be added to fine-tune the color. Dyes are also used to a limited extent.

Ink-jet printing is being increasingly used to produce color filters, replacing the traditional but inefficient photolithography [43].

LCDs are based on subtractive color technology and are therefore not as bright as emissive displays such as CRTs and the newer organic light-emitting devices.

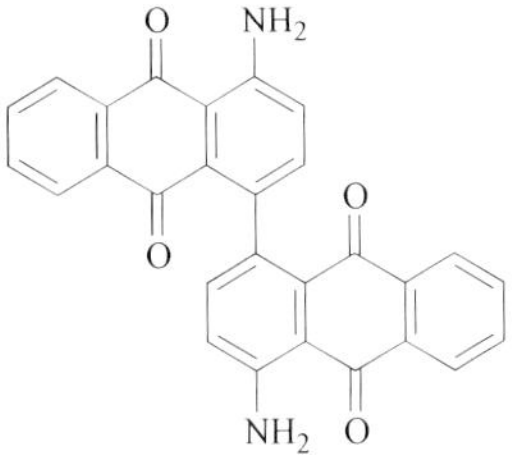

48

CuPc(Br)$_6$(Cl)$_{10}$
49

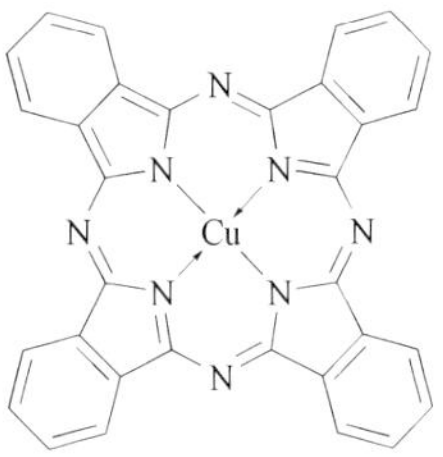

3
Copper phtalocyanine (CuPc)

6.3.3.3 Organic Light-Emitting Devices

Organic light-emitting devices (OLEDs) are one of several technologies competing for the market for next-generation emissive flat panel displays. It is the one most likely to triumph over field emission displays (basically CRT technology in miniature) and plasma display panels.

The basic principle of an OLED is simple. Electricity generates a positive hole and an electron. These are attracted to each other and, upon contact, annihilation occurs generating light (Figure 6.13). The process is essentially the opposite of that in electrophotography, whereby light generates a hole and an electron. The wavelength of the light depends on the band gap between the HOMO and LUMO of the molecules involved. For a full color image, the energy levels have to be selected to produce red, green, and blue light. This can be difficult, and fluorescent dyes which emit at the appropriate wavelength may be added as dopants. In this case, the energy from annihilation is transferred to the dopant, which emits it as light of the desired wavelength. Representative dopants [44] are the red pyran dye (**50**), the green (**51**), and the blue perylene (**52**).

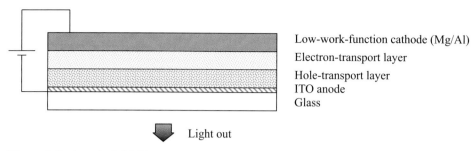

Figure 6.13 A typical OLED

50, **51**, **52**

Two types of OLEDs are being developed: small-molecule [45] and polymer [46]. Most work has been done on small-molecule OLED. Representative small-molecule hole-transport [47] (*p*-type) and electron-transport (*n*-type) organic semiconductors are shown in Figure 6.14. Not surprisingly, the hole-transport molecules are identical or very similar to those used in electrophotography.

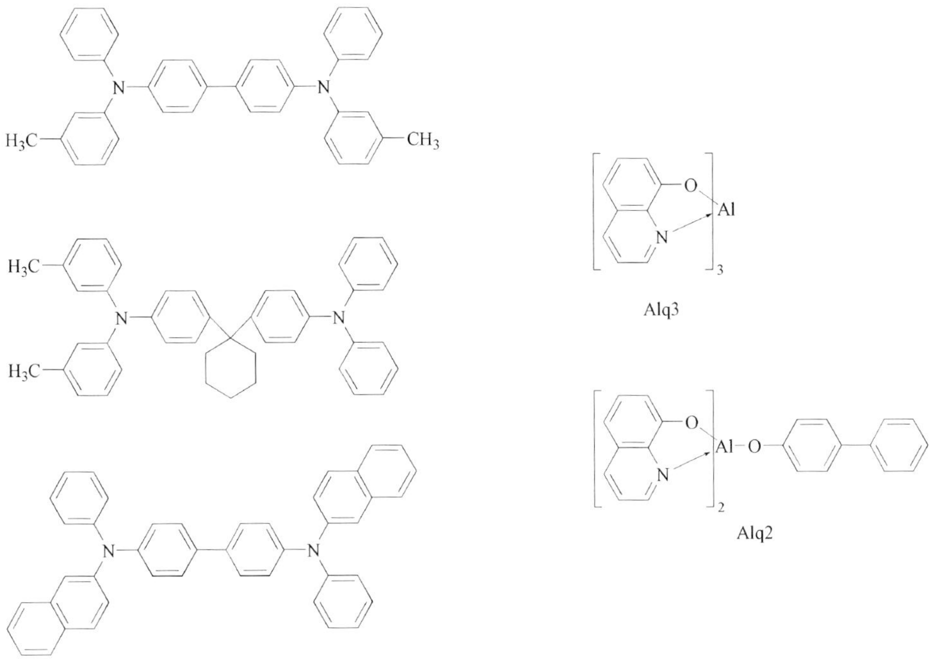

Hole-transport molecules   Electron-transport molecules

Figure 6.14 Typical hole-transport and electron-transport molecules

Light-emitting polymers [46, 48] are a more recent development and may eventually prove superior to small-molecule OLED. Typical polymers are the green poly(*para*-phenylene vinylene) (PPV; **53**), the orange-red dialkoxy derivatives (**54**), and the blue polyfluorene (**55**) [48].

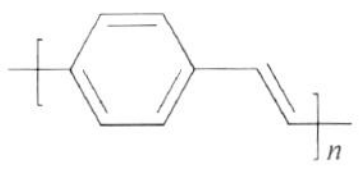

53

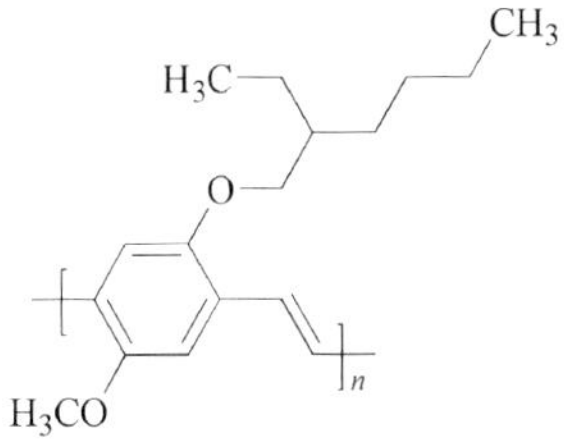

54

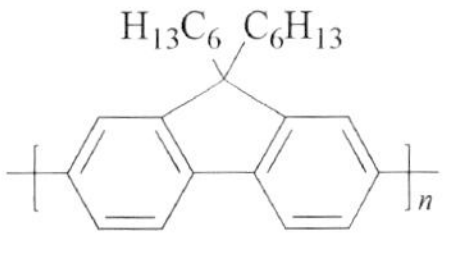

55

6.3.3.4 Electrochromic Displays

Electrochromic displays work by using a dye that changes color, usually from colorless to colored, with electricity [3]. The system commercialized by Gentex [49]

for use in the rear-view mirrors of luxury cars is based on viologens. Viologens are analogues of the weedkiller paraquat (R = Me). In the bis-cationic form (**56**) viologens are colorless; electrical reduction produces the purple-colored radical cation (**57**).

$$R-\overset{+}{N}\underset{56}{\bigcirc}-\bigcirc-\overset{+}{N}-R \quad \underset{-e}{\overset{+e}{\rightleftharpoons}} \quad R-\overset{\cdot}{N}\underset{57}{\bigcirc}=\bigcirc=\overset{\cdot+}{N}-R$$

6.3.4 Electronic Materials

Electronic materials have already been encountered under imaging (e.g., in OPCs) and displays (e.g., in OLEDs). Two areas worthy of further consideration are organic semiconductors and solar cells.

6.3.4.1 Organic Semiconductors

Organic semiconductors can be subdivided into hole-transport materials (*p*-type semiconductors) and electron-transport materials (*n*-type semiconductors). Organic semiconductors are used in numerous devices, including field-effect transistors, light-emitting diodes, Schottky diodes, photovoltaics (solar cells), and the light-emitting electroluminescent devices discussed above. Because of the lack of suitable electron-transport materials, the organic semiconductor devices studied have been mainly unipolar devices based on hole-transport materials.

As with electroluminescent molecules for emissive displays, organic semiconductors can be small molecules or polymers. Since polymers are intrinsically doped almost by definition and cannot be purified to the same degree as discrete molecular compounds, molecular crystalline solids will always be superior to polymeric semiconductors in thin-film transistors in terms of charge mobility and conductivity. Although a few polymers can approach the high levels of mobility needed for thin-film transistors, these materials do not give the high on/off ratios required for a practical device. The two main advantages of polymers are their potential for solution processing, which avoids the use of vacuum deposition for molecular compounds, and their flexibility. High purity and reproducibility will be key to the success of polymeric semiconductors.

Current state-of-the-art hole-transport and electron-transport materials [50] are shown in Figure 6.15. Hole-transport materials include oligo- (**58**) and polythiophenes (**59**) [51], poly(thienylene vinylenes) (**60**) [52], and pentacenes [53] (**61**). Electron-transport materials include copper phthalocyanine (**3**), hexadecafluoro copper phthalocyanine (**62**), naphthalenetetracarboxy dianhydride (**63**), and perylene tetracarboxy dianhydride (**64**).

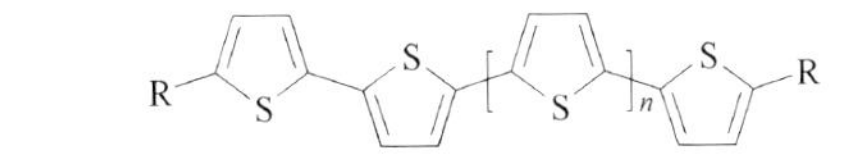

$n = 1 - 5$

58

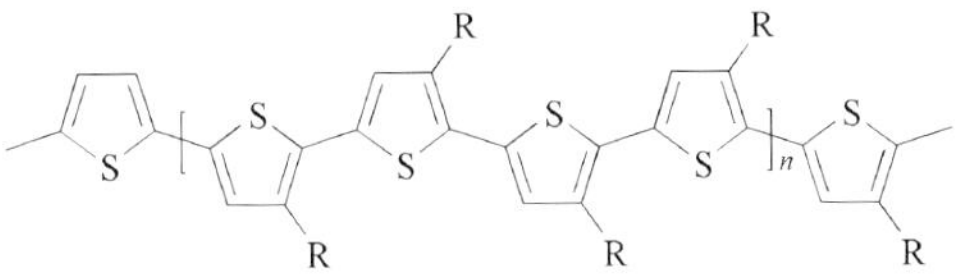

59

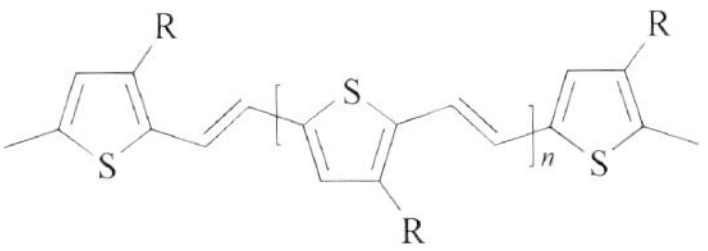

60

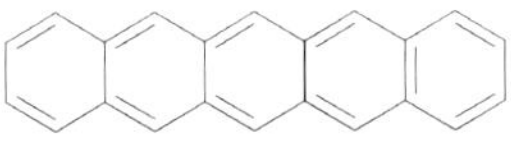

61

Hole-transport materials

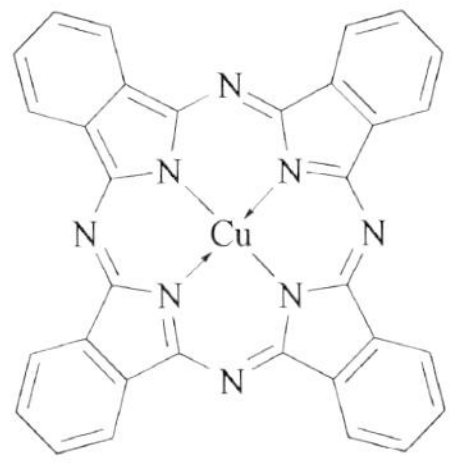

3

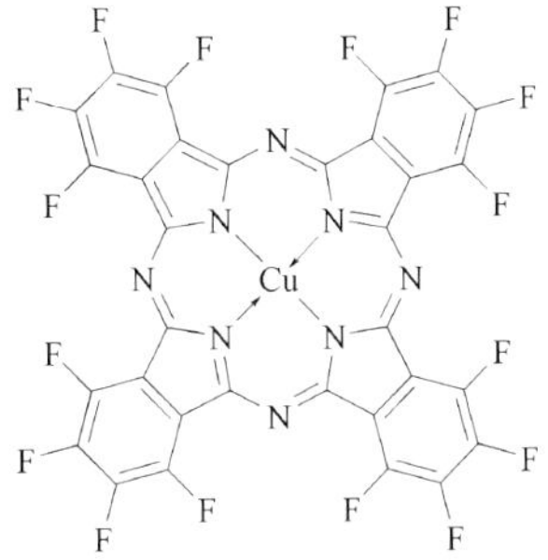

62

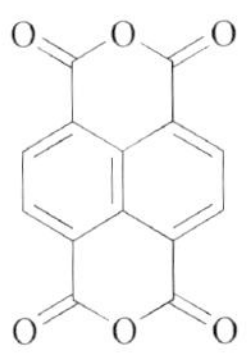

63

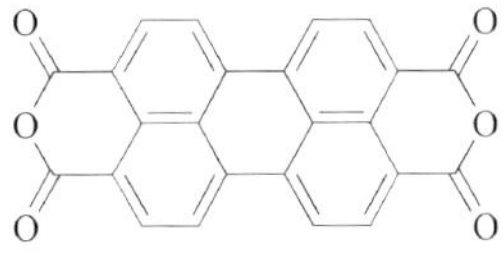

64

Electron-transport materials

Figure 6.15 Typical organic semiconductors

6.3.4.2 Solar Cells

Solar cells, or photovoltaic devices, have been studied for many years [3]. Most of the current work is focused on dye-sensitized nanocrystalline solar cells. These provide a technical and economically viable alternative to present-day photovoltaic devices. In contrast to conventional systems, in which the semiconductor assumes both the task of light absorption and charge carrier transport, the two functions are separated in dye-sensitized nanocrystalline solar cells [54] (cf. OPCs). Light is absorbed by the dye sensitizer, which is anchored to the surface of a wide-band-gap semiconductor. Charge separation takes place at the interface via photoinduced electron injection from the dye into the conduction band of the solid. Carriers are transported in the conduction band of the semiconductor to the charge collector.

The nanocrystalline solids are metal oxides, especially titanium dioxide [54–58]. Various dyes are used. Transition metal complexes such as (**65**) and (**66**) have broad absorption bands and allow the harvesting of a large fraction of sunlight [54, 58]. Fluorescent dyes are also used, such as Eosin-Y (**67**) [57]. Dye-sensitized nanocrystalline solar cells are now giving efficiencies in excess of 10 % [54, 58], compared to just 1 % ten years ago [3].

65 **66** **67**

Solar cells can be used as electric windows to generate low-cost electricity at peak times [59].

6.3.4.3 Nonlinear Optical Dyes

Nonlinear optics is the interaction of laser radiation with a substance to produce new radiation which is altered in phase, frequency, and amplitude from the incident radiation. There are several types of nonlinear effects but the most important are second-order frequency doubling and reverse saturable absorption.

Frequency Doubling. As the name implies, in frequency doubling a substance doubles the frequency of the incident laser radiation. This effect is important in telecommunications and optical data storage. For example, in telecommunications the most efficient way to transmit data is by using infrared radiation, e.g., 1200 nm radiation from an indium phosphide laser [60]. Detection of infrared radiation is inefficient. In contrast, visible radiation is much easier to detect but is an inefficient transmitter of data. Consequently, an important application of nonlinear optical (NLO) materials is to convert infrared radiation into visible and thus enable easier detection of the signal.

Frequency doubling in optical data storage would enable more data to be stored per unit area (ca. fourfold) because of the smaller spot size of the frequency-doubled radiation. However, the availability of low-cost semiconductor lasers, combined with the wavelengths required for CD-R and DVD-R systems (see earlier), has reduced the need for NLO materials in this area.

Molecules capable of high polarizability (hyperpolarizability) give the best NLO effects. Organic molecules such as (**68**), which have powerful donor and acceptor groups conjugated through a delocalized π-electron system, give the largest second-order effects. Fluorinated molecules such as (**69**), covalently

6.3.4 Electronic Materials

bound to an acrylic polymer, possess good NLO properties [39]. However, in all cases, the molecules must align themselves in the macroscopic state for the NLO effect to be realized [3].

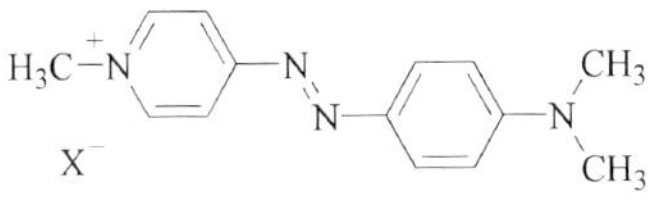

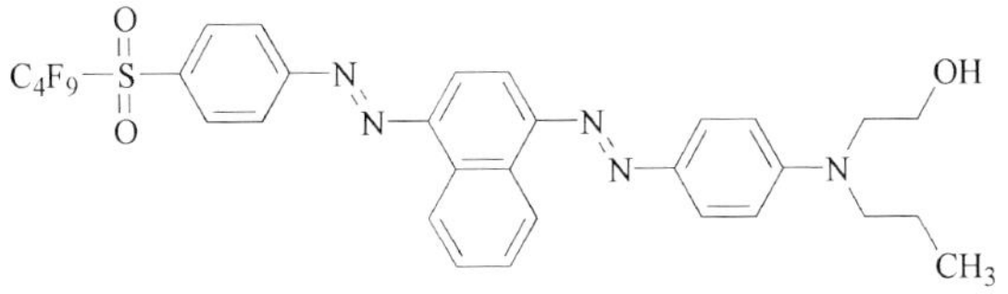

Reverse saturable absorption is another important nonlinear optical effect. Reverse saturable absorbers (RSAs) act as optical limiters by absorbing laser radiation. Therefore, they are particularly useful for laser protection, both civilian and military. Colorless infrared-absorbing RSAs which can absorb green laser radiation are especially important as they safeguard the eyes of pilots, tank commanders, etc. from enemy lasers [61].

The theory of RSAs is complex, involving two-photon absorption and absorption from S_1 to higher excited states [61]. Various dyes function as RSAs, including stilbenes (**70**) and (**71**), and phthalocyanines (**72**) [62].

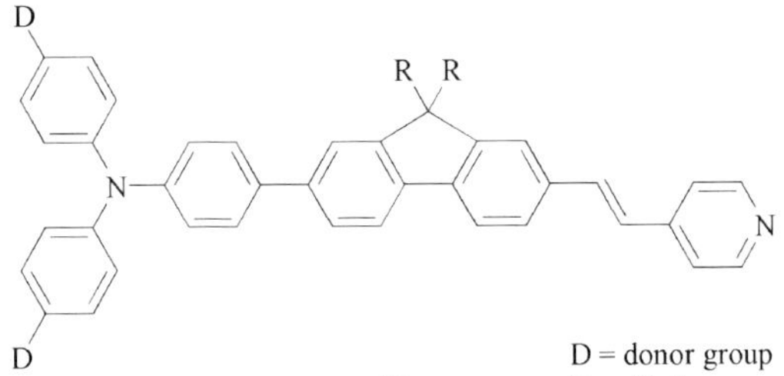

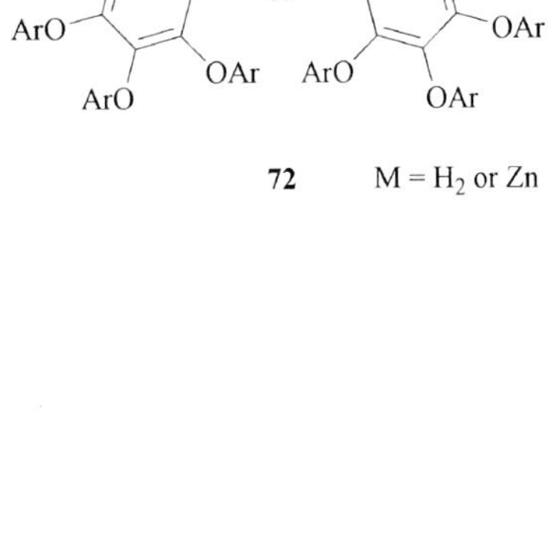

6.3.4.4 Laser Dyes

The main advantages of dye lasers are spectrum coverage (from less than 400 to 1000 nm) and tunability. In general, the laser dyes used nowadays have changed little from those used a decade ago [3]. Coumarins such as (**73**) and xanthenes such as Rhodamine 6G (**74**), are the main types. However, fluorinated coumarin dyes such as (**75**) have higher light stability than their nonfluorinated counterparts [39].

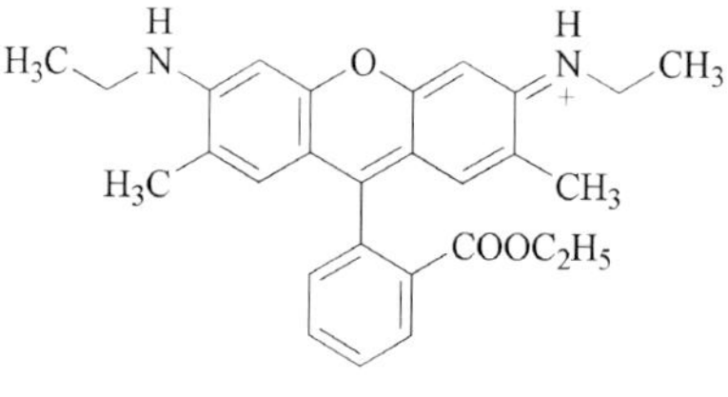

73

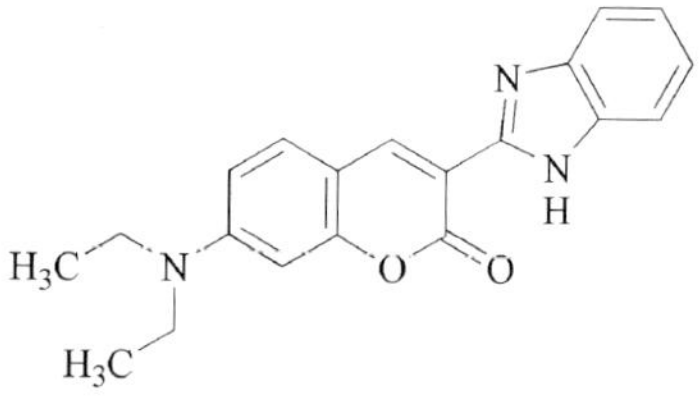

74

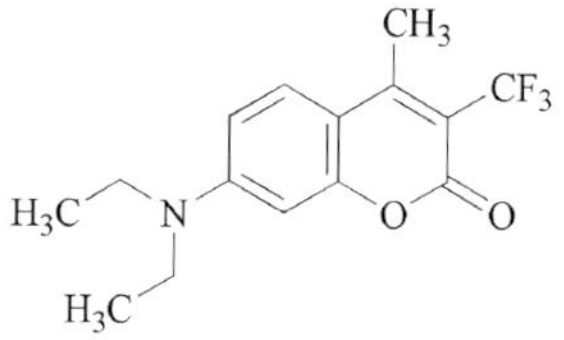

75

Most of the research over the past twenty years has been aimed at improving the performance of dye lasers by incorporating the laser dyes into a solid matrix [63]. The choice of dye has the biggest effect on performance, with pyrromethine dyes having the best performance [64]. Dyes such as Rhodamine 6G and Coumarin 540A have also been incorporated into an acrylic polymer matrix [65].

For more information on dyes for dye lasers, see [66].

6.3.5 Biomedical Applications

Dyes have a long history of use in medicine both for diagnostic and therapeutic purposes. Thus, crystal violet (**76**) is used for staining bacteria (Gram test), and the azo dye Prontosil rubrum (**77**) was the first drug that produced the active agent sulfanilamide (**78**) on reduction in the body [67].

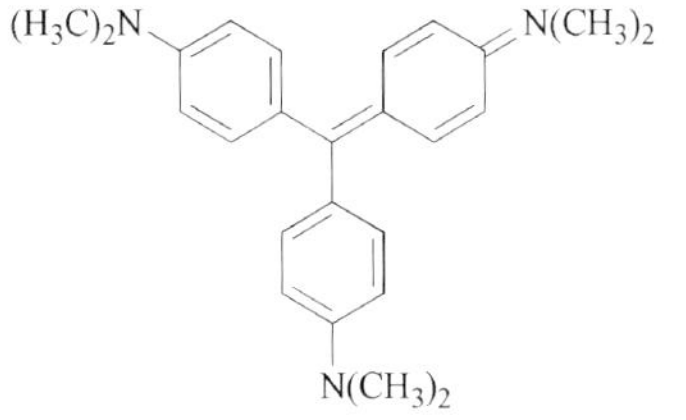

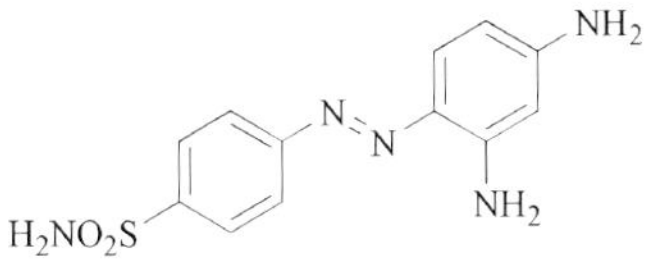

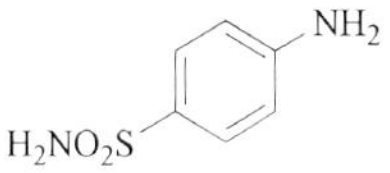

6.3.5 Biomedical Applications

[Structures 76, 77, 78]

More recently, azo dyes (**79**) bearing crown ether functionalities have been made to selectively detect metal cations such as lithium [68] and calcium [69].

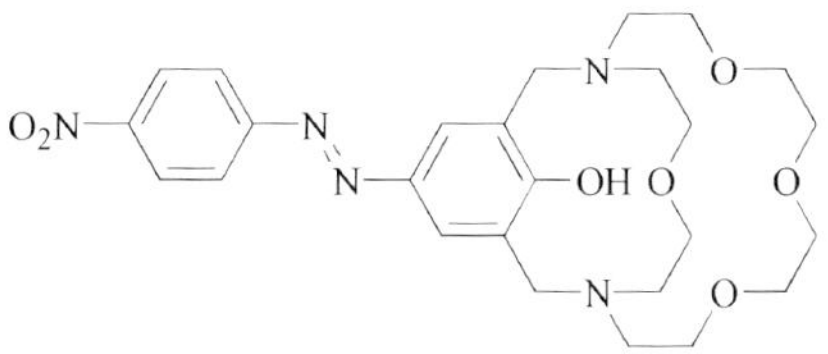

[Structure 79]

Much of the recent activity has focused on fluorescent sensors and probes and on therapy, particularly photodynamic therapy.

6.3.5.1 Fluorescent Sensors and Probes

Over the past 25 years fluorescent dyes, especially cyanines, have found increasing use in biology [70, 71]. Cyanine dyes (**80**) have narrow absorption and emission bands, high extinction coefficients (70 000 for Cy1 to 250 000 for Cy5), good quantum yields, and ready tunability (the absorption maxima are red-shifted by ca. 100 nm for each additional double bond). These attributes have led to use in many biological applications, including DNA sequencing, immunoassays, fluorescence microscopy, and single-molecule detection [70, 71]. For labeling proteins and modified DNA, a reactive group such as a succinidyl ester, is attached to the fluorescent dye. Reactive Cy5 (**81**) is a typical example [70].

X, Y = C(CH₃)₂, O and S

m = 0 Cy1
m = 1 Cy2
m = 2 Cy3
m = 3 Cy4
etc.

80

81

Over the last ten years, there has been increasing interest in the use of near-IR-active compounds, particularly the fluorescent species, for use in biomedical analytical applications. There are several reasons for this activity, perhaps the most significant being the lack of background interference from biological media in the 650–1000 nm region. This imaging window allows highly sensitive detection of analytes at very low levels. Infrared radiation, being of lower energy, has less of a damaging effect than the more commonly used UV or visible wavelengths and as such is preferred in the study of living tissue. Rayleigh scattering diminishes with the fourth power of wavelength, which implies that the penetration of IR light through tissues will be more efficient. This is important for both incident light and that emitted by fluorophores, especially those buried deep within biological tissue. Interest has continued to grow as IR excitation sources, typically semiconductor laser diodes, have become more readily available at low cost.

Infrared-active chromophores used in biological and medical applications are selected for their solubility, quantum yield (in the case of fluorophores), and their resistance to degradation in the system under investigation. It is very important to avoid perturbing the biological environment and thus creating artefacts. Consequently, dyes of a smaller molecular size tend to be preferred, such as the cyanines and squaryliums, over the larger molecules like the phthalocyanines. A reasonable degree of synthetic flexibility is essential in order to be able to introduce appropriate functionality through which to control solubility and other physical properties as well as to allow the formation of conjugates via covalent bond formation with active species such as antibodies. Infrared fluorophores, having a short fluorescence lifetime, generally exhibit better lightfastness than their visible counterparts and a decreased probability of nonradiative quenching.

Cyanines are by far the most popular and extensively researched chromophore for use in biological and medical imaging applications [70, 72]. They are used almost exclusively in fluorescence-related protocols. Their popularity may be attributable to factors such as their excellent synthetic flexibility and fluorescence properties, the latter being very high in some instances. Cyanines are the only synthetic near-infrared fluorophores which are commercially available (for example, Cy5 and Cy7 (**82**) from Amersham) for use in this area [70, 73]. There are a number of different structural types of cyanine for sale, mainly as laser dyes (see above), and several have been adapted for use in biological research [74].

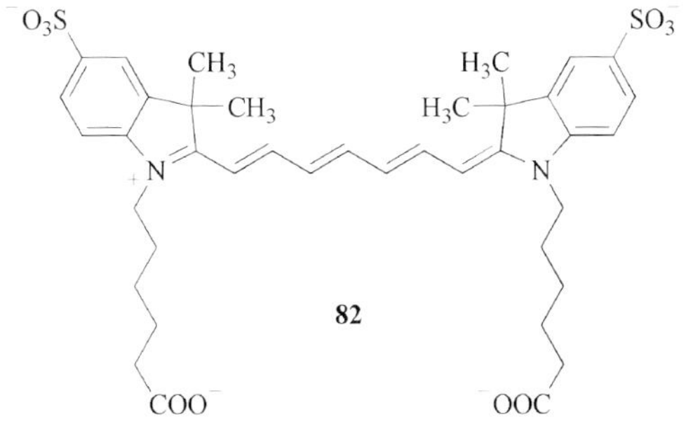

82

Squarylium dyes such as (**83**) [75] have probably received less attention than cyanine dyes due to the fact that the majority of syntheses furnish symmetrical species which are difficult to monofunctionalize in reactions such as the formation of peptide conjugates. The unsymmetrical types have been reported but seem to suffer from about 50 % decrease in extinction coefficient. Squaryliums are also more difficult to handle due to their low solubility. Very few water-soluble systems have been reported. These compounds are also used exclusively as fluorophores, but quantum yields are highly dependent on substituents and environment.

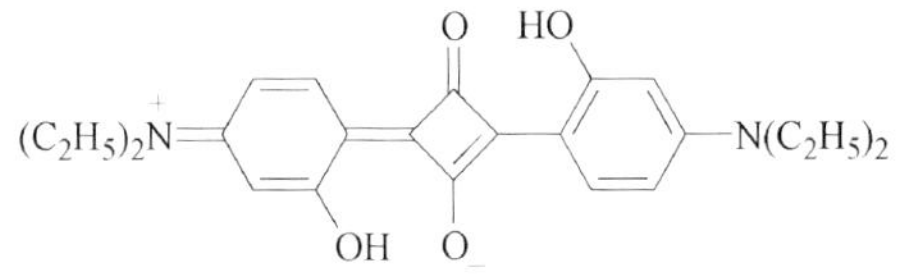

83

Biological and medical imaging at its most sophisticated involves some means of targeting specific features of interest. Typically, this is achieved by using antibodies which have been generated in response to the structure or analyte to be studied. The antibodies are covalently bound to the imaging agent, usually a fluorophore, prior to introduction into the subject. More recent research has shown that peptides which mimic the epitopic region of the antibody can be used as vectors. This is a significant improvement since peptides can be easily synthesized to order at a much lower cost.

6.3.5.2 Photodynamic Therapy

Photodynamic therapy (PDT) is a developing approach to the treatment of cancer and certain other diseases, such as age-related macular degeneration, which uses a combination of a photosensitizing dye and laser light to obtain a therapeutic effect [75]. There is also an absolute requirement for oxygen. If any of the three components, photosensitizer, light, or molecular oxygen is missing, there is no biological effect.

The application of PDT to a patient is simple. The photosensitizing dye is administered to the patient, who is then kept in the dark for up to 48 h, during which the dye becomes optimally located in any tumor tissue. A predetermined dose of laser light is delivered, typically for ca. 15 min. This is believed to produce highly reactive singlet oxygen and/or radicals, which kill the tumor [76].

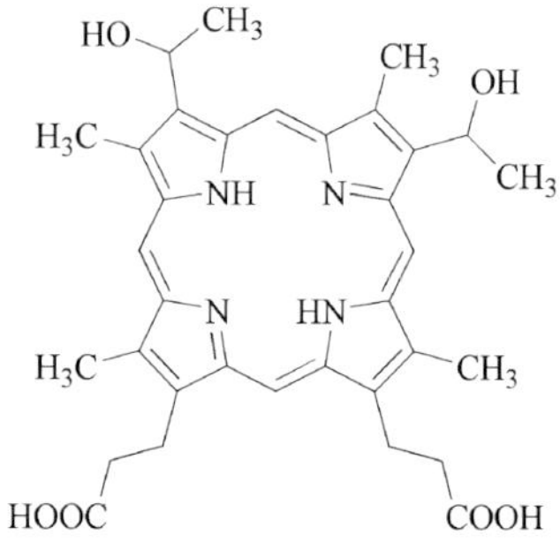

84

Some of the key properties of a photosensitizer are high efficiency of singlet-oxygen generation, strong absorption in the red and particularly near-IR (660–800 nm), preferential affinity for tumor rather than healthy tissue, and rapid clearance from the body. Porphyrin and phthalocyanine dyes fit these criteria best, and much work has been done on these dye types [76]. Photofrin, a hematoporphyrin derivative, was the first photosensitizer to be approved for clinical use. It is a complex mixture of monomeric porphyrins (protoporphyrin, hematoporphyrin (**84**), and hydroxyethylvinyldeuteroporphyrin) and oligomers of these porphyrins.

Second-generation photosensitizers include chlorins such as *meta*-tetrahydroxyphenyl chlorin (**85**) and *N*-aspartylchlorin [76], and phthalocyanines such as the zinc phthalocyanines (**86**) [77] and (**87**) [78].

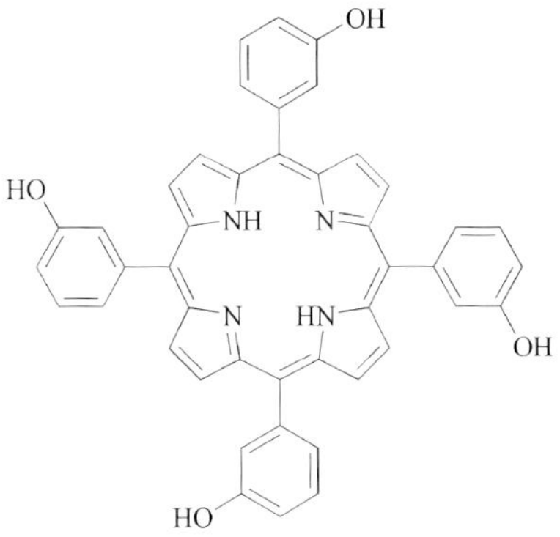

85

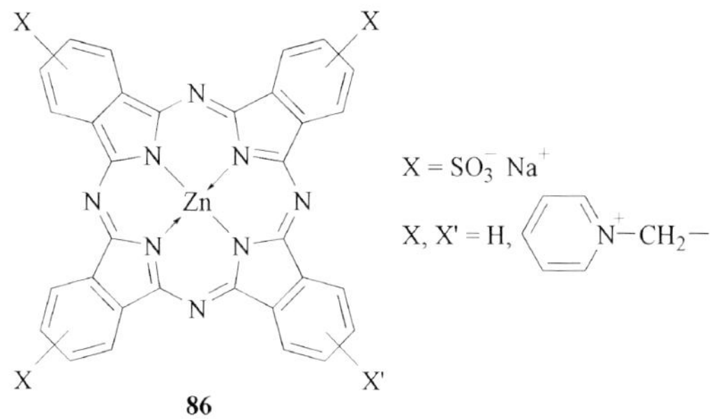

86 X = SO₃⁻ Na⁺

X, X' = H, [pyridinium-CH₂—]

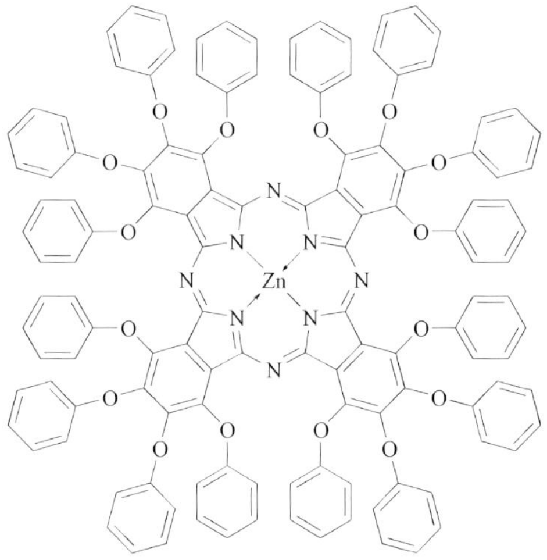

87

6.4 References

[1] P. Gregory, *Rev. Prog. Coloration* **24** (1994) 1–16.
[2] R. S. Gairns, "Electrophotography" in *Chemistry and Technology of Printing and Imaging Systems*, P. Gregory (ed.), Blackie, London, 1996.
[3] P. Gregory, *High Technology Applications of Organic Colorants*, Plenum, New York, 1991.
[4] Avecia, WO 0 032 537, 1999.

6.4 References

[5] W. Bauer, R. Baur, J. Geisenburger, H.-T. Macholdt, W. Zoeller, *Am. Ink Maker* **77** (1999) no. 6, 48.
[6] U. Schlosser et al., *Shikizai Kyokaishi* 70 (1997) no. 2, 92.
[7] R. Bradbury, "Thermal Printing" in *Chemistry and Technology of Printing and Imaging Systems*, P. Gregory (ed.), Blackie, London, 1996.
[8] Y. L. Koch, *Chem. Rev.* **93** (1993) 449.
[9] R. A. Hann in *Chemical Technology in Printing and Imaging Systems*, J. A. G. Drake (ed.), Royal Society of Chemistry, 1993.
[10] K. W. Hutt, I. R. Stephenson, H. C. v. Tran, A. Kaneko, R. A. Hann, Laser Dye Transfer, *Proceedings of IS&T's 8th International Congress on NIP Technologies*, Williamsburg, VA, Oct. 25–30, 1992, p. 367.
[11] P. Gregory, *Chem. Brit.* Jan., 1989, 47.
[12] R. Bradbury in *Modern Colorants—Synthesis and Structures*, A. T. Peters, H. S. Freeman (eds.), Elsevier, London, 1995.
[13] P Gregory, *Chemistry in Britain*, August 2000, p. 39.
[14] S. Hindagolla, D. Greenwood, N. Hughes, "Waterfast Dye Development for Ink Jet Printing", *IS and Ts 6th Int. Congress Advances in Non-Impact Printing*, Orlando, Florida, Oct 21–26, 1990.
[15] J. S. Campbell, P. Gregory, P. M. Mistry, "Novel Waterfast Black Dyes for Ink Jet Printing", *IS and Ts 7th Int. Congress on Advances in Nonimpact Printing*, Portland, Oregon, Oct 6–11, 1991.
[16] R. W. Kenyon in *Chemistry and Technology of Printing and Imaging Systems*, P. Gregory (ed.), Blackie, Glasgow, 1996, chap. 5, p 113.
[17] R. A. Work, H. J. Spinelli, *R and R News*, Sept. 1996, p. 28.
[18] J. R. Provost, *Surface Coating International* **77** (1994) 36.
[19] J. R. Provost, *Color Science '98* **2** (1999) 159.
[20] P. Gregory, "Electrostatic, Ionographic, Magnetographic and Embryonic Printing" in: *Chemistry and Technology of Printing and Imaging Systems*, P. Gregory (ed.), Blackie, London, 1996.
[21] A. Castegnier, *IS&T's NIP 12, International Conference on Digital Printing Technologies*, Montreal, 1996, p. 276.
[22] Meade Corp., US 4 399 209, 1983.
[23] *Chem. Eng. News*, Dec. 9, 1991, p. 29.
[24] Y. Sakuma, B. R. Waterhouse, A. S. Diamond, *Reprographic Chemicals Worldwide*, SRI International, Menlo Park, 1989.
[25] P. Hunt, "Optical Data Storage Systems" in *Chemistry and Technology of Printing and Imaging Systems*, P. Gregory (ed.), Blackie, London, 1996.
[26] R. Hurditch, "Dyes for Optical Storage Disc Media" *Chemichromics '99*, New Orleans, Jan. 27–29, 1999.
[27] V. Novotny, L. Alexandru, "Light Induced Phenomena in Dye-Polymer Systems" *J. Appl. Phys.* **50** (1979) 1215.
[28] M. Emmelius, G. Pawlowski, H. W. Vollman, "Materials for Optical Data Storage", *Angew. Chem. Int. Ed.* **28** (1989) 1445.
[29] J. E. Kuder, *Proc. SPIE 40th Annual Conference*, May 17–22, 1097.
[30] M. Matsui, "Optical Recording Systems" in *Infrared Absorbing Dyes*, M. Matsuoka, (ed.), Plenum, New York, 1990.
[31] ICI, EP 155 780, 1984 (P. J. Duggan, P. F. Gordon).
[32] Ciba-Geigy, US 5 594 128, 1997.
[33] Ciba, Supergreen Product Dossier, Issue 2.0, November 1999.

[34] J. Seto, "Photochromic Dyes" in *Infrared Absorbing Dyes*, M. Matsuoka, (ed.), Plenum, New York 1990.
[35] A.M. Cox, R. D. Blackburn, D. P. West, T. A. King, F. A. Wade, D. A. Leigh, *Appl. Phys. Lett.* **68** (1996) 2801.
[36] Eastman Kodak, US 5 372 907, 1994; Eastman Kodak, US 5 372 915, 1994.
[37] P. Gregory, *J. Porphyrins Phthalocyanines* **3** (1999) 468–476.
[38] G. W. Gray, *Molecular Structure and the Properties of Liquid Crystals*, Academic Press, London, 1962.
[39] M. Matsui, *Color Science '98, Vol. 1: Dye and Pigment Chemistry*, J. Griffiths (ed.), University of Leeds 1999, p. 242.
[40] S. Musa, "Working Knowledge – Active Matrix Liquid Displays", *Scientific American*, Nov., 1997.
[41] R. W. Sabnis, *Displays* **20** (1999) 119.
[42] M. Iwasaki, *Kinosei Ganryo no Gijutsuto Oyo Tenkai*, Tokyo, 1998, 188.
[43] Canon, EP 665 449, 1995; Avecia, WO 0 029 493, 2000; Dai Nippon Printing, EP 1 008 873, 2000.
[44] Kodak, US 4 720 432, 1988.
[45] C. W. Tang, S. A. van Slyke, *Appl. Phys.* **51** (1987) 913.
[46] J. H. Burroughes, P. L. Burns, D. D. C. Bradley, A. R. Brown, R. N. Marks, K. MacKay, R. H. Friend, A. B. Holmes, *Nature* **347** (1990) 539.
[47] Kodak, US 4 539 507, US 4 720 432, US 5 061 569.
[48] B. S. Chuah, A. B. Holmes, S. C. Moretti, J. C. de Mello, J. J. M. Hallas, R. H. Friend, *Color Science '98, Vol. 1: Dye and Pigment Chemistry*, J. Griffiths (ed.), University of Leeds 1999, p. 233.
[49] Gentex, US 4 902 108, 1990; Gentex, US Patent Application, 1 992 000 874 175, 1992.
[50] Z. Bao, J. A. Rogeri, H. E. Katz, *J. Mater. Chem.* **9** (9) 1895 1999.
[51] R. D. McCullough, *Adv. Mater.* **10** (1999) 93.
[52] BP, EP 0 182 548, 1984; Allied Signal, US 5 162 473, 1992; Motorola, US 5 192 930, 1992; M. Matters, D. M. DeLeeuw, *Opt. Mater.* **12** (1999) 89.
[53] M. Matters, D. M. DeLeeuw, *Synth. Metals* **85** (1997) 1403.
[54] M. Gratzel, *Prog. Photovoltaics* **8** (2000) 171.
[55] A. K. Jana, *J. Photochem. Photobiol. A* **132** (2000) 1.
[56] M. Gratzel, *Optoelectron. Prop. Inorg. Compd.* **1999**, 169.
[57] H. Arakawa, K. Sayama, *Res. Chem. Intermed.* **26** (2000) 145.
[58] S. K. Deb, R. Ellingson, S. Ferrere, A. J. Frank, B. A. Gregg, A. J. Nozik, N. Park, G. Schlichtorl, A. Zaban, *AIP Conf. Proc.* **1999**, 473.
[59] K. Kalyanasundaram, M. Gratzel, *Proc. Indian Acad. Sci. Chem. Sci.* **109** (1997) 447.
[60] H. Nakazumi, *J. Soc. Dyers Colourists* **104** (1988) 121.
[61] C. W. Spangler, *J. Mater. Chem.* **9** (1999) 2013.
[62] Zeneca, US 5 486 274, 1996.
[63] K.Iragashi, *Shikai Kyokaishi* **70** (1997) 102.
[64] D. M. Rahn, T. A. King, *Proc. SPIE-Int. Soc. Opt. Eng.* **1999**, 3613.
[65] A. Costela, I. Garcia-Moreno, J. M. Figuera, F. Amat-Guerri, R. Sastre, *Laser Chem.* **18** (1998) 63.
[66] I. D. Watkins, G. E. Tulloch, T. Maine, L. Spiuia, D. R. MacFarlane, J. L. Woolfrey, *Ceram. Trans.* **81** (1998) 223.
[67] A. Albert, *Selective Toxicity: The Physico-Chemical Basis of Therapy*, 5th Ed., Chapman and Hall, London, 1973, pp. 139–141.
[68] A. F. Scholl, I. O. Sutherland, *J. Chem. Soc. Chem. Comm.* **1992**, 1716.
[69] A. Mason, I. O. Sutherland, *J. Chem. Soc. Chem. Comm.* **1994**, 1131.

[70] A. Waggoner, L Ernst, *Color Science '98, Vol 1: Dye and Pigment Chemistry,*, J. Griffiths (ed.), University of Leeds, 1999 p. 298.

[71] R. W. Horobin, *Color Science '98, Vol. 1: Dye and Pigment Chemistry*, p. 242, J. Griffiths (ed.), University of Leeds, 1999.

[72] United States Patent 5 286 846, 1993 (A. Waggoner et al.).

[73] S. Hardwicke et al., poster presented at Drug Discovery '99, Boston, MA, August, 1999.

[74] J. Mama, *Advances Color Sci. Technol.* **2** (1999) no. 3, 162.

[75] T. J. Dougherty, *Photochem. Photobiol.* **58** (1993) 895.

[76] J. E. Brown, S. B. Brown, D. I. Vernon et al., *Color Science '98, Vol 1: Dye and Pigment Chemistry*, J. Griffiths (ed.), University of Leeds, 1999, p. 259.

[77] J. E. Cruse-Sawyer, J. Griffiths, B. Dixon, S. B. Brown, *Color Science '98, Vol 1: Dye and Pigment Chemistry*, J. Griffiths (ed.), University of Leeds, 1999, p. 309.

[78] Zencca, EP 484 027, 1990.

7 Optical Brighteners

7.1 Introduction

Optical brighteners or, more adequately, fluorescent whitening agents (FWAs) are colorless to weakly colored organic compounds that, in solution or applied to a substrate, absorb ultraviolet light (e.g., from daylight at ca. 300–430 nm) and reemit most of the absorbed energy as blue fluorescent light between ca. 400 and 500 nm.

In daylight optical brighteners can thus compensate for the esthetically undesirable yellowish cast found in white industrial substrates, such as textiles, papers, or plastics. Furthermore, since a portion of the daylight spectrum not perceived by the eye is converted to visible light, the brightness of the material is enhanced to give a dazzling white.

Figure 7.1 illustrates the action of optical brighteners [1–3]. White substrates show a yellowish cast because their reflectance for incident visible light is lower at short wavelengths than at long ones (curve a), i.e., they absorb short-wavelength light.

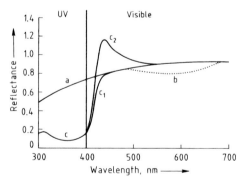

Figure 7.1 Reflectance spectra for a textile material
a) Bleached; b) Blued; c) Treated with optical brightener [25] (c_1 absorption, c_2 fluorescence)

7.1 Introduction

The first method of compensating for the yellowish cast was the use of a blue dye such as ultramarine or indigo. Bluing lowers the reflectance of the sample in the long-wavelength part of the visible spectrum (curve b). As a result, the sample takes on a neutral white appearance, but at the same time it loses brightness so that it looks grayer.

Optical brighteners lower the reflectance mainly in the UV and near-visible by absorption (curve c_1); at visible wavelengths (mostly with a maximum at 435–440 nm), they greatly increase the reflectance through fluorescence (curve c_2). The brightener acts as a supplementary emission source. Optical brighteners are more effective the cleaner and whiter the substrate is. In no case can they replace cleaning or bleaching.

7.1.1 Physical Principles

Figure 7.2 illustrates the processes involved in light absorption and fluorescence by optical brighteners [4–8].

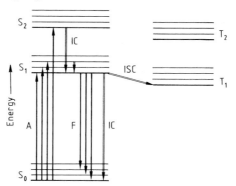

Figure 7.2 Energy diagram of optical brighteners and transitions
A = absorption; F = fluorescence; IC = internal conversion; ISC = intersystem crossing; S = singlet state; T = triplet state

Absorption (A) of light quanta by the brightener molecules induces transitions from the singlet ground state S_0 to vibrational levels of the electronically excited singlet states S_1 and S_2. The electronic state and vibrational level reached depend on the wavelength of the absorbed light. Absorption occurs within ca. 10^{-15} s and results in a gain in energy E and a change in the electron distribution within the brightener molecule.

Brightener molecules that have been excited to a higher electronic state such as S_2 or into a higher vibrational level of S_1 relax by a nonradiative process within ca. 10^{-12} s to the vibrational ground state of S_1 (internal conversion, IC), which has a lifetime of ca. 10^{-9} s. This time is sufficient for the brightener geometry to adapt to the electron distribution in the S_1 state.

Brighteners in the S_1 state are deactivated by several routes. Fluorescence results from radiative transitions to vibrational levels of the ground state. The fluorescence emission has a lower average energy than the absorbed light (Figure 7.2), i.e., fluorescence occurs at longer wavelengths than absorption (Stokes shift).

Deactivation processes competing with fluorescence are mainly nonradiative deactivation to the S_0 state (IC) and nonradiative transition to a triplet state (intersystem crossing, ISC). Photochemical products are often formed from this triplet state. Important photochemical reactions are the $E \rightarrow Z$ isomerization of ethylene, the oxidation of pyrazoline to pyrazole, and the dimerization of coumarins.

The efficiency of fluorescence is measured by the quantum yield Φ:

$$\Phi = \frac{\text{Number of quanta emitted}}{\text{Number of quanta absorbed}}$$

It is determined by the relative rates of fluorescence emission and the competing processes.

In many brighteners deactivation from the S_1 state depends on the local environment. When fixed in (highly viscous) solid substrates, brighteners fluoresce with high quantum yields (Φ ca. 0.9). In (low-viscosity) solvents, such as are used before or during application of the brightener, the quantum yield is often low and the probability of formation of photoproducts is higher.

Requirements for optimally effective brighteners are as follows:

1) Brighteners should absorb as much UV light as possible and should have little inherent color. This means that the absorption maximum should lie in the 350–375 nm range, the extinction coefficient should be high, and the absorption band should decrease steeply near 400 nm.
2) Fluorescence should be as intense as possible and produce maximum whiteness. This means that the quantum yield should be near 1.0. The fluorescence maximum should be between 415 and 445 nm depending on the preferred shade (violet-blue to greenish blue),and the fluorescence band should decrease as rapidly as possible on the long-wavelength side.
3) The brightener distribution in the substrate should be monomolecular, even at high concentrations, because aggregates of brightener molecules show dramatically lowered quantum yields and often different spectra.

Figure 7.3 shows typical absorption (A) and fluorescence (F) spectra for optical brighteners. Fluorescence spectra F_1 are measured in dilute solutions. On brightened substrates at the usual (i.e., higher) brightener concentrations, short-wavelength fluorescence is reabsorbed by the brightener with the result that the fluorescence band becomes narrower (F_2). A further increase in brightener concentration shifts the center of the band towards longer wavelengths corresponding to more greenish effects, up to unwanted shades.

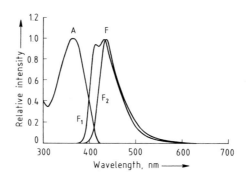

Figure 7.3 Typical absorption (A) and fluorescence (F) spectra of optical brighteners [25]
F_1 = dilute solution; F_2 = applied to substrate

7.1.2 Molecular Structure

The chromophores of optical brighteners are π electron systems in which π–π* transitions occur. The chromophores must be rigid [5], and their conformations should differ only slightly in the electronic ground state and in the first excited state.

Brightener chromophores can be regarded as combinations of building blocks with the following numbers of π electrons:

2 π electrons	vinylene, imino
4 π electrons	2-pyrazolin-1,3-diyl
16 π electrons	1,4-phenylene, 2,5-furanediyl, 2,5-thiophenediyl, phenyl, fur-2-yl, pyrazol-4-yl, pyrazol-1-yl, 1,2,3-triazol-2-yl, 1,2,4-triazol-1-yl, 1,3,5-triazin-2-yl, 1,2,4-oxadiazol-5-yl
10 π electrons	1,4-naphthylene, benzofuran-2-yl, benzoxazol-2-yl, benzimidazol-2-yl
12 π electrons	3,7- coumarindiyl
14 π electrons	naphtho[1,2-*d*]triazol-2-yl
16 π electrons	naphthalimido, pyren-1-yl

The individual building blocks absorb at wavelengths too short for brighteners. Linking the building blocks into a system of conjugated double bonds yields chromophores that absorb at longer wavelengths.

By themselves, the building blocks with two or four π electrons (chiefly vinylene and 2-pyrazolin-1,3-diyl) absorb at very short wavelengths, but they make a significant contribution to shifting the absorption of other building blocks to longer wavelengths. They are usually combined with phenyl and phenylene groups. Examples are compounds **4**, **13**, and **49**.

In contrast, the cyclic building blocks with six π electrons give a very slight bathochromic shift. If several of these building blocks are combined, brighteners of great length and high molecular mass, as well as a wide range of structures, can be obtained. Examples are compounds **10** and **32**.

The absorption and fluorescence of coumarin with an exocyclic π center, naphthalimide, and pyrene are close to the favorable region for brighteners, and so only slight lengthening of these chromophore systems is necessary (**63**) or possible (**64–66, 67**).

The optical properties of chromophores can be varied by means of substituents. Electron donors (e.g., alkyl, alkoxy, or substituted amino groups) and electron acceptors (e.g., cyano, alkylsulfonyl, or carbalkoxy groups) can intensify fluorescence depending on their position on the chromophore.

Substituents that have little influence on the chromophore system affect the application properties and are important for substantivity (see Section 4).

For reviews of optical brightener systems, see [1, 3, 6–21].

7.1.3 History of Whitening

Optical brightening was discovered in 1929 by P. Krais [22]. The whiteness of viscose rayon and semibleached flax yarn was increased by treatment with an aqueous solution of esculin (**1**) [*531-75-9*] and drying. Krais recognized that this effect was due to the strong blue fluorescence of esculin, a β-D-glucoside of 6,7-dihydroxycoumarin that is obtained by extracting horse chestnut bark. (For historical references, see [23].)

The first industrial application in 1935 involved the use of sodium umbelliferone acetate [*19491-88-4*] and starch [24] and, later, chiefly β-methylumbelliferone (**2**) [*90-33-5*] [6]. As with esculin, however, these treatments lacked light- and wetfastness.

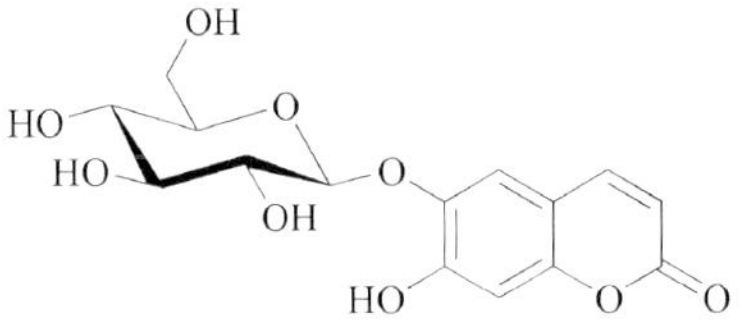

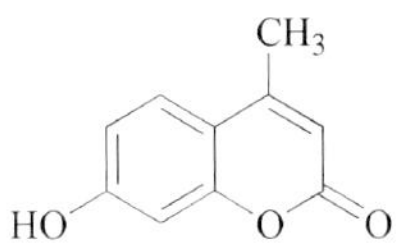

With the 1,3,5-triazin-2-yl derivatives of 4,4′-diaminostilbene-2,2′-disulfonic acid proposed by B. Wendt in 1940 [25], optical brighteners came into full-scale industrial use. In 1941, 4,4′-bis[(4-anilino-6-hydroxy-1,3,5-triazin-2-yl)amino]stilbene-2,2′-disulfonic acid disodium salt was introduced commercially under the trade name Blankophor B (**3**) [*1264-32-0*].

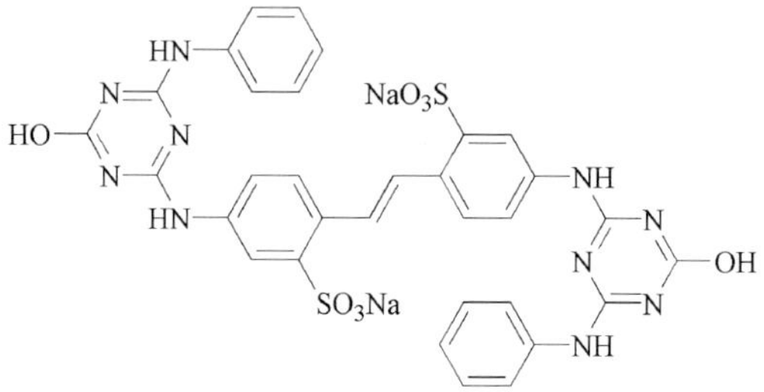

Subsequent development in all industrialized countries was rapid, especially during the period 1945–1975, and led to more than 1000 commercial products by 1989. Almost 400 compounds of distinct chemical structure or composition are registered in the Colour Index. Many of these compounds, including **3** and heterocycles such as imidazolone, dibenzothiophene-5,5′-dioxides, methine derivatives, dihydrocollidine, pyrazine, carbostyrils, pyrazolo[3,4-*b*]quinolines, and 5,6-benzocoumarins are no longer used and have been replaced by others.

7.2 Chemistry of Technical Products

7.2.1 Carbocycles

The first commercial carbocyclic products were derivatives of 4,4′-diaminostilbene-2,2′-disulfonic acid, e.g., 4,4′-bis(phenylureido)stilbene-2,2′-disulfonic acid disodium salt [2606-93-1] [26] and acylamino- or methoxy-substituted benzoylamino derivatives [6]. Because of their poor efficiency (i.e., yield of whiteness) and hue, these products have disappeared from the market.

Carbocyclic brighteners gained new impetus after 1959, when the trialkylphosphite-activated ethylenation process developed by Horner became available [27]. This process, which features a high yield and high selectivity for the *E* isomers needed for optical brighteners, is extremely interesting from an industrial point of view. The principal products are distyrylarenes and divinylstilbenes, which have retained their importance to the present day.

7.2.1.1 Distyrylbenzenes

Cyano-substituted 1,4-distyrylbenzenes (Table 7.1) are carbocyclic brighteners with a very high whiteness and good efficiency for plastics and synthetic fibers, especially polyesters [28]. Asymmetrical compounds, mixtures of two or three different bis(cyanostyryl)benzenes [29–32], or combinations with polyester brighteners differing in structure (e.g., **13**, **35**, **36**, and **38**) improve uptake and diffusion into the fiber [33] and allow the white hue to be varied over a wide range. On polyester substrates, however, the di(cyanostyryl)benzenes have only moderately good lightfastness.

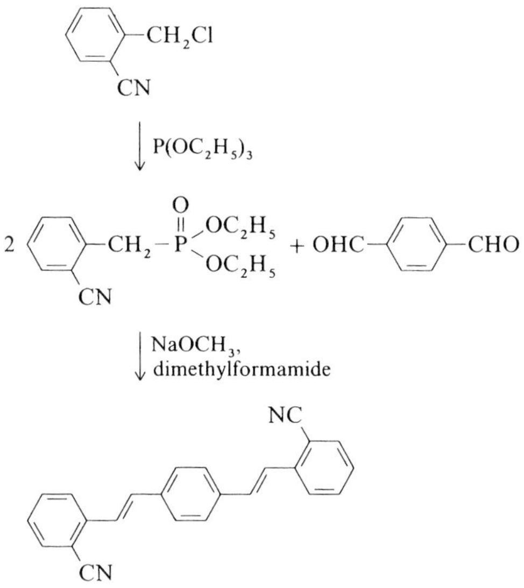

For the synthesis of **4**, 2-cyanobenzyl chloride is first treated with triethyl phosphite to give 2- cyano-(1-diethylphosphonomethyl)benzene [*23973-65-1*] , which reacts with terephthalaldehyde in the presence of sodium methoxide to yield 1,4-bis(2-cyanostyryl)benzene (**4**) [*13001-39-3*] [28]

Table 7.1 1,4-Distyrylbenzenes.

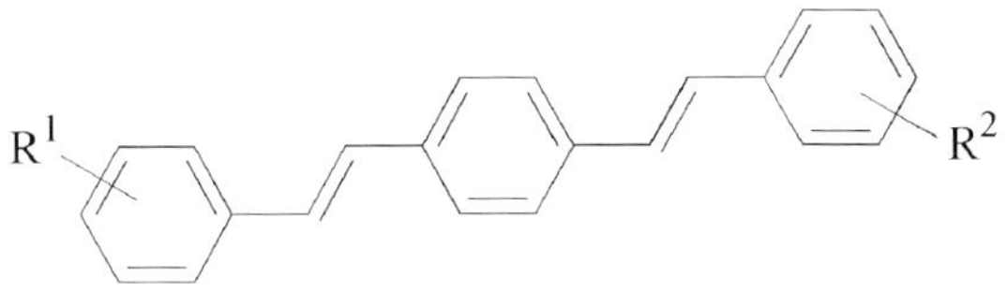

FWA	R^1 (position)	R^2 (position)	CAS registry no.	Reference
5	–CN (2)	–CN (3)	[*79026-03-2*]	
6	–CN (2)	–CN (4)	[*13001-38-2*]	[29]
7	–CN (3)	–CN (3)	[*36755-00-7*]	[28]
8	–CN (3)	–CN (4)	[*79026-02-1*]	[32]
9	–CN (4)	–CN (4)	[*13001-40-6*]	[28]

7.2.1.2 Distyrylbiphenyls

4,4′-Distyrylbiphenyls, discovered in 1967, can be produced in a similar way to distyrylbenzenes [34]. The reaction of 4,4′-bis(chloromethyl)biphenyl with trimethyl phosphite yields 4,4′-bis(dimethoxyphosphonomethyl)biphenyl [27344-43-0]. Reaction with, for example, benzaldehyde-2-sulfonic acid gives 4,4′-di(2-sulfostyryl)biphenyl disodium salt (**10**) [27344-41-8].

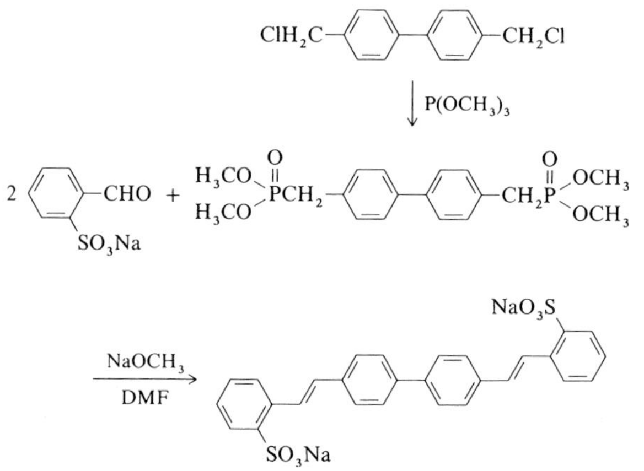

Anionic distyrylbiphenyls have very high efficiencies and are fairly chlorine stable brighteners that are used for cellulosic fibers and especially for detergents (Table 7.2). Nonionic [35] and cationic compounds [36] have also been developed, for example [72796-88-4]: R^1 = –O(CH$_2$)$_2$N$^+$MeEt$_2$, CH$_3$OSO$_3^-$ in the 2-position, R^2 = H (see formula in Table 7.2).

Table 7.2 4,4′-Distyrylbiphenyls

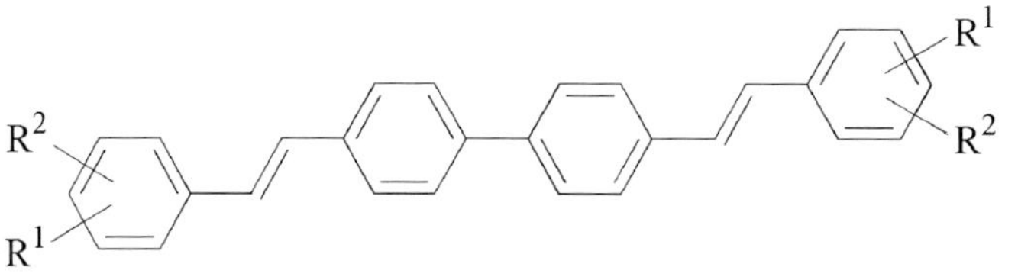

FWA	R^1 (position)	R^2 (position)	CAS registry no.	Reference
11	–SO$_3$Na (3)	Cl (4)	[42380-62-1]	[34]
12	–OCH$_3$ (2)	H	[40470-68-6]	[35]

7.2.1.3 Divinylstilbenes

Polyester divinylstilbene brighteners with a very high efficiency [37] include 4,4′-bis(ethoxycarbonylvinyl)stilbene (**13**) [*60683-03-6*], which can be obtained by the reaction of stilbene-4,4′-dicarbaldehyde with triethyl phosphonoacetate in the presence of sodium methoxide.

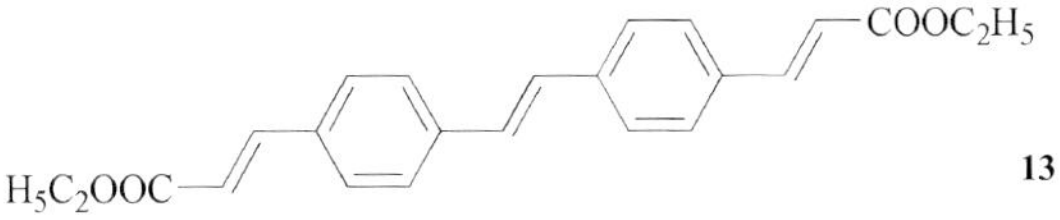

$$H_5C_2OOC-CH=CH-C_6H_4-CH=CH-C_6H_4-CH=CH-COOC_2H_5$$

13

Another divinylstilbene brightener with an even higher efficacy is 4,4′-bis(cyanovinyl)stilbene [*60682-87-3*] [37, 38].

7.2.2 Triazinylaminostilbenes

1,3,5-Triazinyl derivatives of 4,4′-diaminostilbene-2,2′-disulfonic acid have been available since 1941 and remain the most important class of optical brighteners in terms of quantity. Some 75 derivatives have been put on the market [1]. In the textile industry, they find use in brightening cotton, regenerated cellulosic fibers, and polyamides. They can also be included in detergent formulations and used to brighten paper, either in the pulp or in the coating.

The starting compound for all triazinylaminostilbenes is 4,4′-dinitrostilbene-2,2′-disulfonic acid [*128-42-7*] , which is obtained by oxidizing 4-nitrotoluene-2-sulfonic acid with aqueous sodium hypochlorite in the presence of sodium hydroxide or, more recently, by atmospheric oxidation in an aqueous ammoniacal medium [39]. The Béchamps reduction using iron filings etched with hydrochloric acid yields 4,4′-diaminostilbene-2,2′-disulfonic acid [*81-11-8*] (DAS).

In subsequent steps, DAS is treated in acidic aqueous solution at 0–5 °C with cyanuric chloride dissolved in ethyl methyl ketone. The remaining chlorine atoms are then replaced by aliphatic, cycloaliphatic, or aromatic amines at 15–35 °C and then at ca. 60 °C. The addition of electrolytes (e.g., sodium carbonate, sodium hydroxide) or water-soluble aprotic solvents at 90 to 100 °C or above leads to the desired β crystal modification, which is nearly colorless.

Another sequence of reactions starting with DAS is needed for weakly basic amines, such as aniline sulfonic acids, and also for alcohols. These are initially treated with cyanuric chloride, then with DAS in an aqueous medium at 40–50 °C. In this way the number of side reactions is substantially reduced.

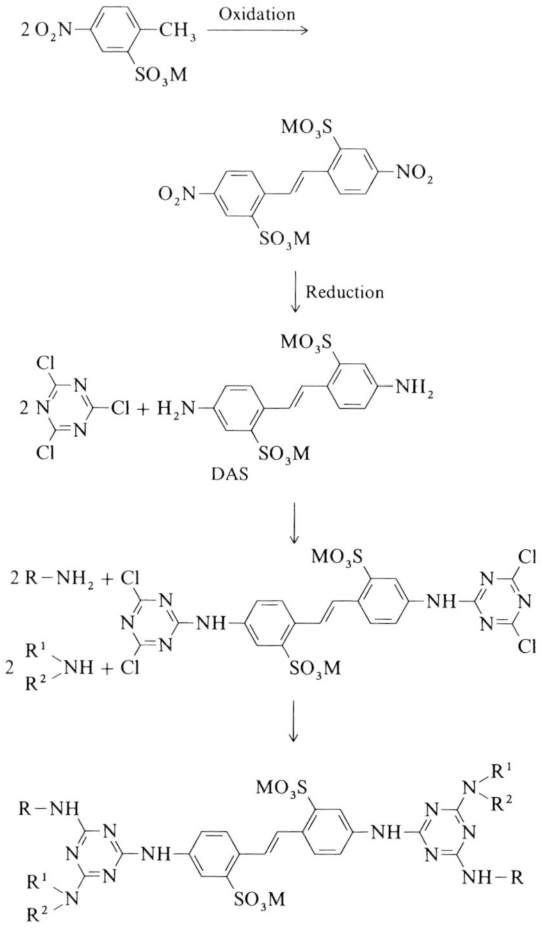

7.2 Chemistry of Technical Products

[Reaction scheme: 2-nitrotoluene-sulfonic acid derivatives undergo oxidation to form a dinitrostilbene disulfonic acid, followed by reduction to form 4,4'-diaminostilbene-2,2'-disulfonic acid (DAS), which reacts with cyanuric chloride and amines to give the final bis(triazinylamino)stilbene disulfonic acid brightener.]

M = H, Na

Some 40 triazinylaminostilbene brighteners are currently marketed; Tables 7.3 and 7.4 respectively list the important anilino and anilinosulfonic acid representatives of bis(4,4'-triazinylamino)stilbene-2,2'-disulfonic acid. The latter can be employed over a wide pH range. All of the listed compounds are distinguished by high whitening effects, good efficiency, and adequate lightfastness, but they have only partial fastness to bleaches and none to chlorite solutions.

7.2.2 Triazinylaminostilbenes

Table 7.3 Anilino derivatives of bis(4,4'-triazinylamino)stilbene-2,2'-disulfonic acid

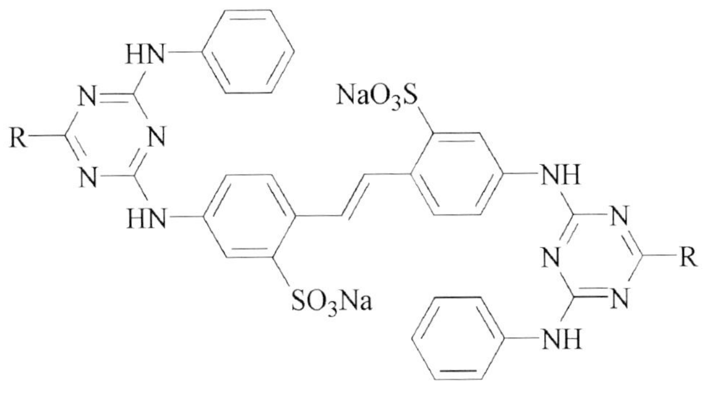

FWA	R	CAS registry no.	Reference
14	–OCH$_2$	[3426-43-5]	[40]
15	–NH–CH$_3$	[35632-99-6]	[41] – [43]
16	–NH–C$_2$H$_5$	[24565-13-7]	[42]
17	–NH–CH$_2$CH$_2$OH	[12224-16-7]	[44]
18	–N(CH$_3$)(CH$_2$CH$_2$OH)	[13863-31-5]	[45]
19	–N(CH$_2$CH$_2$OH)$_2$	[4193-55-9]	[25], [46]
20	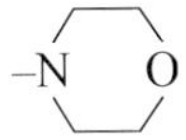 –N(morpholino)	[16090-02-1]	[41]
21	–NH–C$_6$H$_5$	[133-66-4]	[25], [46]
22	–N(CH$_2$CH$_2$OH)(CH$_2$CH$_2$CONH$_2$)	[68444-86-0]	[47]

Table 7.4 Anilinosulfonic acid derivatives of bis(4,4′-triazinylamino)stilbene-2,2′-disulfonic acid

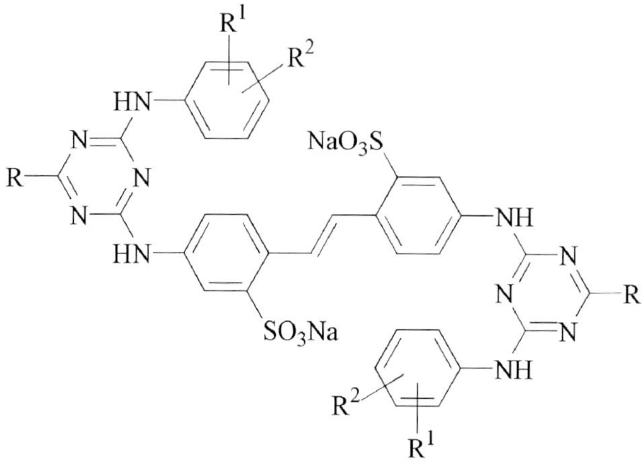

FWA	R	R^1 (position)	R^2 (position)	CAS registry no.	Reference
23	$-NH-CH_2CH_2OH$	$-SO_3Na$ (3)	H	[61968-74-9]	[48]
24	$-N(CH_2CH_2OH)_2$	$-SO_3Na$ (3)	H	[12224-02-1]	[49], [50]
25	$-N(CH_2CHCH_3)_2$ $\vert$ OH	$-SO_3Na$ (4)	H	[99549-42-5]	[51]
26	$-N(CH_2CH_2OH)_2$	$-SO_3Na$ (4)	H	[16470-24-9]	[50], [52]
27	$-N\begin{smallmatrix}CH_3\\CH_2CH_2OH\end{smallmatrix}$	$-SO_3Na$ (4)	H	[74228-28-7]	[53]
28	$-N(C_2H_5)_2$	$-SO_3Na$ (2)	$-SO_3Na$ (5)	[83512-97-4]	[54], [55]
29	$-N(CH_2CH_2OH)_2$	$-SO_3Na$ (2)	$-SO_3Na$ (5)	[76482-78-5]	[55]
30	$-N$(morpholino)	$-SO_3Na$ (2)	$-SO_3Na$ (5)	[55585-28-9]	[55]
31	$-N(CH_2CO_2Na)_2$	$-SO_3Na$ (4)	H	[174305-36-3]	[56]

7.2.3 Stilbenyl-2H-triazoles

Stilbenyl-2H-triazoles are used as highly chlorine and chlorite stable brighteners for cotton and polyamides.

7.2.3.1 Stilbenyl-2*H*-naphtho[1,2-*d*]triazoles

The first important member of this group was 4,4′-bis(6-sulfonaphtho[1,2-*d*]triazol-2-yl)stilbene-2,2′-disulfonic acid. As the tetrasodium salt [*7426-67-7*] [57], however, it produced a greenish white brightening effect on cotton. Removal of one of the naphthotriazole groups resulted in asymmetric compounds giving a neutral white, but at the cost of efficiency, e.g., the sodium salt of 4-(2*H*-naphtho[1,2-*d*]triazol-2-yl)stilbene-2-sulfonic acid [58].

Today, due to the moderate application properties this type of compound is no longer marketed.

7.2.3.2 Bis(1,2,3-triazol-2-yl)stilbenes

Bis(1,2,3-triazol-2-yl)stilbene brighteners have high efficiencies and good light-fastness and are very resistant to chemical bleaches. They are employed as brighteners for cotton and polyamides. An example is 4,4′-bis(4-phenyl-1,2,3-triazol-2-yl)stilbene-2,2′-disulfonic acid dipotassium salt (**32**) [*52237-03-3*] [59].

Compound **34** is obtained by bishydrazone formation from hydroxyiminoacetophenone (**33**) and 4,4′-dihydrazinostilbene-2,2′-disulfonic acid followed by cyclization with urea [60].

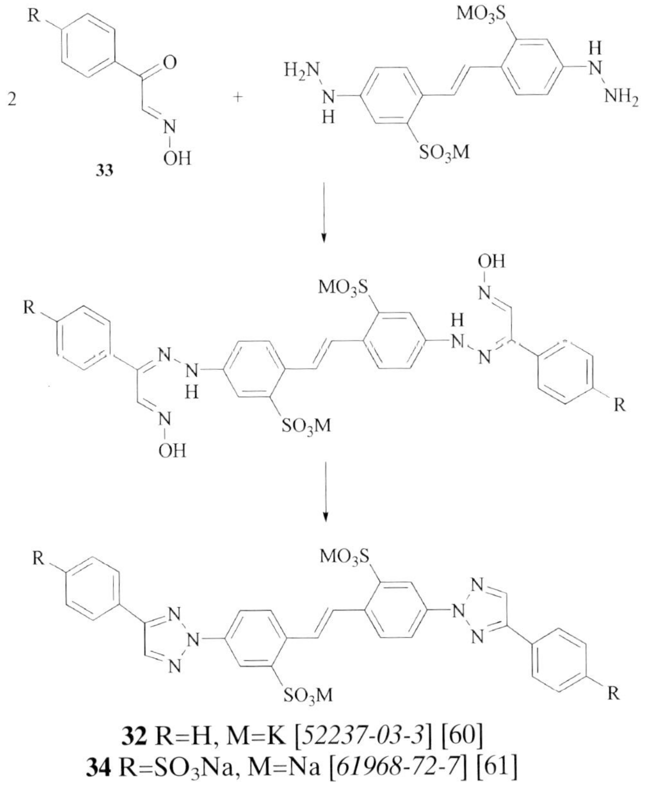

32 R=H, M=K [*52237-03-3*] [60]
34 R=SO$_3$Na, M=Na [*61968-72-7*] [61]

7.2.4 Benzoxazoles

The benzoxazole ring is especially suitable as an end group for lengthening conjugated systems that have the ability to resonate. The ring is easily synthesized and imparts good to excellent lightfastness, especially on polyester substrates. Brighteners are applied in the textile industry or added to the spinning mass. In addition, synthetic fibers such as polyamides and cellulose acetates, as well as plastics such as polyolefins, polystyrene, or poly(vinyl chloride) can be brightened with these compounds.

7.2.4.1 Stilbenylbenzoxazoles

Stilbenylbenzoxazoles are produced by cyclocondensation of stilbene-4-carboxylic acids with 2-aminophenols or by olefination with aromatic anils (anil synthesis); the latter requires activated tolyl compounds as reactants [62].

For example, 5,7-dimethyl-2-(4-tolyl)benzoxazole and the Schiff's base from diphenyl-4-carbaldehyde and aniline react at 60–65 °C to yield 5,7-dimethyl-2-(4'-phenylstilben-4-yl)benzoxazole (**35**) [*40704-04-9*] [63].

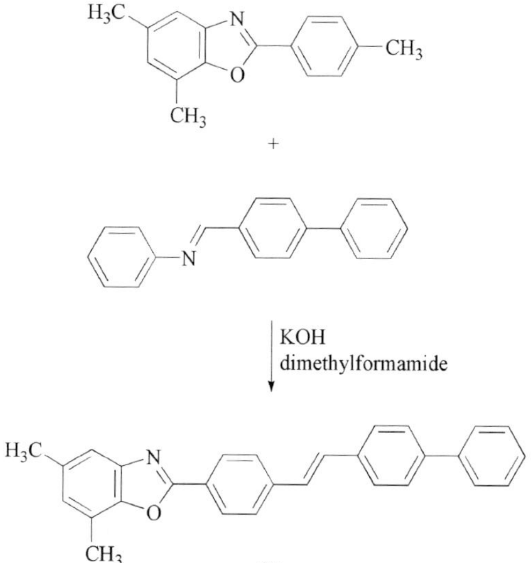

Other stilbenylbenzoxazoles are listed in Table 7.5. Mixtures of **35–37** with polyester brighteners of different structure (e.g., **4**, **6**, and **41**) are commercially available.

7.2.4 Benzoxazoles

Table 7.5 Stilbenylbenzoxazoles

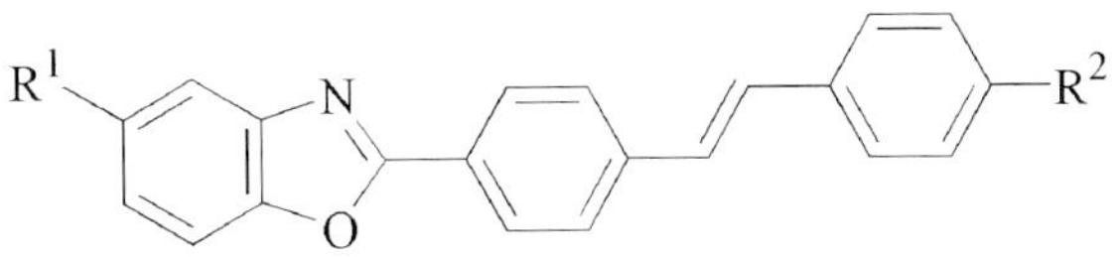

FWA	R^1	R^2	CAS registry no.	Reference
36	–CH$_3$	–COOCH$_3$	[18039-18-4]	[64], [65]
37	H	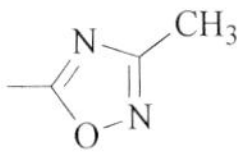 3-methyl-1,2,4-oxadiazol-5-yl	[64893-28-3]	[66]

7.2.4.2 Bis(benzoxazoles)

Table 7.6 lists the most important bis(benzoxazole) brighteners.

Table 7.6 Bis(benzoxazoles)

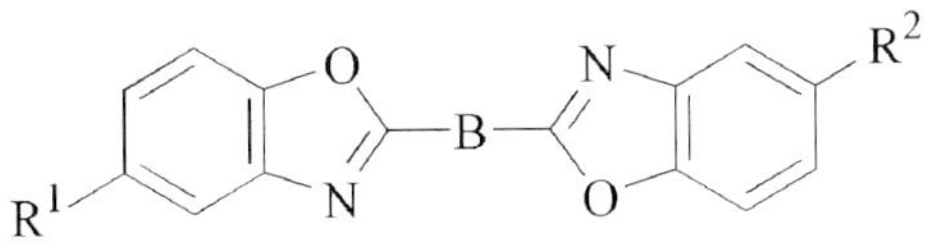

FWA	R^1	R^2	B	CAS registry no.	Reference
38	–CH$_3$	–CH$_3$	 –CH=CH–	[1041-00-5]	[67], [68]
39	H	H	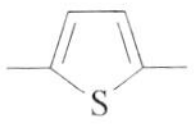 thiophene-2,5-diyl	[2866-43-5]	[69], [70]
40	–C(CH$_3$)$_3$	–C(CH$_3$)$_3$	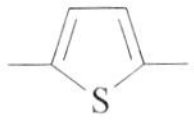 thiophene-2,5-diyl	[7128-64-5]	[71]
41	H	H	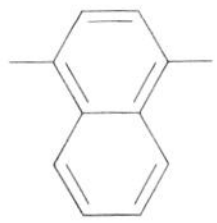 naphthalene-1,4-diyl	[5089-22-5]	[72]
42	–CH$_3$	–CH$_3$	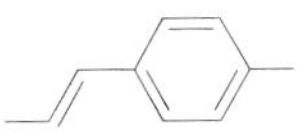 4-styryl-phenyl	[1552-46-1]	[73]

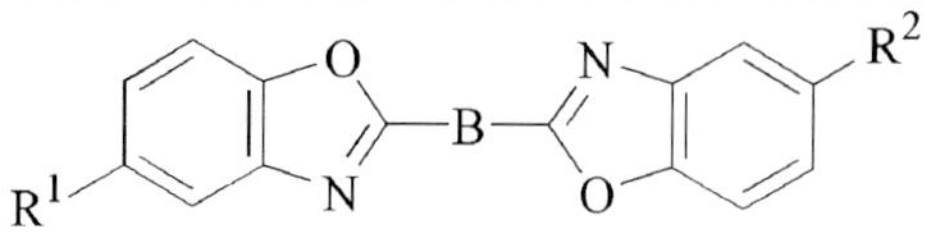

FWA	R¹	R²	B	CAS registry no.	Reference
43	H	H	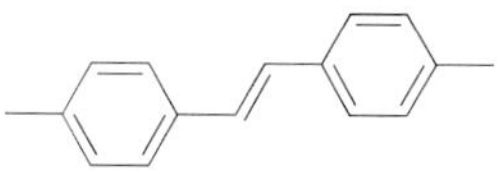 (4,4'-stilbenediyl)	[1533-45-5]	[74], [75]
44	–CH₃	H	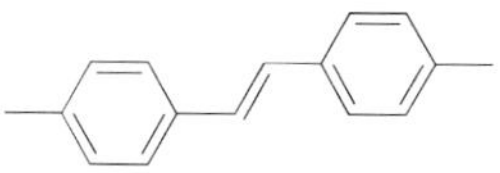 (4,4'-stilbenediyl)	[5242-49-9]	[76]

In addition to the most common synthesis (the reaction of dicarboxylic acids or their chlorides with 2-aminophenols) other industrially interesting processes have been developed recently.

Compound **38** is obtained, for example, by condensing 2-amino-4-methylphenol with maleic acid to yield the diamide. Ring closure and the formation of the ethylene bridge occur on heating in a mixture of diphenyl ether and diphenyl at 135–155 °C in the presence of boric acid [68].

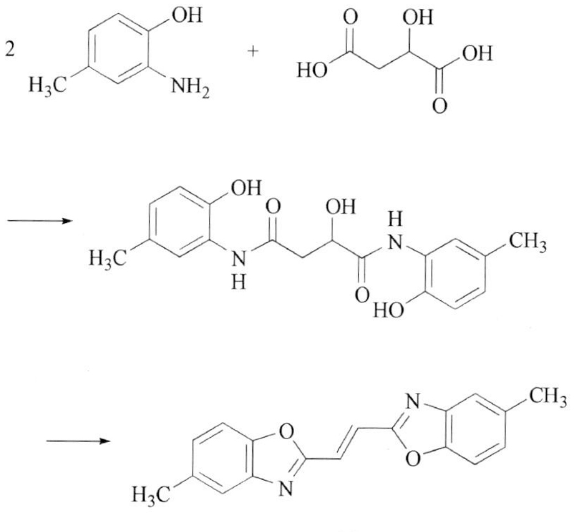

38

To obtain 2,5-bis(benzoxazol-2-yl)thiophene (**39**), two equivalents of 2-chloromethylbenzoxazole are treated with sodium sulfide and the thiophene ring is formed with glyoxal hydrate [70]:

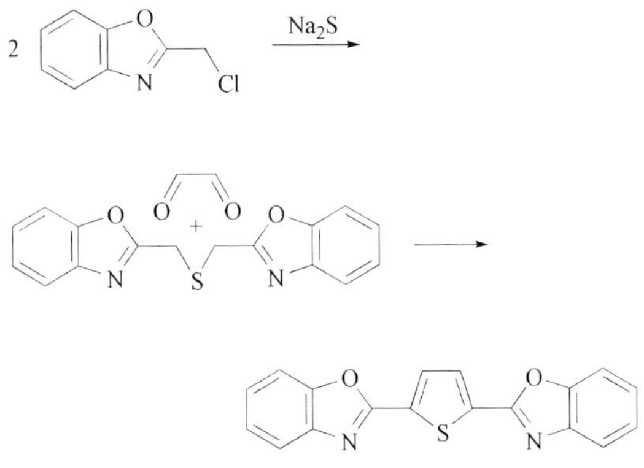

7.2.5 Furans, Benzo[b]furans, and Benzimidazoles

For **43**, a new olefination reaction was discovered. 2-[4-(chloromethyl)phenyl]benzoxazole is treated in polar aprotic solvents with strong bases [76].

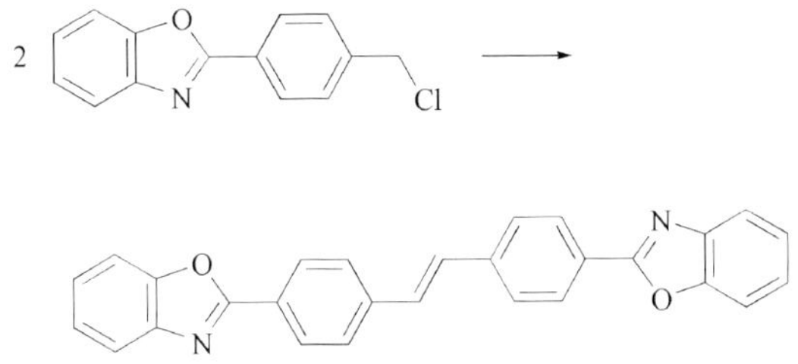

7.2.5 Furans, Benzo[b]furans, and Benzimidazoles

Furans and benzo[b]furans are further building blocks for optical brighteners. They are used, for example, in combination with benzimidazoles and benzo[b]furans as biphenyl end groups.

7.2.5.1 Bis(benzo[b]furan-2-yl)biphenyls

The 4,4′-bis(benzo[b]furan-2-yl)biphenyls can be regarded as distyrylbiphenyls that have had their rings closed and are thus fixed in the *E* configuration. Sulfonated derivatives are described as brighteners for polyamides and cellulosic fibers [77].

Compound **45** is produced by first treating 4,4′-bis(chloromethyl)biphenyl with salicylaldehyde to give 4,4′-biphenyldiyl-bis(methylenoxy-2-benzaldehyde). The formyl groups of this compound are converted to phenyliminomethyl groups with aniline. The double ring closure in dimethylformamide with potassium hydroxide is almost quantitative [78, 79]. The final step is sulfonation. For other syntheses, see [80].

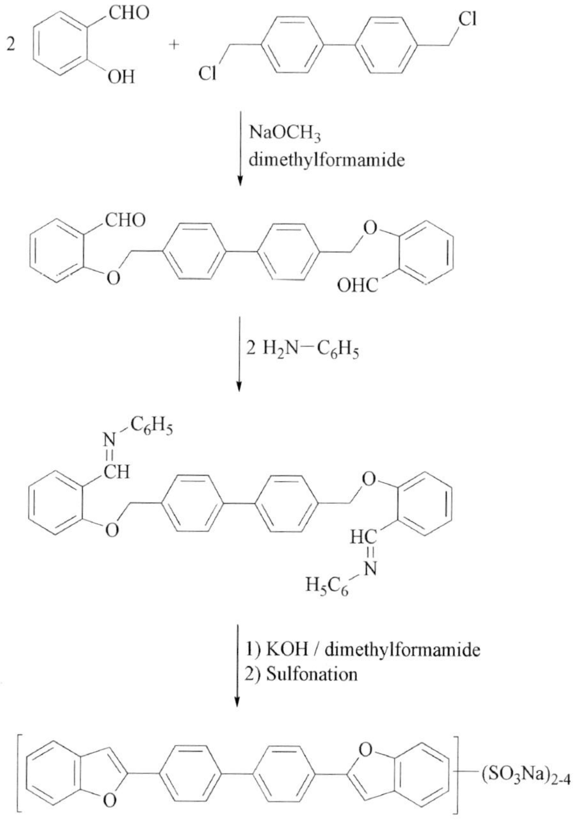

[Reaction scheme showing synthesis of compound 45:

2 salicylaldehyde (CHO, OH) + 4,4'-bis(chloromethyl)biphenyl

↓ NaOCH$_3$, dimethylformamide

bis(ether-aldehyde) intermediate

↓ 2 H$_2$N—C$_6$H$_5$

bis(imine) intermediate with N-C$_6$H$_5$ groups

↓ 1) KOH / dimethylformamide
2) Sulfonation

Final product 45: bis(benzofuran-2-yl)biphenyl with (SO$_3$Na)$_{2-4}$]

45

7.2.5.2 Cationic Benzimidazoles

2,5-Di(1-methylbenzimidazol-2-yl)furan [*4751-43-3*] [81], which has the ability to form salts, was one of the first cationic benzimidazole brighteners. More recently, cationic benzimidazoles have been developed as brilliant, chlorine-fast, lightfast brighteners for polyacrylonitrile and cellulose acetate (Table 7.7). They are marketed in the form of cold- and heat-resistant concentrated aqueous solutions with long shelf lives.

In the synthesis of **47**, 4-methoxysalicylaldehyde is condensed with 2-chloromethyl-1-methyl-5-methylsulfonylbenzimidazole, ring closure is effected, and the product is quaternized with dimethyl sulfate [83]. To enhance water solubility, the methyl sulfate anion is converted to the acetate anion; the benzimidazole ring is

7.2.5 Furans, Benzo[b]furans, and Benzimidazoles

first opened with sodium hydroxide and then reclosed with acetic acid, because direct ion exchange is unfavorable for solubility reasons [84].

Table 7.7: Cationic Benzimidazoles

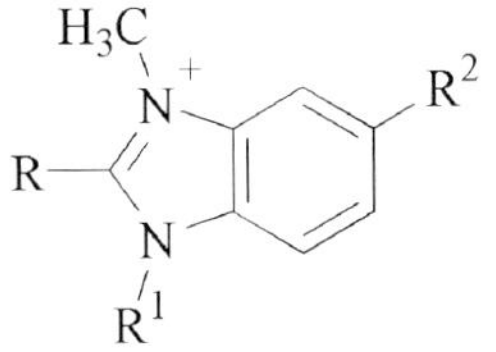

FWA	R	R^1	R^2	A^-	CAS registry no.	Reference
46	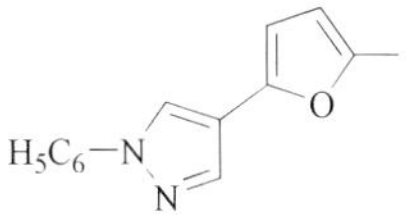 (H₅C₆—N-pyrazole-furan-CH₃)	$CH_2\ COOC_2H_5$	H	Br^-/Cl^-	[72829-17-5] [74878-56-1]	[82]
47	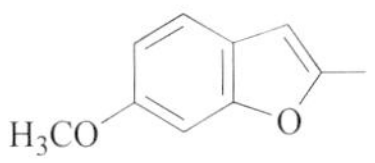 (H₃CO-benzofuran)	CH_3	$-SO_2CH_3$	CH_3COO^-	[74878-48-1]	[83], [84]
48	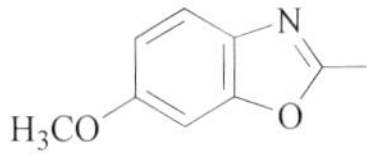 (H₃CO-benzoxazole)	$CH_2\ C_6H_5$	H	$CH_3OSO_3^-$	[66371-25-3]	[85]

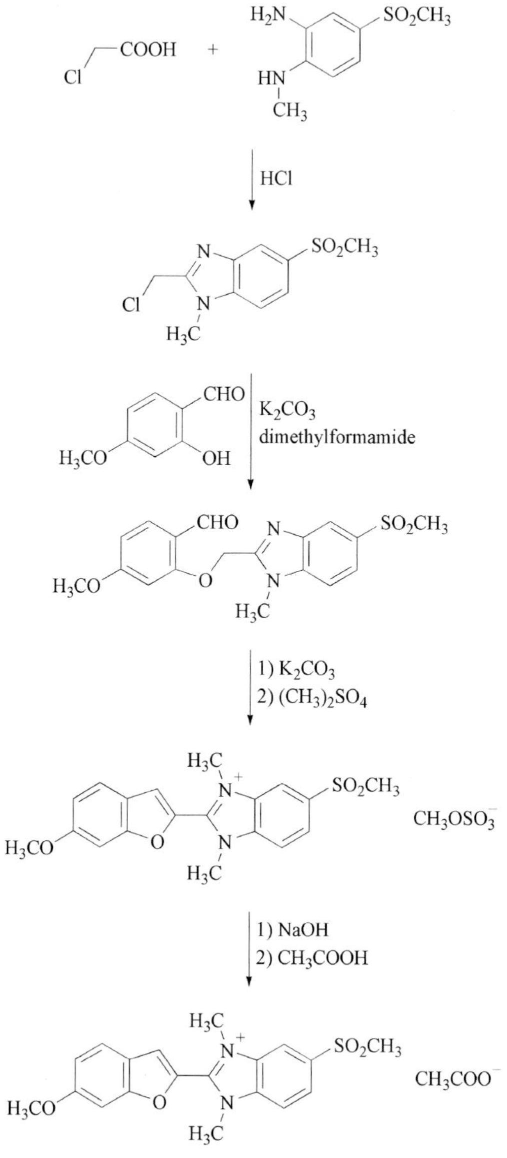

Compound **47** can also be synthesized from 6-methoxybenzo[*b*]furan-2-carboxylic acid and 2-methylamino-5-methylsulfonylaniline by cyclocondensation [83].

7.2.6 1,3-Diphenyl-2-pyrazolines

Although the intense blue fluorescence of 1,3-diphenyl-2-pyrazolines had long been known [86], the usefulness of these compounds as optical brighteners was not recognized until 1949 [87]. The most valuable brighteners are the 3-(4-chlorophenyl)-1-(4-phenylsulfonic acid) derivatives because of their ready manufacturing ability, excellent optical properties, and versatility in application.

Industrial synthesis involves the reaction of β-chloro-4-chloropropiophenones with phenylhydrazines containing substituents at the 4-position, such as 4-hydrazinobenzenesulfonamide. The β-chloro-4-chloropropiophenone needed as starting product can be obtained from the Friedel–Crafts reaction of chlorobenzene with 2-chloropropionyl chloride.

1-(4-Amidosulfonylphenyl)-3-(4-chlorophenyl)-2-pyrazoline (**49**) [*2744-49-2*] [88] was used for the brightening of polyamides in the laundry.

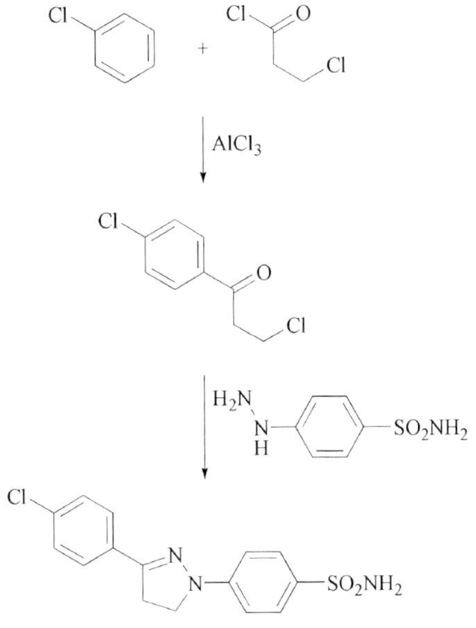

The compounds listed in Table 7.8 and **55** are industrially important nonionic and anionic 1,3-diphenyl-2-pyrazoline derivatives. The anionic compounds are employed to brighten polyamides but do not always have adequate lightfastness. The poor wet lightfastness can be improved by the insertion of substituents in the 3-phenyl and pyrazoline rings, as in **54** and **55** [*81209-71-4*] [94]. The pyrazoline ring system is not stable against oxidative bleaches, which dehydrogenate it to the corresponding pyrazole.

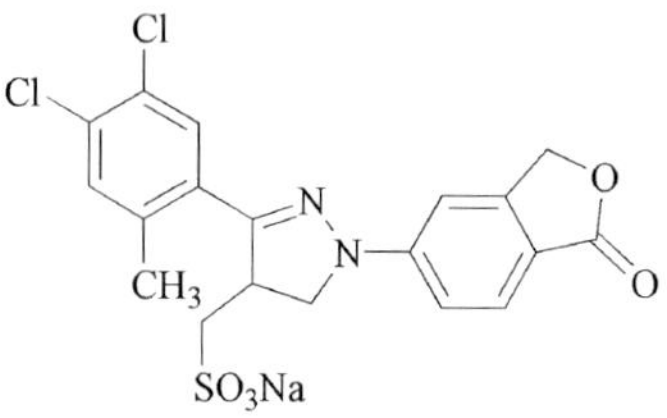

7.2 Chemistry of Technical Products

[Structure 55: a 1,3-diphenyl-2-pyrazoline derivative with 4,5-dichloro-2-methyl substitution, CH₂SO₃Na side chain, and an isobenzofuranone (phthalide) N-aryl group]

The tertiary and quaternary amine salts of 1,3-diphenyl-2-pyrazoline derivatives (Table 7.9) are brilliant, moderately lightfast brighteners for (modified) polyacrylonitriles. Some can also be used to brighten cellulose acetate. Depending on the structure and nature of the anion, concentrated aqueous solutions with long shelf lives can be obtained; such products are needed, for example, in gel brightening.

The side chains R in Table 7.8 and SO_2R in Table 7.9 can be added before or after cyclocondensation, for example, by the attachment of $NaHSO_3$, amines, or alcohols to vinyl or allylsulfonyl groups.

Table 7.8 Nonionic and anionic 1,3-diphenyl-2-pyrazolines

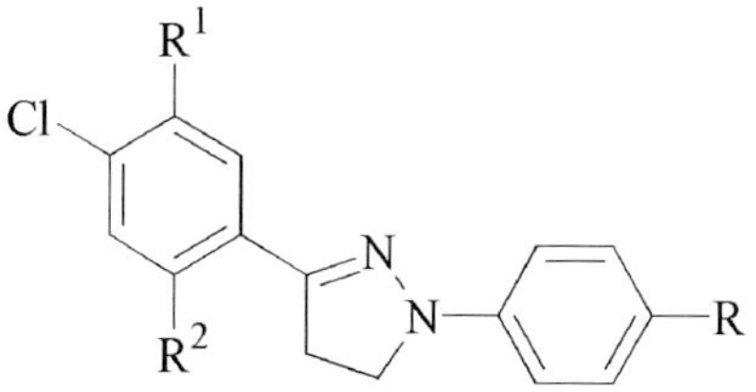

FWA	R^1	R^2	R	CAS registry no.	Reference
50	H	H	$-SO_2CH_3$	[60650-43-3]	[89]
51	H	H	$-SO_2-CH_2CH_2OH$	[3656-22-2]	[90]
52	H	H	$-SO_2-(CH_2)_2-SO_3Na$	[27441-70-9]	[91]
53	H	H	$-COONa$	[32020-25-0]	[92]
54	Cl	CH_3	$-SO_2-(CH_2)_2-SO_3Na$	[61931-42-8]	[93]

Table 7.9 Cationic 1,3-diphenyl-2-pyrazolines

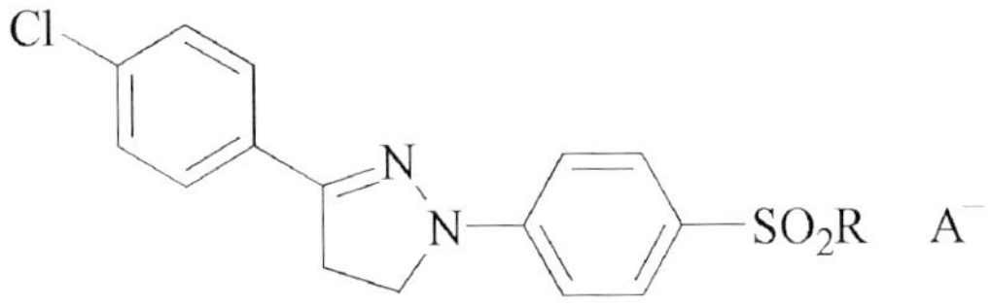

FWA	R	A⁻	CAS registry no.	Reference
56	–(CH$_2$)$_2$–$\overset{+}{\text{N}}$H(CH$_3$)$_2$	H–P(=O)(OH)–O⁻	[106359-93-7]	[95]
57	–(CH$_2$)$_2$–$\overset{+}{\text{N}}$H(CH$_3$)$_2$	HCOO⁻		[95]
58	–CH$_2$–CH(CH$_3$)–$\overset{+}{\text{N}}$H(CH$_3$)$_2$	CH$_3$CH(OH)COO⁻	[85154-08-1]	[96]
59	–(CH$_2$)$_2$–CONH–(CH$_2$)$_3$–$\overset{+}{\text{N}}$H(CH$_3$)$_2$	Cl⁻	[42952-22-7]	[97]
60	–(CH$_2$)$_2$–O–CH(OH)–CH$_2$–$\overset{+}{\text{N}}$H(CH$_3$)$_2$	Cl⁻	[63310-12-3]	[98]
61	–NH–(CH$_2$)$_3$–$\overset{+}{\text{N}}$(CH$_3$)$_3$	CH$_3$OSO$_3^-$	[12270-54-1]	[99]
62	–NH–(CH$_2$)$_3$–$\overset{+}{\text{N}}$H(CH$_3$)$_2$(CH$_2$CH$_2$OH)	CH$_3$COO⁻	[36086-26-7]	[100]

7.2.7 Coumarins

Further modification of the coumarin ring system in esculin (**1**) led to the development of 7-amino-4-methylcoumarins with an *N*-alkyl group at the 7-position. The sole representative of this series still on the market is 7-diethylamino-4-methylcoumarin (**63**) [*91-44-1*] [101]. Despite its poor lightfastness, the good optical properties of this compound have kept it in use for the brightening of wool, cellulose acetate rayon, and polyamides. Compound **63** is produced by cyclocondensation of 3-diethylaminophenol with ethyl acetoacetate in the presence of zinc chloride [102].

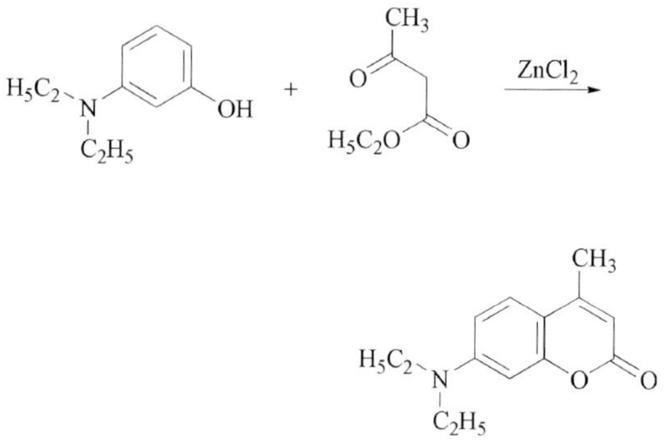

Substituted 3-phenylcoumarins with methoxy [103] and triazinylamino groups [104] at the 7-position were developed later. These have been superseded in industry by the 3,7-bis[(hetero)aryl]coumarins [105–110].

7.2.8 Naphthalimides

The 4-aminonaphthalimides and their *N*-alkylated derivatives are brilliant greenish yellow fluorescent dyes. Acylation of the amino group at the 4-position of the naphthalimide ring shifts the fluorescence toward blue [111], yielding compounds suitable for use as optical brighteners, such as 4-acetylamino-*N*-(*n*-butyl)naphthalimide [*3353-99-9*] [112].

Later workers, chiefly in Japan, used naphthalimides with alkoxy substituents at the 4- or 4,5-positions and obtained brighteners with good lightfastness for polyester substrates and good chlorite fastness for polyacrylonitriles. The first commercial product was 4-methoxy-*N*-methylnaphthalimide (**64**) [*3271-05-4*] [113]. Table 7.10 lists the most important compounds.

7.2.8 Naphthalimides

Table 7.10 Naphthalimides

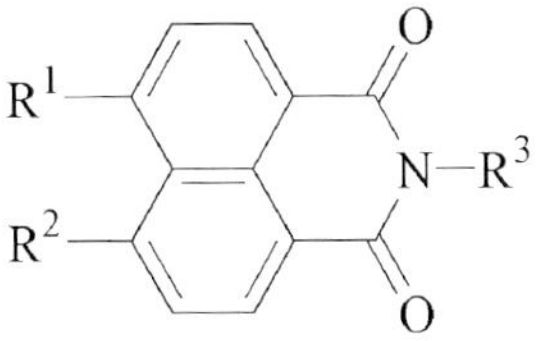

Structure: naphthalimide with R^1, R^2 substituents on the naphthalene, N–R^3 on the imide nitrogen, A$^-$ (anion).

FWA	R^1	R^2	R^2	A$^-$	CAS registry no.	Reference
64	–OCH$_3$	H	–CH$_3$		[3271-05-4]	[113]
65	–OC$_2$H$_5$	–OC$_2$H$_5$	–CH$_3$		[22330-48-9]	[114]
66	–OCH$_3$	H	–O–(CH$_2$)$_3$–CH$_3$		[22826-31-7]	[115]

The naphthalimide ring system has the drawback of low effectiveness, which is mainly attributable to the low molar extinction coefficient. The industrial synthesis of alkoxynaphthalimides begins with acenaphthene. Chlorination and subsequent oxidation with dichromate give the corresponding naphthalic acids [116], which are converted to the anhydrides on drying. Mild reaction with methylamine, followed by reaction with sodium methoxide or sodium ethoxide, gives, e.g., **64** or **65**.

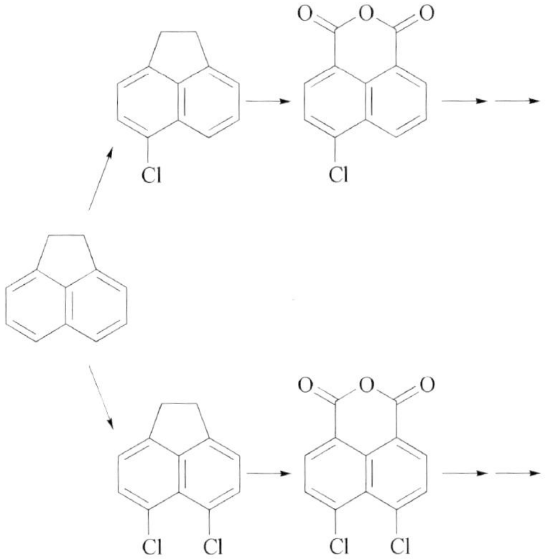

7.2.9 1,3,5-Triazin-2-yl Derivatives

The sole representative of this class is 1-(4,6-dimethoxy-1,3,5-triazin-2-yl)pyrene (**67**), [*3271-22-5*], which gives a bluish white brightening effect on polyester substrates; it is employed chiefly in combination with other polyester brighteners such as **38**. Compound **67** is obtained by the Friedel–Crafts reaction of cyanuric chloride with pyrene with subsequent replacement of the remaining chlorines with methoxyl groups [117].

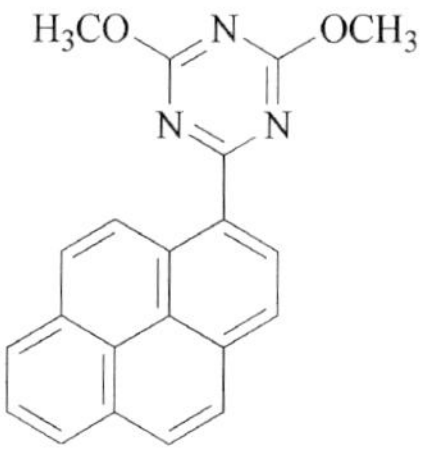

67

7.3 Commercial Forms and Brands

Optical brighteners are formulated specifically according to structure, substrate, and mode of application. Nondusting, easily meterable forms are preferred: free-flowing granules with high crushing strength, solutions, or slurries (concentrated dispersions).

Nonionogenic optical brighteners must be as finely divided as possible so that they can be quickly and optimally distributed in apolar substrates. They are marketed as powders (e.g., for plastics or spinning masses) or as low-viscosity dispersions (e.g., for the textile industry).

Formulations are optimized with additives to control granule strength, dissolution, chemical stability, viscosity, stability on storage, or fungal growth. They can also contain toning dyes and auxiliaries to improve their efficiency in a given application.

Several thousand products with more than 100 trade names are on the market worldwide.

7.4 Uses

7.4.1 General Requirements

Optical brighteners (fluorescent whitening agents, FWAs) are used primarily in the textile, detergent, and paper industries and are also added to plastics [118]. The mode of application and performance requirements differ greatly from one substrate to another, but there are five common basic principles:

1) FWAs can only improve whiteness effectively if the substrate does not absorb too strongly in their excitation and fluorescence range. Objects with a poor basis white, such as unbleached textiles or poorly cleaned articles, cannot be brightened to a high degree of whiteness even with high concentrations of FWAs [119].
2) FWAs only act effectively if they are monomolecularly dispersed in the substrate. This depends on the substrate, the FWA type, and the mode of application [4, 5, 120].
3) Whitening effects increase with FWA concentration up to a saturation limit. For this reason, the applied concentrations of FWA must be kept low: ca. 0.002–0.2 wt % relative to the substrate, or roughly one-tenth of the usual concentration of dyes [4].
4) Since the use of FWAs in industrial applications should not create any problems, specific products have been developed for each type of application.
5) The required shades and fastnesses needed are also selection criteria for a suitable FWA.

7.4.2 Textile Industry

Brighteners can be applied to any substrate by batch or continuous processes [9, 17, 121–125]. The processes are the same as those used for dyeing.

Cellulosic Fibers. Substantivity for cellulosic fibers is offered mainly by anionic FWAs, preferably with sulfonic acid groups. Brighteners such as **14**, **19**, **24**, and **26** with moderate to high affinities are suitable for batch processes. If simultaneous bleaching takes place in the same bath, the brightener must be stable with respect to bleaching chemicals such as peroxide or chlorite [126].

Continuous processes allow high throughputs of woven and knitted fabrics [127]. To achieve a level brightening effect from the beginning to the end of the process and between the center and the edges of the material, FWAs with low to moderate affinity for the fiber are used; examples of principal types are **23**, **25**, **30**, **31**, and **34**.

Brightening is often combined with a high-grade finishing process. The goods are impregnated with a synthetic resin in the padding machine, usually under strongly acidic conditions (Lewis acids can also be employed as catalysts). Brighteners must therefore meet additional stability requirements [128]; types such as **10** and **31** are especially suitable for such a process.

Polyamides. Batch brightening usually takes place in combination with reductive bleaching (dithionite). Continuous processes (padding) are followed by steam, heat, or acid treatment (acid shock) [129]. Anionic FWAs are absorbed similarly to acid dyes, or insoluble dispersion brighteners are suitable. Examples of products listed in order of increasing fastness are diaminostilbenes (**14** and **19**); pyrazolines (**50** and **54**); distyrylbiphenyls (**10**); and bis(triazolyl)stilbenes (**32**).

Polyacrylonitriles. An increasing proportion of the polyacrylonitrile fibers produced have a good basis white and do not require further bleaching [130, 131]. The use of FWAs that can be applied in the chlorite bleaching bath is therefore steadily declining. Suitable types include benzimidazoles such as **46–48**.

If chlorite stability is not required, other less expensive, highly effective products can be used: pyrazoline derivatives such as **49** and **50** and cationic products such as **58–60** and **62**. The optical brighteners are being increasingly applied immediately after the polyacrylonitrile leaves the spinning nozzle while it is still in the gel state. Gel application has already become the dominant process in some countries (e.g., UK). Increasingly stringent fastness requirements relating to, e.g., perborate washing and exposure to light are satisfied by types **58** and **60**.

Polyesters. Dispersions of nonionogenic FWAs are suitable for hydrophobic fibers such as polyesters. Penetration into the fibers and optimal distribution are achieved by batchwise treatment at 110–130 °C under pressure, or at the boiling point with a carrier (a substance that improves diffusion of the brightener by swelling the fiber); the continuous process in the padding machine involves subsequent heat treatment at 180–210 °C, which simultaneously results in good dimensional stability of the textile [132]. The FWA molecules become oriented during thermal fixation [133]. Batch processes can be combined with bleaching, which is necessary for many polyester qualities and polyester–cellulosic blends.

The brighteners should be adapted to the heat treatment; they should represent an optimal compromise between diffusion into the polyester material and sublimation fastness [32, 33]. High to very high lightfastness and chlorite stability are offered by naphthalimides such as **64** and **65**; and benzoxazoles such as **35–38**, **41**, **42**, and especially **39**. Compounds such as **4–6**, **13**, and **67** are very effective but less stable to chlorite and light. Boosters (e.g., alkyl esters of stilbene dicarboxylic acids) have been suggested as a way of intensifying the action of FWAs [134]. Brighteners that sublime can be applied by transfer printing [135, 136].

Other Fibers. The main products used for cellulose acetate or triacetate fibers are pyrazoline derivatives such as **49–51**, and for wool typical polyamide brighteners.

7.4.3 Detergent Industry

Regular replacement of brighteners degraded during use improves the acceptance of textile articles and thus helps extend their service lives [137, 138]. Optical brighteners are therefore used in laundry detergents and are required above all for cotton fabrics [139].

Heavy-Duty Powders. Appropriate brighteners for regions with low washing temperatures include **17** and especially **10** with its better lightfastness. FWAs such as **18** and especially **20** are used in countries where typical washing temperatures are moderate to high [140].

The preferred FWAs lend a bright, fresh, clean appearance to the detergent powder. Bleaching systems in modern detergents are effective even at low temperature, so that stringent demands are placed on FWA stability [141]. For example, triazinylaminostilbenes must be as free as possible from nonfluorescing triazine byproducts, which lead to discoloration and an unpleasant odor during storage of the detergent powder [142]. The active ingredient may also be degraded; preference is therefore given to **10**.

Heavy-Duty Liquid Detergents. The ease of incorporation and affinity of FWAs are crucial in such formulations. Heavy-duty liquid detergents can also be employed as "prespotters" that are applied in concentrated form to heavily soiled areas before the normal washing process. Bright spots can be formed, especially on colored fabrics, due to washing out of the color; they are intensified by the fluorescence of the locally high FWA concentration. For this reason, FWAs have been developed which cause less pronounced spotting that cannot be seen. The main products in use are **10** and **21** [143].

The addition of bleaches to aqueous or nonaqueous heavy-duty liquid detergents will impose even more stringent stability requirements on FWAs [144–146].

Other Laundry Products. Laundry boosters are bleaching systems that are added separately to the wash. They contain **10** as a brightener; this compound is easy to incorporate and shows good compatibility and stability during storage [147].

Light-duty detergents contain coumarins (**63**) as brighteners for polyamide and wool. At the washing temperatures acceptable for polyesters, FWAs cannot penetrate into the fibers. Special products containing water-soluble polymers and a lightfast brightener such as **10** have therefore been developed for use in the rinse bath after the washing of, for example, curtains [148].

7.4.4 Paper Industry

White paper must satisfy increasingly high standards of whiteness, which can be obtained by means of the quality and degree of bleaching of the raw materials as well as optical brightening. Brighteners can be added at three stages of papermaking: in the pulp, to the paper surface, and during coating [2, 149]. Application in the pulp has leveled off in recent years, while brightening of the paper surface and especially coating have made strong gains.

Brightening in Pulp. The brightener is added in an aqueous solution to the paper pulp during preparation in the hollander or in mixing chests. Inexpensive, highly soluble triazinylaminostilbenes such as **19**, **22**, **24**, **26**, and **30** are used. For acid-sized paper qualities, good stability toward acid and alum are necessary. There is, however, a trend toward neutral sizing.

Impregnation of Paper Web. The paper web is impregnated from one or both sides with an FWA solution that often contains other additives. In the size press, the paper is therefore brightened only on the surface, thus economizing on brightener material. Examples of products employed in this way are **24**, **26**, and **28–31**.

Addition to Coating. The paper is often coated with a material that contains binder and/or pigment, which usually reduces the whiteness [150]. To make up for this loss of whiteness or to provide additional brightness, FWAs such as **15**, **19**, **22**, **23**, **28**, and **29** are added to the coating.

7.4.5 Plastics and Synthetic Fibers

Brighteners for plastics and synthetic fibers must be "dissolved" in the material. They are added dry to the polymer powder, granulate, or chips, either before or during polymerization or else later before high-temperature forming (e.g., immediately prior to spinning in the case of synthetic fibers). Very good chemical and thermal stabilities are prerequisites for such FWAs. The brightener can also be added to the polymer solution if this is appropriate [132, 151, 152]. Low FWA concentrations are often used (e.g., 0.005 wt %) just enough to offset the yellowish cast of the material. In poly(vinyl chloride) articles, higher degrees of whiteness are achieved mainly with compounds such as **12**, **44**, and especially **40**; in polyester fibers, with products such as **43**, and especially **44**.

7.4.6 Other Uses

Optical brighteners are also used to enhance the whiteness of various inorganic and organic substrates, such as tripolyphosphates, carbonates, and photographic papers [152–154].

Markings visible under UV light can be made with FWAs, for example, to identify petroleum products such as fuel oil or to verify the uniform distribution of sprayed products.

Compounds such as **40** are valuable as scintillators that convert the radiation emitted by radioactive isotopes into visible light, which can easily be measured [155]. Substances such as **24** are preferentially absorbed by fungi, making them easily detectable in the skin or other types of cells allowing accurate clinical diagnosis and control [156].

Some FWAs, especially **10**, are of interest as laser dyes [5, 157].

7.5 Analytical Methods and Whiteness Assessment

7.5.1 Analytical Methods

Identification. Chromatographic methods (interpretation of fluorescing spots in thin layer chromatography or of UV absorption peaks and fluorescence light emission peaks in high performance liquid chromatography) allow fairly rapid and certain identification of FWAs [158–164]. The same methods can be used for qualitative and quantitative characterization of the FWA purity in terms of fluorescent or nonfluorescent byproducts.

Determination in Solutions. The content of active substance is determined by photometric measurement of the extinction (absorbance) in solution [158]. Both the wavelength and the height of the absorption maximum depend on the solvent. Concentrations of FWAs in application baths can be monitored quantitatively by comparative fluorescence measurements against calibrated standards [165]. When dilute FWA solutions are handled, their photosensitivity must be kept in mind. Lighting that includes UV wavelengths can lead to rearrangement (e.g., $E{\rightarrow}Z$ photo-isomerization) and/or degradation of the FWAs to give molecules with different spectra and chromatographic properties [18, 168].

Determination in Substrates. The FWA on a substrate can be directly determined by reflectance measurements. Alternatively, FWAs may be extracted with appropriate solvents and determined in solution; in the case of plastics, this involves dissolving the substrate [166, 167, 169].

Determination in the Aquatic Environment. Identification and quantification of FWAs in sewage water and sludges as well as in surface waters and sediments require special methods of extraction and sensitive determination of photo-isomers [170].

7.5.2 Assessment of Whitening Effect

Whitening can be assessed either visually or instrumentally (by colorimetry). The quantities evaluated are the *degree of whiteness* and the *tint*. Special attention must be paid to the spectral distribution of the incident light in the UV and visible ranges, since this controls the fluorescent contribution of the brightener to the whiteness [171].

Visual assessment is performed by comparison with references in diffuse daylight [172]. White scales consisting of stages with well-defined whiteness values (e.g., the cotton or plastic white scale) can be used [173].

For instrumental evaluation spectrophotometers are preferred. The colorimetric values obtained depend on the instrument and its instantaneous state (chiefly the sample illumination conditions) and must be controlled by suitable methods [174]. The influence of the instrument can be eliminated by conversion to the standard illumination D 65, representing daylight with a correlated color temperature of 6504 K [175]. The degree of whiteness and the tint values can then be calculated from these colorimetric data with appropriate formulas. A selection of currently used whiteness formulas can be found in [176, 177]. For recent attempts at standardizing the assessment methods for white objects, see [9, 178–185].

7.6 Environmental Aspects

Both the toxicology and the ecotoxicology of commercial FWAs for cotton/paper have been extensively studied [7, 19, 186].

Acute toxicity by the oral and dermal routes places these substances outside any poison class. The oral median lethal doses (LD_{50}) are generally above 2000 mg/kg. There are no indications of carcinogenic, mutagenic, or teratogenic, sensitizing, phototoxic effects for the optical brighteners so far tested. Chronic rat studies with optical brighteners produced a favorable no observable adverse effect level (NOAEL) [7]. FWAs for cotton/paper do not penetrate the skin in measurable amounts, and when fed to animals they are eliminated unchanged. Extensive work has been done to characterize aquatic toxicity and photo- and biodegradation. Human and environmental risk assessments of FWAs for cotton/paper show no adverse effects to humans and the environment [186].

As industrial chemicals, FWAs are regulated by national laws such as the Toxic Substance Control Act in the United States and the relevant classification, packaging, and labeling directives in the European Community. For specific applications such as food packaging, the national regulatory requirements must be complied with.

7.7 References

[1] A. E. Siegrist, C. Eckhardt, J. Kaschig, E. Schmidt: "Optical Brighteners" in *Ullmann's Encyclopedia of Industrial Chemistry*, vol. A18, VCH, Weinheim 1991, pp. 153–176.
[2] A. E. Siegrist: "Die Anwendung optischer Aufheller in der Papierindustrie", *Papier (Darmstadt)* **8** (1954) 109–120.
[3] R. Zweidler: "Einführung in die Chemie der optischen Aufheller", *Textilveredlung* **4** (1969) 75–87.
[4] T. Förster: *Fluoreszenz organischer Verbindungen*, Vandenhoek und Ruprecht, Göttingen, 1951.
[5] F. P. Schäfer: *Dye Lasers*, Springer Verlag, Berlin 1973.
[6] H. Gold: "Fluorescent Brightening Agents" in K. Venkataraman (ed.): *The Chemistry of Synthetic Dyes*, vol. V, Academic Press, New York–London 1971, pp. 535–679.
[7] R. Anliker, G. Müller: "Fluorescent Whitening Agents" in F. Coulston, F. Korte (eds.): *Environmental Quality and Safety*, vol. IV, Thieme Verlag, Stuttgart, 1975.
[8] M. Zahradnik: *The Production and Application of Fluorescent Brightening Agents*, John Wiley & Sons, Chichester 1982.
[9] R. Levene, M. Lewin: "The Fluorescent Whitening of Textiles" in M. Lewin, S. B. Sello (eds.): *Handbook of Fiber Science and Technology*, vol. 1, part B, Marcel Dekker, New York–Basel, 1984, pp. 257–304.
[10] A. K. Sarkar: "Fluorescent Whitening Agents" in J. G. Cook (ed.): *Textile Technology*, Merrow Publishing Co., Watford, England, 1971.
[11] D. Barton, H. Davidson: "Fluorescent Brighteners, 1967–1973", *Rev. Prog. Color. Relat. Top.* **5** (1974) 3–11.
[12] A. Dorlars, C. W. Schellhammer, J. Schroeder: "Heterocyclen als Bausteine neuer optischer Aufheller", *Angew. Chem.* **87** (1975) 693–707.

[13] E. Siegel: "Organic Dyes and Optical Brightening Agents", *Int. Rev. Sci. Org. Chem. Ser. Two*, 1976, vol. 2, 259–297.
[14] E. A. Kleinheidt, H. Theidel: "Optische Aufheller" in A. Chwala, W. Anger, C. Chwala (eds.): *Handbuch Textilhilfsmittel*, Verlag Chemie, Weinheim 1977, pp. 645–666.
[15] *Kirk-Othmer*, 3rd. ed., **4**, pp. 213–226.
[16] Y. Hashida, K. Matsui: "Recent Development in Fluorescent Whitening Agents", *Senryo to Yakuhin* **25** (1980) 128–137.
[17] R. Williamson: "Fluorescent Brightening Agents" in *Textile Science and Technology*, vol. 4, Elsevier, Amsterdam, 1980.
[18] I. H. Leaver, B. Milligan: "Fluorescent Whitening Agents, A Survey (1974–1982)", *Dyes Pigm.* **5** (1984) 109–144.
[19] A. E. Siegrist, H. Hefti, H. R. Meyer, E. Schmidt: "Fluorescent Whitening Agents 1973–1985", *Rev. Prog. Color. Relat. Top.* **17** (1987) 39–55.
[20] H. Zollinger: "Fluorescent Brighteners" in *Color Chemistry*, VCH, Weinheim, 1987, pp. 203–213.
[21] B. M. Krasovitskii, B. M. Bolotin: *Organic Luminescent Materials*, VCH, Weinheim 1988.
[22] P. Krais: "Ueber ein neues Schwarz und ein neues Weiss", *Melliand Textilber.* **10** (1929) 468–469.
[23] R. Anliker in [7] pp. 12–18.
[24] Ultrazell, GB 472 473, 1935 (L. Mellersch-Jackson).
[25] IG Farbenind., DE 752 677, 1940 (B. Wendt).
[26] IG Farbenind., DE 746 569, 1940 (S. Petersen, O. Bayer, B. Wendt).
[27] Hoechst, GB 981 178, 1958 (L. Horner, H. Hoffmann, H. G. Wippel), US 3 156 739, 1960 (L. Horner, H. Hoffmann, W. Klink).L. Horner, H. Hoffmann, H. G. Wippel, G. Klahre, *Chem. Ber.* **92** (1959) 2499.
[28] BASF, GB 920 988, 1959 (W. Stilz, H. Pommer, K. H. Koenig), US 3 177 208, 1959 (W. Stilz, H. Pommer).
[29] Hoechst, US 4 336 155, 1979 (T. Martini et al.).
[30] Ciba-Geigy, US 4 778 623, 1979 (L. Guglielmetti).
[31] BASF, US 4 380 514, 4 464 284, 1980 (G. Seybold).
[32] Ciba-Geigy, EP-A 238 446, 1986; GB 2 200 660, 1987 (L. Guglielmetti, H. R. Meyer, D. Reinehr, K. Weber).
[33] T. Martini, H. Probst: "Synergismen bei optischen Aufhellern", *Melliand Textilber.* **65** (1984) 327–329.
[34] Ciba-Geigy, US 3 984 399, 1967 (K. Weber, P. Liechti, H. R. Meyer, A. E. Siegrist).
[35] Ciba-Geigy, GB 1 360 142, 1971 (K. Weber, C. Lüthi).
[36] Ciba-Geigy, US 4 339 393, 4 468 352, 4 602 087, 1979 (C. Lüthi, H. R. Meyer, K. Weber).
[37] Sandoz, US 4 108 887, 4 196 229, 1975 (F. Fleck, J. Heller).
[38] Ciba-Geigy, US 4 708 558, 1985 (K. Burdeska, K. Weber, D. Reinehr); US 4 533 505, 1981 (A. Spencer).
[39] Ciba-Geigy, EP-A 305 646, 1987 (L. Guglielmetti).
[40] Ciba, US 2 846 496, 1949 (F. Ackermann). General Aniline & Film Corp., US 2 713 046, 1951 (W. W. Wilson, H. B. Freyermuth).
[41] ICI, US 2 612 501, 1947 (R. H. Wilson).
[42] Ciba, US 2 763 650, 1951 (F. Ackermann).
[43] Nippon, JP-Kokai 59 5 8 80, 1956 (T. Noguchi, M. Sumitami), *Chem. Abstr.* **54** (1960) 14707g.

[44] Nisso, JP-Kokai 6 325 833, 1961 (G. Hayakawa, M. Orita, M. Hayashi, T. Obizu), *Chem. Abstr.* **60** (1964) 5524h.
[45] Geigy, US 2 762 801, 1953 (H. Häusermann); GB 1 116 619, 1966 (H. Häusermann).
[46] IG Farbenind., FR 874 939, 1940 (B. Wendt).
[47] Sandoz, GB 1 243 276, 1968 (F. Fleck).
[48] Nippon, JP-Kokai 6 3 21 636, 1961 (S. Ono, T. Noguchi).
[49] Bayer, DE 1 044 758, 1956 (B. Dach); GB 841 189, 1957 (J. Hegemann, A. Mitrowski, H. Roos).
[50] Bayer, GB 1 114 021, 1965 (E. A. Kleinheidt, H. Gold). Nippon, JP-Kokai 58 7 6 38, 1958 (S. Ono, T. Noguchi), *Chem. Abstr.* **54** (1960) 2381i.
[51] Sandoz, GB 896 533, 1958 (F. Fleck).
[52] Du Pont, US 3 025 242, 1961 (R. C. Seyler). Amer. Cyanamid, US 3 211 665, 1962 (W. Allen, R. F. Gerard).
[53] Geigy, GB 930 393, 1959 (H. Häusermann).
[54] Wolfen, DD 55 668, 1966 (B. Noll, G. Chmell, G. Knöchel, L. Zölch).
[55] Geigy, US 3 479 349, 1967 (R. C. Allison, F. Fischer, H. Häusermann).
[56] Sandoz, WO 9 600 221, 1996 (J. Cowmann, J. Farrar, J. Martin, M. Graham, N. Mackinnon).
[57] Bayer, US 2 668 777, 1948 (H. Gold, S. Petersen).
[58] Geigy, US 2 784 183, 1951 (E. Keller, R. Zweidler, H. Häusermann). Ciba-Geigy, US 4 791 211, 1982 (R. B. Lund, L. W. Bass).
[59] Bayer, US 3 485 831, 1965 (A. Dorlars, O. Neuner, R. Pütter).
[60] Bayer, US 3 666 758, 1967 (A. Dorlars, O. Neuner).
[61] Bayer, GB 1 172 657, 1967 (A. Dorlars, O. Neuner).
[62] A. E. Siegrist, *Helv. Chim. Acta* **50** (1967) 906–957. I. J. Fletcher, A. E. Siegrist, *Adv. Heterocycl. Chem.* **23** (1978) 171–261.
[63] Ciba-Geigy, US 3 850 914, 1974 (C. Lüthi).
[64] Nippon, JP-Kokai 7 6 40 090, 1968 (S. Mitsukuni, S. Kiyoshi), *Chem. Abstr.* **86** (1977) 122956n.
[65] Hoechst, US 4 129 412, 1976 (D. Günther, E. Schinzel, R. Erckel, G. Rösch).
[66] Hoechst, US 4 142 044, 1979 (D. Günther, R. Erckel, E. Schinzel, G. Rösch); US 4 231 741, 4 363 744, 1977 (D. Günther, R. Erckel, G. Rösch, H. Probst).
[67] Ciba, US 2 483 428, 1945 (J. Meyer, F. Ackermann); US 2 488 289, 1945 (J. Meyer, C. Gränacher, F. Ackermann); US 2 875 089, 1955 (F. Ackermann, A. E. Siegrist). Intracolor, US 3 649 623, 1955 (F. Ackermann, M. Dünnenberger, A. E. Siegrist).
[68] Ciba-Geigy, EP 31 296, 1979 (W. Schreiber).
[69] Ciba, US 2 995 564, 1958 (M. Dünnenberger, A. E. Siegrist, E. Maeder); US 3 095 421, 1960 (P. Liechti, A. E. Siegrist, M. Dünnenberger, E. Maeder); US 3 127 416, 1960 (P. Liechti, A. E. Siegrist, E. Maeder).
[70] Ciba-Geigy, GB 1 561 031, 1976 (L. Guglielmetti).
[71] Ciba, US 3 135 762, 1961 (E. Maeder, P. Liechti, A. E. Siegrist, M. Dünnenberger).
[72] Hoechst, US 3 709 896, 3 743 625, 1966 (H. Frischkorn, U. Pintschovius, H. Behrenbruch).
[73] Nippon, US 3 629 246, 1962 (T. Tanaka).
[74] Kodak, US 3 260 715, 1962 (D. G. Saunders); US 3 322 680, 1962 (D. G. Hedberg, M. S. Bloom, M. V. Otis).
[75] Hoechst, US 4 585 875, 1983 (L. Heiss).
[76] Nippon, JP-Kokai 7 2 36 861, 1963 (K. Ohkawa, T. Kinoshita, T. Tanaka), *Chem. Abstr.* **78** (1973) 85946w; JP-Kokai 6 6 20 225, 1964 (T. Tanaka), *Chem. Abstr.* **67** (1967) 55173x; JP-Kokai 6 9 06 978, 1964 (T. Tanaka), *Chem. Abstr.* **71** (1969) 4505g; JP-Kokai

6 923 028, 1966 (T. Tanaka), *Chem. Abstr.* **72** (1970) 12705g. Kodak, US 3 586 673, 1968 (M. S. Bloom, J. A. Hill). Hoechst, US 4 282 355, 1978 (R. Erckel, P. Jürges).
[77] Hoechst, US 3 559 350, 1971 (W. Sahm, E. Schinzel, G. Rösch). Nippon, JP-Kokai 7 5 40 627, 1973 (S. Mitsukuni, S. Hajime), *Chem. Abstr.* **83** (1975) 116988w.
[78] Hoechst, US 3 864 333, 1971 (W. Sahm, A. Horn); US 3 892 807, 3 994 879, 1972 (W. Sahm).
[79] W. Sahm, E. Schinzel, P. Juerges, *Justus Liebigs Ann. Chem.* **1974**, 523–538.
[80] Hoechst, US 4 133 953, 1977, US 4 309 536, 1979 (E. Schinzel).
[81] Ciba, US 2 488 094, 1942 (C. Gränacher, F. Ackermann); US 3 103 518, 1963 (M. Dünnenberger, A. E. Siegrist).
[82] Sandoz, US 4 018 789, 1974 (P. S. Littlewood, A. V. Mercer).
[83] Ciba-Geigy, GB 1 331 670, 1970 (H. Schläpfer); EP 10 063, 1978 (L. Guglielmetti, C. Lüthi).
[84] Ciba-Geigy, US 4 313 846, 4 455 436, 1978 (H. R. Meyer, M. Siegrist).
[85] Sandoz, GB 1 577 127, 1976 (P. S. Littlewood).
[86] L. Knorr, P. Duden, *Ber. Dtsch. Chem. Ges.* **26** (1893) 115. K. v. Auwers, H. Voss, *Ber. Dtsch. Chem. Ges.* **42** (1909) 4411–4427.
[87] Ilford, US 2 639 990, 1949 (J. D. Kendall, G. F. Duffin).
[88] Bayer, US 3 135 742, 1956 (A. Wagner, A. Schlachter, H. Marzolph).
[89] Bayer, US 3 378 389, 1962 (C. W. Schellhammer, A. Wagner).
[90] Hoechst, US 3 255 203, 1962 (E. Schinzel, K. H. Lebkücher).
[91] Hoechst, US 3 619 234, 1968 (J. Weihsbach, O. Smerz, G. Rösch).
[92] Bayer, US 2 879 174, 1955 (A. Wagner, S. Petersen).
[93] Hoechst, GB 1 360 490, 1970 (H. Mengler, E. Schinzel, G. Rösch); US 3 865 816, 1971 (H. Mengler); DE-OS 2 702 635, 1977 (O. Smerz, E. Schinzel, G. Rösch, J. Weihsbach).
[94] Sandoz, US 3 997 556, 1974 (I. J. Bolton, A. V. Mercer).
[95] Ciba-Geigy, US 4 816 590, 1985 (H. R. Meyer).
[96] Bayer, EP 73 996, 1983 (H. Theidel).
[97] Sandoz, US 3 849 406, 1971 (H. Aebli, F. Fleck, P. S. Littlewood, A. V. Mercer).
[98] Hoechst, US 3 560 485, 1964 (E. Schinzel, S. Bildstein, K. H. Lebkücher); US 3 690 947, 1969 (G. Rösch, W. Linke, O. Smerz, E. Schinzel).
[99] Bayer, US 3 131 079, 1960 (A. Wagner, S. Petersen).
[100] Bayer, US 3 189 772, 3 925 367, 1970 (G. Boehmke, H. Theidel).
[101] Ciba, US 2 610 152, 1946 (F. Ackermann). Sandoz, US 2 654 713, 1951, US 2 791 564, 1952 (F. Fleck).
[102] H. von Pechmann, M. Schaal, *Ber. Dtsch. Chem. Ges.* **32** (1899) 3690.
[103] Bayer, US 3 351 482, 1960 (R. Raue).
[104] Geigy, US 2 945 033, 1957; US 3 429 880, 1965 (H. Häusermann). Bayer, US 3 242 177, 1963 (C. W. Schellhammer, E. Degener, H. G. Schmelzer, A. Wagner); US 3 244 711, 1964 (O. Berendes, H. Gold, C. W. Schellhammer).
[105] Geigy, US 3 123 617, 1961 (H. Häusermann).
[106] Bayer, US 3 646 052, 1966 (O. Neuner, A. Dorlars).
[107] Sandoz, US 3 288 801, 1962 (F. Fleck, H. Balzer, H. Aebli).
[108] Bayer, GB 1 313 253, 1970 (A. Dorlars, C. W. Schellhammer).
[109] Bayer, US 4 005 098, 1968 (A. Dorlars, C. W. Schellhammer, W. D. Wirth); US 3 271 412, 1961 (R. Raue, H. Gold).
[110] Bayer, EP 74 590, 1981 (J. Koll, H. Theidel, H. H. Mölls).
[111] General Aniline & Film Corp., US 2 600 080, 1946 (M. O. Shrader).
[112] BASF, GB 741 798, 1952 (E. Tolksdorf, F. Schubert, E. Kern).

[113] Mitsubishi, US 3 310 564, 1961 (T. Kasai).
[114] Mitsubishi, JP-Kokai 71 05 596, 1968 (T. Kasai), *Chem. Abstr.* **75** (1971) 7435c; JP-Kokai 71 13 953, 1968 (S. Kasai), *Chem. Abstr.* **76** (1972) 114764b.
[115] Mitsubishi, JP-Kokai 70 03 671, 1967 (H. Okada, M. Kaneko), *Chem. Abstr.* **73** (1970) 16310u.
[116] Nippon, JP-Kokai 74 05 333, 1969 (T. Noguchi, S. Matsunaga).
[117] ICI, US 3 157 651, 1961 (J. R. Atkinson, S. Hartley). Bayer, US 4 278 795, 1979 (H. Harnisch). Hoechst, US 4 675 402, 1984 (E. Schinzel).
[118] M. Lüscher: "Psychological Aspects of White", in [7] pp. 1–11.
[119] I. Soljacic, K. Weber: "Die Wirkung organischer Fremdstoffe auf die Fluoreszenz optischer Aufheller", *Textilveredlung* **14** (1979) no. 3, 97–101.
[120] I. Soljacic, K. Weber: "Ueber die Konzentrationslöschung der Fluoreszenz optischer Aufheller", *Textilveredlung* **6** (1971) no. 12, 796–802. "Washing Powders containing FWA", *Res. Discl.* **151** (1976) no. 11, 18–20.
[121] H. Hefti: "Whitening of Textiles", in [7], pp. 51–58.
[122] H. K. Banerjee, K. G. Pai: "Textile Optical Whiteners", *Colourage* **22** (1975) no. 7, 25–35.
[123] R. Williamson: "Fluorescent Brightening Agents", *Int. Dyer Text. Printer Bleacher Finish.* **157** (1977) no. 8, 359–364; no. 9, 408–416.
[124] R. Levene, M. Lewin: "Some Aspects of the Fluorescent Whitening of Textiles", *Colour. Annu.* 1982–83, A44–A49.
[125] J. J. Donze: "L'azurage optique", *Ind. Text. (Paris)* **1177** (1987) no. 5, 519–524.
[126] W. Schuerings: "Weissausrüsten von Polyester-Cellulosetextilien", *Text.-Prax. Int.* **29** (1974) no. 11, 1570–1584.
[127] W. Schuerings: "Optische Aufheller für Baumwolle in Semi- und Vollkontinue-Verfahren", *Textilveredlung* **15** (1980) no. 7, 236–241.
[128] W. Schoelermann, H. Mantz: "Der Einfluss von Reaktant und Katalysator auf den Weissgrad in der Hochveredlung", *Textilveredlung* **8** (1973) no. 10, 506–514.
[129] G. Rösch: "L'azurage optique des fibres synthétiques", *Text. Chim.* **28** (1972) no. 4, 43–50.
[130] A. Decorte: "La blancheur des fibres acryliques", *Text. Chim.* **27** (1971) no. 4, 4–10.
[131] R. Anliker, H. Hefti, A. Räuchle, H. Schläpfer: "Das Aufhellen von Polyacrylnitril-Fasern", *Textilveredlung* **11** (1976) no. 9, 369–375.
[132] C. Eckhardt, H. Hefti: "A Comparison of Fluorescent Brightening in the Production of Synthetic Polymer Fibers and in the Textile Industry", *J. Soc. Dyers Colour.* **87** (1971) no. 11, 365–370.
[133] H. Hefti: "Optische Aufheller für Polyesterfaserstoffe", *Textilveredlung* **4** (1969) no. 2, 94–101.
[134] Ciba-Geigy, EP-A 0 280 652, 1988 (H. Hefti, U. Lehmann, D. Reinehr, J. Kaschig).
[135] Storey Br, GB 38 491/73, 1973.
[136] R. Toex: "Optisches Aufhellen von Polyesterfasern im Transferdruck", *Chemiefasern Textilind.* **30** (1980) no. 8, 642–643.
[137] P. Carter: "FWA for the Soap and Detergents Industry", *MVC Rep. Miljoevardscentrum, Stockholm*, **2** (1973) 51–56; *Chem. Abstr.* **82** (1975) 32767 q.
[138] C. Eckhardt, R. von Rütte: "FWA in Detergents", in [7]. pp. 59–64.
[139] A. Siegrist: "The Requirements of Present-Day Detergent FWA", *J. Am. Oil Chem. Soc.* **55** (1978) 114. P. Stensby: "FWA" in W. G. Cutler (ed.): Detergency, Part III, Marcel Dekker, New York 1981, pp. 730–806. U. Schüssler: "FWA for Detergents,

2nd World Conf. on Detergents", *J. Am. Oil Chem. Soc.* **63** (1986) 187. W. Findley: "FWA for Modern Detergents", *J. Am. Oil Chem. Soc.* **65** (1988) no. 4, 679–683.
[140] Ciba-Geigy, US 4 311 605, 1982 (C. Eckhardt, R. von Rütte).
[141] G. Becker: "Das Waschen bei niedrigen Waschtemperaturen", Dok. Chemiefasertagung Dornbirn, 1977, 1–29; 1980, 1–13.
[142] Unilever, GB 1 286 459, 1972 (J. Evans).
[143] Procter & Gamble, EP-A 0 237 119, 1987 (J. Wevers, J. Walker). Bayer, DE 3 726 266 A1, 1987 (U. Schüssler, F. Seng). Sandoz, EP-A 0 314 630, 1988 (J. Farrar, M. Graham).
[144] Ciba-Geigy, DE-OS 2 730 246, 1978 (F. Günter, C. Eckhardt).
[145] Clorox, US 4 764 302, 1988 (G. Baker, J. Farr, D. Carty). Albright & Wilson, EP-A 0 273 775, 1988 (E. Messenger, A. Martin). Procter & Gamble, EP-A 0 293 040, 1988 (F. de Buzzaccarini, J. P. Boutique).
[146] Colgate-Palmolive, GB 2 158 838 A, 1985 (M. Julemont, L. Laitem, M. Marchal); US 4 781 856, 1988 (T. Ouhadi, L. Dehan); DE 3 704 876 A1, 1987 (T. Ouhadi, L. Dehan); US 4 772 413, 1988 (J. Massaux, N. Mineo, L. Laitem).
[147] Procter & Gamble, US 4 179 390, 1979 (G. Spadini, I. Tolliday).
[148] Ciba-Geigy, US 4 008 172, 1977, Re. 30 227, 1980 (R. von Rütte, W. Gasser). Bayer, DE-OS 3 046 482 A1, 1982 (U. Schimmel, U. Schüssler, A. Brüggemann).
[149] W. Mischler: "FWA in the Paper Industry", *Ciba-Geigy Rev.* 1973, no. 1, 35–38.
[150] H. G. Oesterlin: "Optische Aufhellung gestrichener Papiere", *Papier (Darmstadt)* **36** (1982) no. 2, 66–72.
[151] A. Wieber, C. Eckhardt: "Mass Whitening of Synthetic Fibers and Plastics", *Ciba-Geigy Rev.* 1973, no. 1, 26–29. A. Wieber: "Mass Whitening of Synthetic Fibers and Plastics", in [7], pp. 65–82.
[152] Weifang S. D. P., Chin. FZSGS 87 108 249, 1988 (Y. Ding, X. Cao, Z. Zhang).
[153] Weifang S. D. P., Chin. FZSGS 87 108 248, 1988 (Y. Ding, X. Cao, M. Zhang).
[154] Kodak, GB 47 527/62, 1962 (D. Saunders, E. Knott, M. Mijovic, H. Wealdstone).
[155] E. Kowalski, R. Anliker, K. Schmid: "Criteria for the Selection of Solutes in Liquid Scintillation Counting", *Int. J. Appl. Radiat. Isot.* **18** (1967) 307–323. C. T. Peng: "Sample Preparation in Liquid Scintillation Counting", *Rev. Amersham Corp.* **17** (1977) 9. A. Dyer: *Liquid Scintillation Counting Practice*, Heyden & Son, East Kilbride, Scotland, 1980.
[156] H. H. Koch, M. Pimsler: "A Nonspecific Fluorescent Stain for Detecting and Identifying Fungi and Algae in Tissue", *Lab. Med.* **18** (1987) no. 9, 603–606. E. D. Wachsmuth: "A Comparison of the Highly Selective Fluorescence Staining of Fungi in Tissue Sections", *Histochem. J.* **20** (1988) 215–221.
[157] H. Telle, U. Brinkmann, R. Raue: "Laser Properties of Triazinylstilbene Compounds", *Opt. Commun.* **24** (1978) no. 1, 33–37; "Laser Properties of Bis-styryl Compounds", *Opt. Commun.* **24** (1978) no. 3, 248–250. M. Ringe, H. Güsten, *Ber. Bunsen-Ges. Phys. Chem.* **90** (1986) 439–444.
[158] J. Lanter: "Properties and Evaluation of FWA", *J. Soc. Dyers Colour.* **82** (1966) 125–132.
[159] G. Anders: "Limits of Accuracy Obtainable in the Direct Determination by Fluorimetry of FWA on TLC", in [7], pp. 104–110.
[160] H. Theidel: "Qualitative TLC of FWA", in [7], pp. 94–103.
[161] H. Theidel: "Direct Determination of FWA by Absorption Measurement in situ on TLC", in [7], pp. 111–114.
[162] G. R. Weis: "Säulenchromatographische Auftrennung von Aufhellergemischen", *Tenside Deterg.* **12** (1975) no. 1, 43–44.

[163] D. Kirkpatrick: "Separation of Optical Brighteners by Solid-Liquid Chromatography," *J. Chromatogr.* **139** (1977) 168–173.

[164] B. P. McPherson, N. Omelczenko: "The Determination of Optical Brighteners in Laundry Detergents by Reverse Phase and Ion Pair High Performance Liquid Chromatography," *J. Amer. Oil Chem. Soc.* **57** (1980), 388–391. A. Nakae, M. Morita, M. Yamanaka: "Separation and Determination of Fluorescent Whitening Agents in Detergents by High Speed Liquid Chromatography," *Bunseki Kagaku* **29** (1980) 69–73. K. Tsuji, S. Setsuda, S. Naito, A. Abe: "Determination of Fluorescent Whitening Agents in Detergents by High Performance Liquid Chromatography," *Bull. Kanagawa P. H. Lab.* **1981**, 65–67. G. Micali, P. Curro, G. Calabro: "High Performance Liquid Chromatographic Separation and Determination of Fluorescent Whitening Agents in Detergents," *Analyst* **109** (1984) 155–158. J. L. Jasperse, P. H. Steiger: "A System of Determining Optical Brighteners in Laundry Detergents by TLC and HPLC," *J. Amer. Oil Chem. Soc.* **69** (1992) 621–625.

[165] C. Eckhardt: "Examen del blanqueo optico de los textiles tenidos o preblanqueados", *Invest. Inf. Text. Tensioactivos* **16** (1973) no. 4, 465–485.

[166] B. Werthmann, R. Borowki: "Möglichkeiten zur quantitativen Bestimmung optischer Aufheller für Papier", *Papier (Darmstadt)* **30** (1976) no. 6, 243–249.

[167] W. Archwal, A. Narkar: "Die Bestimmung optischer Aufheller auf Baumwolle mittels Cadoxen als Lösungsmittel", *Textilveredlung* **7** (1972) no. 1, 19–24.

[168] H. Theidel: "Zur Stereochemie optischer Aufheller vom Stilben-Typ", *Melliand Textilber.* **45** (1964) no. 5, 514–519. W. Weller: "The Preparation of the Pure cis-Isomers of Four FWA of the Sulfonated Stilbene Type", *J. Soc. Dyers Colour.* **95** (1979) no. 5, 187–190. K. Smit, K. Ghiggino: "Fluorescence and Photoisomerisation of Two Stilbene-Based Dyes", *Dyes Pigm.* **8** (1987) no. 2, 83–97.

[169] J. Schulze, T. Polcaro, P. Stensby: "Analysis of FWA in US Home Laundry Detergents", *Soap Cosmet. Chem. Spec.* **50** (1974) no. 11, 46–52. H. Bloching, W. Holtmann, M. Otten: "Beitrag zur Analytik von Weisstönern in Waschmitteln", *Seifen Oele Fette Wachse* **105** (1979) no. 2, 33–38.

[170] T. Poiger, J. A. Field, T. M. Field, W. Gieger: "Determination of Detergent Derived Fluorescent Whitening Agents in Sewage Sludges by Liquid Chromatography," *Analytical Methods and Instrumentation* **1** (1993) 104–113. T. Poiger, J. A. Field, T. M. Field, W. Gieger: "Occurrence of Fluorescent Whitening Agents in Sewage and River Water Determined by Solid-phase Extraction and High-performance Liquid Chromatography", *Environmental Science and Technology* **30** (1996) 2220–2226.

[171] G. Anders, E. Ganz: "Metameric White Samples for Testing the Relative UV Content of Light Sources and of Daylight", *Appl. Opt.* **18** (1979) 1067–1072.

[172] C. Eckhardt: "Visual Assessment of Whiteness", *Ciba-Geigy Rev.* 1973, no. 1, 10–13.

[173] G. Anders: "Control of Whites: The Cibanoid White Scale", *J. Soc. Dyers Colour.* **84** (1968) no. 2, 125–132. J. Rieker, D. Gerlinger: "Neuerungen auf dem Gebiet der farbmetrischen Weissbewertung", *Melliand* **65** (1984) 631–632.

[174] F. Gärtner, R. Griesser: "Eine Vorrichtung zur Messung von optischen Aufhellern mit konstanter UV-Anregung", *Farbe* **24** (1975) 199–207.

[175] D. Eitle, E. Ganz: "Eine Methode zur Bestimmung von Normfarbwerten für fluoreszierende Proben", *Textilveredlung* **3** (1968) 389–392.

[176] G. Wyszecki, W. S. Stiles: *Color Science, Concepts and Methods, Quantitative Data and Formulas*, J. Wiley & Sons, New York–London–Sydney, 1967.

[177] R. Hunter, *J. Opt. Soc. Am.* **48** (1958) 597. A. Berger, *Farbe* **8** (1959) 187. P. Stensby, *Soap & Chem. Spec.* **43** (1967) 80.

[178] A. S. Stenius et al., *Farbe* **26** (1977) 3–104.

[179] S. V. Vaeck: "Some New Experiments on the Colorimetric Evaluation of Whiteness", *J. Soc. Dyers Colour.* **95** (1979) 262–269.
[180] F. Grum: "Whiteness and Fluorescence", *Text. Chem. Color.* **13** (1981) 187–189.
[181] E. Ganz: "Whiteness: Photometric Specification and Colorimetric Evaluation", *Appl. Opt.* **15** (1976) 2039–2058. E. Ganz: "Whiteness Formulas: A Selection", *Appl. Opt.* **18** (1979) 1073–1078. E. Ganz, R. Griesser: "Whiteness: Assessment of Tint", *Appl. Opt.* **20** (1981) 1395–1396. R. Griesser: "Stand der instrumentellen Weissbewertung unter besonderer Berücksichtigung der Beleuchtung", *Textilveredlung* **18** (1983) 157–162.
[182] A. Brockes: "The Evaluation of Whiteness", *CIE-J.* **1** (1982) no. 2, 38–39.
[183] B. Mensak: "Exakte Weißmetrik mit Spektralphotometern", *Maschen Industrie* **3** (2000) 40–42.
[184] R. Griesser: "Assessment of Whiteness and Tint of Fluorescent Substrates with Good Inter-Instrument Correlation", *Color Res. Appl.* **19** (1994) no. 6, 446–460.
[185] R. Hayhurst, K. Smith: "Instrumental Evaluation of Whiteness", *J. Soc. Dyers Colourists* **111** (1995) 263–266.
[186] E. J. van de Plassche, P. F. H. Bont, J. M. Hesse: "Exploratory Report Fluorescent Whitening Agents (FWAs)", Dutch National Institute of Public Health and Environment, Report 601503013 (1999). J. Plautz, P. Richner: "Human & Environmental Risk Assessment on Ingredients of European Household Cleaning Products: Fluorescent Brightener FWA-5", HERA, International Association for Soaps, Detergents and Maintenance Products, A.I.S.E., 2001; http://www.heraproject.com/RiskAssessment.cfm.

8 Health and Safety Aspects

8.1 Introduction

The analysis of a product which is to be marketed nowadays requires, apart from the identification of its technical and application properties, comprehensive toxicological and ecological considerations. Dyes are no exception. Manufacturers, processors, and customers are always faced with the question of whether the production or use of a specific dye might present an unacceptably high toxicological or ecological risk for humans or the environment. Eliminating or at least reducing these risks to legally acceptable levels is very important. For acceptability, criteria must be developed for each individual case with regard to legal provisions.

The evaluation and knowledge of physical, chemical, toxicological, and ecological properties of a dye is essential in order to assess its influence on the environment and to estimate whether or not a given product presents a hazard potential. It is important also to consider the type of handling and use that a dye is likely to undergo, the extent of human exposure that might be expected, and the amount involved.

Primary considerations in dye manufacture and use, as with any other chemical, include personal safety, atmospheric emissions, wastewater quality, and appropriate waste disposal.

Brochures on safe handling of dyes [1] as well as on product stewardship of dyes [2] have been published. An ETAD online database on publications of toxicological and ecological effects concerning dyes (and pigments), as well as on appropriate legislation, is available for ETAD member companies [3].

8.2 Toxicology and Toxicity Assessments

More than 30 years ago and long before chemical and environmental regulations existed, the colorants manufacturing industry of Western Europe began to investigate toxicological and ecological properties of dyes (and pigments). Today numerous laws and regulations require manufacturers to assess the hazard potential of each of their substances [4].

Toxicology studies are concerned with a variety of aspects, primarily with (1) Acute toxicity, (2) Irritation of skin and eyes, (3) Toxicity after repeated application, (4) Sensitization, (5) Mutagenicity, and (6) Cancerogenicity.

8.2.1 Acute Toxicity

The first step to determine whether a dye is hazardous is the appropriate evaluation or testing of the acute toxicity of dyes, as defined by the EU Directive 67/548/EEC (with numerous amendments). A comprehensive review on such data including skin and eye irritation of numerous commercial dyes, derived from Safety Data Sheets, showed that the potential for these acute toxic effects ("harmful" or "toxic") was very low. Although the review stems from an early date, it can be asumed that the results are still valid today [5].

8.2.2 Sensitization

Dermatologists have reported cases of skin reactions suspected to be caused by textile dyes [6–10]. There is evidence that some reactive dyes cause contact dermatitis, allergic conjunctivitis, rhinitis, occupational asthma, or other allergic reactions in textile workers. The reason is believed to be the capability of reactive dyes to combine with human serum albumin (HSA) to give a dye–HSA conjugate which acts as an antigen [11]. The antigen in turn gives rise to specific immunoglobulin E (IgE) antibodies and, through the release of intermediates such as histamine, produces allergic reactions. In 1985, a study was undertaken of 414 workers who were exposed to dye powders, e.g., as dye-store workers, mixers, weighers, dyehouse operators, and laboratory staff. Twenty-one of them were identified as having allergic symptoms, including occupational asthma, due to one or more reactive dyes [12].

Reactive dyes that caused respiratory or skin sensitization in workers on occupational exposure were compiled by ETAD [13]. In the EU these dyes, which are all listed in the European Inventory of Existing Chemical Substances (EINECS), should be labeled accordingly (Table 8.1).

Table 8.1 List of reactive dyes to be classified as respiratory/skin sensitizers

C.I. name	C.I. no.	CAS no.	R phrase
Reactive Black 5	20505	[17095-24-8] (4 Na)	R42/43
Reactive Blue 114		[72139-17-4] (2 Na)	R42/43
Reactive Blue 204		[85153-92-0] (6 Na)	R42/43
Reactive Orange 4	18260	[70616-90-9] (3 Na)	R42/43
Reactive Orange 12	13248	[70161-14-7] (3 Na)	R42/43
		[93658-87-8] (x Na)	R42/43
Reactive Orange 14		[12225-86-4] (acid)	R42
Reactive Orange 16		[20262-58-2] (2 Na)	R42
		[106027-83-2] (2 Li)	R42
Reactive Orange 64		[83763-57-9] (x Na)	R42/43
Reactive Orange 67		[83763-54-6] (x Na)	R42/43
Reactive Orange 86		[57359-00-9] (3 Na)	R42/43
Reactive Orange 91		[63817-39-0] (3 Na)	R42/43
Reactive Red 29		[94006-25-4] (5 Na)	R42
		[70865-39-3] (4 Na)	R42
Reactive Red 65		[70210-40-1] (2 Na)	R42/43
Reactive Red 66	17555	[70210-39-8] (2 Na)	R42/43
Reactive Red 123		[85391-83-9] (x Na)	R42/43
		[68959-17-1] (2 Na)	R42/43
Reactive Red 219		[149057-72-7](4 Na)	R42
Reactive Red 225		[83399-95-5] (x Na)	R42/43
Reactive Violet 33		[69121-25-1] (3 Na)	R42/43
Reactive Yellow 25		[72139-14-1] (3 Na)	R42/43
Reactive Yellow 39	18971	[70247-70-0] (2 Na)	R42/43
Reactive Yellow 175		[111850-27-2] (2 Na)	R42

To minimize the risk, the exposure to dye dust must be avoided either by using liquid dyes, dyes with low-dusting formulations, and by appropriate personal protective equipment. After dyeing and fixation, reactive dyes have different toxicological properties, because the reactive group is no longer present, and the high fastness properties mean no exposure to the skin of the wearer. Therefore, no cases of allergic reactions are reported in consumers wearing textiles dyed with reactive dyes.

It is estimated that 1–2 % of the total allergic diseases which were treated in German hospitals were caused by contact with textiles, and most of them were due to dyes. The most conspicuous dyes were disperse dyes, especially when they were used for skintight, close-fitting clothes made from synthetic fibers. The sweatfastness properties of the dyes on the different textiles are an important factor in whether an allergic reaction is caused or not. Disperse dyes with a sensitizing potential may cause allergic skin reaction if dyed on polyamide or semi-acetate, on which the low wetfastness could allow the dyes to migrate to the skin [14]. In the 1980s some severe cases of allergic diseases were described [15] related to stockings and pantyhoses made of polyamide [16] and in the 1990s to sportswear (leggings) made of semi-acetate.

The German Federal Institute for Consumer Protection and Veterinary Medicine (BgVV) evaluated the available literature and came to the conclusion that above all the following disperse dyes represent a health risk to consumers and should therefore not be used anymore for clothes (Table 8.2).

Table 8.2 Disperse dyes representing a health risk to consumers (BgVV)

C.I. Name	C.I. no.	CAS no.
C.I. Disperse Blue 1	64500	[2475-45-8]
C.I. Disperse Blue 35		[12222-75-2]
C.I. Disperse Blue 106	11935	[68516-81-4]
C.I. Disperse Blue 124	111938	[15141-18-1]
C.I. Disperse Yellow 3	11855	[2832-40-8]
C.I. Disperse Orange 3	11005	[730-40-5]
C.I. Disperse Orange 37/76		[12223-33-5]
C.I. Disperse Red 1	1110	[2872-52-8]

Up to now there is no legal prohibition in any country, but some organizations, e.g., the International Association for Research and Testing in the Field of Textile Ecology (Öko-Tex), which bestows eco-labels on environmentally and toxicologically proven textiles, refuses eco-labels for some dyes [17].

8.2.3 Mutagenicity

Some dyes exhibit a mutagenic potential. The Ames test is commonly used as a first screening for the prediction of mutagenicity of a substance. It is a bacterial point mutation test inducing activity, which uses special strains of the bacteria *Salmonella typhimurium* with growth-dependence on the amino acid histidine. The dose-dependent reversion to histidine-independent growth is the marker for a point mutation.

For the mutagenicity testing of azo dyes the Prival test, a modification of the Ames test, was found to be superior. The reductive enzymatic cleavage of the azo bond occuring in mammals is simulated by this test.

It is generally accepted that carcinogenesis is a multiphase process in which genotoxic effects can be important contributors in the initiation phase. Since the bacterial reverse mutation test is a highly sensitive assay for the induction of point mutations in bacteria, rather than a test for the complex multiple step process of carcinogenesis in mammals, a close correlation between Ames test results and rodent cancer bioassays cannot be expected [18]. Validation studies (see [19] and references cited therein) demonstrate a relatively low degree of correlation between the end points of mutagenicity in bacteria and carcinogenicity in rodents.

To establish a meaningful assessment of mutagenicity (and possible carcinogenicity), positive results in the Ames/Prival test should be followed by further in vitro gene mutatation tests using mammalian cells, e.g., the HPGRT test or the mouse lymphoma test. Provided a mutagenicity potential has been found in these studies, a chromosome aberration assay, such as the cytogenetic test in vitro, would be a reasonable next step. After demonstrating a genotoxic potential in vitro, and in order to determine a possible mutagenic potential in animals, corresponding in vivo tests, e.g., the cytogenetic in vivo test or the micronucleus test should subsequently be carried out. The test results might also enable the prediction of a carcinogenic potential.

8.2.4 Carcinogenicity

8.2.4.1 Introduction

The toxic nature of some dyes and intermediates has long been recognized. Acute, or short-term effects are generally well known. They are controlled by keeping the concentration of the chemicals in the workplace atmosphere below prescribed limits and avoiding physical contact with the material. Chronic effects, on the other hand, frequently do not become apparent until after many years of exposure. Statistically higher incidences of benign and malignant tumors, especially in the bladders of workers exposed to certain intermediate and dye manufacturing processes, were recorded in dye-producing countries during the period 1930–1960. The specific compounds involved were 2-naphthylamine [*91-59-8*], 4-aminobiphenyl [*92-67-1*], benzidine (4,4′-diaminobiphenyl) [*92-87-5*], fuchsine (*C.I. Basic Violet 14*, 42510 [*632-99-5*]), and auramine (*C.I. Solvent Yellow 2*, 11020 [*60-11-7*]). There is considerable evidence that metabolites of these compounds are the actual carcinogenic agents [20, 21]. Strict regulations concerning the handling of known carcinogens have been imposed in most industrial nations. These regulations [22] have caused virtually all the dye companies to discontinue use of these compounds.

According to animal tests some dyes are proven carcinogens and are probably carcinogenic to humans. Table 8.3 lists dyes which are recognized to cause cancer in animals and are therefore classified as potential human carcinogens.

Table 8.3 Dyes classified as potential carcinogens

C.I. name	C.I. no.	Chemical class	Diazo component	Classification		
				IARC[a]	NTP[b]	67/548/EEC[c]
Acid Dye	16155	azo	pseudocoumidine	2B		
Acid Red 26	16150	azo	xylidine	2B		
Acid Violet 49	42640	triphenylmethane		2B		
Basic Yellow 2	42100	ketoniminc		2B		
Basid Red 9	42500	triphenylmethane		2B	B	Cat. 2
Basic Violet 14	42510	triphenylmethane		2B		
Disperse Orange 11	60700	anthraquinone			B	
Disperse Blue 1	64500	anthraquinone		2B	B	Cat. 2
Solvent Yellow 1	11000	azo	aniline	2B		Cat. 2
Solvent Yellow 2	11020	azo	aniline	2B	B	
Solvent Yellow 34	41001:1	diphenylmethane		2B[d]		

[a] International Agency for Research on Cancer (IARC): 2B: possibly carcinogenic to humans [23].
[b] U.S. National Toxicology Program, 9th NTP Report, B: reasonably anticipated to be a human carcinogen [24].
[c] 67/548/EEC Annex 1: Cat. 2: sufficient evidence to be carcinogenic to humans.
[d] Production is classified in group 1 (carcinogenic to humans).

8.2.4.2 Metabolism of Azo Dyes

With regard to metabolism, azo dyes are the most widely investigated class of dyes. According to metabolic pathway two groups of azo dyes can be distinguished: (1) water-soluble dyes, mostly bearing sulfo groups, and (2) solvent-soluble dyes with nonpolar substituents.

By far the most predominant metabolic pathway for water-soluble azo dyes is cleavage of the azo linkage by azoreductase of the liver and extrahepatic tissue or by intestinal microflora in the body [25, 26]. Oxidative metabolism occurs for lipid-soluble dyes, e.g., solvent dyes. Three oxidation pathways are known for such dyes (1): *C*-(ring-)hydroxylation, (2) *N*-hydroxylation at a primary or secondary amino group or (3) by stepwise oxidation of the methyl groups of dimethylamino compounds (demethylation). All three oxidative degradation ways leave the azo bond intact. For further details of the mechanisms, see [27, 28].

8.2.4 Carcinogenicity

The metabolism of the benzidine-based dyes *C.I. Direct Red 28*, *C.I. Direct Blue 6*, *C.I. Direct Brown 95*, and *C.I. Direct Black 38* was studied in Rhesus monkeys. After ingestion of the dyes, benzidine and monoacetylbenzidine could be detected as metabolites in the urine. This indicated that the dyes had been converted to benzidine [29]. Recent in vitro studies on *C. I. Direct Blue 14* show that bacteria isolated from healthy human skin have reductase activity and are able to cleave the dye into the corresponding arylamine, in this case 3,3′-dimethylbenzidine [30].

The role of certain aromatic amines in the etiology of bladder cancer has long been known, and a carcinogenic effect of benzidine dyes on the human bladder is very likely according to epidemiologic studies. In case of poor working practice and the use of dusting products, workers can be exposed to high levels of dye dust which can exceed the permissible limit. It could be demonstrated that workers exposed to the dust of benzidine-based dyes, excreted benzidine and the related metabolites *N*-acetyl- and *N,N*-diacetylbenzidine [31]. In the blood serum from female textile workers employed in dye printing, warehouse, and color room shops, benzidine was detected [32]. There is evidence that dyes based on the benzidine congeners *o*-dianisidine and *o*-tolidine may also be metabolized to their parent compounds [33].

o-Dianisidine is also used as the starting material in the manufacture of metallized copper-complex dyes. These dyes no longer contain the methoxyl groups and it is therefore likely that metallized *o*-dianisidine dyes would be metabolized not to *o*-dianisidine but to 3,3′-dihydroxybenzidine, which is evidently noncarcinogenic.

The possible release of the aromatic amines, which are carcinogenic in rodents, and subsequent metabolic activation in the organism is assumed to be the reason for the carcinogenic effect of some members of this dye class in animal testing. These findings support the conclusion that dyes which could be metabolized to a carcinogenic aromatic amine should be considered to be carcinogenic.

In Germany bladder cancer is recognized as an existing occupational disease for textile workers [34]. This knowledge is the reason for the recommendation of the German MAK Kommission to handle the dyes in the same way as the amines, which can be released under reducing conditions. In the next step the German, Austrian, and Dutch authorities prohibited their use in some consumer articles [35] (see Section 8.4.3).

Table 8.4 lists the carcinogenic amines according to TRGS 614 (Limitation of use of dyes which are likely to cleave into carcinogenic aromatic amines) (March 2001) [36], which are classified as carcinogenic by TRGS 905 and 67/548/EEC Annex 1.

Table 8.4 Categories of listed carcinogenic amines

Name	CAS-No.	TRGS 905[a, b]	67/548/EEC
4-Aminobiphenyl	[92-67-1]		Carc. Cat. 1
Benzidine	[92-87-5]		Carc. Cat. 1
4-Chloro-*o*-toluidine	[95-69-2]	Carc. Cat. 1	
2-Naphthylamine	[91-59-8]		Carc. Cat. 1
o-Aminoazotoluene	[97-56-3]		Carc. Cat. 2
5-Nitro-*o*-toluidine	[99-55-8]	Carc. Cat. 3	
p-Chloroaniline	[106-47-8]		Carc. Cat. 2
4-Methoxy-*m*-phenylenediamine	[615-05-4]	Carc. Cat. 2	
4,4'-Diaminodiphenylmethane	[101-77-9]		Carc. Cat. 2
3,3'-Dichlorobenzidine	[91-94-1]		Carc. Cat. 2
3,3'-Dimethoxybenzidine	[119-90-4]		Carc. Cat. 2
3,3'-Dimethylbenzidine	[119-93-7]		Carc. Cat. 2
4,4'-Methylendi-*o*-toluidine	[838-88-0]		Carc. Cat. 2
6-Methoxy-*m*-toluidine	[120-71-8]	GefStV	
4,4'-Methylenebis(-2-chloroaniline)	[101-14-4]		Carc. Cat. 2
4,4'-Oxydianiline	[101-80-4]	Carc. Cat. 2	
4,4'-Thiodianiline	[139-65-1]	Carc.Cat. 2	
o-Toluidine	[95-53-4]		Carc. Cat. 2
4-Methyl-*m*-phenylendiamine	[95-80-7]		Carc. Cat. 2
2,4,5-Trimethylaniline	[137-17-7]	Carc. Cat. 2	
o-Anisidine[c]	[90-04-0]		Carc. Cat. 2
4-Aminoazobenzene	60-09-3	Carc. Cat. 2	
4-Amino-3-fluorophenol[d]	399-95-1	Carc. Cat.2	
6-Amino-2-ethoxynaphthalene[d]		GefStV	

[a] Technische Regeln für Gefahrstoffe (German Technical Rules for Dangerous Substances).
[b] TRGS 905 lists only CMT substances which are not correspondingly regulated by other laws.
[c] Application is banned on carpets, articles that may come into contact with the oral cavity, and toys containing leather or textiles.
[d] Azo dyes cleaving off these amines are not known. No analytical proof is necessary.

ETAD has compiled a list of azo dyes which upon reduction of the azo bond would form the aromatic amines listed in Table 8.4. The list covers more than 500 azo dyes, of which at least 142 are still available on the world market [37].

8.3 Environmental Assessment/Fate

8.3.1 Introduction

Within the EU the chemical control directives are based on three steps: (1) Hazard assessment, (2) Risk assessment, and (3) Risk management.

Hazard assessment includes the classification of substances and preparations that are hazardous to the environment. Dyes are preparations, and therefore ecotoxicological and ecological data have to be generated for classification and possible labeling of the correponding preparation. Major factors to be considered are the biodegradability and toxicity to aquatic organisms (fish, bacteria, Daphnia, algae). Dyes in general are not readily biodegradable. The criteria for classification and labeling of dyes have been described in Directive 67/548/EEC as last amended (2001/59/EC) and in Directive 99/45/EC, to which ETAD has issued a Guidance Document for the dye-using industries [38].

8.3.2 Treatment of Dye-Containing Wastewater

Dyes, because they are intensely colored, present special problems in wastewater; even a very small amount is noticeable. However, the effect is more aesthetically displeasing rather than hazardous, e.g., red dyes discharged into rivers and oceans. Of more concern is the discharge of toxic heavy metals such as mercury and chromium.

Wastewaters from both dye manufacturing plants and dyehouses are treated before leaving the plant, e.g., by neutralization of acidic and alkaline liquors and removal of heavy metals, and in municipal sewage works. Various treatments are used [39].

Biological treatment is the most common and most widespread technique used in effluent treatment, having been employed for over 150 years. There are two types of treatment, aerobic and anaerobic. The aerobic system needs air (oxygen) for the bacteria to perform the degradation process on the activated sludge, whereas anaerobic bacteria operate in the absence of air. Activated sludge usually removes only a moderate amount (10–20 %) of the color.

Removal of color by adsorption on activated carbon is also employed. Activated carbon is very effective in removing low concentrations of soluble chemicals, including dyes. Its main drawback is its limited capacity. Consequently, activated carbon is best for removing color from dilute effluent.

Chemical treatment of the effluent with a flocculating agent is generally the most efficient and most robust way to remove color. The process involves adding a flocculating agent, such as ferric (Fe^{3+}) or aluminum (Al^{3+}) ions, to the effluent.

This induces flocculation. A coagulant may also be added to assist the process. The final product is a concentrated sludge that is easy to dispose of.

Chemical oxidation is a more recent method of effluent treatment, especially chemical effluent. This procedure uses strong oxidizing agents such as ozone, hydrogen peroxide, chlorine, or potassium permanganate to force degradation of even some of the more resistant organic molecules. At present, these treatments remain very expensive and are of limited scale, though they may have some promise for the future.

Due to their poor water solubility (less than 1 mg/L), disperse dyes have low acute ecological impact. Especially the acute toxicity to aquatic life is generally low. Nevertheless, according to the European chemicals legislation disperse dyes are classified as substances, that "may cause long-term adverse effects in the aquatic environment" [40, 41], because they are not easily biodegradable and suspected of being potentially bioaccumulative due to their hydrophobicity.

In practice, unfixed disperse dyes discharged from spent dyebaths to wastewater treatment plants are easily eliminated by coprecipitation with the sewage sludge and may be anaerobically degraded in the digestion process.

Most commercial disperse dyes are formulated with naphthalenesulfonate or lignosulfonate dispersing agents. These dispersants are completely discharged with the spent dyebath. They have poor biodegradability and will pass the treatment plants to a notable extent. Although not harmful to aquatic life, these dispersants contribute to the persistent COD (Chemical Oxygen Demand) load of the wastewater. Recently, disperse dyes containing highly effective biodegradable dispersants were brought onto the market [42].

Further strategies being implemented to minimize dye and related chemical effluent include designing more environmentally friendly chemicals, more efficient (higher yielding) manufacturing processes, and more effective dyes, for example, reactive dyes with higher fixation rate.

8.4 Legislation

In the last decade the number of laws and regulations providing health and environmental protection during the manufacture, use, and storage of chemicals has further risen worldwide.

8.4.1 Registration/Notification of New Substances

At first, for new substances a very comprehensive registration procedure has to be carried out for the EU and several industrialized countries in the world, provided the substance is not listed in the corresponding inventory of existing chemical substances. Filing a notification in the EU requires numerous data determined by testing. Physicochemical properties are required including melting/boiling point, vapor pressure, solubility, and flammability/explosion characteristics. The toxicological studies include acute toxicity tests, oral, by inhalation, and dermal; skin and eye irritation; skin sensitization; subacute toxicity, oral, by inhalation, and dermal; and mutagenicity tests, in vitro: reverse mutation assay (Ames test) on *Salmonella typhimurium* and/or *Escherichia coli* and mammalian cytogenic test; and in vivo: mouse micronucleus test.

Finally, ecotoxicological studies, designed to assess the impact of the substance on the environment, include acute toxicity tests to fish, daphnia, and algae and a battery of tests for the biodegradability of the substance and its biological oxygen demand characteristics.

Registration of a new chemical substance in the USA, Canada, Japan, Australia, Korea, China, or the Philippines requires similar comprehensive sets of data, although there are some differences. Obtaining all the data for a full registration can be time-consuming and costly. In 2000 it cost approximately € 150 000 to 200 000 and took about two years to register a new substance in Europe.

To expedite the launch of a new chemical and allow further time to complete the toxicological package for full registration, at first a "limited announcement" can be filed with a notification time of 30 d. By this procedure, however, not more than 1 t/a of the chemical is allowed to be sold in the EU.

In the EU and several other countries the legislation regulates the notification of new substances, classification, packaging, labeling and possible prohibition of dangerous substances, the safety of the work place, and the use of chemicals for more sensitive purposes, such as in foodstuffs, drugs, cosmetics, and consumer goods.

8.4.2 Principal Chemical Legislation also Relevant to Dyes

Dyes are affected by numerous regulations, the most important of which are listed in Table 8.5. Since the States of the EU are obliged to implement the EU regulations into national law, with the exception of the German specific regulation on water-endangering classes, national regulations are not listed within the scope of this book.

Important chemical and environmental laws worldwide relevant to dyes are as follows:

European Union:

– Directive for the Classification, Packaging, and Labelling of Dangerous Substances (67/548/EEC), and several amendments and adaptions, as last amended by Dir. 2001/59/EC.

- Directive for the Classification, Packaging, and Labelling of Dangerous Preparations (88/379/EEC), and amendments.
- Restrictions on the Marketing and Use of Certain Dangerous Substances and Preparations 76/769/EEC, and amendments, as last amended by Dir. 2002/61/EC.
- Directive on the Control of Major Accident Hazards involving Dangerous Substances (96/82/EC)
- Protection of Workers from the Risk to Exposure to Carcinogens at Work (89/391/EEC, 90/394/EEC)
- Directive concerning Integrated Pollution Prevention (96/61/EC)
- MSDS Directive 91/155/EEC

Germany:

- List of water-endangering substances under the Federal Water Act (2001)

Switzerland:

- Poison Law (1969).
- Environmental Protection Act (1985).

USA:

- Toxic Substance Control Act (1976).
- OSHA Hazard Communication Standard (1985).

Canada:

- Canadian Environmental Protection Act (1994).

Japan:

- Chemical Substance Control Law (1973).
- Industrial Safety and Health Law (1972).

Australia:

- National Industrial Chemicals Notification and Assessment Scheme (1990)

The legal requirements are supplemented by numerous other more or less specific health and environmental acts, not only in the above-mentioned countries, but also in most other industrialized nations.

8.4.3 Special Regulations for Dyes (Colorants)

Legislation covers all chemicals, including dyes. Only the use of chemicals and colorants in foodstuffs, food-packaging materials, or pharmaceuticals is mentioned here. The exposure level of dyes is generally very low, but people are inadvertently exposed to dyes and other synthetic chemicals for these applications through dermal contact. Therefore, the use of colorants is especially regulated in many countries. General requirements on dyes for the incorporation into packaging materials for foodstuff are: (1) The dyes must not be dangerous according to the regulation; (2) Limits of impurities, such as aromatic amines and heavy metals (see below) have to be met; and (3) The nonmigration principle, that is, the dye must not migrate from the packaging material into the foodstuff in detectable traces.

The migration tendency of colorants depends on the dye structure, the plastic or paper material in which it is incorporated, and whether the foodstuff is oil- or water-based.

The limit concentrations of extractable primary aromatic amines not bearing carboxyl or sulfonyl groups are 500 ppm (as aniline); the threshold limits for metals are required (Table 8.6).

Table 8.6 Threshold concentration of metals in dyes used in packaging material for foodstuffs

Metal	ppm	Metal	ppm
Antimony	500	Cadmium	100
Arsen	100	Chromium	1000
Barium	100	Mercury	50
Lead	100	Selenium	100

The given limits refer to the amount which are extractable with 0.1 N hydrochloric acid [43].

Colorless additives of dye preparations must also be listed if used for the above-mentioned purposes. For example, in Germany they must be listed in various BgVV recommendations for dyes, paper, and plastics used for articles coming into contact to foodstuffs.

Dyes such as benzidine-based dyes which are classified as carcinogens, mutagens, or teratogens according to the Directive 67/548/EEC (CMR substances, Table 8.3), are subject of the Council Directive 76/769/EEC (Restrictions on the Marketing and Use of Certain Dangerous Substances and Preparations) and must not be marketed as such or in preparations if they contain more than the maximum concentration of 0.1 %. In principle, the directive does not ban the import of "articles", e.g., textiles dyed with benzidine-based dyes, because "articles" are not covered by the directive.

This gap was closed by the German, Austrian, and Dutch Consumer Goods Ordinances, banning the use of articles dyed with azo dyes that may release carcinogenic aromatic amines (see Table 8.4) by reductive cleavage of azo groups. In 2002 the Directive 2002/61/EC which is the 19th amendment of the Council Directive 76/769/EEC, prohibits the use of these azo dyes on the EC level. The dyes may not be used for textile or leather or other articles which have the potential for coming into direct and prolonged contact with the human skin, e.g., clothing, bedding sheets, blankets, towels, wigs, bracelets, baby napkins, covers for baby chairs, etc. The ban also covers the import and marketing of the above-mentioned articles dyed with these dyes.

The presence and amount of the carcinogenic amines are to be tested according to an official German test method [44]. The test result leads to banning of the article if the trace amount of a listed amine exceeds 30 ppm.

Because of their ingestion a different situation exists for *food dyes*, and their toxicological investigation is the most comprehensive one (see Section 5.5). The colorants approved for food coloration in the EEC are regulated in 94/36/EC (On Colours for Use in Foodstuff). Purity criteria for these colors for use in foodstuff are regulated under 95/45/EC. The regulation is continually updated.

The food dyes permitted in the USA are compiled in the Code of Federal Regulations, Vol. 21, Food, Drugs and Cosmetics, Parts 70, 71, 73, 74, 80–82; amendments are published in the Federal Register.

8.4.4 Material Safety Data Sheets

Already in 1974 the member companies of ETAD (Ecological and Toxicological Association of Dyes and Organic Pigments Manufacturers) voluntarily obliged themselves to develop a Safety Data Sheet with appropriate information on the hazardous potential of colorants. Customers were supplied with these health hazard data long before any official legislation existed. Nowadays Safety Data Sheets have spread all over the world.

Material Safety Data Sheets (MSDS) for dyes and other chemicals are customarily used to provide the necessary information for safety handling to the user. Although in Europe they must only be legally provided for hazardous substances and preparations according to the EU Directive 91/155/EEC, the majority of dye producers provide MSDS for all products, including those which are not classified as hazardous.

The Safety Data Sheet contains information, such as the identity of the dye, possible hazardous components and physicochemical, toxicological, and ecological data, first aid and emergency measures, occupational exposure limits, and information on personal protective equipment [45].

The toxicological data should provide evidence of adverse effects to humans. The acute oral, dermal or inhalation toxicity is characterized by the LD_{50} or LC_{50} (lethal dose or lethal concentration). In addition to the acute toxicity, possible long-term effects are reported, such as mutagenic, carcinogenic, or teratogenic effects.

8.5 References

8.5.1 General References

ETAD has published numerous Position Papers, Guidelines, lectures and studies on specific problems of colorants concerning toxicology, ecology, and legislation: ETAD General Secretariat, Clarastr.4, CH-4005 Basel, Switzerland, Tel. (+41) 61-690-9966; www.etad.com

ETAD has developed an online database comprising sources of toxicological, environmental, and legal publications on colorants with almost 12 900 documents concerning more than 2000 different dyes (and pigments); it is available for ETAD member companies only.

A comprehensive "Survey of Azo Colorants in Denmark" comprising consumption, use, health, and environmental aspects has been compiled by the Danish Environmental Protection Agency: H. Ollgard et al., Survey of Azo Colorants in Denmark, Danish Technological Institute, Environment, Danish Environmental Protection Agency, November 1998.

C.I. Disperse Blue 79 and *79:1* are produced on a large scale. American ETAD member companies managed a testing program as part of a negotiated consent agreement with the U.S. Environmental Protection Agency to study the health and environmental effects. The toxicological testing program included studies on mutagenicity, 90-d subchronic toxicity, developmental toxicity, and the metabolism/pharmacokinetics in rats. The $ 600 000 program provided evidence for the safety of *C.I. Disperse Blue 79* under the conditions of the health and environmental studies conducted. The dye seems to be safe for the user and the consumer wearing textiles dyed with it [46, 47].

C.I. Reactive Black 5, an important type of reactive dye, was comprehensively studied for its toxicological and ecological profile. It proved to be of low acute toxicity and is non-irritant, a weak sensitizer, and has no genotoxic potential. Even in its hydrolyzed form it is not hazardous to the effluent water [4].

8.5.2 Specific References

[1] USOC Safe Handling of Dyes. A Guide for the Protection of Workers Handling Dyes, 1995 (USOC: US Committee of ETAD, Washington, D.C.).
[2] C. T. Helmes, "Dye Care: A Product Stewardship Program for Dyes", *Am. Dyest. Rep.* **83** (1994) no. 8, 40–41.
[3] ETAD: Ecological and Toxicological Association of Dyes and Organic Pigments Manufacturers, Basel; www.etad.com.
[4] K. Hunger, R. Jung, *Chimia* **45** (1991) 297–300.
[5] R. Anliker, *J. Soc. Dyers Colour.* **95** (1979) 317–326.
[6] K. L. Hatch, *Text. Research J.* **54** (1984) 664–682.
[7] K. L. Hatch, H. I. Maibach, *J. Amer. Academy Dermatol.* **12** (1985) 1079–1092.
[8] K. L. Hatch, H. I. Maibach, *Textile Chem. Colorist* **30** (1998) 22–29.
[9] K. L. Hatch, H. I.Maibach, *Contact Dermatitis* **42** (2000) 187–195.
[10] M. Pratt, V. Taraska, *Amer. J. Contact Dermatitis* **11** (2000) 30–41.
[11] C. M. Luczynska, M. D. Topping, *J. Immunol. Methods* **95** (1986) 177–186.

[12] T. Platzek, *Bundesgesundhbl.* **40** (1997) 239–241.
[13] H. Motschi, *J. Soc. Dyers Colour.* **116** (2000) 251–252.
[14] J. M. Wattie, *J. Soc. Dyers Colour.* **103** (1987) 304–307.
[15] B. M. Hausen, A. Kleinheinz, H. Mensing, *Allerg. J.* **2** (1993) 13–16.
[16] B. M. Hausen, K.-H. Schulz, *Dtsch. Med. Wochenschr.* **109** (1984) 1469–1475.
[17] Öko-Tex Standard 100, Edition 01/2000.
[18] ETAD Information No.7, Significance of the Bacterial Reverse Mutation Test as Predictor for Rodent and Human Carcinogenity, Basle, 1998.
[19] J. Ashby, R. W. Tennant, E. Zeiger, S. Stasiewicz, *Mutat. Res.* **223** (1989) 73–103.
[20] W. C. Heuper, *Occupational and Environmental Cancers of the Urinary System*, Yale University Press, New Haven, Conn., 1969, p. 216.
[21] P. Gregory, *High Technology Applications of Organic Colorants*, Chap. 12, Plenum, New York, 1991.
[22] Fed. Reg. 1973, 38, 10929.
[23] International Agency for Research on Cancer (IARC) Monographs on the Evaluation of Carcinogenic Risks to Humans, Lyon.
[24] U.S. Dept. of Health and Human Services, National Toxicology Program, Technical Report Series.
[25] C. E. Cerniglia, Z. Zhuo, B. W. Manning, T. W. Federle, R. H. Heflich, *Mutat. Res.* **175** (1986) 11–16.
[26] M. U. Ellis, G. J. McPherson, N. Ashcraft, *J. Soc. Leather Technol. Chem.* **81** (1997) 52–56.
[27] K. Hunger, *Chimia 48* (1994) 520–522.
[28] M. A. Brown, S. C. DeVito, *Critical Rev. Environm Sci. Technol.* **23** (1993) 249–324.
[29] E. Rinde, W. Troll, *J. Nat. Cancer Inst.* **55** (1975) 181–182.
[30] T. Platzek, C. Lang, G. Grohmann, U.-S. Gi, W. Baltes, *Hum. Experim. Toxicol.* **18** (1999) 552–559.
[31] R. Anliker, D. Steinle, *J. Soc. Dyers Colour.* **104** (1988) 377–384.
[32] T. Korosteleva, A. Skachkov, I. Shvaidetskii, *Gig. Tr. Prof. Zabol.* **18** (1974) no. 5, 21–24.
[33] R. K. Lynn, D. W. Danielson, A. M. Ilias, K. Wong, J. M. Kennish, H. B. Matthews, *Toxicol. Appl. Pharmacol.* **56** (1980) 248–258.
[34] Z. W. Myslak, H.M. Bolt, *Zbl. Arbeitsmed.* **38** (1988) 310–321.
[35] ETAD Information No.6, German Ban of Use of Certain Azo Compounds in some Consumer Goods, revised version, Basle, 1998.
[36] TRGS 614: Verwendungsbeschränkungen für Azofarbstoffe, die in krebserzeugende aromatische Amine gespalten werden können (AGS, March 2001).
[37] IFOP im Verband der Chemischen Industrie e.V.: Azofarbstoffe, die in krebserzeugende Amine gemäß TRGS 614 gespalten werden können, Frankfurt, March 2001.
[38] Guidance for the User Industry on the Environmental Hazard Labelling of Dyestuffs, ETAD, Basel.
[39] K. Socha, *Textile Month*, **52**, Dec. 1992.
[40] EU Directive 93/21/EEC.
[41] EU Common Position Paper No 54/98.
[42] J. Gruetze, "Dianix ECO Dyes – a New Approach to Dispersing Dyestuffs" 25th, Aachen Textile Conference, Nov. 25/26, 1998.
[43] Kunststoffe im Lebensmittelverkehr, Empfehlungen des Bundesinstituts für gesundheitlichen Verbraucherschutz und Veterinärmedizin, H. Wiezcorek (Hrsg.), Carl Heymanns Verlag, Köln.

[44] Amtliche Sammlung von Untersuchungsverfahren nach § 35 des Lebensmittel- und Bedarfsgegenständegesetzes, Gliederungsnummer B 82.02-2 bis B 82.02-4, Beuth-Verlag Köln.
[45] U. Sewekow, A. Weber, *Melliand Textilber. Int.* **75** (1994) 656–659, E165–E167.
[46] C. T.Helmes, "Disperse Blue 79", *Textile Chem. Color.* **25** (1993) 15–17.
[47] E. A. Clarke, C. T. Helmes, "Cooperation between Authorities and Industry: Disperse Blue 79:1", *Chimia* **48** (1994) no. 11, 519–520.

List of Examples of Commercially Available Dyes

C.I. Color Index Name	C.I. Const. No	CAS No.	Page
Acid Dyes			
Acid Black 1	20470	1064-48-8	293
Acid Black 2	50420	8005-03-6	435, 499
Acid Black 24	26370	3071-73-6	293
Acid Black 26	27070	6262-07-3	285
Acid Black 48	65005	1328-24-1	199
Acid Black 180	13710	11103-91-6	317
Acid Black 194	–	61931-02-0	294
Acid Black 210	300825	99576-15-5	293
Acid Black 234	30027	157577-99-6	293
Acid Blue 9	42090	3844-45-9	501
Acid Blue 25	62055	6408-78-2	196, 203, 294, 436, 454
Acid Blue 40	62125	6424-85-7	196, 203
Acid Blue 43	63000	2150-60-9	197
Acid Blue 62	62045	4368-56-3	196, 203
Acid Blue 74	73015	860-22-0	214
Acid Blue 92	13390	3861-73-2	291
Acid Blue 93	42780	28983-56-4	435
Acid Blue 113	26360	3351-05-1	285, 290, 293
Acid Blue 117	17055	10169-12-7	281
Acid Blue 129	62058	6397-02-0	196, 203
Acid Blue 158	14880	6370-08-7	294
Acid Blue 193	15707	12392-64-2	294, 438, 457
Acid Blue 193	15707	75214-58-3	309
Acid Brown 14	20195	5850-16-8	293
Acid Brown 20	17640	6369-33-1	281
Acid Brown 123	35030	6473-04-7	437
Acid Brown 349	–	72827-73-7	294
Acid Green 1	10020	19381-50-1	110
Acid Green 16	–	12768-78-4	294
Acid Green 25	61570	4403-90-1	198, 203, 294
Acid Green 27	61580	6408-57-7	294

List of Examples of Commercially Available Dyes

C.I. Color Index Name	C.I. Const. No	CAS No.	Page
Acid Green 28	–	12217-29-7	294
Acid Green 41	62560	4430-16-4	198
Acid Orange 3	10385	6373-74-6	294, 455
Acid Orange 7	15510	633-96-5	279, 469
Acid Orange 10	16230	1936-15-8	292
Acid Orange 19	14690	3058-98-8	279
Acid Orange 65	14170	6408-90-8	293
Acid Orange 67	14172	12220-06-3	292
Acid Orange 74	18745	10127-27-2	294
Acid Red 1	18050	3734-67-6	282
Acid Red 7	14895	5858-61-7	436
Acid Red 13	16045	2302-96-7	279
Acid Red 14	14720	3567-69-9	291
Acid Red 32	17065	6360-10-7	281, 289
Acid Red 37	17045	6360-07-2	280
Acid Red 42	17070	6245-60-9	280, 289
Acid Red 52	45100	3520-42-1	500
Acid Red 68	17920	6369-40-0	281
Acid Red 88	15620	1658-56-6	291, 455
Acid Red 119	–	70210-06-9	293
Acid Red 131	–	70210-37-6	292
Acid Red 134	24810	6459-69-4	284
Acid Red 138	18073	15792-43-5	282
Acid Red 154	24800	6507-79-5	284
Acid Red 183	18800	6408-31-7	437
Acid Red 186	18810	52677-44-8	457
Acid Red 249	18134	6416-66-6	292
Acid Red 299	–	67674-28-6	293
Acid Violet 14	17080	4404-39-1	280
Acid Violet 17	42650	4129-84-4	469
Acid Violet 42	62026	6408-73-7	198
Acid Violet 43	60730	4430-18-6	199
Acid Yellow 1	10316	846-70-8	110
Acid Yellow 11	18820	6359-82-6	436
Acid Yellow 17	18965	6359-98-4	292, 500
Acid Yellow 23	19140	1934-21-0	282, 500
Acid Yellow 25	18835	6359-85-9	289
Acid Yellow 36	13065	587-98-4	279
Acid Yellow 38	25135	13390-47-1	284
Acid Yellow 42	22910	6375-55-9	293

List of Examples of Commercially Available Dyes 645

C.I. Color Index Name	C.I. Const. No	CAS No.	Page
Acid Yellow 44	23900	2429-76-7	283
Acid Yellow 56	24825	6548-24-9	284
Acid Yellow 65	14170	6408-90-8	290
Acid Yellow 76	18850	6359-88-2	282
Acid Yellow 99	13900	10343-58-5	294
Acid Yellow 127	18888	73384-78-8	293
Acid Yellow 151	13906	12715-61-6	294, 438, 456
Acid Yellow 194	–	85959-73-5	294
Basic Dyes			
Basic Blue 3	51005	2787-91-9	470
Basic Blue 41	11105	12270-13-2	266
Basic Blue 54	11052	1500-59-6	265
Basic Blue 64	–	12217-44-6	466
Basic Blue 140	–	61724-62-7	468
Basic Green 4	42000	569-64-2	470
Basic Orange 1	11320	4438-16-8	248
Basic Orange 2	11270	532-82-1	248
Basic Orange 21	–	47346-66-7	273
Basic Red 1	45160	989-38-8	470
Basic Red 12	48070	6320-14-5	108
Basic Red 14	–	12217-48-0	264, 273
Basic Red 18	11085	25198-22-5	248
Basic Red 29	–	42373-04-6	249
Basic Red 46	110825	12221-69-1	249, 267
Basic Red 51	–	12270-25-6	268
Basic Red 111	–	118658-98-3	467
Basic Violet 14	42510	632-99-5	629
Basic Violet 16	48013	6359-45-1	264, 273
Basic Yellow 1	49005	2390-54-7	261
Basic Yellow 11	–	4208-80-4	272
Basic Yellow 13	–	12217-50-4	272
Basic Yellow 15	11087	72208-25-4	248
Basic Yellow 21	48060	6359-50-8	272
Basic Yellow 28	48054	54060-92-3	57, 272
Basic Yellow 29	–	38151-74-5	272
Basic Yellow 40	–	12221-86-2	108
Basic Yellow 51	480538	83949-75-1	273
Basic Yellow 70	–	71872-36-1	273

List of Examples of Commercially Available Dyes

C.I. Color Index Name	C.I. Const. No	CAS No.	Page
Direct Dyes			
Direct Black 19	35255	6428-31-5	176, 498-499
Direct Black 22	35435	6473-13-8	176
Direct Black 51	27720	3442-21-5	176
Direct Black 150	32010	6897-38-7	176
Direct Black 166	30026	75131-19-8	176
Direct Black 168 (3Li)	–	108936-08-9	498
Direct Black 168 (3Na)	30410	85631-88-5	440, 499
Direct Blue 71	34140	4399-55-5	176, 464
Direct Blue 78	34200	2503-73-3	176, 456
Direct Blue 86	74180	1330-38-7	440
Direct Blue 93	22810	13217-74-8	312
Direct Blue 106	51300	6527-70-4	112, 121, 439
Direct Blue 109	–	33700-25-3	120
Direct Blue 199	74190	12222-04-7	501
Direct Blue 218	24401	28407-37-6	465
Direct Blue 273	–	76359-37-0	466
Direct Green 26	34045	6388-26-6	465
Direct Green 28	14155	6471-09	172
Direct Orange 102	29156	6598-63-6	176, 464
Direct Orange 34	40215	1325-54-8	111
Direct Red 23	29160	3441-14-3	176
Direct Red 75	25380	2829-43-8	176, 500
Direct Red 81	28160	2610-11-9	176, 462
Direct Red 239	–	28706-25-4	176, 464
Direct Red 253	–	142985-51-1	176, 462
Direct Violet 51	27855	6227-10-7	464
Direct Yellow 11	40000	1325-37-7	176
Direct Yellow 12	24895	2870-32-8	176, 461
Direct Yellow 22	13925	10190-69-9	175
Direct Yellow 27	13950	10190-68-8	175
Direct Yellow 28	19555	8005-72-9	175, 463
Direct Yellow 29	19556	6537-66-2	175
Direct Yellow 51	29030	6420-29-7	176, 463
Direct Yellow 86	29325	50925-42-3	500
Direct Yellow 132	–	61968-26-1	500
Direct Yellow 137	–	71838-47-6	175, 462
Direct Yellow 147	–	35294-62-3	175, 462

List of Examples of Commercially Available Dyes 647

C.I. Color Index Name	C.I. Const. No	CAS No.	Page
Disperse Dyes			
Disperse Blue 1	64500	2475-45-8	628
Disperse Blue 7	62500	3179-90-6	185, 203
Disperse Blue 14	61500	2475-44-7	185, 203
Disperse Blue 26	63305	3860-63-7	185, 203
Disperse Blue 31	64505	1328-23-0	185, 203
Disperse Blue 56	63285	12217-79-7	184, 203
Disperse Blue 72	60725	81-48-1	183, 202
Disperse Blue 73	63265	12222-75-2	184, 203
Disperse Blue 79	11345	3956-55-6	147
Disperse Blue 106	11935	68516-81-4	628
Disperse Blue 124	111938	15141-18-1	628
Disperse Blue 148	11124	52239-04-0	149
Disperse Blue 165	–	41642-51-7	147
Disperse Blue 354	48480	74239-96-6	140
Disperse Blue 365	–	108948-36-3	142
Disperse Blue 366	–	84870-65-5	147
Disperse Green 9	110795	58979-46-7	149
Disperse Orange 3	11005	730-40-5	628
Disperse Orange 29	26077	19800-42-1	155
Disperse Orange 44	–	4058-30-4	146
Disperse Orange 56	–	67162-11-2	152
Disperse Red 1	1110	2872-52-8	628
Disperse Red 17	11210	3179-89-3	453
Disperse Red 60	60756	17418-58-5	148, 181, 185, 202
Disperse Red 72	11114	12223-39-1	147
Disperse Red 177	11122	68133-69-7	137
Disperse Red 338	111430	63134-15-6	148
Disperse Violet 27	60724	19286-75-0	183, 202
Disperse Violet 28	61102	81-42-5	182, 202
Disperse Yellow 3	11855	2832-40-8	137, 628
Disperse Yellow 5	12790	6439-53-8	152
Disperse Yellow 9	10375	6373-73-5	110
Disperse Yellow 14	10340	961-68-2	110
Disperse Yellow 42	10338	5124-25-4	143
Disperse Yellow 54	47020	7576-65-0	139
Disperse Yellow 64	42023	10319-14-9	155
Disperse Yellow 82	–	27425-55-4	140
Disperse Yellow 99	–	25857-05-0	139
Disperse Yellow 160	–	75216-43-2	156

C.I. Color Index Name	C.I. Const. No	CAS No.	Page
Disperse Yellow 211	12755	70528-90-4	153
Disperse Yellow 241	128450	83249-52-9	153
Disperse Yellow 332	55165	35773-43-4	141
Fluorescent Brightener 130	501101	87-01-4	457
Food Dyes			
Food Black 1	28440	2519-30-4	490
Food Red 1	14700	4548-53-2	489
Food Red 3	14720	3567-69-9	490
Food Red 7	16255	2611-82-7	490
Food Red 9	16185	915-67-3	490
Food Red 14	45430	16423-68-0	491
Food Yellow 3	15985	2783-94-0	490
Food Yellow 4	19140	1934-21-0	489
Mordant Dyes			
Mordant Black 3	14640	3564-14-5	286
Mordant Black 9	16500	2052-25-7	292
Mordant Black 11	14645	1787-61-7	292
Mordant Blue 7	17940	3819-12-3	287
Mordant Blue 13	16680	1058-92-0	292
Mordant Brown 33	13250	3618-62-0	287
Mordant Brown 48	11300	6232-53-7	287
Mordant Red 7	18760	3618-63-1	286
Mordant Red 19	18735	1934-24-3	293
Mordant Red 30	19360	6359-71-3	287
Mordant Yellow 1	14025	584-42-9	286
Mordant Yellow 5	14130	6054-98-4	286
Mordant Yellow 30	18710	10482-43-6	287
Oxidation Dyes			
Oxidation Base 1	50440	13007-86-8	452
Oxidation Base 10	76060	13007-86-8	452
Pigment Violet 23	51319	6358-30-1	468
Reactive Dyes			
Reactive Black 5	20505	17095-24-8	117
Reactive Black 8	18207	79828-44-7	311
Reactive Blue 15	74459	12225-39-7	122
Reactive Blue 70	–	61968-92-1	130
Reactive Blue 83	–	12731-65-8	130
Reactive Blue 84	–	12731-66-7	130

List of Examples of Commercially Available Dyes 649

C.I. Color Index Name	C.I. Const. No	CAS No.	Page
Reactive Blue 104	–	61951-74-4	130
Reactive Blue 160	137160	71872-76-9	130
Reactive Blue 163	–	72847-56-4	122
Reactive Blue 182	–	68912-12-9	130
Reactive Blue 209	–	110493-61-3	130
Reactive Blue 212	–	86457-82-1	130
Reactive Blue 216	137155	89797-01-3	130, 319
Reactive Blue 220	–	128416-19-3	130
Reactive Blue 221	–	93051-41-3	130
Reactive Blue 235	–	106404-06-2	130
Reactive Green 15	–	61969-07-1	130
Reactive Red 180	181055	98114-32-0	500
Reactive Violet 4	17965	12769-08-3	443
Reactive Yellow 4	13190	1222-45-8	442
Solubilised Sulphur Dyes			
Solubilised Sulphur Black 1	53186	1326-83-6	499
Solubilised Sulphur Red 11	–	61969-41-3	83, 225
Solvent Dyes			
Solvent Black 3	26150	4197-25-5	301
Solvent Black 5	50415	11099-03-9	434, 499
Solvent Black 7	50415:1	8005-02-5	434, 499
Solvent Blue 44	–	61725-69-7	501
Solvent Blue 45	–	37229-23-5	501
Solvent Blue 66	42799	58104-34-0	299
Solvent Blue 122	60744	67905-17-3	299
Solvent Brown 53	48525	64969-98-6	299
Solvent Green 3	61565	128-80-3	301
Solvent Green 28	625580	71839-01-5	299
Solvent Orange 5	518745:1	13463-42-8	300
Solvent Orange 56	–	12227-68-8	296, 300
Solvent Orange 60	564100	61969-47-9	299
Solvent Red 8	12715	33270-70-1	296
Solvent Red 23	26100	85-86-9	298
Solvent Red 24	26105	85-83-6	298
Solvent Red 49	–	509-34-2	108
Solvent Red 52	68210	81-39-0	301
Solvent Red 91	–	61901-92-6	501
Solvent Red 109	–	53802-03-2	296
Solvent Red 111	60505	82-38-2	301

C.I. Color Index Name	C.I. Const. No	CAS No.	Page
Solvent Red 135	564120	71902-17-5	299
Solvent Red 212	48530	61300-98-9	299
Solvent Yellow 2	11020	60-11-7	629
Solvent Violet 13	60725	81-48-1	301
Solvent Yellow 14	12055	842-07-9	300
Solvent Yellow 16	12700	4314-14-1	301
Solvent Yellow 21	18690	5601-29-6	300, 444, 458
Solvent Yellow 32	48045	61931-84-8	296
Solvent Yellow 44	56200	2478-20-8	108
Solvent Yellow 56	11021	2481-94-9	300
Solvent Yellow 82	–	12227-67-7	296, 300
Solvent Yellow 83:1	–	61116-27-6	501
Solvent Yellow 94	45350:1	518-45-6	108
Solvent Yellow 163	58840	13676-91-0	299
Sulphur Dyes			
Sulphur Black 1	53185	1326-82-5	226, 499
Sulphur Black 6	53295	1327-16-8	82, 215
Sulphur Black 7	53300	1327-17-9	82, 215
Sulphur Black 11	53290	1327-14-6	82, 215, 371
Sulphur Blue 9	53430	1327-56-6	82
Sulphur Brown 52	53320	1327-18-0	81
Sulphur Orange 1	53050	1326-49-4	81
Sulphur Red 14	–	81209-07-6	83, 225
Sulphur Yellow 4	53160	1326-75-6	80
Vat Dyes			
Vat Black 25	69525	4395-53-3	193, 203
Vat Black 27	69005	2379-81-9	191
Vat Black 29	65225	6049-19-0	195, 203
Vat Blue 1	73000	482-89-3	205, 213
Vat Blue 4	69800	81-77-6	195, 203, 367
Vat Blue 5	73065	7475-31-2	213
Vat Blue 14	69810	1324-27-2	367
Vat Blue 20	59800	116-71-2	193, 203
Vat Blue 21	67920	6219-97-2	193
Vat Blue 25	70500	6247-39-8	193, 367
Vat Blue 26	60015	4430-55-1	193
Vat Blue 30	67110	6492-78-0	190
Vat Blue 43	53630	1327-79-3	82, 371
Vat Blue 64	66730	15935-52-1	190

C.I. Color Index Name	C.I. Const. No	CAS No.	Page
Vat Brown 1	70800	2475-33-4	191, 203
Vat Brown 3	69015	131-92-0	191, 203
Vat Brown 45	59500	6424-51-7	195
Vat Brown 55	70905	4465-47-8	367
Vat Green 1	59825	128-58-5	193, 203
Vat Green 3	69500	3271-76-9	193, 203
Vat Green 8	71050	14999-97-4	191, 203
Vat Green 11	69850	1328-41-2	195
Vat Green 12	70700	6661-46-7	193
Vat Orange 1	59105	1324-11-4	195, 203
Vat Orange 2	59705	1324-35-2	195
Vat Orange 3	59300	4378-61-4	195
Vat Orange 9	59700	128-70-1	195, 203
Vat Orange 15	69025	2379-78-4	191, 203
Vat Orange 17	65415	6370-77-0	187
Vat Orange 26	–	12237-49-9	367
Vat Red 10	67000	2379-79-5	152, 190
Vat Red 13	70320	4203-77-4	190
Vat Red 18	60705	409-68-3	190
Vat Red 28	65710	6370-82-7	187
Vat Violet 1	60010	1324-55-6	193, 203
Vat Violet 15	63355	6370-58-7	187
Vat Violet 16	65020	4003-36-5	190
Vat Yellow 5	56005	4370-55-2	330
Vat Yellow 10	65430	2379-76-2	187
Vat Yellow 12	65405	6370-75-8	187
Vat Yellow 20	68420	4216-01-7	187
Vat Yellow 28	69000	4229-15-6	191
Violet 51	27905	5489-77-0	176

Index

A

Acid and Direct Dyes 454
Acid and Metal-Complex Dyes on Polyamide 386
– 1:2 Metal-Complex Dyes 390
– Acid Dyes 388
– Chemical and Physical Structure of the Fiber 386
– Interactions Between Dye and Fiber 387
Acid Dyes 195, 276
– Chrome Dyes 285
– Leveling dyes 277
– Milling dyes 277
– Polyamide Dyes 289
– Primary disazo dyes of the type $D_1 \rightarrow K \leftarrow D_2$ 283
– Secondary disazo dyes 285
– Silk Dyes 291
– Wool Dyes, Acid Monoazo Dyes 278
Acid Dyes on Plant Fibers 378
Acid Dyes on Wool 382
Acrylic Fibers (PAC) 412
Afterchroming 286
Alcohol- and Ester-Soluble Dyes 295
Ames test 628
Aminoanthraquinones 201
– 1,4-Diaminoanthraquinones with External Sulfonic Acids Groups 198
– 1-Aminoanthraquinone-2-sulfonic Acids 196
– Diaminodihydroxyanthraquinone-sulfonic Acids 197
Anionic Azo Dyes 276
– see also Acid Dyes
Anthraquinone Chromophore 35
Anthraquinone Dyes 178
– Dyes for Cellulose Ester and Synthetic Polyamide Fibers 184
– Dyes for Polyester Fibers 180
– Transfer Dyes 185
– Vat Dyes 187
  – Acylaminoanthraquinones 187
  – Anthraquinoneazoles 189
  – Anthrimide Carbazoles 191
  – Benzanthrone Dyes 193
  – Highly Condensed Ring Systems 195
  – Indanthrones 195
  – Linked Anthraquinones 190
  – Phthaloylacridone 193
  – Vat Dyes on Cellulosic Fibers 362
Anthraquinonesulfonic Acids 200
Azo/Azomethine Complex Dyes 86
Azo Chromophore 14
– Coupling Components 20
– Diazo Components 16
– Diazotization Methods 19
Azo Coupling in Practice 28
Azo Dyes
– Metabolism 630
– Tautomerism 29
Azo/hydrazone tautomerism 29

B

Basic Dyes, see Cationic Dyes
Basic Dyes on Cellulose 378
Bath Dyeing Technology 342
– Circulating Machines 342
– Process Control 342
Benzidine-based Dyes 637
Benzodifuranone Dyes 37
BgVV 628

C

Carbocyclic Azo Dyes 33
Carcinogenic amines 632
Cationic Azo Dyes 227
– Cationic Charge in the Coupling Component 227
  – Coupling Components With Alkylammonium Groups 229
  – Coupling Components with two different cationic groups 236
  – Polyamines as Coupling Components 228
– Cationic Charge in the Diazo Component

- Diazo components with aminoalkyl groups 237
- Diazo components with Cycloammonium Groups 242
- Diazo Components with Trialkylammonium Residues 238
- Diazo components with two different cationic residues 241
- Cationic Dyes with Sulfur or Phosphorus as Charge-Carrying Atoms 246
- Different Cationic Charges in Both the Coupling and the Diazo Component 244
- Dyes with Releasable Cationic Groups 247
- Introduction of Cationic Substituents into Preformed Azo Dyes 245

Cationic Dyes
- as Chromophores 44
- for Paper, Leather, and Other Substrates 53
- for Synthetic Fibers 52
- in Fiber Blends 419
- on Acrylic Fibers 412
- on Aramide Fibers 418
- with Delocalized Charge 45
- with Localized Charge 49
- Continuous Processes with Cationic Dyes 417
- Dyeing of Special Fiber Types by the Exhaustion Process
  - High-Bulk Material 416
  - Microfibers 416
  - Modacrylic Fibers 417
  - Pore Fibers 416
- Exhaustion Process 414
- Mixtures of Acrylic with other Synthetic Fibers 420
- PAC-CEL Blends 419
- PAC-Wool Blends 419
- Retarders and Auxiliaries 413

Cationic Methine Dyes 254
- Cyanine Dyes 268
  - Diazadimethinecyanine 270
  - Monomethinecyanine 269
  - Pentamethinecyanine 271
  - Trimethinecyanine 271
- Hemicyanine Dyes 255
- Higher Vinylogues of Hemicyanine Dyes 260
- Phenylogous Hemicyanine Dyes 261
- Streptocyanine Dyes 254

- Styryl dyes 263

Cellulose molecule 459
Chrome Dyes on Wool 384
Chromophores of Dye Classes 13
Chrysoidines 227
C.I. Disperse Blue 79 and 79:1 147, 639
C.I. Reactive Black 5 117, 639
CIELAB 34
Classification of Dyes by Use or Application Method 3
- Acid Dyes 5
- Cationic (Basic) Dyes 5
- Direct Dyes 5
- Disperse Dyes 3
- Reactive Dyes 3
- Solvent Dyes 5
- Sulfur Dyes 5
- Vat Dyes 5

Colorfastness of Textiles 348
Colour Index 6, 427, 429
Continuous and Semicontinuous Dyeing 343
Cotton Dyeing 362, 368 f., 375 f., 403 ff.
Coupling and Diazotization Dyes on Cellusosic Fibers 379
Coupling Components 20
- Aminonaphtholsulfonic Acids 25
- Aminophenols 25
- Aminophenolsulfonic Acids 25
- Anilines, diaminobenzenes 21
- Compounds with Reactive Methylene Groups 26
- Naphtholsulfonic Acids 24
- Phenols, Naphthols 22

D

Di- and Triarylcarbenium and Related Chromophores 59
Diazo Components 16
- Diamines H_2N-A-NH_2 18
- Heteroyclic amines and diamines 19
- Naphthylaminesulfonic Acids 16

Diazotization Methods 19
Diphenylmethane and Triphenylmethane Dyes 47
Direct Dyes 158
- Aftertreatment with Cationic Auxiliaries
  - Cationic Formaldehyde Condensation Resins 174
  - Quaternary Ammonium Compounds 173

- Aftertreatment with Formaldehyde 174
- Aftertreatment with Metal Salts 175
- Anthraquinone Direct Dyes 172
- Condensation Dyes 166, 170
- Conventional Direct Dyes 161
- Copper Complexes of Azo Dyes 167
- Diazotization Dyes 174
- Direct Dyes with Aftertreatment 172
- Direct Dyes with a Urea Bridge 170
- Disazo Dyes 162, 169
- Monoazo Dyes 161, 168
- Precursors 159
- Structural Characteristics
  - Coplanarity 159
  - Long Chain of Conjugated Double Bonds 159
- Synthesis 168
- Tetrakisazo Dyes 165
- Triazinyl Dyes 167
- Trisazo Dyes 164, 170

Direct Dyes for Fiber Blends 361
Direct Dyes on Cellulosic Fibers
- Aftertreatment 361
- Dyeing Parameters 359
- Dyeing Principle 358
- Dyeing Techniques
  - Exhaustion Process 360
  - High-Temperature Dyeing Process 360
  - Pad Processes 360
  - Printing with Direct Dyes 360

Dispensing Dyes and Chemicals 345
Disperse Dyes 134
- Aftertreatment 145
- Anthraquinone Dyes 138
- Azo Dyes
  - Disazo Dyes 138
  - Monoazo Dyes 135
- Other Chromophores
  - Methine Dyes 139
  - Nitro Dyes 143
  - Quinophthalone Dyes 139
  - Thioindigo 143
- Synthesis
  - Disazo Dyes 145
  - Monoazo Dyes 144

Disperse Dyes on Cellulose Acetate 409
- Dyeing Cellulose 2.5 Acetate 409
- Dyeing Cellulose Triacetate 410

Disperse Dyes on other Fibers 411
- Polypropylene fibers 411
- Poly(vinyl chloride) fibers 411

Disperse Dyes on Polyamide Fibers 410
Disperse Dyes on Polyester and other Man-Made Fibers 392
- Aftertreatment 403
- Dyeing Blends Containing Polyester Fibers 403
  - Disperse and Direct Dyes 406
  - Disperse and Reactive Dyes 405
  - Disperse and Sulfur Dyes 406
  - Disperse and Vat Dyes 405
  - Disperse Dyes and Leuco Esters of Vat Dyes 406
  - Dyeing Processes 404
  - Polyester-Cellulose Blends 403
  - Polyester-Wool Blends 407
- Dyeing in Aqueous Liquor 392
- Dyeing Processes for Polyester Fibers 396
  - Continuous and Semicontinuous Dyeing Processes 399
  - Disperse Dyes for Different Applications 396
  - Dyeing from Aqueous Dye Baths 397
  - Dyeing of Modified PES Fibers 401
  - Dyeing of PES Microfibers 400
  - Printing wth Disperse Dyes on Man-Made Fibers 401
  - Special Dyeing Processes 398
- Thermosol Process 395

Drew-Pfitzner type 32
Dye Classes for Reactive Dyes 118
Dye Fixation 344
Dyeing Process with Different Classes of Dyes 388
Dyeing Techniques for Cellulose 349
Dyeing Technology 340
Dyeing with Indigo
- Dyeing Technique on Cotton 368
- Indigo on Wool 369

Dyes, General Survey 1
Dyes Soluble in Polymers 298
Dyewood 432

E

Economic Aspects 10
Environmental Assessment

- Treatment of Dye-Containing Wastewater 633
  - Biological treatment 633
  - Chemical oxidation 634
  - Chemical treatment 633

Equipment and Manufacture of Dyes 7
ETAD 638

F

Fat- and Oil-Soluble Dyes 297
Fluorescent Dyes 107
Fluorescent Whitening Agents (FWAs) 585
Food and Drug Administration (FDA) 491
Food Dyes 486
- Legal Aspects 492
  - Codex Alimentarius 492
  - EU and other European Countries 492
  - Japan 494
  - USA 493
- Purity Requirements 491
- Synthetic Dyes Approved for Coloring of Foodstuffs 487
  - European Union 487
  - Japan 489
  - USA 488
- Uses and Individual Substances 487
Formazan Dyes 97
Formazan Dyes (Other Chromophores) 111
Functional Dyes
- Interactions 543
  - Electricity 545
  - Electromagnetic Radiation 544
  - Frictional Forces 545
  - Heat 544
  - Pressure 545
Functional Dyes by Application 545
- Biomedical Applications 576
  - Fluorescent Sensors and Probes 577
- Displays 566
  - Cathode Ray Tube 566
  - Electrochromic Displays 571
  - Liquid Crystal Displays 566
  - Organic Light-Emitting Devices 569
- Elastomeric Urethane Fibers 411
- Electronic Materials
  - Laser Dyes 576
  - Nonlinear Optical Dyes 574
  - Organic Semiconductors 572
  - Solar Cells 573

- Imaging
  - Black Dyes 554
  - Cyan Dyes 554
  - Dyes for Ink-Jet Printing 555
  - Ink-Jet Dyes for Paper 557
  - Ink-Jet Dyes for Special Media 555
  - Ink-Jet Printing 555
  - Laser Printing and Photocopying 545
  - Magenta Dyes 553
  - Other Imaging Technologies 558
  - Thermal Printing 551
  - Yellow Dyes 553
- Invisible Imaging 559
  - Optical Data Storage 560
  - Other Technologies 564
- Photodynamic Therapy 579
Fur Dyeing 451
Fur Dyes 446
- Acid Dyes 454
  - Direct Dyes 455
  - Milling dyes 455
- Animal Rights 447
- Disperse Dyes 453
- Final Stage 449
- Fur Dressing 448
- Fur Finishing 448
- Fur Hair and Classification 448
- Garments and Fashion 450
- Labels 450
- Metal-Complex Dyes 456
- Origin of Fur 446
- Oxidation Bases 452
- Vegetable Dyes 451
FWAs, Uses 611
- Analytical Methods 615
- Assessment of Whitening Effect 616
- Detergent Industry 613
  - Heavy-Duty Liquid Detergents 613
  - Heavy-Duty Powders 613
  - Other Laundry Products 613
- Environmental Aspects 616
- General Requirements 611
- Other Uses 615
- Paper Industry 614
  - Addition to Coating 614
  - Brightening in Pulp 614
  - Impregnation of Paper Web 614
- Plastics and Synthetic Fibers 614
- Textile Industry 611
  - Cellulosic Fibers 611

- Other Fibers   612
- Polyacrylonitriles   612
- Polyamides   612
- Polyesters   612

H

H Acid   281
Hair Dyes
- Bleaching   473
  - Bleaches   474
  - Chemistry   474
- Dye Classes
  - Anionic Dyes   480
  - Cationic Dyes   480
  - Direct Dyes   479
  - Disperse Dyes   480
  - Dyeing with Inorganic Compounds   480
  - Nitro Dyes   479
- Dyeing with Oxidation Dyes
  - Couplers   477
  - Oxidation   475
  - Primary Intermediates   477
- Dye-Removal Preparations   483
- Product Forms   481
- Testing Hair Dyes   483
Haloanthraquinones   201
Health and Safety Aspects   625
Heterocyclic Azo Dyes   34
Hydroxyanthraquinones   202

I

Indicator Dyes   526 ff
- Chelate-Forming Chemical Structures   539
- Classes of Indicators   526 ff, 527
- General Principles   526
- Indicator Papers   540
  - Bleeding Indicator Papers   540
  - Nonbleeding Indicator Papers   541
- Metal Indicators   537
- Redox Indicators   537
Indigo   40, 205, 213
Indigo, Halogen Derivatives   213
Indigoid Chromophore   40
- Color   41
- Solvatochromism   42
Indigoid Dyes   204
- Biotechnological Synthesis   211
  - Bacterial De-Novo Synthesis   212
  - Biotransformation of Indoles   211
  - Microbiological Synthesis   211
- Chemical Properties   206
- Chemical Synthesis   207
  - Heuman I Process   208
  - Heuman II Process   209
- Leuco Indigo   210
- Physical Properties   205
Ink Dyes
- Dye Classes
  - Dyes for Ink-Jet Application   497
  - Dyes for Writing, Drawing, and Marking   501
  - Fields of Application for Ink Jet Printing   502
- Ink-Jet Technology   495
- Inks
  - Ink-Jet Inks   503
  - Writing, Drawing, and Marking Inks   505
- Properties of Ink-Jet Prints   507
- Writing, Drawing and Marking Materials   497
IR-Absorbing Dyes   561

L

Laboratory Dyeing Techniques   349
Leather   427
Leather Dyes   427
- Acid Dyes   434
- Amphoteric Dyes   435
- Aniline Leather   428
- Anthraquinone Dyes   435
- Azo Metal-Complex Dyes   437
- Basic Dyes   433
  - Azine Dyes   433
  - Other Cationic Dyes   434
- Condensation Dyes   439
- Direct Dyes   439
- Dyewood   432
- Low-Molecular Azo Dyes   436
- Natural and Mordant Dyes, Tanning Agents   431
- Phthalocyanine Dyes   440
- Pigments   444
- Polyazo Dyes   440
- Reactive Dyes   442
- Resorcinol Azo Dyes   437
- Solvent Dyes   443
- Sulfur Dyes   441
Legislation

- Material Safety Data Sheets 638
- Notification of New Substances 634
- Principal Chemical Legislation 635
    - Chemical and Environmental Laws 635
- Special Regulations for Dyes 636
Leuco Esters of Vat Dyes on Cellulosic Fibers 367

M

1:1 Metal-Complex Dyes 288
1:2 Metal-Complex Dyes 288
Metal-Complex Dyes 302
- Azo Metal-Complex Dyes
    - Anodized aluminum 321
    - Color Photography 322
    - Electrophotography 322
    - Ink Jet Dyes 320
    - IR-Absorbing Dyes 323
    - Redox Catalysts 321
    - Solvent Dyes 319
- Chromium and Cobalt Complexes for Wool and Polyamides 304
    - 1:1 Chromium Complexes Containing Sulfonic Acid Groups 304
    - 1:2 Chromium Complexes with One Sulfonic Acid Groups 307
    - 1:2 Chromium Complexes with Two or More Sulfonic Acid Groups 309
    - 1:2 Metal Complexes without Water-Solubilizing Groups 305
    - Sulfonated Cobalt Complexes 310
- Copper Complexes for Paper (see also 5.3) 315
- Formazan Dyes 316
    - Analytical Reagents 324
    - Bioindicators 325
    - Cotton Dyes 319
    - Ink-Jet Dyes 326
    - Optical Recording Dyes 326
    - Photoreagents 325
    - Wool and Polyamide Dyes 317
- Metal-Complex Dyes for Polypropylene 316
- Metal Complexes for Cotton 311
    - Demethylative Coppering 312
- Metal Complexes for Leather (see also 5.1) 313
- PVC fastness 314
Metal-Complex Dyes on Wool 385
Metal Complexes as Chromophores 85
- Chromium Complexes 87
- Cobalt Complexes 90
- Copper Complexes 86
- fac configuration 94
- Formazan Dyes 97
- Iron Complexes 91
- mer configuration 95
- Metal-Containing Bidentate Formazans 102
- Metal-Containing Tetradentate Formazans 104
- Metal-Containing Tridentate Formazans 102
- Sterochemistry and Isomerism 94
Metallized Azo Dyes 32
Methine Dyes 56
Mordant Dyes on Cellulosic Fibers 377

N

Naphthoquinone and Benzoquinone Dyes 329
- 1,4-Naphthoquinone Dyes 331
- 1,5-Naphthoquinones 335
- Benzoquinone Dyes 330
- Heteroannelated 1,4-Naphthoquinones 332
- Naphthoxidine 336
- Simple 1,4-Naphthoquinones 331
Napthol AS Dyes on Cellulosic Fibers 375
- Application 375
- Dyeing Processes on Cellulosic Fibers 376
- Printing on Cellulosic Fibers 377
Nitro and Nitroso Dyes 110
Nitroanthraquinones 201
Nomenclature of Dyes 6
Nontextile Dyeing 427

O

Optical Brighteners 585
- 1,3,5-Triazin-2-yl Derivatives 610
- 1,3-Diphenyl-2-pyrazolines 605
- Benzoxazoles, Bis(benzoxazoles) 599
    - Stilbenylbenzoxazoles 598
- Carbocycles 590
- Chemistry of Technical Products
    - Carbocycles 590
    - Distyrylbenzenes 590
    - Distyrylbiphenyls 592
    - Divinylstilbenes 593
- Commercial Forms 610

- Coumarins 607
- Furans, Benzo[b]furans, and Benzimidazoles
  - Bis(benzo[b]furan-2-yl)biphenyls 601
  - Cationic Benzimidazoles 602
- History of Whitening 589
- Molecular Structure 588
- Naphthalimides 608
- Physical Principles 586
- Stilbenyl-2H-triazoles 596
  - Bis(1,2,3-triazol-2-yl)stilbenes 597
  - Stilbenyl-2H-naphtho[1,2-d]triazoles 597
- Triazinylaminostilbenes 593
Organic Photoconductors 547
Oxidation Dyes on Cellulosic Fibers 378

P

Paper Dyes 459
- Acid Dyes 469
- Catonic Dyes 470
- Direct Dyes 460
  - Anionic Direct Dyes 461
  - Cationic Direct Dyes 466
- Organic Pigments 472
- Special Requirements 472
- Sulfur Dyes 471
Pfeiffer-Schetty type 32
Photographic Dyes
- Anthraquinone Dyes 523
- Azo Dyes
  - Color Masking 520
  - Diffusion-Transfer Imaging Systems 519
  - Silver Dye Bleach Processes 520
- Azomethine and Indoaniline Image Dyes 516
  - Color Developers 517
  - Cyan Indoaniline Dyes 518
  - Magenta Azomethine Dyes 518
  - Yellow Azomethine Dyes 517
- Cyanine Dyes 509
- Merocyanine Dyes 515
- Metallized Dyes 521
- Oxonol Dyes 516
- Sensitizing Dyes
  - Application 511
  - Cyanine Dyes 512
  - Production 512
- Triarylmethane Dyes 522
- Xanthene Dyes 521
Phthalocyanine Chromophore 68

- Copper Phthalocyanine 73
- Diiminoisoindolenine Process 74
- Phthalocyanine Derivatives 74
- Pthalocyanine Sulfonic Acids 75
- Synthesis 70
Phthalogen Dyes on Cellulosic Fibers 379
Polyamide Dyeing 391
Polyamide Dyes 289
Polycyclic Aromatic Carbonyl Dyes 38
Polyester Fibers 134, 340 ff., 390 ff.
Polymethine and Related Chromophors 56
- Azacarbocyanines 57
- Diazahemicyanines 58
- Hemicyanines 57
- Styryl Dyes 58
Polypropylene 316
Polypropylene Fibers 411
Poly(vinyl chloride) Fibers 411
Principles of Dyeing 341
- Exhaustion dyeing 341
- Mass Dyeing 341
- Pigment Dyeing 341
Principles of Dyeing of Wool and Silk
- Bonding Forces 381
- Dyeing Silk 382
- Dye selection 382
- Technological Aspects of Wool Dyeing 381
Printing 345
Prival test 629

Q

Quinophthalone Dyes 109

R

Reactive Dyes 113
- Anthraquinone Dyes 119
- Azo Dyes 119
- Double-Anchor Dyes 117
- Formazan Dyes 122
- Metal-Complex Azo Dyes 119
- Mixed-anchor systems 117
- Mono-Anchor Dyes 114
- Multiple-Anchor Dyes 118
- Phthalocyanine Dyes 122
- Synthesis
  - Azo Dyes 123
  - Dioxazine Dyes 124
  - Metal-Complex (Formazan) Dyes 123
- Triphenodioxazine Dyes 120

Reactive Dyes for Printing on Cellulose 357
Reactive Dyes on Cellulose and Other Fibers 349
- Diffusion Rate 352
- Exhaustion Dyeing 353
- Hydrolysis and Reactivity 350
- Pad Dyeing Processes 354
- Special Processes and Development Trends 355
- Stability of the Dye-Fiber Bond 352
- Substantivity 351
Reactive Dyes on Polyamide 357
Reactive Dyes on Wool and Silk 356
Registration of New Substances 634

S

Silk Dyeing 356, 381 ff.
Silk Dyes 291
Sodium dithione (hydrosulfite) 363
Solvent Dyes
- Alcohol- and Ester-Soluble Dyes 295
- Dyes Soluble in Polymers 298
- Fat- and Oil-Soluble Dyes 297
- Solvent Dyes for Other Applications 299
Standardization of Textile Dyes 346
- Color Strength, Hue, Chroma 346
- Dusting 347
- Solution and Dispersion Behavior 347
Stilbene Dyes (Other Chromophores) 111
Substantivity 158 f.
Sulfur Compounds as Chromophores 78
- Polysulfide Melt Dyes 81
- Pseudo Sulfur Dyes 83
- Sulfur Bake and Polysulfide Bake Dyes 79
- Sulfur Dyes 78
Sulfur Dyes 215
- C.I. Leuco Sulphur Dyes 224
- C.I. Solubilised Sulphur 224
- C.L. Sulphur Dyes 224
- Indophenols 219
- Polysulfide Melt, Solvent Reflux, or Reflux Thionation Process 219
- Pseudo Sulfur Dyes 225
- Sulfur and Polysulfide Bake or Dry Fusion 216
Sulfur Dyes on Cellulosic Fibers 370
- Additives to the Dye Bath
  - Reducing Agents 371
  - Water Softeners 372
  - Wetting Agents 372
- Combination with Other Dyes 375
- Dyeing Process
  - Absorption 372
  - Alkylation 373
  - Oxidation 372
  - Rinsing or Washing 373
- Dyeing Techniques
  - Batch (Exhaustion) Processes 373
  - Continuous Pad Dyeing Processes 374
- Types and Mode of Reaction 370
Supramolecular Structure 387

T

Tanning 427
Textile Dyeing 339
Toxicology 626
- Acute Toxicity 626
- Azoreductase 630
- Carcinogenicity 629
- Metabolism of Azo Dyes 630
- Mutagenicity 628
- Sensitization 626
Triphenodioxazine Dyes (Other Chromophores) 112

V

Vat Dyeing Process 363
- Aftertreatment (Soaping) 365
- Dye Absorption in the Exhaustion Process 364
- Dyeing Techniques 365
  - Batch Processes 365
  - Continuous Processes 366
  - Printing with Vat Dyes 366
- Oxidation 364
- Principles of Vat Dyeing 362
- Vatting 363
Vat Dyes 187
Vat Dyes for Fiber Blends 367
Vesuvin 227

W

Wastewater 633
Wool Dyeing 369, 381 ff.
World fiber consumption 339
World polyester fiber production 392

Wicks, Zeno W. / Jones, Frank N. / Pappas, S. Peter

Organic Coatings

Science and Technology
2nd edition

From the reviews of the First Edition:

"Excellently written in a clear and vivid style ... a valuable source of information."
Progress in Organic Coatings.

"[This book] does an excellent job of connecting the theory of polymer chemistry to the practical facts of organic coatings ... an extremely useful reference."
Choice.

Substantially reorganized in this accessible, self-contained volume, Organic Coatings: Science and Technology, Second Edition provides a systematic, up-to-date survey of the principles underlying the production and use of organic coatings and paints. Complete with 250 figures, this immensely useful text/reference includes:

- New developments in the field since the publication of the First Edition
- Concise descriptions of raw materials, physical concepts, formulation, applications, and properties
- Troubleshooting guidance for coatings scientists and technologists
- Precise definitions of coatings industry terminology for newcomers to the field
- Extensive references reflecting current literature
- An appendix listing useful sources.

CONTENTS: What are Coatings?/ Polymerization and Film Formation/ Flow/ Mechanical Properties/ Exterior Durability/ Adhesion/ Corrosion Protection by Coatings/ Latexes/ Amino Resins/ Binders Based on Isocyanates: Polyurethanes/ Epoxy and Phenolic Resins/ Acrylic Resins/ Polyester Resins/ Drying Oils/ Alkyd Resins/ Other Resins and Cross-Linkers/ Solvents/ Color and Appearance/ Pigments/ Pigment Dispersion/ Pigment Volume Relationships/ Application Methods/ Film Defects/ Solvent-Borne and High Solids Coatings/ Water-Borne Coatings/ Electrodeposition Coatings/ Powder Coatings/ Radiation Cure Coatings/ Product Coatings for Metal Substrates/ Product Coatings for Nonmetallic Substrates/ Architectural Coatings/ Special Purpose Coatings/ Perspectives on Coatings Design/ Appendix/ Index

ISBN 0471 24507 0 1999 654 pp Hbk
€ *159.00*/£ 89.50/ US $ 125.00***

John Wiley & Sons, Ltd.
Baffins Lane · Chichester, West Sussex · PO 19 1UD, UK
Fax: +44 (0)1243-775878
e-mail: cs-books@wiley.co.uk · http://www.wiley.co.uk

WILEY

Freitag, Werner / Stoye, Dieter (eds.)
W. Freitag, Creanova Spezialchemie GmbH; D. Stoye, Hüls Aktiengesellschaft, Marl, Germany

Paints, Coatings and Solvents

2nd edition

"This book is a valuable read for anyone interested in this field"
(Composites in Science and Technology)

CONTENTS: Introduction/ Historical Development/ Composition of Paints/ Film Formation/ Future Outlook/ Types of Paints and Coatings/ Paint Systems/ Pigments and Extenders/ Paint Additives/ Paint Removal/ Production Technology/ Paint Application/ Properties and Testing/ Analysis/ Uses/ Environmental Protection and Toxicology/ Solvents/ References

ISBN 3527 28863 5 1998 431 pp Hbk
€ *149.00*/£ 95.00/US $ 150.00***

Gierenz, Gerhard / Karmann, Werner (eds.)
W. Karmann, Beiersdorf AG, Hamburg, Germany

Adhesives and Adhesive Tapes

CONTENTS: ADHESION THEORIES/ RAW MATERIALS FOR ADHESIVES/ CLASSIFICATION FOR ADHESIVES/ INDIVIDUAL ADHESIVE SYSTEMS: Adhesives that set without a chemical reaction; Adhesives setting by chemical reaction/ BONDING TECHNIQUES/ TESTING OF ADHESIVES/ APPLICATIONS OF ADHESIVES: Bookbinding; Gluing of Wood and Wooden Materials; Footwear Adhesives; Bonding of Plasics; Bonding of Elastomers; Bonding of Metals; Adhesives for Wallcoverings; Floorcovering Adhesives; Building Construction Adhesives; Adhesives for Bonding Textile Fabrics; Flocking Adhesives; Adhesives for Bonding Glass; Adhesives in Automobile Manufacture; Adhesives in Aircraft Construction and Sealants in Electronics; Medical Adhesives; Household Adhesives; Applications of Anaerobic Adhesives/ ECONOMIC ASPECTS

ISBN 3527 30110 0 2001 146 pp Hbk
€ *79.00*/£ 50.00/US $ 75.00***

* Euro-price valid only for Germany and Austria
** All prices are approx prices and subject to change

WILEY-VCH
P.O. Box 10 11 61 · 69451 Weinheim, Germany
Fax: +49 (0) 62 01-60 61 84
e-mail: service@wiley-vch.de · http://www.wiley-vch.de/

WILEY-VCH

Zollinger, Heinrich
H. Zollinger, Eidgenössische TH Zürich, CH

Color Chemistry

Syntheses, Properties and Applications of Organic Dyes and Pigments
2nd edition

The well-received monograph Color Chemistry, now revised and updated in its 2nd edition, provides a thorough treatment of the synthesis, properties, and industrial applications of organic dyes and pigments.

This is what the reviewers had to say about Color Chemistry:

"Recommended as essential reading not only to color chemists in all stages of their careers, but to chemists unilaterally. They will find it interesting, informative, stimulating and very readable."
Dyes and Pigments

"By confining the discussion to topics of current technical importance and using a mechanistic organic approach, an informative overall balance is achieved..."
Chemistry in Britain

"This book will stand as the definitive treatment of the subject for years to come...Professor Zollinger's important contribution to the scientific literature belongs in every serious collection."
Textile Research Journal

ISBN 3527 28352 8   1991   512 pp   Hbk
€ 169.00*/£ 105.00/US $ 230.00**

Kuehni, Rolf G.

Color — An Introduction to Practice & Principles

FROM THE CONTENTS: Color and its Sources; What is Color?; From Light to Color; Color Perception Phenomena; Orderly Arrangements of Color; Putting Numbers on Color; Orderly Arrangements of Color Revisited; Colorants and Their Mixture; Color Reproduction; Color and Man; Color (Theory) in Art; Harmony of Colors; Timetable of Color in Science and Art.

ISBN 0471 14566 1   1996   198 pp   Hbk
€ 99.00*/£ 57.50/US $ 79.95**

Berns, Roy S.

Billmeyer and Saltzman's Principles of Color Technology

3rd edition

Fully updated-the classic comprehensive introduction to color technology

Supplemented with copious numerical examples, graphs, and illustrations that clarify and explain complex material, as well as side bars that present technical details in a well-organized, accessible manner, this excellent and exciting introduction for newcomers to the field is also a valuable reference for experienced color technologists, color specialists, chemical and industrial engineers, computer scientists, research scientists, and mathematicians interested in color.

CONTENTS: Defining Color; Describing Color; Measuring Color; Measuring Color Quality; Colorants; Producing Colors; Back to Principles; Appendix; Bibliography; Index

ISBN 0471 19459 X   2000   264 pp   Hbk
€ 129.00*/£ 71.50/US $ 99.95**

Tilley, Richard

Colour and the Optical Properties of Materials

An Exploration of the Relationship Between Light, the Optical Properties of Materials & Colour

CONTENTS: Light and Colour/ Colours Due to Refraction and Dispersion/ Crystals and Light/ The Production of Colour by Reflection/ Colour Due to Scattering/ Colour Due to Diffraction/ Colour from Atoms and Ions/ Colour from Molecules/ Colour from Charge Transfer and Luminescence/ Colour in Metals, Semiconductors and Insulators/ Fibre Optics and Data Transmission/ Displays/ Lasers and Holograms/ Appendix/ Index

ISBN 0471 85197 3   1999   348 pp   Hbk
€ 129.00*/£ 70.00/US $ 150.00**

ISBN 0471 85198 1   1999   348 pp   Pbk
€ 44.90*/£ 24.95/US $ 49.95**

*   Euro-price valid only for Germany and Austria
** All prices are approx prices and subject to change

John Wiley & Sons, Ltd.
Baffins Lane · Chichester, West Sussex · PO 19 1UD, UK
Fax: +44 (0)1243-775878
e-mail: cs-books@wiley.co.uk · http://www.wiley.co.uk

WILEY-VCH
P.O. Box 10 11 61 · 69451 Weinheim, Germany
Fax: +49 (0) 62 01-60 61 84
e-mail: service@wiley-vch.de · http://www.wiley-vch.de/

Buxbaum, Gunter (ed.)
G. Buxbaum, Bayer AG, Krefeld

Industrial Inorganic Pigments

2nd complete new revised edition

"Everything there is to know about inorganic pigments"

CONTENTS: GENERAL CHEMICAL AND PHYSICAL PROPERTIES/ COLOR PROPERTIES/ STABILITY TOWARDS LIGHT, WEATHER, HEAT, AND CHEMICALS/ BEHAVIOR OF PIGMENTS IN BINDERS/ WHITE PIGMENTS: Titanium Dioxide; Zinc Sulfide; Zinc Oxide/ COLORED PIGMENTS: Oxides and Hydroxides; Cadmium Pigments; Bismuth Pigments; Chromate Pigments; Ultramarine Pigments; Iron Blue Pigments/ BLACK PIGMENTS: Properties; Raw Materials; Production Processes; Testing and Analysis; Transportation and Storage; Uses; Economic Aspects; Toxicology and Health Aspects/ SPECIALTY PIGMENTS: Magnetic Pigments; Anticorrosive Pigments; Luster Pigments; Transparent Pigments; Luminescent Pigments

ISBN 3527 28878 3 1998 302 pp Hbk
€ 139.00*/£ 85.00/US $ 205.00**

Herbst, Willy / Hunger, Klaus
W. Herbst, Hoechst AG; K. Hunger, Hoechst AG

Industrial Organic Pigments

Production, Properties, Applications
2nd completely revised edition

"Everything there is to know about organic pigments"

CONTENTS: Definition: Pigments – Dyes/ Classification of Organic Pigments/ Chemical and Physical Characterization of Pigments/ Commercial Properties and Terminology/ Particle Size Distribution and Application Properties/ Areas of Application for Organic Pigments/ Azo Pigments/ Polycyclic Pigments/ Aluminum Pigment Lakes/ Pigments With Known Chemical Structure Which Cannot Be Assigned to Other Chapters/ Pigments With Hitherto Unknown Chemical Structure/ Ecology, Toxicology, Legislation

ISBN 3527 28836 8 1997 668 pp Hbk
€ 219.00*/£ 140.00/US $ 325.00**

Smith, Hugh Macdonald (ed.)
H. M. Smith, Sun Chemical Corporation, Cincinnati, USA

High Performance Pigments

CONTENTS: Introduction/ Global Market for High Performance Organic Pigments/ Overview of High Performance Inorganic Pigments/ Cadmiums as High Performance Inorganic Pigments/ Cerium Pigments, a New Class of High Performance Inorganic Pigments/ Complex Metal Oxides, High Performance Inorganic Pigments/ Titanates as High Performance Inorganic Pigments/ High Performance Photoluminescent Pigments/ Special Effect Pigments/ Yellow High Performance Inorganic and Organic Pigments/ Isoindolinone Pigments/ High Performance Azo Pigments/ Benzimidazolone Pigments/ Quinacridone Pigments/ PyrroloPyrrole Pigments/ Perylene Pigments/ High Performance Thiazine and Oxazine Pigments/ Dioxazine High Performance Pigments/ High Performance Phthalocyanine Pigments/ Crystal Design of High Performance Pigments/ Analytical and Physical Characterization of High Performance Pigments/ Regulatory Affairs for High Performance Pigments: Europe, North America/ Ecology and Toxicology of High Performance Pigments/ Index

ISBN 3527 30204 2 Due December 2001
approx 400 pp Hbk
Approx € 159.00*/£ 95.00/US $ 125.00**

Schwertmann, Udo / Cornell, Rochelle M.

Iron Oxides in the Laboratory

Preparation and Characterization
2nd completely revised and enlarged edition

CONTENTS: Introduction/ The Iron Oxides and Hydroxides/ General Preparative Techniques/ Methods of Characterization/ Methods of Synthesis/ Goethite/ Lepidocrocite/ Feroxyhyte/ Ferrihydrite/ Akaganeite/ Hematite/ Magnetite/ Maghemite/ Iron Hydroxy Salts/ Index

ISBN 3527 29669 7 2000 204 pp Hbk
€ 89.00*/£ 55.00/US $ 79.95**

* Euro-price valid only for Germany and Austria
** All prices are approx prices and subject to change

John Wiley & Sons, Ltd.
Baffins Lane · Chichester, West Sussex · PO 19 1UD, UK
Fax: +44 (0)1243-775878
e-mail: cs-books@wiley.co.uk · http://www.wiley.co.uk

WILEY-VCH
P.O. Box 10 11 61 · 69451 Weinheim, Germany
Fax: +49 (0) 62 01-60 61 84
e-mail: service@wiley-vch.de · http://www.wiley-vch.de/

Paul, Swaraj (ed.)
Surface Coatings
Science and Technology
2nd edition

CONTENTS: Synthesis of Polymeric Binders (S. Paul); Industrial Resins (S. Paul); Pigments (S. Paul); Principles of Solvent Selection in Paint Formulations (J. Klein & D. Wu); Pigment Dispersion Formulation (H. Jakubauskas); Surface Preparation and Paint Application (S. Paul); Paint Properties and Their Evaluation (S. Paul); Types of Coatings (S. Paul); New Technologies (S. Paul, et al.); Index.

ISBN 0471 95818 2 1995 948 pp Hbk
€ 329.00*/£ 185.00/US $ 425.00**

Gutoff, Edgar B. / Cohen, Edward D.
Coating & Drying Defects – Troubleshooting Operating Problems

Using a non-mathematical approach, this book offers all the tools needed to troubleshoot defects as well as the means to eliminate them. It contains brief descriptions of key coating processes and presents a methodology to guide readers from the start of the troubleshooting procedure, when the defect is first discovered, to the finish where the mechanism for the formation of the flaw is defined and then eliminated.

CONTENTS: Troubleshooting or Problem-Solving Procedure/ Coater Diagnostic Tools/ Problems Associated with Feed Preparation/ Problems Associated with Roll Coating and Related Processes/ Problems in Slot, Extrusion, Slide, and Curtain Coating/ Coating Problems Associated with Coating Die Design/ Surface-Tension-Driven Defects/ Problems Associated with Static Electricity/ Problems Associated with Drying/ Problems Associated with Web Handling (G. Kheboian)/ Coating Defects Catalog/ Index

ISBN 0471 59810 0 1995 304 pp Hbk
€ 129.00*/£ 71.50/US $99.95**

Fettis, Gordon (ed.)
G. Fettis, Dept.of Chemistry,Univ.of York,U.K.
Automotive Paints and Coatings

Dedicated wholly to automotive coatings, this book is the first of its kind. It provides an in-depth coverage of the subject and in keeping with the international nature of the automotive business the book has a truly multinational flavour with authors selected from Australia, Japan, Europe and the USA. An authoritative and informative treatment of all aspects of coatings formulation are presented together with their manufacture and application. Numerous chapters written by experts in the field deal with substrate pretreatment, undercoats, surfacers and topcoats. Finishes for both metals and non-metallics are described as well as speciality coatings such as sealers, antichip and underbody paints. Further valuable information on commercial support for the sale of finishes in the automotive industry and the licensing of technology is also given.

Specialists involved in a wide range of disciplines in the coatings industry including chemists, chemical engineers and commercial staff will find this up-to-date source of exceptional interest.

CONTENTS: Pretreatment/ Undercoats/ Surfacers/ Topcoats/ Speciality coatings/ Underbody paints/ Paints for non-metallics/ Technology licensing/ Technical service/ Market support

ISBN 3527 28637 3 1994 255 pp Hbk
€ 149.00*/£ 95.00/US $ 235.00**

Bieleman, Johan (ed.)
J. Bieleman, Serva Delden B. V.
Additives for Coatings

CONTENTS: Introduction/ Fundamentals/ Thickener/ Surface-Active Compounds/ Additives for the Modification of Surfaces/ Flow-Control Agents and Film-Forming Components/ Catalytic-Active Compounds/ Special Additives/ Security Measures and Disposal/ Quality Assurance

ISBN 3527 29785 5 2000 390 pp Hbk
€ 159.00*/£ 95.00/US $ 185.00**

* Euro-price valid only for Germany and Austria
** All prices are approx prices and subject to change

John Wiley & Sons, Ltd.
Baffins Lane · Chichester, West Sussex · PO 19 1UD, UK
Fax: +44 (0)1243-775878
e-mail: cs-books@wiley.co.uk · http://www.wiley.co.uk

WILEY-VCH
P.O. Box 10 11 61 · 69451 Weinheim, Germany
Fax: +49 (0) 62 01-60 61 84
e-mail: service@wiley-vch.de · http://www.wiley-vch.de/

Zollinger, Heinrich
Color
A Multidisciplinary Approach

FROM THE CONTENTS: INTRODUCTION: What do we Mean by Color?; Historical Survey/ PHYSICS OF LIGHT AND COLOR: The Nature (Theory) of Light; Color by Refraction: Newton's Experiments; Color of the Rainbow; Peacock's Colors, a Phenomenon of Interference; How Many Causes of Color do we Know?/ CHEMISTRY OF COLOR: History of Colorants; Inorganic Pigments; Organic Colorants; Correlations between Chemical Structure and Color of Chemical Compounds/ COLORIMETRY: Color Measurements; Color – Harmony or Contrasts?/ HOW DO WE SEE COLORS: Perception and Cognition of Color; Anatomy of the Human Eye; Photochemistry of the Retina; What does the Eye tell the Brain?; Psychophysical Investigations on Color Vision; Color Vision of Animals/ HOW DO WE NAME COLORS?: From Color Chemistry to Color Linguistics; The Phenomenon (Prodigy) of Human Language; Categorization of the Color Space by Color Naming; Color and Phonological Universals; Cultural Influence on Color Naming/ COLOR IN ART AND IN OTHER CULTURAL ACTIVITIES: Color in European Art from Antiquity to Gothic; From Renaissance to Neo-Impressionism; Art in the 20th Century; Color in the Art of Non-European Culture: The Case of Japan; Color in Psychology; Goethe's "Farbenlehre"; Sound-Color Synesthesia; Epilogue; Acknowledgements; Author Index; Subject Index

ISBN 3906390 18 7 1999 268 pp Hbk
€ 109.00*/£ 65.00/US $ 120.00**

Wyszecki, Günther / Stiles, W. S.
Color Science
Concepts and Methods, Quantitative Data and Formulae
2nd edition

CONTENTS: Physical Data/ The Eye/ Colorimetry/ Photometry/ Visual Equivalence and Visual Matching/ Uniform Color Scales/ Visual Thresholds/ Theories and Models of Color Vision/ Appendix/ References/ Author and Subject Indexes

ISBN 0471 39918 3 2000 968 pp Pbk
€ 75.00*/£ 42.95/US $ 59.95**

Völz, Hans G.
H. G. Völz, Krefeld, Germany
Industrial Color Testing
Fundamentals and Techniques

This book is the first complete treatment focusing on theoretical and practical aspects of testing pigments, dyes, and pigmented and dyed coatings. It provides basic knowledge for newcomers in the field and serves as a reference work for experts.

Part 1 explains the dependence of color on spectra, of spectra on scattering and absorption, and of scattering and absorption on the content of coloring matter.

Part 2 deals with the significance of color measurement and the acceptability of color differences. It describes the determination of hiding power and transparency, tinting strength and lightening power.

The book provides the answers to questions arising in the production, processing, and application of coloring matter in vehicles. It is a fundamental resource for engineers in industry, scientists in research and development, educators, and students.

CONTENTS: Color Properties/ How Colors Depend on Spectra/ How Spectra Depend on the Scattering and Absorption of Light/ How Light Scattering and Absorption Depend on the Content of Coloring Material/ How Light Scattering and Absorption Depend on the Physics of the Pigment Particle/ Measurement and Evaluation of Nonluminous Perceived Colors/ Determination of Hiding Power and Transparency/ Determination of Tinting Strength and Lightening Power

ISBN 3527 28643 8 1994 390 pp Hbk
€ 139.00*/£ 85.00/US $ 195.00**

* Euro-price valid only for Germany and Austria
** All prices are approx prices and subject to change

John Wiley & Sons, Ltd.
Baffins Lane · Chichester, West Sussex · PO 19 1UD, UK
Fax: +44 (0)1243-775878
e-mail: cs-books@wiley.co.uk · http://www.wiley.co.uk

WILEY-VCH
P.O. Box 10 11 61 · 69451 Weinheim, Germany
Fax: +49 (0) 62 01-60 61 84
e-mail: service@wiley-vch.de · http://www.wiley-vch.de/

TP
893
I43
2003
CHEM

LIBRARY USE ONLY